ANALOG ELECTRONICS-II

[For E.C.E. & E.E.E. Students)

By

J.B. Gupta
B.Sc. (Elec. Engg.); C. Eng. (I);
F.I.E. (India); M.I.S.T.E.

Rajeev Manglik
B.E. (Elec.) Pb. University

Rohit Manglik
B.E. (Elec.) Univ. of Roorkee

S.K. KATARIA & SONS®

Publisher of Engineering Books

4885/109, Prakash Mahal, Dr. Subhash Bhargav Lane,
Opposite Delhi Medical Association, Daryaganj, New Delhi–110002 (INDIA)
Phone: +91-11-43551243; *Mobile:* +91-9871775858, 7678425209
e-mail: katariabook@gmail.com, katariabooks@yahoo.com
Website: www.skkatariaandsons.com

(e-books, subscription and bulk buying also available on our website)

Published by:

S.K. Kataria & Sons®

4885/109, Prakash Mahal, Dr. Subhash Bhargav Lane,
Opposite Delhi Medical Association, Daryaganj, New Delhi-110002 (INDIA)
Phone: +91-11-43551243; *Mobile:* +91-9871775858, 7678425209
e-mail: katariabook@gmail.com, katariabooks@yahoo.com
Website: www.skkatariaandsons.com

Edited By : **Shalini Manglik**

First Edition: January, 2012; Reprint: 2026

Laser Typesetting by :
AISA ENTERPRISES

Printed at :
New A.S. Offset Printers

PREFACE

The primary aim of presenting this book is to meet the needs of Second Year (Fourth Semester) students of B.Tech. (E.C.E. and E.E.E.) studying in Engineering Colleges affiliated to G.G.S.I.P. University, Delhi. The book has been written keeping in view the latest revised syllabus. The subject matter has been presented in an easily comprehensible manner, logical sequence using simple language and to provide a detailed exposition to the fundamentals. A large number of workedout examples, theoretical exercises, short answer type questions with answers and numerical problems included in the text are from latest question papers of Universities/Deemed Universities/Technical Universities including G.G.S.I.P. University or are of equivalent standard.

The book consists of 15 chapters divided into 4 Units according to the latest syllabus prescribed by G.G.S.I.P. University, Delhi for the subject of Analog Electronics-II.

The solution to illustrative examples covering almost all topics/subtopics of different chapters makes the entire presentation easy to follow. Selected exercises and numerical problems with answers given at the end of each chapter help the students in evaluating their understanding of the subject. Short answer type questions with answers help the students in preparing for theory and viva voce/ practical examinations.

Every care has been taken to eliminate misprints, omissions and errors but it is too much to expect that no inaccuracy/misprint/error has crept in and the authors would be grateful to the readers for bringing to their notice any such errors/misprints/omissions they may come across while going through the book.

The authors will feel satisfied if the book meets the needs of the students for whom it is meant. Suggestions and criticism for the improvement of the book shall be welcomed.

The authors will like to thank all the Scientists, Engineers, Authors and Publishers whose works and text have been the source of enlightenment, inspiration and guidance to us in the presentation.

Lastly, but not the least, the authors are thankful to their publisher, Sanjeev Kumar Kataria for bringing out the book in such a short time and for pricing it moderately despite heavy cost of paper and printing.

December 23, 2011 **AUTHORS**

CONTENTS

UNIT IV : ACTIVE RC FILTERS AND APPLICATIONS OF IC ANALOG MULTIPLIER .... 315—390

Unit-I

BUILDING BLOCKS OF ANALOG ICs

CONTENTS OF THIS UNIT

CHAPTER

1

Differential Amplifiers

1.1. INTRODUCTION

INSIDE THIS CHAPTER

The differential amplifier is a fundamental building block of analog circuits. It is the input stage of virtually every operational amplifier (op-amp), and is the basis of a high-speed digital logic circuit family, called **emitter-coupled logic (ECL).**

Initially invented for use with vacuum tubes, the basic differential-amplifier configuration was subsequently implemented with discrete BJTs. However, it was the advent of ICs that has made the differential amplifier extremely popular in both bipolar and MOS technologies. First, as we shall shortly see, the performance of the differential amplifier depends critically on matching between the two sides of the circuits. IC fabrication is capable of providing matched devices whose parameters track over wide ranges of changes in environmental conditions. Second, by their nature, differential amplifiers make use of more components (approaching twice as many) than single-ended circuits and we know that a significant advantage of IC technology is the availability of large number of transistors at relatively low cost.

Basically, there are two reasons for using differential amplifiers in preference to single-ended amplifiers. First, the differential amplifiers are much less sensitive to noise and interference than single-ended circuits. The second reason for preferring differential amplifiers is that the differential configuration enables us to bias the amplifier and to couple amplifier stages together without the necessity of bypass and coupling capacitors such as those used in the designs of discrete-circuit amplifiers.

Differential amplifier at the input stage of an op-amp provides very high input impedance, noise rejection and differential amplification capabilities. Since the differential amplifier is the basic building block of an op-amp, the analysis of differential amplifier will not only clarify the operation of an op-amp but also set a ground work to follow the performance characteristics of the op-amp.

The differential amplifier shown in Fig. 1.2, as its name implies, amplifies the difference between the two input signals V_{in_1} and V_{in_2}. The differential amplifier is also referred to as a **difference amplifier.**

Figure 1.2 shows the basic BJT differential amplifier pair configuration. It consists of two identical bipolar junction transistors (BJTs) shown in Fig. 1.1 with their emitters E_1 and E_2 coupled together. The collectors of transistors Q_1 and Q_2 are connected through resistors R_C to a positive supply voltage $+V_{CC}$. Emitter E_1 of transistor Q_1 is connected to emitter E_2 of transistor Q_2 (this connection places R_{E_1} in parallel with R_{E_2}) and the combination is connected to negative supply voltage $-V_{EE}$. The parallel combination of R_{E_1} and R_{E_2} is denoted by R_E. Since both positive and negative bias voltages are used in the circuit, the need for coupling capacitors and voltage divider biasing resistors at the inputs of Q_1 and Q_2 has been eliminated. By design, transistors Q_1 and Q_2 are to remain biased in forward active region. It is assumed that the two collector resistors R_C are equal.

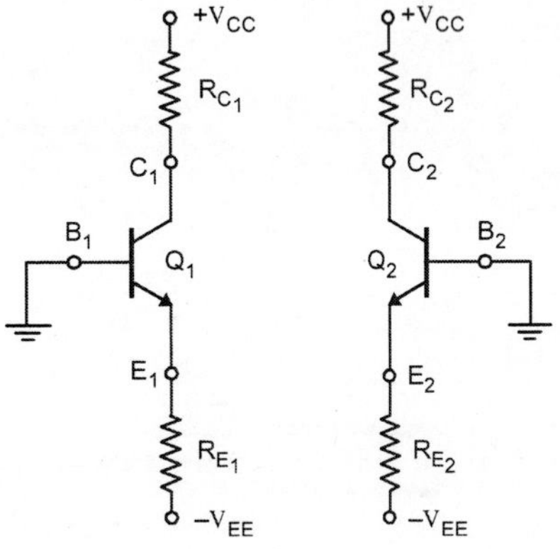

Fig. 1.1. ***Two Identical Emitter-Biased Transistors***

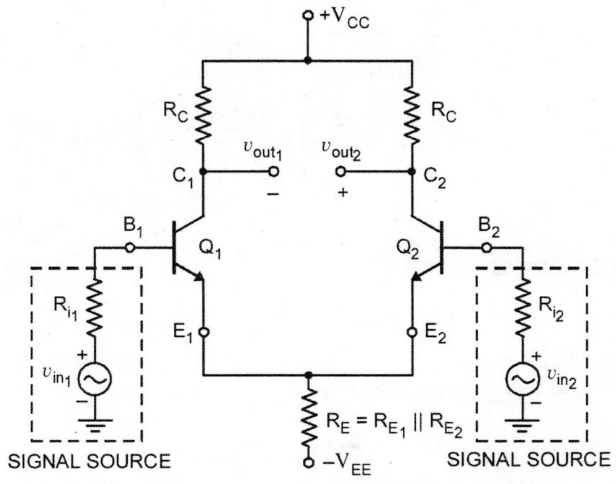

Fig. 1.2. ***Basic BJT Differential Amplifier Circuit***

The configurations are categorised on the basis of the number of inputs and the way an output voltage is measured. Accordingly, differential amplifier configurations may be classed as follows :

1. Dual-input, balanced-output differential amplifier configuration.
2. Dual-input, unbalanced-output differential amplifier configuration.
3. Single-input, balanced-output differential amplifier configuration.
4. Single-input, unbalanced-output differential amplifier configuration.

When both the inputs are used, the configuration is said to be dual input, otherwise single input. On the other hand, when the output voltage is measured between two collectors, it is referred to as a **balanced output,** because both collectors C_1 and C_2 are at the same dc potential w.r.t. ground. In case the output is measured across any one collector (either C_1 or C_2) w.r.t. ground, the configuration is known as an **unbalanced output.**

For a differential amplifier we need two matched semiconductors of the same type (BJTs or FETs). All the components in the two emitter biased circuits constituting a differential amplifier, must be matched in all respects for proper operation. Also, the magnitude of the supply voltages $+V_{CC}$ and $-V_{EE}$ must be equal. Fortunately, a variety of matched transistor pairs are readily available for differential amplifier applications.

Among the four configurations, single input configurations are rarely used. Hence, our discussion will be limited to dual input configurations.

1.2. BJT DIFFERENTIAL PAIR

The basic BJT differential pair configuration is shown in Fig. 1.3. Two identical transistors, Q_1 and Q_2, whose emitters are connected together, are biased by a constant-current source I_Q, which is connected to a negative supply voltage $-V_{EE}$. The collectors of transistors Q_1 and Q_2 are connected through resistors R_C to a positive supply voltage $+V_{CC}$. By design, transistors Q_1 and Q_2 are to remain biased in the forward active region. Two collector resistors are assumed to be equal and v_{B_1} and v_{B_2} are assumed to be ideal sources (*i.e.*, output resistances of these sources are negligibly small).

Since both positive and negative bias voltages are used in the circuit, coupling capacitors and voltage divider biasing resistors at the inputs of transistors Q_1 and Q_2 are not required. If the input signal voltages v_{B_1} and v_{B_2} in the circuit shown in Fig. 1.3 are both zero, transistors Q_1 and Q are still biased in the active region by the constant-current source I_Q. The common-emitter voltage v_E would be of the order of –0.7 V. This circuit, then, is referred to as a dc-coupled differential amplifier, so the differences in dc input voltages can be amplified. Although the difference amplifier contains two transistors, it is considered a single stage amplifier. Its characteristics are similar to those of the common-emitter amplifier.

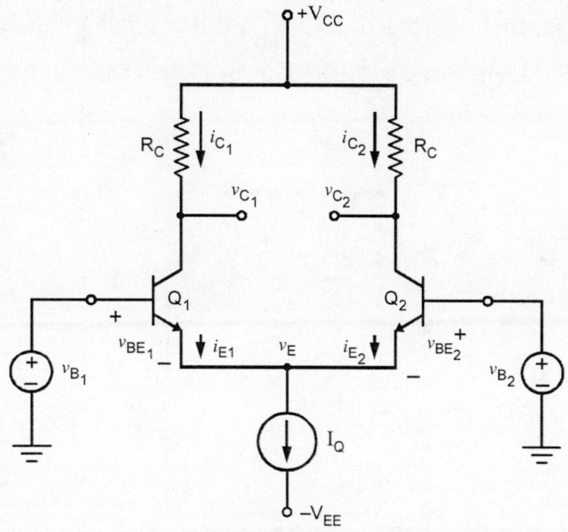

Fig. 1.3. ***Basic BJT Differential-Pair Configuration***

1.2.1. Basic Operation

For understanding the operation of the BJT differential pair, let us consider the circuit shown in Fig. 1.4 in which two bases are connected together and to a common-mode voltage v_{CM}.

$$v_{B_1} = v_{B_2} = v_{CM}$$

Since transistors Q_1 and Q_2 are perfectly matched, the bias current source being assumed to be ideal with infinite output resistance, it follows that the current I_Q will remain constant and from symmetry I_Q will divide equally between the two transistors. Thus emitter currents $i_{E_1} = i_{E_2} = \frac{1}{2} I_Q$ and the voltage at the emitters will be $v_{CM} - V_{BE}$ where V_{BE} is the base-emitter voltage assumed to be approximately 0.7 V corresponding to an emitter current of $\frac{1}{2} I_Q$. The voltage at each collector will be $V_{CC} - I_C R_C = V_{CC} - \frac{1}{2} \alpha I_Q R_C$ and difference in voltage between two collectors will be zero.

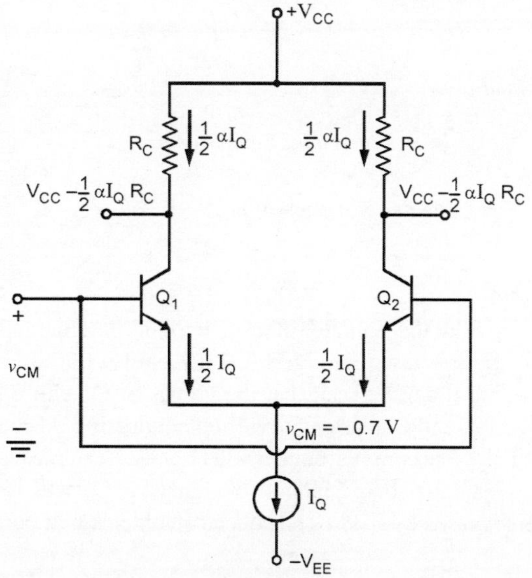

Fig. 1.4. ***Differential BJT Pair with a Common-Mode Input Signal v_{CM}***

Now, let the common-mode input signal v_{CM} be varied. As long as transistors Q_1 and Q_2 remain in the active region, the current I_Q will still divide equally between transistors Q_1 and Q_2, and the voltages at the collectors will remain unchanged. Thus the differential pair does not respond to (*i.e.*, it rejects) common-mode input signals.

Now let the voltage v_{B_2} be made 0 (by grounding base terminal B_2) and let v_{B_1} be + 1 V, as shown in Fig. 1.5 (*a*). With a bit of reasoning, it can be understood that transistor Q_1 will be *on* and conducting all the current I_Q and that transistor Q_2 will be *off*. For transistor Q_1 to be *on* (with $V_{BE_1} = 0.7$ V), the emitter has to be at approximately +0.3 V, which keeps the emitter-base junction of transistor Q_2 reverse biased.

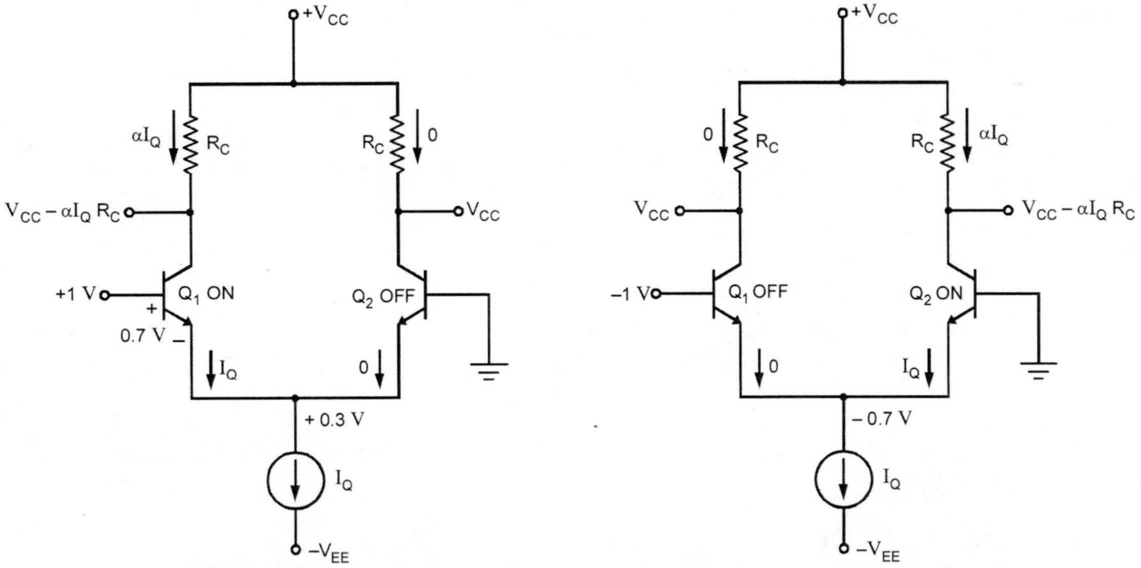

(a) BJT Differential Pair With a Large Differential Input Signal

(b) BJT Differential Pair With a Large Differential Input Signal of Polarity Opposite to That in (a)

Fig. 1.5

The collector voltages will be

$$v_{C_1} = V_{CC} - \alpha I_Q R_C$$

and $$v_{C_2} = V_{CC}$$

Now let the voltage of base terminal B_1 changed to −1 V, as shown in Fig. 1.5 (*b*), Again with some reasoning it can be seen that the transistor Q_1 will be turned *off*, and transistor Q_2 will be conducting all the current I_Q. The common emitter will be at −0.7 V. It means that the emitter-base junction of transistor Q_1 will be reverse biased by 0.3 V. The collector voltages will be

$$v_{C_1} = V_{CC}$$

and $$v_{C_2} = V_{CC} - \alpha I_Q R_C$$

From the above it can be well understood that the differential pair certainly responds to large difference-mode signals. In fact, with relatively small difference voltages entire bias current can be steered from one side of the pair to the other. This property of current steering of the differential pair plays vital role in logic circuits. It can be easily seen that the differential pair can be used to implement the single-pole double-throw switch that can be used in the realization of the current-mode inverter.

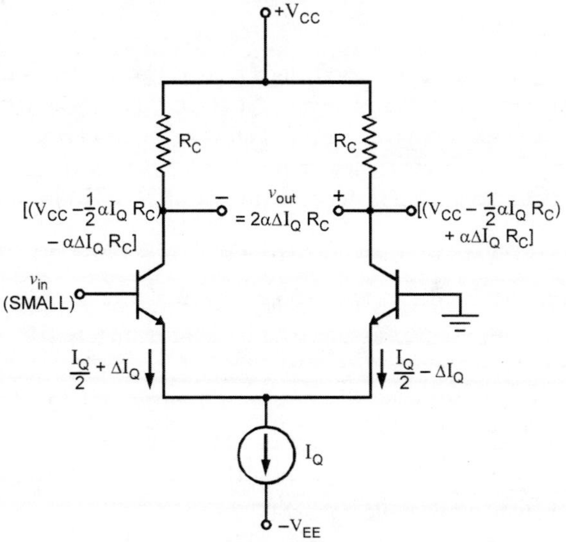

Fig. 1.6. ***BJT Differential Pair With a Small Differential Input Signal, v_{in}***

For using the BJT differential pair as a linear amplifier a very small differential signal (a few mV) is applied to the base of one of the transistors (say, Q_1). Now the transistor Q_1 will conduct current $\frac{1}{2} I_Q + \Delta I_Q$

and the transistor Q_2 will conduct current $\frac{1}{2}I_Q - \Delta I_Q$, with ΔI_Q being proportional to the difference input voltage. The output voltage taken between the two collectors will be $2\alpha\Delta I_Q R_C$, which is proportional to the differential input signal v_{in}.

Example 1.1. Determine the quiescent collector current i_{C_1} and collector-emitter voltage v_{CE_1} for the difference amplifier shown in Fig. 1.3.

The circuit parameters are : $V_{CC} = 10$ V, $-V_{EE} = -10$ V, $I_Q = 1.2$ mA and $R_C = 8.2$ kΩ.
The transistor parameters are : $V_{BE} = 0.7$ V, $V_A = \infty$, $\alpha = 1$ (neglect base currents) when

(i) $v_{B_1} = v_{B_2} = v_{CM} = 0$ V

(ii) $v_{B_1} = v_{B_2} = v_{CM} = 5$ V

and *(iii)* $v_{B_1} = v_{B_2} = v_{CM} = -5$ V

Solution : Emitter current, $I_{E_1} = I_{E_2} = \frac{1}{2}I_Q = \frac{1}{2} \times 1.2 = 0.6$ mA

Collector currents, $I_{C_1} = I_{C_2} = I_{E_1} = I_{E_2} = 0.6$ mA **Ans.** $\because$ base currents are negligible.

Collector voltage, $v_{C_1} = v_{C_2} = V_{CC} - I_{C_1} R_C = 10 - 0.6 \times 8.2 = 5.08$ V

(i) When $v_{CM} = 0$ V, $v_E = -0.7$ V

$v_{CE_1} = v_{C_1} - v_E = 5.08 - (-0.7) = 5.78$ V **Ans.**

(ii) When $v_{CM} = 5$V, $v_E = 5 - 0.7 = 4.3$ V

and $v_{CE_1} = v_{C_1} - v_E = 5.08 - 4.3 = 0.78$ **Ans.**

(iii) When $v_{CM} = -5$ V, $v_E = -5 - 0.7 = -5.7$ V

and $v_{CE_1} = v_{C_1} - v_E = 5.08 - (-5.7) = 10.78$ V **Ans.**

Example 1.2. Determine emitter voltage and collector voltages for the circuit given in Fig. 1.7. Assume $|v_{BE}|$ of a conducting transistor to be approximately 0.7 V and neglect base currents *i.e.*, $\beta = \infty$.

Solution : The given circuit is a differential amplifier consisting of two PNP transistors Q_1 and Q_2.

With a bit of reasoning, it can be seen that PNP transistor Q_1 will not conduct (*i.e.*, off) and PNP transistor Q_2 will carry all the current i_E.

Emitter voltage, $v_E = +0.7$ V **Ans.**

Emitter current, $i_E = \frac{V_{EE} - v_E}{R_E} = \frac{10 - 0.7}{2.2 \text{ k}\Omega} = 4.23$ mA

Collector current, $i_{C_1} = 0$

Collector voltage, $v_{C_1} = -10$ V

Collector current, $i_{C_2} = i_E = 4.23$ mA

Collector voltage, $v_{C_2} = -10 + 4.23 \times 2.2 = -0.7$ V **Ans.**

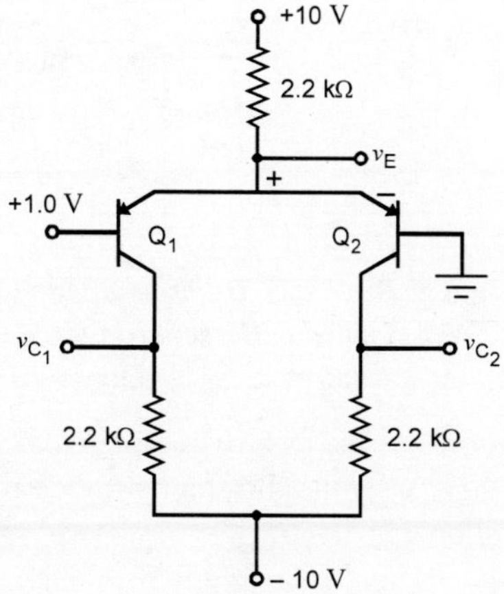

Fig. 1.7

1.2.2. Large-Signal Operation

General analysis of a differential-pair configuration can be performed by using the exponential relationship between collector current and base-emitter voltage. We know that

$$i_{C_1} = I_S e^{v_{BE_1}/V_T} \quad \text{...(1.1)}$$

and

$$i_{C_2} = I_S e^{v_{BE_2}/V_T} \quad \text{...(1.2)}$$

Transistors Q_1 and Q_2 having being assumed to be perfectly matched and operating at the same temperature, so the coefficient I_S is the same in each expression.

Neglecting base currents and assuming I_Q as an ideal constant current source, we have

$$i_{C_1} + i_{C_2} = I_Q \quad \text{...(1.3)}$$

where i_{C_1} and i_{C_2} are the total instantaneous currents, which may include the signal currents. So, we have

$$I_S\left[e^{v_{BE_1}/V_T} + e^{v_{BE_2}/V_T}\right] = I_Q \quad \text{...(1.4)}$$

Taking the ratios of i_{C_1} to I_Q and i_{C_2} to I_Q, we have

$$\frac{i_{C_1}}{I_Q} = \frac{1}{1 + e^{(v_{BE_2} - v_{BE_1})/V_T}} \quad \text{...(1.5)}$$

$$\frac{i_{C_2}}{I_Q} = \frac{1}{1 + e^{-(v_{BE_2} - v_{BE_1})/V_T}} \quad \text{...(1.6)}$$

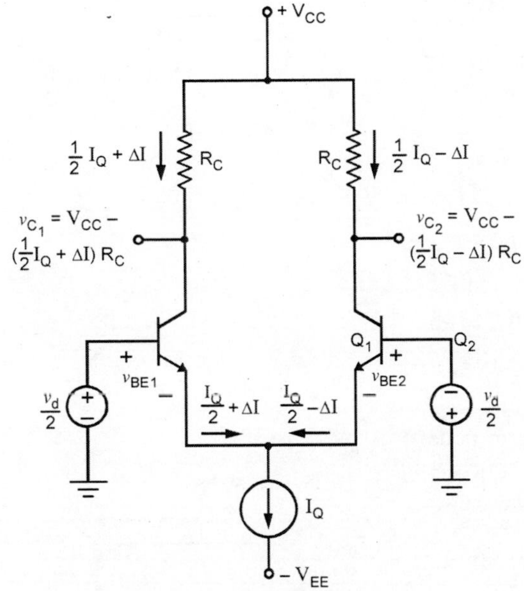

Fig. 1.8. ***Basic BJT Differential Amplifier With Applied Differential-Mode Voltage***

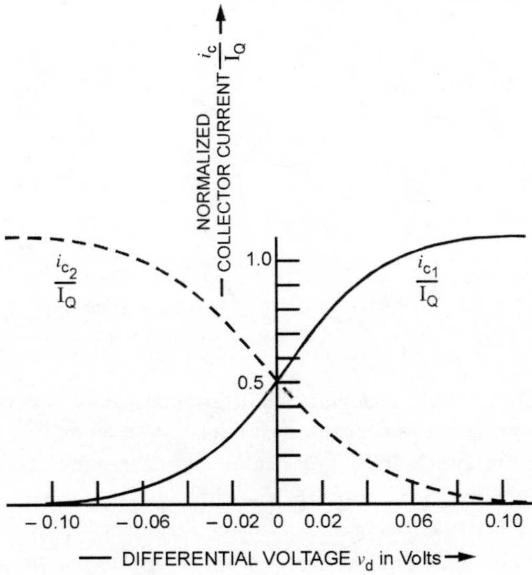

Fig. 1.9. ***Normalized DC Transfer Characteristics For BJT Differential Amplifier***

From Fig. 1.8, we see that

$$v_{BE_1} - v_{BE_2} \cong v_d \quad \text{...(1.7)}$$

where v_d is the differential-mode input voltage. Eqs. (1.5) and (1.6) can then be written in terms of v_d as follows :

$$i_{C_1} = \frac{I_Q}{1 + e^{-v_d/V_T}} \quad \text{...(1.8)}$$

and

$$i_{C_2} = \frac{I_Q}{1 + e^{+v_d/V_T}} \quad \text{...(1.9)}$$

Equations (1.8) and (1.9) describe the basic current-voltage characteristics of the differential amplifier. If the differential-mode input voltage is zero, then the bias current I_Q divides equally between the two transistors. However, when a differential-mode signal v_d is applied, a difference occurs between collector currents i_{C_1} and i_{C_2}, which in turn results in a change in the collector terminal voltage. This is the fundamental operation of the differential amplifier. If a common-mode signal $v_{CM} = v_{B_1} = v_{B_2}$ is applied, the bias current I_Q still divides equally between the two transistors.

The normalized plot of the *dc transfer characteristics* for a differential amplifier is shown in Fig. 1.9. There are two basic observations which can be made. First, the gain of the differential amplifier is proportional to the slopes of the transfer curves about the point $v_d = 0$. In order to have a linear amplification, the excursion of v_d about zero must be kept small.

Differential input signal is limited to less than about $V_T/2$ in order to operate it in linear mode.

Second, as the magnitude of v_d becomes sufficiently large, essentially all of the current I_Q goes to one transistor, and the other transistor effectively turns *off*. It may be noted that a difference voltage of about $4V_T$ ($\simeq$ 100 mV) is sufficient to switch the current almost entirely to one side of the BJT pair.

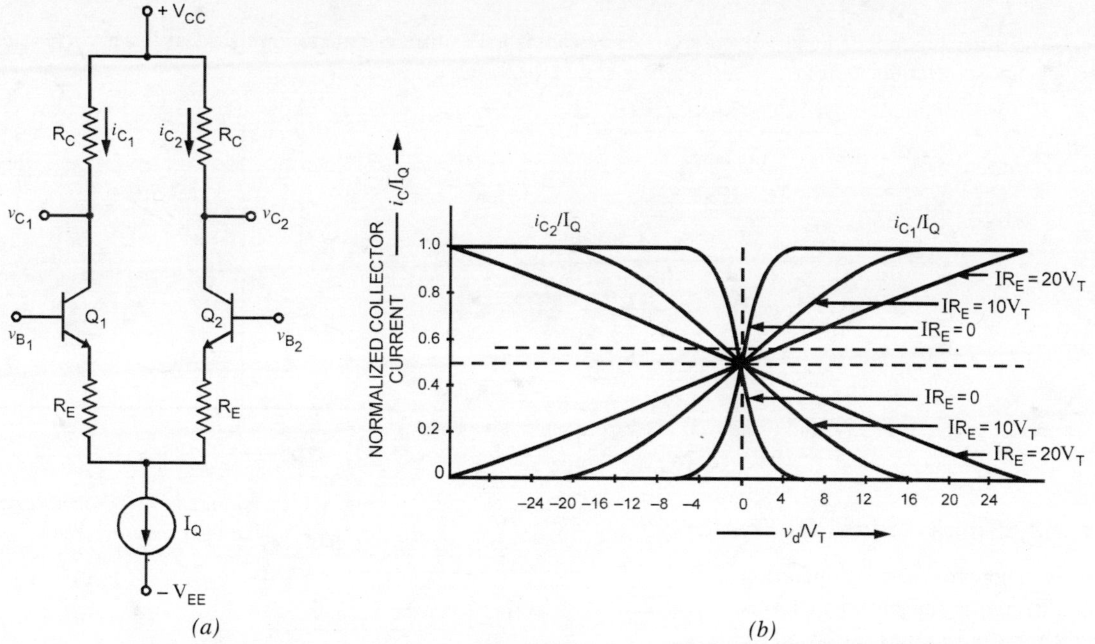

Fig. 1.10

The fact that such a small signal can switch the current from one side of the BJT differential pair to the other means that it can be used as a fast current switch. Another reason for the high speed of operation of the differential device as a switch is that neither of the transistors saturates. The absence of saturation in the normal operation of the BJT differential pair makes the logic family based on it the fastest form of logic circuits available.

Another noteworthy point is that the linear range of operation of the BJT differential amplifier can be extended by including two equal resistors R_E in series with the emitters of Q_1 and Q_2, as shown in Fig. 1.10 (*a*).

The resulting transfer characteristics for three different values of R_E are given in Fig. 1.10 (*b*). From Fig. 1.10 (*b*) it is observed that the linear range expansion is obtained at the cost of reduction in transconductance g_m (which is the slope of the transfer curve at $v_d = 0$) and, therefore, reduction in gain. This is what expected, R_E is performing in exactly the same way as the emitter resistance R_E does in the CE amplifier with emitter degeneration.

Example 1.3. Determine the maximum differential-mode input signal that can be applied and still maintain linearity in the differential amplifier. An expanded view of the normalised i_{C_1} versus v_d characteristic is shown in Fig. 1.11. A linear approximation that corresponds to the slope at $v_d = 0$ is superimposed on the curve. Determine $v_{d\,(max)}$ such that the difference between the linear approximation and the actual curve does not exceed 1 per cent.

Solution : The collector current i_{C_1} in terms of v_d can be written as

$$i_{C1\,(actual)} = \frac{I_Q}{1 + e^{-v_d/V_T}} \qquad \text{Refer to Eq. (1.8)}$$

The slope of $v_d = 0$ is found to be

$$g_f = \left.\frac{di_{C_1}}{dv_d}\right|_{v_d=0} = I_Q(-1)\left[(1+e^{-v_d/V_T})\right]^{-2}\left(\frac{-1}{V_T}\right)(e^{-v_d/V_T})\Big|_{v_d=0} = \frac{I_Q}{4V_T} \quad ...(i)$$

where g_f is the forward transconductance.

The linear approximation $i_{C_1} - v_d$ can be given as

$$i_{C_1}\text{(linear)} = 0.5\,I_Q + g_f v_d = 0.5\,I_Q + \frac{I_Q}{4V_T}v_d \quad ...(ii)$$

The differential-mode input voltage $v_{d\,(max)}$ that causes a 1% difference between the ideal linear curve and the actual curve is determined from

$$\frac{i_{C_1\,(linear)} - i_{C_1\,(actual)}}{i_{C_1\,(linear)}} = 0.01$$

or

$$\frac{0.5\,I_Q + \dfrac{I_Q}{4V_T}v_{d(max)} - \dfrac{I_Q}{1+e^{-v_{d(max)}/V_T}}}{0.5I_Q + \dfrac{I_Q}{4V_T}v_{d(max)}} = 0.01$$

$$\text{or}\quad 0.99\left[0.5 + \frac{v_{d(max)}}{4V_T}\right] = \frac{1}{1+e^{-v_{d(max)}/V_T}}$$

Assuming $V_T = 26$ mV, and using trial and error method, we have

$$v_{d(max)} \cong 18\text{ mV} \quad \textbf{Ans.}$$

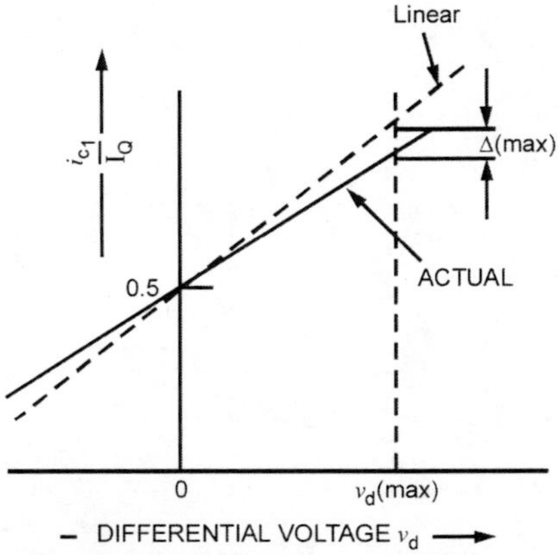

Fig. 1.11. ***Expanded View, Normalized $i_{C_1} - v_d$ Transfer Characteristic***

1.2.3. Small-Signal Operation

We can now consider the operation of differential amplifier in terms of the small-signal parameters. Fig. 1.12 depicts the BJT differential amplifier with a difference voltage v_d applied between the two bases. It means that the dc level at the input (common-mode input voltage) has been somehow established. For example, one of the two input terminals can be grounded and v_d applied to the other input terminal. Alternatively, the differential amplifier may be supplied from the output of another differential amplifier. In the latter case, the voltage at one of the input terminals will be $V_{CM} + \frac{1}{2}v_d$ and that at the other input terminal will be $V_{CM} - \frac{1}{2}v_d$.

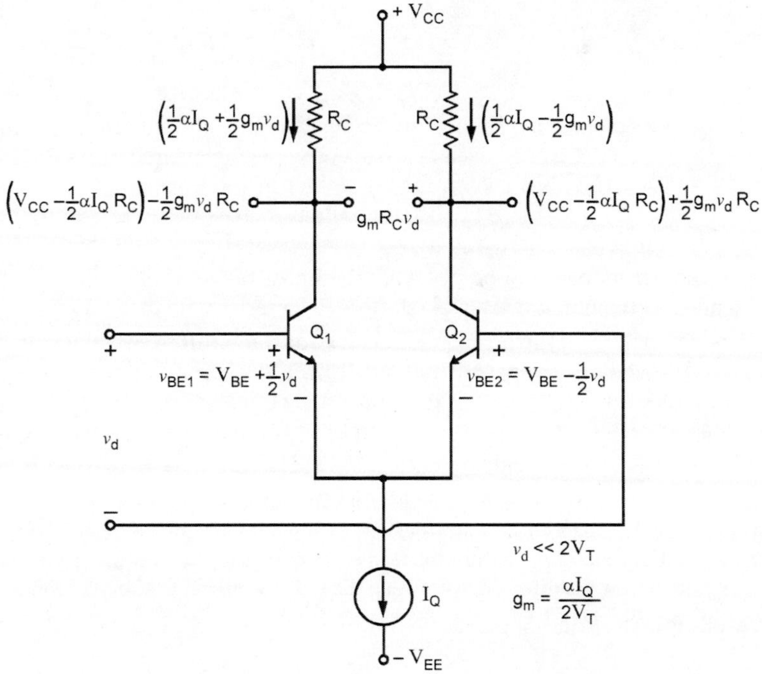

Fig. 1.12. ***BJT Differential Amplifier With Differential Mode Input Signal***

For the circuit shown in Fig. 1.12, the collector currents with v_d applied, may be written as

$$i_{C_1} = \frac{\alpha I_Q}{1+e^{-v_d/V_T}} \qquad ...(1.10)$$

$$i_{C_2} = \frac{\alpha I_Q}{1+e^{v_d/V_T}} \qquad ...(1.11)$$

Multiplying the numerator and the denominator of the RHS of Eq. (1.10) by $e^{v_d/2V_T}$, we have

$$i_{C_1} = \frac{\alpha I_Q e^{v_d/2V_T}}{e^{v_d/2T} + e^{-v_d/2V_T}}$$

Assuming $v_d << 2V_T$, the exponential term may be expanded in series and retaining only first two terms, we have

$$i_{C_1} = \frac{\alpha I_Q\left(1+\frac{v_d}{2V_T}\right)}{1+\frac{v_d}{2V_T}+1-\frac{v_d}{2V_T}} = \frac{1}{2}\alpha I_Q + \frac{1}{2}\alpha I_Q \frac{v_d}{2V_T} \qquad ...(1.12)$$

Similarly we have expression for collector current i_{C_2}

$$i_{C_2} = \frac{1}{2}\alpha I_Q - \frac{1}{2}\alpha I_Q \frac{v_d}{2V_T} \qquad ...(1.13)$$

Equations (1.12) and (1.13) reveal that when $v_d = 0$, the bias current I_Q splits evenly between the two transistors of the differential pair. Thus each transistor is biased at an emitter current of $\frac{I_Q}{2}$. With small-signal differential voltage v_d applied differentially (*i.e.,* between the bases of transistors Q_1 and Q_2), the collector current of transistor Q_1 increases by an increment i_c and that of transistor Q_2 reduces by an equal amount. The incremental (or signal) current component i_c is given by

$$i_c = \frac{1}{2}\alpha \frac{I_Q}{V_T} \times \frac{v_d}{2} \qquad ...(1.14)$$

From the symmetry of the circuit (Fig. 1.12), the differential voltage v_d should split evenly between base-emitter junctions of the two transistors. Thus, the total base-emitter voltage will be

$$v_{BE_1} = V_{BE} + \frac{1}{2}v_d \qquad ...(1.15)$$

$$\text{and } v_{BE_2} = V_{BE} - \frac{1}{2}v_d \qquad ...(1.16)$$

where V_{BE} is the dc base-emitter voltage corresponding to an emitter current of $\frac{I_Q}{2}$. So the collector current of transistor Q_1 will increase by $\frac{1}{2}g_m v_d$ and that of Q_2 will reduce by $\frac{1}{2}g_m v_d$ where g_m is the transconductance of transistors Q_1 and Q_2 which are equal and given by

$$g_m = \frac{I_C}{V_T} = \frac{\alpha I_Q/2}{V_T} \qquad ...(1.17)$$

Substituting $\frac{\frac{1}{2}\alpha I_Q}{V_T} = g_m$ from Eq. (1.17) into Eq. (1.14), we have

$$i_C = \frac{1}{2}g_m v_d$$

An Alternative Technique of Analysis : Since the current source I_Q has been assumed ideal, its incremental resistance is infinite. Thus the differential voltage v_d appears across a total resistance of $2r_e$ where

$$r_e = \frac{V_T}{I_E} = \frac{V_T}{I_Q/2} \quad ...(1.18)$$

Correspondingly, there will be a signal current i_e, as shown in Fig. 1.13, given by

$$i_e = \frac{v_d}{2r_e} \quad ...(1.19)$$

Thus the collector of transistor Q_1 will exhibit a current increment i_c and that of transistor Q_2 will exhibit a current decrement i_c, where

$$i_c = \alpha i_e = \frac{\alpha v_d}{2r_e} = \frac{1}{2} g_m v_d \quad ...(1.20)$$

This method of analysis is particularly useful when resistances are included in the emitter leads, as depicted in Fig. 1.14. For this circuit, we have

$$i_e = \frac{v_d}{2(r_e + R_E)} \quad ...(1.21)$$

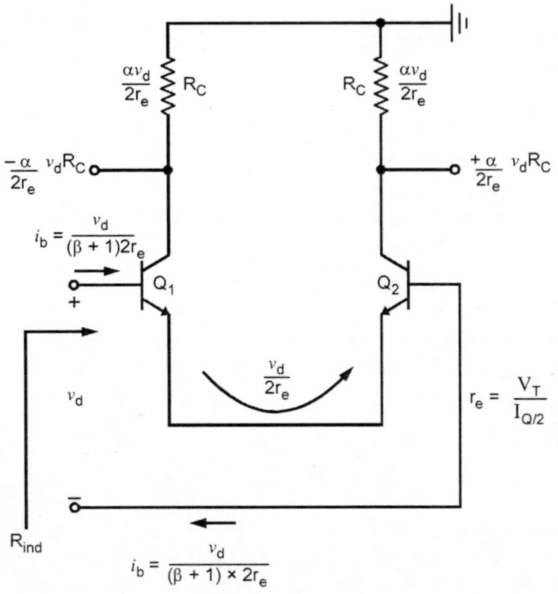

Fig. 1.13. *A Simple Technique For Determination of Signal Currents in a Differential Amplifier Excited by v_d, (DC Quantities not shown)*

Input Differential Resistance. Unlike the MOSFET differential amplifier, which has an infinite input resistance, the BJT differential pair has a finite input resistance because of finite value of β of the BJT.

The input differential resistance is the resistance seen from the two bases *i.e.,* the resistance seen by the differential input signal v_d. From the circuits of BJT differential amplifier given in Figs. 1.12 and 1.13 it can be seen that the base current of transistor Q_1 shows an incremental current i_b and that of transistor Q_2 shows an equal decrement.

$$i_b = \frac{i_e}{\beta+1} = \frac{v_d/2r_e}{(\beta+1)} = \frac{v_d}{2(\beta+1)r_e} \quad ...(1.22)$$

Thus the differential input resistance is given as

$$R_{in} = \frac{v_d}{i_b} = 2(\beta+1)r_e = 2r_\pi \quad ...(1.23)$$

The above result is just a restatement of the familiar resistance-reflection rule; namely, the resistance seen between the two bases is equal to the total resistance in the emitter circuit multiplied by (β + 1).

The input differential resistance for the circuit given in Fig. 1.14, using resistance-reflection rule, is given as

$$R_{ind} = (\beta+1)(2r_e + 2R_E) = 2(\beta+1)(r_e + R_E) \quad ...(1.24)$$

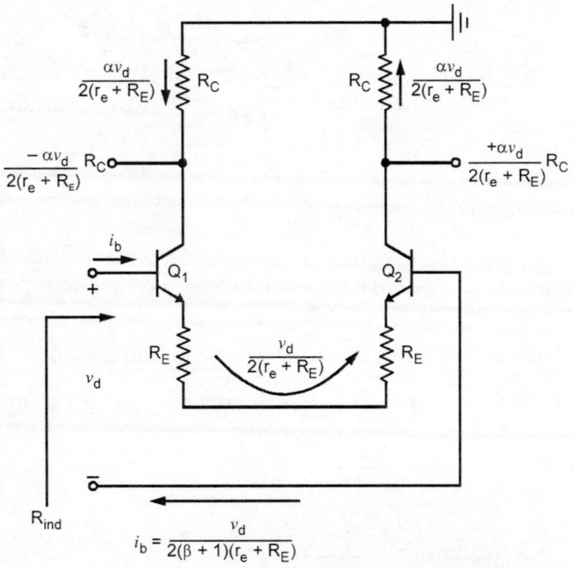

Fig. 1.14. *A BJT Differential Amplifier With Emitter Resistors R_E*

Differential Voltage Gain. It has been established that for small difference input voltages ($v_d << 2V_T$) *i.e.,* $v_d <<$ (about 20 mV), the collector currents are given by

$$i_{C_1} = I_C + \frac{1}{2} g_m v_d \quad \text{...(1.25)}$$

$$i_{C_2} = I_C - \frac{1}{2} g_m v_d \quad \text{...(1.26)}$$

where $$I_C = \alpha I_E = \frac{1}{2} \alpha I_Q \quad \text{...(1.27)}$$

Thus the total voltage at the collectors will be

$$v_{C_1} = (V_{CC} - I_C R_C) - \frac{1}{2} g_m R_C v_d \quad \text{...(1.28)}$$

and $$v_{C_2} = (V_{CC} - I_C R_C) + \frac{1}{2} g_m R_C v_d \quad \text{...(1.29)}$$

The quantities in parentheses are simply the dc voltages at each of the two collectors.

The output voltage singal of a BJT differential amplifier can be taken either differentially (*i.e.,* between the two collectors), or single-endedly (*i.e.,* between one collector and ground). If the output is taken differentially, then the differential gain of the differential amplifier will be

$$A_d = \frac{v_{C_1} - v_{C_2}}{v_d} = - g_m R_C \quad \text{...(1.30)}$$

On the other hand if output voltage is taken single-endedly (say, between the collector of transistor Q_1 and ground), then the differential gain will be

$$A_d = \frac{v_{C_1}}{v_d} = -\frac{1}{2} g_m R_C \quad \text{...(1.31)}$$

For the differential amplifier with resistances in the emitter leads (circuit shown in Fig. 1.14), the differential gain with the output taken differentially is given by

$$A_d = -\frac{\alpha(2R_C)}{2r_e + 2R_E} \simeq -\frac{R_C}{r_e + R_E} \quad \text{...(1.32)}$$

This equation is quite familiar. It states that the voltage gain is equal to the ratio of the load resistance in the collector circuit ($2R_C$) and the total resistance in the emitter circuit ($2r_e + 2R_E$).

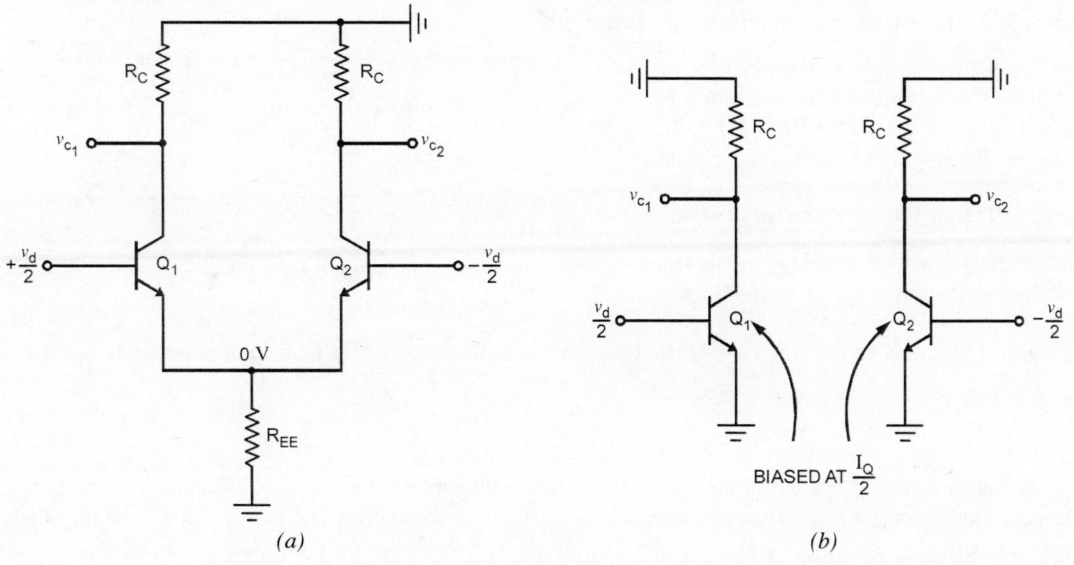

Fig. 1.15. ***Equivalence of BJT Differential Amplifier in (a) To The Two CE Amplifiers in (b)***

Equivalence of The Differential Amplifiers To a CE Amplifier. The analysis and the results on the previous page are quite similar to those obtained in case of a CE amplifier stage. The differential amplifier is in fact equivalent to a CE amplifier, as depicted in Fig. 1.15. Figure 1.15 (*a*) depicts a differential amplifier supplied by a differential signal v_d which is applied in a *complementary* (*push-pull* or balanced) manner. Output resistance R_{EE} of the bias current source has also been included. From symmetry, it is obvious that the signal voltage at the emitters will be zero. Thus, the circuit is equivalent to the two CE amplifiers given in Fig. 1.15 (*b*), where each of the transistor is biased at an emitter current of $\frac{I_Q}{2}$. It is noteworthy point that the finite output resistance R_{EE} of the current source will not affect the operation. The equivalent circuit shown in Fig. 1.15 (*b*) is valid for differential operation.

In some applications, the differential amplifier is not supplied in a complementary manner; rather, the input signal may be applied to one of the input terminals and the other terminal is grounded, as illustrated in Fig. 1.16. In such a case, the signal voltage at the emitters will not be zero, and, therefore, the resistance R_{EE} will affect the circuit operation. Nevertheless, if R_{EE} is sufficiently large ($R_{EE} >> r_e$), as usual, then the differential voltage will still split evenly (approximately) between the two junctions, as depicted in Fig. 1.16. Thus the operation of differential amplifier in this case will be almost identical to that in the case of symmetric supply and the CE equivalence can be used. Since in the circuits shown in Fig. 1.15, $v_{c_2} = -V_{c_1}$, the two CE transistors shown in Fig. 1.15 (*b*) give the similar results regarding the differential amplifier performance. So, only one is sufficient for analysis of the differential small-signal operation of the differential amplifier, and it is known as the *differential half-circuit.*

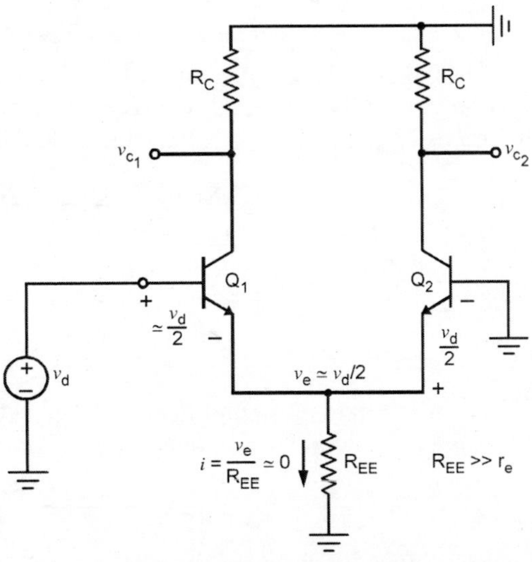

Fig. 1.16. ***BJT Differential Amplifier Supplied in a Single-Ended Manner***

In case of a CE transistor excited by $+\frac{v_d}{2}$ as the half-differential circuit, the BJT is replaced by its low-frequency equivalent circuit, and the resulting circuit is given in Fig. 1.17. In evaluation of the model parameters g_m, r_o and r_π, note that the half-circuit is biased at $\frac{I_Q}{2}$.

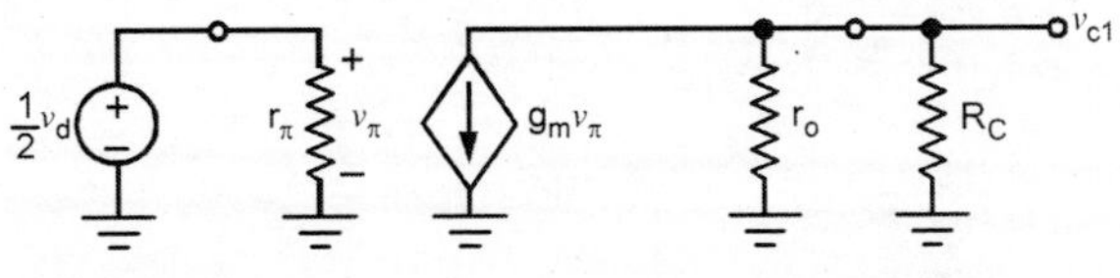

(*a*) *BJT Differential Half Circuit* (*b*) *Equivalent Circuit Model*

Fig. 1.17.

The voltage gain of the DA (with the output taken differentially) is equal to the voltage gain of the half circuit *i.e.*, $\frac{v_{c_1}}{\frac{v_d}{2}}$. Inclusion of r_o will modify the gain expression expressed by Eq. (1.30) as follows

$$A_d = -g_m (R_C \,||\, r_o) \qquad ...(1.33)$$

The input differential resistance of the DA is twice that of the half-circuit—*i.e.*, $2r_\pi$. Finally, it is noted that the differential half-circuit of the DA given in Fig. 1.14 is a CE transistor with a resistor R_E in the emitter lead.

Common-Mode Gain and CMRR. A differential amplifier excited by a common-mode voltage v_{cm} is shown in Fig. 1.18 (*a*). The resistance R_{EE} is the incremental output resistance of the bias current source.

From symmetry it can be seen that the circuit is equivalent to that given in Fig. 1.18 (*b*), in which each of the two transistors Q_1 and Q_2 is biased at an emitter current of $\frac{I_Q}{2}$ and has a resistance $2R_{EE}$ in the emitter lead. So the common-mode output voltage v_{c_1} will be

$$v_{c_1} = -v_{cm} \times \frac{\alpha R_C}{2R_{EE} + r_e} \simeq -v_{cm} \times \frac{\alpha R_C}{2R_{EE}} \quad \text{...(1.34)} \because r_e << R_{EE}$$

Similarly,
$$v_{c_2} \simeq -v_{cm} \times \frac{\alpha R_C}{2R_{EE}} \quad \text{...(1.35)}$$

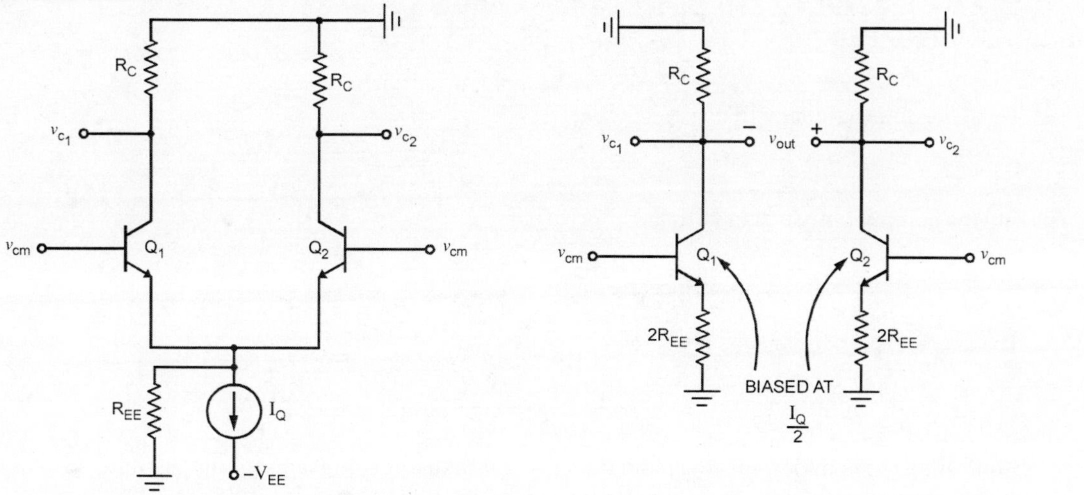

(a) BJT Differential Amplifier Supplied at a Common-Mode Voltage Signal, v_{cm}

(b) Equivalent Half-Circuits For Common-Mode Calculations

Fig. 1.18

For differential output, the output common-mode voltage will be $(v_{c_1} - v_{c_2})$ resulting in 0 volt and the common-mode gain A_{cm} also will be zero.

On the other hand, for single-ended output, the common mode gain A_{cm} will be finite and is given by*

$$A_{cm} = -\frac{\alpha R_C}{2R_{EE}} \quad \text{...(1.36)}$$

Since in this case the differential gain is

$$A_d = \frac{1}{2} g_m R_C \quad \text{...(1.37)}$$

So the common-mode rejection ratio will be

$$\text{CMRR} = \left| \frac{A_d}{A_{cm}} \right| \simeq g_m R_{EE} \quad \text{...(1.38)} \because \alpha \simeq 1$$

---

* The expressions in Eq. (1.34) and (1.35) are obtained by neglecting r_o. A detailed derivation will provide $\frac{v_{c_1}}{v_{cm}}$ and $\frac{v_{c2}}{v_{cm}}$ are approximately

$$\frac{-\alpha R_C}{2R_{EE}}\left(1 - \frac{2R_{EE}}{\beta r_o}\right)$$

where it is assumed that $R_C << \beta r_o$ and $2R_{EE} >> r_\pi$. This expression reduces to those in Eq. (1.34) and (1.35) when $2R_{EE} << \beta r_o$.

Normally CMRR is expressed in decibels,

$$\text{CMRR} = 20\log_{10}\left|\frac{A_d}{A_{cm}}\right| \qquad \text{...(1.39)}$$

Each of the circuit in Fig. 1.18 (*b*) is known as the *common-mode half-circuit.*

Practical circuits are not perfectly symmetrical, as assumed, and therefore A_{cm} will not be zero even when the output is taken differentially. For illustration, consider the case of perfect symmetry except for mismatch ΔR_C in the collector resistances *i.e.,* let the collector resistance of Q_1 be R_C and that of Q_2 be $R_C + \Delta R_C$. It means that

$$v_{c_1} = -v_{cm}\frac{\alpha R_C}{2R_{EE} + r_e}$$

and $$v_{c_2} = -v_{cm}\frac{\alpha(R_C + \Delta R_C)}{2R_{EE} + r_e}$$

Thus, the output signal will be

$$v_{out} = v_{c_1} - v_{c_2} = v_{cm}\frac{\alpha \Delta R_C}{2R_{EE} + r_e}$$

and the common-mode gain will be

$$A_{cm} = \frac{\alpha \Delta R_C}{2R_{EE} + r_e} \simeq \frac{\Delta R_C}{2R_{EE} + r_e}$$

The above expression can be rewritten as

$$A_{cm} = \frac{R_C}{2R_{EE} + r_e} \times \frac{\Delta R_C}{R_C} \qquad \text{...(1.40)}$$

Comparing the common-mode gain in Eq. (1.40) with that for single ended output in Eq. (1.36), we see that the common-mode gain is much smaller in case of differential output.

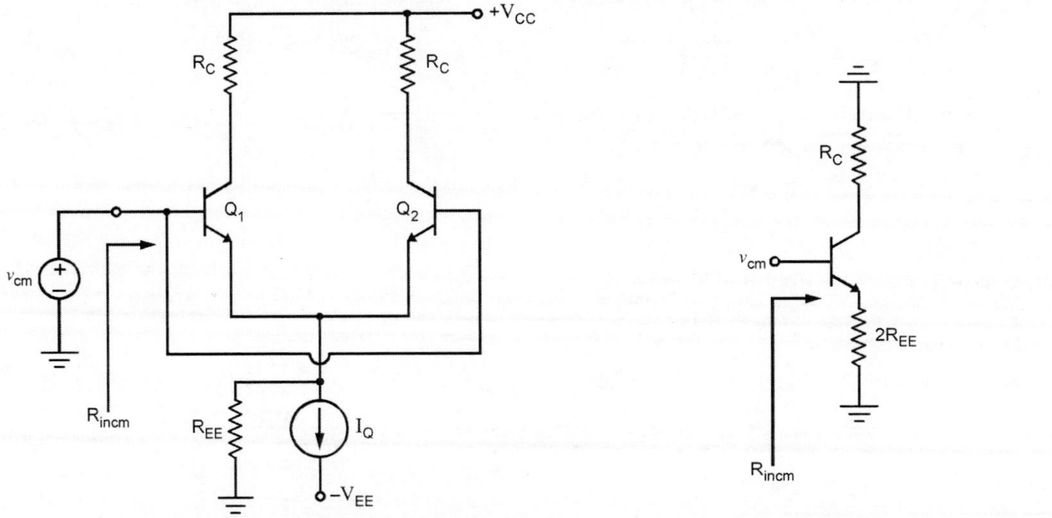

(*a*) *Definition of The Common-Mode Input Resistance* $R_{in\ cm}$ (*b*) *Equivalent Common-Mode Half Circuit*

Fig. 1.19

The input signals v_1 and v_2 to a differential amplifier usually contain a common-mode component

$$v_{cm} \equiv \frac{v_1 + v_2}{2} \qquad \text{...(1.41)}$$

and a differential component $v_d \equiv v_1 - v_2$ ...(1.42)

Thus, the output signal will be given, in general, by

$$v_{out} = A_d(v_1 - v_2) + A_{cm}\frac{v_1 + v_2}{2} \quad ...(1.43)$$

Common-Mode Input Resistance. The definition of the common-mode input resistance $R_{in\ cm}$ is illustrated in Fig. 1.19 (*a*). Figure 1.19 (*b*) depicts the equivalent common-mode half circuit, the input resistance is 2 $R_{in\ cm}$.

The approximate expression for $R_{in\ cm}$ is given by

$$2R_{in\ cm} \simeq (\beta + 1)(2R_{EE} \,||\, r_o)$$

Thus,
$$R_{in\ cm} \simeq (\beta + 1)\left(R_{EE} \,||\, \frac{r_o}{2}\right) \quad ...(1.44)$$

Equation (1.44) indicates that since R_{EE} is typically of the order of r_o and $R_{in\ cm}$ will be very large.

Example 1.4. For the differential amplifier shown in Fig. 1.20, determine the following :

(*i*) The input differential resistance.

(*ii*) The overall differential voltage gain, neglecting the effect of r_o.

(*iii*) The worst-case common-mode gain, if two collector resistances are accurate to within ±1%.

(*iv*) The input common-mode resistance, assuming Early voltage to be 200 V.

(*v*) The CMRR in dB. Assume β to be 125.

Solution : Each transistor is biased at an emitter current

$$I_E = \frac{I_Q}{2} = \frac{1.25}{2} = 0.625 \text{ mA}$$

So,
$$r_{e_1} = r_{e_2} = \frac{V_T}{I_E} = \frac{25 \text{ mV}}{0.625 \text{ mA}} = 40\ \Omega$$

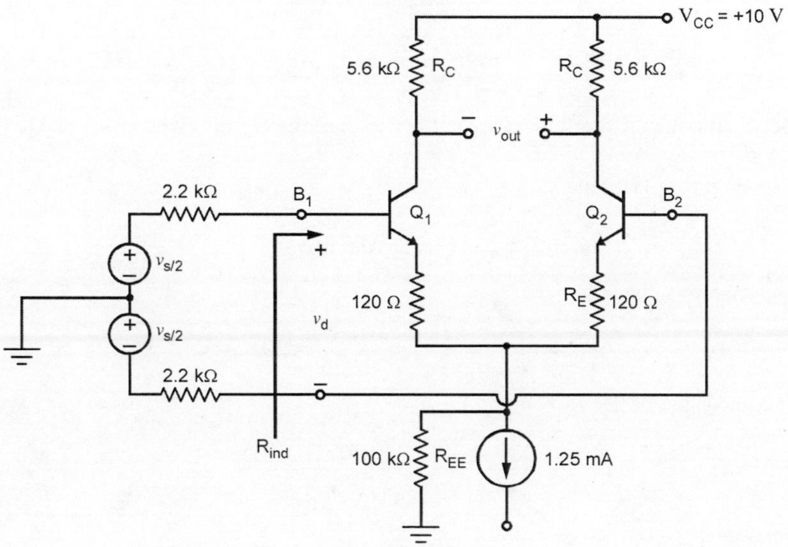

Fig. 1.20

(*i*) The input differential resistance is given as

$$R_{ind} = 2(\beta + 1)(r_e + R_E)$$

$$= 2(125 + 1)(40 + 120)\ \Omega = 40{,}320\ \Omega \text{ or } 40.32 \text{ k}\Omega \textbf{ Ans.}$$

(*ii*) The voltage gain from the signal source to the bases of transistors Q_1 and Q_2,

$$\frac{v_d}{v_s} = \frac{R_{ind}}{R_S + R_{ind}} = \frac{40.32}{2.2 + 40.32 + 2.2} = 0.9 \text{ V/V} \quad \textbf{Ans.}$$

The voltage gain from the bases of the two transistors Q_1 and Q_2 to their output is

$$\frac{v_{out}}{v_d} = \frac{\text{Total resistance in the collectors}}{\text{Total resisance in the emitters}}$$

$$= \frac{2R_C}{2(r_e + R_E)} = \frac{2 \times 5.6}{2 \times (40 + 120) \times 10^{-3}} = 35 \text{ V/V}$$

Overall differential voltage is given as

$$A_d = \frac{v_{out}}{v_s} = \frac{v_d}{v_s} \times \frac{v_{out}}{v_d} = 0.9 \times 35 = 31.5 \text{ V/V} \quad \textbf{Ans.}$$

(*iii*) The worst-case common-mode gain,

$$A_{cm} = \frac{R_C}{(2R_{EE} + r_e)} \times \frac{\Delta R_C}{R_C} \qquad \text{Refer to Eq. (1.40)}$$

$$= \frac{5.6}{2 \times 100 + 0.04} \times 0.02 = 5.6 \times 10^{-4} \text{ V/V} \quad \textbf{Ans.}$$

(*iv*)

$$r_o = \frac{V_A}{\frac{I_Q}{2}} = \frac{200}{0.625} = 320 \text{ k}\Omega$$

Input common-mode resistance is

$$R_{in\,cm} = (\beta + 1)\left(R_{EE} \,||\, \frac{r_o}{2}\right) \qquad \text{Refer to Eq. (1.44)}$$

$$= (125 + 1)\left(100 \text{ k}\Omega \,||\, \frac{320}{2} \text{ k}\Omega\right) = 7{,}754 \text{ k}\Omega \text{ or } 7.754 \text{ M}\Omega \quad \textbf{Ans.}$$

(*v*)

$$\text{CMRR} = 20 \log_{10} \frac{A_d}{A_{cm}} = 20 \log_{10} \frac{31.5}{5.6 \times 10^{-4}} = 95 \text{ dB} \quad \textbf{Ans.}$$

Example 1.5. Determine the differential-mode voltage gain of the circuit given in Fig. 1.21. When (*i*) $R_E = 0$ and (*ii*) $R_E = 330\ \Omega$.

Solution : The transistor is biased at

$$I_E = \frac{1}{2} I_Q = \frac{1}{2} \times 1 = 0.5 \text{ mA}$$

The transconductance is given as

$$g_m = \frac{I_{CQ}}{V_T} \simeq \frac{I_E}{V_T} = \frac{0.5}{0.025} = 20 \text{ mA/V}$$

(*i*) Differential-mode gain when $R_E = 0$

$$A_d = \frac{1}{2} g_m R_C = \frac{1}{2} \times 20 \times 6.8$$

$$= 68 \text{ V/V} \quad \textbf{Ans.}$$

(*ii*) Differential-mode gain when $R_E = 330\ \Omega$

$$A_d = \frac{g_m R_C}{2(1 + g_m R_E)} = \frac{20 \times 6.8}{2(1 + 20 \times 0.33)}$$

$$= 8.95 \text{ V/V} \quad \textbf{Ans.}$$

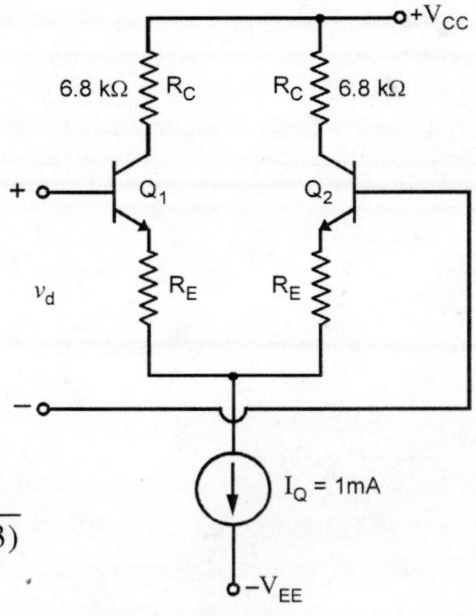

Fig. 1.21

1.3. MOSFET DIFFERENTIAL AMPLIFIER

Many differential amplifiers are constructed with FETs because of the large impedance they present to the input signals—an exceptionally important property in many applications, including operational amplifiers and instrumentation amplifiers. A large voltage gain is also important in such applications. Although the FET does not provide much voltage gain, an FET differential amplifier is usually the first stage in a multistage amplifier whose overall gain is very large. FET differential amplifiers are widely found in linear integrated circuits (LICs), as FETs are easily fabricated in IC form.

Op-amps using FETs in the input stage offer some very significant advantages over BJT op-amps, especially in the areas as input impedance, input bias and offset currents and slewing rate as illustrated below (Table 1.1).

TABLE 1.1

Parameter	BJT	JFET	MOSFET
Input impedance	kΩ	Giga ohms (10^9 Ω)	Tera ohms (10^{12} Ω)
Input Gate Current	μA	1 nA	1 pA
Input Offset Current	20 nA	2 pA	0.5 pA
Slewing Rate	1 V/μs	3 V/μs	10 V/μs

The basic MOS differential-pair configuration is shown in Fig. 1.22. It consists of two identical (matched) MOS, M_1 and M_2, whose sources are joined together and biased by a constant current source I_Q (current source is assumed to be ideal and supposed to have infinite output resistance). Like the basic bipolar configuration, the basic MOSFET differential amplifier uses both positive and negative bias voltages, thereby eliminating the need for coupling capacitors and voltage divider biasing resistors at the gate terminals. Even with $V_{G_1} = V_{G_2} = 0$, the MOSFETs M_1 and M_2 can be biased in saturation region by the current source I_Q. The circuit, then is also a *dc coupled differential amplifer.*

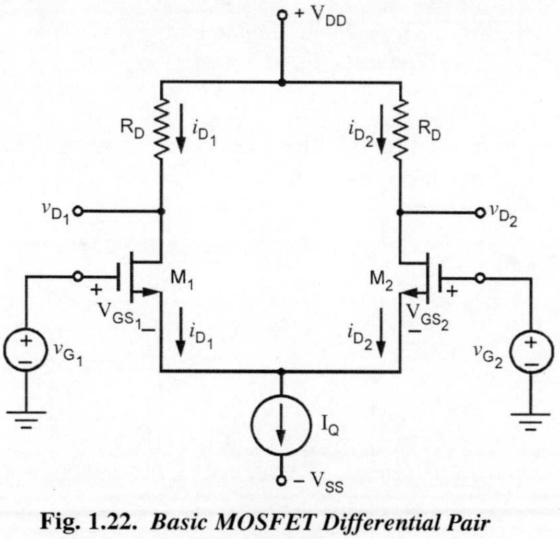

Fig. 1.22. ***Basic MOSFET Differential Pair Configuration***

1.3.1. Small-Signal Operation of MOS Differential Pair

Figure 1.23 (*a*) depicts the MOS differential amplifier with input voltages

$$v_{G_1} = V_{CM} + \frac{1}{2} v_d \quad ...(1.45)$$

$$v_{G_2} = V_{CM} - \frac{1}{2} v_d \quad ...(1.46)$$

where V_{CM} denotes a common-mode dc voltage within the input common-mode range of the differential amplifier. It is required for setting the dc voltage of the MOSFET gates. Typically V_{CM} is the middle value of power supply. Thus, in our case, where two complementary supplies are used, V_{CM} is typically zero.

The differential input signal v_d is applied in a *complementary* (or *balance*) way, that is, gate voltage v_{G_1} is increased by $\frac{v_d}{2}$ and gate voltage v_{G_2} is reduced by $\frac{v_d}{2}$. This would be the case, for example, if the differential amplifier were supplied from the output of another differential amplifier. Sometimes, however, the differential input is applied in a single-ended fashion, as shown in Fig. 1.23. The difference in the performance resulting is too subtle a point for our current requirements.

As shown in Fig. 1.24 (*a*), the output from the amplifier can be taken either between one of the drain terminal and ground or between the two drain terminals. In the formal case, the resulting single-ended outputs v_{out_1} and v_{out_2} will be riding on the top of dc voltages at the drain terminals $\left(V_{DD} - \frac{1}{2} I_Q R_D\right)$. In the latter case, the resulting *differential* output v_{out} (having a 0 V dc component) will be entirely a signal component. There are also other significant advantages of taking the output voltage differentially.

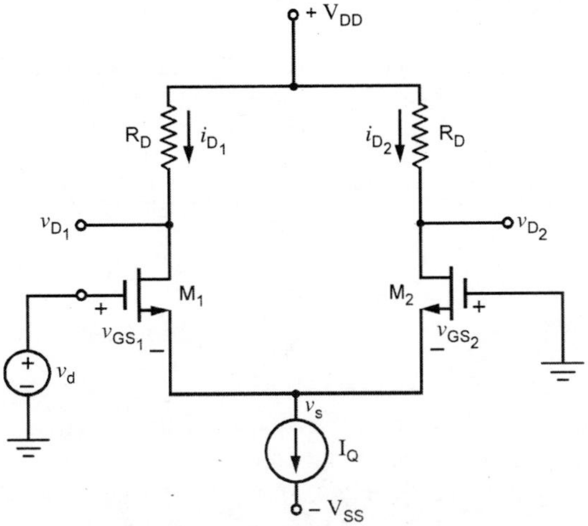

Fig. 1.23. ***MOSFET Differential Pair With a Differential Input Signal v_d Applied***

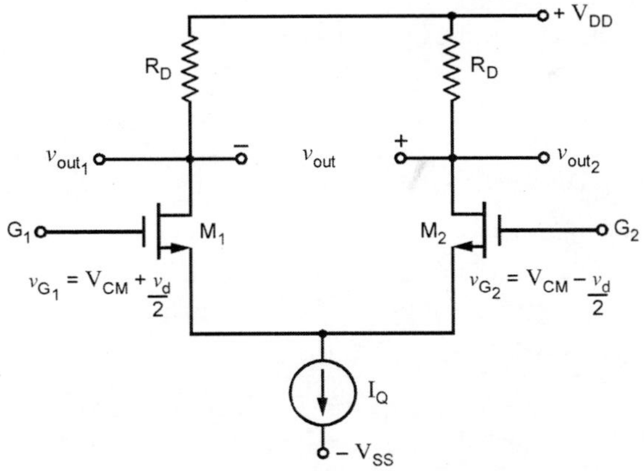

Fig. 1.24. *(a) The Circuit With a Common-Mode Voltage Applied To Set The DC Bias Voltage At The Gates and With v_d Applied in a Complementary (or Balanced) Manner*

The small-signal operation of differential amplifier is analyzed mainly for the determination of its voltage gain in response to the differential input signal v_d. Toward that end, the circuit with power supplies removed and V_{CM} eliminated is depicted in Fig. 1.24 (*b*).

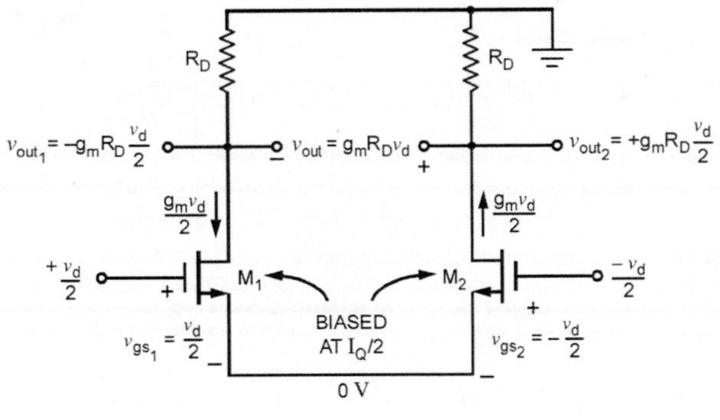

(b) Circuit For Small-Signal Analysis

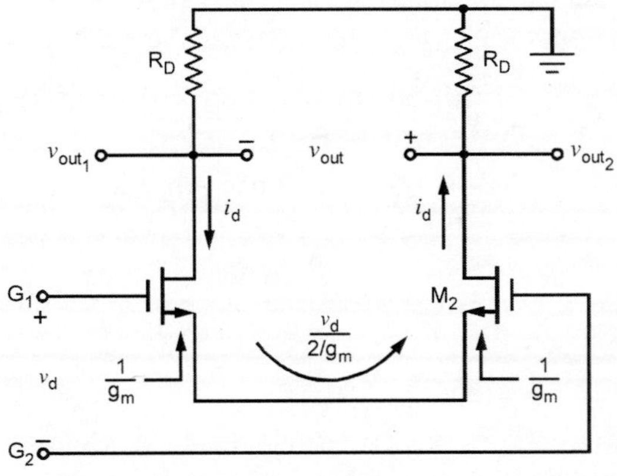

(c) An Alternative Way of Looking at The Small-Signal Operation of The Circuit

Fig. 1.24. ***Small-Signal Analysis of The MOS Differential Amplifier***

For the time being the effect of channel-length modulation and body effect are neglected. Finally note that each of the MOSFETs M_1 and M_2 is biased at a dc current of $\frac{1}{2} I_Q$ and is operating at an overdrive voltage V_{OV}.

The signal voltage at the joint source connection is seen to be zero (acting as a sort of virtual ground) because of symmetry of the circuit and because of the balanced way in which differential input signal v_d is applied. Thus, the MOSFET M_1 has a gate-to-source voltage signal $v_{gs_1} = \frac{v_d}{2}$ and the MOSFET M_2 has $v_{gs_2} = -\frac{v_d}{2}$. Assuming $\frac{v_d}{2} << V_{OV}$, the condition for small-signal approximation, the changes causing drain currents of MOSFETs M_1 and M_2 will be proportional to v_{gs_1} and v_{gs_2} respectively. Thus the MOSFET M_1 will have a drain current increment $\frac{1}{2} g_m v_d$ and the MOSFET M_2 will have a drain current decrement $\frac{1}{2} g_m v_d$, where g_m denotes the transconductance of the two devices, which is equal.

So the transconductance, $$g_m = \frac{2I_D}{V_{OV}} = \frac{2 \times \frac{I_Q}{2}}{V_{OV}} = \frac{I_Q}{V_{OV}} \quad ...(1.47)$$

The observation that a signal ground is established at the source terminals of the MOSFETs without making use of a large bypass capacitor, is very significant advantage of the differential pair configuration, as already mentioned in Art. 1.1.

The necessity of differential-pair operation is that it gives complementary current signals in the drains. The two current signals are simply passed through a pair of matched drain resistors, R_D, and drain voltage signals are obtained.

$$v_{out_1} = -g_m \frac{v_d}{2} . R_D \quad ...(1.48)$$

and $$v_{out_2} = +g_m \frac{v_d}{2} R_D \quad ...(1.49)$$

If the output is taken in a single-ended manner, the resulting gain is

$$\frac{v_{out_1}}{v_d} = -\frac{1}{2} g_m R_D \quad ...(1.50)$$

$$\frac{v_{out_2}}{v_d} = \frac{1}{2} g_m R_D \quad ...(1.51)$$

In case, the output is taken differentially, the gain is

$$A_d \equiv \frac{v_{out_2} - v_{out_1}}{v_d} = g_m R_D \quad ...(1.52)$$

On comparing Eqs. (1.50), (1.51) and (1.52), it is observed that the advantage of taking the output differentially is an increase in gain by a factor of 2. Although the differential outputs are preferred, but there are some applications in which single-ended output is required.

An alternative and useful way of looking the operation of differential pair in response to a differential input signal is shown in Fig. 1.24 (*c*). Here, advantage is taken of the fact that the resistance between the gate and source of a MOSFET, looking into the source, is $\frac{1}{g_m}$. Consequentlly, a total resistance, between G_1 and G_2, in the source circuit is $\frac{2}{g_m}$. It means that drain current i_d can be obtained simply by dividing different input signal v_d by $\frac{2}{g_m}$, as shown in the figure.

Effect of the MOSFET's r_0. Now we will consider the effect of the finite output resistance of each of MOSFETs M_1 and M_2 and realistic assumption of the bias current source I_Q having a finite output resistance R_{SS}. The resulting MOS differential-pair circuit for small-signal analysis, is depicted in Fig. 1.25 (*a*). Observe that the circuit is perfectly symmetrical, and consequently, the voltage signal at the common source connection will be zero. So, the signal current through R_{SS} will be zero and R_{SS} plays no role in determination of differential gain.

Because of the virtual ground on the common source connection, the circuit shown in Fig. 1.25 (*a*) is reduced to that shown in Fig. 1.25 (*b*). It consists of two identical common-source amplifiers, one supplied with $\frac{v_d}{2}$ and the other with $\frac{-v_d}{2}$. It means that only one of the two circuits needs to be analyzed. Thus, either of the two common-source circuits is known as the *differential half circuit*.

From the equivalent circuit shown in Fig. 1.25 (*b*)

$$v_{\text{out}_1} = -\frac{1}{2} g_m (R_D || r_o) v_d \quad ...(1.53)$$

$$v_{\text{out}_2} = \frac{1}{2} g_m (R_D || r_o) v_d \quad ...(1.54)$$

$$v_{\text{out}} = v_{\text{out}_2} - v_{\text{out}_1} = g_m (R_D || r_o) v_d \quad ...(1.55)$$

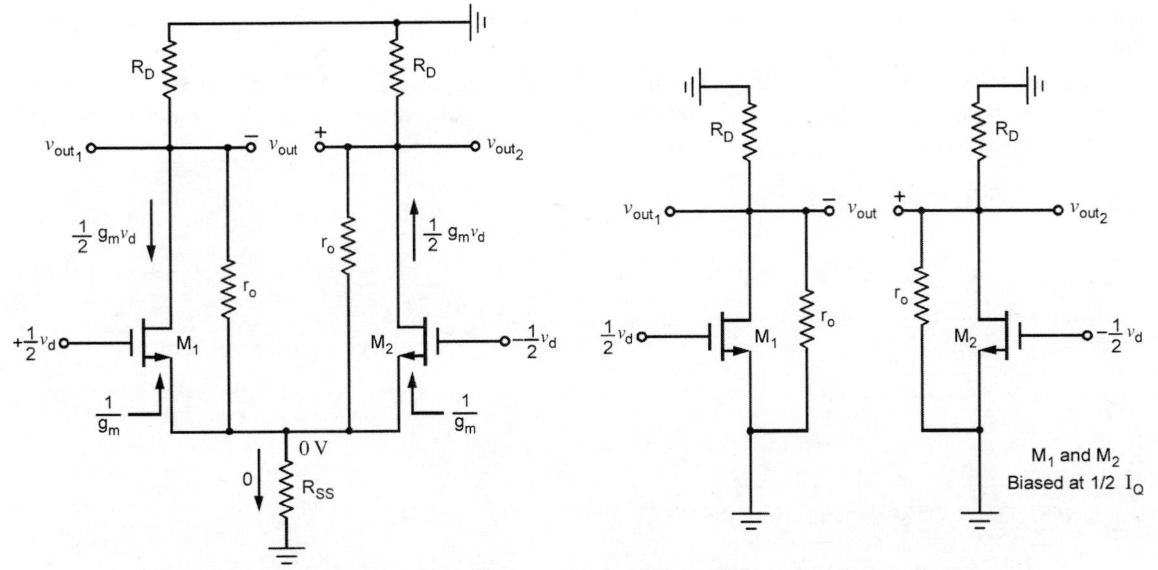

(a) MOS Differential Amplifier With r_0 and R_{SS} Taken Into Consideration

(b) Equivalent Circuit For Determination of Differential Gain

Fig. 1.25

1.3.2. Common-Mode Gain and Common-Mode Rejection Ratio (CMRR)

Now, we will consider the operation of the MOS differential pair with a common-mode input signal v_{cm}, as depicted in Fig. 1.26 (*a*). Here v_{cm} represents a disturbance, or interference signal that is coupled somehow to both input terminals. Although not indicated, the dc voltage of the input terminals must still be defined by a voltage V_{CM} as we have seen earlier.

Because of the symmetry of the circuit, it can be split up into two identical halves, as depicted in Fig. 1.26 (*b*). Each of the two halves, called the *common-mode half-circuit,* is a MOSFET biased at $\frac{I_Q}{2}$ and having a source degeneration resistance $2R_{SS}$. Neglecting the effect of r_o, the voltage gain of each of the two identical half circuits can be expressed as

$$\frac{v_{\text{out}_1}}{v_{cm}} = \frac{v_{\text{out}_2}}{v_{cm}} = \frac{-R_D}{\frac{1}{g_m} + 2R_{SS}} \quad ...(1.56)$$

Above equation can be approximated as

$$\frac{v_{out_1}}{v_{cm}} = \frac{v_{out_2}}{v_{cm}} = \frac{-R_D}{2R_{SS}} \quad ...(1.57) \text{ because usually } R_{SS} >> \frac{1}{g_m}$$

Case 1 : When the output is taken at single end,

$$|A_{cm}| = \frac{R_D}{2R_{SS}} \quad ...(1.58)$$

$$|A_d| = \frac{1}{2} g_m R_D \quad ...(1.59)$$

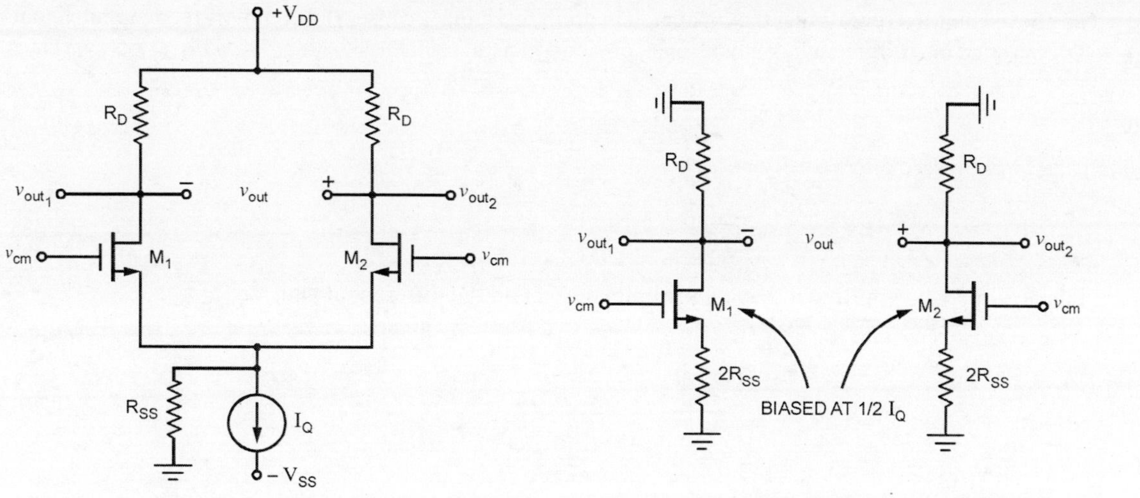

(a) MOS Differential Amplifier With a Common Mode Input Signal v_{cm}

(b) Equivalent Circuit For Determination of Common Mode Gain (r_o Being Ignored)

Fig. 1.26

Thus, the CMRR is given by

$$\text{CMMR} = \frac{|A_d|}{|A_{cm}|} = \frac{\frac{1}{2} g_m R_D}{\frac{\frac{1}{2} R_D}{R_{SS}}} = g_m R_{SS} \quad ...(1.60)$$

Case 2. When the output is taken differentially

$$A_{cm} = \frac{v_{out_2} - v_{out_1}}{v_{cm}} = 0 \quad ...(1.61)$$

$$A_d = \frac{v_{out_2} - v_{out_1}}{v_d} = g_m R_D \quad ...(1.62) \text{ Refer to Eq. (1.52)}$$

Thus, $$\text{CMRR} = \frac{|A_d|}{|A_{cm}|} = \frac{g_m R_D}{0} = \infty \quad ...(1.63)$$

Thus, even though RSS is finite, taking the output differentially results in an infinite CMRR. However, this is true only when the circuit is perfectly matched.

Effect of R_D Mismatch on CMRR. When the two drain resistances differ, say by ΔR_D, as usually, the CMRR will be finite even if the output is taken differentially. This can be shown as follows :

Consider the circuit given in Fig. 1.26 (*b*). Let the drain resistance for MOSFET M_1 be R_D and that for MOSFET M_2 be $(R_D + \Delta R_D)$. The drain signal voltages arising from v_{cm} can be given as

$$v_{out_1} \cong \frac{-R_D}{2R_{SS}} v_{cm} \quad ...(1.64)$$

$$v_{out_2} \cong -\frac{R_D + \Delta R_D}{2R_{SS}} v_{cm} \quad ...(1.65)$$

$$\text{Thus, } v_{out_2} - v_{out_1} = -\frac{\Delta R_D}{2R_{SS}} v_{cm} \quad ...(1.66)$$

The above equation reveals that the mismatch in drain resistances causes the common-mode input signal v_{cm} to be converted into a differential output signal, obviously an unavoidable situation. From Eq. (1.66), we have

$$A_{cm} = -\frac{\Delta R_D}{2R_{SS}} \quad ...(1.67)$$

which can be expressed alternatively as

$$A_{cm} = \frac{-R_D}{2R_{SS}} \left(\frac{\Delta R_D}{R_D} \right) \quad ...(1.68)$$

Since the mismatch in R_D will have a negligible effect on the differential gain, so

$$A_d \cong -g_m R_D \quad ...(1.69)$$

$$\text{and CMRR} = \left| \frac{A_d}{A_{cm}} \right| = \frac{2g_m R_{SS}}{\dfrac{\Delta R_D}{R_D}} \quad ...(1.70)$$

Effect of g_m Mismatch on CMRR. Consider the circuit shown in Fig. 1.27 for determination of effect of a mismatch between the values of the transconductance g_m of the two MOSFETs on the CMRR of the differential pair. Since the circuit is mismatched, we cannot use the common-mode half-circuit.

From the circuit shown in Fig. 1.27, we can write

$$i_{d_1} = g_{m_1} v_{gs_1} \quad ...(1.71)$$

$$i_{d_2} = g_{m_2} v_{gs_2} \quad ...(1.72)$$

Dividing Eq. (1.71) by Eq. (1.72), we have

$$\frac{i_{d_1}}{i_{d_2}} = \frac{g_{m_1}}{g_{m_2}} \cdot \frac{v_{gs_1}}{v_{gs_2}} = \frac{g_{m_1}}{g_{m_2}} \quad ...(1.73) \quad \because v_{gs_1} = v_{gs_2}$$

The two drain currents add together in R_{SS} to give

$$v_s = (i_{d_1} + i_{d_2}) R_{SS}$$

$$\text{or } i_{d_1} + i_{d_2} = \frac{v_s}{R_{SS}} \quad ...(1.74)$$

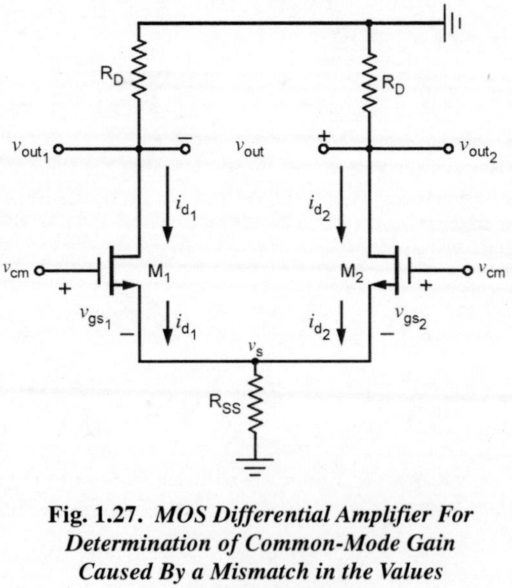

Fig. 1.27. ***MOS Differential Amplifier For Determination of Common-Mode Gain Caused By a Mismatch in the Values of g_m of M_1 and M_2***

Since MOSFETs M_1 and M_2 are in effect operating as a source follower with a source resistance R_{SS} that is typically larger than $\frac{1}{g_m}$. So,

$$v_s \cong v_{cm} \quad ...(1.75)$$

and therefore, Eq. (1.74) can be written as

$$i_{d_1} + i_{d_2} \cong \frac{v_{cm}}{R_{SS}} \quad ...(1.76)$$

Combining Eqs. (1.73) and (1.76), we have

$$i_{d_1} = \frac{g_{m_1} v_{cm}}{(g_{m_1} + g_{m_2}) R_{SS}} \quad ...(1.77)$$

$$\text{and } i_{d_2} = \frac{g_{m_2} v_{cm}}{(g_{m_1} + g_{m_2}) R_{SS}} \quad ...(1.78)$$

If the mismatch between g_{m_1} and g_{m_2} is very small, say, Δg_m (*i.e.* $g_{m_1} - g_{m_2} = \Delta g_m$), it can be assumed that $g_{m_1} + g_{m_2} \cong 2g_m$ where g_m is the nominal value of g_{m_1} and g_{m_2}. Thus

$$i_{d_1} = \frac{g_{m_1} v_{cm}}{2g_m R_{SS}} \quad ...(1.79)$$

and

$$i_{d_2} = \frac{g_{m_2} v_{cm}}{2g_m R_{SS}} \quad ...(1.80)$$

The differential output voltage is now given by

$$v_{out_2} - v_{out_1} = -i_{d_2} R_D + i_{d_1} R_D = R_D(i_{d_1} - i_{d_2}) = \frac{(g_{m_1} - g_{m_2}) v_{cm} R_D}{2g_m R_{SS}} = \frac{\Delta g_m v_{cm} R_D}{2g_m R_{SS}}$$

From which the common-mode gain can be obtained as

$$A_{cm} = \frac{R_D}{2R_{SS}} \times \frac{\Delta g_m}{g_m} \quad ...(1.81)$$

Since the g_m mismatch will have a negligible effect on differential gain

$$A_d = -g_m R_D \quad ...(1.82)$$

and the resulting CMRR will be

$$\text{CMRR} \cong \frac{A_d}{A_{cm}} = \frac{2g_m R_{SS}}{\dfrac{\Delta g_m}{g_m}} \quad ...(1.83)$$

The similarity between Eqs. (1.70) and (1.83) may be noted.

Example 1.6. A MOS differential pair is operated at a bias current of 1 mA. The MOSFET has g_m = 5 mA/V; R_D = 4 kΩ and R_{SS} = 20 kΩ.

Find the differential gain, the common mode gain, and CMRR in dB when

(*a*) the output is taken single-endedly and the circuit is perfectly matched.

(*b*) the output is taken differentially and the circuit is perfectly matched.

(*c*) the output is taken differentially but the drain resistances have a mismatch of 1%.

(*d*) the output is taken differentially but there is mismatch of 1% in g_m.

Solution : Transconductance, g_m = 5 mA/V

(*a*) Differential gain, $A_d = \frac{1}{2} g_m R_D = \frac{1}{2} \times 5 \times 4 = 10$ **Ans.**

Common-mode gain, $A_{cm} = \frac{R_D}{2R_{SS}} = \frac{4}{2 \times 20} = 0.1$ **Ans.**

$\text{CMMR} = \frac{A_d}{A_{cm}} = \frac{10}{0.1} = 100$ or $20 \log_{10} 100$ dB = 40 dB **Ans.**

(*b*) Differential gain, $A_d = g_m R_D = 5 \times 4 = 20$ **Ans.**

Commom-mode gain, $A_{cm} = 0$ **Ans.**

$$\text{CMMR} = \frac{A_d}{A_{cm}} = \frac{20}{0} = \infty \text{ or } \infty \text{ dB} \quad \textbf{Ans.}$$

(*c*) Differential gain, $A_d = g_m R_D = 5 \times 4 = 20$ **Ans.**

$$\text{Common-mode gain, } A_{cm} = \frac{R_D}{2R_{SS}} \times \frac{\Delta R_D}{R_D} = \frac{4}{2 \times 20} \times \frac{1}{100} = 0.001 \quad \textbf{Ans.}$$

$$\text{CMMR} = \frac{A_d}{A_{cm}} = \frac{20}{0.001} = 20{,}000 \text{ or } 86 \text{ dB} \quad \textbf{Ans.}$$

(*d*) Differential gain, $A_d = g_m R_D = 5 \times 4 = 20$ **Ans.**

$$\text{Common-mode gain, } A_{cm} = \frac{R_D}{2R_{SS}} \times \frac{\Delta g_m}{g_m} = \frac{4}{2 \times 20} \times 0.01 = 0.001$$

$$\text{CMMR} = \frac{A_d}{A_{cm}} = \frac{20}{0.001} = 20{,}000 \text{ or } 86 \text{ dB} \quad \textbf{Ans.}$$

1.4. CONSTANT CURRENT BIAS

In a dual-input, balanced-output differential amplifier, we have used the combination of R_E and V_{EE} to produce emitter dc bias current. Alternatively, constant current bias circuit can also be employed to set up the emitter dc bias current. Actually, the constant current bias circuit is considered better as it provides current stabilization and therefore, assures a stable operating point for the differential amplifier.

Dual-input, balanced-output differential amplifier with a resistive constant current bias is depicted in Fig. 1.28 (*a*). By comparing the two circuits (circuits shown in Figs. 1.2 and 1.28 (*a*)) it is noticed that the emitter resistor R_E has been replaced by a constant current transistor Q_3 circuit. The dc collector current in transistor Q_3 is established by resistors R_1, R_2 and R_3 and may be determined as follows :

Voltage at the base of transistor Q_3,

$$V_{B_3} = -V_{EE} \times \frac{R_2}{R_1 + R_2} \quad \text{...(1.84)}$$

By voltage divider rule, neglecting base loading effect

Voltage at the emitter of transistor Q_3

$$V_{E_3} = V_{B_3} - V_{BE_3} = \frac{-R_2 V_{EE}}{R_1 + R_2} - V_{BE_3} \quad \text{...(1.85)}$$

Collector current,

$$I_{C_3} \cong I_{E_3} = \frac{V_{E_3} - (-V_{EE})}{R_3} = \frac{V_{EE} + V_{E_3}}{R_3}$$

Substituting the value of V_{E_3} from Eq. (1.85) in above equation, we have

$$I_{C_3} = \frac{V_{EE} - \left[\frac{R_2 V_{EE}}{R_1 + R_2} + V_{BE_3}\right]}{R_3}$$

Because the two halves of the differential amplifier are symmetrical, each has half of the current I_{C_3}. Thus,

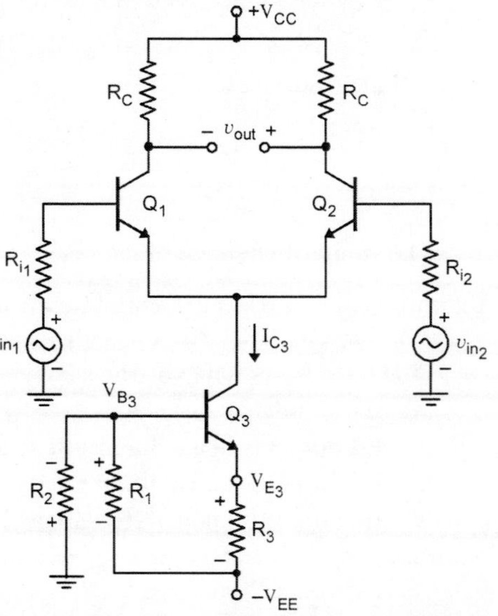

Fig. 1.28 (*a*). *Dual-Input Balanced-Output Differential Amplifier With Constant Current Bias*

$$I_{E_1} = I_{E_2} \cong \frac{I_{C_3}}{2} = \frac{V_{EE} - \left[\frac{R_2 V_{EE}}{R_1 + R_2} + V_{BE_3}\right]}{2R_3} \quad \text{...(1.86)}$$

The collector current of transistor $Q_3 (I_{C_3})$ is fixed and must be invariant because there is no input signal into either the base or emitter of transistor Q_3. Thus the transistor Q_3 is a source of constant emitter current for transistors Q_1 and Q_2 of the differential amplifier [Fig. 1.28 *(a)*]

The constant current bias, in addition to supply of constant emitter current, also provides a very high resistance. This is because the ac equivalent of the dc current source is ideally an open circuit.

The thermal stability of the constant current transistor Q_3 can be improved if resistor R_1 is replaced by diodes D_1 and D_2 as depicted in Fig. 1.28 *(b)*. The base of transistor Q_3 is biased with the voltage divider consisting of components resistor R_2, and diodes D_1 and D_2. Diodes D_1 and D_2 assist in holding the emitter current I_{E_3} constant even though there is large variation in temperature.

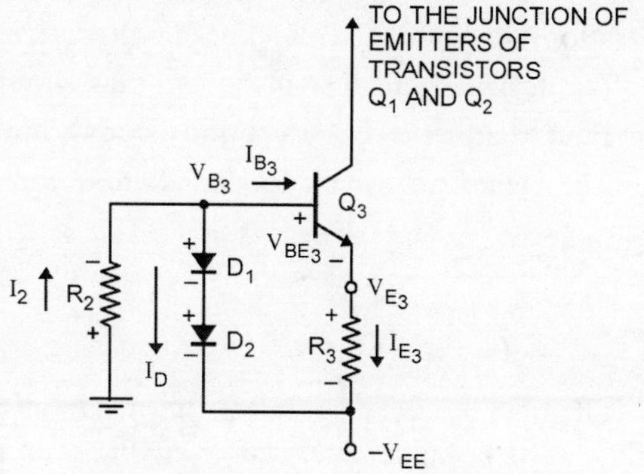

Fig. 1.28 *(b)*. ***Constant Current Bias With Diode Compensation For Variation in V_{BE}***

From the circuit shown in Fig. 1.28 *(b)*, it is seen that current I_2 flows to the node at the base of transistor Q_3 and then divides into diode current I_D and base current I_{B_3} of transistor Q_3. In case the temperature of transistor Q_3 goes up, its base-emitter voltage V_{BE_3} reduces (in silicon transistors, V_{BE} reduces by 2 mV/°C, and in germanium transistors, V_{BE} reduces by 1.6 mV/°C). Because of reduction in V_{BE_3}, the voltage drop across R_3 tends to rise and therefore, emitter current I_{E_3}. On the other hand, the voltage drops across diodes D_1 and D_2 also reduce, causing a greater portion of current I_2 to contribute to diode current I_D. This causes the base current I_3 to fall, which prevents any significant increase in emitter current I_{E_3}.

The transistor Q_3 emitter current I_{E_3} may be determined as follows :

Voltage at the base of transistor Q_3,

$$V_{B_3} = -V_{EE} + 2V_D$$

Assuming equal voltage drop across the diodes and denoting it by V_D, voltage at the emitter of transistor Q_3,

$$V_{E_3} = V_{B_3} - V_{BE_3} = -V_{EE} + 2V_D - V_{BE_3}$$

and emitter current through transistor Q_3,

$$I_{E_3} = \frac{V_{E_3} - (-V_{EE})}{R_3} = \frac{V_{E_3} + V_{EE}}{R_3} = \frac{-V_{EE} + 2V_D - V_{BE_3} + V_{EE}}{R_3} = \frac{2V_D - V_{BE_3}}{R_3}$$

Assuming same characteristics for transistor Q_3 and diodes D_1 and D_2 *i.e.,* $V_D = V_{BE_3}$, we have

$$I_{E_3} = \frac{V_D}{R_3} \quad \text{...(1.87)}$$

Thus the emitter current through transistor Q_3 depends upon V_D and resistor R_3. The voltage drop across the diodes depends upon the diode current I_D and diode current is a part of current I_2 which depends upon the value of resistance R_2. Thus emitter current I_{E_3} can be changed by changing either R_2 or R_3.

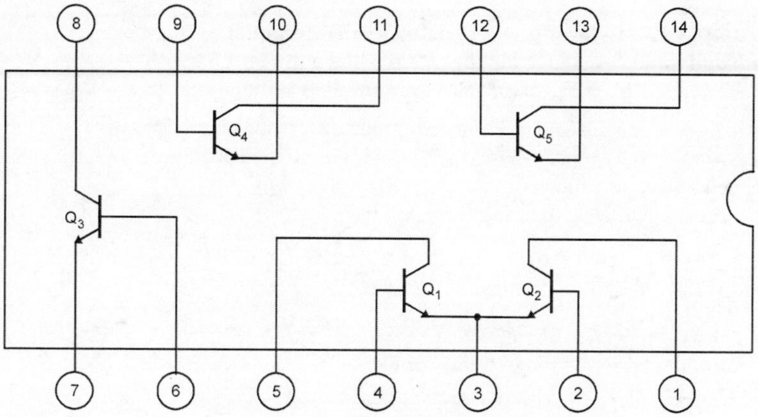

Fig. 1.29. ***Functional Diagram of CA3086***

Performance can be made better, if we use a transistor array like CA3086 as a constant current source [Fig. 1.29]. Here an isolated transistor is employed, and required diodes are formed by emplying transistors connected for diode operation. No doubt, discrete diodes will work as well. Figure 1.29 shows the functional diagram of CA3086.

For designing a constant current bias circuit depicted in Fig. 1.28 (*b*), we may proceed as follows :

First of all let us choose a transistor Q_3 emitter current I_3.

The values of resistor R_3 and R_2 now can be determined from Eqs. (1.87) and (1.88) respectively by assuming $V_{D_1} = V_{D_2} = 0.7$ V and $I_2 = I_{E_3}$

$$R_2 = \frac{V_{EE} - 1.4\text{ V}}{I_{E_3}} \qquad ...(1.88)$$

Often we use a zener diode in place of diodes as depicted in Fig. 1.30, because zeners are available over a wide range of voltages and can have a matching temperature coefficient of voltage to those of transistors.

From circuit diagram given in Fig. 1.30.

$$V_{B_3} = -V_{EE} + V_Z$$

and voltage at the emitter of transistor Q_3,

$$V_{E_3} = V_{B_3} - V_{BE_3} = -V_{EE} + V_Z - V_{BE_3}$$

and so the current through resistor R_3,

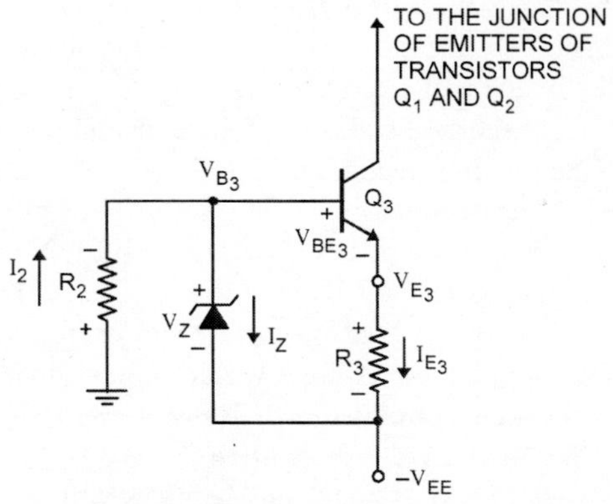

Fig. 1.30. ***Constant Current Bias With Zener Compensation For Variation in V_{BE}***

$$I_3 = \frac{V_{E_3} - (-V_{EE})}{R_3} = \frac{V_{E_3} + V_{EE}}{R_3} = \frac{-V_{EE} + V_Z - V_{BE_3} + V_{EE}}{R_3} = \frac{V_Z - V_{BE_3}}{R_3} \qquad ...(1.89)$$

and $$R_2 = \frac{V_{EE} - V_Z}{I_2} \qquad ...(1.90)$$

The value of R_2 should be selected so as to give $I_2 = 1.2\, I_{ZT}$ where I_{ZT} is the minimum current required to cause the zener diode to conduct in the reverse region (*i.e.,* to block the rated voltage V_Z) and its value is specified on the data sheet of a zener diode.

The zener diode is quite useful for maintaining a constant base voltage and therefore, the constant emitter current in a constant current bias circuit.

Example 1.7. Determine (*i*) Q-point, (*ii*) Differential voltage gain and input resistance for the circuit shown in Fig. 1.31 with parameters $V_{D_1} = V_{D_2} = 0.7$ V, $V_{BE} = 0.7$ V, $\beta = 100$.

Solution : For a dual-input, balanced-output differential amplifier with constant current bias, shown in Fig. 1.31, the required quantities are determined as follows :

From circuit shown in Fig. 1.31, we have

$$V_{B_3} = -V_{EE} + V_{D_1} + V_{D_2} = -15 + 0.7 + 0.7 = -13.6\text{ V}$$

Voltage at emitter terminal of transistor Q_3,

$$V_{E_3} = V_{B_3} - V_{BE_3} = -13.6 - 0.7 = -14.3\text{ V}$$

Emitter current through transistor Q_3

$$I_{E_3} = \frac{V_{E_3} - (-V_{EE})}{R_3} = \frac{V_{E_3} + V_{EE}}{R_3} = \frac{-14.3 + 15}{180} = 4\text{ mA}$$

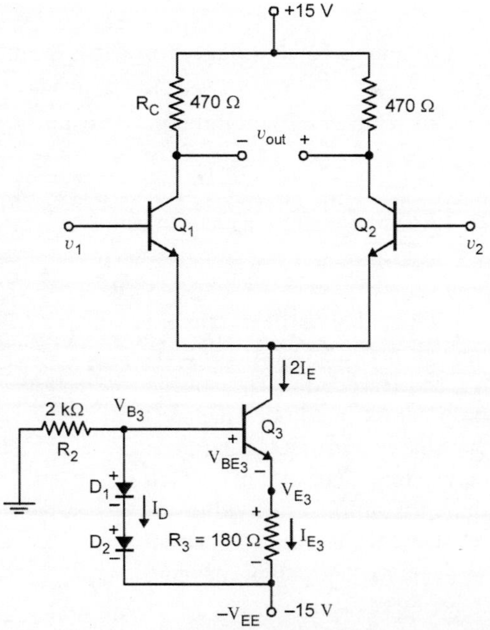

Fig. 1.31

(*i*) Quiescent current, $I_{CQ} = I_E = \frac{I_{E3}}{2} = \frac{4 \text{ mA}}{2} = 2 \text{ mA}$ **Ans.**

Quiescent collector-emitter voltage, $V_{CEQ} = V_{CC} + V_{BE} - I_{CQ}R_C = 15 + 0.7 - 2 \times 10^{-3} \times 470 = 14.76 \text{ V}$ **Ans.**

AC emitter resistance, $r'_e = \frac{26 \text{ mV}}{I_E \text{ in mA}} = \frac{26 \text{ mV}}{2 \text{ mA}} = 13\ \Omega$

(*ii*) Differential voltage gain, $A_d = \frac{R_C}{r'_e} = \frac{470}{13} = 36.15$ **Ans.**

(*iii*) Input resistance, $R_{in_1} = R_{in_2} = 2\beta r'_e = 2 \times 100 \times 13 = 2.6 \text{ k}\Omega$ **Ans.**

Example 1.8. For the circuit shown in Fig. 1.32 (*a*), $\beta_{ac} = \beta_{dc} = 100$, $V_{BE} = V_{D_1} = 0.715$ V, $V_Z = 6.2$ V, $I_{ZT} = 41$ mA. Calculate :

(*a*) Voltage gain of differential amplifier,
(*b*) Input resistance,
(*c*) Operating point values I_{CQ} and V_{CEQ}.

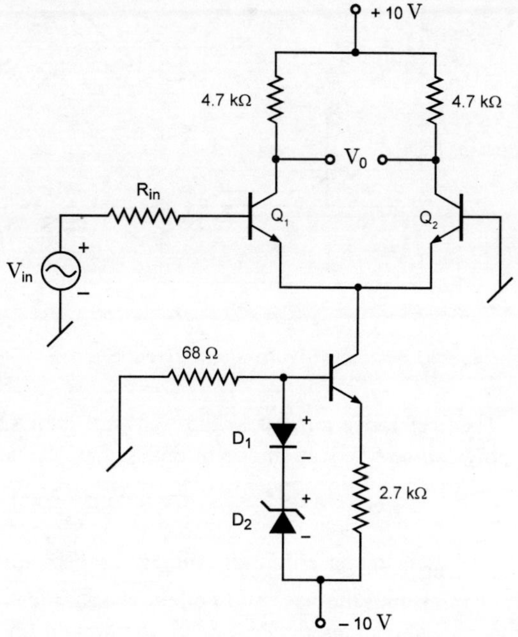

(*a*)

Solution : The given configuration is a single-input, balanced output differential amplifier with a constant current bias. For determination of operating point (Q-point), we will analyse constant current bias first.

Voltage at the base of transistor Q_3,

$$V_{B_3} = -V_{EE} + V_Z + V_{D_1} = -10 + 6.2 + 0.715 = -3.085 \text{ V}$$

Voltage at the emitter of transistor Q_3,

$$V_{E_3} = V_{B_3} - V_{BE_3} = -3.085 - 0.715 = -3.8 \text{ V}$$

Emitter current through transistor Q_3,

$$I_{E_3} = \frac{V_{E_3} - (-V_{EE})}{R_3} = \frac{-3.8 - (-10)}{2.7 \text{ k}\Omega} = 2.296 \text{ mA}$$

Since, the differential amplifier is symmetrical, the current I_3 splits into two equal parts and therefore,

$$I_{CQ_1} = I_{CQ_2} = I_{CQ} = \frac{I_{E_3}}{2} = \frac{2.296}{2} = 1.148 \text{ mA}$$

Quiescent collector-emitter voltage,

$$V_{CEQ} = V_{CC} + V_{BE} - I_{CQ} R_C$$
$$= 10 + 0.715 - 1.148 \times 10^{-3} \times 4.7 \times 10^3 = 5.32 \text{ V}$$

(*c*) Thus $Q_{point} = (I_{CQ}, V_{CEQ}) = (1.148 \text{ mA}, 5.32 \text{ V})$ **Ans.**

Dynamic emitter resistance,

$$r'_e = \frac{26 \text{ mV}}{1.148 \text{ mA}} = 22.65\ \Omega$$

(*a*) Voltage gain, $A_d = \frac{R_C}{r'_e} = \frac{4.7 \times 1{,}000}{22.65} = 207.5$ **Ans.**

(*b*) Differential input resistance,

$R_{in} = 2\beta_{ac} r'_e = 2 \times 100 \times 22.65 = 4.53 \text{ k}\Omega$ **Ans.**

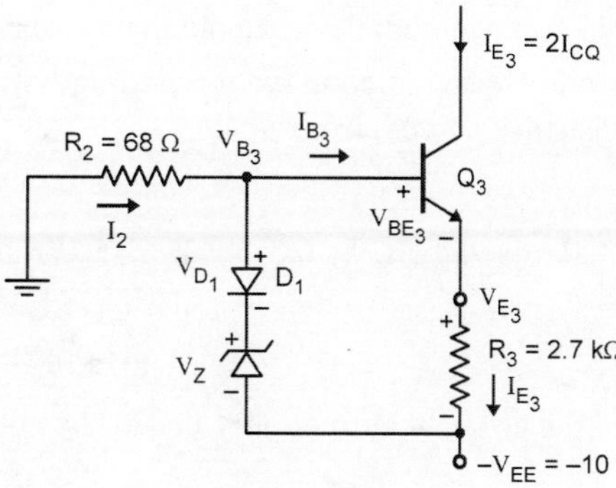

Fig. 1.32 (*b*)

1.5. CURRENT MIRROR

The **two-transistor current source**, also called a **current mirror,** is the basic building block in the design of integrated circuit current source. Figure 1.33 (*a*) depicts the basic current-source circuit, which consists of two matched or identical transistors Q_1 and Q_2, operating at the same temperature, with their base terminals and emitter terminals connected together. The base-emitter voltage, V_{BE} is, therefore, the same in the two transistors. Transistor Q_1 is connected as a diode; consequently, when the supply voltage is applied, the base-emitter junction of transistor Q_1 is forward biased and a reference current I_{REF} is established. Although there is a specific relationship between I_{REF} and V_{BE_1}, V_{BE_1} can be thought of as the result of I_{REF}. Once V_{BE_1} is established, it is applied to the base-emitter junction of transistor Q_2. The applied V_{BE_2} turns Q_2 *on* and generates the output current, which is employed to bias a transistor or transistor circuit.

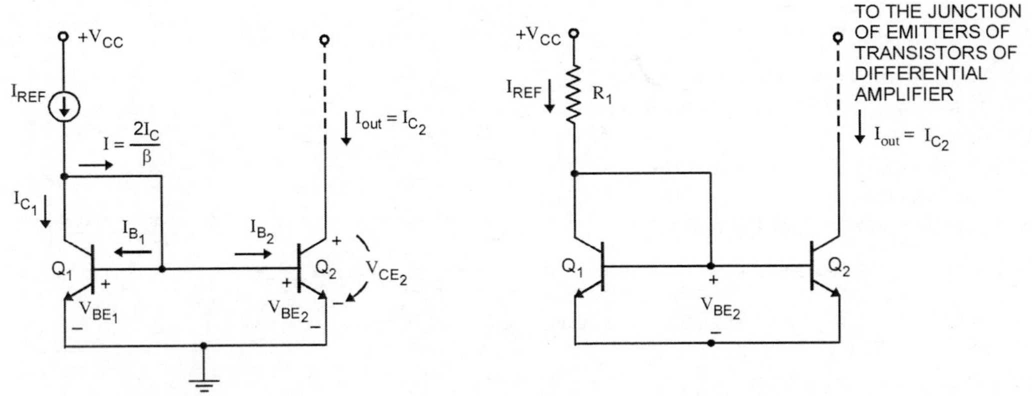

(a) Basic Two-Transistor Current Source *(b) Two Transistor Current Source With Resistor R_1*

Fig. 1.33

The reference current in the two-transistor source can be established by connecting a resistor to the supply source $+V_{CC}$ as shown in Fig. 1.33 (*b*). The reference current is then

$$I_{REF} = \frac{V_{CC} - V_{BE}}{R_1} \qquad \text{...(1.91)}$$

where V_{BE} is the base-emitter voltage corresponding to collector current (I_{REF}).

Connecting the base and collector terminals of a BJT effectively provides a two-terminal device with ampere-volt characteristics that are identical to the i_C versus v_{BE} characteristic of the BJT.

Current Relationship. Fig. 1.33 (*a*) depicts the current in the two transistor current source. Since V_{BE} is the same in both devices, and the transistors are identical, then $I_{B_1} = I_{B_2}$ and $I_{C_1} = I_{C_2}$. Transistor Q_2 is assumed to be biased in the forward-active region. Applying Kirchhoff's current law to the collector node of transistor Q_1, we have

$$I_{REF} = I_{C_1} + I = I_{C_1} + I_{B_1} + I_{B_2}$$

Replacing I_{C_1} by I_{C_2} ($\because I_{C_1} = I_{C_2}$) and noting that $I_{B_2} = \frac{I_{C_2}}{\beta}$, above equation becomes

$$I_{REF} = I_{C_2} + \frac{2I_{C_2}}{\beta} = \frac{\beta + 2}{\beta} I_{C_2} \cong I_{C_2} \qquad \text{...(1.92)}$$

i.e., the constant current provided at the collector of transistor Q_2 mirrors that of transistor Q_1

Since $$I_{REF} = \frac{V_{CC} - V_{BE}}{R_1}$$

The current I_{REF} set by V_{CC} and R_1 is mirrored in the current into the collector of transistor Q_2. Transistor Q_1 is referred to as a diode-connected transistor because its base and collector are shorted together. The operation of the current mirror may be described as follows:

1. Q_1 acts like a diode and a current is established through it.
2. Q_1 will develop a voltage drop V_{BE}, in response to the current.

3. The base-emitter of Q_2 is in parallel with Q_1 ($V_{BE_1} = V_{BE_2} = V_{BE}$).
4. Q_2's collector current will be established in response to its V_{BE}.
5. Since the two transistors are matched, Q_2's collector current will be approximately equal to Q_1's collector current. Q_2's collector current is said to "mirror" the current through Q_1.

Q_1 is often called a *compensating diode* because it automatically compensates for variations in temperature. Matching of diode curve means matching at all temperatures as well as voltages. When the temperature increases, the voltage across the emitter diode falls approximately 2 mV per degree. Since the voltage across the compensating diode also decreases by 2 mV per degree, the collector current is little affected with temperature increase.

1.5.1. Modified Current Mirror

Equation (1.92) reveals that the collector current in the transistor Q_2 differs from the reference current I_{REF} by a factor $\left(\frac{\beta + 2}{\beta}\right)$. For $\beta >> 2$, $I_{C_2} = I_{REF}$, but for low-current transistors I_{C_2} can differ significantly from I_{REF}. However, this error can be reduced by introducing another transistor as an emitter follower. The resulting circuit, as shown in Fig. 1.34, is a modified current mirror circuit or **three-transistor current source.** We again assume that all transistors are identical; therefore, since the V_{BE} is the same for transistors Q_1 and Q_2, $I_{B_1} = I_{B_2}$ and $I_{C_1} = I_{C_2}$.

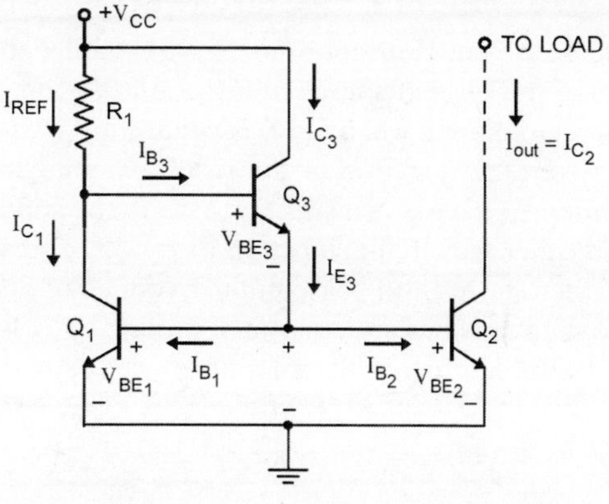

Fig. 1.34. ***Basic Three-Transistor Current Source***

Transistor Q_3 supplies the base currents to Q_1 and Q_2, so these base currents should be less dependent on the reference current. Also, since the current in transistor Q_3 is substantially smaller than that in either Q_1 or Q_2, the current gain is expected to be less than those of transistors Q_1 and Q_2. We may define the current gain of transistors Q_1 and Q_2 as $\beta_1 = \beta_2 = \beta$, and current gain of the transistor Q_3 as β_3.

Applying Kirchhoff 's current law to the collector node of transistor Q_1, we have

$$I_{REF} = I_{C_1} + I_{B_3} \quad \text{...(1.93)}$$

$$\text{Since} \quad I_{B_1} + I_{B_2} = 2I_{B_2} = I_{E_3} \quad \text{...(1.94)}$$

$$\text{and} \quad I_{E_3} = (1 + \beta_3)I_{B_3} \quad \text{...(1.95)}$$

Combining above Eqs. (1.93), (1.94) and (1.95), we have

$$I_{REF} = I_{C_1} + \frac{I_{E_3}}{(1 + \beta_3)} = I_{C_1} + \frac{2I_{B_2}}{(1 + \beta_3)} \quad \text{...(1.96)}$$

Replacing I_{C_1} by I_{C_2} and noting that $I_{B_2} = \frac{I_{C_2}}{\beta}$, the above Eq. (1.96) may be rewritten as

$$I_{REF} = I_{C_2} + \frac{2I_{C_2}}{\beta(1 + \beta_3)} = I_{C_2}\left[1 + \frac{2}{\beta(1 + \beta_3)}\right] \quad \text{...(1.97)}$$

The output or bias current is then

$$I_{C_2} = I_{out} = \frac{I_{REF}}{1 + \frac{2}{\beta(1 + \beta_3)}} \quad \text{...(1.98)}$$

The reference current is given by

$$I_{REF} = \frac{V_{CC} - V_{BE_3} - V_{BE}}{R_1} \simeq \frac{V_{CC} - 2V_{BE}}{R_1} \quad \text{...(1.99)}$$

As a first approximation base-emitter voltages of transistors Q_1 and Q_3 are assumed to be equal (*i.e.* $V_{BE_3} = V_{BE_1}$) as indicated in Eq. (1.99).

A comparison of Eq. (1.98) for the three-transistor current source and Eq. (1.92) for the two-transistor current source reveals that the approximation of $I_{out} \cong I_{REF}$ is better for three-transistor circuit.

Note : For a two-transistor current source (Fig. 1.33) and for a three-transistor current source (Fig. 1.34), a large valued resistor R_1 is required for a small biasing current, which is not feasible because it becomes costly in terms of chip area. Usually, these current sources are employed for producing a current of about 0.3 mA–0.5 mA.

1.5.2. Multiple Current Source

In the previous current sources, we established a reference current I_{REF} and one load current I_{out}. In the two-transistor current source in Fig. 1.33 (*a*), the base-emitter junction of the diode-connected transistor Q_1 is forward biased, when the bias voltage $+V_{CC}$ is applied. Once V_{BE} is established, the voltage is applied to the base-emitter junction of transistor Q_2, which turns Q_2 *on* and generates the output current I_{out}. The base-emitter voltage of transistor Q_1 can also be applied to additional transistors, to generate multiple load currents. Consider the circuit shown in Fig. 1.35. Transistor Q_R; which is the reference transistor, is connected as a diode. The resulting base-emitter voltage of reference transistor Q_R, established by reference current I_{REF}, is applied to N output transistors, producing N load currents. The relationship between each load current and the reference current, assuming all transistors are matched, is

$$I_{out_1} = I_{out_2} = \ldots I_{out_N} = \frac{I_{REF}}{1 + \dfrac{1+N}{\beta}} \qquad \ldots(1.100)$$

The collectors of multiple output transistors can be connected together, changing the load current versus reference current relationship. For an example, the circuit depicted in Fig. 1.36 has three output transistors with common collectors and a load current I_{out}. Assuming that transistors Q_R, Q_1, Q_2 and Q_3 are all matched if the current gain β is very large, the base currents can be neglected and $I_1 = I_2 = I_3 = I_{REF}$, and the output current $I_{out} = 3I_{REF}$. This process is not recommended for discrete devices, since a mismatch between devices will usually cause one device to carry more current than the other devices. Connecting transistors in parallel increases the effective base-emitter area of the device. In actual IC fabrication, the base-emitter area would be doubled or tripled to provide an output current twice or three times the value of I_{REF}.

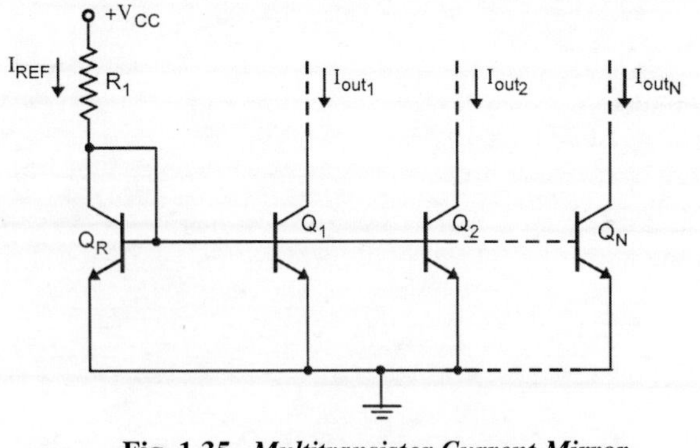

Fig. 1.35. *Multitransistor Current Mirror*

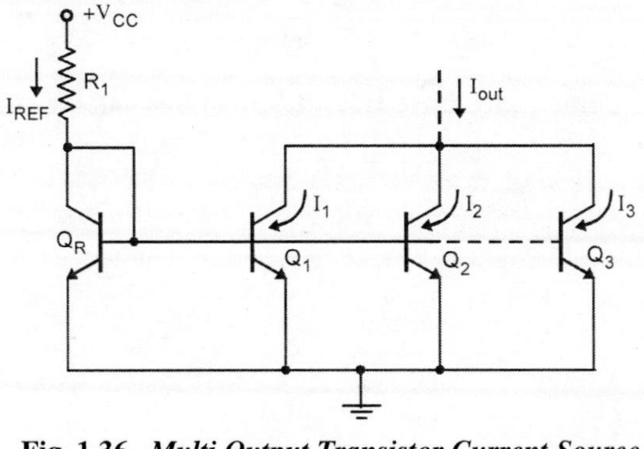

Fig. 1.36. *Multi Output Transistor Current Source*

1.5.3. Wilson Current Source

This is another configuration of a three-transistor current source, known as a **Wilson current source.** The circuit is shown in Fig. 1.37. This circuit has a large output resistance. We again assume that all the transistors are identical, with $I_{B_1} = I_{B_2} = I_B$; $I_{C_1} = I_{C_2}$ and $V_{BE_1} = V_{BE_2}$.

Applying KCL at the emitter of transistor Q_3, we have

$$I_{E_3} = I_{C_2} + 2I_B = I_{C_2}\left(1+\frac{2}{\beta}\right) \quad ...(1.101)$$

Also
$$I_{E_3} = I_{C_3} + I_{B_3} = I_{C_3}\left(1+\frac{1}{\beta}\right) \quad ...(1.102)$$

Comparing Eqs. (1.101) and (1.102), we have

$$I_{C_3}\left(1+\frac{1}{\beta}\right) = I_{C_2}\left(1+\frac{2}{\beta}\right)$$

$$\text{or } I_{C_3} = I_{out} = \left(\frac{\beta+2}{\beta+1}\right)I_{C_2} \quad ...(1.103)$$

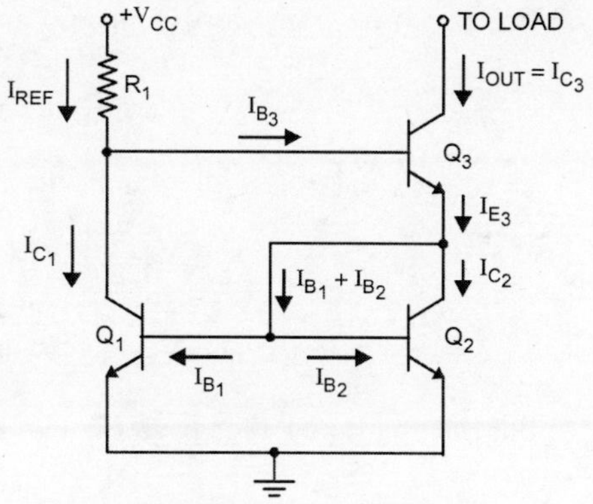

Fig. 1.37. ***Wilson Current Source***

Since $I_{C_1} = I_{C_2}$

$$I_{out} = \left(\frac{\beta+2}{\beta+1}\right)I_{C_1} \quad ...(1.104)$$

Applying KCL at the collector of transistor Q_1

$$I_{REF} = I_{C_1} + I_{B_3} = \left(\frac{\beta+1}{\beta+2}\right)I_{out} + \frac{I_{out}}{\beta} = \left(\frac{\beta^2+2\beta+2}{\beta^2+2\beta}\right)I_{out}$$

$$\text{or } I_{out} = \frac{\beta^2+2\beta}{\beta^2+2\beta+2}I_{REF} \quad ...(1.105)$$

where
$$I_{REF} = \frac{V_{CC} - 2V_{BE}}{R_1} \quad ...(1.106)$$

The difference
$$I_{out} - I_{REF} = \frac{2}{\beta^2+2\beta+2}I_{REF} \quad ...(1.107)$$

is extremely small error for modest values of β.

In the Wilson current source, the output resistance looking into the collector of Q_3 is $R_{out} \cong \frac{\beta r_{03}}{2}$, which is approximately a factor β/2 larger than that of either the two-transistor source or the basic three-transistor source. This means that, in the Wilson current source the change in bias current I_{out} with a change in output collector voltage is much smaller.

1.5.4. Widlar Current Source

The basic current mirror shown in Fig. 1.33 (*b*) has a limitation that for low value current source the resistance R_1 required is sufficiently large and cannot be fabricated economically in IC circuit. For a two-transistor current shown in Fig. 1.33 (*b*), if a load current of $I_0 = 10$ μA is required, then for $V_{CC} = 10$ V, the required resistance value is

$$R_1 = \frac{V_{CC} - V_{BE}}{I_{REF}} \cong \frac{10-0.7}{10 \times 10^{-6}} = 930 \text{ k}\Omega.$$

The solution to above problem is a circuit shown in Fig. 1.38, called a **Widlar current source.** This circuit is particularly suitable for low value of current. The circuit differs from the basic current mirror [Fig. 1.33(*b*)] only in the resistance R_E that is included in the emitter lead of transistor Q_2. A voltage difference is caused across emitter resistor R_{E_1} so that the V_{BE_2} is lesser than V_{BE_1}. A smaller V_{BE} produces a smaller collector current. It means that the load current I_{out} is lesser than reference current I_{REF}.

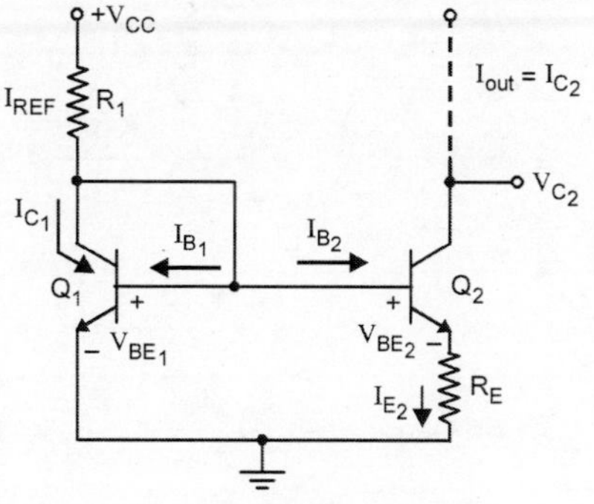

Fig. 1.38

For $\beta >> 1$ for transistors Q_1 and Q_2 and for transistors to be identical

$$I_{REF} \cong I_{C_1} = I_s e^{V_{BE_1}/V_T} \quad ...(1.108)$$

$$\text{and } I_{out} = I_{C_2} = I_s e^{V_{BE_2}/V_T} \quad ...(1.109)$$

$$\text{and } V_{BE_1} = V_T \log_e \frac{I_{REF}}{I_s} \quad ...(1.110)$$

$$\text{and } V_{BE_2} = V_T \log_e \frac{I_{out}}{I_s} \quad ...(1.111)$$

Subtracting Eq. (1.111) from Eq. (1.110), we have

$$V_{BE_1} - V_{BE_2} = V_T \log_e \frac{I_{REF}}{I_{out}} \quad ...(1.112)$$

From the circuit, we see that

$$V_{BE_1} - V_{BE_2} = I_E R_E = \left(\frac{\beta+1}{\beta}\right) I_{out} R_E \cong I_{out} R_E \quad ...(1.113)$$

Comparing Eqs. (1.112) and (1.113), we have

$$I_{out} R_E = V_T \log_e \frac{I_{REF}}{I_{out}} \quad ...(1.114)$$

$$\text{where } I_{REF} = \frac{V_{CC} - V_{BE}}{R_1} \quad ...(1.115)$$

Equation (1.114) provides the relationship between the reference and bias currents. Sometimes, emitter resistances are used in both the transistors Q_1 and Q_2 as depicted in Fig. 1.39.

From the circuit shown in Fig. 1.39.

$$V_{BE_2} - V_{BE_1} = V_T \log_e \frac{I_{C_2}}{I_{C_1}} \quad ...(1.116)$$

Applying KVL in the base-emitter loop, we have

$$V_{BE_2} - V_{BE_1} = I_{C_1} R_{E_1} - I_{C_2} R_{E_2} \quad ...(1.117)$$

Neglecting base current

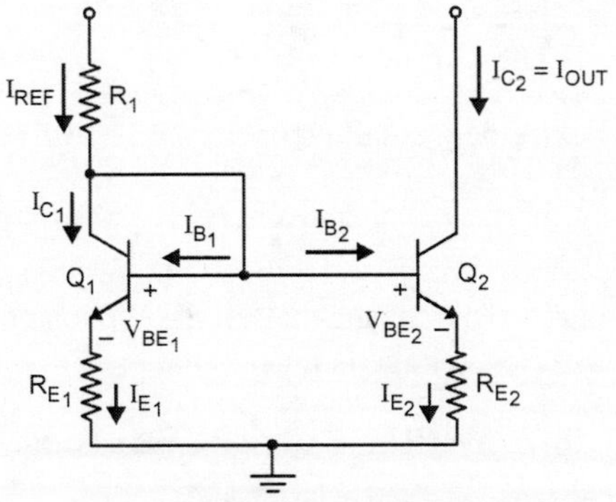

Fig. 1.39

Comparing Eqs. (1.116) and (1.117), we have

$$I_{C_1} R_{E_1} - I_{C_2} R_{E_2} = V_T \log_e \frac{I_{C_2}}{I_{C_1}} \quad ...(1.118)$$

$$\text{or } \frac{I_{C_2}}{I_{C_1}} \times \frac{R_{E_2}}{R_{E_1}} = 1 - \frac{V_T}{I_{C_1} R_{E_1}} \log_e \frac{I_{C_2}}{I_{C_1}} \quad ...(1.119)$$

$$\text{or } \frac{I_{C_2}}{I_{C_1}} = \frac{R_{E_1}}{R_{E_2}} \left[1 - \frac{V_T}{I_{C_1} R_{E_1}} \log_e \frac{I_{C_2}}{I_{C_1}}\right] \quad ...(1.120)$$

For the range $0.1 < \frac{I_{C_2}}{I_{C_1}} < 10$ it can be assumed that $\frac{I_{C_2}}{I_{C_1}} \cong \frac{R_{E_1}}{R_{E_2}}$. Thus even large ratio is obtained by the modified circuit (Fig. 1.39).

Example 1.9. Determine the mirrored current I in the circuit shown in Fig. 1.33 (*b*), if R_1 = 25 kΩ and V_{CC} = 12 V.

Solution : Mirrored current, $I = I_{REF} = \dfrac{V_{CC} - V_{BE}}{R_1}$ Refer to Eq. (1.91)

$$= \frac{12\text{ V} - 0.7\text{ V}}{25\text{ k}\Omega} = 0.452\text{ mA} \quad \textbf{Ans.}$$

Example 1.10. Determine reference current and output current for the two-transistor current source [Fig. 1.33 (*b*)] with circuit parameters +V_{CC} = 10 V, R_1 = 15 kΩ. The transistor parameters are : $V_{BE(ON)}$ = 0.7 V, β = 100.

Solution : Reference current, $I_{REF} = \dfrac{V_{CC} - V_{BE}}{R_1} = \dfrac{10 - 0.7}{15 \times 10^3} = 0.62\text{ mA}$ **Ans.**

Output current, $I_{out} = I_{REF}\dfrac{\beta}{\beta + 2} = 0.62 \times \dfrac{100}{102} = 0.608\text{ mA}$ **Ans.** Refer to Eq. (1.92)

Example 1.11. Calculate the current I in the circuit shown in Fig. 1.40.

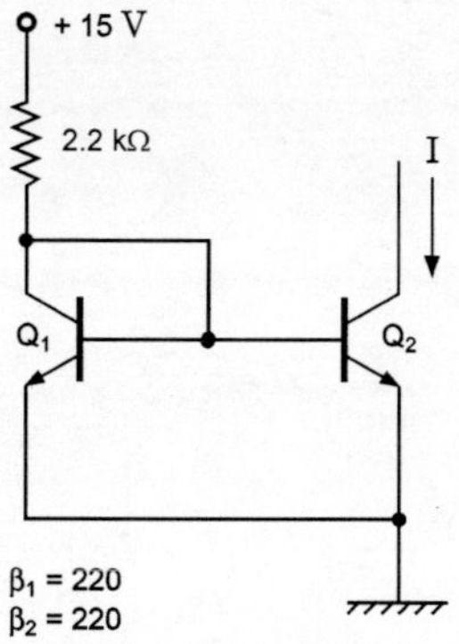

Fig. 1.40

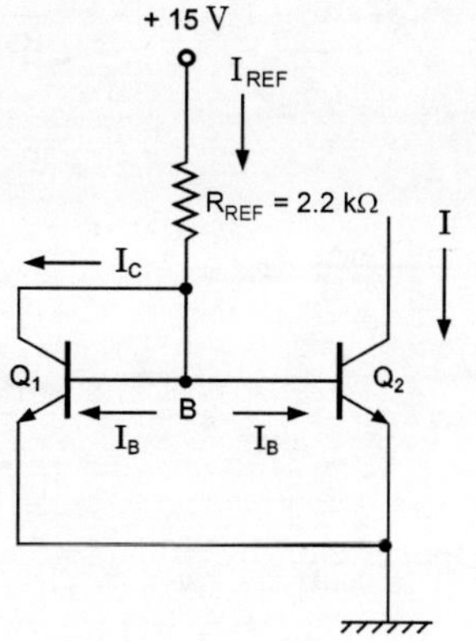

Fig. 1.41

Solution : Redrawing the circuit, as shown in Fig. 1.41, and applying Kirchhoff's current law at point B, we have

$$2I_B + I_C - I_{REF} = 0$$

$$\text{and } I_C = \beta I_B$$

$$\therefore\ I = I_C = \left(\frac{\beta}{2+\beta}\right)I_{REF} = \left(\frac{\beta}{2+\beta}\right)\left(\frac{V_{CC} - V_{BE}}{R_{REF}}\right) = \left(\frac{220}{2+220}\right)\left(\frac{15 - 0.7}{2.2 \times 10^3}\right) = 6.44$$

mA **Ans.**

Example 1.12. Calculate current I in the circuit of Fig. 1.42.

Solution : Applying Kirchhoff's second law to base-emitter loop of the transistor, we have

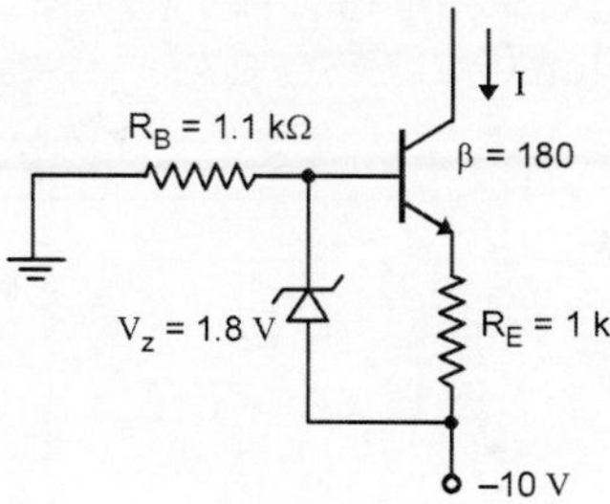

Fig. 1.42

$$V_Z - V_{BE} - I_E R_E = 0$$

$$\text{or } I_E R_E = V_Z - V_{BE} = 1.8 - 0.7 = 1.1\text{ V}$$

$$\text{Emitter current, } I_E = \frac{I_E R_E}{R_E} = \frac{1.1\text{ V}}{1\text{ k}\Omega} = 1.1\text{ mA}$$

$$\text{Current, } I = I_C = \frac{\beta}{\beta + 1}I_E = \frac{180}{181} \times 1.1 = 1.094\text{ mA} \quad \textbf{Ans.}$$

Example 1.13. Explain the working of a current source suitable for generating low values of constant currents. Design the current source for I_0 = 10 μA. Assume V_{CC} = 10 V, V_{BE} = 0.7 V and β = 100.

[G.G.S.I.P. Univ. Analog Electronics-II May 2011]

Solution : Referring to Widlar current source shown in Fig. 1.38, let us decide the value of I_{REF}, say 1 mA. Then

$$R_1 = \frac{V_{CC} - V_{BE}}{I_{REF}} = \frac{10 - 0.7}{1 \times 10^{-3}} = 9.3\ \text{k}\Omega$$

From Eqs. (1.112) and (1.113), we have

$$R_E = \frac{V_T}{I_{out} \times \left(\dfrac{\beta + 1}{\beta}\right)} \log_e \frac{I_{REF}}{I_{out}}$$

$$= \frac{0.025}{10 \times 10^{-6} \times \dfrac{100 + 1}{100}} \log_e \left(\frac{1\text{ mA}}{10\ \mu\text{A}}\right) = \frac{0.025}{10.1 \times 10^{-6}} \log_e 100 = 11.4\ \text{k}\Omega \quad \textbf{Ans.}$$

Example 1.14. Determine the value of each current depicted in Fig. 1.34 with parameters V_{CC} = 9 V, R_1 = 12 kΩ. The transistor parameters for all transistors are $V_{BE\,(ON)}$ = 0.7 V, β = 100.

Solution : Reference current, $I_{REF} = \dfrac{V_{CC} - 2V_{BE}}{R_1}$ Refer to Eq. (1.99)

$$= \frac{9 - 2 \times 0.7}{12 \times 10^3} = 0.6333\text{ mA} \quad \textbf{Ans.}$$

Output current, $I_{out} = \dfrac{I_{REF}}{1 + \dfrac{2}{\beta(1 + \beta_3)}}$ Refer to Eq. (1.98)

$$= \frac{0.6333}{1 + \dfrac{2}{100(1 + 100)}} = 0.6332\text{ mA} \quad \textbf{Ans.} \qquad \because\ \beta_1 = \beta_2 = \beta_3 = \beta = 100$$

Collector current, $I_{C_2} = I_{out} = 0.6332$ mA **Ans.**

Base current of transistor Q_3, $I_{B_3} = I_{REF} - I_{C_1} = I_{REF} - I_{C_2} \quad \because\ I_{C_1} = I_{C_2}$

$$= 0.6333 - 0.6322 = 0.0001\text{ mA} = 0.1\ \mu\text{A} \quad \textbf{Ans.}$$

Emitter current of transistor Q_3, $I_{E_3} = (1 + \beta_3) I_{B_3} = (1 + 100) \times 0.1 \times 10^{-6} = 10.1\ \mu\text{A}$ **Ans.** Refer to Eq. (1.95)

Base currents, $I_{B_1} = I_{B_2} = \dfrac{I_{E_3}}{2} = \dfrac{10.1}{2} = 5.05\ \mu\text{A}$ **Ans.**

Example 1.15. In Fig. 1.43, a transistor array CA3086 is used and $\beta_{dc} = \beta_{ac}$ = 100 and V_{BE} = 0.715 V, find collector current of each transistor.

Solution : Since Q_1 has its collector base shorted, it is working as a diode. Thus

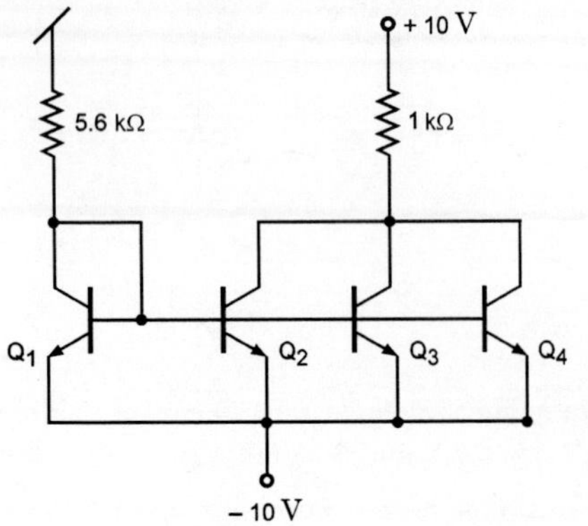

Fig. 1.43

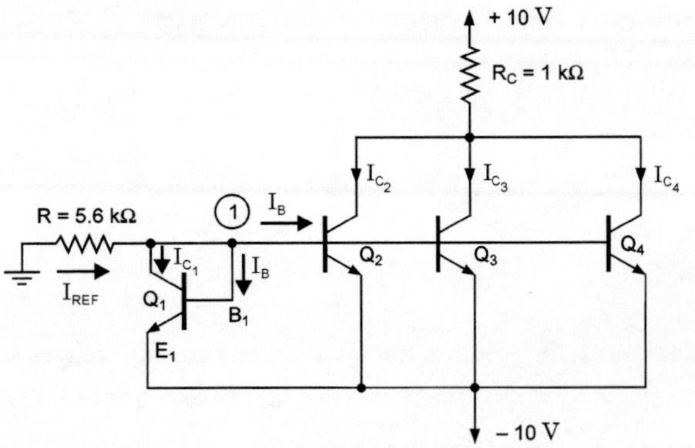

Fig. 1.44

$$V_{BE_1} = 0.715 \text{ V}$$

Current through resistor of 5.6 kΩ,

$$I_{REF} = \frac{V_{EE} - V_{BE}}{R} = \frac{10 - 0.715}{5.6 \times 10^3} = 1.658 \text{ mA}$$

Applying Kirchhoff's current law at node (1), we have

$$2I_B + I_{C_1} - I_{REF} = 0$$

$$\text{or} \quad \frac{2I_{C_1}}{\beta} + I_{C_1} = I_{REF} \qquad \because I_{C_1} = \beta I_B$$

$$\text{or} \quad I_{C_1}\left(\frac{2}{\beta} + 1\right) = I_{REF}$$

$$\text{or} \quad I_{C_1} = \frac{I_{REF}}{(2+\beta)} \times \beta = \frac{1.658 \times 100}{102} = 1.626 \text{ mA}$$

By symmetry $I_{C_1} = I_{C_2} = I_{C_3} = 1.626$ mA

So current through $R_C = 3I_C = 3 \times 1.626 = 4.88$ mA **Ans.**

1.6. ACTIVE LOAD

Differential voltage gain, A_d can be increased by using large collector resistance values as gain is proportional to load resistor R_C. However, there are limitations to the maximum value of R_C to be employed because of the following two reasons :

1. A large value of collector resistance R_C needs a large chip area.
2. For large valued R_C, quiescent drop across it increases and therefore a large power supply will be required to maintain a particular value of quiescent collector current.

However, this problem can be solved by using an **active load.*** Figure 1.45 shows a current mirror used as a load resistor (active load). Transistors Q_1 and Q_2 are the differential pair biased with a constant current, I_T and transistors Q_3 and Q_4 form the active load, the transistor Q_3 being diode connected.

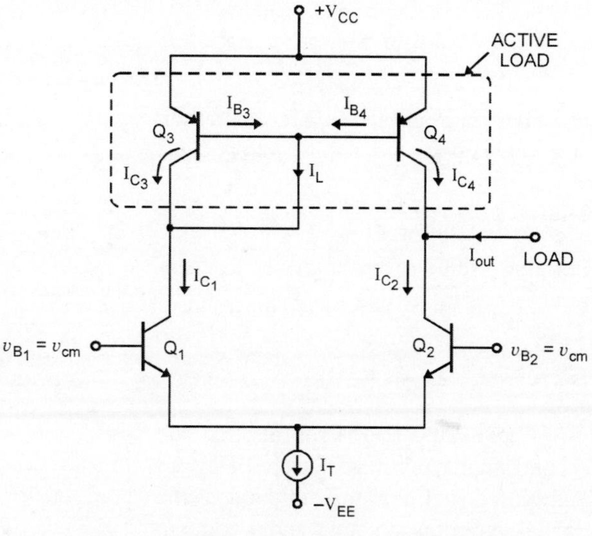

Fig. 1.45. ***BJT Differential Amplifier With Active Load***

If all the transistors are assumed identical, then a pure applied common-mode voltage means that $v_{B_1} = v_{B_2} = v_{cm}$, and the tail current I_T splits evenly between transistors Q_1 and Q_2. Neglecting base currents, $I_{C_3} = I_{C_4}$ through the current-source circuit and currents $I_{C_1} = I_{C_2} = I_{C_3} = I_{C_4} = \frac{I_T}{2}$ with no load connected at the output.

* In amplifier ICs, a constant current source replaces the load resistance R_C. A constant current source provides a high voltage gain because the dynamic resistance of an ideal current source is infinite and so a constant current source offers a very high load resistance. Such a load is referred to as an active load.

The quiescent load current may be determined as follows :

$$I_L = I_{B_3} + I_{B_4} = \frac{1}{\beta}(I_{C_3} + I_{C_4})$$

$$= \frac{I_{C_1} + I_{C_2}}{\beta} = \frac{I_T}{\beta} \quad ...(1.121)$$

Since $\beta >> 1$, I_L will be very small and thus there will be no loading problem.

In actual differential amplifier circuits, base currents are not zero. In addition, a second amplifier stage is connected at the differential amplifier output. Fig. 1.46 depicts a differential amplifier with an active load circuit, corresponding to a three-transistor current source, as well as a second amplifying stage. Usually, the common emitter current gain β is a function of collector current. However, for simplicity, all transistor current gains are assumed to be equal, even though current level in transistor Q_5 is much smaller than in the other transistors. Current I_{out} is the dc bias current from the gain stage. Assuming all transistors are matched and $v_{B_1} = v_{B_2} = v_{cm}$, current I_T splits evenly and $I_{C_1} = I_{C_2}$. To ensure that transistors Q_2 and Q_4 are biased in the forward active mode, dc currents must be balanced *i.e.,* $I_{C_3} = I_{C_4}$. We see that

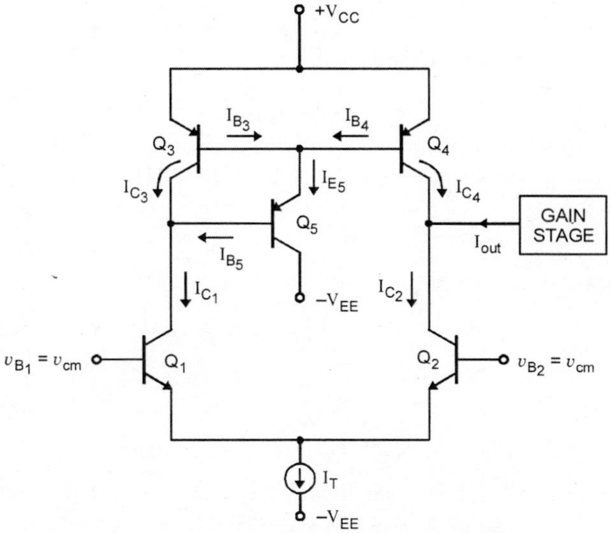

Fig. 1.46. ***BJT Differential Amplifier With Three-Transistor Active Load and Second Gain Stage***

$$I_{E_5} = I_{B_3} + I_{B_4} = \frac{I_{C_3}}{\beta} + \frac{I_{C_4}}{\beta} \quad ...(1.122)$$

Then $$I_{B_5} = \frac{I_{E_5}}{1+\beta} = \frac{I_{C_3} + I_{C_4}}{\beta(1+\beta)} \quad ...(1.123)$$

If the base currents and I_{out} are small, then

$$I_{C_3} + I_{C_4} \cong I_T \quad ...(1.124)$$

Thus $$I_{B_5} \cong \frac{I_T}{\beta(1+\beta)}$$

For the circuit to be balanced *i.e.,* for $I_{C_1} = I_{C_2}$ and $I_{C_3} = I_{C_4}$, we must have

$$I_{out} = I_{B_5} = \frac{I_T}{\beta(1+\beta)} \quad ...(1.125)$$

Above Eq. (1.125) reveals that the second amplifying stage must be designed and biased such that the direction of the dc bias current is as shown and is equal to the result of Eq. (1.124).

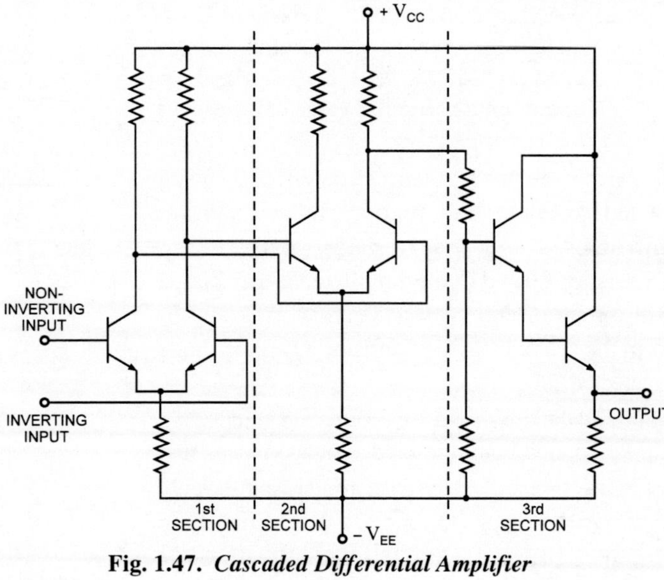

Fig. 1.47. ***Cascaded Differential Amplifier***

1.7. CASCADED DIFFERENTIAL AMPLIFIER STAGES

Figure 1.47 gives an idea of how differential amplifiers can be cascaded to give an op-amp. Here the output of first section differential amplifier is fed to the second section, which is again a differential amplifier. This provides the best common mode rejection ratio and differential voltage gain. The third section of op-amp is a common collector or common emitter follower stage which has low output impedance. The output is single ended, as shown in Fig. 1.47. It can provide only one phase with respect to ground and that is why two inputs are referred to *inverting* and *noninverting* inputs. The noninverting input will be in phase with the output terminal. The inverting input will be 180° out of phase with the output terminal.

In many cases the matching of transistors as well as resistor values is necessary for proper functioning of the differential stages. That is why the use of transistor array like CA3086 is recommended. This transistor array have inherent advantages unique to IC circuits such as compactness, ease of physical handling and electrical and thermal matching.

1.8. LEVEL TRANSLATOR (LEVEL SHIFTER)

This is usually an emitter follower circuit which is used to shift the dc level at the output of the intermediate stage downward to zero volt w.r.t. ground.

The level translator should be included to shift the output dc level because of following reasons:

1. The direct coupling in cascaded differential amplifiers produces rise in dc level at the emitter (from stage to stage). This increase in dc level alters the operating point of coming stages and thus, limits the output voltage swing and sometimes distorts the output signal.

2. The voltage at the output terminal of second stage (Fig. 1.47) is well above the ground. This additional voltage circuits the peak-to-peak output voltage swing without distortion and produces distortions (errors) in the dc output signal.

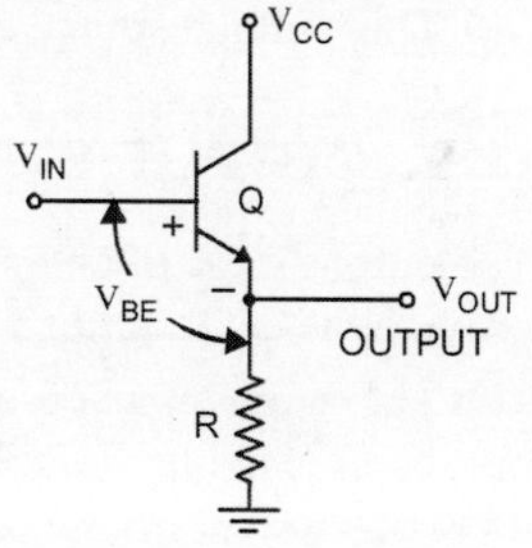

(a)

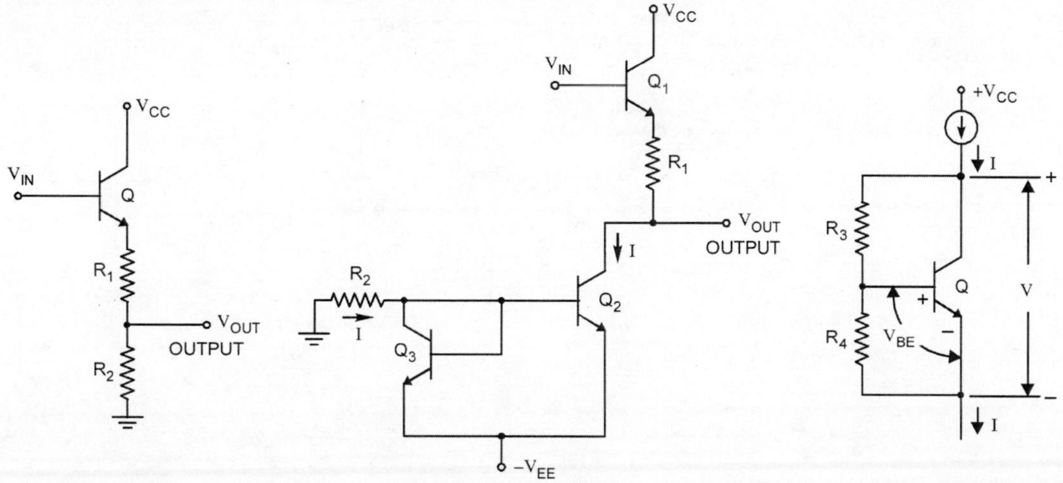

(b) With Voltage Divider *(c) With Current Mirror* *(d) A Voltage Source V which is Multiple of V_{BE}*

Fig. 1.48. ***Level Shifters Using Emitter Follower Buffer***

There are many circuits available to work as level translator. The simplest one is depicted in Fig. 1.48 (*a*) which is basically an emitter follower. Thus, the level shifter also acts as a buffer for isolation of the high gain stage from the output stage. Available amount of shift is

$$V_{out} = V_{in} = -V_{BE} = -0.7\text{ V} \qquad ...(1.126)$$

If the shift given by Eq. (1.126) is not sufficient, then we can use an emitter follower with a voltage divider as shown in Fig. 1.48 (*b*). The voltage shift is now increased by the voltage drop across R_1. However, this arrangement has the drawback of attenuation of signal voltage by $\frac{R_2}{R_1 + R_2}$. This can be easily overcome by replacing R_2 by a current mirror I as depicted in Fig. 1.48 (*c*).

Now available amount of shift is

$$V_{out} - V_{in} = -(V_{BE} + IR_1) \qquad ...(1.127)$$

In this arrangement ac attenuation is avoided due to high resistance of current source.

Another voltage commonly used in μA 741 op-amp is depicted in Fig. 1.48 (*d*). Neglecting base current in comparison to the current in voltage divider R_3, R_4, the circuit behaves as a multiplier of V_{BE} as

$$V = V_{BE} \times \frac{R_3 + R_4}{R_4} = V_{BE}\left(1 + \frac{R_3}{R_4}\right) \qquad ...(1.128)$$

This voltage source can also be used to replace R_1 in Fig. 1.48 (*b*).

1.9. OUTPUT STAGES

The output stage of an op-amp is another requirement which should have very small output impedance and provide the external load current so that large output voltage swing is obtained.

The output of an op-amp is the voltage measured at the output of this final stage which is usually a push-pull complementary amplifier, as shown in Fig. 1.49.

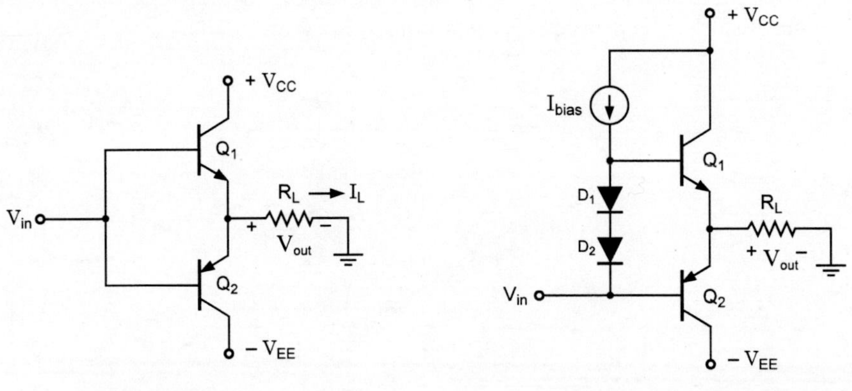

(a) Complementary Push-Pull Amplifier For The Output Stage

(b) Series Connected P-N Diodes To Remove Cross-over Distortion

Fig. 1.49

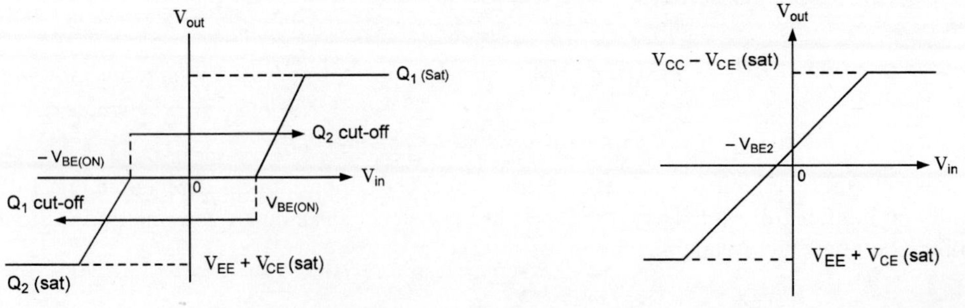

(a) Voltage Transfer Characteristics of Fig 1.49 (a) (Results in Cross-over Distortion)

(b) Voltage Transfer Characteristics of Fig 1.49 (b) (Cross-over Distortion Eliminated By Using Pair of Diodes)

Fig. 1.50

When positive cycle of input voltage V_{in} arrives, Q_1 behaves as current supplier to load R_L and Q_2 is off.

When negative cycle of V_{in} arrives, Q_1 is off, Q_2 acts as a sink to remove current from the load, (it decreases the load current).

But the circuit bears cross-over distortion when $V_{in} = 0$ (when voltage swings from positive to negative). To reduce this problem of getting $V_{in} = V_{BE} = 0.7$ V when it should be zero, we provide a constant dc bias voltage between the two bases so that small current exists in quiescent condition. For this, two diodes D_1 and D_2 are connected, as shown in Fig. 1.49 (*b*).

The transfer characteristics shows that if $V_{in} = 0$, $V_{out} \neq 0$. Here, V_{in} is obtained from level shifter, we can have $V_{out} = 0$ with $V_{in} = 0$ by making quiescent value of $V_{in} = -V_{BE2}$. Figure 1.51 can also reduce the cross-over distortion. The output voltage is about 1.1 V between two bases. Here both transistors Q_1 and Q_2 conduct slightly under quiescent condition.

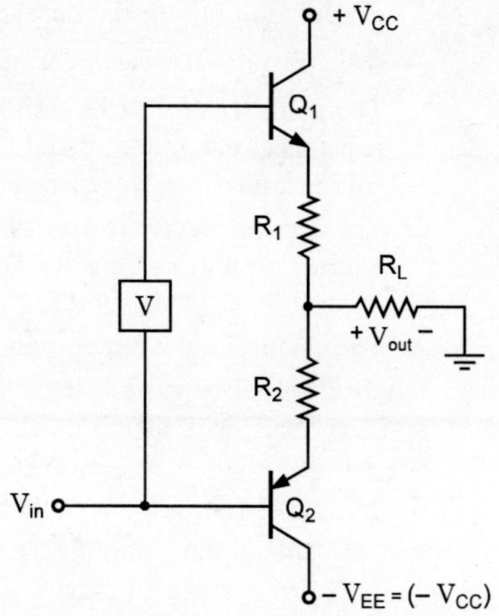

Fig. 1.51. ***Using The B-E Junction To Reduce The Cross-over Distortion***

HIGHLIGHTS

1. An amplifier, which is designed to amplify the difference between two input signals is called the differential amplifier. The differential amplifier is also referred to as a difference amplifier. The main requirements of differential amplifiers are matched devices and use of more components and IC fabrication is capable of providing matched devices and is also capable of providing large number of transistors at relatively low cost. Basically there are two reasons for using differential amplifiers in preference to single ended amplifiers. First, the differential amplifiers are much less sensitive to noise and interference than single-ended circuits. The second reason for preferring differential amplifiers is that the differential configuration enables us to bias the amplifier and to couple amplifier stages together without the necessity of bypass and coupling capacitors.
2. The configurations are categorised on the basis of the number of inputs and the way an input voltage is measured. Accordingly, differential amplifier configurations may be classed as follows :
   (*i*) Dual-input, balanced-output differential amplifier configuration.
   (*ii*) Dual-input, unbalanced-output differential amplifier configuration.
   (*iii*) Single-input, balanced-output differential amplifier configuration.
   (*iv*) Single-input, unbalanced-output differential amplifier configuration.
   Among the four configurations, single-input configurations are rarely used.
3. With the two input terminals connected to a suitable dc voltage V_{CM}, the bias current of a perfectly symmetrical differential pair divides equally between the two transistors of the pair, resulting in a zero voltage difference between the two collectors/drains. To steer the current completely to one side of the pair, a differential input voltage v_d of at least $4V_T$ in case of bipolar and $\sqrt{2}V_{OV}$ is required.
4. Superimposing a differential input voltage v_d on the dc common-mode input voltage V_{CM} such that $v_{in_1} = V_{CM} + v_d$ and $v_{in_2} = V_{CM} - v_d$ causes a virtual signal ground to appear on the common emitter/source connection. In response to differential input voltage v_d, the current in transistor Q_1 increases by $\frac{1}{2}g_m v_d$ and the current in transistor Q_2 decreases by $\frac{1}{2}g_m v_d$. Thus voltage signals of $\pm\frac{1}{2}g_m(R_C \| r_o)v_d$ develop at the two collectors in case of BJT differential amplifier $\left[\pm\frac{1}{2}g_m(R_D \| r_o)v_d \text{ in case of MOS differential amplifier}\right]$. If the output is taken single-endedly, *i.e.*, between one of the collectors/drains and ground, a differential gain of $\frac{1}{2}g_m(R_C \| r_o)$ or $\frac{1}{2}g_m(R_D \| r_o)$

is realized. Taking the output differentially, *i.e.,* between the two collectors/drains, the differential gain realized is twice as large; $g_m(R_C || r_o)$ or $g_m(R_D || r_o)$.

5. The analysis of a differential amplifier for determination of differential voltage gain, differential input resistance, etc. is facilitated by using the differential half-circuit which is a common-emitter or common-source transistor biased at I/2.

6. The constant current bias circuit is considered better than the combination of R_E and V_{EE} to produce emitter dc bias current. Performance can be improved further by using transistor array like CA3086 as a constant current source.

7. The differential voltage gain A_d is improved and as a result, the differential gain A_d is much higher with an active load than with an ordinary resistor.

8. Level shifter is usually an emitter follower circuit which is used to shift the dc level at the output of the intermediate stage downward to zero volt w.r.t. ground.

The level translator should be included to shift the output dc level because of following reasons:

1. The direct coupling in cascaded differential amplifiers produces rise in dc level at the emitter (from stage to stage). This increase in dc level alters the operating point of coming stages and thus, limits the output voltage swing and sometimes distorts the output signal.
2. The voltage at the output terminal of second stage (Fig. 1.47) is well above the ground. This additional voltage circuits the peak-to-peak output voltage swing without distortion and produces distortions (errors) in the dc output signal.

Thus, the above undesirable effects can be reduced by adding a level translator (or shifter) to shift the output dc level down to zero volt to ground.

9. The output stage of an op-amp is another requirement which should have very small output impedance and provide the external load current so that large output voltage swing is obtained.

EXERCISES

1. What is differential amplifier ? Why are differential amplifiers preferred over single-ended amplifiers ? Give the classification of differential amplifier configurations.

2. Draw and explain the basic BJT differential pair configuration. Explain the operation of this circuit. Show that the BJT differential pair does not respond to input signals in ideal conditions.

3. Explain the large-signal operation of a BJT differential amplifier. How can the linear range of operation of a BJT differential amplifier be increased ?

4. Explain the small-signal operation of a BJT differential pair. Derive the expression for the input differential resistance of a BJT differential pair.

5. Show that, in case of BJT differential amplifier, the differential gain is the ratio of the load resistance in the collector circuit and the total resistance in the emitter circuit.

Explain the equivalence of the BJT differential amplifier to a common-emitter amplifier.

6. Derive the expressions for CMRR in case of BJT differential amplifier when the output is taken (*i*) differentially (*ii*) single-ended. Derive the expression for common-mode input resistance.

7. Derive and plot the transfer characteristics of an emitter-coupled differential amplifier and explain the various regions of operation. **[G.G.S.I.P. Univ. Analog Electronics-II May 2011]**

8. Derive an expression for the CMRR of an emitter-coupled differential amplifier. Suggest one method to increase CMRR. **[G.G.S.I.P. Univ. Analog Electronics-II May 2011]**

9. Derive the expression for the differential mode gain, differential input impedance, common-mode gain of a bipolar differential stage and discuss the common-mode rejection ratio.

[G.G.S.I.P. Univ. Analog Electronics-II February 2009]

10. Draw the circuit of a MOSFET source coupled amplifier. Derive an expression for its CMRR.

[G.G.S.I.P. Univ. Analog Electronics-II May 2010]

11. Explain the operation of a MOS differential amplifier with differential input voltage. How the bias current is steered from one MOSFET to another ? Show that the differential amplifier corresponds to differential input signals.

12. Explain the small-signal operation of a MOS differential amplifier. What is the objective of analysis of the small-signal operation of a MOS differential pair ? State clearly the assumptions made in the analysis. Show that the differential output is double of single-ended output in a MOS differential amplifier.

13. Explain the small-signal operation of a MOS differential amplifier taking the effect of MOSFETs r_o into consideration.

14. Explain the (*i*) effect of R_D mismatch and (*ii*) g_m mismatch on CMRR of a differential amplifier.

15. What is constant current bias ? How is it advantageous over emitter? How is it obtained in a differential amplifier ? Explain with circuit diagram ?

16. Discuss the current-mirror circuit with proper labelling diagram.

17. What is importance of current mirror ? Draw the circuit of the basic current mirror and explain its working.
[G.G.S.I.P. Univ. Analog Electronics-II May 2010]

18. Discuss the operation and significance of a multiple output transistor current mirror.

19. Draw the circuit of BJT three transistor current source and give its advantage.

20. Draw a neat circuit of a bias-compensated current mirror, derive the expression for the ratio of I_0/I_{ref} and discuss its superiority over the simple current mirror.
[G.G.S.I.P. Univ. Analog Electronics-II February 2009]

21. Draw the circuit of a Wilson current source and show that its output current is almost equal to the reference current. **[G.G.S.I.P. Univ. Analog Electronics-II May 2010]**

22. Explain the working of a current source suitable for generating low values of constant currents.
[G.G.S.I.P. Univ. Analog Electronics-II May 2011]

23. What is an active load ? Where is it used ? **[G.G.S.I.P. Univ. Analog Electronics-II May 2011]**

24. Discuss the operation of an active load in a differential amplifier. What is primary advantage of using an active load ?

25. What is a level translator circuit ? Why is it used with the cascaded differential amplifier ? Explain with circuit diagram.

26. What do you mean by a complementary emitter follower circuit employed at output stage of an op-amp ? Discuss.

27. Describe the operation and characteristics of a BJT complementary push-pull output stage. What are the advantages of this circuit ?

28. Write short notes on the following :

(*i*) BJT differential pair.
(*ii*) MOS differential pair.
(*iii*) Constant current bias.
(*iv*) Current mirror.
(*v*) Wilson current source.
(*vi*) Active load.
(*vii*) Level translator.
(*viii*) Output stages.

SHORT ANSWER TYPE QUESTIONS WITH ANSWERS......................

Q. 1. What is differential amplifier ?

Ans. An amplifier, which is designed to amplify the difference between two input signals is called the differential amplifier. The differential amplifier is also referred to as a difference amplifier.

Q. 2. Advent of ICs made the differential amplifier extremely popular in both BJT and MOS technologies. Why ?

Ans. The main requirements of differential amplifiers are matched devices and use of more components and IC fabrication is capable of providing matched devices and is also capable of providing large number of transistors at relatively low cost.

Q. 3. Why are differential amplifiers preferred over single-ended amplifiers ?

Ans. Basically there are two reasons for using differential amplifiers in preference to single-ended amplifiers. First, the differential amplifiers are much less sensitive to noise and interference than single-ended circuits. The second reason for preferring differential amplifiers is that the differential configuration enables us to bias the amplifier and to couple amplifier stages together without the necessity of bypass and coupling capacitors.

Q. 4. What are differential configurations of differential amplifier ?

Ans. The configurations are categorised on the basis of the number of inputs and the way an output voltage is measured. Accordingly, differential amplifier configurations may be classed as follows :

1. Dual-input, balanced-output differential amplifier configuration.
2. Dual-input, unbalanced-output differential amplifier configuration.
3. Single-input, balanced-output differential amplifier configuration.
4. Single-input, unbalanced-output differential amplifier configuration.

When both the inputs are used, the configuration is said to be dual input, otherwise single input. On the other hand, when the output voltage is measured between two collectors, it is referred to as a **balanced output,** because both collectors C_1 and C_2 are at the same dc potential w.r.t. ground. In case the output is measured across any one collector (either C_1 or C_2) w.r.t. ground, the configuration is known as an **unbalanced output.**

Q. 5. Why dc analysis of a differential amplifier circuit is required ?

Ans. DC analysis of a differential amplifier circuit is required for determination of operating point (quiescent collector current I_{CQ} and quiescent collector-emitter voltage V_{CEQ}).

Q. 6. What is the value of differential voltage which is sufficient to switch the current almost entirely one side of the BJT pair ?

Ans. A difference voltage of about 4 V_T ($\simeq$ 100 mV) is sufficient to switch the current almost entirely to one side of the BJT pair.

Q. 7. How can the linear range of operation of the BJT differential amplifier be extended ?

Ans. The linear range of operation of the BJT differential amplifier can be extended by including two equal resistors in series with the emitters of BJTs Q_1 and Q_2 but this will reduce the gain.

Q. 8. What is meant by differential input resistance of a BJT differential amplifier ?

Ans. The input differential resistance of a BJT differential amplifier is defined as the resistance seen from the two bases *i.e.,* resistance seen by the differential input signal v_d.

Q. 9. What is meant by input common-mode range of a MOS differential pair ?

Ans. Input common-mode range of a MOS differential pair is defined as the range of common-mode voltage over which the differential pair operates properly.

Q. 10. What for small-signal operation of differential pair is analyzed ?

Ans. The small-signal operation of a differential pair is analyzed mainly for the determination of its voltage gain in response to differential input signal v_d.

Q. 11. What is CMRR ?

Ans. CMRR is defined as the ratio of differential voltage gain to common-mode voltage gain and it is given by the equation

$$\text{CMRR} = \frac{A_d}{A_{cm}}.$$

Q. 12. Why R_E is replaced by a constant current source in a differential amplifier ?

Ans. The resistor R_E in the differential amplifier is replaced by a constant current source in order to improve CMRR without affecting the operating point of the amplifier and without increasing the negative supply voltage $- V_{EE}$.

Q. 13. What is the current mirror ?

Ans. The circuit in which the output current is forced to be equal to the input current is called current mirror circuit. Figure 1 shows block diagram and circuit diagram.

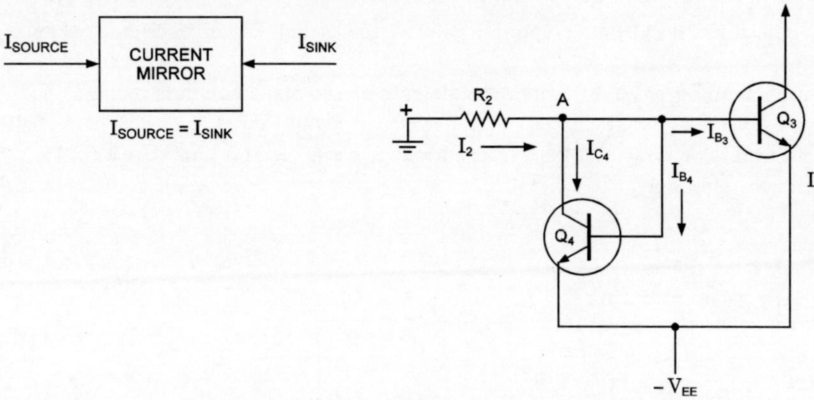

Fig. 1

Q. 14. Explain the difference between constant current bias and current mirror.

[G.G.S.I.P. Univ. Analog Electronics-II May 2011]

Ans. The constant current bias is a circuit that provides current stabilization and therefore assures a stable operating point for the differential amplifier. The circuit in which output current is forced to be equal to be the input current is said to be a current mirror circuit. Thus, in current mirror the output current is a mirror image of input current. The current mirror is a special case of constant current bias and, therefore, can be used to provide constant current in different amplifier stages.

Q. 15. What is level translator ? Draw any level translator circuit.

Ans. Level translator circuit is usually an emitter follower circuit which is used to shift the dc level at the output of the intermediate stage downward to zero volt w.r.t. ground in a cascaded differential amplifier.

Level translator circuit is shown in Fig. 1.48.

Q. 16. Why level translator is used with cascaded differential amplifier ?

Ans. The level translator is used with cascaded differential amplifier because of the following reasons :

1. The direct coupling in cascaded differential amplifiers produces rise in dc level at the emitter (from stage to stage). This increase in dc level alters the operating point of coming stages and thus, limits the output voltage swing and sometimes distorts the output signal.
2. The voltage at the output terminal of second stage (Fig. 1.47) is well above the ground. This additional voltage circuits the peak-to-peak output voltage swing without distortion and produces distortions (errors) in the dc output signal.

Thus, the above undesirable effects can be reduced by adding a level translator (or shifter) to shift the output dc level down to zero volt to ground.

Q. 17. Fill up the blank in the following :

(*i*) In a differential amplifier, if the output voltage is measured at one of the collector with respect to ground, the configuration is called as _________.

(*ii*) A differential amplifier _________ the difference between the two input signals.

[**Ans.** (*i*) unbalanced output, (*ii*) amplifies]

PROBLEMS ....

1. Determine the quiescent collector currents i_{C_1} and collector emitter voltage v_{CE} for a difference amplifier shown in Fig. 1.3. The circuit parameters are : V_{CC} = 10 V, $-V_{EE}$ = –10 V, I_Q = 1 mA,

$R_C = 10\ k\Omega$. The transistor parameters are : $V_{BE} = 0.7$ V, $V_A = \infty$. Neglect base currents when $v_{B_1} = v_{B_2} = 4.5$ V. [**Ans.** 0.5 mA, 5 V, 1.2 V]

2. Determine the differential-mode voltage gain of the circuit given in Fig. 1 when (*i*) $R_E = 0$ and $R_E = 400\ \Omega$. [**Ans.** (*i*) 80 V/V (*ii*) 10.81 V/V]
3. A differential amplifier has a differential mode gain of 100 and a common mode gains 0.01. Determine CMMR. [**Ans.** 10,000 or 80 dB]
4. In Fig. 2, a transistor array CA3086 is used and $\beta_{dc} = \beta_{ac} = 100$ and $V_{BE} = 0.715$ V, find collector current of each transistor. [**Ans.** 1.626 mA]

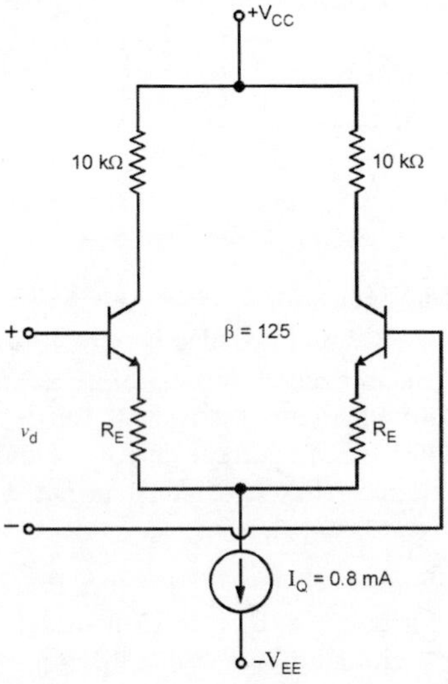

Fig. 1

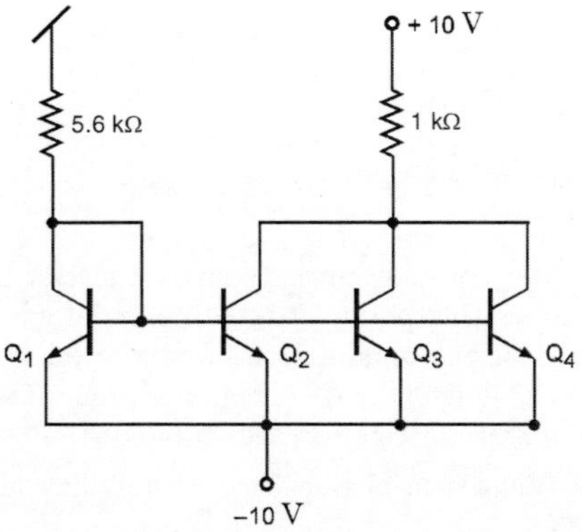

Fig. 2

5. Determine the current I in the circuit shown in Fig. 3. [**Ans.** 3.377 mA]

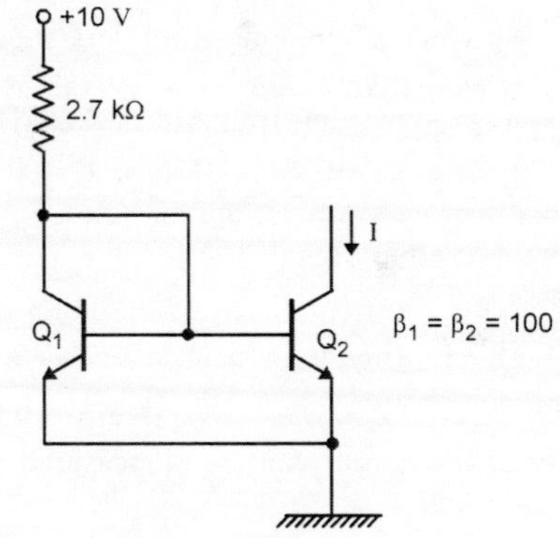

Fig. 3

❑❑❑

CHAPTER

2

Operational Amplifiers (Op-Amps)

INSIDE THIS CHAPTER

2.1. INTRODUCTION

An *operational amplifier*, abbreviated as op-amp, is basically a multistage, very high gain (typically 200,000), direct-coupled, negative feedback amplifier that provides a stabilized voltage gain. An op-amp has high input impedance (exceeding 100 kΩ) and low output impedance (less than 100 Ω) and has capability of amplifying signals having frequency ranging from zero Hz to 1 MHz *i.e.,* op-amp can be used to amplify dc as well as ac input signals. Although discrete op-amps are built, designers of industrial electronic circuits now use integrated circuit operational amplifiers (IC op-amps) almost exclusively. They are made with different internal configurations in linear ICs. An op-amp is so named as it was originally designed to perform mathematical operations such as summation, subtraction, multiplication, differentiation and integration etc. Nowadays, the op-amps are put to a variety of other uses such as sign changing, scale changing, phase shifting, voltage regulation, analog computer operations, in instrumentation and control systems, oscillator circuits, pulse generators, square-wave generators, triangular-wave generators, comparators, analog-to-digital and digital-to-analog converters (ADC and DAC); voltage-to-current converters, current-to-voltage converters, sample-and-hold circuits etc.

The op-amp manufactured with integrated transistors, diodes, resistors and capacitors, is an extremely versatile device that is found doing countless tasks, as enumerated above and many more. The op-amp is a complete amplifier having the advantages of an IC such as low cost, small size, high reliability, temperature stability and low value of offset voltage and current. Furthermore, it is designed in such a way that external components like resistors, capacitors etc. can be connected to its terminals and external characteristics can be changed. Various amplifier characteristics can be tailored to fit a particular application merely by changing a few components, without redesigning the entire amplifier. This is because the voltage gain, input impedance, output impedance, and frequency bandwidth depend almost solely on stable external resistors and capacitors. The voltage gains of IC op-amps are extremely large, typically 200,000. Because of this large voltage gain, externally connected resistors must be employed to provide negative feedback for most of the applications.

The operational amplifier was first built with vacuum tubes. In the early 1960s the tubes were replaced by transistors. A single op-amp then required only one printed circuit card full of components. This reduction in size and power consumption broadened its applications. Op-amps were used in many signal conditioning areas, test and measuring equipments, and industrial controls.

With the advent of integrated circuit techniques, it became possible to place the entire op-amp into a single eight-lead mini-DIP package. The first generation 709 was introduced by Fairchild in 1965, with the still popular, second generation, 741 op-amp available in 1968. Surprisingly, the cost of these op-amp ICs is comparable to a single discrete transistor. Most small signal analog circuits designed today use the operational amplifier as the basic active device, and fall back on discrete transistors only when the op-amp will fail to solve the problem.

An op-amp referred to as the IC741 has become an industry standard, Fairchild's original μA741 is also manufactured by various other manufacturers under their own designation, as listed below

Manufacturer	**IC designation**
National Semiconductor	LM741
Motorola	MC1741
RCA	CA3741
Texas Instruments	SN52741
Signetics	N5741

The last three digits in each manufacturers designation are 741 indicating that all these op-amps have same specifications and, therefore, behave the same way.

Some linear ICs are available in different classes such as 741, 741A, 741C, 741E, 741S and 741SC showing different versions of same op-amp. These differ in their voltage gain, temperature range, noise level and other characteristics. The 741 is a military-grade op-amp having operating temperature range –55° to 125°C and the 741C is a commercial grade op-amp (the C stands for commercial grade) with operating temperature range 0° to 70°/75 °C. 741 C is the least expensive and most widely used linear IC. It has an input impedance of 2 MΩ, a voltage gain of 100,000, and an output impedance of 75 Ω. On the other hand, the 741A and 741E are improved versions of the 741 and 741C, respectively, in that they have improved electrical specifications over their counterparts. The 741C and 741E are identical to the 741 and 741A except that the former have their performance guaranteed over a 0° to 70/75°C temperature range, instead of – 55° to 125°C. The 741S and 741SC are military- and commercial-grade op-amps respectively, with a higher slew rate than the 741 and 741C.

2.2. BLOCK DIAGRAM OF A TYPICAL OPERATIONAL AMPLIFIER (OP-AMP)

An operational amplifier being a multistage amplifier, consists of some basic building blocks as shown in block diagram (Fig. 2.1).

The block diagram of an op-amp shown in Fig. 2.1 consists of a four stage direct-coupled amplifier in cascade. In summarized form, it can be explained as shown in Fig. 2.1.

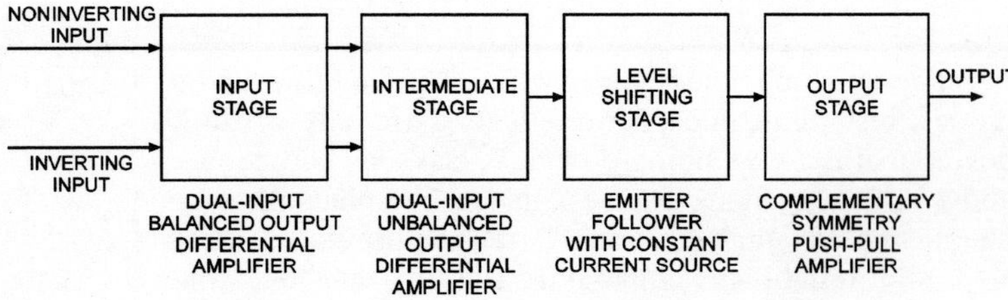

Fig. 2.1. ***Block Diagram of a Typical Op-amp***

The **first stage** is double-ended high-gain (60 dB) differential amplifier *i.e.,* dual-input balanced output differential amplifier with a constant current source (in order to increase CMRR). In this stage high gain is desirable so that there would be a negligible effect on the output of any shortcoming in the following stages. This is the reason why this stage is generally responsible for most of the gain of op-amp. Also this stage determines the input resistance of the op-amp. Output of this stage is taken between collectors of two emitter biased circuits so that output remains balanced and the dc voltage at output in quiescent condition maintains zero level.

The second stage, called the **intermediate stage**, is usually another differential amplifier, which is driven by the output of first stage. In most amplifiers the intermediate stage is dual input, unbalanced (single-ended) output differential amplifier in order to increase the gain. The differential mode voltage gain of such an amplifier is half of the gain of the dual-input, balanced output differential amplifier. In this amplifier output is measured at the collector of only one of the two transistors w.r.t. ground. In the quiescent condition some dc voltage exists at the output terminal and there is no other collector voltage at output to balance or nullify this output dc voltage. This is the reason why this type of amplifier is called unbalanced output type differential amplifier. This unbalanced dc voltage present at the output acts as an error voltage in the desired output signal.

The third stage, known as **level shifting stage**, is usually an emitter follower circuit in order to shift the dc level at the output of the intermediate stage downward to zero volt with respect to ground. It is noteworthy here that error signal is developed in the intermediate stage due to direct coupling and gets amplified in the succeeding stages. This increase in dc level tends to shift the operating point of the succeeding stages which also limits the output voltage swing or may distort the output signal. To overcome these problems, use of a level translator (shifter) circuit becomes necessary to bring this dc level to zero volt (ground potential).

The final stage, called **output stage,** is usually a push-pull complementary amplifier. This stage increases the output voltage swing and the current supplying capability of the amplifier. A well designed output stage also provides low output resistance. The output of the op-amp is the voltage measured at output of this final stage push-pull amplifier with respect to ground.

2.3. SCHEMATIC SYMBOL OF AN OP-AMP

The schematic symbol of an op-amp is shown in Fig. 2.2 (*a*). A is the voltage gain, v_1 is noninverting input and v_2 is inverting input. The differential input is

$$v_{in} \text{ or } v_d = v_1 - v_2 \qquad ...(2.1)$$

and output voltage, $v_{out} = A\, v_{in} = A\,(v_1 - v_2)$ ...(2.2)

Note that voltages v_1, v_2 and v_{out} are node voltages. This means that they are always measured with respect to ground. The differential input voltage, v_{in} is the difference of two node voltages v_1 and v_2.

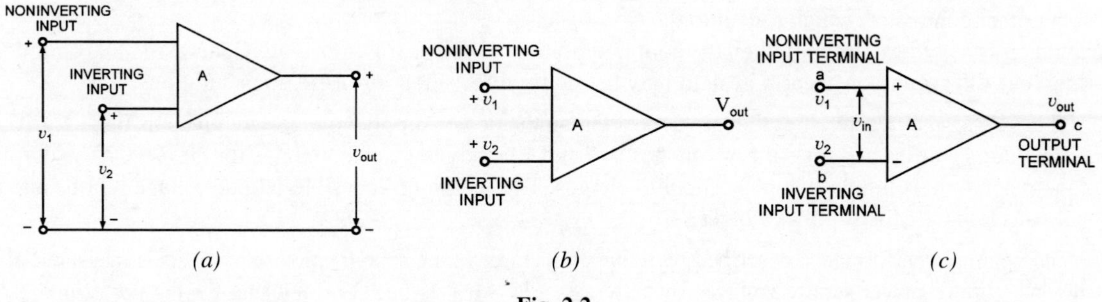

Fig. 2.2

Often, the common ground line is omitted and the schematic symbol becomes, as illustrated in Fig. 2.2 (*b*).

Most widely used circuit symbol for an op-amp is shown in Fig. 2.2 (*c*). It essentially consists of two input terminals and one output terminal. Here the inputs are marked with plus (+) and minus (–) to indicate noninverting and inverting inputs, respectively. A signal applied to the *plus* input appears with the same polarity and amplified at the output, while an input applied to the *minus* terminal appears amplified but inverted at the output.

It should be clearly understood that *plus and minus signs never mean that the voltages v_1 and v_2 are positive and negative respectively*. It also does not imply that *positive voltage is to be applied to the terminal marked positive and negative voltage to the terminal marked negative.*

The output voltage is directly proportional to the input voltage which is the difference of v_1 and v_2 (*i.e.*, $v_{in} = v_1 - v_2$) and is given by Eq. (2.2). The constant of proportionality is the voltage gain of the amplifier and is denoted by English letter A.

The op-amp's input can be single-ended or double-ended or differential input depending whether the input voltage is applied to one input terminal only or to both. Similarly output can also be either single-ended or double-ended. Most commonly used configuration is with two input terminals and one output terminal.

Op-amps have five basic terminals namely two input terminals (noninverting input terminal and inverting input terminal), one output terminal and two power supply terminals (–ve bias supply terminal and +ve bias supply terminal). The significance of other terminals varies with the type of the op-amp.

2.4. POWER SUPPLY REQUIREMENTS OF OP-AMPs

For operation of an op-amp, power supply voltages are required to be provided. Normally, bipolar voltages are required *i.e.*, two voltages are required, one positive and the other equal in magnitude, but negative. Often, these supply voltages are omitted from the schematics. However, proper schematic and power supply connections are shown in Fig. 2.3.

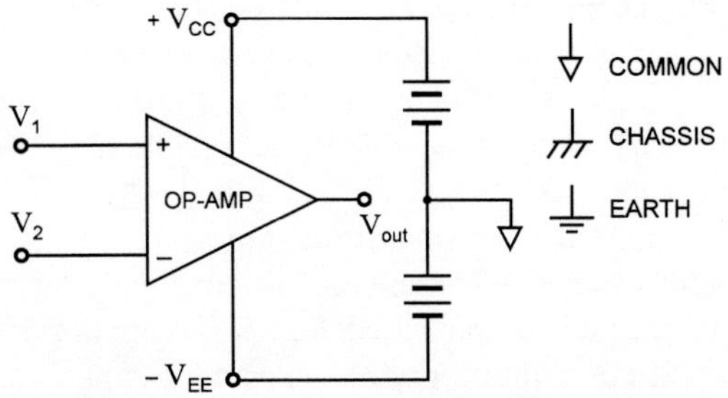

Fig. 2.3. *Power Supply Connections For an Op-amp*

The point at which these two supplies are connected is called *analog common* or just *common*. It is given by the schematic symbol of a triangle, point down. Usually all sources, loads and measurements are referred to this point. If this point is connected to the metal chassis in which the circuits are housed, then the point is at chassis ground. If common point is tied to the third wire of the 230 V ac line or to some other conductor, which runs into the earth (such as a cold water pipe), then the common point is called *earth* or *ground.* Connecting this common point to chassis ground or to earth ground may be required for safety or noise considerations.

TABLE 2.1. *Power Supply Dependent Specifications For The 741 C Op-amp*

Absolute Maximum Ratings	
Power Supply Voltage	± 18 V
Differential Input Voltages	± 30 V
Common Mode Input Voltages	± 15 V

Three op-amp characteristics that are dependent on the power supply voltages are given in Table 2.1. The manufacturer specifies that the supply voltages (V^+ and V^-) may not exceed ±18 V for the 741C. The output voltage magnitude is limited by these supply voltages. The maximum possible output voltage will be about 2 or 3 volts less than the supply voltages.

The op-amp's performance described by the manufacturer in the specification data sheets is measured at a particular value of power supply voltages (usually it is ±15 V). It is not at all unusual for the performance to change considerably as the supply voltage is changed, often getting worse at lower supply voltages. So, if there is no major reason to do otherwise, or if achieving rated performance is important, then use of power supply voltages prescribed by the manufacturer must be ensured.

It is specified that the difference in potential between the two inputs (differential input voltage) may not exceed 30 V for the 741C. Also, neither input may be ±15 V away from circuit common (common mode input voltage). In addition, though not specifically mentioned, should either input exceed the power supply voltage, damage may be caused. If the op-amp has a ±12 V supply, neither input should exceed 12 V, even though the specifications indicate ±15 V.

Care should be taken that power supply connections are not reversed otherwise the IC may get damaged.

2.5. SCHEMATIC DIAGRAM OF 741 IC OPERATIONAL AMPLIFIER

The schematic diagram of 741 IC operational amplifier is shown in Fig. 2.4 (*a*). It is not necessary to understand every detail about the circuit design and its operation in order to be able to use the IC. What must be understood

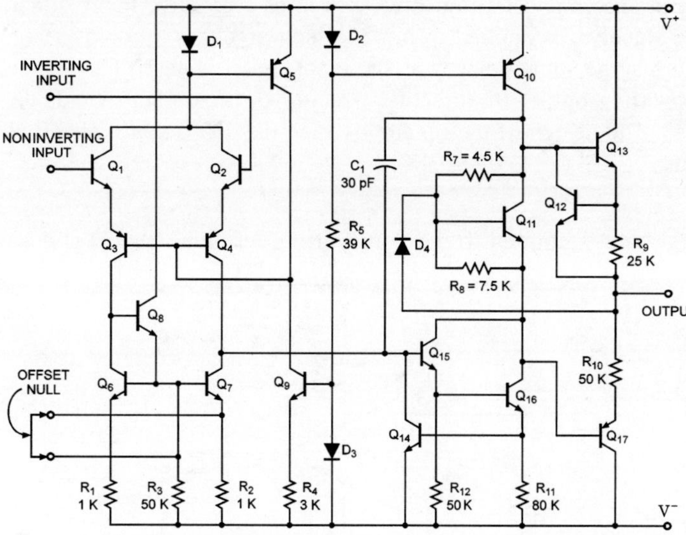

(a) Schematic Diagram of 741 IC Operational Amplifier

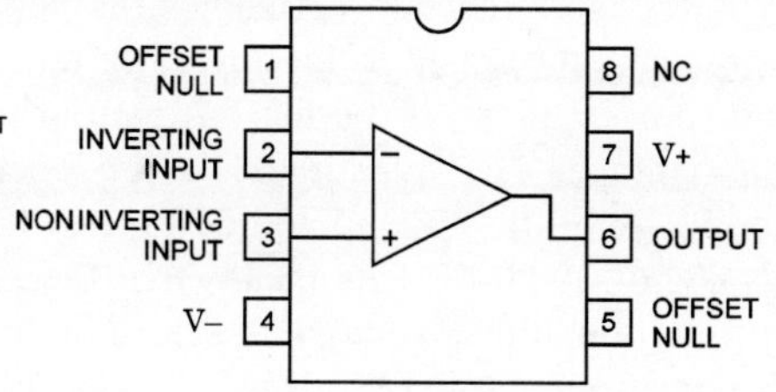

(b) Top View of Dual-in-Line Package

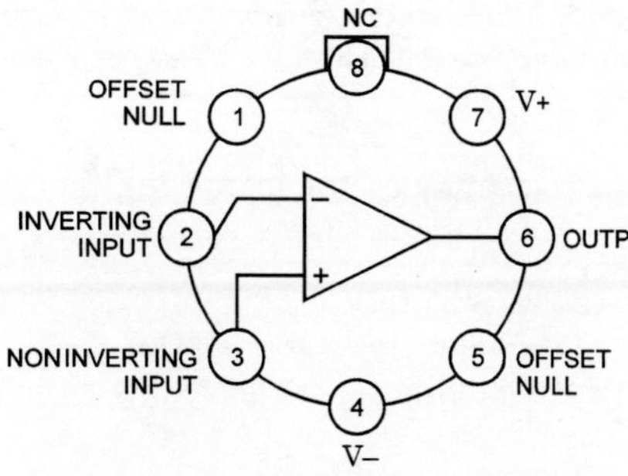

(c) Top View of Metal Can Package

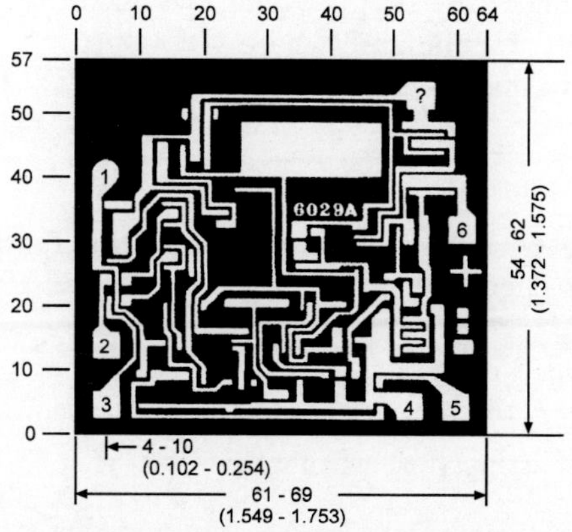

(d) Metallization Mask Layout of CA 741CH

Fig. 2.4. ***Dimensions in parenthesis are in mm and are derived from the basic inch dimensions as indicated. Grid graduations are in mils (10^{-3} inch). Schematic Diagram of 741 IC Operational Amplifier, Together with Terminal Connections for a 741 Dual-in-Line Package and for a 741 in a Metal Can and Metallization Mask Layout of CA741CH [Courtesy Harris Semiconductor]***

is the correct terminal connection and biasing for the input terminal, as well as the limitations on supply voltages, signal levels, and external load. In the schematic diagram shown in Fig. 2.4 (*a*), the base leads of

Q_1 and Q_2 are connected to the pins on the IC unit and serve as the two inputs of the op-amp. Q_1 and Q_2 transistors form a differential amplifier circuit. Transistors Q_{12} and Q_{13} constitute a complementary emitter follower output stage. The output is taken at the emitter of transistors Q_{12} and Q_{13}. This push-pull configuration allows the op-amp to have a very low output impedance, which is analogous to a voltage source having a very small internal resistance.

Figure 2.4 (*b*) shows the connnection terminals for a single 741 op-amp in a dual-in-line package, and Fig. 2.4 (*c*) shows the connection terminals for a 741 enclosed in a metal can package. From Figs. 2.4 (*b*) and 2.4 (*c*) it is obvious that there are two input terminals (pins 2 and 3) and one output terminal (pin 6). Both a positive (pin 7) and negative (pin 4) supply voltage must be furnished.

The schematic arrangement showing the output and input signals and the positive and negative power supplies is shown in Fig. 2.5. The op-amp can accommodate a positive as well as a negative input signal simultaneously or either one separately. The signal v_1 is applied to pin 3, which produces a noninverting output *i.e.,* output voltage appearing at pin 6 has the same polarity as the input signal voltage. The signal v_2 is applied to pin 2, which produces an inverting output (the internal circuitry of the op-amp yields an output signal of inverted or opposite polarity). The design of the op-amp is such that the output voltage is related to the two input voltages by equation

$$v_{out} = A\,(v_1 - v_2) = Av_d$$

Keep in mind that v_1 denotes the voltage that is applied to the noninverting terminal (pin 3) and v_2 is the signal applied to the inverting terminal (pin 2) and v_d is the net differential input voltage. The quantity A is the open-loop gain of the op-amp and customary has values of 100,000. This means that the net differential input voltage is raised at the output terminal (pin 6) by this factor. Clearly, then, if overloads are to be avoided, it is essential to limit v_d to those values that will keep output voltage v_{out} at least below the magnitude of the supply voltage V_{CC}. Accordingly, for the case where $A = 1 \times 10^5$ it follows that v_d must be limited to

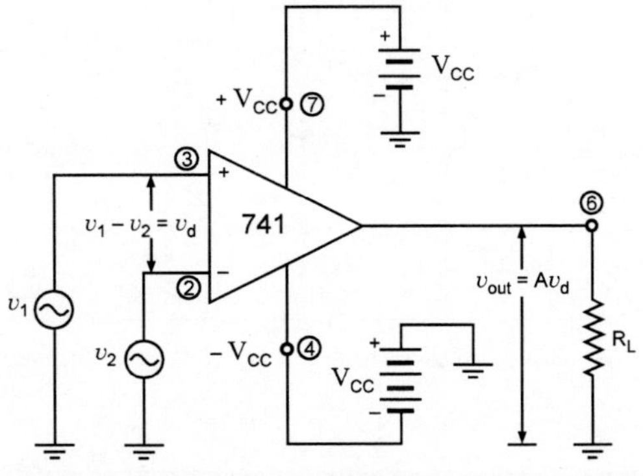

Fig. 2.5. ***Schematic Diagram of Op-Amp Showing Supply Voltages, Output and Signal Inputs With Connection to Ground Reference***

$$v_d = \frac{15\text{ V}}{1 \times 10^5} = 150\ \mu\text{V}$$

which is obviously a very small quantity. Fortunately, there are techniques for increasing this limit to more practical values.

To make the amplifier free from common-mode undesired signals, like from pickups etc., neither of its input terminals should be grounded. This can be achieved by using differential input mechanism in which both the input terminals are at same voltage level with respect to ground. One terminal is noninverting terminal and the other is inverting terminal.

2.6. IDEAL OP-AMP

The op-amp is said to be *ideal* if it has the following characteristics :

1. Its open-loop gain A is infinite. When an op-amp is operated without any connection between the output and any of the inputs (*i.e.,* without feedback), it is said to be in the *open-loop condition.* Infinite voltage gain means the voltage difference required between the two inputs to produce any output voltage is zero.

2. Its input resistance (*i.e.,* the resistance measured between inverting and noninverting terminals) R_{in} is infinite. It means that the input current (current drawn from the source) is zero and so it does not load the source. It also means that an ideal op-amp is a voltage controlled device.
3. Its output impedance R_{out} is zero *i.e.,* the output voltage v_{out} does not depend on the load resistance connected between the output terminals *i.e.,* output voltage v_{out} is independent of the current drawn by the load. The output thus can drive an infinite number of other devices.
4. Perfect balance. Because of infinite voltage gain, the voltage between the inverting and noninverting terminals of input *i.e.,* differential input voltage $v_d = v_2 - v_1$ is essentially zero (*i.e.,* $v_1 = v_2$) for finite output voltage v_{out}. This implies that v_1 and v_2 track each other *i.e.,* a virtual short circuit exists between the two input terminals but with no current flowing between the two terminals, as R_{in} is infinite.
5. Infinite frequency bandwidth *i.e.,* it has flat frequency response from dc to infinity so that any frequency signal from zero to infinity Hz can be amplified without attenuation.
6. Drift of characteristics with temperature is nil.
7. Common-mode rejection ratio (CMRR) is infinite so that amplifier is free from undesired common-mode signals such as pickups, thermal noise etc.
8. Slew rate is infinite so that output voltage changes occur simultaneously with input voltage changes.
9. Output voltage is zero when input voltage is zero *i.e.,* offset voltage is zero.

Because of having infinite input resistance and zero output resistance, ideal op-amp behaves as an *ideal voltage controlled voltage source* (VCVS).

To amplify the output of passive waveshaping networks without loading the network and also not affecting the nature of waveform a high gain VCVS with infinite bandwidth is required. An ideal op-amp bears all these characteristics, so it is suitable for such a purpose.

But none of these parameters given in ideal characteristics can be achieved by any practical op-amp. However, by use of negative feedback a practical op-amp can be brought close to some of the above ideal characteristics, such as input resistance, output resistance and bandwidth.

Actually ideal characteristics are only for assumption so that analysis for determination of performance of an op-amp for any application can be simplified.

2.7. OP-AMP EQUIVALENT CIRCUIT (OR MODEL)

Figure 2.6 shows an op-amp equivalent circuit, which is useful in analyzing the basic operating principles of op-amps and in observing the effects of feedback arrangements.

The voltage source Av_d is an equivalent Thevenin voltage source, and R_{out} is the Thevenin equivalent resistance looking back into the output terminal of the op-amp.

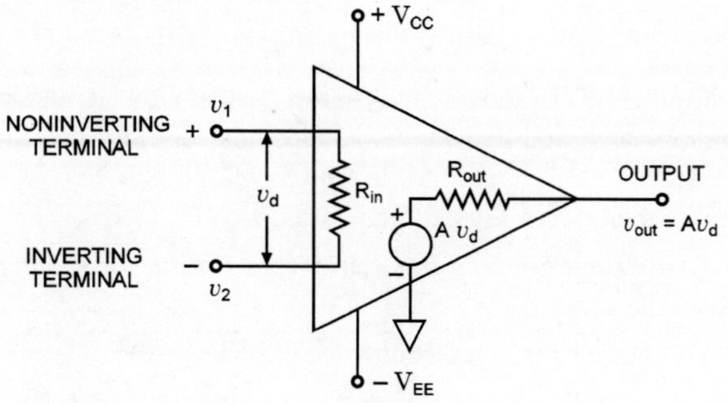

Fig. 2.6

The output voltage of the circuit shown in Fig. 2.6 is given as

$$v_{out} = Av_d = A\,(v_1 - v_2) \qquad ...(2.3)$$

where A is large signal voltage gain, v_d is differential input voltage, v_1 and v_2 are the input voltages at noninverting and inverting terminals respectively with respect to ground.

Equation (2.3) shows that output voltage is directly proportional to the algebraic difference of the two input voltages which means that the op-amp amplifies the difference between the two input voltages, not the input voltages themselves. For this reason the polarity of the output voltage depends on the polarity of the difference voltage v_d.

2.8. IDEAL VOLTAGE TRANSFER CURVE

Equation (2.3) is the basic op-amp equation in which it has been assumed that output offset voltage is zero. This equation is quite useful in studying the characteristics of an operational amplifier and in the analysis of different circuit configurations employing feedback.

The curve drawn between output voltage v_{out} and input differential voltage v_d keeping gain A constant (Fig. 2.7) represents the Eq. (2.3) graphically. The curve shown in Fig. 2.7 is known as an *ideal voltage transfer curve,* ideal because in drawing this curve output offset voltage is assumed to be zero. In normal op-amp use (with negative feedback), the output offset voltage is nearly zero and is ignored for simplifying the calculations. It is noteworthy here that the curve shown in Fig. 2.7 is not drawn to scale : if it were drawn to scale, it would have been an almost vertical line because of very large voltage gain A.

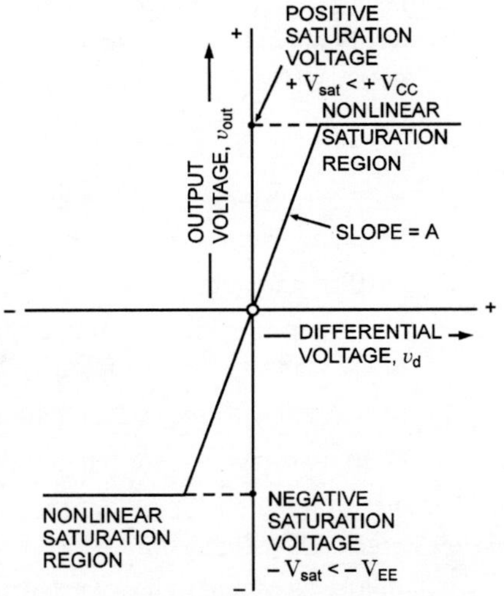

Fig. 2.7. ***Ideal Voltage Transfer Curve***

The curve shown in Fig. 2.7 reveals that the output voltage v_{out} cannot exceed the positive and negative saturation voltages. Positive and negative saturation voltages are specified by an output voltage swing rating of the op-amp for given values of supply voltages. In other words it can be stated that the output voltage v_{out} increases in direct proportion to the input differential voltage v_d only until it attains the saturation value and then it becomes constant, as shown in Fig. 2.7.

2.9. BASIC OP-AMP CIRCUIT

The basic circuit of an op-amp is shown in Fig. 2.8. An input signal V_2 is applied through a resistor R_1 to the inverting or minus input terminal. The output voltage is fedback through resistor R_f to the same input terminal. The noninverting or plus input terminal is connected to earth. For determination of overall gain of the circuit *i.e.,* $\frac{V_{out}}{V_2}$, some more details of the op-amp unit will have to be considered.

Figure 2.8 (*b*) shows the op-amp replaced by an equivalent circuit having input resistance R_{in} and output voltage source and resistance. As an ideal op-amp has infinite input resistance (*i.e.,* $R_{in} = \infty$), zero output resistance (*i.e.,* $R_{out} = 0$) and infinite voltage gain (*i.e.,* $A = \infty$), the connection for the ideal amplifier illustrated is redrawn in Figs. 2.8 (*c*) and 2.8 (*d*).

Using superposition theorem we can solve for the input voltage V_{in} in terms of the components due to each of the sources.

For source V_2 only (assuming $-A V_{in}$ to be zero)

$$V_{1\,in} = \frac{R_f}{R_1 + R_f} V_2$$

For source $-A V_{in}$ only (assuming V_2 to be zero)

$$V_{2\,in} = \frac{R_1}{R_1 + R_f} (-A V_{in})$$

$$\text{Total input voltage } V_{in} = V_{1\,in} + V_{2\,in} = \frac{R_f V_2}{R_1 + R_f} - \frac{AR_1 V_{in}}{R_1 + R_f}$$

$$\text{or } V_{in} = \frac{R_f V_2}{R_f + R_1(1 + A)} \quad \text{...(2.4)}$$

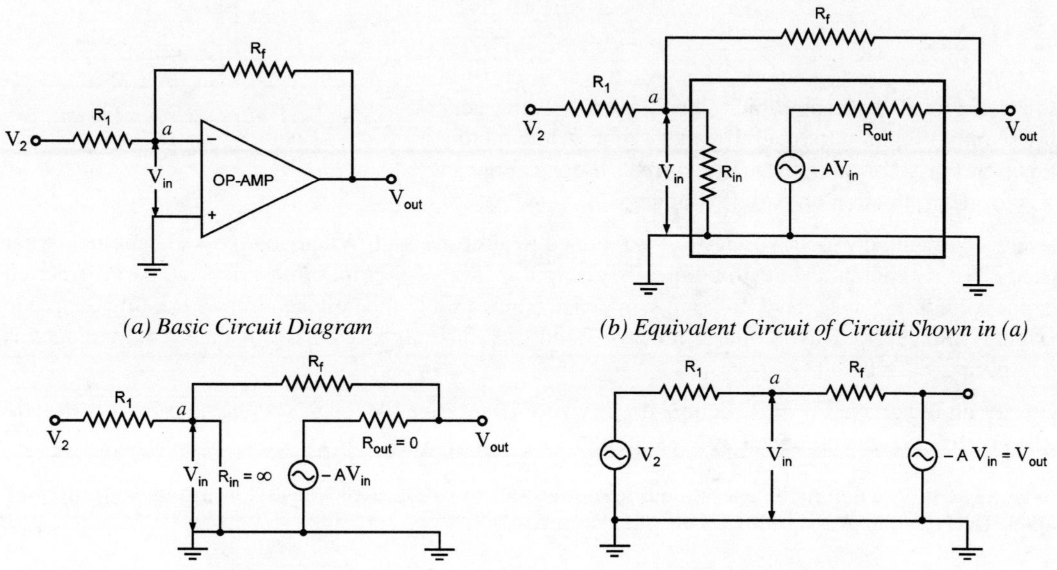

(a) Basic Circuit Diagram

(b) Equivalent Circuit of Circuit Shown in (a)

(c) Ideal Op-amp

(d) Ideal Equivalent Circuit

Fig. 2.8

If A >> 1 and A R_1 >> R_f as is usually true, then

$$V_{in} \simeq \frac{R_f}{AR_1} V_2$$

$$\frac{V_{out}}{V_2} = \frac{-AV_{in}}{V_2} = \frac{-A}{V_2} \frac{R_f V_2}{AR_1} = \frac{-R_f}{R_1} \quad \text{...(2.5)}$$

From above Eq. (2.5) it is obvious that the ratio of overall output to the input voltage is dependent only on the values of resistors R_1 and R_f provided A is very large.

By selecting precise resistor values for R_f and R_1, a wide range of gains can be obtained. The gain will be as accurate as the resistors used and only slightly affected by temperature and other circuit factors.

2.9.1. Virtual Ground

As already mentioned in Art. 2.4, the output voltage V_{out} is to be limited to the supply voltage V_{CC}, which is of the order of 10 or 15 V then the input voltage

$$V_{in} = \frac{V_{out}}{A} = \frac{10 \text{ or } 15 \text{ V}}{10^5} = 100 \text{ or } 150\ \mu\text{V} \quad \text{if A is assumed to be } 10^5$$

The value of V_{in} compared to all other voltages, is very small and may be considered to be zero volt. Note that although V_{in} is almost equal to zero but not exactly zero because output voltage is V_{in} times the amplifier gain. Of course, amplifier gain is very high, of the order of 10^5.

The fact that $V_{in} \cong 0$ volt leads to the concept that at the input to the amplifier there exists a virtual short circuit or virtual ground. The concept of a virtual short implies that although the input voltage is nearly zero, there is no current through the amplifier input to ground.

The concept of virtual ground is depicted in Fig. 2.9. The thick line indicates that a short may be considered to exist with $V_{in} \cong 0$ V, but this is a virtual short in that no current passes through the short to the ground. Current is through resistors R_1 and R_f, as illustrated.

Using the virtual ground concept, equations for the current I may be written as

$$I = \frac{V_2}{R_1} = -\frac{V_{out}}{R_f} \quad \text{or} \quad \frac{V_{out}}{V_2} = -\frac{R_f}{R_1}, \quad \text{the same as Eq. (2.5)}$$

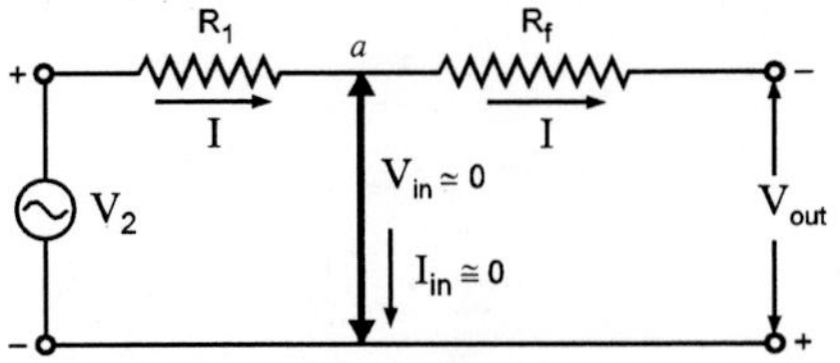

Fig. 2.9. ***Virtual Ground in an Op-amp***

The virtual ground concept, which depends on A being very large, allows simple solution for overall voltage gain. Although the circuit shown in Fig. 2.9 is not physically correct, it does allow an easy means for determination of overall voltage gain.

Now let us explain how input voltage V_{in} is reduced to almost 0 volt. When voltage V_2 is applied, point *a* attains some positive potential but simultaneously output voltage V_{out} comes into existence. Some fraction of this output voltage V_{out} is fedback to point *a* in phase opposition to the voltage already existing there due to input signal voltage V_2. This is due to negative feedback. The algebraic sum of these two voltages is almost zero giving $V_{in} \cong 0$ volt.

Obviously input voltage V_{in} will become exactly zero if negative feedback voltage at point *a* is exactly equal to the positive voltage produced by V_2 at point *a*.

Example 2.1. The differential input operational amplifier, shown in Fig. 2.10, consists of a base amplifier of infinite gain. Derive an expression for its output voltage.

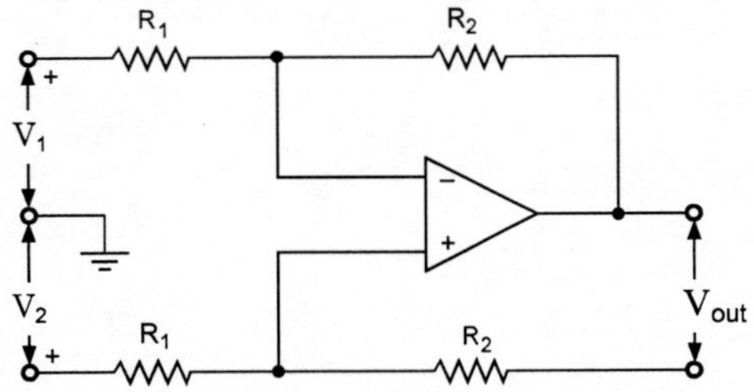

Fig. 2.10

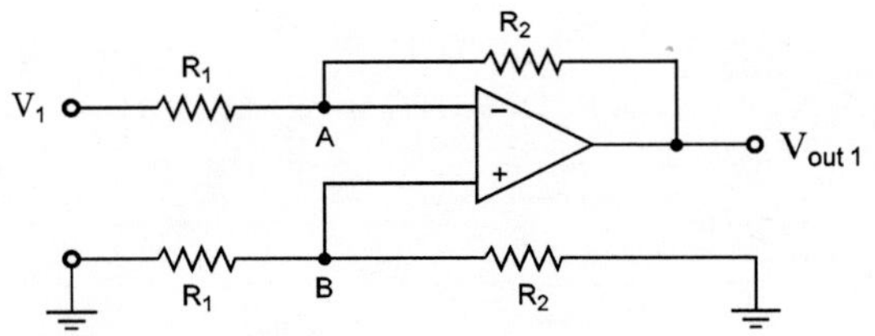

Fig. 2.11

Solution : Applying superposition theorem for V_{out},

Let $V_{out} = V_{out1} + V_{out2}$

Applying Kirchhoff's current law at node A (Fig. 2.11) and solving the equation we have

$$V_{out1} = \left(-\frac{R_2}{R_1}\right)V_1$$

Applying Kirchhoff's current law at node B (Fig. 2.12) and solving the equation we have

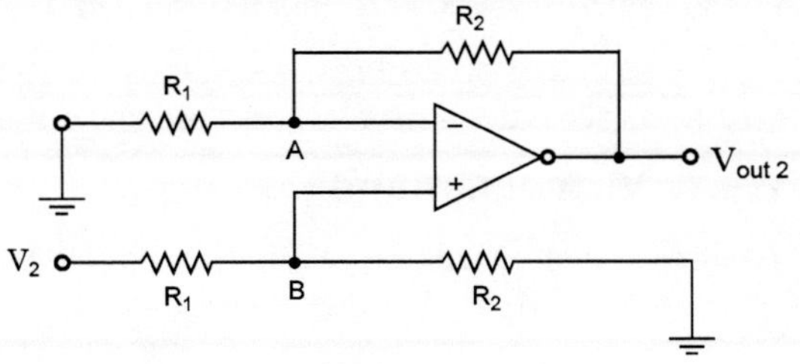

Fig. 2.12

$$V_{out_2} = \left(\frac{R_2}{R_1 + R_2}\right)\left(1 + \frac{R_2}{R_1}\right)V_2 = \frac{R_2}{R_1}V_2$$

$$\therefore V_{out} = V_{out_1} + V_{out_2} = \frac{R_2}{R_1}(V_2 - V_1)$$

Example 2.2. Determine the output voltage in each of the following cases for the open-loop differential amplifier shown in Fig. 2.13.

(*i*) $V_{in_1} = 5\ \mu V$ dc, $V_{in_2} = -7\ \mu V$ dc

(*ii*) $V_{in_1} = 10$ m V_{rms}, $V_{in_2} = 20$ m V_{rms}

The op-amp is 741 with following specifications:

$A = 2 \times 10^5$, $R_{in} = 2\ M\Omega$, $R_{out} = 75\ \Omega$, $+V_{CC} = +15$ V, $-V_{EE} = -15$ V, output voltage swing = 14 V.

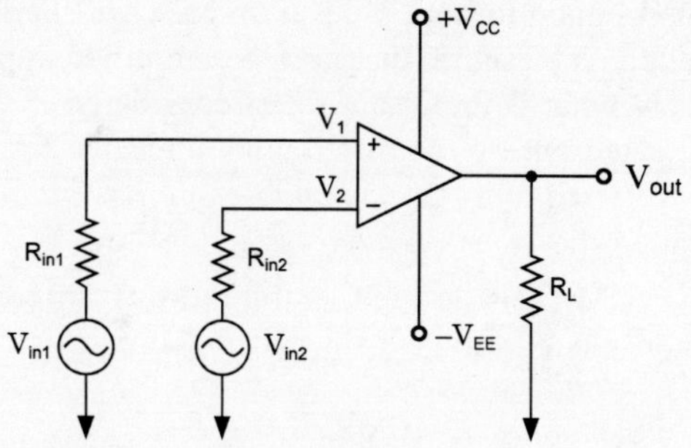

Fig. 2.13

Solution : (*i*) $V_{out} = (V_{in_1} - V_{in_2})\,A_V$

$= (5 + 7)\,10^{-6} \times 2 \times 10^5 = 2.4$ volts **Ans.**

(*ii*) $V_{out} = 2 \times 10^5\,(10\sqrt{2}\ \sin \omega t - 20\sqrt{2}\ \sin \omega t) \times 10^{-3}$

$= -2{,}828 \sin \omega t$ volts **Ans.**

In this case, the output will fluctuate between +7 V and –7 V.

Example 2.3. For the op-amp circuit shown in Fig. 2.14, determine the output voltage V_{out}. Assume that the op-amps are ideal.

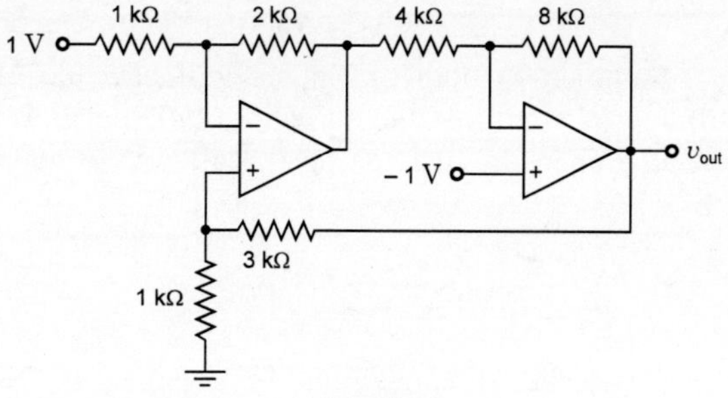

Fig. 2.14

Solution : The circuit is labelled, as shown in Fig. 2.15

Voltage at node A, $V_A = \dfrac{V_{out} \times R_5}{R_5 + R_6}$

$= \dfrac{V_{out} \times 1}{1+3} = \dfrac{V_{out}}{4}$

Voltage at node C, $V_C = -1$ V

Applying Kirchhoff 's current law at node B and node A respectively, we have

$$\frac{V_B + 1}{4} + \frac{V_B - V_A}{2} = 0$$

$$\text{or} \quad \frac{3}{4} V_B + 0.25 = \frac{V_A}{2} = \frac{V_{out}}{8} \qquad ...(i)$$

$$\text{and} \quad \frac{V_A - 1}{1} + \frac{V_A - V_B}{2} = 0$$

$$\text{or} \quad \frac{V_B}{2} + 1 = \frac{3}{2} V_A = \frac{3}{8} V_{out} \qquad ...(ii)$$

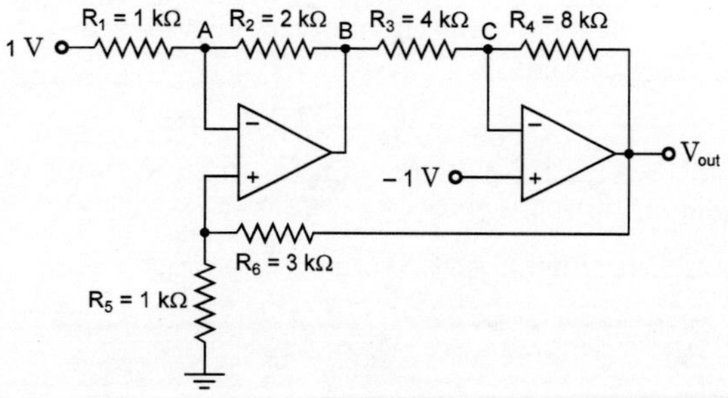

Fig. 2.15

Solving Eqs. (*i*) and (*ii*) we have

$$V_{out} = \frac{20}{7} \text{ volts} \quad \textbf{Ans.}$$

2.10. COMMON-MODE REJECTION RATIO (CMRR)

One of the more common features of a differential amplifier is its ability to cancel out or reject certain types of undesired voltage signals. Such undesired signals are referred to as "noise" and may occur owing to voltages induced by stray magnetic fields in the ground or signal wires, as voltage variations in the voltage supply. Here the important point is that these noise signals are not the signals that are desired to be amplified in the differential amplifier. Their distinguishing feature is that the noise signal appears equally at both inputs of the circuit. It means that any undesired (or noise) signals that appear in polarity, or common to both input terminals, will be largely rejected, or cancelled out at the differential amplifier output. The signal that is to be amplified appears at only one input or opposite in polarity at both inputs. Now what is important to be considered is that, if undesirable noise does occur, up to what extent it is rejected out by the differential amplifier ? A measure of this rejection of signals common to both inputs is referred to as the *common-mode rejection* of the amplifier and a numerical value is assigned, which is called the *common-mode rejection ratio* (CMRR).

CMRR is defined as the ratio of differential voltage gain to common-mode voltage gain and it is given as

$$\text{CMRR} = \frac{A_d}{A_{cm}}$$

Fig. 2.16

If a differential amplifier is perfect, CMRR would be infinite because in that case common-mode voltage gain A_{cm} would be zero.

Figure 2.16 represents a linear active device with two input signals V_1 and V_2 and one output signal V_{out}, each measured with respect to ground. In an ideal differential amplifier the output signal V_{out} should be given as

$$V_{out} = A\,(V_1 - V_2) \quad ...(2.6)$$

where A is the gain of the differential amplifier.

Common to both the inputs will have no effect on the output voltage. However, in a practical differential amplifier, the output not only depends upon the difference signal V_d of the two input signals, but also upon the average level, called the *common-mode signal* V_{cm} where

$$V_d = V_1 - V_2 \quad ...(2.7)$$

$$\text{and } V_{cm} = \frac{1}{2}(V_1 + V_2) \quad ...(2.8)$$

For example the output in the two cases having input signals of (V_1 = +100 μV and V_2 = – 50 μV) and (V_1 = + 1,200 μV and V_2 = + 1050 μV) will not be exactly the same, even though the algebraic difference of two input signals V_d in both cases is the same *i.e.,* 150 μV.

Now let the output voltage V_{out} in Fig. 2.16 be expressed as a linear combination of two input voltages as below

$$V_{out} = A_1V_1 + A_2V_2 \quad ...(2.9)$$

where A_1 is the voltage gain for input V_1 with V_2 grounded and A_2 is the voltage gain for the input V_2 with V_1 grounded.

But from Eqs. (2.7) and (2.8)

$$V_1 = V_{cm} + \frac{1}{2}V_d \quad ...(2.10)$$

$$\text{and } V_2 = V_{cm} - \frac{1}{2}V_d \quad ...(2.11)$$

Substituting the values of V_1 and V_2 from above Eqs. (2.10) and (2.11) in Eq. (2.9) we have

$$V_{out} = A_1\left(V_{cm} + \frac{1}{2}V_d\right) + A_2\left(V_{cm} - \frac{1}{2}V_d\right) = \frac{1}{2}(A_1 - A_2)V_d + (A_1 + A_2)V_{cm} \quad ...(2.12)$$

$$= A_d\,V_d + A_{cm}\,V_{cm} \quad ...(2.13)$$

$$\text{where } A_d = \frac{A_1 - A_2}{2} \quad ...(2.14)$$

$$\text{and } A_{cm} = (A_1 + A_2) \quad ...(2.15)$$

Here A_d is the voltage gain for the difference signal while A_{cm} is the voltage gain for the common-mode signal.

The voltage gain A_d may be measured directly by selecting $V_1 = -V_2 = 0.5$ V so that $V_d = 1$ V and $V_{cm} = 0$ V. Under such conditions output voltage is $A_d \times (1 \text{ V})$ *i.e.,* the output voltage equals A_d.

Similarly for measuring A_{cm}, select $V_1 = V_2 = 1$ V, so that $V_d = 0$ and $V_{cm} = 1$ V. Under such conditions output voltage is $A_{cm} \times (1 \text{ V})$ *i.e.,* the output voltage equals A_{cm}.

Thus, values of differential voltage gain A_d and common-mode voltage gain A_{cm} can be determined simply by measurement of output voltage.

Having measured A_d and A_{cm} for the amplifier, CMMR can be determined from the following relation

$$\text{CMMR} = \frac{A_d}{A_{cm}} \qquad \text{...(2.16)}$$

The value of CMMR can also be expressed in logarithmic terms as

$$\text{CMMR in dB} = 20 \log_{10} \frac{A_d}{A_{cm}} \qquad \text{...(2.17)}$$

From Eqs. (2.13) and (2.16) we have an expression for the output voltage in the following form

$$V_{out} = A_d V_d \left(1 + \frac{1}{\text{CMMR}} \frac{V_{cm}}{V_d}\right) \qquad \text{...(2.18)}$$

From above Eq. (2.18) it is seen that the amplifier should be designed so that CMRR is large in comparison to the ratio of common-mode signal to the difference signal (V_{cm}/V_d).

Even if both components of voltage V_{cm} and V_d exist at the inputs, the value of $\left(\frac{1}{\text{CMMR}} \frac{V_{cm}}{V_d}\right)$ will be very small, for CMRR very large, and the output voltage will be almost equal to $A_d V_d$. It means that the output will be almost completely due to the difference signal with the common-mode input signals rejected or cancelled out.

Example 2.4. A differential dc amplifier has a differential mode gain of 100 and a common-mode gain 0.01. What is its CMMR in dB ?

Solution : Differential mode gain, $A_d = 100$

Common-mode gain, $A_{cm} = 0.01$

$$\text{CMRR} = \frac{A_d}{A_{cm}} = \frac{100}{0 \cdot 01} = 10^4$$

$$\text{CMRR in dB, CMRR}' = 20 \log_{10} \text{CMRR} = 20 \log_{10} 10^4 = 20 \times 4 = 80 \text{ dB} \quad \textbf{Ans.}$$

Example 2.5. For a given op-amp, CMRR = 10^4 and differential gain $A_d = 10^5$. What is the common-mode gain ?

Solution : Differential voltage gain, $A_d = 10^5$

CMRR $= 10^4$

$$\text{Common-mode gain } A_{cm} = \frac{A_d}{\text{CMRR}} = \frac{10^5}{10^4} = 10 \quad \textbf{Ans.} \qquad \because \text{CMRR} = \frac{A_d}{A_{cm}}$$

Example 2.6. The two input terminals of an op-amp are connected to voltage signals of strength 745 μV and 740 μV respectively. The gain of the op-amp in differential mode is 5 × 10^5 and its CMRR is 80 dB. Calculate the output voltage and percentage error due to common-mode.

Solution : Input voltage $V_1 = 745$ μV and $V_2 = 740$ μV

Differential voltage, $V_d = V_1 - V_2 = 745 - 740 = 5$ μV

Common mode signal, $V_{cm} = \frac{V_1 + V_2}{2} = \frac{745 + 740}{2} = 742.5\ \mu V$

Differential voltage gain, $A_d = 5 \times 10^5$

$\text{CMRR} = 80 \text{ dB} = \text{Antilog } \frac{80}{20} = 10{,}000$

Output voltage, $V_{out} = A_d V_d \left[1 + \frac{1}{\text{CMRR}} \frac{V_{cm}}{V_d}\right]$

$$= 5 \times 10^5 \times 5 \times 10^{-6} \left[1 + \frac{1}{10{,}000} \times \frac{742.5 \times 10^{-6}}{5 \times 10^{-6}}\right] V = 2.537125 \text{ V} \quad \textbf{Ans.}$$

So error voltage $= V_{out} - A_d V_d = 2.537125 - 5 \times 10^5 \times 5 \times 10^{-6} = 0.037125$ V

Percentage error $= \frac{\text{Error voltage}}{V_{out}} \times 100 = \frac{0.037125}{2.537125} \times 100 = 1.4633\%$ **Ans.**

2.11. OPEN-LOOP OP-AMP CONFIGURATIONS

Open-loop means that there is no connection between input and output terminals either direct or via another network. It means that output signal is not fed back in any form as part of the input signal, and the loop that would have been formed with feedback is open. An op-amp in open-loop configuration acts as a high-gain amplifier. The three open-loop op-amp configurations are :

1. Differential amplifier
2. Inverting amplifier
3. Noninverting amplifier

2.11.1. Differential Amplifier

The circuit of open-loop op-amp differential amplifier is shown in Fig. 2.17. In this circuit, inputs are applied to both the inverting and noninverting terminals. Since, in this configuration, the difference between two input signals is amplified, the configuration is called the *differential amplifier*.

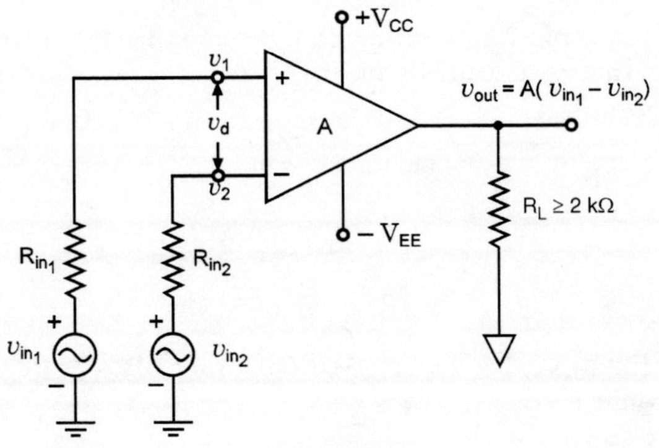

Fig. 2.17

Input signals v_{in_1} and v_{in_2} may be either ac or dc voltages as op-amp is capable of amplifying both dc and ac input voltages. Source resistances R_{in_1} and R_{in_2} are usually negligibly small as compared to input resistance of op-amp (R_{in}). Neglecting voltage drop across source resistors

$$v_1 = v_{in_1} \text{ and } v_2 = v_{in_2}$$

and output voltage is given as

$$v_{out} = A(v_1 - v_2) = A(v_{in_1} - v_{in_2}) \quad ...(2.19)$$ where A is open-loop gain.

From above Eq. (2.19), it is obvious that the output voltage is A times the difference between two input voltages and the polarity of the output voltage depends upon the polarity of the input difference voltage $(v_{in_1} - v_{in_2})$.

2.11.2. Inverting Amplifier

In inverting configuration, the input signal is applied to the inverting (–) input terminal and noninverting (+) input terminal is grounded, as illustrated in Fig. 2.18.

In Fig. 2.18, since $v_1 = 0$, and $v_2 = v_{in}$, output voltage,

$$v_{out} = Av_d = A(v_1 - v_2) = -A\,v_{in} \quad ...(2.20)$$

The negative sign indicates that the output voltage is out of phase w.r.t. input voltage by 180°. Thus output voltage is A times the input voltage and is of opposite polarity.

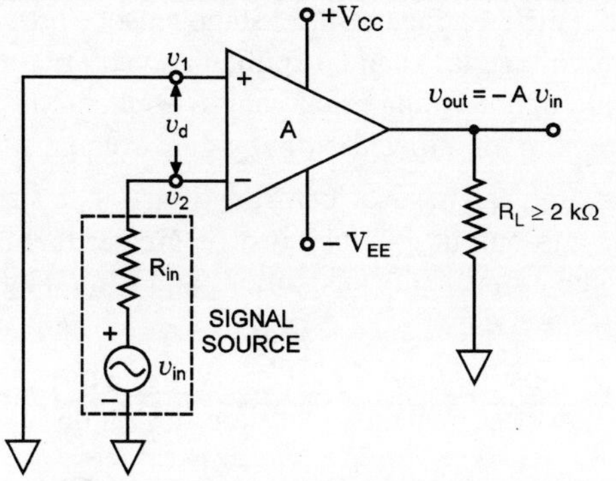

Fig. 2.18. *Inverting Amplifier*

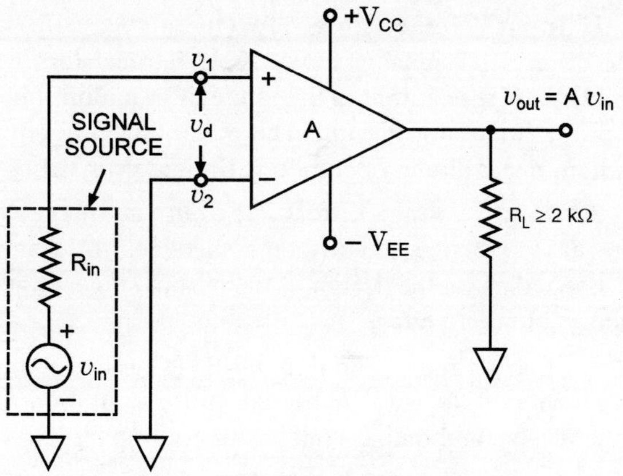

Fig. 2.19. *Noninverting Amplifier*

2.11.3. Noninverting Amplifier

In this configuration, the input signal is applied to the noninverting (+) input terminal and inverting (–) input terminal is grounded, as illustrated in Fig. 2.19.

In Fig. 2.19, since $v_1 = v_{in}$ and $v_2 = 0$, output voltage,

$$v_{out} = Av_d = A(v_1 - v_2) = Av_{in} \quad ...(2.21)$$

The above Eq. (2.21) reveals that the output voltage v_{out} is A times the input voltage v_{in} and is of the same polarity (*i.e.,* v_{out} and v_{in} are in phase).

2.11.4. Limitations of Open-Loop Op-Amp Configurations

In all the three open-loop op-amp configurations any input signal (differential or single), which even slightly exceeds zero, drives the output into saturation because of very high gain (nearly infinity) of op-amp. Thus, when an op-amp is operated in the open-loop configuration, the output either goes to positive saturation or negative saturation levels or switches between positive and negative saturation levels and thus clips the output above these levels. So open-loop configurations of op-amp cannot be used in linear applications. However, open-loop op-amp configurations are used in certain nonlinear applications such as square-wave generation.

The op-amp can be effectively employed in linear applications if feedback is introduced *i.e.,* an output signal is fed back to the input either directly or via another network.

Example 2.7. Find the output voltage of an open-loop op-amp when the differential input voltage is ± 25 μV. Assume open-loop gain A to be 200,000.

Solution : Output voltage, $V_{out} = A \times V_d = 200{,}000 \times (\pm 25) \times 10^{-6} = \pm\ 5$ V **Ans.**

2.12. PRACTICAL OP-AMP PARAMETERS

In ideal op-amps, we assume, for example, that the differential voltage gain is infinite, the input resistance is infinite and the output resistance is zero. In practical op-amp circuits, these ideal parameter values are not realized. There are several reasons for it, such as biasing currents, variations in transistor parameters (β and V_{BE} values), change in operating point due to variation in supply voltage and thermal drifts.

Here we define some of the practical op-amp parameters.

2.12.1. Practical Op-Amp Parameter Definitions

1. Input Voltage Limits. Two input voltage limitations are to be considered—a dc input voltage limit and a differential signal input voltage. All transistors in the input differential amplifier stage must be properly biased, so there is a limit in the range of common-mode input voltage that can be applied and still maintain the proper transistor biasing. The maximum differential input signal voltage that can be applied and still maintain linear circuit operation is limited primarily by the maximum allowed output signal voltage.

2. Output Voltage Limits. The output voltage of the op-amp can never exceed the limits of dc supply voltages. In practice, the difference between the bias voltage and output voltage must be greater than 1 to 4 V, depending on the design of the output stage. Otherwise, the output voltage saturates and is no longer a function of input voltage.

For linear circuit operation, all BJTs in an op-amp circuit must be biased in the forward active region and all MOSFETs must be biased in the saturation region. For these reasons, there are limitations to the range of input and output voltages in op-amp circuit.

3. Output Current Limitation. The maximum current out of (or into) the op-amp is determined by the current ratings of the output transistors. Practical op-amp circuits cannot source or sink an infinite amount of current.

4. Finite Open-Loop Voltage Gain. The open-loop voltage gain of an ideal op-amp is assumed to be infinite. In practice, the open-loop gain of any op-amp circuit is always finite. The nonideal parameter value affects the circuit performance.

5. Input Resistance. The input resistance R_{in} is the small signal resistance between the inverting and noninverting terminals when a differential voltage is applied. Ideally, this parameter is infinite, but especially for BJT circuits, this parameter is finite.

A nonzero closed-loop input resistance in conjunction with a finite open-loop input resistance implies that the signal current into the op-amp is not zero, as assumed in the ideal case.

6. Output Resistance. The output resistance is the Thevenin equivalent small-signal resistance looking back into the output terminal of the op-amp measured with respect to ground. The ideal output resistance is zero, which means that there is no loading effect at the output. In practice, this value is not zero.

Since the ideal op-amp has a zero output resistance, the output voltage is independent of the load impedance. The op-amp acts as an ideal voltage source and there is no loading effect. An actual op-amp circuit has a nonzero output resistance, which means that the output voltage, and, therefore, the closed-loop gain is a function of load impedance.

7. Finite Bandwidth. In the ideal op-amp, the bandwidth is infinite. In practical op-amps, the bandwidth is finite because of capacitances within the op-amp circuit.

8. Slew Rate. The slew rate is defined as the maximum rate of change in output voltage per unit of time. The maximum rate at which the output voltage can change is also a function of capacitances within the op-amp circuit.

9. Input Offset Voltage. In an ideal op-amp, the output voltage is zero for zero differential input signal voltage. However, mismatches between input devices, for instance, may create an output voltage with zero input. The input offset voltage is the applied differential input voltage required to induce a zero output voltage.

10. Input Bias Currents. In an ideal op-amp, the input current to the op-amp is assumed to be zero. However, in practical op-amps, especially with BJT input devices, the input bias currents are not zero.

2.13. OP-AMP CHARACTERISTICS

Because an op-amp is a dc amplifier, we have to consider both dc and ac characteristics in troubleshooting, analysis and in designing of op-amp circuits.

2.13.1. DC Characteristics

The ideal op-amp draws no current from the source driving it. The circuit response does not vary with temperature. Actual op-amps do not operate in this way. Current is drawn from the source into the op-amp's inputs. There are slight differences in the way the two inputs respond to current and voltage. An actual op-amp will shift its operation with temperature. These nonideal dc characteristics and compensation techniques will be described here.

1. Input Offset Voltage. When the inputs of the op-amp are grounded, there is almost always an *output offset voltage**, as illustrated in Fig. 2.20 (*a*), because the input transistors have different V_{BE} values. For instance, the data sheet of a typical 741C lists a maximum input offset voltage of ±2 mV. This difference of 2 mV is an unwanted input signal that is amplified to give an *output offset voltage*. The output offset voltage can be eliminated by applying a voltage of 2 mV to the inverting input as shown in Fig. 2.20 (*b*). Then the output offset voltage is reduced to zero. Since the offset can have either polarity, it might be necessary to reverse the polarity of the 2 mV.

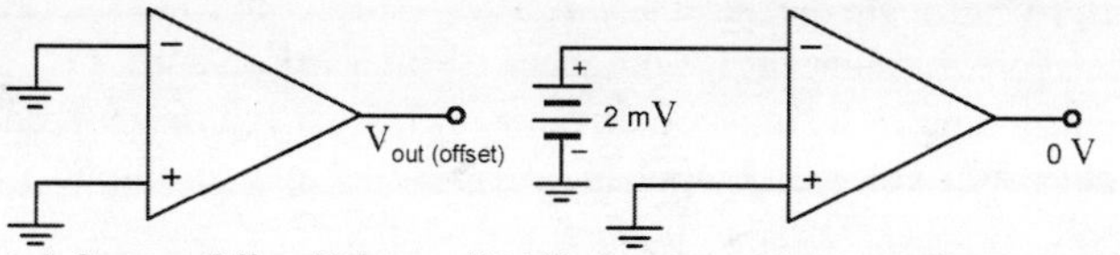

(a) Output Offset Voltage (b) Elimination of Output Offset Voltage

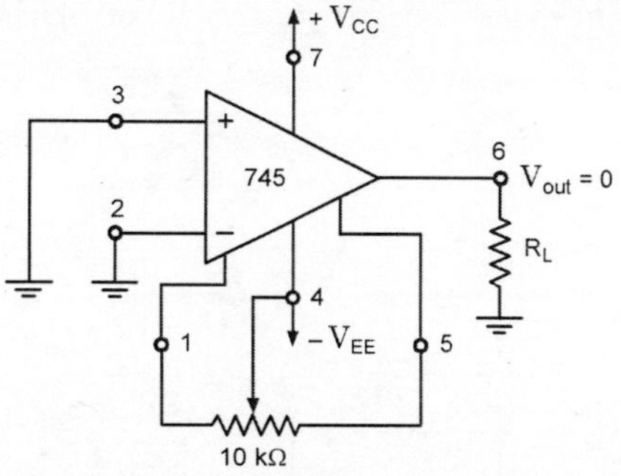

(c) Voltage Offset Null Circuit For The 741

Fig. 2.20

Thus the input offset voltage $V_{in\ (offset)}$ may be defined as that voltage, which is to be applied between the input terminals to balance the amplifier.

The adjustment explained above [Fig. 2.20 (*b*)] could be made to eliminate offset errors if the offset voltages were constant. But these voltages are not constant and depend upon temperature and other parameters such as supply voltage.

Therefore, in most of the practical op-amps extra terminals are provided where an external potentiometer can be connected and adjusted to reduce the offset voltage to zero. For instance, for the 741-type op-amp the manufacturer recommends that a 10 kΩ potentiometer be placed across offset null pins 1 and 5 and a wiper is connected to the negative supply pin 4, as shown in Fig. 2.20 (*c*).

2. Input Bias Current. The op-amp's input is a differential amplifier. It may be made of BJTs or FETs. In either case, these transistors are required to be biased, and this takes current. For the BJTs this bias current, supplied by the external circuit, is the base bias current, biasing the input transistors into their linear region. The

*Output offset voltage is the dc voltage present at the output terminal when the two input terminals are grounded.

bias current for the FET input op-amp is actually the leakage current across the reverse-biased gate to channel junction.

Manufacturers specify input bias current as the average of the bias current at each input

$$i.e., \ I_B = \frac{I_{B_1} + I_{B_2}}{2} \text{ for } V_{out} = 0 \quad ...(2.22)$$

The value of input bias current I_B is very small, in the range of a few to a few hundred nanoamperes. It is a dc current and is specified on the op-amp data sheets. For example, I_B is 500 nA maximum at supply voltages of ±15 V dc for the μA741. The value of I_B has been significantly reduced in later generation op-amps.

Even though very small, the input bias current I_B can cause a significant output offset voltage in circuits using relatively large feedback resistors. This output offset voltage may not be as large as that caused by the input offset voltage.

Figure 2.21 shows a basic inverting amplifier. For 0 input voltage, output voltage V_{out} must be zero. However, both the inverting and noninverting inputs take bias current from the external circuit. Current I_B^+ comes directly from ground, and develops no voltage. Ignoring the input offset voltage, current I_B^- comes entirely through the feedback resistor (since both ends of R_{in} are at 0 V, no current can flow through it). Therefore, bias cirrent I_B^- produces a voltage drop across R_f. Since the left end of R_f is held at virtual ground, the right end of R_f must be driven to $I_B^- R_f$ by the op-amp output. So, instead of a 0 output voltage (ignoring input offset voltage), output is $I_B^- R_f$. So to reduce output offset voltage due to input bias current I_B, use of small feedback resistors is recommended.

To eliminate or reduce the output offset voltage due to input bias current I_B some scheme has to be devised to make voltages V_1 and V_2 at the inputs equal.

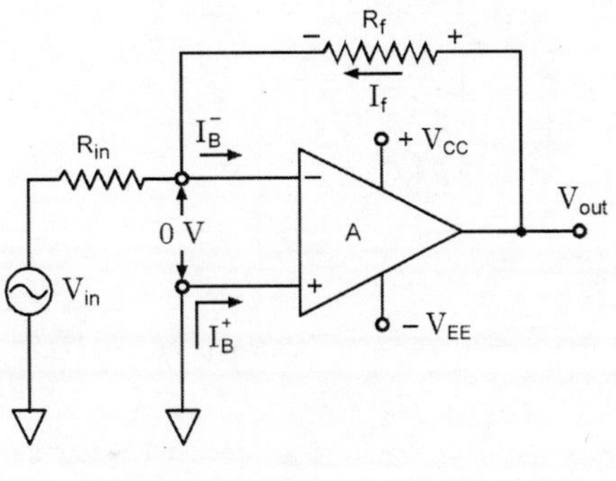

Fig. 2.21

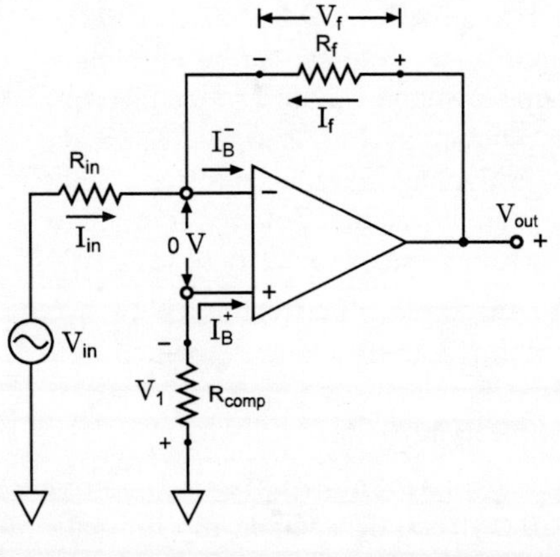

Fig. 2.22. ***Bias Current Compensation Circuit***

The compensating circuit is given in Fig. 2.22. A compensating resistor has been added between ground and the noninverting input. Current I_B^+ flowing through this resistor develops a voltage V_1.

Summing the voltages around the input feedback-output loop, we have

$$-V_1 + 0 + V_f - V_{out} = 0$$

$$\text{or} \quad V_{out} = V_f - V_1$$

If the R_{comp} is selected of proper size, V_1 will cancel V_f and output will be zero. The value of R_{comp} is derived below

$$V_1 = I_B^+ R_{comp}$$

$$\text{or } I_B^+ = \frac{V_1}{R_{comp}}$$

$$I_{in} = \frac{V_1}{R_{in}} \text{ and } I_f = \frac{V_f}{R_f}$$

For compensation $V_f = V_1$

$$\text{So } I_f = \frac{V_f}{R_f} = \frac{V_1}{R_f}$$

The current $I_B^- = I_f + I_{in} = \dfrac{V_1}{R_f} + \dfrac{V_1}{R_{in}} = V_1\left[\dfrac{1}{R_f} + \dfrac{1}{R_{in}}\right] = V_1\left(\dfrac{R_{in} + R_f}{R_f R_{in}}\right)$

Assuming $I_B^- = I_B^+$

$$V_1\left(\frac{R_{in} + R_f}{R_{in}\,R_f}\right) = \frac{V_1}{R_{comp}}$$

$$\text{or } R_{comp} = \frac{R_{in}\,R_f}{R_{in} + R_f} = R_{in} \,||\, R_f \qquad \text{...(2.23)}$$

So for compensating for bias currents, R_{comp} (total resistance seen from the noninverting input to ground) is made equal to the total resistance seen from the inverting input to ground. With FET-input op-amps the effect is usually negligible.

3. Input Offset Current. In case of an op-amp having input transistors of equal V_{BE} values, though the input offset voltage would be zero, but the problem may arise because of the bias currents. If either input to the op-amp has a large resistance in the return path, an output offset voltage may exist. For example, a resistance R_B exists between the noninverting input and ground, as shown in Fig. 2.23 (*a*). In such a condition a voltage will appear at the noninverting input due to flow of bias current I_{B_1} through R_B and is given as

$$V_1 = I_{B_1} R_B \qquad \text{...(2.24)}$$

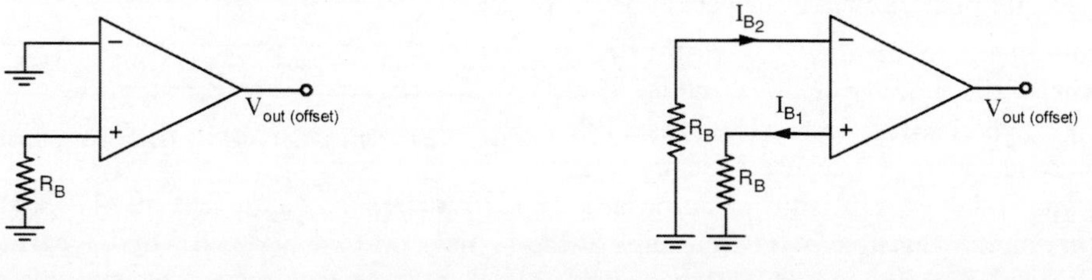

(a) Output Offset Voltage Due to Return-Path Resistance *(b) Reduced Offset Voltage by Equal Return-Path Resistors*

Fig. 2.23

Since the inverting input is grounded, $V_2 = 0$ and $V_{in} = V_1 = I_{B_1} R_B$.

The unwanted input voltage is amplified to produce an output offset voltage. For small values of R_B, the resulting output offset voltage may be small enough to ignore.

The output offset voltage produced due to return path resistance for one input may be eliminated if we add an equal resistance to the other input, as illustrated in Fig. 2.23 (*b*). In this case

$$V_{in} = I_{B_1} R_B - I_{B_2} R_B$$

$$\text{or } V_{in} = I_{in\ (offset)}\ R_B \qquad \text{...(2.25)}$$

The unwanted input voltage, because of smaller value of $I_{in\ (offset)}$ is quite small.

Input offset current $I_{in\,(offset)}$ is defined as the difference between the two currents entering the input terminals of a balanced amplifier for $V_{out} = 0$

$$i.e.,\ I_{in\,(offset)} = I_{B_1} - I_{B_2} \text{ for } V_{out} = 0 \quad ...(2.26)$$

In general, $I_{in(offset)}$ is smaller than I_{bias} by a factor of 2 to 20 (with bipolar op-amps generally showing better matching than FET op-amps).

4. Thermal Drift. Offset current, and offset voltage change with temperature. We may have carefully nulled a circuit shortly after switching on power supply at an ambient temperature of 25°C. Later on, we may find that temperature has risen to 40°C, and the circuit is no longer nulled. This is called *thermal drift.*

2.13.2. AC Characteristics

Although the op-amp is used in some industrial circuits as strictly a dc amplifier, many op-amp applications are ac. Bias currents, offset currents, offset voltage, and drift all affect the steady-state (dc) response of an op-amp. How the op-amp's output responds to variations in its input must also be studied.

1. Gain Bandwidth Product. For an ideal op-amp, it has been assumed that the open-loop gain is arbitrarily large, and this does not vary significantly with frequency. Neither of these assumptions hold good for an actual op-amp.

The frequency response curve of a 741C is shown in Fig. 2.24. The dc and very low frequency open-loop gain of 741C is about 2×10^5, which is generally large enough to be considered arbitrarily large. However, with the increase in frequency, the gain decreases proportionally. The 741C has a critical frequency f_c of 10 Hz. Above the critical frequency, the voltage gain decreases at a rate of 20 dB per decade until it reaches 1 at a frequency of 1 MHz. Expressed another way, increasing the frequency by a factor of 10 decreases the gain by 10.

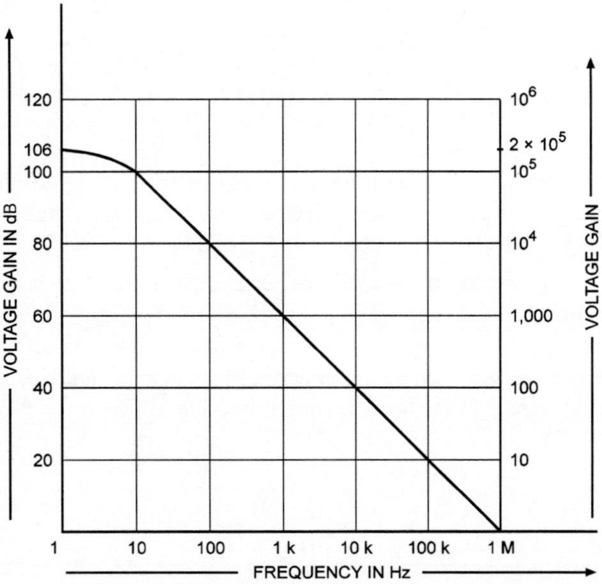

Fig. 2.24. ***Open-Loop Frequency Response of a 741C***

The *unity-gain frequency*, f_{unity} is the frequency where the voltage gain is unity. In Fig. 2.24, f_{unity} is 1 MHz. Data sheets usually specify the value of f_{unity} because it represents the upper limit of the useful gain of an op-amp. For instance, the data sheet of 741C lists an f_{unity} of 1 MHz. This means that the 741C can amplify the signals up to 1 MHz. Beyond 1 MHz, the voltage gain is less than unity and the 741C is useless. If higher f_{unity} is required, other op-amps are available. For instance, the LM 318 has an f_{unity} of 15 MHz, which means it can provide usable voltage gain all the way to 15 MHz.

The higher f_{unity} op-amps are costlier. An alternative to buying more expensive higher f_{unity} op-amps is to build the circuit with several stages, each stage with a lower gain (and therefore, higher frequency response).

2. Rise Time. For small signal ($V_{peak} < 1$ V) sinusoidal waves, the bandwidth provides all of the frequency response information, that is usually required. However, for small signal rectangular waves, the rise time is a more critical parameter.

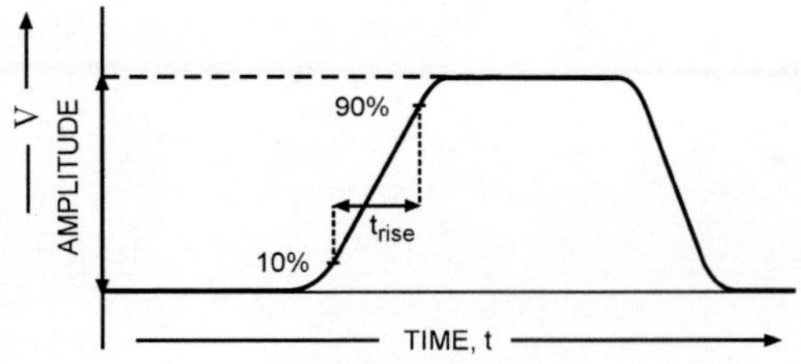

Fig. 2.25

Rise time is illustrated in Fig. 2.25. It is the time required for the output signal to rise from 10% of the amplitude to 90% of the amplitude. The faster the op-amp is, the shorter the rise time, and the output signal appears more like an ideal rectangular wave.

For small signals (V_{peak} < 1 V) the rise time is directly related to the high frequency cutoff (f_H) and is given as

$$t_{rise} = \frac{0.35}{f_H} \qquad ...(2.27)$$

For determination of GBW, the op-amp is configured as a voltage follower ($A_{closed\text{-}loop}$ = 1) and small signal rise time is measured. Having measured t_{rise} with $A_{closd\text{-}loop}$ = 1

$$GBW = A_{closed\text{-}loop} \times f_H = 1 \times \frac{0.35}{f_{rise}} \qquad ...(2.28)$$

It is noteworthy here that open-loop gain is very difficult to measure directly, because $V_{in\ (offset)}$ usually drives the op-amp's output into saturation. So, gain bandwidth (GBW) is not typically measured by completing a frequency response plot such as shown in Fig. 2.24.

3. Slew Rate. Gain bandwidth (GBW) and rise time are both characteristics of small signals, where peak output voltage V_{peak} is less then 1 V. For many front-end applications this may be adequate, but most power or driver stages must output larger signals. For larger signal outputs, V_{peak} > 1 V, the op-amp's speed is limited by the slew rate.

The slew rate of an op-amp is defined as the maximum rate at which the output voltage can change, no matter how large an input signal is applied.

$$i.e., \; SR = \left.\frac{dV_{out}}{dt}\right|_{max} \qquad ...(2.29)$$

This is usually measured in V/μs. As illustrated in Fig. 2.26, the instantaneous change in input voltage does not appear instantaneously but slews so many volts in a given period of time.

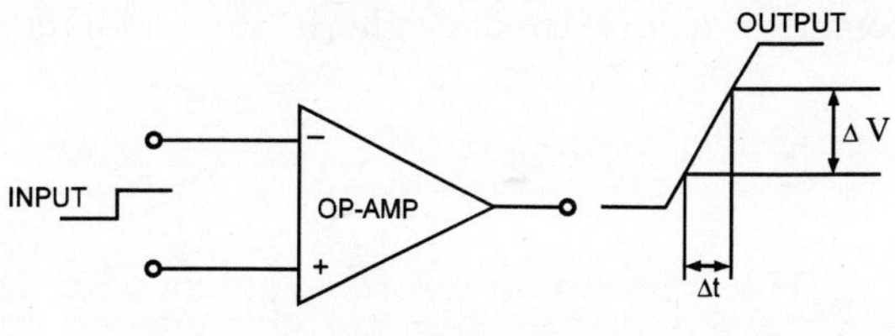

Fig. 2.26

The slew rate provides a parameter that specifies the maximum rate of change of the output voltage when driven by a large step-input signal. If one tried to drive the output at a rate of voltage change greater than slew rate, the output would not be able to change fast enough and would not vary over the full range expected, resulting in signal clipping or distortion. In any case the output would not be an amplified duplicate of the input signal if the op-amp slew rate is exceeded.

Practical op-amps are available with slew rates from 0.1 V/μs to well above 1,000 V/μs. The National semiconductor LH 0063 C has a slew rate of 6,000 V/μs. The slew rate of a 741 C is 0.5 V/μs.

Slew Rate Equation. Since the slew rate on a data sheet is generally listed for unity gain, let us consider the voltage follower circuit shown in Fig. 2.39. Further more let us assume that input is a large-amplitude and high-frequency sinusoidal wave. The equation of the input and output are as given below:

$$V_{in} = V_{P\,in} \sin \omega t \text{ and } V_{out} = V_{P\,out} \sin \omega t \qquad ...(2.30)$$

Rate of change of the output is

$$\frac{dV_{out}}{dt} = V_{P\,out}\, \omega \cos \omega t$$

and the maximum rate of change of the output occurs when cos ωt = 1 *i.e.,*

$$\left.\frac{dv_{out}}{dt}\right|_{max} = V_{P\,out}\, \omega$$

or slew rate, $SR = 2\pi f \, V_{P\,out}$ volts/second ...(2.31)

where f is input frequency in Hz and $V_{P\,out}$ is the peak value of the output sinusoidal wave in volts.

4. Full-Power Response. From Eq. (2.31)

$$f_{max} = \frac{SR_{max}}{2\pi V_{P\,out}}$$

This f_{max} is called the *full-power response.* It is the maximum frequency of a large-amplitude sinusoidal wave that the op-amp can amplify without distortion. The noteworthy point is that it is entirely separate from GBW product (which limits the output frequency because of a drop in gain). Above the full-power response frequency, the op-amp cannot charge the compensation capacitor fast enough to cause the output signal to swing to $V_{P\,out}$.

5. AC Noise. Any unwanted signals at the output of an amplifier are often referred to as *noise.* Under this broad definition, variation in output voltage V_{out} caused by bias current, offset current, or input offset voltage are examples of dc noise. We have seen in Art. 2.13.1 how these effects are reduced.

AC noise can be divided into two major categories; those caused by external effects (or interference), and those caused by effects internal to the amplifier circuit itself. We are concerned here with internally generated noise.

Noise voltage is primarily *Johnson* or *thermal* noise. Every resistive element, inside and outside of the op-amp IC, generates Johnson noise. Noise voltage is given as

$$V_n = \sqrt{4kTR\Delta f}$$

where k is the Boltzmann's constant, T is temperature in degrees absolute, R is resistance of the conductor and Δf is bandwidth of the measurement.

Simplifying this for room temperature

$$v_n = \frac{4nV}{\sqrt{Hz}} \times \sqrt{\frac{R}{1\,k\Omega}}$$

Thus, a resistor of 100 kΩ, at room temperature will produce a noise of

$$v_n = \frac{4nV}{\sqrt{Hz}} \times \sqrt{\frac{100\,k\Omega}{1\,k\Omega}} = \frac{40\,nV}{\sqrt{Hz}}$$

The *noise current* is often referred to as *shot noise.* The exact numbers of electrons flowing past a point at any instant is random. The average rate is the dc current. This random variation of current, about I_{dc}, is the noise current. It is directly proportional to I_{dc} and to the circuit's bandwidth *i.e.,*

$$i_n \propto I_{dc} \times \Delta f$$

For determination of the overall output noise, the effects of the Johnson noise from the resistors around the op-amp, the op-amp's own Johnson noise, and its shot noise must all be considered. Taking all these effects into consideration the output noise from an op-amp is given as

$$V_{noise\,out} = v_{n\,out} \times \sqrt{BW_n} \quad ...(2.32) \quad \text{where } BW_n = 1.57 f_H$$

2.13.3. Power Supply Rejection Ratio (PSRR)

The power supply rejection ratio (PSRR) is defined as the ratio of the change in input offset voltage to the corresponding change in one power supply voltage, with all remaining power supply voltages kept constant.

$$\textit{i.e.,}\ PSRR = \frac{\text{Change in input offset voltage}}{\text{Change in supply voltage}} = \frac{\Delta V_{in\,(offset)}}{\Delta V_{supply}} \quad ...(2.33)$$

It is also called the *supply voltage rejection ratio (SVRR)* or power supply sensitivity (PSS). It is expressed in μV per volt or in decibels. Lower value of PSRR in μV/V is an indication of better performance of the op-amp. It ranges from 10^{-5} to 7×10^{-5}.

Example 2.8. The output voltage of a certain op-amp changes by a 20 V in 4 μs. What is slew rate ?

Solution :

$$\text{Slew rate, SR} = \frac{dV_{out}}{dt} = \frac{20\ \text{V}}{4\ \mu\text{s}} = 5\ \text{V}/\mu\text{s} \quad \textbf{Ans.}$$

Example 2.9. Determine input bias current and input offset current if I_{B_1} = 10 μA and I_{B_2} = 7.5 μA.

Solution :

$$\text{Input bias current, } I_{in\,(bias)} = \frac{I_{B_1} + I_{B_2}}{2} = \frac{10 + 7.5}{2} = 8.75\ \mu\text{A} \quad \textbf{Ans.}$$

$$\text{Input offset current, } I_{in\,(offset)} = I_{B1} - I_{B2} = 10 - 7.5 = 2.5\ \mu\text{A} \quad \textbf{Ans.}$$

Example 2.10. The slew rate of an op-amp is 6 V/μs when closed-loop gain is unity. The amplified output signal is observed to be a pure sinusoid, $v_{out} = V_{max} \cos \omega t$ provided the frequency of this signal does not exceed a certain limit. Find the value of this limiting frequency before the output signal is distorted by the slew rate limit if (*i*) V_{max} = 1 V, (*ii*) V_{max} = 10 V.

Solution : Slew rate of op-amp, $SR_{max} = 6\ \text{V}/\mu\text{s}$

From Eq. (2.31), we have

$$f_{max} = \frac{SR_{max}}{2\pi V_{max\,out}}$$

Substituting V_{max} = 1 V and V_{max} = 10 V in above equation, we have

(*i*) For V_{max} = 1 V

$$f_{max} = \frac{6\ \text{V}/\mu\text{s}}{2\pi \times 1\ \text{V}} = \frac{6 \times 10^6\ \text{V/s}}{2\pi \text{V}} = 0.955\ \text{MHz} \quad \textbf{Ans.}$$

(*ii*) For V_{max} = 10 V

$$f_{max} = \frac{6 \times 10^6}{2\pi \times 10} = 95.5\ \text{kHz} \quad \textbf{Ans.}$$

Example 2.11. A peak-to-peak input signal of 0.5 V has to produce a peak-to-peak undistorted output voltage of 3 V with a rise time of 4 μs. Can 741 op-amp be used for such an application. Explain your answer.

Solution : Since the magnitude of the output exceeds 1 V, the slew rate will have to be considered.

The time interval given is the rise time and during this time interval the signal is to change from 10 per cent of the peak-to-peak amplitude to 90 per cent of the peak-to-peak amplitude. Thus

$$\Delta V = 90\% \text{ of } V_{PP} - 10\% \text{ of } V_{PP} = (90\% - 10\%)\ V_{PP} = 0.8\ V_{PP}$$

The required slew rate,

$$SR = \frac{\Delta V}{\Delta t} = \frac{0.8\ V_{PP}}{\text{Rise time}} = \frac{0.8 \times 3}{4\ \mu\text{s}} = 0.6\ \text{V}/\mu\text{s}$$

It means the op-amp used must have a slew rate equal to or faster than 0.6 V/μs.

The 741 op-amp has a SR of 0.5 V/μs and, therefore, it is too slow and cannot be used.

2.14. FREQUENCY COMPENSATION

In op-amps the variations in gain and phase shift with respect to frequency are due to the internal capacitors and stray capacitances, which are due to physical characteristics of semiconductor devices (BJTs and FETs) and the internal construction of the op-amp.

The rate of change of gain as well as the phase shift of an op-amp can be controlled by adding specific components, most common of which are resistors and capacitors. The network formed by such components and employed for modifying the rate of change of gain and phase shift is known as *compensating network.* The main purpose of a compensating network is to improve the performance of an op-amp circuit over the desired frequency range by controlling its gain and phase shift.

First generation op-amps such as 301, 709 have no internal frequency compensation, whereas later-generation op-amps such as 741, 351 and 318 have internal compensation. Thus op-amps are of two types viz internally compensated and externally compensated.

The internally compensated op-amps use a metal oxide capacitor built within the IC. The circuit configuration is given in Fig. 2.27. Although this works well, internal compensation does not allow us any control over the op-amp frequency response.

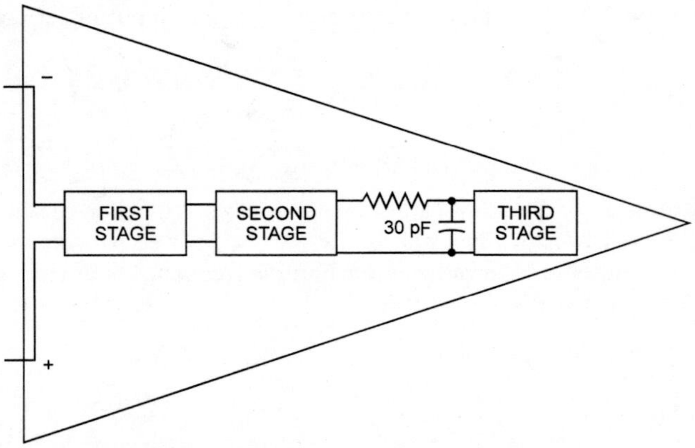

Fig. 2.27. ***Internally Compensated Op-Amp***

The 301 and 709 op-amps have no internal frequency compensation capacitor. Instead, frequency compensation terminals are provided, and compensation capacitors are to be connected externally. Failure to connect these external compensation capacitors will practically guarantee that the op-amp will oscillate. However, the op-amp's frequency response can be tailored as per needs by the value and type of compensation network used.

The compensating networks used in op-amps for modifying the rate of change of gain and the phase shift are of two types viz the phase lag and phase lead. These two networks are indicative of their functions, since phase lag contributes a negative phase angle and phase lead a positive phase angle. In each case an R – C network is connected into the circuit to form as a part of loop. Usually, as recommended by the device manufacturer, the R – C network is connected to specified points within the op-amp. For this purpose, external terminals are provided for access to points inside the IC.

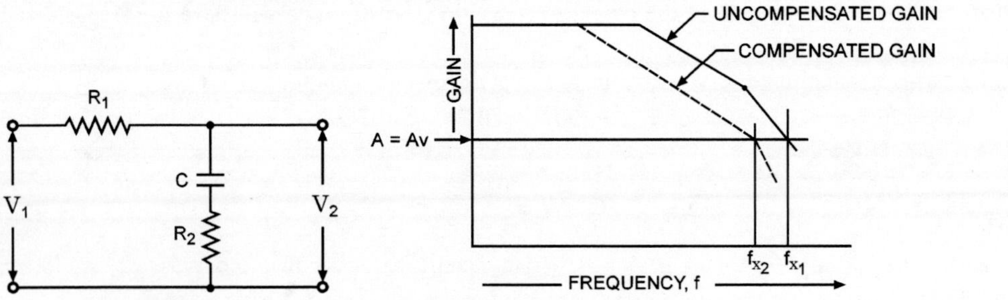

(a) Network *(b) Effect of Lag Compensation on Gain-Frequency Response*

Fig. 2.28. ***Phase-Lag Compensation***

A *phase-lag network* stabilizes an amplifier by reducing the open-loop gain, so that the frequency at which $A\beta = 1$ occurs when the phase shift is too small for instability. The components of the phase-lag network are calculated to introduce additional phase lag at some low frequency where op-amp phase shift θ is still so small that additional shift has no effect. For the circuit shown in Fig. 2.28 (*a*), it is obvious that at frequencies for $X_C >> R_2$, the voltage v_2 lags behind v_1. A phase lag as great as – 90° might be introduced. However, at higher frequencies where $X_C << R_2$, no significant phase lag occurs, and the network merely introduces some attenuation. The effect of this attenuation is to move the frequency f_{x_1} at which $\beta A = 1$ (for a given closed-loop gain A_v) to a lower frequency f_{x_2}, as illustrated in Fig. 2.28 (*b*). Because f_{x_2} is smaller than f_{x_1} the phase shift at f_{x_2} is smaller than that at f_{x_1}, and the circuit is likely to be stable.

A *phase-lead network* cancels some of the phase lag. In the phase-lead network shown in Fig. 2.29 (*a*) when $X_C >> R_1$, v_2 leads v_1. This phase lead cancels some of the unwanted phase lag in the op-amp, as illustrated in Fig. 2.29 (*b*), thus making the circuit more stable.

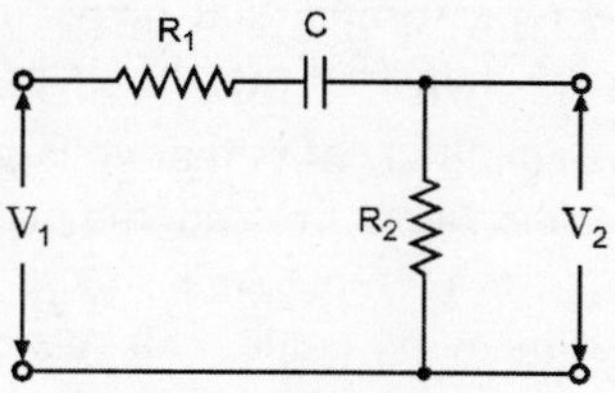

(a) Network

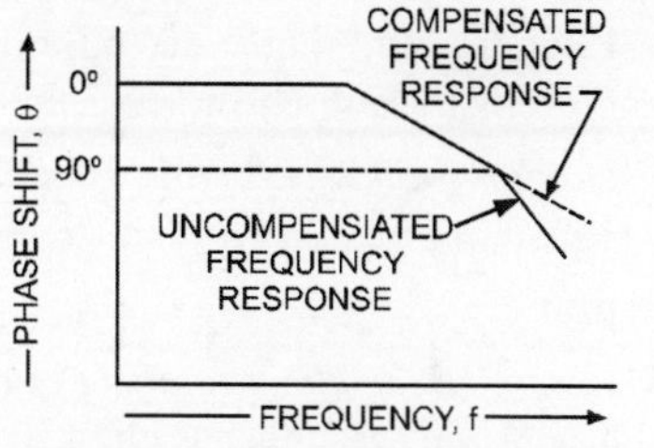

(b) Effect of Lead Compensation on Phase-Frequency Response

Fig. 2.29. ***Phase-Lead Compensation***

Calculation of component values for an op-amp compensation network may be a tedious job, and the frequency response plot is required for the particular op-amp used. For simplifying the compensation work, IC manufacturers publish recommended compensation networks and the component values on op-amp data sheets. The recommended component values for the amplifier closed-loop gain are noted from the data sheet and then connected to the op-amp, as illustrated on the data sheet.

Single capacitor compensation is the simplest compensation and is shown in Fig. 2.30. The minimum value of compensation capacitance is dependent on the resistor feedback network.

$$C_{min} = \frac{R_{in} \times 30\text{ pF}}{R_{in} + R_f} \qquad ...(2.34)$$

The worthnoting point is that the same equation holds good for inverting as well as noninverting op-amps, so use of noninverting amplifier will allow you to have lower values of R_{in} and C without sacrificing the input impedance.

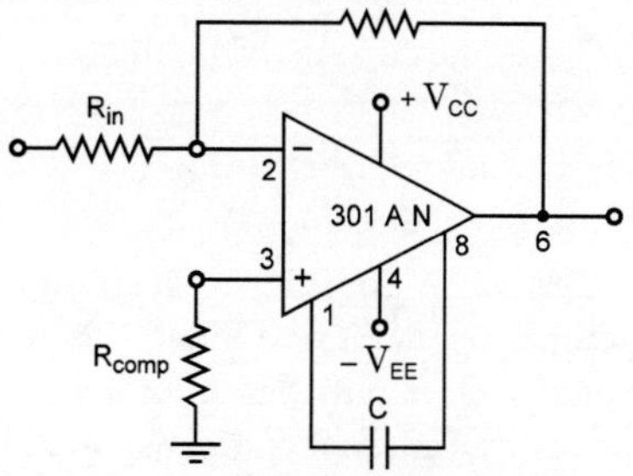

Fig. 2.30. ***301 Single Capacitor Compensation***

2.15. BASIC SPECIFICATIONS OF OP-AMPS

The ideal op-amp cannot be had in practice. But, for a remarkably low cost, monolithic op-amps can be manufactured which, when properly used, closely approximate the ideal. Here key nonideal characteristics will be discussed, along with simple techniques and precautions for minimizing their effects.

The ideal op-amp characteristic values and typical characteristic values for the 741C are given in Table 2.2.

TABLE 2.2. ***Ideal and Typical Characteristics of a Monolithic Op-amp IC***

Particulars	Ideal	Typical Values For 741C
Voltage gain (open-loop)	∞	2×10^5
Output impedance	0	75 Ω
Input impedance	∞	2 MΩ
Offset voltage	0	2 mV
Offset current	0	20 nA
Bandwidth (BW)	∞	1 MHz

Although the *open-loop gain* of the op-amp is not infinite, generally an error of less than 10% will be introduced by assuming it to be infinity. This will be true if for any circuit using op-amps the *overall (closed-loop) gain of the circuit is limited to 100 or less.*

The typical input and output impedances, though not ideal, need to be considered only for a few demanding industrial measurement applications.

The amplitude of output voltage for most op-amps is not restricted by the output impedance, typically 75 Ω for the 741 C. Instead, the magnitude of current the op-amp can deliver to the load limits the output voltage. In an attempt to make these ICs as foolproof as possible, a short-circuit current limit circuit has been added to many op-amps. This means that the load impedance can be shorted without overheating the IC. Instead, the output current will be limited to a safe level (25 mA for the 741 C). The output voltage drops appropriately to keep the output current below the safe limit as the load impedance decreases. This is very handy. Keep in mind that every op-amp is not protected against short circuit. So it is necessary to read the manufacturer's literature. Shorting the output of an unprotected op-amp will probably cause damage to it.

Use an op-amp which can withstand a short circuit indefinitely such as the 741 C.

Interestingly, the last three parameters given in Table 2.2 along with several others limit the op-amp's application.

To minimize the effect of offset currents, the size of resistance seen from each input terminal is reduced as much as practical. This has already been discussed in Art. 2.13.1(3). Usually, keeping R_B in the kΩ range is quite adequate.

The bandwidth parameter is actually a measurement of the maximum possible gain at a given frequency. For the 741 C, at no point GBW (gain bandwidth product) should exceed 1 MHz. The open-loop gain for 741 C becomes unusably small above about 10 kHz $\left(\dfrac{1 \times 10^6}{10 \times 10^3}\ i.e.\ 100\right)$. For this reason *use 741 C only at dc to upper-audio frequencies. Use a wideband op-amp above* 10 kHz. The LM 318 has a bandwidth of 15 MHz, providing useful gain up to at least 150 kHz.

Certain op-amps have exceptionally high gain at high frequencies (for example, 301). These are usually uncompensated. At first glance, they may seem advantageous. However, high frequency noise is amplified, coupled back into the input through interwiring capacitance, amplified, coupled back to the input ... oscillations result. External compensation network is to be provided to make such an op-amp usable.

Generally, to simplify circuits, *use internally compensated op-amps (741 C, LM 318).* Avoid uncompensated op-amps (such as 301).

2.16. COMPARISON OF TYPICAL VALUES FOR IC OP-AMPS

Table 2.3 compares several IC op-amps on the basis of some of the parameters already defined. The prefixes that appear on these devices are associated with specific manufacturers. For example, μA is used by Fair child, LF/LM and LH by National, OP by Precision Monolithics, NE/SA/SE by Signetics, MC by Motorola, RC by Raytheon and TL/TLC by Texas Instruments.

A quick review of Table 2.3 can serve to establish typical parameter ranges. The open-loop gain, A ranges from 6×10^3 to 2×10^5, with the upper figure much more typical. CMRR ratings fall in the range of 86 and 123 dB. Input bias currents are typically from 10^{-11} to 10^{-7} A, depending on the way in which the circuit has been optimized. Input offset voltages usually range from 1 to 3 mV except for those op-amps designed for ultra-low values. The frequency response can be indicated by the unity-gain bandwidth or the rise time for a step input. Typical supply voltages are ±15 V and standby supply currents are of the general magnitude of mA.

2.17. NEGATIVE FEEDBACK IN OP-AMPS

An open-loop op-amp offers very high voltage gain [ranging from 10^5 to 10^6 at dc (or 0 Hz)] and narrower bandwidth (band of frequencies for which the gain remains constant)–almost zero. Besides being large, the

TABLE 2.3. *Comparison of Typical Values For IC Op-Amps*

Device Code	Description	A (× 10^3)	CMMR in db	Input Bias Current in n A	Input Offset Voltage in m V	Frequency Response		Slew Rate in V/μ s	Supply	
						Unity GBW in MHz	Rise Time in μ s		Voltage in Volts	No-load Current in m A
RC 741	General purpose	200	90	80	1		0.3	0.5	± 15	1.7
NE 5230	Low supply voltage	300	97	40	0.3	0.6		0.25	± 0.75 min	0.5
LM 124	Quad Op-amp/single supply	100	85	45	2	1		0.3	± 15 30, 0	1.5
NE 5512	High input impedance	200	100	6	1	3		1	± 15	3.4
NE 530	High slew rate	200	90	65	2		0.06	35	± 18	2
NE 5535	Dual op-amp/high slew rate	500	90	65	2		0.25	15	± 18	1.8
LM 218	High speed	200	100	120	2	15		50	± 15	5
OP-07 E	Ultra-low offset voltage	400	123	± 1.2	0.03	0.6		0.3		
TL 060 M	Low-power JFET input	6	86	0.03	3	1		3.5	± 15	0.2
LF 155 A	JFET	200	100	0.03	1	5		5	± 15	2
MC 1539	Uncompensated	120	110	200	1		0.08-0.19	6-34	± 15	3
MC 1776	Programmable	200	86	2	2		0.6	0.03-0.35	± 3	0.013-0.13

* Q-point currents set by an external resistor.

voltage gain of the open-loop op-amp is not a constant. The voltage gain varies with variations in temperature and power supply voltages as well as mass production techniques. Because of these features the open-loop op-amps are not suitable for many linear applications as explained below :

1. Because of the very high voltage gain of open-loop op-amp, only the smaller signals (of the order of microvolts or less) having very low frequency may be amplified accurately without distortion. For instance, if the input signal is of 1 mV, then output voltage will be equal to $1 \times 10^5 \times 1$ mV *i.e.,* 100 V (assuming open-loop gain = 10^5). Such a large output cannot be obtained because the output from the op-amp can never exceed the saturation levels $\pm V_{sat}$. However, the signals of small amplitudes are very susceptible to noise and almost impossible to obtain in the laboratory.
2. The large variations in voltage gain of an open-loop op-amp makes it unsuitable for many linear applications.
3. Because of narrower bandwidth, the open-loop op-amp is impractical in ac applications. For instance, the open-loop bandwidth of the 741C op-amp is roughly 5 Hz but in almost all ac applications a bandwidth exceeding 5 Hz is required.

However, the gain of the op-amp can be controlled and stabilized and the bandwidth can be increased if necessary modification is introduced in the op-amp basic circuit. This modification involves the use of feedback. The process of combining a fraction of output energy (voltage or current) back to the input is called **feedback.** When feedback voltage (or current) is so applied as to increase the input signal (*i.e.,* when the feedback is in phase with input signal), it is called the *positive* or *direct* or *regenerative feedback.* Since positive feedback causes excessive distortion and instability, it is seldom used in amplifiers. However, because of its capability of increasing the power of the original signal it is used in oscillator circuits.

When feedback voltage (or current) is so applied as to weaken the input signal (*i.e.,* when feedback is in phase opposition to the input signal), it is called *negative, inverse* or *degenerative feedback.* Negative feedback reduces the amplifier gain but it has numerous advantages (such as gain stability, reduction in nonlinear distortion, reduction in noise, increase in bandwidth or improvement in frequency response, increase in input impedance and decrease in output impedance) and is frequently used in amplifier circuits. Negative feedback also reduces the effects of input offset voltage and variations in temperature and supply voltages on the output of the op-amp.

2.18. BLOCK DIAGRAM REPRESENTATION OF FEEDBACK CONFIGURATIONS

An op-amp using the feedback is known as a **feedback amplifier** and sometimes referred to as a **closed-loop amplifier** because the feedback forms a closed loop between the input and the output. A feedback amplifier essentially consists of two parts viz. an amplifier (an op-amp) and a feedback network as illustrated in Fig. 2.31. The function of the feedback network is to return a fraction of the output energy (voltage or current) to the input. The feedback network may be made of either passive components (resistors, inductors and capacitors), active components or combination of both. The feedback network is usually in the form of a passive two port network and is usually formed of resistors, inductor and capacitors.

The feedback may also be classified as *voltage feedback* and *current feedback.* In voltage feedback the energy fedback to the input terminals is proportional to the output voltage while in current feedback, the energy fedback to the input terminals is proportional to the current through the load.

Both voltage and current can be fedback to the input either in series or in parallel resulting in four basic feedback connections, given below :

1. Voltage-series feedback, as shown in Fig. 2.31 (*a*).
2. Voltage-shunt feedback, as shown in Fig. 2.31 (*b*).
3. Current-series feedback, as shown in Fig. 2.31 (*c*).
4. Current-shunt feedback, as shown in Fig. 2.31 (*d*).

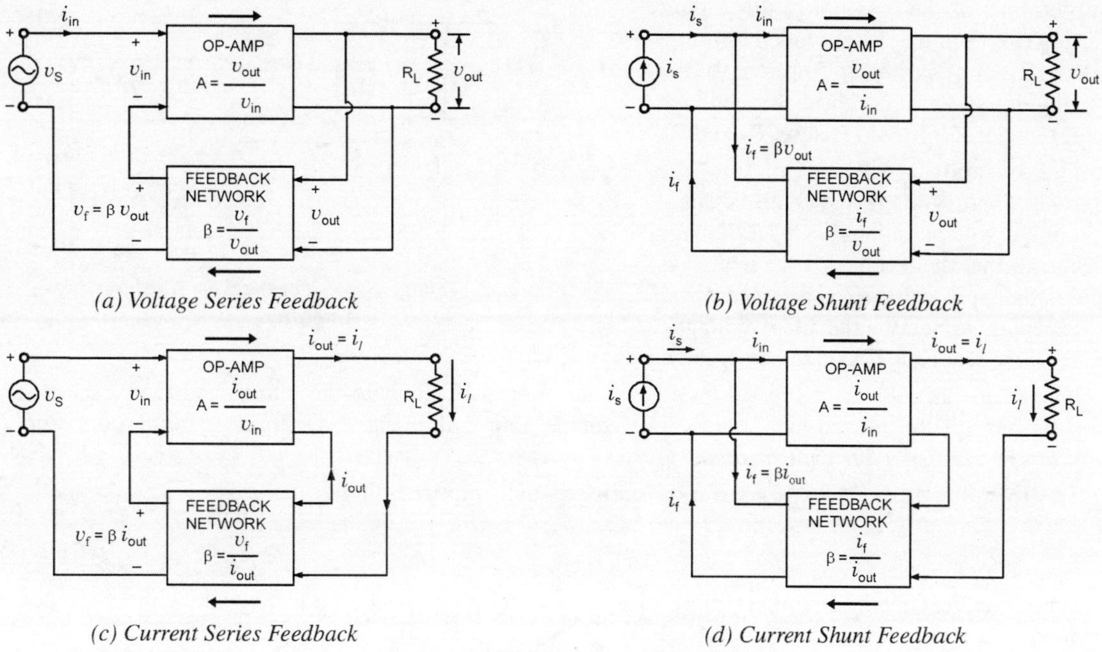

Fig. 2.31. ***Feedback Configurations***

In the list above, voltage refers to connecting the output voltage as input to the feedback network and current refers to tapping off some output current through the feedback network. Series refers to connecting the feedback signal in series with the input signal voltage; and shunt refers to connecting the feedback signal in parallel (shunt) with an input current source.

Series feedback connections tend to increase the input impedance of the amplifier. This is because in this case feedback signal is returned to the input in series to oppose the applied voltage causing input current to fall and consequently makes the input impedance to increase. In case of shunt or parallel connections, the current drawn from the signal source is increased by an amount equal to feedback current I_f and therefore, input impedance falls. Voltage feedback tends to reduce the output impedance while current feedback tends to increase the output impedance.

In all four of the configurations depicted in Fig. 2.31, the direction of signal through the op-amp is from the input to output while in an ideal case, the direction of signal through the feedback circuit is exactly opposite *i.e.,* from the output to input.

The most widely used closed configurations are :

1. Voltage-series feedback configuration.
2. Voltage-shunt feedback configuration.

Hence only these two configurations will be discussed here in details.

2.19. VOLTAGE-SERIES FEEDBACK OR NONINVERTING AMPLIFIER

Figure 2.32 depicts the diagram of a voltage-series feedback amplifier. The op-amp is represented by a schematic symbol, including its large-signal voltage gain A and the feedback circuit is composed of two resistors R_1 and R_f.

The circuit depicted in Fig. 2.32 is commonly known as *noninverting amplifier with feedback* or *closed-loop noninverting amplifier* because it uses a feedback and input signal is applied to the noninverting input terminal of the op-amp. In this circuit inverting (–) terminal is grounded through resistor R_1 and output is applied to the inverting input terminal through feedback circuit composed of two resistors R_1 and R_f.

The op-amp always amplifies the difference input voltage v_d which is equal to the input voltage minus feedback voltage v_f. It means that the feedback voltage always opposes the input voltage (or is out of phase by 180° with respect to the input voltage); therefore, the feedback is said to be *negative.*

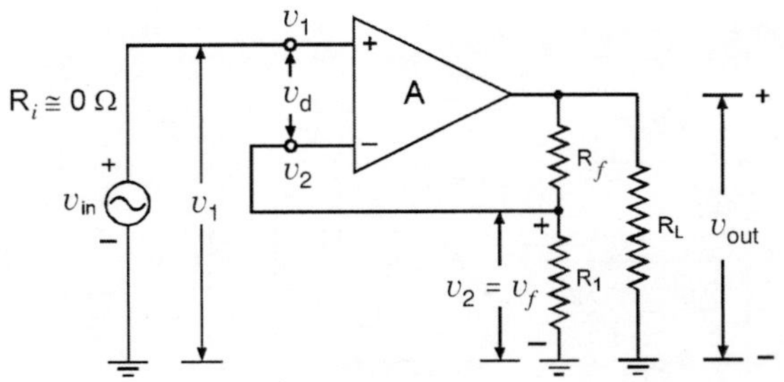

Fig. 2.32. ***Voltage-Series Feedback or Noninverting Amplifier***

It is seen that there is a *closed loop* from the amplifier input to the output and then back to the input. Because of this closed loop, the overall voltage gain with negative feedback is termed the *closed-loop voltage gain* or *gain with feedback,* A_f.

The overall voltage gain (closed-loop gain) of the amplifier in Fig. 2.32 is

$$A_f = \frac{v_{out}}{v_{in}} \qquad \text{...(2.35)}$$

When the feedback network is disconnected, the loop is opened and the gain without feedback is referred to as the *open-loop gain A.*

Open-loop voltage gain (or gain without feedback) is

$$A = \frac{v_{out}}{v_d} \qquad \text{...(2.36)}$$

and gain of feedback circuit is

$$B = \frac{v_f}{v_{out}} \qquad \text{...(2.37)}$$

2.19.1. Closed-Loop Voltage Gain

Voltage appearing at noninverting terminal, $v_1 = v_{in}$

Voltage appearing at inverting terminal, $v_2 = v_f$, feedback voltage

and feedback voltage, $v_f = \dfrac{R_1}{R_1 + R_f} v_{out}$ ...(2.38) Using the voltage divider rule

From Eq. (2.38)

$$\text{Output voltage, } v_{out} = Av_d = A(v_1 - v_2)$$

$$= A\left[v_{in} - \frac{R_1}{R_1 + R_f} v_{out}\right] \qquad \because v_1 = v_{in},\ v_2 = v_f = \frac{R_1}{R_1 + R_f} v_{out}$$

or $\quad v_{out}(R_1 + R_f) + AR_1 v_{out} = Av_{in}(R_1 + R_f)$

or $\quad v_{out}(R_1 + R_f + AR_1) = A(R_1 + R_f)v_{in}$

$$\text{or } v_{out} = \frac{A(R_1 + R_f)v_{in}}{R_1 + R_f + AR_1}$$

$$\text{Thus, } A_f = \frac{v_{out}}{v_{in}} = \frac{A(R_1 + R_f)}{R_1 + R_f + AR_1}$$

Generally A is very large (typically 10^5). So $AR_1 >> (R_1 + R_f)$ and $R_1 + R_f + AR_1 \cong AR_1$ and closed-loop gain is

$$A_f = \frac{v_{out}}{v_{in}} = \frac{A(R_1 + R_f)}{AR_1} = \left(1 + \frac{R_f}{R_1}\right) \quad ...(2.39)$$

From Eq. (2.39) it is obvious that the magnitude of output voltage can be controlled by the proper selection of resistors R_1 and R_f. In fact, this circuit arrangement readily provides us with a *voltage-controlled voltage source* (VCVS). Accordingly if R_f is set equal to 24 kΩ and R_1 is 1 kΩ, the output voltage of the op-amp will be 25 times the value of input voltage. The noteworthy point is that with the arrangement shown in Fig. 2.32, the input voltage can be made to take on values much larger than the highly restrictive values associated with V_d when used without resistors R_1 and R_f. The circuit model used to represent a voltage-controlled voltage source is shown in Fig. 2.33. The factor K is a numeric which for the circuit of Fig. 2.32 clearly takes on the value of $\left(1 + \frac{R_f}{R_1}\right)$. The input circuit feeds into an open circuit in recognition of the fact that the input resistance is assumed to be very high.

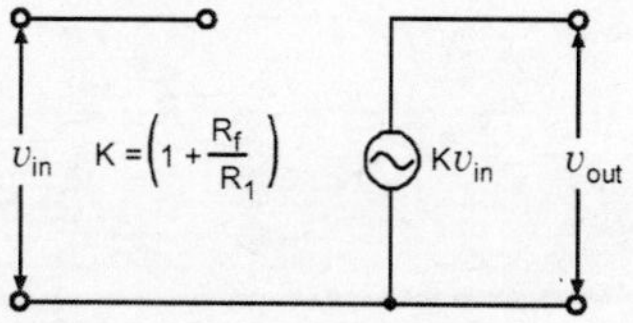

Fig. 2.33. ***Circuit Mode For VCVS***

Since this circuit provides the output voltage K times the input voltage with same polarity, it is known as *positive scaler.*

Another interesting result can be obtained from Eqs. (2.37) and (2.38)

$$B = \frac{v_f}{v_{out}} = \frac{R_1 v_{out}}{(R_1 + R_f) v_{out}} = \frac{R_1}{R_1 + R_f} \quad ...(2.40)$$

Comparing Eqs. (2.39) and (2.40), we have

$$A_f = \frac{1}{B} \text{ (ideal)} \quad ...(2.41)$$

The above equation shows that the gain of the feedback circuit is the reciprocal of the closed-loop voltage gain. It means that for given values of R_1 and R_f the values of A_f and B are fixed.

Finally the closed-loop voltage gain A_f can be expressed in terms of open-loop gain A and feedback circuit gain B as follows :

From Eq. (2.36)

$$v_{out} = Av_d = A(v_1 - v_2) = A(v_{in} - v_f)$$

Substituting $v_f = Bv_{out}$ from Eq. (2.37) in above equation, we have

$$v_{out} = Av_{in} - ABv_{out}$$

$$\text{or } v_{out}(1 + AB) = Av_{in}$$

and closed-loop gain,

$$A_f = \frac{v_{out}}{v_{in}} = \frac{A}{1 + AB} \quad ...(2.42)$$

Amplifier **gain stabilization** is the most important advantage of negative feedback. Amplifiers that are required to have stable voltage gain are always designed as negative feedback amplifiers.

From Eq. (2.41), it is seen that to design a negative feedback amplifier with a particular closed-loop gain, it is only necessary to design the feedback network to give $A_f = \frac{1}{B}$.

2.19.2. Input Resistance With Feedback

An amplifier with noninverting voltage feedback is shown in Fig. 2.34. The op-amp has an open-loop impedance of approximately R_{in}. The amplifier has an overall closed-loop input impedance (or input impedance with feedback) of R_{inf}. The input resistance with feedback is given by

$$R_{in\,f} = \frac{v_{in}}{i_{in}} = \frac{v_{in}}{\frac{v_d}{R_{in}}} \qquad \because i_{in} = \frac{v_d}{R_{in}}$$

and $$v_d = \frac{v_{out}}{A} \qquad \text{Refer to Eq. (2.36)}$$

$$v_{out} = \frac{A}{1+AB} v_{in} \qquad \text{Refer to Eq. (2.42)}$$

Hence $$R_{inf} = \frac{v_{in}}{\frac{v_{out}}{A}} R_{in}$$

$$= AR_{in} \frac{v_{in}}{\frac{A v_{in}}{1+AB}}$$

$$= R_{in}(1 + AB) \qquad \text{...(2.43)}$$

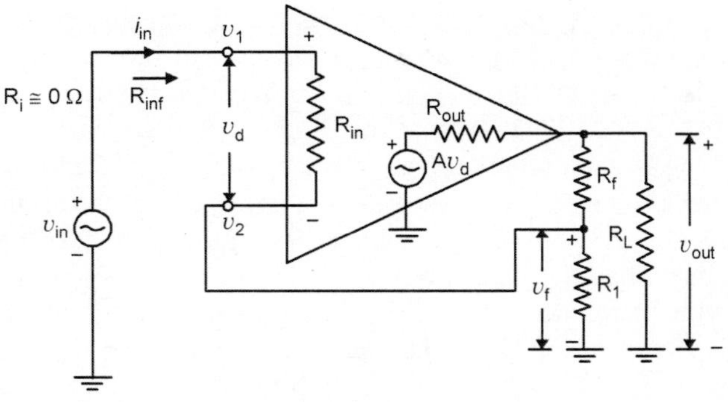

Fig. 2.34

The above equation reveals that series voltage feedback increases the input impedance of an amplifier by a factor (1 + AB).

This is the same factor involved in gain reduction.

2.19.3. Output Resistance With Feedback

The op-amp shown in Fig. 2.35 has an open-loop output resistance of R_{out}. The output resistance with feedback R_{outf} is defined as the resistance looking back into the amplifier from the output terminal as shown in Fig. 2.35. This resistance may be determined as follows :

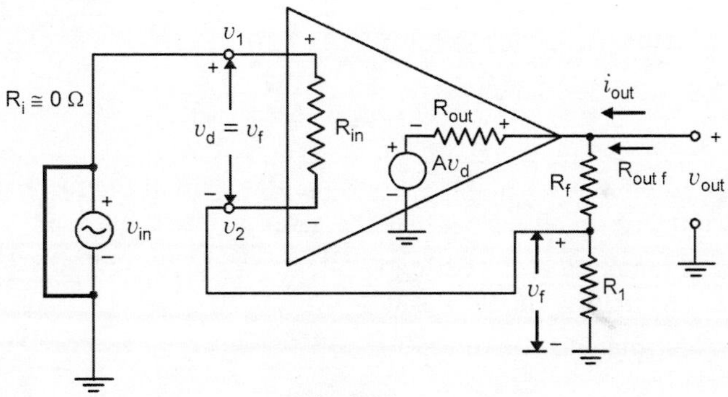

Fig. 2.35

The input terminals are shorted, so that feedback voltage v_f is the only voltage at the amplifier input. The output voltage v_{out} develops the feedback voltage v_f and this generates a voltage $Av_d = Av_f$ in series with R_{out} as illustrated in Fig. 2.35. Assuming an instantaneous positive polarity for v_{out}, the output current i_{out} flows in the direction indicated.

Writing an equation for voltage drops around the output circuit of the negative feedback amplifier,

$$v_{out} = i_{out}R_{out} - Av_d = i_{out}R_{out} - Av_f$$

or $$i_{out}R_{out} = v_{out} + Av_f = v_{out} + ABv_{out}$$

$$= v_{out}(1 + AB) \qquad \because \text{From Eq. (2.37) } v_f = Bv_{out}$$

or $$\frac{v_{out}}{i_{out}} = \frac{R_{out}}{1+AB}$$

or Output resistance, $$R_{outf} = \frac{v_{out}}{i_{out}} = \frac{R_{out}}{1+AB} \qquad \text{...(2.44)}$$

Above equation reveals that the series voltage feedback reduces the output impedance of an amplifier by a factor of (1 + AB).

This is the same factor involved in the reduction of amplifier gain and the increase of the input impedance.

2.19.4. Nonlinear Distortion With Negative Feedback

The final stage of an op-amp has **nonlinear distortion** when the signal swings over most of the ac load line. Large swings in the current cause the dynamic emitter resistance r'_e of the transistor to vary during the cycle. This means that the open-loop voltage gain varies throughout the cycle because of variations in r'_e. It is this variation in voltage gain which causes nonlinear distortion.

Noninverting voltage feedback reduces nonlinear distortion because the feedback stabilises the closed-loop voltage gain, making it almost independent of variations in open-loop voltage gain. As long as the loop gain is much larger than unity, the output equals 1/B times the input voltage. This implies that the output will be a faithful reproduction of the input.

Figure 2.36 illustrates how distortion is reduced under large-signal conditions, the op-amp produces a distortion voltage, designated v_{dist} which may be visualized as a new voltage source in series with the original source Av_d. Without negative feedback, all the distortion voltage v_{dist} would appear at the output. But with negative feedback a fraction of the distortion voltage is fed back to the inverting input. This is amplified and arrives at the output with inverted phase, almost completely neutralizing the original distortion caused by the output voltage.

In Fig. 2.36, the output voltage is

$$v_{out} = Av_d + v_{dist} = A(v_{in} - Bv_{out}) + v_{dist} \qquad \because v_d = v_{in} - v_f = v_{in} - Bv_{out}$$

or $$v_{out} = \frac{A}{1+AB}v_{in} + \frac{v_{dist}}{1+AB}$$

The first term is what we desire because it represents the amplified input voltage. The second term is the closed-loop distortion that appears at the final output. This term can be written as

$$v_{distf} = \frac{v_{dist}}{1+AB} \qquad ...(2.45)$$

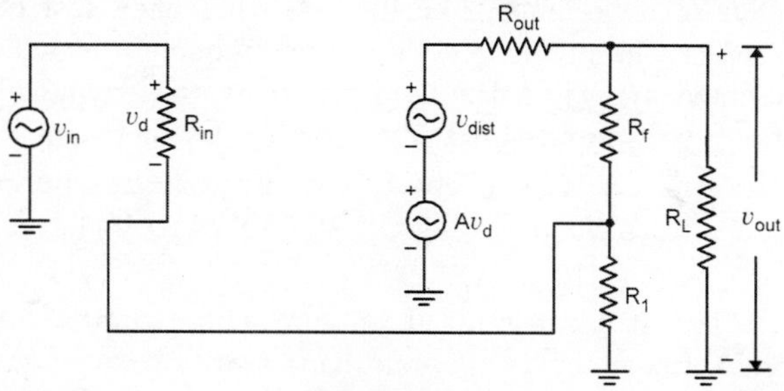

Fig. 2.36

When the loop gain is much larger than unity, the closed-loop distortion is much smaller than the open-loop distortion.

2.19.5. Noise With Negative Feedback

Noise in an amplifier is reduced in the same way as distortion. However, distortion is usually introduced in the last stage of the amplifier and, therefore, is always within the feedback loop whereas noise in an amplifier may be generated or introduced from outside source at any stage that may not be within the feedback loop. The noise can be reduced by negative feedback only if its source lies within the feedback loop after the first stage.

As nonlinear distortion generated within the amplifier is reduced with negative feedback by a factor (1 + BA), the noise introduced within amplifier also gets reduced by the same factor (1 + BA) on using negative feedback and the performance of the amplifier is thus much improved.

2.19.6. Output Offset Voltage With Negative Feedback

We know that an output offset voltage can exist even when the input voltage is zero. Figure 2.37 shows a feedback amplifier with an output offset voltage in series with the original voltage source Av_d. This new voltage source represents the output offset voltage without feedback. The actual output offset voltage with negative feedback is much smaller. The reasoning is similar to that given for nonlinear distortion. Some of the output offset voltage is fed back to the inverting input, cancelling most of the original output offset voltage. The output offset voltage with feedback may be given by

$$v_{oof} = \frac{v_{oo}}{1 + AB} \quad ...(2.46)$$

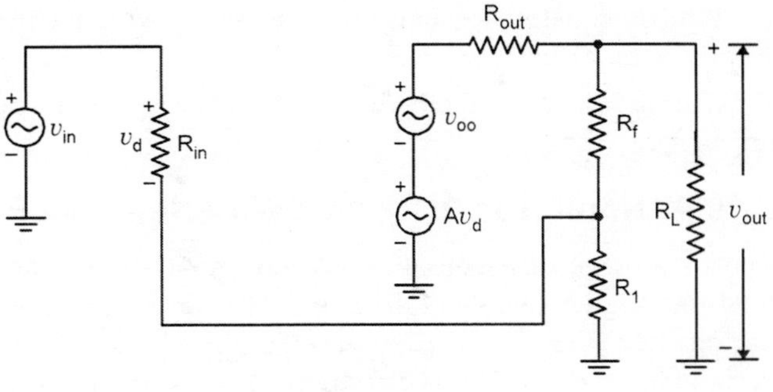

Fig. 2.37

2.19.7. Bandwidth With Feedback

The bandwidth (BW) of an amplifier without feedback is equal to the separation between 3-dB frequencies f_1 and f_2. If A is the gain then gain-bandwidth product is A × BW. With the negative feedback the amplifier gain is reduced and since gain bandwidth product has to remain constant in both cases, so obviously the bandwidth will increase to compensate for the reduction in gain. This is explained as follows :

The open-loop gain-frequency curve of the 741 C op-amp is depicted in Fig. 2.38. This curve indicates that for a gain of 2×10^5, the bandwidth is roughly 5 Hz. It means the gain-bandwidth product (GBW) is $2 \times 10^5 \times 5$ Hz = 1 MHz. On the other extreme, the bandwidth is roughly 1 MHz for unity gain. In other words, gain-bandwidth product can be said constant. However, this is true only in case of op-amps, like the 741, that have just one break frequency below unity gain-bandwidth. For 741, the break frequency (the frequency at which gain A is 3 dB down from its value at 0 Hz) denoted by f_o is 5 Hz. Contrary to it, the frequency at which the gain is unity is referred to as the **unity gain-bandwidth.** The relationship between the break frequency f_o, open-loop voltage gain A, bandwidth with feedback f_f and closed-loop gain A_f may be derived as follows :

Unity gain-bandwidth (UGB) may be written as

$$\text{UGB} = Af_o \quad ...(2.47)$$

This is because for an op-amp with a single break frequency f_0, the gain-bandwidth product is constant and equals unity gain-bandwidth.

Alternatively, only for a single break frequency op-amp,

$$\text{UGB} = A_f f_f \quad ...(2.48)$$

Equating Eqs. (2.47) and (2.48), we have

$$Af_0 = A_f f_f$$

or Bandwidth with feedback,

$$f_f = \frac{Af_o}{A_f} \quad ...(2.49)$$

However, for the noninverting amplifier with feedback,

$$A_f = \frac{A}{1 + AB} \quad \text{Refer to Eq. (2.42)}$$

Substituting the value of A_f in Eq. (2.49), we have

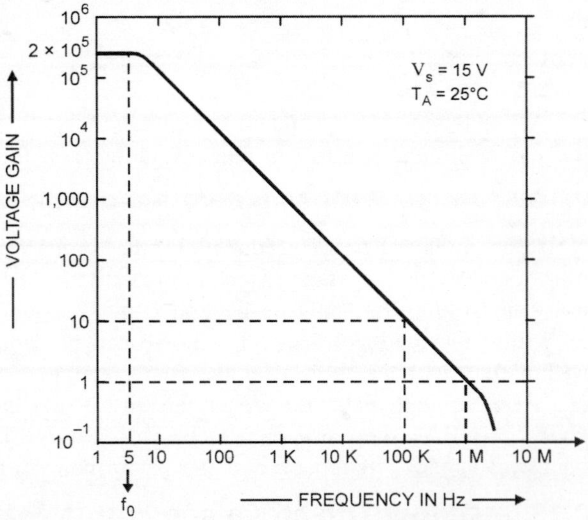

Fig. 2.38. *Open-Loop Gain-Frequency Curve For 741 C*

$$f_f = \frac{Af_0}{\frac{A}{(1+AB)}} = f_0(1+AB) \qquad ...(2.50)$$

The above equation reveals that the series voltage feedback increases the bandwidth of an amplifier by a factor (1 + AB).

This is the same factor involved in the reduction of amplifier gain and output impedance and the increase of the input impedance.

2.19.8. Voltage Follower

An interesting result is obtained, when the resistor R_f is set equal to zero, as shown in Fig. 2.39 or R_1 is made ∞ by keeping it open circuited. A glance at Eq. (2.39) makes it apparent that for such a circuit, we have

$$\text{Output voltage, } v_{out} = v_{in} \qquad ...(2.51)$$

i.e., the output voltage of the op-amp exactly tracks the input voltage both in sign and magnitude. This is the reason that this circuit is called a *voltage follower.*

The result displayed by Eq. (2.51) can also be obtained from direct application of Eq. (2.36). From Fig. 2.39 it is obvious that v_2 is equal to v_{out} and v_1 is equal to v_{in}, so

$$v_{out} = Av_d = A(v_1 - v_2) = A(v_{in} - v_{out})$$

$$\text{or} \quad \frac{v_{out}}{v_{in}} = \frac{A}{1+A} \cong 1 \text{ for } A >> 1 \qquad ...(2.52)$$

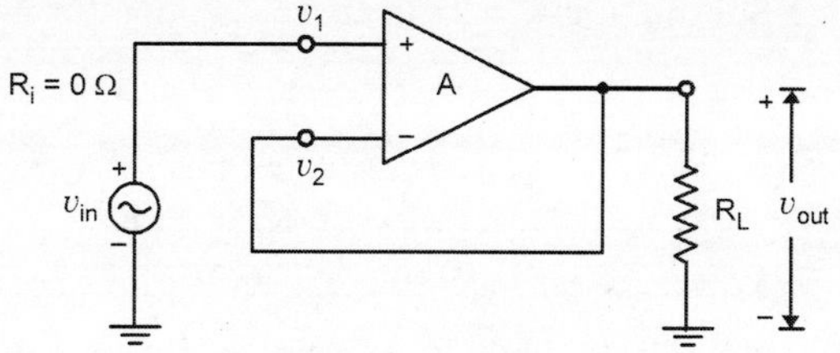

Fig. 2.39. *Voltage Follower*

Thus for voltage follower, $A_f = 1$

Since the voltage follower is a special case of the noninverting amplifier, all the formulas derived for the latter are applicable to the former (voltage follower) except that the gain of the feedback circuit is unity (*i.e.,* B = 1). Thus for a voltage follower

$$\text{Closed-loop gain, } A_f = 1$$

$$\text{Closed-loop input impedance, } R_{inf} = AR_{in}$$

$$\text{Closed-loop output impedance, } R_{outf} = \frac{R_{out}}{A}$$

$$\text{Bandwidth with feedback, } f_f = Af_0$$

The direct connection of the output voltage to the inverting terminal of the op-amp represents the case of 100% negative feedback of the output to the input. In such cases the output quantity is always related to the input quantity by the intermediate Eq. (2.52). Of course, when the open-loop gain A is assumed to be having very large value (of the order of 10^5), the corresponding closed-loop gain becomes approximately unity. However, a unity transmission gain is not the only feature of the voltage (or unity) follower. A more detailed analysis of the equivalent circuit of the op-amp reveals that the resistance looking into the input terminals takes on the value of AR_{in}. We know that when the op-amp is used in the open-loop mode (*i.e.,* without a feedback connection from the output) R_{in} already has a high value of 1 or 2 MΩ. Hence for a typical value of A = 10^5, the input resistance for the voltage follower becomes $1 \times 10^6 \times 10^5$ = 100,000 MΩ or more. It reveals that the op-amp produces virtually no effect on the input signal voltage source even if the latter should be characterized by a high internal resistance.

Detailed analysis of the equivalent circuit of the op-amp also reveals that the output resistance R_{out} looking into the output terminals, which is normally 75 Ω, is reduced to $\frac{R_{out}}{A}$, when 100% feedback is used. As a result the output resistance of the op-amp becomes $\frac{75}{10^5}$ *i.e.,* of the order of 0.75 mΩ. This means that on the output side, the op-amp behaves like a voltage source with negligible internal resistance.

Because of three unique characteristics of extremely high input resistance (of the order of 10^5 MΩ), unity transmission gain, and extremely low output resistance (of the order of few mΩ), the voltage follower circuit is an ideal circuit device to serve as a *buffer* (or isolation unit). Thus, it can be made to prevent the disturbance of one part of a circuit on another as might be encountered, for example, in filter design.

Example 2.12. Design an amplifier to have a voltage gain of +21.

Solution : From Eq. (2.39) $A_f = 1 + \frac{R_f}{R_1}$

Substituting $A_f = 21$ in above equation for A_f, we have

$$21 = 1 + \frac{R_f}{R_1}$$

$$\text{or } R_f = 20\, R_1$$

R_1 may be taken as 10 kΩ, then R_f comes out to be 20 × 10 = 200 kΩ **Ans.**

Example 2.13. What is the range of the voltage gain adjustment in the following circuit.

Solution : From Fig. 2.40

$$R_f = 500 \text{ k}\Omega$$

$$R_{1(min)} = 10 \text{ k}\Omega \text{ and } R_{1(max)} = 10 + 10 = 20 \text{ k}\Omega$$

From Eq. (2.39)

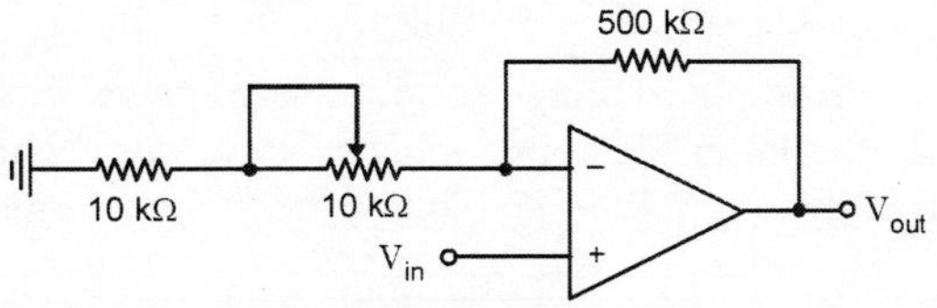

Fig. 2.40

$$\text{Closed-loop voltage gain, } A_f = \left(1 + \frac{R_f}{R_1}\right)$$

$$\text{So } A_{f(min)} = 1 + \frac{R_f}{R_{1(max)}} = 1 + \frac{500}{20} = 26$$

$$\text{So } A_{f(max)} = 1 + \frac{R_f}{R_{1(min)}} = 1 + \frac{500}{10} = 51$$

So range of voltage gain adjustment is 26 to 51. **Ans.**

Example 2.14. Determine voltage gain, input resistance, output resistance and bandwidth for the circuit given in Fig. 2.32 with R_f = 100 kΩ and R_1 = 100Ω. The 741C op-amp connected in the circuit has the following parameters :

A = 2 × 10^5; R_{in} = 2 MΩ, R_{out} = 75 Ω and f_0 = 5 Hz.

Solution : Feedback fraction, $B = \frac{R_1}{R_1 + R_f} = \frac{100}{100 + 100 \times 10^3} = \frac{1}{1{,}001}$ ...Refer to Eq. (2.40)

$$\text{Feedback factor, } AB = 2 \times 10^5 \times \frac{1}{1{,}001} = 199.8$$

$$1 + AB = 1 + 199.8 = 200.8$$

$$\text{Voltage gain, } A_f = \left(1+\frac{R_f}{R_1}\right) = 1+\frac{100\times10^3}{100} = 1{,}001 \qquad \text{...Refer to Eq. (2.39)}$$

Input resistance, $R_{inf} = R_{in}(1 + AB) = 2 \times 10^6 \times 200.8 = 401.6\ M\Omega$ **Ans.** ...Refer to Eq. (2.43)

$$\text{Output resistance, } R_{outf} = \frac{R_{out}}{1+AB} = \frac{75}{200.8} = 0.3735\ \Omega \ \textbf{Ans.}$$

Bandwidth, $f_f = f_0(1 + AB) = 5 \times 200.8 = 1{,}004$ Hz **Ans.** ...Refer to Eq. (2.50)

Example 2.15. Determine the voltage gain, input resistance, output resistance and bandwidth for the circuit shown in Fig. 2.39. The 741C op-amp connected in the circuit has the following parameters :

$A = 2 \times 10^5$, $R_{in} = 2\ M\Omega$, $R_{out} = 75\ \Omega$, $f_0 = 5$ Hz

Solution : For voltage follower, $B = 1 \because R_f = 0$

Voltage gain, $A_f = 1$ **Ans.**

Input resistance, $R_{inf} = AR_{in} = 2 \times 10^5 \times 2 \times 10^6 = 4 \times 10^{11}\ \Omega$ or $400\ G\Omega$ **Ans.**

$$\text{Output resistance, } R_{outf} = \frac{R_{out}}{A} = \frac{75}{2\times10^5} = 0.000375\ \Omega \ \textbf{Ans.}$$

Bandwidth, $f_f = Af_o = 2 \times 10^5 \times 5 = 1$ MHz **Ans.**

Example 2.16. The variable resistance varies from 0 to 100 kΩ. Find out the maximum and the minimum closed-loop voltage gains.

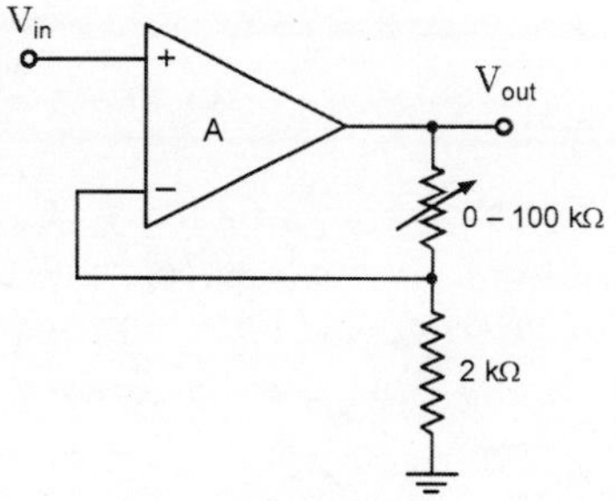

Fig. 2.41

Solution : From Fig. 2.41,

$R_1 = 2\ k\Omega$

$R_{f(min)} = 0$

and $R_{f(max)} = 100\ k\Omega$

From Eq. (2.39)

$$\text{Closed-loop voltage gain, } A_f = \left(1+\frac{R_f}{R_1}\right)$$

$$\text{So } A_{f(max)} = 1 + \frac{R_{f(max)}}{R_1} = 1 + \frac{100}{2} = 51 \ \textbf{Ans.}$$

$$\text{and } A_{f(min)} = 1 + \frac{R_{f(min)}}{R_1} = 1 + 0 = 1 \ \textbf{Ans.}$$

2.20. VOLTAGE-SHUNT FEEDBACK OR INVERTING AMPLIFIER

Figure 2.42 shows the voltage-shunt feedback amplifier using an op-amp. Its noninverting or plus input terminal is grounded whereas the external input signal v_{in} is applied to the inverting or minus input terminal through resistance R_1. A feedback resistor R_f is connected from the output to the inverting input terminal of the op-amp, as illustrated in Fig. 2.42.

This arrangement provides a negative feedback because any increase in output signal results in a feedback signal into the inverting input, causing a decrease in the output signal.

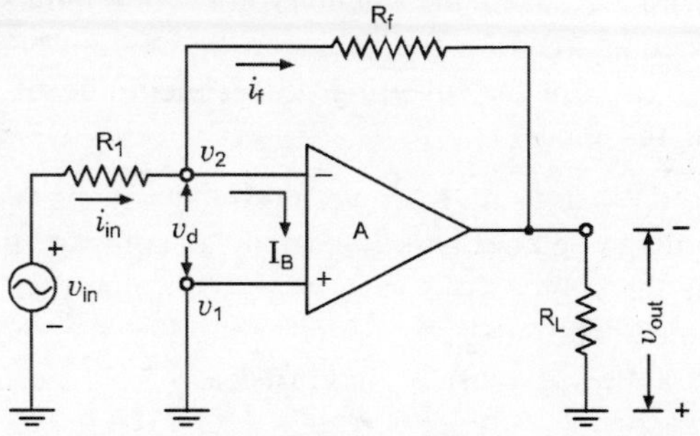

Fig. 2.42. ***Voltage-Shunt Feedback Amplifier or Inverting Amplifier With Feedback***

Due to zero v_d (*i.e.,* zero potential difference between input terminals) no current actually flows from inverting input terminal to noninverting terminal *i.e.,* input bias current I_B is zero.

Applying Kirchhoff's current law at input node v_2 in Fig. 2.42 we have

$$i_{in} = i_f + I_B$$

$$\text{and } i_{in} = i_f \text{ since } I_B = 0$$

$$\text{or } \frac{v_{in} - v_2}{R_1} = \frac{v_2 - v_{out}}{R_f} \quad ...(2.53)$$

$$\text{Since } v_1 - v_2 = \frac{v_{out}}{A} \text{ and } v_1 = 0 \text{ being grounded,}$$

$$v_2 = \frac{-v_{out}}{A}$$

Substituting the value of $v_2 = \frac{-v_{out}}{A}$ in Eq. (2.53) we have

$$\frac{v_{in} - \frac{(-v_{out})}{A}}{R_1} = \frac{\frac{-v_{out}}{A} - v_{out}}{R_f}$$

$$\text{or } A_f = \frac{v_{out}}{v_{in}} = -\frac{AR_f}{R_1 + R_f + AR_1} \quad ...(2.54)$$

The –ve sign in above equation indicates that the input and output voltages are out of phase by 180° (*i.e.,* of opposite polarities). Because of this phase inversion, this configuration is commonly called an *inverting amplifier* with feedback.

Since the open-loop gain A is very large (of the order of 10^5), A $R_1 >> (R_1 + R_f)$ and Eq. (2.54) becomes

$$A_f = \frac{v_{out}}{v_{in}} = -\frac{AR_f}{AR_1} = -\frac{R_f}{R_1} \quad ...(2.55)$$

Because the magnitude of the output voltage can be controlled by the selection of R_f and R_1 $\left(\because v_{out} = \frac{-R_f}{R_1} v_{in}\right)$, we have another version of a voltage-controlled voltage source (VCVS). Thus the output of an op-amp in the inverting mode can also be made to serve as *dependent voltage source.* The corresponding model for this mode is drawn in Fig. 2.43. A comparison of the model displayed in Fig. 2.33 for the noninverting mode indicates that the difference lies mainly in the input resistance encountered. For the noninverting mode it is so high that the model on the input side indicates an open circuit. For the inverting mode, the value of the input resistance is finite and is accordingly included in the model.

When using the inverting mode of the op-amp as a dependent voltage source, it is important to compare the internal resistance of the source voltage v_{in} with the value of R_1 in Fig. 2.42. If the source resistance, R_s is not negligible, then equation for the output voltage v_{out} must be modified as

$$\frac{v_{out}}{v_1} = -\frac{R_f}{R_1 + R_s} \quad ...(2.56)$$

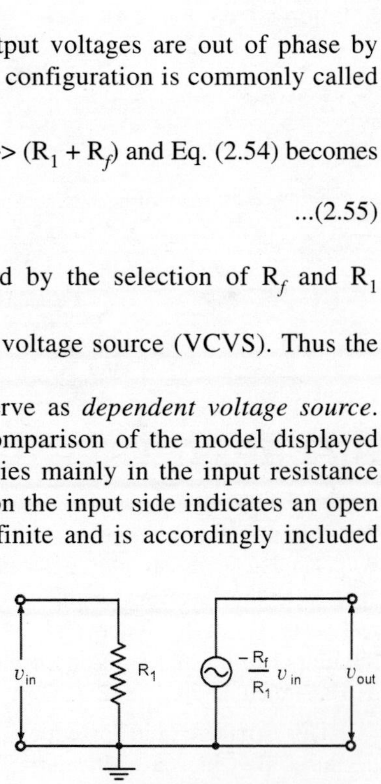

Fig. 2.43. ***Circuit Model For VCVS (Op-Amp in Inverting Mode)***

Inverting op-amp is a very versatile component and can be employed for performing a number of mathematical simulation in control system design by proper choice of R_f and R_1 (or Z_f and Z_1).

2.20.1. Analog Inverter and Paraphase Amplifier

In Fig. 2.42 if $R_f = R_1$, then amplifier gain A_f, being equal to $\frac{v_{out}}{v_{in}}$ or equal to $\frac{-R_f}{R_1}$, is –1 and the $v_{out} = -v_{in}$. Thus the sign of the input signal has simply been reversed. Hence such a circuit is called a *sign changer* or a *phase inverter*. If two such amplifiers are cascaded, the output from the second stage exactly equals the signal input in magnitude and is in phase. Thus the output from the two stages is the same as input—both in phase and magnitude, and as such a system is an excellent *paraphase amplifier*.

2.20.2. Negative Scaler

If the ratio $\frac{R_f}{R_1}$ = K, a real constant, then amplifier gain $A_f = -K$. Thus the input voltage scale has been multiplied by a factor –K to give the output voltage scale. The circuit, therefore, can act as *negative scaler* or *scale changer*. In scale changer, the precision resistors are used to obtain accurate value of the scale factor K.

2.20.3. Phase Shifter

If resistors R_f and R_1 in circuit shown in Fig. 2.42 are replaced by impedances Z_f and Z_1 respectively so that Z_f and Z_1 are equal in magnitude but differ in phase angle, the inverting op-amp shifts the phase of the sinusoidal input voltage without making any change in its amplitude. Thus any phase shift from 0° to 360° (or ±180°) can be obtained.

2.20.4. Input Resistance With Shunt Feedback

The input resistance with shunt feedback can be easily determined by applying Miller's theorem. According to Miller's theorem the feedback resistance of an inverting amplifier can be split into an input Miller resistance and an output Miller resistance.

The input Miller resistance is

$$R_{in\ (Miller)} = \frac{R_f}{A+1} \quad ...(2.57)$$

and the output Miller resistance is

$$R_{out\ (Miller)} = \frac{AR_f}{A+1} \quad ...(2.58)$$

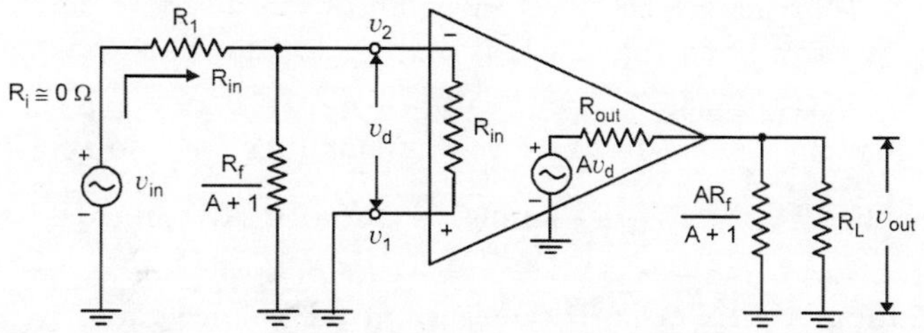

Fig. 2.44. ***Inverting Amplifier With Millerized Feedback Resistance***

Whenever there is a feedback resistor between the input and output of an inverting amplifier, it can be split into an input Miller resistance and output Miller resistance (Fig. 2.44). Because A is extremely high, the input Miller resistance approaches zero and the output Miller resistance approaches R_f.

In the circuit shown in Fig. 2.44, the input resistance with feedback R_{inf} is

$$R_{inf} = R_1 + \frac{R_f}{A+1} \parallel R_{in} \quad ...(2.59)$$

Since R_{in} and A are very large,

$$\frac{R_f}{A+1} \parallel R_{in} \cong 0$$

$$\text{and} \quad R_{inf} = R_1 \quad ...(2.60)$$

2.20.5. Other Benefits of Shunt Feedback

Inverting voltage feedback differs from noninverting voltage feedback primarily because of the way it affects the input characteristics of the amplifier. The input impedance is ideally equal to external resistance connected in series with the input signal. But on the output side, inverting voltage feedback has the same effects as noninverting voltage feedback. That is, it decreases the output impedance, distortion, and the output offset voltage by the same factor as in case of noninverting voltage feedback. It means Eqs. (2.44), (2.45) and (2.46) are also applicable to inverting voltage feedback amplifiers.

2.20.6. Bandwidth With Shunt Feedback

As mentioned in Art. 2.19.7, the gain-bandwidth product of a single break frequency op-amp is always constant. Since the gain of the amplifier with feedback is always less than that without feedback, the bandwidth of the amplifier with feedback f_f must be greater than that without feedback and is given by

$$f_f = f_0(1 + AB) \qquad ...(2.61)$$

where break frequency of the op-amp,

$$f_0 = \frac{\text{Unity gain bandwidth}}{\text{Open-loop voltage gain}}$$

$$= \frac{UGB}{A} \qquad \text{(True only in case of break frequency op-amp like 741)}$$

Substituting the value of $f_0 = \frac{UGB}{A}$ in Eq. (2.61), we have

Bandwidth of the amplifier with feedback

$$f_f = \frac{UGB}{A}(1 + AB) = \frac{UGB}{A_f} \times K \qquad ...(2.62) \quad \text{where } K = \frac{R_f}{R_1 + R_f} \text{ and } A_f = \frac{AK}{1 + AB}$$

Thus the closed-loop bandwidth of the inverting amplifier can be determined using Eq. (2.61) if f_0 is known and Eq. (2.62) if unity gain bandwidth is known.

Note : Since inverter explained in Art. 2.20.1 is a special case of the inverting amplifier, all the equations derived for the inverting amplifier are also applicable to the inverter. The equations can be applied by merely substituting $\frac{A}{2}$ for (1 + AB) since $B = \frac{1}{2}$.

Example 2.17. What is the range of the voltage gain adjustment in the following circuit ?

[U.P. Technical Univ. Electronics Engineering First Semester 2006-07]

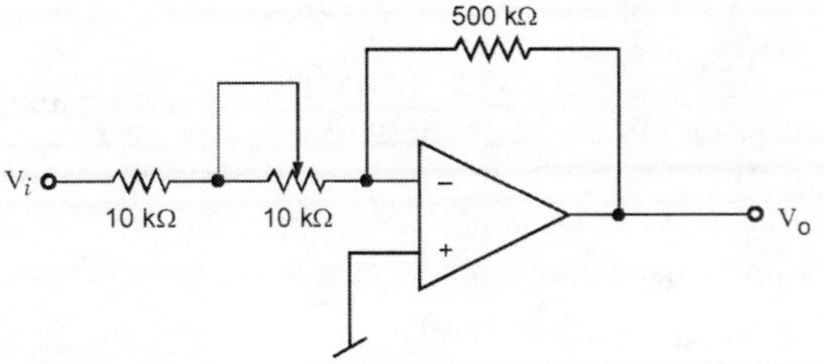

Fig. 2.45

Solution : From Eq. (2.55)

$$\text{Closed-loop voltage gain, } A_f = \frac{-R_f}{R_1}$$

$$\text{So } A_{f(\min)} = \frac{-R_f}{R_{1(\max)}} = \frac{-500\text{ k}\Omega}{(10 + 10)\text{k}\Omega} = -25$$

$$A_{f(\max)} = \frac{-R_f}{R_{1(\min)}} = \frac{-500\text{ k}\Omega}{10\text{k}\Omega} = -50$$

Thus the voltage gain adjustment is –25 to –50 **Ans.**

Example 2.18. Determine the voltage gain, input resistance, output resistance and bandwidth for the circuit shown in Fig. 2.42 if R_1 = 330 Ω and R_f = 3.3 kΩ. The 741C op-amp connected in the circuit has the following parameters :

$A = 2 \times 10^5$, R_{in} = 2 MΩ, R_{out} = 75 Ω and f_0 = 5 Hz.

Solution :

$$\text{Feedback fraction, } B = \frac{R_1}{R_1 + R_f} = \frac{330}{330 + 3.3 \times 10^3} = \frac{1}{11}$$

$$\text{Feedback factor, } AB = 2 \times 10^5 \times \frac{1}{11} = 18{,}181.8$$

$$(1 + AB) = 1 + 18{,}181.8 = 18{,}182.8$$

$$\text{Closed-loop voltage gain, } A_f = \frac{-R_f}{R_1} = \frac{3.3 \times 10^3}{330} = -10 \quad \textbf{Ans.}$$

$$\text{Input resistance with feedback, } R_{inf} = R_1 = 330\ \Omega$$

$$\text{Output resistance with feedback, } R_{outf} = \frac{R_{out}}{1 + AB} = \frac{75}{18{,}182.8} = 0.00412\ \Omega \quad \textbf{Ans.}$$

$$\text{Closed-loop bandwidth, } f_f = f_0(1 + AB) = 5 \times 18{,}182.8 = 90.92 \text{ kHz} \quad \textbf{Ans.}$$

Example 2.19. Determine the voltage gain, input resistance, output resistance and bandwidth of the inverter (Fig. 2.42). The 741C op-amp connected in the circuit has the following parameters :

$A = 2 \times 10^5$, R_{in} = 2 MΩ, R_{out} = 75 Ω and f_0 = 5 Hz.

Solution : For the inverter circuit shown in Fig. 2.42,

$$\text{Feedback fraction, } B = \frac{R_1}{R_1 + R_f} = \frac{1}{2} \qquad \because R_1 = R_f$$

$$\text{Voltage gain, } A_f = -1$$

$$\text{Input resistance, } R_{inf} = 330\ \Omega \quad \text{Assuming } R_1 = 330\ \Omega$$

$$\text{Output resistance, } R_{outf} = \frac{R_{out}}{\frac{A}{2}} = \frac{75}{\frac{2 \times 10^5}{2}} = 0.00075\ \Omega \quad \textbf{Ans.}$$

$$\text{Bandwidth, } f_f = \frac{A}{2} f_0 = \frac{2 \times 10^5}{2} \times 5 \text{ Hz} = 0.5 \text{ MHz} \quad \textbf{Ans.}$$

Example 2.20. 741C op-amp having the following parameters is connected as an

(*i*) Noninverting amplifier with R_1 = 1 kΩ and R_f = 10 kΩ.

(*ii*) Inverting amplifier with R_1 = 470 Ω and R_f = 4.7 kΩ

A = 200,000, R_{in} = 2 MΩ, R_0 = 75 Ω, f_0 = 5 Hz, supply voltages = ±15 V,

Output voltage swing = ±13 V

Calculate the values of A_f, R_{1f}, R_{of}, f_f and V_{ooT} for (*i*) and (*ii*) configurations.

Solution : Given that

$$\text{Open-loop gain, } A = 200{,}000$$

$$\text{Input resistance, } R_{in} = 2\ M\Omega$$

$$\text{Output resistance, } R_{out} = 75\ \Omega$$

$$f_0 = 5 \text{ Hz}$$

$$\text{Supply voltages, } V_{CC} = 15 \text{ V}$$

$$\text{and } V_{EE} = -15 \text{ V}$$

$$\text{Output voltage swing} = \pm 13 \text{ V}$$

(*i*) **Noninverting Amplifier :**

$$\text{Feedback fraction, B} = \frac{R_1}{R_1 + R_f} = \frac{1 \times 10^3}{1 \times 10^3 + 10 \times 10^3} = \frac{1}{11}$$

$$\text{Feedback factor, AB} = 2 \times 10^5 \times \frac{1}{11} = 18{,}181.8$$

$$1 + AB = 1 + 18{,}181.8 = 18{,}182.8$$

$$\text{Voltage gain, } A_f = \left(1 + \frac{R_f}{R_1}\right) = 1 + \frac{10 \times 10^3}{1 \times 10^3} = 11 \textbf{ Ans.}$$

$$\text{Input resistance, } R_{inf} = R_{in}(1 + AB) = 2 \times 10^6 \times 18{,}182.8 = 36.3656 \text{ G}\Omega \quad \textbf{Ans.}$$

$$\text{Output resistance, } R_{outf} = \frac{R_{out}}{(1 + AB)} = \frac{75}{18{,}182.8} = 0.00412\ \Omega \textbf{ Ans.}$$

$$\text{Bandwidth, } f_f = f_0\,(1 + AB) = 5 \text{ Hz} \times 18{,}182.8 = 90.92 \text{ kHz} \quad \textbf{Ans.}$$

Total output offset voltage with feedback,

$$V_{ooT} = \frac{\text{Output voltage swing}}{1 + AB} = \frac{\pm 13 \text{ V}}{18{,}182.8} = \pm 0.000715 \text{ V or } \pm 0.715 \text{ mV} \textbf{ Ans.}$$

(*ii*) **Inverting Amplifier :**

$$\text{Feedback fraction, B} = \frac{R_1}{R_1 + R_f} = \frac{470}{470 + 4.7 \times 10^3} = \frac{1}{11}$$

$$1 + AB = 1 + 2 \times 10^5 \times \frac{1}{11} = 18{,}182.8$$

$$\text{Closed loop gain, } A_f = \frac{-R_f}{R_1} = \frac{4.7 \times 10^3}{470} = -10 \quad \textbf{Ans.}$$

$$\text{Input resistance with feedback, } R_{inf} = R_1 = 470\ \Omega$$

$$\text{Output resistance with feedback, } R_{out\,f} = \frac{R_{out}}{1 + AB} = \frac{75}{18{,}182.8} = 0.00412\ \Omega \quad \textbf{Ans.}$$

$$\text{Closed-loop bandwidth, } f_f = f_0(1 + AB) = 5 \text{ Hz} \times 18{,}182.8 = 90.92 \text{ kHz} \quad \textbf{Ans.}$$

$$\text{Total output offset voltage with feedback, } V_{ooT} = \frac{\pm 13 \text{ V}}{18{,}182.8} = \pm 0.715 \text{ mV} \textbf{ Ans.}$$

Example 2.21. An inverting amplifier has R_f = 500 kΩ and R_1 = 5 kΩ. Determine the amplifier circuit voltage gain, input resistance and output resistance. Determine also the output voltage and input current if the input voltage is 0.1 V. Assume op-amp to be ideal one.

Solution : Resistance R_1 = 5 kΩ

Feedback resistance, R_f = 500 kΩ

Input voltage, V_{in} = 0.1 V

$$\text{Amplifier circuit voltage gain, } A_f = -\frac{R_f}{R_1} = -\frac{500 \text{ k}\Omega}{5 \text{ k}\Omega} = -100 \textbf{ Ans.}$$

Amplifier circuit input resistance, R_{in} = R_1 = 5 kΩ **Ans.**

Amplifier circuit output resistance, R_{out} = Output resistance of op-amp = 0 Ω **Ans.**

Output voltage, $V_{out} = A_f V_{in} = -100 \times 0.1 = -10$ V **Ans.**

$$\text{Input current, } I_{in} = \frac{V_{in}}{R_1} = \frac{0.1}{5 \times 1{,}000} = 0.02 \text{ mA} \textbf{ Ans.}$$

Example 2.22. Compute the voltage gain and power gain of the amplifier circuit shown in Fig. 2.46. The op-amp shown is ideal.

Solution : (*i*) Given amplifier is an inverting amplifier

$$\text{where } V_{out} = -\frac{R_f}{R_{in}}V_{in} = -\frac{1\,M\Omega}{1\,M\Omega} \times V_{in} = -V_{in}$$

$$\text{so } A_v = \frac{V_{out}}{V_{in}} = -1 \quad \textbf{Ans.}$$

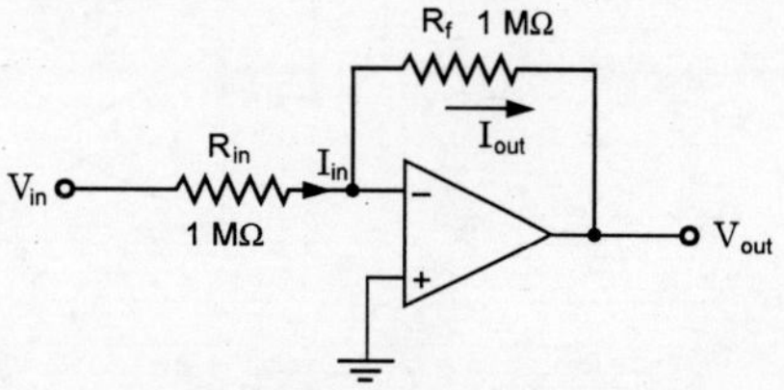

Fig. 2.46

(*ii*) Because it is a unity gain inverter,

$$\text{so } I_{in} = I_{out}$$

$$\text{or } A_{in} = \frac{I_{out}}{I_{in}} = 1 \quad \textbf{Ans.}$$

(*iii*) Power gain of op-amp circuit,

$$A_p = |A_V \times A_i| = |1 \times 1| = 1 \quad \textbf{Ans.}$$

Example 2.23. Design an inverting op-amp circuit with a voltage gain of $A_V = \frac{V_0}{V_i} = -30$ and an input resistance that is largest value possible but under constraint that the largest resistance value is limited to 1 MΩ.

[*U.P. Technical Univ. Analog Integrated Circuits 2004-05*]

Solution : Voltage gain, $\frac{V_0}{V_i} = \frac{-R_f}{R_1} = -30$

or $R_f = 30\,R_1$

Obviously R_f will be larger resistance, being 30 times of R_1 and largest resistance value is limited to 1 MΩ.

So $R_f = 1$ MΩ **Ans.**

$$\text{and } R_1 = \frac{R_f}{30} = \frac{1 \times 10^6}{30} = 33\text{ k}\Omega \quad \textbf{Ans.}$$

Example 2.24. Design an inverting op-amp circuit with a voltage gain of $A_V = \frac{V_{out}}{V_{in}} = -8$. When the input voltage is $V_{in} = -1$ V, the maximum current in R_1 and R_2 must be no larger than 15 mA. Determine the minimum values of R_1 and R_2.

Solution : Voltage gain, $A_V = \frac{V_{out}}{V_{in}} = -8.$

Because of the virtual ground on the inverting input,

$$\text{Resistance } R_1 = \frac{V_{in}}{I_1} = \frac{1\text{ V}}{15 \times 10^{-6}\text{ A}} = 66.66\text{ k}\Omega \quad \text{(use 68 k}\Omega\text{ standard value)}$$

$$\text{and Resistance } R_f = -A_f R_1 \quad \text{...[Equation (2.55)]}$$

$$= -(-8) \times 68 = 544\text{ k}\Omega \quad \text{[Use 560 k}\Omega\text{ standard value]}$$

Example 2.25. For the circuit shown in Fig. 2.47, find out voltage V_{out}.

Solution : $I_1 = \frac{V_{in}}{R_1}$ ...(*i*)

Applying KVL to loops, we have

$I_1R_2 + (I_1 - I_2)R_4 = 0$ ...(*ii*)

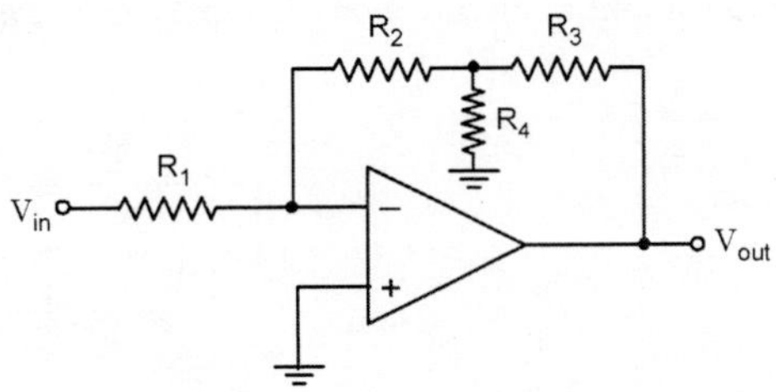

Fig. 2.47

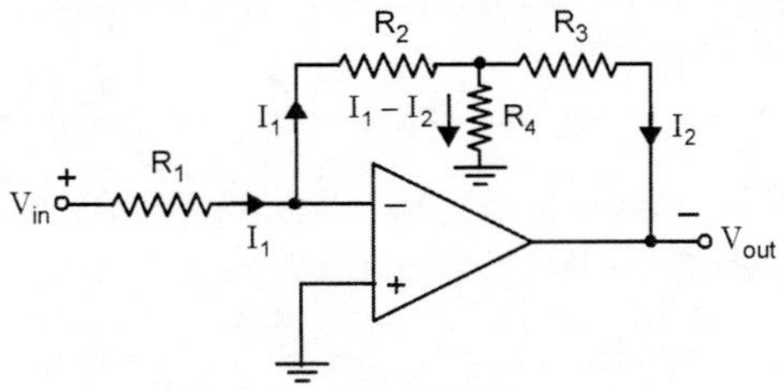

Fig. 2.48

$$\text{and } V_{out} = (I_1 - I_2)R_4 - I_2R_3 = I_1R_4 - I_2(R_3 + R_4) \quad ...(iii)$$

From Eq. (*ii*), we have

$$I_2 = \frac{I_1(R_2 + R_4)}{R_4} \quad ...(iv)$$

Substituting $I_2 = I_1 \dfrac{(R_2 + R_4)}{R_4}$ from Eq. (*iv*) in Eq. (*iii*), we have

$$V_{out} = I_1R_4 - \frac{I_1(R_2 + R_4)(R_3 + R_4)}{R_4}$$

$$= I_1R_4 - I_1\left[\frac{R_2R_3 + R_2R_4 + R_3R_4 + R_4^2}{R_4}\right]$$

$$= I_1R_4 - I_1\left(\frac{R_2R_3}{R_4} + R_2 + R_3 + R_4\right) = -I_1\left(\frac{R_2R_3}{R_4} + R_2 + R_3\right) \quad ...(v)$$

From Eqs. (*i*) and (*v*), we have

$$\frac{V_{out}}{V_{in}} = -\frac{\dfrac{R_2R_3}{R_4} + R_2 + R_3}{R_1} \quad \textbf{Ans.}$$

2.21. OP-AMP DIFFERENTIAL AMPLIFIER

Sometimes it is necessary to amplify the voltage difference between two input lines neither of which is grounded. In this case, the amplifier is called a *differential amplifier*. The circuit of an op-amp differential amplifier is shown in Fig. 2.49.

As already known, in a differential amplifier, neither input is grounded. This reduces the amount of noise injected into the amplifier, because any noise appears simultaneously on both input terminals and the amplifying circuitry rejects it being a common mode signal.

A close examination of the circuit given in Fig. 2.49 reveals that it is a combination of inverting and noninverting amplifiers. When v_{in_1} is zero, the circuit appears as an inverting amplifier while when v_{in_2} is zero the circuit becomes a noninverting amplifier.

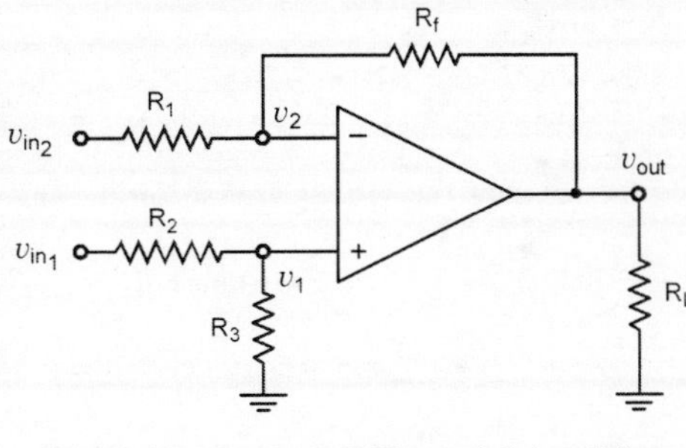

Fig. 2.49. ***Op-Amp Differential Amplifier***

Since the circuit has two inputs v_{in_1} and v_{in_2}, superposition theorem will be used for determination of voltage gain of the amplifier.

When v_{in_1} is zero volt, circuit becomes an inverting amplifier and, therefore, output voltage due to v_{in_2} is

$$v_{out2} = \frac{-R_f}{R_1} v_{in_2} \qquad ...(2.63)$$

Now assuming $v_{in_2} = 0$, the circuit is a noninverting amplifier having a voltage-divider network consisting of resistor R_2 and R_3 at the noninverting input. Therefore,

$$v_1 = \frac{R_3}{R_2 + R_3} v_{in_1}$$

and output due to v_{in_1} then is

$$v_{out_1} = \left(1 + \frac{R_f}{R_1}\right) v_1 = \left(1 + \frac{R_f}{R_1}\right)\left(\frac{R_3}{R_2 + R_3}\right) v_{in_1}$$

If $R_1 = R_2$ and $R_f = R_3$

$$v_{out_1} = \left(\frac{R_f(R_2 + R_3)}{R_1(R_2 + R_3)}\right) v_{in_1} = \frac{R_f}{R_1} v_{in_1} \qquad ...(2.64)$$

The net output voltage, $v_{out} = v_{out_1} + v_{out_2} = \frac{R_f}{R_1} v_{in_1} - \frac{R_f}{R_1} v_{in_2} = \frac{-R_f}{R_1}(v_{in_2} - v_{in_1})$ ...(2.65)

Differential voltage gain, $A_d = \frac{v_{out}}{(v_{in_2} - v_{in_1})} = \frac{-R_f}{R_1}$ ...(2.66)

The above equation resembles Eq. (2.55) and hence the gain of the differential amplifier is same as that of inverting amplifier.

The differential amplifier can be operated in three different modes :

1. Single Ended Mode. The differential amplifier is in a single ended mode when input signal is applied to only one of the two input terminals *i.e.,* either of the two inputs is zero. Thus in this mode, the differential amplifier will act either as an inverting amplifier or as a noninverting amplifier depending on whether the signal is applied to the inverting or noninverting terminal.

2. Differential Mode. In this mode of operation, magnitude of the two input signals may be same or different but their polarity is opposite at every instant of time *i.e.,*

$$v_{in_2} = -v_{in_1}$$

So input voltage, $v_{in} = v_{in_2} - v_{in_1} = v_{in_2} - (-v_{in_2}) = 2\, v_{in_2}$

Hence $v_{out} = \frac{-R_f}{R_1}[v_{in_2} - v_{in_1}] = \frac{-2R_f}{R_1} v_{in_2}$

3. Common Mode. In this mode the two input signals are same in magnitude as well as in phase at every instant of time

So input voltage, $v_{in} = v_{in_2} - v_{in_1} = 0$ $\qquad \because v_{in_2} = v_{in_1}$

and output voltage is zero.

Thus there is no output voltage when the differential amplifier is connected in common mode.

2.21.1. Input Resistance

The input resistance, R_{in} of the differential amplifier is the resistance worked out looking into either one of the two input terminals with the other grounded.

Hence with $v_{in_1} = 0$ V, the circuit given in Fig. 2.49 becomes that of an inverting amplifier, the input resistance of which is

$$R_{inf_1} \simeq R_1 \quad ...(2.67)$$

Similarly with $v_{in_2} = 0V$, the circuit given in Fig. 2.49 becomes that of noninverting amplifier whose input resistance is given as

$$R_{inf_2} = R_2 + R_3 \quad ...(2.68)$$

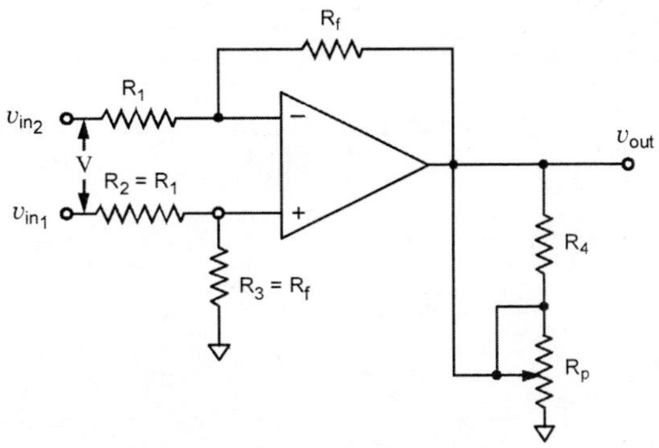

Fig. 2.50. ***Differential Amplifier With Variable Gain***

Equations (2.67) and (2.68) reveal that the input resistance looked into by the signal sources v_{in1} and v_{in2} is not the same. For proper functioning of amplifier circuit given in Fig. 2.49, it is necessary that resistances R_1 and $(R_2 + R_3)$ are made much larger than source resistances so as to avoid loading of signal sources. In cases where a variable gain is required, the circuit given in Fig. 2.49 is modified by introducing a potentiometer as depicted in Fig. 2.50. The voltage gain can be varied from the closed-loop gain of $\frac{-2R_f}{R_1}$ to the open-loop gain of A.

Example 2.26. For the circuit shown in Fig. 2.51 find out voltage gain A_V.

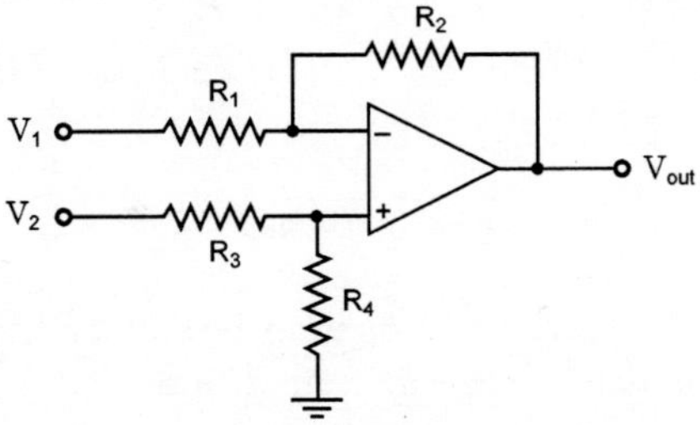

Fig. 2.51

Solution : Output voltage due to V_1

$$V_{out_1} = \frac{-R_2}{R_1} V_1 \quad ...\text{Refer to Eq. (2.63)}$$

Output voltage due to V_2,

$$V_{out_2} = \left(1 + \frac{R_2}{R_1}\right)\left(\frac{R_4}{R_3 + R_4}\right) V_2 \quad ...\text{Refer to Eq. (2.64)}$$

$$\text{So } V_{out} = V_{out_1} + V_{out_2} = \frac{-R_2}{R_1} V_1 + \left(\frac{R_1 + R_2}{R_1}\right)\left(\frac{R_4}{R_3 + R_4}\right) V_2$$

$$\text{Differential voltage gain, } A_{Vd} = \frac{V_{out}}{V_{in}} = \frac{\left(\frac{R_1 + R_2}{R_1}\right)\left(\frac{R_4}{R_3 + R_4}\right) V_2 - \frac{R_2}{R_1} V_1}{V_1 - V_2} \quad \textbf{Ans.}$$

Example 2.27. For the op-amp shown in Fig. 2.52, find the values of R_1 and R_2 for the output to be $v_0 = -5v_a + 3v_b$.

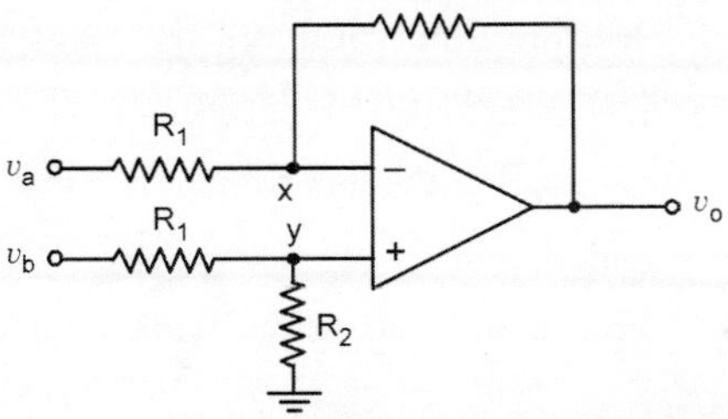

Fig. 2.52

Solution : *Output voltage due to v_a*

$$v_{out_1} = \frac{-250}{R_1} v_a \quad ...\text{Refer to Eq. (2.63)}$$

Output voltage due to v_b

$$v_{out_2} = \left(1 + \frac{250}{R_1}\right)\left(\frac{R_2}{R_1 + R_2}\right) v_b$$

$$\text{so } v_{out} = v_{out_1} + v_{out_2} = \frac{-250}{R_1} v_a + \left(1 + \frac{250}{R_1}\right)\left(\frac{R_2}{R_1 + R_2}\right) v_b$$

$$= \frac{-250}{R_1} v_a + \left(\frac{R_1 + 250}{R_1}\right)\left(\frac{R_2}{R_1 + R_2}\right) v_b \quad ...(i)$$

and $v_{out} = -5v_a + 3v_b$ ...(*ii*) (given)

Comparing Eqs. (*i*) and (*ii*), we have

$$\frac{-250}{R_1} = -5 \quad \text{or} \quad R_1 = 50 \text{ k}\Omega \; \textbf{Ans.}$$

and $$\left(\frac{R_1 + 250}{R_1}\right)\left(\frac{R_2}{R_1 + R_2}\right) = 3$$

Substituting $R_1 = 50$ kΩ in above equation, we have

$$\left(\frac{50 + 250}{50}\right)\left(\frac{R_2}{50 + R_2}\right) = 3$$

$$\text{or } R_2 = 50 \text{ k}\Omega \; \textbf{Ans.}$$

Example 2.28. Obtain CMRR for the circuit shown in Fig. 2.53.

Solution : Output voltage, as worked out in Example 2.26 above, is

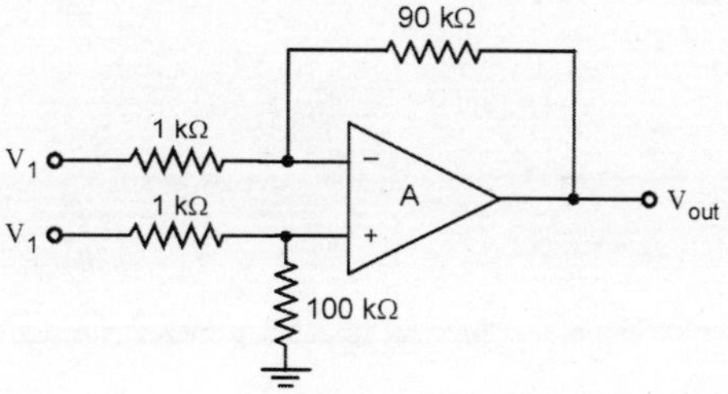

Fig. 2.53

$$V_{out} = \frac{-R_2}{R_1}V_1 + \left(\frac{R_1 + R_2}{R_1}\right)\left(\frac{R_4}{R_3 + R_4}\right)V_2 \quad ...(i)$$

Substituting R_2 = 90 kΩ, R_1 = 1 kΩ, R_3 = 1 kΩ and R_4 = 100 kΩ in above Eq. (*i*), we have

$$V_{out} = -90V_1 + 91 \times \frac{100}{101}V_2$$

$$= \frac{9100}{101}V_2 - 90V_1 = 90.1V_2 - 90V_1 \quad ...(ii)$$

Differential voltage gain, $$A_d = \frac{V_{out}}{V_1 - V_2} = \frac{90.1\,V_2 - 90\,V_1}{V_1 - V_2} \quad ...(iii)$$

Substituting $V_1 = V_2 = V_{cm}$ in Eq. (*ii*), we have

Common-mode output voltage, $V_{out\ cm} = 90.1\ V_{cm} - 90\ V_{cm} = 0.1\ V_{cm}$

Common-mode voltage gain, $$A_{cm} = \frac{V_{out\ cm}}{V_{cm}} = \frac{0.1V_{cm}}{V_{cm}} = 0.1 \quad ...(iv)$$

CMRR for the given circuit $$= \frac{A_d}{A_{cm}} = \frac{\dfrac{90.1V_2 - 90V_1}{V_1 - V_2}}{0.1} = \frac{901V_2 - 900V_1}{V_1 - V_2} \; \textbf{Ans.}$$

Example 2.29. If $R_1 = R_2 = 1$ kΩ, $R_F = R_3 = 10$ kΩ, v_d = 5 mV sinewave at 1 kHz and v_{n_i} = 2 mV at 60 Hz. Calculate :

(*a*) **Output voltage at 1 kHz.**

(*b*) **Magnitude of the induced 60 Hz noise at the output.**

The op-amp is μA 741 with CMRR (dB) = 90 dB.

[*Pb. Technical Univ. Linear Integrated Circuits 2006-07*]

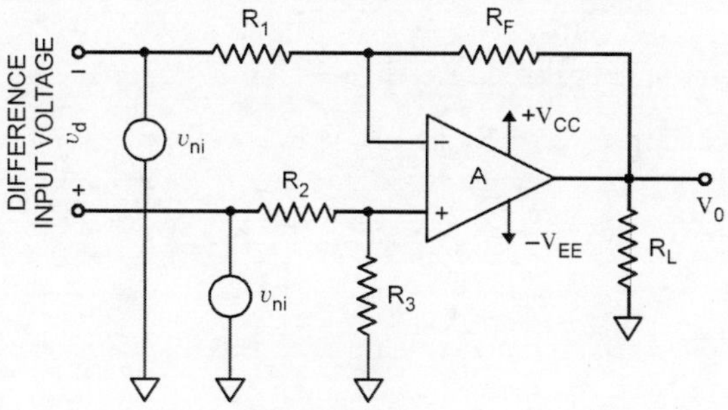

Fig. 2.54

Solution : Differential voltage gain,

$$A_d = \frac{-R_f}{R_1} \quad ...\text{Refer to Eq. (2.66)}$$

$$= \frac{-10 \text{ k}\Omega}{1 \text{ k}\Omega} = -10$$

(*i*) Output voltage, $v_{out} = A_d v_d = -10 \times 5\text{ mV} = 50\text{ mV}$ **Ans.**

$$\text{CMRR (dB)} = 90\text{ dB}$$

$$\text{CMRR} = \text{Antilog}\frac{\text{CMRR (dB)}}{20} = \text{Antilog}\frac{90}{20} = 31{,}623$$

$$\text{Common mode gain, } A_{cm} = \frac{A_d}{\text{CMMR}} = \frac{10}{31{,}623} = 0.000316$$

(*ii*) Magnitude of the induced 60 Hz noise at the output

$$= A_{cm} v_{cm} = 0.000316 \times 2\text{ mV} = 0.632\ \mu\text{V} \quad \textbf{Ans.}$$

Example 2.30. Find the output voltage of the following op-amp circuit shown in Fig. 2.55.

Solution : Output voltage, $V_{out} = A_{vd} \times V_{in}$

$$= A_{vd}(V_1 - V_2)$$

$$= \frac{R_1 + R_2}{R_1} \times \frac{R_4}{R_3 + R_4} \times V_2 - \frac{R_2}{R_1} V_1$$

$$= \frac{5+10}{5} \times \frac{1}{5+1} V_2 - \frac{10}{5} V_1$$

$$= \frac{1}{2} V_2 - 2V_1 \quad \textbf{Ans.}$$

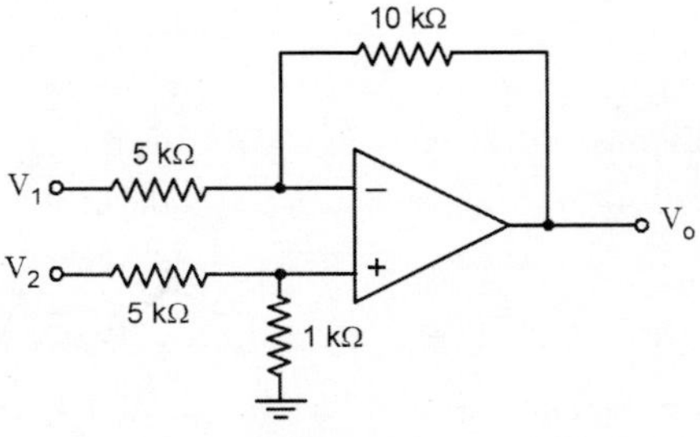

Fig. 2.55

2.22. DIFFERENTIAL AMPLIFIER WITH TWO OP-AMPS

The gain as well as input resistance of the differential amplifier can be increased by using two op-amps. The characteristics of such an amplifier are identical to those of a noninverting amplifier. A practical circuit using two-op-amps is given in Fig. 2.56.

The circuit is composed of a two stage amplifier, the first stage being of a noninverting amplifier and the second being of a differential amplifier, both having gains of different values.

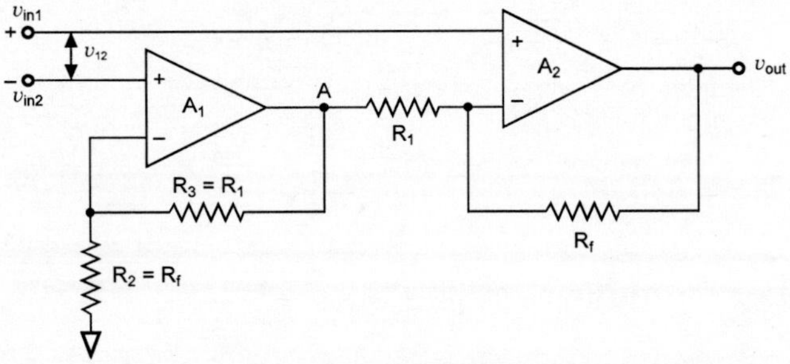

Fig. 2.56

The output of first stage,

$$v_A = \left(1 + \frac{R_3}{R_2}\right) v_{in_2} \quad \text{...(2.69)}$$

Output from second stage,

$$v_{out} = \frac{-R_f}{R_1} v_A + \left(1 + \frac{R_f}{R_1}\right) v_{in_1} \quad \text{...(2.70)}$$

$$= \frac{-R_f}{R_1}\left(1 + \frac{R_3}{R_2}\right) v_{in_2} + \left(1 + \frac{R_f}{R_1}\right) v_{in_1} \quad \text{...(2.71)}$$

Since $R_3 = R_1$ and $R_f = R_2$, we have

$$v_{out} = \left(1 + \frac{R_f}{R_1}\right) v_{in_1} - \frac{R_f}{R_1}\left(1 + \frac{R_1}{R_2}\right) v_{in_2}$$

$$= \left(1 + \frac{R_f}{R_1}\right)(v_{in_1} - v_{in_2}) \quad \text{...(2.72)}$$

$$\text{Voltage gain, } A_d = \frac{v_{out}}{v_{in_1} - v_{in_2}} = \left(1 + \frac{R_f}{R_1}\right) \quad \text{...(2.73)}$$

Input Resistance. Input resistance of first stage, $R_{inf_2} = R_{in}(1 + AB)$ ...(2.74)

where R_{in} is the open-loop resistance of op-amp, as first stage is noninverting amplifier.

$$B = \frac{R_2}{R_2 + R_3} \qquad ...(2.75)$$

With v_{in_2} grounded *i.e.*, $v_{in_2} = 0$, the second stage also becomes a noninverting amplifier whose input resistance may be given as

$$R_{inf_1} = R_{in}(1 + AB) \qquad ...(2.76)$$

where R_{in} is open-loop resistance of op-amp

$$B = \frac{R_1}{R_1 + R_f} \qquad ...(2.77)$$

$$\text{As } R_1 = R_3,\ R_f = R_2,\ R_{inf_2} \neq R_{inf_1}$$

The output signal may be of smaller magnitude than expected. This is a drawback of such an amplifier.

Example 2.31. Determine (*i*) the voltage gain (*ii*) output voltage and internal resistance for the differential amplifier given in Fig. 2.56. Given that

$$R_1 = R_3 = 540\ \Omega\ ;\ R_f = R_2 = 5.4\ k\Omega;\ v_{in_1} = -2.5\ V,\ v_{in_2} = -3.5\ V$$

Input impedance R_i = 2 MΩ and open-loop voltage gain = 2 × 10^5.

Solution :

$$\text{Voltage gain, } A_d = \left(1 + \frac{R_f}{R_1}\right) = \left(1 + \frac{5.4 \times 10^3}{540}\right) = 11 \textbf{ Ans.}$$

$$\text{Output voltage, } v_{out} = A_d(v_{in_1} - v_{in_2}) = 11(-2.5 + 3.5) = 11\text{ V } \textbf{Ans.}$$

$$R_{inf_1} = R_{in}\left[1 + \frac{AR_1}{R_1 + R_f}\right] = 2 \times 10^6 \left[1 + \frac{2 \times 10^5 \times 540}{540 + 5,400}\right] = 36.37 \times 10^9\ \Omega \textbf{ Ans.}$$

$$R_{inf_2} = R_{in}\left[1 + \frac{AR_2}{R_2 + R_3}\right] = 2 \times 10^6 \left[1 + \frac{2 \times 10^5 \times 5,400}{5,400 + 540}\right] = 363.7 \times 10^9\ \Omega \textbf{ Ans.}$$

2.23. DIFFERENTIAL AMPLIFIER WITH THREE OP-AMPS

The differential amplifier with three op-amps, depicted in Fig. 2.57 consists of two stages—one formed by op-amps A_1 and A_2 and second by op-amp A_3. Hence for determination of overall voltage gain of the circuit shown in Fig. 2.57, it becomes imperative to determine the voltage gain of each stage.

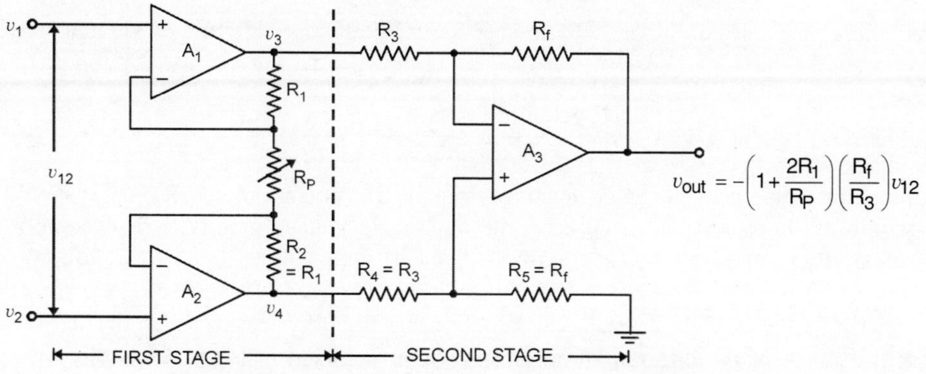

Fig. 2.57. ***Differential Amplifier With Three Op-amps***

The first stage may be viewed as two differential amplifiers, as depicted in Fig. 2.58.

The output voltages of the two differential amplifiers shown in Fig. 2.58 are determined by using superposition theorem.

For circuit depicted in Fig. 2.58 (*a*)

$$v_3 = \frac{2R_1 + R_P}{R_1 + R_P} v_1 - \frac{R_1}{R_1 + R_P} v_4 \qquad ...(2.78)$$

and for circuit depicted in Fig. 2.58 (*b*).

$$v_4 = \frac{2R_1 + R_P}{R_1 + R_P} v_2 - \frac{R_1}{R_1 + R_P} v_3 \qquad ...(2.79)$$

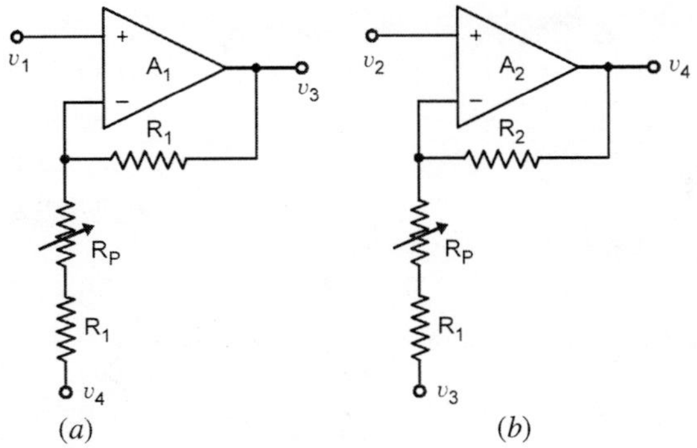

Fig. 2.58

The output voltage of the first stage may be given as

$$v_{34} = v_3 - v_4$$

Substituting the values of v_3 and v_4 respectively from Eqs. (2.78) and (2.79) in above equation, we have

$$v_{34} = \frac{2R_1 + R_P}{R_1 + R_P}(v_1 - v_2) + \frac{R_1}{R_1 + R_P}(v_3 - v_4) = \frac{2R_1 + R_P}{R_1 + R_P} v_{12} + \frac{R_1}{R_1 + R_P} v_{34}$$

or $$v_{34}\left(1 - \frac{R_1}{R_1 + R_P}\right) = \frac{2R_1 + R_P}{R_1 + R_P} v_{12}$$

or $$v_{34} \times \frac{R_P}{R_1 + R_P} = \frac{2R_1 + R_P}{R_1 + R_P} v_{12}$$

or $$\frac{v_{34}}{v_{12}} = \frac{2R_1 + R_P}{R_P} \qquad ...(2.80)$$

The gain of the second stage may be determined using Eq. (2.66) and is given by

$$\frac{v_{out}}{v_{34}} = \frac{-R_f}{R_3} \qquad ...(2.81)$$

The overall voltage gain is given by

$$A_d = \frac{v_{out}}{v_{12}} = \frac{v_{34}}{v_{12}} \times \frac{v_{out}}{v_{34}}$$

Substituting the values of $\frac{v_{34}}{v_{12}}$ and $\frac{v_{out}}{v_{34}}$ respectively from Eqs. (2.80) and (2.81) in above equation, we have

$$A_d = \frac{2R_1 + R_P}{R_P} \times \frac{-R_f}{R_3} = -\left(1 + \frac{2R_1}{R_P}\right)\frac{R_f}{R_3} \qquad ...(2.82)$$

Thus the obtainable gain can be changed by adjusting the potentiometer (R_P) but it should never be set at zero or infinity. In case R_P is made zero, the obtainable voltage gain A_d will become indeterminant, whereas in case R_P is made infinity (open-circuit), the first stage will act as a voltage follower.

2.23.1. Input Resistance

The input resistance of the differential amplifier circuit depicted in Fig. 2.57 is the same as the input resistance of the first stage (the resistance determined at input v_1 and v_2) looking into the circuit with other terminal grounded.

In Fig. 2.58 (*a*), for example, when v_4 is reduced to zero (*i.e.*, when v_2 is grounded), the circuit is a noninverting amplifier and the input impedance is given as

$$R_{inf} = R_{in}(1 + AB) \qquad \text{where } B = \frac{R_1 + R_P}{(R_1 + R_P + R_1)} = \frac{R_1 + R_P}{2R_1 + R_P}$$

$$= R_{in}\left[1 + \frac{R_1 + R_P}{2R_1 + R_P}A\right] \qquad ...(2.83)$$

Similarly, the input resistance determined at input v_2 will be the same as determined above [Eq. (2.83)].

If the differential amplifier circuit depicted in Fig. 2.57 is slightly modified by replacing the first stage by voltage followers, as depicted in Fig. 2.59, we have a circuit that has the main advantage of possessing highest input resistance of three amplifiers and this resistance is given by

$$R_{inf} = R_{in}(1 + A) \qquad ...(2.84)$$

where R_{in} is the input resistance of the op-amp and A is the open-voltage gain of the op-amp.

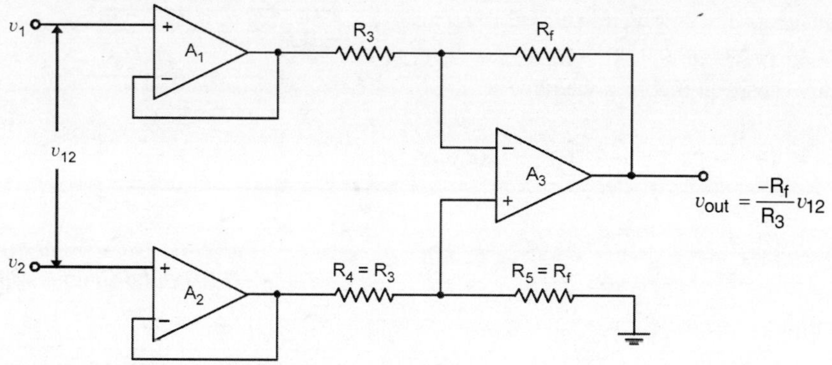

Fig. 2.59

However, the voltage gain of the circuit given in Fig. 2.59 is the same as the voltage gain of the differential amplifier with one op-amp *i.e.*,

$$A_d = \frac{v_{out}}{v_{12}} = \frac{-R_f}{R_3} \qquad ...(2.85) \quad \text{[Refer to Eq. (2.66)]}$$

2.24. DUAL INPUT BALANCED OUTPUT DIFFERENTIAL AMPLIFIER

The analysis of such an amplifier is detailed in Art. 2.21. Since the output considered was a balanced output, the following assumptions were made :

$$R_1 = R_2 \text{ and } R_f = R_3$$

In the applications, the op-amp is used with a single ended or unbalanced output. In such circuits the performance of the differential amplifier is the minimum dc and ac differential voltages that can be detected. Component mismatches and drifts of component values with temperature produce dc differential voltages at the output which are indistinguishable from the dc component of the signal being amplified.

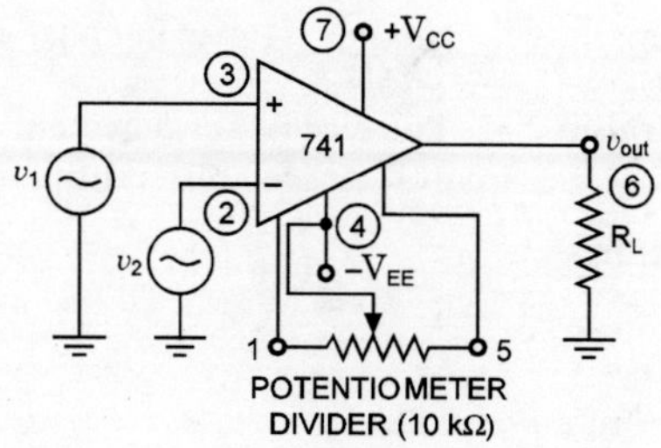

Fig. 2.60

Such mismatches and drifts cause non-zero common-mode-to-differential mode gain as well as differential-to-common-mode gain. Non-zero $(A_{cm} - A_d)$ is important as it converts common-mode input to differential outputs resulting in error in subsequent stages.

In order to obtain a controlled and balanced output a voltage offset null circuit has to be added to the op-amp circuit.

In most of the practical op-amps extra terminals are provided where a potentiometer can be connected for adjustment/reduction of offset voltage to zero. For the 741 op-amp manufacturers recommend a 10 kΩ potentiometer to be connected.

2.25. DIFFERENTIAL INPUT AND DIFFERENTIAL OUTPUT AMPLIFIER

In almost all the applications, the op-amp is employed with a single-ended or unbalanced output. However, there are certain applications in which a differential output is desired. The differential input and differential output amplifier is widely employed as a pre-amplifier and in driving push-pull arrangements.

One possible arrangement of a differential input and differential output amplifier is given in Fig. 2.61. This circuit makes use of two identical op-amps.

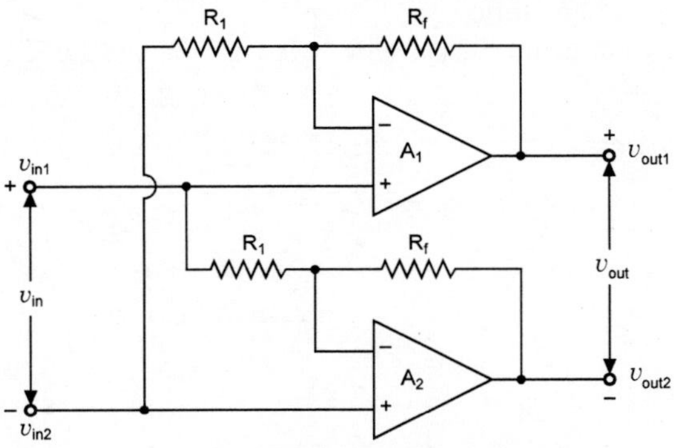

Fig. 2.61. ***Differential Input and Differential Output Amplifier***

Using the superposition theorem we have

$$v_{out_1} = \left(1+\frac{R_f}{R_1}\right) v_{in_1} - \left(\frac{R_f}{R_1}\right) v_{in_2} \quad ...(2.86)$$

Similarly
$$v_{out_2} = \left(1+\frac{R_f}{R_1}\right) v_{in_2} - \left(\frac{R_f}{R_1}\right) v_{in_1} \quad ...(2.87)$$

Differential output voltage, $v_{out} = v_{out_1} - v_{out_2}$

$$= \left(1+\frac{R_f}{R_1}\right) v_{in_1} - \left(\frac{R_f}{R_1}\right) v_{in_2} - \left(1+\frac{R_f}{R_1}\right) v_{in_2} + \left(\frac{R_f}{R_1}\right) v_{in_1}$$

$$= \left(1+\frac{2R_f}{R_1}\right) (v_{in1} - v_{in2}) = \left(1+\frac{2R_f}{R_1}\right) v_{in} \quad ...(2.88)$$

The above equation reveals that the input and output are in phase or are of the same polarity provided that $v_{in} = v_{in_1} - v_{in_2}$ and $v_{out} = v_{out_1} - v_{out_2}$.

The circuit given in Fig 2.61 is very useful in noisy environments, especially if the input signal is comparatively smaller. This is because it rejects the common-mode noise voltages.

Example 2.32. The circuit arrangement given in Fig. 2.61 is to be employed as a pre-amplifier. Determine the gain of the circuit if desired differential output is of at least 4.25 V with an differential input of 100 mV.

Solution :

$$v_{in} = 100 \text{ mV} = 0.1 \text{ V}$$

$$v_{out} = 4.25 \text{ V}$$

From Eq. (2.88)
$$v_{out} = \left(1+\frac{2R_f}{R_1}\right) v_{in}$$

or
$$\frac{2R_f}{R_1} = \frac{v_{out}}{v_{in}} - 1 = \frac{4.25}{0.1} - 1 = 41.5$$

or $R_f = 20.75\, R_1$

Taking $R_1 = 100\ \Omega$; $R_f = 2{,}075\ \Omega \simeq 2.1\text{ k}\Omega$ **Ans.** [Use 2.2 kΩ standard value]

2.26. OUTPUT RESISTANCE AND BANDWIDTH OF DIFFERENTIAL AMPLIFIERS WITH FEEDBACK

We know that the closed-loop op-amp configurations—noninverting, inverting and differential make use of negative feedback and the same type of connections on the output. Thus, the expressions for output resistance of all the three configurations must be same. The output resistance of differential amplifier, is therefore, given by

$$R_{out\,f} = \frac{R_{out}}{1 + \frac{A}{A_d}} \qquad ...(2.89) \text{ Refer to Eq. (2.44)}$$

The exception is that $\qquad B = \frac{1}{A_d}$

where A_d is closed-loop gain of the differential amplifier, R_{out} is the output resistance of the op-amp and A is open-loop voltage gain of the op-amp. No doubt A_d has different values for each of differential amplifier configurations.

The bandwidth of the differential amplifiers, like noninverting or inverting amplifiers, also depends on the closed-loop gain of the amplifier and it is given by

$$f_f = \frac{\text{Unity gain bandwidth}}{\text{Closed-loop voltage gain, } A_d} \qquad ...(2.90)$$

or

$$f_f = \frac{Af_0}{A_d} \qquad ...(2.91)$$

where f_0 is the open-loop break frequency of the op-amp.

Example 2.33. Determine voltage gain, input resistance, output resistance and bandwidth of the differential amplifier circuit given in Fig. 2.57 if $R_1 = R_2 = 3.3$ kΩ, R_P is a 5 kΩ potentiometer, $R_3 = R_4 = 1.2$ kΩ and $R_f = R_5 = 3.9$ kΩ. The op-amps are 741C, the supply voltages are ±15 V and R_P is set at 2.5 kΩ.

Solution : Voltage gain,

$$A_d = -\left(1 + \frac{2R_1}{R_P}\right)\frac{R_f}{R_3} \qquad \text{Refer to Eq. (2.82)}$$

$$= -\left(1 + \frac{2 \times 3.3}{2.5}\right) \times \frac{3.9}{1.2} = -11.83 \text{ Ans.}$$

Input resistance,

$$R_{inf} = R_{in}\left[1 + \frac{R_1 + R_P}{2R_1 + R_P} A\right] \qquad \text{Refer to Eq. (2.83)}$$

$$= 2 \times 10^6 \left[1 + \frac{3.3\text{ k}\Omega + 2.5\text{ k}\Omega}{2 \times 3.3\text{ k}\Omega + 2.5\text{ k}\Omega} \times 2 \times 10^5\right] = 254.947\text{ G}\Omega \text{ Ans.}$$

Output resistance,

$$R_{out\,f} = \frac{R_{out}}{1 + \frac{A}{A_d}} = \frac{75\ \Omega}{1 + \frac{2 \times 10^5}{11.83}} = 0.004436\ \Omega \text{ Ans.}$$

Bandwidth,

$$f_f = \frac{Af_0}{A_d} = \frac{2 \times 10^5 \times 5\text{ Hz}}{11.83} = 84.53\text{ kHz} \text{ Ans.}$$

HIGHLIGHTS

1. An operational amplifier, abbreviated as op-amp, is basically a multistage, very high gain, direct-coupled, negative feedback amplifier that provides a stabilized voltage gain.

2. The op-amp's input can be single-ended or double-ended or differential input depending whether the input voltage is applied to one input terminal only or to both. Similarly output can also be either single-ended or double-ended. Most commonly used configuration is with two input terminals and one output terminal.

   Op-amps have five basic terminals namely two input terminals (noninverting input terminal and inverting input terminal), one output terminal and two power supply terminals (–ve bias supply terminal and +ve bias supply terminal). The significance of other terminals varies with the type of the op-amp.

3. The op-amp is said to be ideal if it has (*i*) infinite open-loop gain, (*ii*) infinite input impedance, (*iii*) zero output impedance, (*iv*) perfect balance, (*v*) infinite frequency bandwidth, (*vi*) infinite slew rate, (*vii*) infinite CMRR and (*viii*) nil drift of characteristics with temperature.

4. An ideal op-amp has an inverting input (–), a noninverting input (+), and a single output referenced to circuit common (ground). The value of this output depends on the difference between the two inputs *i.e.*, $v_{out} = A(v_{in_1} - v_{in_2})$.

5. The common-mode rejection ratio (CMRR) is a figure of merit of a differential amplifier, since it is the ratio of differential voltage gain A_d (the desired gain), to the common-mode voltage gain A_{cm}.

$$\text{CMRR} = \frac{A_d}{A_{cm}}$$

   Higher CMRR ensures that the common-mode signals such as noise are rejected successfully and the output voltage is proportional only to the differential input voltage.

6. Open-loop means that there is no connection between input and output terminals either direct or via another network. An op-amp in open-loop configuration acts as a high-gain amplifier. The open-loop op-amp configurations are : (*i*) Differential amplifier, (*ii*) Inverting amplifier and (*iii*) Noninverting amplifier. The output in the three configurations are given by

   $v_{out} = A(v_{in_1} - v_{in_2})$ in case of differential amplifier.

   $v_{out} = -A v_{in}$ in case of inverting amplifier

   $v_{out} = A\ v_{in}$ in case of noninverting amplifier

   When an op-amp is operated in the open-loop configuration, the output either goes to positive saturation or negative saturation levels or switches between positive and negative saturation levels and thus clips the output above these levels. So open-loop op-amp configurations are not used in linear applications.

7. The parameters to be considered for dc applications are input offset voltage, input offset current, input bias current and drift.

   The parameters to be considered for ac applications are gain bandwidth product (GBW), rise time, slew rate, ac noise and full-power response.

8. Input offset voltage may be defined as the voltage which is to be applied between the input terminals to balance the amplifier.

9. Slew rate of an op-amp is defined as the maximum rate of change of output voltage per unit time and is expressed in V/μs. It indicates how fast the output voltage can change when op-amp is driven by a large step input signal.

10. Power supply rejection ratio (PSSR) is defined as the ratio of the change in input offset voltage to the corresponding change in one power supply voltage, with all remaining power supply voltages remaining constant.

11. There are two basic op-amp circuits known as noninverting amplifier and inverting amplifier circuits.

    Noninverting amplifier makes use of voltage series feedback and always amplifies the difference input voltage v_d which is equal to the input voltage minus feedback voltage v_f. In this amplifier circuit, the input signal is applied to the noninverting input terminal. The overall voltage gain (closed-loop gain) of the amplifier is given by

$$A_f = \frac{v_{out}}{v_{in}} = \frac{A}{1+AB}$$

where A is open-loop voltage gain and B is the gain of feedback circuit given by $B = \frac{v_f}{v_{out}}$.

12. Series-voltage feedback increases the input impedance of an amplifier by a factor (1 + AB) and reduces the output impedance by the same factor *i.e.,* (1 + AB)

$$\text{i.e., } R_{in\,f} = (1 + AB)R_{in}$$

$$\text{and } R_{out\,f} = \frac{R_{out}}{(1+AB)}.$$

13. Nonlinear distortion generated within the amplifier, noise introduced within the amplifier, and output offset voltage are reduced with negative feedback by a factor (1 + AB).

The series-voltage feedback increases the bandwidth of the amplifier by a factor (1 + AB).

14. Voltage follower is an electronic circuit in which output voltage tracks the input voltage both in sign and magnitude.

Because of three unique characteristics of extremely high input resistance (of the order of 10^5 MΩ), unity transmission gain, and extremely low output resistance (of the order of few mΩ), the voltage follower circuit is an ideal circuit device to serve as a *buffer* (or isolation unit). Thus, it can be made to prevent the disturbance of one part of a circuit on another as might be encountered, for example, in filter design.

15. Inverting amplifier makes use of voltage shunt feedback. In this amplifier circuit, the input signal is applied to the inverting (–) input terminal and feedback rersistor is connected from the output to the inverting input terminal of the op-amp. In an inverting amplifier, $A_f = -\frac{R_f}{R_1}$.

16. Inverting amplifier can be used as an analog inverter, negative scaler and phase shifter respectively by making $R_f = R_1$, $R_f = KR_1$ and replacing resistances R_f and R_1 by impedances Z_f and Z_1 in the circuit given in Fig. 2.42. Paraphase amplifier can be had simply by cascading the two analog inverters.

17. Inverting voltage feedback differs from noninverting voltage feedback primarily because of the way it affects the input characteristics of the amplifier. The input impedance is ideally equal to external resistance connected in series with the input signal. But on the output side, inverting voltage feedback has the same effects as noninverting voltage feedback. That is, it decreases the output impedance, distortion, and the output offset voltage by the same factor as in case of noninverting voltage feedback.

18. Since inverter is a special case of the inverting amplifier, all the equations derived for the inverting amplifier are applicable to the inverter. The equations can be applied by merely substituting A/2 for (1 + AB) because B = 1/2.

19. Differential amplifier is a combination of inverting and noninverting amplifiers and amplifies the voltage difference between two input lines neither of which is grounded.

20. Negative feedback is necessary to make an op-amp to operate linearly. This also, to a large degree, removes the dependence of circuits on the op-amp's open-loop gain. The closed-loop gain of the circuit can be strictly determined by the characteristics of the negative feedback network (usually fixed, stable, linear resistors).

Voltage followers, noninverting amplifiers, inverting amplifiers and difference amplifiers can be built with op-amp and negative feedback networks. Analysis of each show that the relationship between input and output voltages is linear. It is set solely by the resistors providing the negative feedback.

There is a wide variety of applications for these linear op-amp circuits. They allow us to build up circuits which directly implement mathematical equations.

21. The gain as well as input resistance of the differential amplifier can be increased by using two op-amps. The characteristics of such an amplifier are identical to those of a noninverting amplifier.

22. In almost all the applications, the op-amp is employed with a single-ended or unbalanced output. However, there are certain applications in which a differential output is desired. The differential input and differential output amplifier is widely employed as a pre-amplifier and in driving push-pull arrangements.

EXERCISES

1. Explain building blocks of an op-amp and their function.
[G.G.S.I.P. Univ. Delhi Analog Electronics-II May-June 2009]
2. Draw the block schematic of an op-amp and explain the function of each block.
[G.G.S.I.P. Univ. Analog Electronics-II May 2011]
3. Explain the functions of all the building blocks of an op-amp.
[G.G.S.I.P. Univ. Analog Electronics-II, May 2010]
4. Explain the block diagram of op-amp with all its stages.
5. Draw and label the pin diagram of IC 741.
6. Explain the working of op-amp 741 schematic diagram with a neat sketch.
7. Describe an ideal op-amp and its electrical characteristics. Draw equivalent circuit and ideal voltage transfer curve. **[G.G.S.I.P. Univ. Analog Electronics-II, May-June 2009]**
8. Draw equivalent circuit of an op-amp and discuss ideal voltage transfer curve.
9. What should be the ideal characteristics of an operational amplifier ? Name them. Distinguish between virtual ground and actual ground. Define the term "common mode rejection ratio". How will you measure it ?
10. Define common mode rejection ratio and explain its significance with respect to differential amplifier.
11. Define the term "common-mode rejection ratio". Explain how to measure it ?
[G.G.S.I.P. Univ. Analog Electronics-II May 2011]
12. Describe the applications of op-amp in open-loop configuration.
13. Explain the terms :
(*i*) Input bias current
(*ii*) Input offset current
(*iii*) Input offset voltage, total output offset voltage
(*iv*) Slew rate. **[G.G.S.I.P. Univ. Analog Electronics-II May 2011]**
14. Explain what do you understand by "offset voltage" and "offset current" of an operational amplifier. Discuss with a neat circuit diagram the techniques used for minimizing offset voltage and offset current in an inverting amplifier.
15. What are practical characteristics of op-amp. Explain.
16. What are the factors that affect the input offset voltage, input bias and input offset currents.
17. What is the effect of variation in power supply voltages on offset voltage ?
18. Define the following electrical properties related to op-amps : Input offset voltage, input resistance, CMMR, slew rate.
19. What are the specifications of an ideal op-amp ? Discuss the meaning of the following terms :
(*i*) Slew rate (*ii*) CMRR (*iii*) Virtual ground.
20. Define CMRR and slew rate. **[G.G.S.I.P. Univ. Delhi Analog Electronics-II May-June 2009]**
21. Discuss the external compensating techniques used in op-amps. Compare their performance.
[G.G.S.I.P. Univ. Delhi Analog Electronics-II May 2010]
22. Define all parameters that are used to specify an op-amp. List their significance.
23. What are practical characteristics of op-amp. Explain.

24. Draw the circuit diagram of a noninverting amplifier. Explain its working and derive the expression for voltage gain.

25. Explain how an op-amp can be used as voltage follower.

26. Draw the circuit of an op-amp voltage follower. Explain its working. Why is it called a noninverting buffer ? **[G.G.S.I.P. Univ. Analog Electronics-II, May-June 2009]**

27. Discuss voltage shunt feedback amplifier with a circuit diagram.

28. What are the characteristics of an ideal op-amp ? Draw the circuit for op-amp as an inverting amplifier and derive the formula for the output.

29. Explain the working of an op-amp based scale changer and obtain expression for its output voltage.

30. Draw the schematic diagram of an ideal inverting op-amp with voltage shunt feedback Z and Z′. Indicate the virtual ground model for calculating the gain.

31. Draw the circuit diagram of an op-amp used as inverting amplifier with negative feedback and derive the expression for its gain.

32. Draw the circuit diagram of a difference amplifier. Select the appropriate values of resistances such that output is given by

$$v_0 = v_1 - v_2$$

where v_0 = output voltage and v_1 and v_2 are input signals.

33. Differentiate between the three closed-loop differential amplifier configurations with negative feedback. **[G.G.S.I.P. Univ. Delhi Analog Electronics-II May-June 2009]**

34. Draw the circuit and explain the differential amplifier with two op-amps. Give its drawback.

35. Draw the circuit and explain the differential amplifier with three op-amps. How does it combine the advantages of the two differential amplifiers (one op-amp and two op-amp) in a single amplifier of having variable voltage gain and very high input resistance.

36. Explain dual input, balanced output differential amplifier with the required circuit diagram. What is its necessity? Explain.

37. Explain differential input and differential output amplifier giving its circuit diagram. Give its applications.

38. Write short notes on the following :

(*i*) Power supply requirements of op-amps.

(*ii*) Schematic diagram of 741 IC op-amp.

(*iii*) Ideal op-amp.

(*iv*) Equivalent circuit of an op-amp

(*v*) CMRR and slew rate in op-amps.

(*vi*) Limitations of open-loop op-amp configurations.

(*vii*) Advantages of negative feedback.

(*viii*) Voltage follower.

(*ix*) Differential amplifier with one op-amp.

(*x*) Differential amplifier with two op-amps.

(*xi*) Differential amplifier with three op-amps.

(*xii*) Differential input and differential output amplifier.

SHORT ANSWER TYPE QUESTIONS WITH ANSWERS....................

Q. 1. What is an op-amp ?

Ans. The name *operational amplifier* stems from its original use for mathematical operations (such as addition, subtraction, multiplication, differentiation and integration) and is abbreviated as *op-amp*. It is a direct-coupled high gain amplifier usually consisting of one or more differential amplifiers and usually followed by a level translator and an output stage (usually a push-pull or push-pull complementary-symmetry pair). With the addition of suitable external feedback components, the modern day op-amp can be employed for a variety of applications.

Q. 2. What are the characteristics of ideal op-amp ?

Ans. The characteristics of an ideal op-amp are: (*i*) infinite open-loop gain, (*ii*) infinite input impedance, (*iii*) zero output impedance, (*iv*) perfect balance, (*v*) infinite frequency bandwidth, (*vi*) infinite slew rate, (*vii*) infinite CMRR and (*viii*) nil drift of characteristics with temperature.

Q. 3. Draw the equivalent circuit of an ideal op-amp.

Ans. Equivalent circuit of an ideal op-amp is shown in Fig. 1.

Q. 4. Define the concept of virtual ground.

Ans. The concept of virtual ground is depicted in Fig. 2. The thick line indicates that a short may be considered to exist with $V_{in} \cong 0$ V, but this is a virtual short in that no current passes through the short to the ground. Current is through resistors R_1 and R_f, as illustrated.

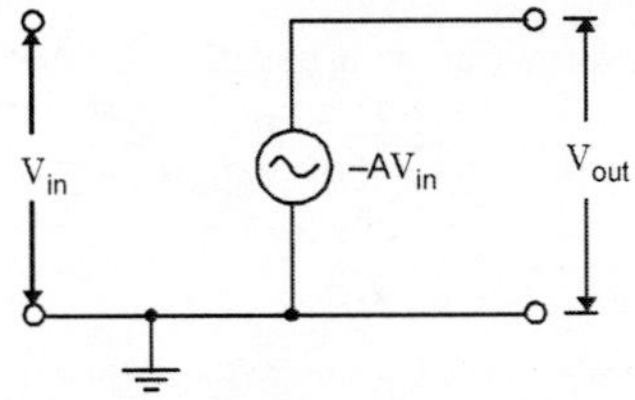

Fig. 1

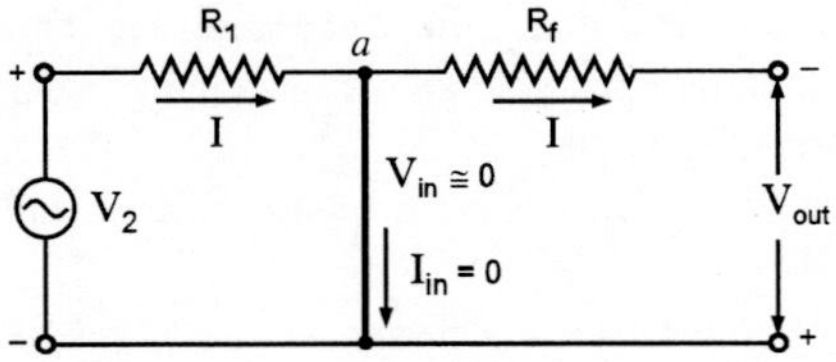

Fig. 2. ***Virtual Ground in an Op-amp***

The virtual-ground concept, which depends on A being very large, allows simple solution for overall voltage gain. Although the circuit shown in Fig. 2 is not physically correct, it does allow an easy means for determination of overall voltage gain.

Obviously input voltage V_{in} will become exactly zero if negative feedback voltage at point *a* is exactly equal to the positive voltage produced by V_2 at point *a*.

Q. 5. What are differential gain and common-mode gain of a differential amplifier ?

Ans. When the difference of the two inputs applied to the two terminals of a differential amplifier is amplified, the resultant gain is termed as differential gain. But when the two input terminals are connected to the same input source then the gain established by the differential amplifier is called the common mode gain.

Q. 6. Define CMRR. Explain with example.

Ans. CMRR is defined as the ratio of differential voltage gain to common-mode voltage gain and it is given as $\text{CMRR} = \dfrac{A_d}{A_{cm}}$. High CMRR ensures that the common mode signals such as noise are rejected successfully and the output voltage is proportional only to the differential input voltage.

Q. 7. What are the applications of op-amp in open-loop configuration ?

Ans. Applications of op-amp in open-loop configuration are :

1. Differential amplifier, 2. Inverting amplifier, 3. Noninverting amplifier.

Q. 8. Why open-loop op-amp configuration is unsuitable for linear applications ?

Ans. When an op-amp is operated in the open-loop configuration, the output either goes to positive saturation or negative saturation levels or switches between positive and negative saturation levels and thus clips the output above these levels. So open-loop op-amp configurations are not used in linear applications.

Q. 9. What is offset voltage ?

Ans. Input offset voltage may be defined as that voltage which is to be applied between the input terminals to balance the amplifier.

Q. 10. Give the typical value of bias current for CA741 operational amplifier.

Ans. 80 nA.

Q. 11. If the currents flowing towards the inverting and noninverting terminals of an op-amp are 3 μA and 1 μA respectively, what is the bias current and the input offset current ?

Ans. 2 μA and 2 μA.

Q. 12. What is slew rate ? List the causes of slew rate and explain its significance in applications.

Ans. Slew rate of an op-amp is defined as the maximum rate of change of output voltage per unit time and is expressed in V/μs.

Slew rate is caused by current limiting and saturation of internal stages of an op-amp when a high frequency large amplitude signal is applied.

Slew rate decides capability of op-amp to change its output rapidly hence it decides high frequency of operation. It changes with change in voltage gain.

The slew rate has important effects on both open-loop and closed-loop op-amp circuits. The slew rate limitations can also affect the performance of the circuit when handling large sinusoidal signals at higher frequencies.

Q. 13. Define CMRR and slew rate. **[G.G.S.I.P. Univ. May-June 2009]**

Ans. CMRR is defined as the ratio of differential voltage gain to common-mode voltage gain and it is given as CMRR = $\frac{A_d}{A_{cm}}$. High CMRR ensures that the common mode signals such as noise are rejected successfully and the output voltage is proportional only to the differential input voltage.

Slew rate of an op-amp is defined as the maximum rate of change of output voltage per unit time and is expressed in V/μs.

Q. 14. What are the different methods of offset compensation ?

Ans. There are two methods of offset compensation. These methods are : Phase-lag network and phase-lead network.

Q. 15. List one advantage of using compensated networks in op-amps.

Ans. The main purpose of providing a compensating network in op-amps is to improve the performance of an op-amp over the desired frequency range by controlling its gain and phase shift.

Q. 16. Define CMRR and PSRR.

Ans. For definition of CMRR refer to Answer of Question 6.

The power supply rejection ratio (PSRR) is defined as the ratio of the change in input offset voltage to the corresponding change in one power supply voltage, with all remaining power supply voltages kept constant.

$$i.e.,\ \text{PSRR} = \frac{\text{Change in input offset voltage}}{\text{Change in supply voltage}} = \frac{\Delta V_{in\,(offset)}}{\Delta V_{supply}}$$

It is also called the *supply voltage rejection ratio* (*SVRR*) or power supply sensitivity (PSS). It is expressed in μV per volt or in decibels. Lower value of PSRR in μV/V is an indication of better performance of the op-amp. It ranges from 10^{-5} to 7×10^{-5}.

Q. 17. What are the input offset current and supply voltage rejection ratio ?

Ans. *Input offset current* $I_{in\,(offset)}$ is defined as the difference between the two currents entering the input terminals of a balanced amplifier for $V_{out} = 0$

i.e., $I_{in\,(offset)} = I_{B_1} - I_{B_2}$ for $V_{out} = 0$

For explanation of supply voltage rejection ratio, refer to Answer of Question 16.

Q. 18. Differentiate between ideal and practical op-amp.

Ans. The characteristics of ideal and practical op-amps are given in Table 2.2.

Q. 19. Name three important specifications of IC op-amps.

Ans. The three important specifications of IC op-amps are high voltage gain, low output impedance and high input impedance.

Q. 20. What is voltage gain of a practical op-amp ?

Ans. Voltage gain of a practical op-amp is of the order of 10^5.

Q. 21. Draw the complete circuit diagram of a voltage series feedback. **[G.G.S.I.P. Univ. December 2009]**

Ans. Refer to Fig. 2.32.

Q. 22. Why voltage series feedback is most commonly used in cascaded amplifiers ?

Ans. In cascaded amplifiers, high input impedance and low output impedance are usually the main requirements and both of these requirements can be met by voltage series feedback, so it is most commonly used in cascaded amplifiers.

Q. 23. What kind of negative feedback is present in a noninverting op-amp ?

Ans. Negative voltage series feedback.

Q. 24. What is effect of negative feedback in a noninverting amplifier on input impedance and bandwidth ?

Ans. Negative feedback in a noninverting amplifier increases each of the input impedance and bandwidth by a factor (1 + AB) where AB is feedback factor.

Q. 25. What is an voltage follower ?

Ans. Voltage follower is an electronic circuit in which output voltage tracks the input voltage both in sign and magnitude.

Q. 26. What are the advantages of using a voltage follower amplifier ?

Ans. Voltage follower has three unique characteristics viz extremely high input impedance, extremely low output impedance and unity transmission gain and is, therefore, an ideal circuit device for use as a buffer amplifier.

Q. 27. In what way is the voltage follower a special case of the noninverting amplifier ?

Ans. If feedback resistor is made zero or R_1 is made ∞ (by keeping it open circuited) in a noninverting amplifier circuit, voltage follower is obtained.

Q. 28. What is an inverting amplifier ?

Ans. In an inverting amplifier, the input is connected to the minus or inverting terminal of op-amp.

Q. 29. How does negative feedback affect the performance of an inverting amplifier ?

Ans. Negative feedback decreases the voltage gain, nonlinear distortion, noise, output impedance and total output offset voltage and increases the input impedance and bandwidth.

Q. 30. What are the applications of an inverting amplifier ?

Ans. Inverting amplifier is a very versatile component and can be used for performing number of mathematical stimulation such as analog inverter, paraphase amplifier, phase shifter, adder, integrator, differentiator.

Q. 31. What is a differential amplifier ?

Ans. Differential amplifier is a combination of inverting and noninverting amplifiers and amplifies the voltage difference between two input lines neither of which is grounded.

Q. 32. Are the characteristics of dual op-amp differential amplifier similar to those of inverting amplifier ?

Ans. No. The characteristics of a dual op-amp differential amplifier are identical to those of a noninverting amplifier.

Q. 33. What is the drawback of dual op-amp differential amplifier ?

Ans. The possible reduction in the amplitude of the output signal is the drawback of dual op-amp differential amplifier.

Q. 34. What is the effect on input resistance if two op-amps are used as differential amplifier ?

Ans. **Input Resistance.** Input resistance of first stage, $R_{inf_2} = R_{in}\ (1 + AB)$

where R_{in} is the open-loop resistance of op-amp, as first stage is noninverting amplifier.

$$B = \frac{R_2}{R_2 + R_3}$$

With v_{in_2} grounded *i.e.*, $v_{in_2} = 0$, the second stage also becomes a noninverting amplifier whose input resistance may be given as

$$R_{inf_1} = R_{in}\ (1 + AB)$$

where R_{in} is open-loop resistance of op-amp

$$B = \frac{R_1}{R_1 + R_f}$$

$$\text{As } R_1 = R_3,\ R_f = R_2,\ R_{inf_2} \neq R_{inf_1}$$

Q. 35. What is the effect on input resistance if three op-amps are used as differential amplifier ?

Ans. The input resistance of the differential amplifier circuit depicted in Fig. 2.57 is the same as the input resistance of the first stage (the resistance determined at input v_1 and v_2 looking into the circuit with other terminal grounded).

In Fig. 2.58 (*a*), for example, when v_4 is reduced to zero (*i.e.*, when v_2 is grounded), the circuit is a noninverting amplifier and the input impedance is given as

$$R_{inf} = R_{in}(1 + AB) \qquad \text{where } B = \frac{R_1 + R_P}{(R_1 + R_P + R_1)} = \frac{R_1 + R_P}{2R_1 + R_P}$$

$$= R_{in}\left[1 + \frac{R_1 + R_P}{2R_1 + R_P}A\right]$$

Similarly, the input resistance determined at input v_2 will be the same as determined above.

If the differential amplifier circuit depicted in Fig. 2.57 is slightly modified by replacing the first stage by voltage followers, as depicted in Fig. 2.59, we have a circuit that has the main advantage of possessing highest input resistance of three amplifiers and this resistance is given by

$$R_{inf} = R_{in}(1 + A)$$

where R_{in} is the input resistance of the op-amp and A is the open-voltage gain of the op-amp.

Q. 36. Why are there variations in gain and phase-shift with respect to frequency in op-amps ?

Ans. In op-amps the variations in gain and phase shift with respect to frequency are due to the internal capacitors and stray capacitances, which are due to physical characteristics of semiconductor devices (BJTs and FETs) and the internal construction of the op-amp.

Q. 37. How can the rate of change of gain and phase-shift of an op-amp be controlled ?

Ans. The rate of change of gain as well as phase shift of an op-amp can be controlled by adding specific components, most common of which are resistors and capacitors.

Q. 38. What is meant by compensating network provided in an op-amp ?

Ans. The network formed by specific components like resistors and capacitors provided in an op-amp to improve its performance over desired frequency range by controlling its gain and phase shift is known as compensating network.

Q. 39. What are the different methods of offset compensation ?

Ans. There are two methods of offset compensation. These methods are : Phase-lag network and phase-lead network.

Q. 40. List one advantage of using compensated networks in op-amps.

Ans. The main purpose of providing a compensating network in op-amps is to improve the performance of an op-amp over the desired frequency range by controlling its gain and phase shift.

Q. 41. What is the difference between internally compensated op-amps and externally compensated op-amps ?

Ans. The internally compensated op-amps like 741, 351, 318 use a metal oxide capacitor built within the IC. The circuit configuration is given in Fig. 2.27. Externally compensated or non-compensated op-amps like 301, 709 have no internal frequency compensation capacitor. Instead, frequency terminals are provided for connecting compensation capacitors externally.

Q. 42. Fill in the blanks :

(*i*) In an ideal operational amplifier the output impedance is __________ (*a*) 50 Ω, (*b*) Infinity.

(*ii*) The intermediate stage of operational amplifier consists of __________ differential amplifier.

(*iii*) Negative feedback is also known as ________ feedback.

(*iv*) Positive feedback is also known as ________ feedback.

(*v*) Negative feedback ________ voltage gain. (increases/decreases)

(*vi*) Negative feedback ________ voltage gain. (stabilizes/destabilizes)

Ans. (*i*) 50 Ω, (*ii*) dual input, unbalanced output, (*iii*) degenerative (*iv*) regenerative (*v*) decreases (*vi*) stabilizes.

PROBLEMS

1. A differential dc amplifier has a differential mode gain of 100 and a common mode gain 0.01. What is its CMMR in dB ? [**Ans.** 80 dB]
2. For a given op-amp, CMMR = 10^5 and differential gain $A_d = 10^5$. Determine common-mode gain A_{CM} of the op-amp. [**Ans.** 1]
3. What is slew rate of an op-amp whose output voltage changes by 30V in 4 μs ? [**Ans.** 7.5 V/μs]

4. In the circuit of Fig. 3, $R_S = 2\ k\Omega$ and $R_L = 5\ k\Omega$. For the op-amp $A = 10^5$, $R_i = 100\ k\Omega$ and $R_o = 50\ \Omega$. For $V_o = 10$ V, calculate V_S and V_o/V_S and estimate the input resistance of the circuit.

[**Ans.** 10 V; 1; 10^7 kΩ]

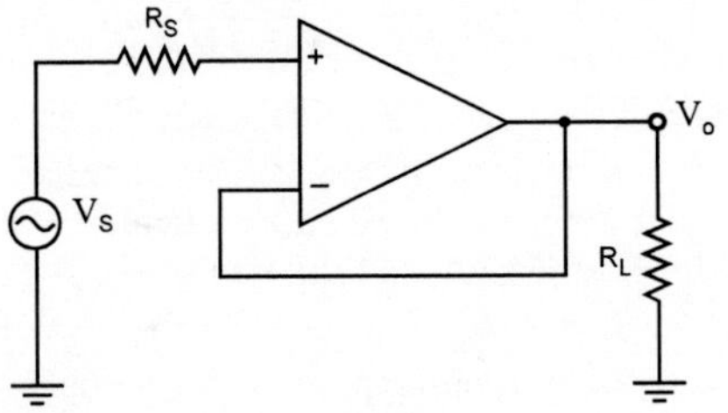

Fig. 3

5. Obtain the output voltage $V_o(t)$ for the circuit of Fig. 4 using ideal op-amp.

$$\left[\textbf{Ans. } -R_3\left(\frac{V_1(t)}{R_1}+\frac{V_2(t)}{R_2}\right)\right]$$

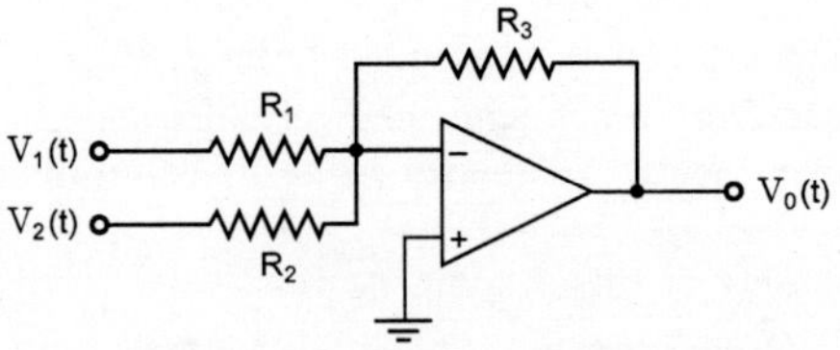

Fig. 4

6. Derive an expression for the closed-loop gain v_o/v_i of the circuit shown in Fig. 5. Assume ideal op-amp.

$$\left[\textbf{Ans. } \frac{-R_2}{R_1}\left(1+\frac{R_4}{R_3}+\frac{R_4}{R_2}\right)\right]$$

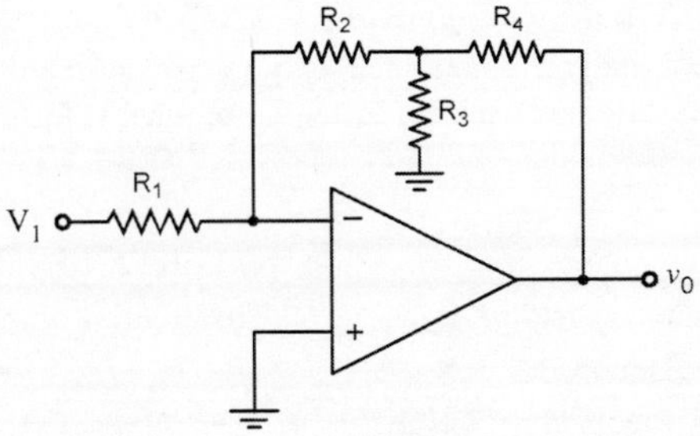

Fig. 5

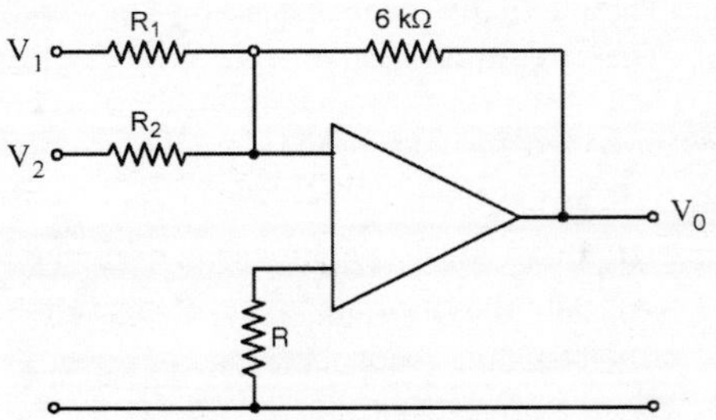

Fig. 6

7. The circuit shown in Fig. 6, stimulates the relation $V_0 = -2V_1 - 3V_2$. Determine the values of R_1 and R_2. [**Ans.** $R_1 = 3\ k\Omega$, $R_2 = 2\ k\Omega$]

8. The 741C operational amplifier having the following parameters is connected as a noninverting amplifier with $R_1 = 1\ k\Omega$, and $R_f = 10\ k\Omega$, $A = 200{,}000$, $R_i = 2\ M\Omega$, $R_0 = 75\ \Omega$, $f_0 = 5$ Hz. Supply voltages = ±15 V, Output voltage swing = ±13 V. Calculate the value of R_{oF} and V_{ooT}. [**Ans.** 4.125 mW; ±0.715 mV]

❑❑❑

CHAPTER

3

Linear Applications of Op-Amps

3.1. INTRODUCTION

We have already discussed the electronics of op-amp, its dc and ac characteristics, parameter limitations and various configurations. Now we will take a look at the op-amp applications. Op-amp may be viewed as a basic building block of linear and nonlinear analog systems. Typical uses of the op-amps are to provide voltage amplitude changes (amplitude and polarity), oscillators, filter circuits and many types of instrumentation circuits. In linear circuits (such as adder, subtractor, integrator, differentiator, voltage-to-current converter, current-to-voltage converter, instrumentation amplifier), the output signal varies directly with the input signal.

INSIDE THIS CHAPTER

There is another class of circuits with highly nonlinear input to output characteristics. The nonlinear circuits include logarithmic and antilogarithmic amplifiers, precision rectifiers, peak detector, sample and hold circuit, clipper and clamper, zero-crossing detector, regenerative comparator, logarithmic multiplier etc. and will be discussed later on (Chapter 11). These nonlinear circuits are very useful in industrial instrumentation, communication and general signal processing.

3.2. SUMMING, SCALING AND AVERAGING AMPLIFIER

The most useful of the op-amp circuits employed in analog computers is the summing amplifier circuit. This circuit can be used to add ac or dc signals. This circuit provides an output voltage proportional to or equal to the algebraic sum of two or more input voltages each multiplied by a constant gain factor. Three-input summing circuit is shown in Fig. 3.1 (*a*).

Using the virtual equivalent circuit, as shown in Fig. 3.1 (*b*), the output voltage can be given in terms of inputs as

$$V_{out} = -R_f\left[\frac{V_1}{R_1}+\frac{V_2}{R_2}+\frac{V_3}{R_3}\right] \qquad ...(3.1)$$

$$\text{If } R_1 = R_2 = R_3 = R_f$$

$$\text{then } V_{out} = -(V_1 + V_2 + V_3) \qquad ...(3.2)$$

Op-amp summing amplifiers are also called *mixers*. One of the advantages of inverting op-amp mixers is that there is no interaction between the inputs. The inverting input is a virtual ground. This prevents one input signal from appearing at the other inputs.

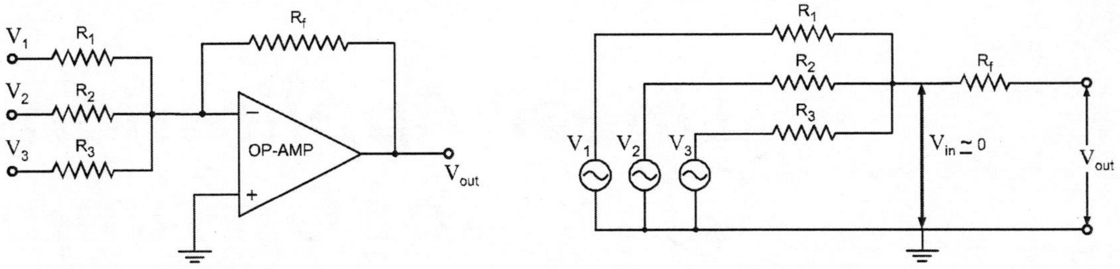

(a) Summing Operational Amplifier *(b) Virtual-Ground Equivalent Circuit*

Fig. 3.1

The circuit shown in Fig. 3.1 (*a*) can be used as a *summing amplifier, scaling amplifier* or *averaging amplifier.*

Equation (3.2) reveals that the output voltage is equal to the negative sum of all input voltages, hence the circuit acts as a *summing amplifier.*

If each input voltage is amplified by a different factor *i.e.,* weighted differently at the output, the circuit becomes a *scaling* or *weighted amplifier.*

The output voltage of the scaling amplifier is given by Eq. (3.1) where

$R_1 \# R_2 \# R_3$.

The circuit shown in Fig. 3.1 (*a*) can be used as an *averaging circuit*, which gives output voltage equal to average of all the input voltages. The modification required in the circuit are of equalising all the input resistors R_1, R_2 and R_3 *i.e.,* $R_1 = R_2 = R_3 = R$ (say) and making gain equal to one over the number of inputs *i.e.,* $= \frac{R_f}{R} = \frac{1}{n}$ where n is the number of inputs.

For example, for three inputs the output voltage,

$$V_{out} = \frac{V_1 + V_2 + V_3}{3} \qquad ...(3.3)$$

Summing or averaging amplifier circuit (Fig. 3.2) can be designed in *noninverting configuration* by selecting appropriate values of resistors *i.e.,* R_f and R_1.

Using the superposition theorem, the voltage V_1 at the noninverting terminal is

$$V_1 = \frac{R/2}{R + R/2} V_a + \frac{R/2}{R + \frac{R}{2}} V_b + \frac{R/2}{R + \frac{R}{2}} V_c$$

$$= \frac{V_a + V_b + V_c}{3} \qquad ...(3.4)$$

Hence the output voltage is

$$V_{out} = \left(1 + \frac{R_f}{R_1}\right) \frac{(V_a + V_b + V_c)}{3} \qquad ...(3.5)$$

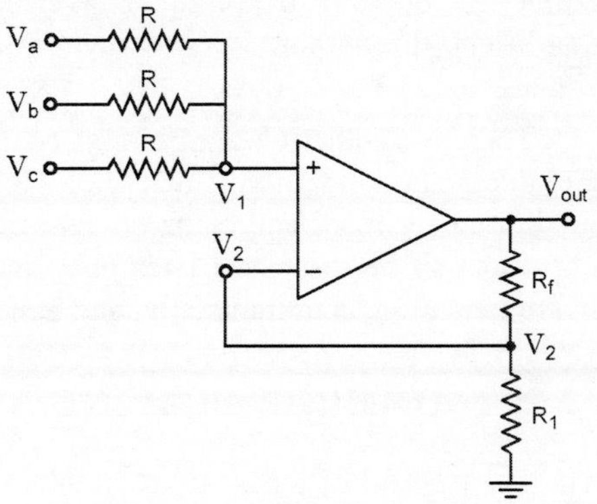

Fig. 3.2. ***Noninverting Configuration For Use as Summing or Averaging Amplifier***

Equation (3.5) reveals that the output voltage is equal to the product of average of all input voltages and the gain of the circuit $\left(1+\frac{R_f}{R_1}\right)$. Hence it is referred to as an *averaging amplifier.* The gain $\left(1+\frac{R_f}{R_1}\right)$ can be set to any value as per application need. For unity gain *i.e.,* when $\left(1+\frac{R_f}{R_1}\right)$ = 1, the circuit will give average of all the inputs.

In case the gain of the circuit (*i.e.,* $1+\frac{R_f}{R_1}$) is made equal to the number of inputs, the output voltage will become equal to the sum of all the input voltages

$$i.e., \; V_{out} = V_a + V_b + V_c \qquad \text{...(3.6)}$$

Hence the circuit is called a *noninverting summing amplifier.*

Summing Amplifier in Differential Configuration. A four-input summing amplifier (Fig. 3.3) may be constructed using the basic differential amplifier shown in Fig. 2.49 by adding two input sources, one each to the inverting and noninverting input terminals through resistor R.

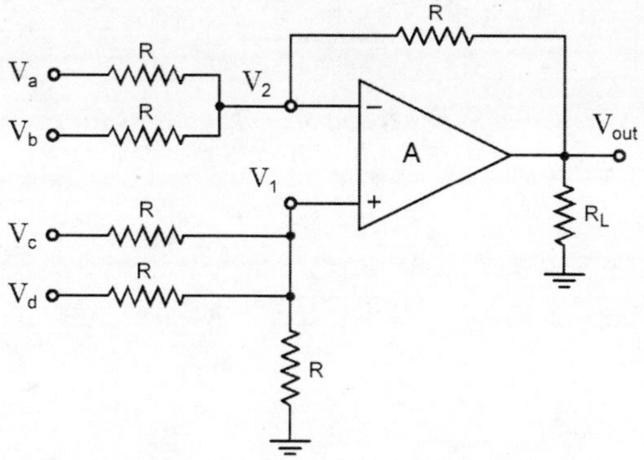

Fig. 3.3. *Summing Amplifier in Differential Configuration*

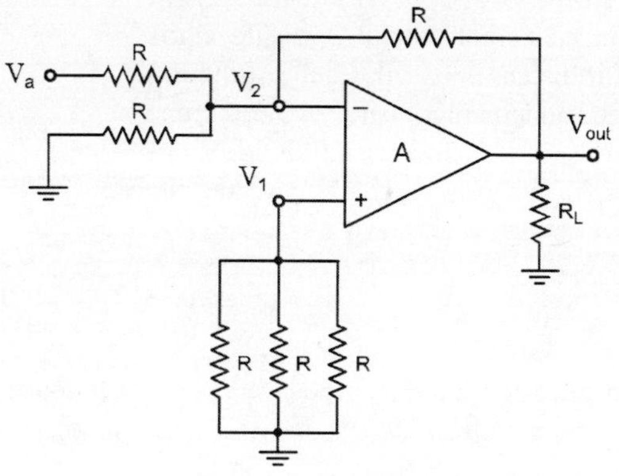

Fig. 3.4

The circuit output can be determined by using the superposition theorem. For instance, for determination of output voltage due to V_a alone, reduce all other input voltages V_b, V_c and V_d to zero, as shown in Fig. 3.4. Infact, this circuit is an inverting amplifier in which the inverting input terminal is at virtual ground (*i.e.,* $V_2 = 0$).

So the output voltage, $V_{out\,a} = \frac{-R}{R} V_a = -V_a$

Similarly the output voltage due to V_b alone, $V_{out\,b} = -V_b$

Now if input voltages V_a, V_b and V_d are made zero, the circuit shown in Fig. 3.3 becomes a noninverting amplifier (Fig. 3.5) in which the voltage V_1 at the noninverting input terminal is

$$V_1 = \frac{R/2}{R+R/2} V_c = \frac{V_c}{3}$$

So the output voltage due to V_c alone,

$$V_{out\,c} = \left(1+\frac{R}{R/2}\right) V_1 = 3\left(\frac{V_c}{3}\right) = V_c$$

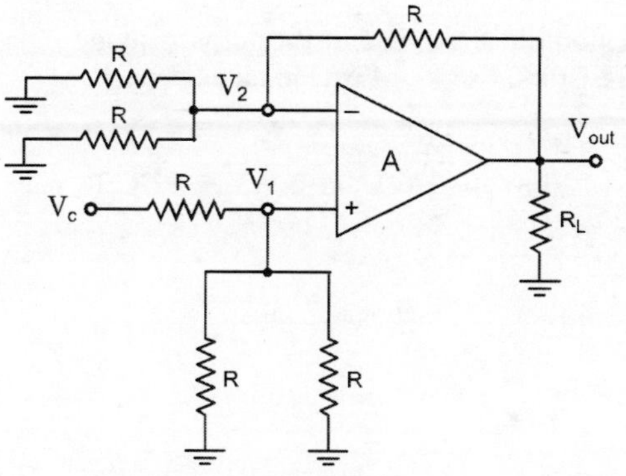

Fig. 3.5

Similarly, the output voltage due to V_d alone,

$$V_{out\,d} = V_d$$

The net output voltage is thus

$$V_{out} = V_{out\,a} + V_{out\,b} + V_{out\,c} + V_{out\,d} = -V_a - V_b + V_c + V_d \quad ...(3.7)$$

The above equation reveals that the output voltage is equal to the sum of the input voltages applied to the noninverting input terminal plus the negative sum of the input voltages applied to the inverting input terminal. Even though in the circuit shown in Fig. 3.3 the gain of the summing amplifier is unity, any scale factor can be used for the inputs by selecting proper external resistors.

Weighted Summer Capable of Implementing Summing Coefficients of Both Signs. The weighted summing amplifier shown in Fig. 3.1 (*a*) has the limitation that all the summing coefficients are of the same sign. Sometimes signals with opposite signs need to be summed up. Such a function can be implemented using two op-amps, as shown in Fig. 3.6. Assuming ideal op-amps, it can be easily shown that the output voltage is given by

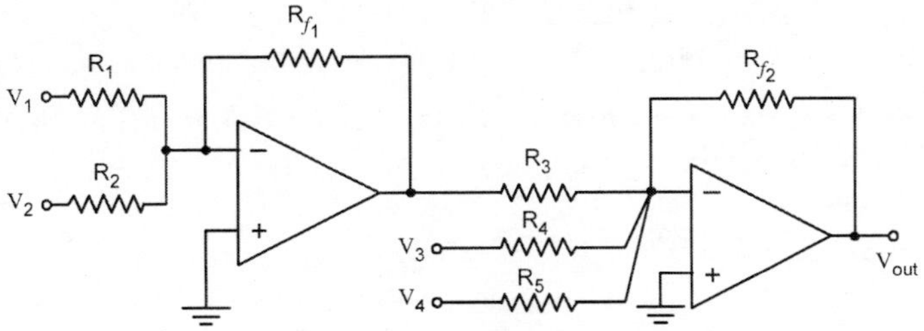

Fig. 3.6. ***A Weighted Summer Capable of Implementing Summing Coefficients of Both Signs***

$$V_{out} = V_1\left(\frac{R_{f1}}{R_1}\right)\left(\frac{R_{f2}}{R_3}\right) + V_2\left(\frac{R_{f1}}{R_2}\right)\left(\frac{R_{f2}}{R_3}\right) - V_3\left(\frac{R_{f2}}{R_4}\right) - V_4\left(\frac{R_{f2}}{R_5}\right) \quad ...(3.8)$$

Example 3.1. Find the output voltage of the following circuit (Fig. 3.7).

[*U.P. Technical Univ. Electronics Engineering First Semester 2006-07*]

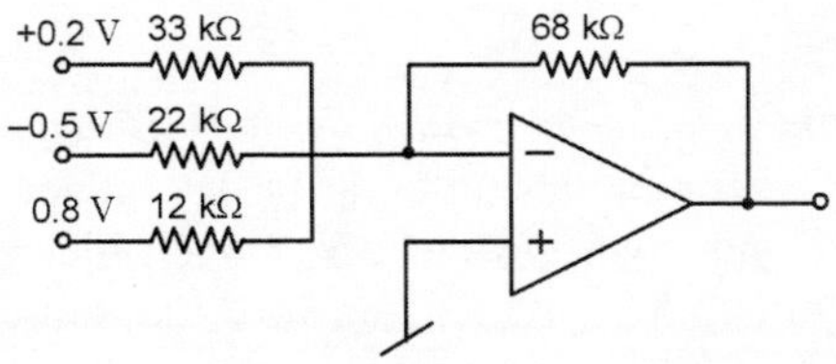

Fig. 3.7

Solution :

$$V_{out} = -\left[\frac{R_f V_1}{R_1} + \frac{R_f V_2}{R_2} + \frac{R_f V_3}{R_3}\right]$$

$$= -\left[\frac{68}{33} \times 0.2 + \frac{68}{22} \times (-0.5) + \frac{68}{12} \times 0.8\right]$$

$$= -[0.412 - 1.545 + 4.533] = -3.40 \text{ V. } \textbf{Ans.}$$

Example 3.2. Calculate the output voltage of an op-amp summing amplifier for the following sets of voltages and resistors. Use $R_f = 1$ MΩ in all cases.

(*a*) $V_1 = +1$ V, $V_2 = +2$ V, $V_3 = +3$ V; $R_1 = 500$ kΩ, $R_2 = 1$ MΩ, $R_3 = 1$ MΩ.

(*b*) $V_1 = -2$ V, $V_2 = +3$ V, $V_3 = +1$ V; $R_1 = 200$ kΩ, $R_2 = 500$ kΩ, $R_3 = 1$ MΩ.

[*U.P. Technical Univ. Electronics Engineering IInd Semester 2006-07*]

Solution : (*a*) Output voltage $= -R_f\left(\frac{V_1}{R_1} + \frac{V_2}{R_2} + \frac{V_3}{R_3}\right) = -\frac{1 \times 10^6}{1 \times 10^3}\left[\frac{1}{500} + \frac{2}{1,000} + \frac{3}{1,000}\right]$

$$= -(2 + 2 + 3) = -7 \text{ V } \textbf{Ans.}$$

(*b*) Output voltage $= -\frac{1 \times 10^6}{1 \times 10^3}\left[\frac{-2}{200} + \frac{3}{500} + \frac{1}{1,000}\right] = -(-10 + 6 + 1) = 3$ V **Ans.**

Example 3.3. Realize a circuit to obtain $V_{out} = -2V_1 + 3V_2 + 4V_3$ using an operational amplifier. Use minimum value of resistance as 10 kΩ.

Solution : For an operational amplifier,

$$V_{out} = -\left[\frac{R_f}{R_1}V_1 + \frac{R_f}{R_2}V_2 + \frac{R_f}{R_3}V_3\right]$$

Comparing the above expression with the given expression for the output

i.e., $V_{out} = -2\,V_1 + 3\,V_2 + 4\,V_3 = -[2\,V_1 - 3\,V_2 - 4\,V_3]$

we have $\frac{R_f}{R_1} = 2$; $\frac{R_f}{R_2} = 3$ and $\frac{R_f}{R_3} = 4$

Resistance R_3 will be of minimum value of 10 kΩ

$R_f = 4\,R_3 = 4 \times 10 = 40$ kΩ **Ans.**

$$R_2 = \frac{R_f}{3} = \frac{40}{3} = 13.33 \text{ k}\Omega \textbf{ Ans..}$$

$$R_1 = \frac{R_f}{2} = \frac{40}{2} = 20 \text{ k}\Omega \textbf{ Ans.}$$

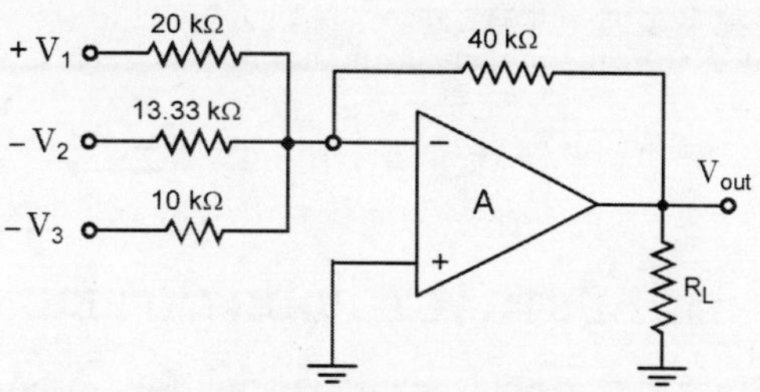

Fig. 3.8

The circuit is shown in Fig. 3.8.

Example 3.4. Figure 3.9 below shows a noninverting op-amp summer with $V_1 = 2$ V and $V_2 = -1$ V. The output voltage V_0 = — [*GATE 1994*]

Solution : According to superposition theorem

$$V_{out} = V'_0 + V''_0$$

where V'_0 is the output produced by V_1 of + 2 V and V''_0 is the output produced by $V_2 = -1$ V

When $V_2 = -1$ V is made zero

Input at noninverting input terminal,

$$V_{s1} = V_1 \times \frac{R\,||\,R}{R + (R\,||\,R)}$$

$$= 2 \times \frac{R/2}{R + \frac{R}{2}} = \frac{2}{3}\text{ V}$$

$$V'_0 = V_{s1}\left(1 + \frac{R_f}{R}\right) = \frac{2}{3}\left(1 + \frac{2R}{R}\right) = 2\text{ V}$$

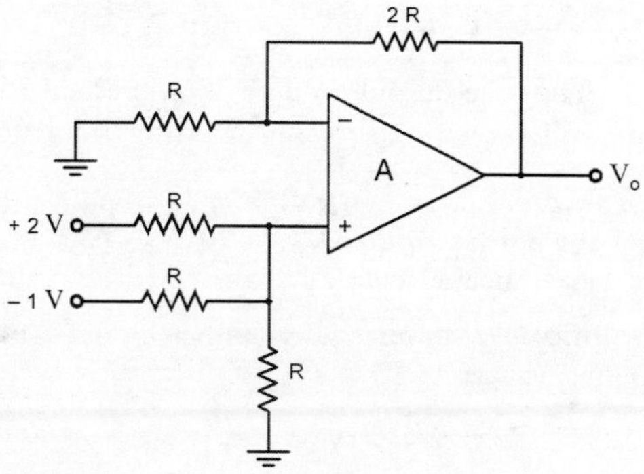

Fig. 3.9

When $V_1 = 2$ V is made zero

$$V_{s2} = V_2 \times \frac{R\,||\,R}{R + (R\,||\,R)} = -1 \times \frac{R/2}{R + \frac{R}{2}} = \frac{-1}{3}\text{ V}$$

and $V''_0 = V''_{s2}\left(1 + \frac{R_f}{R}\right) = \frac{-1}{3}\left(1 + \frac{2R}{R}\right) = -1$ V

Output voltage, $V_{out} = V'_0 + V''_0 = 2 + (-1) = 1$ V **Ans.**

Example 3.5. Find the output of a given circuit in Fig. 3.10.

[G.G.S.I.P. Univ. Analog Electronics-II May 2011]

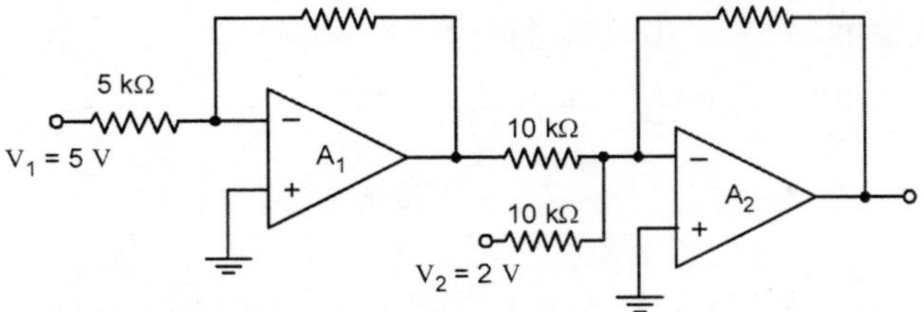

Fig. 3.10

Solution : Output voltage of op-amp A_1,

$$V_{out} = -V_1 \times \frac{R_f}{R_1}$$

$$= -5 \times \frac{10}{5} = -10 \text{ V}$$

For op-amp A_2, there are two inputs : one of −10 V, the output of op-amp A_1 and second 2 V, so the output from op-amp A_2

$$V'_{out} = -10 \times \frac{20}{10} + 2 \times \frac{20}{10} = -20 + 4 = -16 \text{ V} \quad \textbf{Ans.}$$

3.3. SUBTRACTOR OR DIFFERENCE AMPLIFIER

Since differential amplifier amplifies the difference of two input signals applied to the inverting and noninverting terminals of an op-amp, it can be used as subtractor circuit. The circuit is shown in Fig. 3.11. This circuit provides an output equal to the difference of two input signals.

By selecting appropriate values for the external resistors, the input signals can be scaled (or attenuated) to the desired value. The circuit so obtained is called the *scaling amplifier.*

In circuit shown in Fig. 3.11 all external resistors are of equal value, and, therefore, gain of the amplifier is unity

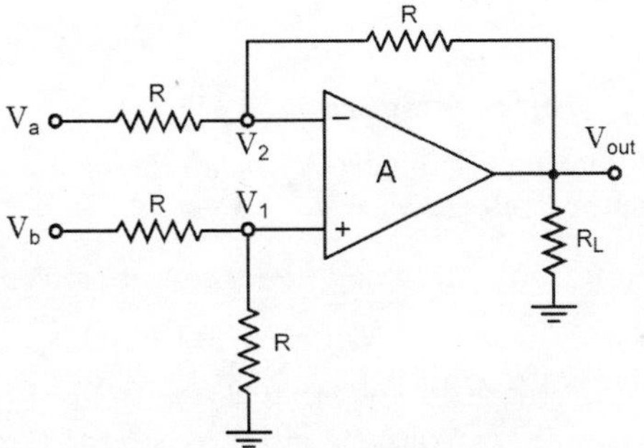

Fig. 3.11. ***Op-Amp Subtractor***

So output voltage, $V_{out} = \frac{-R}{R}[V_a - V_b] = V_b - V_a$ ...(3.9)

Thus the output voltage V_{out} is equal to the voltage applied to the noninverting input terminal less the voltage applied to the inverting input terminal. Hence the circuit is called a *subtractor.*

Example 3.6. Design a difference amplifier using an op-amp to get the output voltage $V_{out} = 10\ (V_2 - V_1)$. Use minimum value of resistance as 10 kΩ.

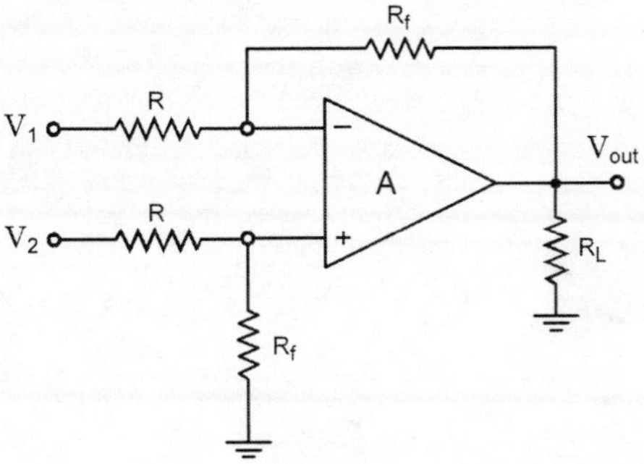

Fig. 3.12

Solution : A differential amplifier circuit is shown in Fig. 3.12. Output voltage is given as

$$V_{out} = \frac{R_f}{R}(V_2 - V_1)$$

Comparing the above expression with the given expression for output *i.e.*, $V_{out} = 10\ (V_2 - V_1)$ we have $\frac{R_f}{R} = 10$

Selecting R = 10 kΩ (minimum value of resistance to be used),

R_f = 10 R = 10 × 10 = 100 kΩ **Ans.**

3.4. VOLTAGE-TO-CURRENT CONVERTER (TRANSCONDUCTANCE AMPLIFIER)

Occasionally in industrial electronics, it is necessary to provide a current proportional to certain voltage, even though the load impedance may vary. A circuit which can perform this job, is called a *voltage-to-cur-*

rent converter. The circuit of voltage-to-current converter is shown in Fig. 3.13. The circuit is similar to that of positive scaler shown in Fig. 2.32 except that the feedback resistor R_f is replaced by a load impedance Z_L. For a single input, the current in the load impedance Z_L is given by

$$i_L = i = \frac{v_{in}}{R} \quad ...(3.10)$$

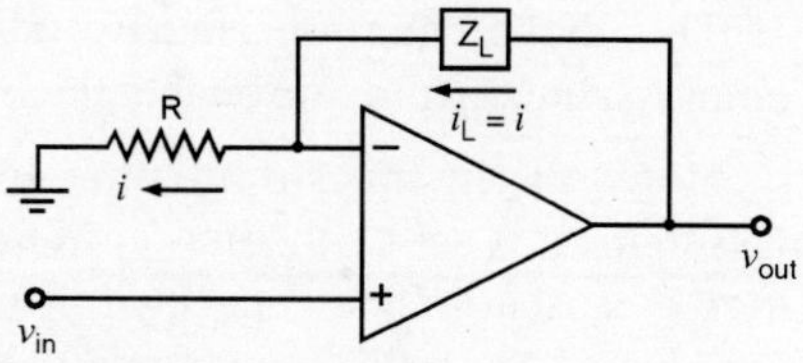

Fig. 3.13. ***Voltage-To-Current Converter***

From above Eq. (3.10) it is obvious that the output current i_L is independent of load impedance Z_L and is proportional to the input voltage v_{in}. This is because of the virtual ground at the inverting input terminal of the op-amp. Such a circuit is employed in analog-to-digital converter (ADC).

One good thing about the op-amp voltage-to-current converter is that it can be driven by a voltage source which is itself not capable of supplying the load current called for by Eq. (3.10). This is because the voltage source only has to drive a noninverting op-amp, whose input impedance is very high (many megohms). The load current itself is supplied by the op-amp.

Another version of the voltage-to-current converter is given in Fig. 3.14. In this circuit, the load is grounded, and an input voltage controls the load current.

Let v_1 be the voltage at noninverting input terminal (Fig. 3.14).

Applying Kirchhoff 's current law at node 1 we have

$$i_1 + i_2 = i_L$$

$$\text{or} \quad \frac{v_{in} - v_1}{R} + \frac{v_{out} - v_1}{R} = i_L$$

$$\text{or} \quad v_{in} - v_1 + v_{out} - v_1 = i_L R$$

$$\text{or} \quad v_1 = \frac{v_{in} + v_{out} - i_L R}{2} \quad ...(3.11)$$

Since the op-amp is in the noninverting mode,

$$\text{Circuit gain} = 1 + \frac{R}{R} = 2$$

Then output voltage, $v_{out} = 2v_1$

Substituting $v_1 = \frac{v_{out}}{2}$ in Eq. (3.11) we have

$$v_{out} = v_{in} + v_{out} - i_L R$$

$$\text{or} \quad \text{load current, } i_L = \frac{v_{in}}{R} \quad ...(3.12)$$

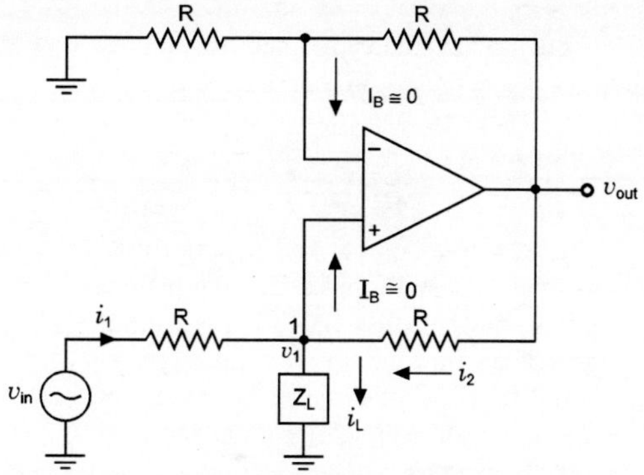

Fig. 3.14. ***Voltage-To-Current Converter With Grounded Load***

The above Eq. (3.12) shows that the load current depends upon the input voltage v_{in} and resistor R. It is an ideal circuit for low-voltage dc and ac voltmeters, LED and zener diode tester.

3.5. CURRENT-TO-VOLTAGE CONVERTER (TRANSRESISTANCE AMPLIFIER)

A device that produces a voltage proportional to input signal current is called a *current-to-voltage converter*. The circuit arrangement is shown in Fig. 3.15. In this circuit, a *photocell* or *photomultiplier tube* that provides output current proportional to the light flux but independent of the load impedance, is connected to the inverting input terminal of the op-amp. R_s is the shunt source resistance. As there is a virtual ground at the inverting input terminal, current flowing through R_s is zero and, therefore, the entire input current i_s flows through the feedback resistor R_f resulting in the output voltage given as

$$v_{out} = -i_s R_f \quad ...(3.13)$$

The above equation clearly indicates that the output voltage is directly proportional to the input (source) current i_s.

However, in this circuit, the lower limit on current measurement is set by the bias current of the inverting input. Often a capacitor C_f is connected in parallel with resistor R_f for reducing the high frequency noise and the possibility of oscillations.

One of the most common uses of the current-to-voltage converter is in sensing current from photodetectors and in digital-to-analog converter applications.

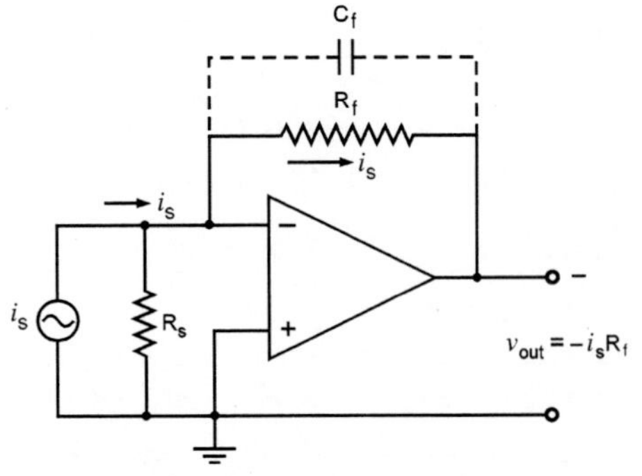

Fig. 3.15. *Current-To-Voltage Converter*

Example 3.7. For the circuit shown in Fig. 3.16, find the output voltage waveform for a sine wave input signal.

[G.G.S.I.P. Univ. Analog Electronics-II May 2010]

Solution : The given circuit is of a full-wave rectifier or absolute value circuit.

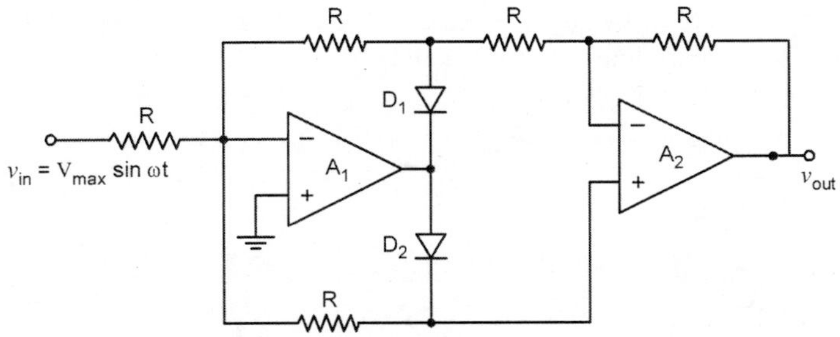

Fig. 3.16

For positive input, *i.e.*, when v_{in} exceeds zero, diode D_1 is *on* and diode D_2 is *off*. Both the op-amps A_1 and A_2 act as inverter as depicted in equivalent circuit (Fig. 3.17). From the equivalent circuit, it is obvious that

$$v_{out} = v_{in} \qquad ...(i)$$

For negative output, *i.e.*, $v_{in} < 0$, diode D_1 is *off* and diode D_2 is *on*. The equivalent circuit is shown in Fig. 3.18. If the output voltage of op-amp A_1 is v volts, the inverting input terminal of op-amp A_2 is also at voltage v, because the differential input to op-amp A_2 is zero.

Applying KCL at node 1, we have

$$\frac{v_{in}}{R} + \frac{v}{2R} + \frac{v}{R} = 0$$

$$\text{or} \quad v = -\frac{2}{3} v_{in} \qquad ...(ii)$$

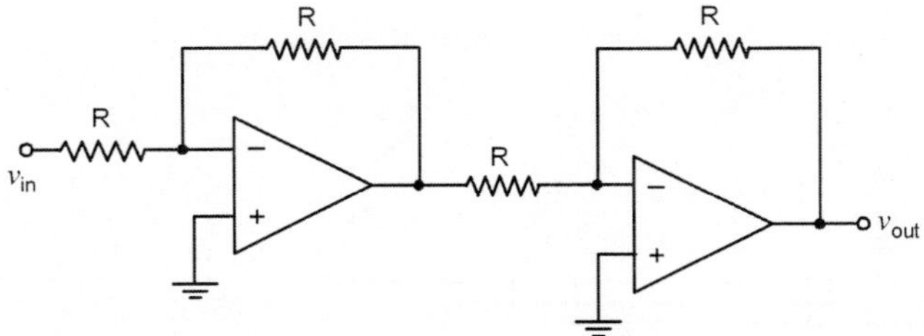

Fig. 3.17. *Equivalent Circuit For Positive Half-Cycle of Input*

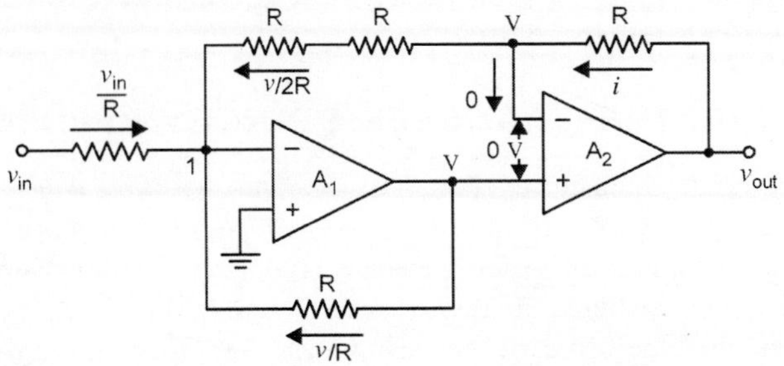

Fig. 3.18. *Equivalent Circuit For Negative Half Cycle*

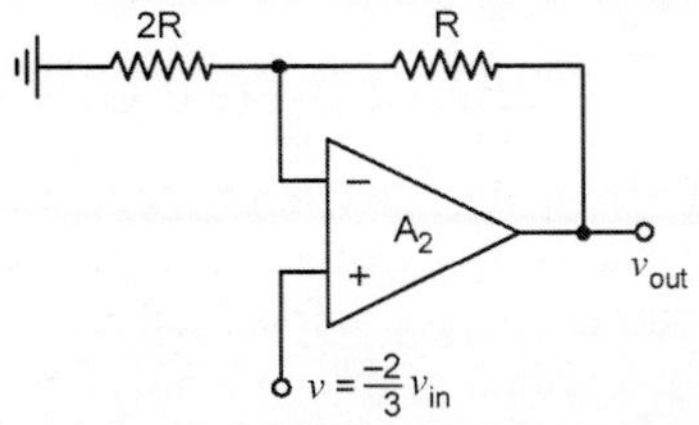

Fig. 3.19. *Equivalent Circuit of Circuit Given in Fig. 3.18.*

The equivalent circuit shown in Fig. 3.18 is a noninverting amplifier as depicted in Fig. 3.19.

The output voltage

$$v_{out} = \left(1 + \frac{R}{2R}\right)\left(-\frac{2}{3}v_{in}\right) = v_{in} \quad ...(iii)$$

The input and output waveform are shown in Fig. 3.20. This circuit is called an absolute value circuit as output is positive even when input is negative. It is possible to have negative outputs for either polarity of input simply by reversing the diodes.

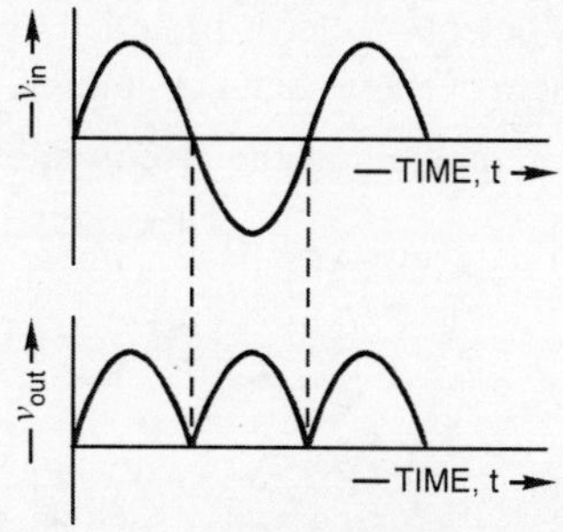

Fig. 3.20. ***Input and Output Waveforms***

3.6. INSTRUMENTATION AMPLIFIER

The low-level signal outputs of electrical transducers often need to be amplified before further processing, and this is done with instrumentation amplifiers. The important features of instrumentation amplifiers are given below:

1. Selectable gain with high gain accuracy and gain linearity.
2. Differential input capability with high common mode rejection, even with sources having unbalanced high output impedances.
3. High stability of gain with low temperature coefficient.
4. Low dc offset and drift errors referred to input.
5. Low output impedance.

Obviously, all kinds of amplifiers are used in instrumentation; however, a particular class and configuration conventionally is given the name *instrumentation amplifier*, mainly to distinguish from the simpler op-amp circuits.

Instrumentation amplifiers differ from the ordinary op-amp in the following respects :

1. The instrumentation amplifier is often a package comprising op-amps, wired up with accurate and stable resistive feedback network to provide a desired gain. The gain is often selectable to a precise value by a single external resistance.
2. The instrumentation amplifier, because of its closed-loop configuration, can be directly used to amplify the signal, by a fixed amplification factor. There is no user effort required in selecting the input and feedback configurations or in choosing suitable components. The gain accuracy, gain stability and drift performance are normally specified by the manufacturer.
3. Depending upon the configuration used, the instrument amplifier will often yield high common mode rejection even with source impedances exceeding 1 MΩ and with the source unbalance of a large value (1 to 10 kΩ). The dc op-amp has a CMMR specified for it which is only applicable for very low source impedances.

In practice, an instrumentation amplifier is a specific combination of dc op-amps wired up with feedback, and needs very little design effort from the user.

The schematic diagram of an instrumentation amplifier possessing many of the general features desired and constructed with dc op-amps is shown in Fig. 3.21.

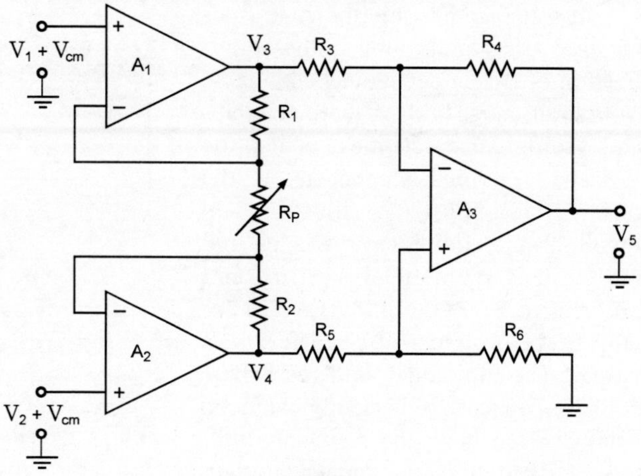

Fig. 3.21

A potentiometer is provided to permit adjusting the scale factor of the circuit. While three op-amps are used here, a single-quad op-amp IC, is all that is required (other than the resistor components.)

An analysis of the circuit shown in Fig. 3.21 gives the following equations if

$$R_3 = R_4 = R_5 = R_6.$$

$$V_3 = \left(1 + \frac{R_1}{R_p}\right) V_1 - \frac{R_1}{R_p}(V_{cm} + V_2) \quad \text{...(3.14)}$$

$$V_4 = \left(1 + \frac{R_2}{R_p}\right) V_2 - \frac{R_2}{R_p}(V_{cm} + V_1) \quad \text{...(3.15)}$$

where $V_1 + V_{cm}$ is the input applied to amplifier A_1 and $V_2 + V_{cm}$ is the input applied to amplifier A_2

$$\text{Output voltage, } V_5 = V_4 - V_3 = \left(1 + \frac{2R}{R_P}\right)(V_2 - V_1) \quad \text{...(3.16) if } R_1 = R_2 = R \text{ (say)}$$

The input amplifiers A_1 and A_2 act as input buffers with unity gain for common mode signals V_{cm} and with a gain $\left(1 + \frac{2R}{R_P}\right)$ for differential signals. A high input impedance is ensured by the noninverting configurations in which they operate. The common-mode (*cm*) rejection is achieved by the following stage which is connected in differential mode. The optimum common-mode rejection can be obtained by adjusting R_5 or R_6 ensuring that

$$\frac{R_4}{R_3} = \frac{R_6}{R_5}.$$

The amplifier A_3 can also be made to have some nominal gain for the whole amplifier by an appropriate selection of R_3, R_4, R_5 and R_6.

The drift problem of instrumentation amplifiers can be improved if amplifiers A_1 and A_2 have offset voltages which tend to track the temperature.

A high gain accuracy can be achieved by using precision metal film resistors for all the resistances. Because of large negative feedback employed, the amplifier has good linearity, typically about 0.01% for a gain less than 10. The output impedance is also low, being in the range of milli-ohms. The input bias current of the instrumentation amplifier is determined by that of the amplifiers A_1 and A_2.

Instrumentation amplifiers are finding increasing application in the amplification of the output signals obtainable from strain gauge bridges, thermocouples and biological electrodes. When the Wheatstone bridge is connected to the differential amplifier, as illustrated in Fig. 3.22, the equivalent voltages and resistances from the inverting and noninverting input terminals of the op-amp to the ground will be different in values. The equivalent voltages will be equal only in case of perfectly balanced bridge. Even then, the resistance values are likely to differ from each other.

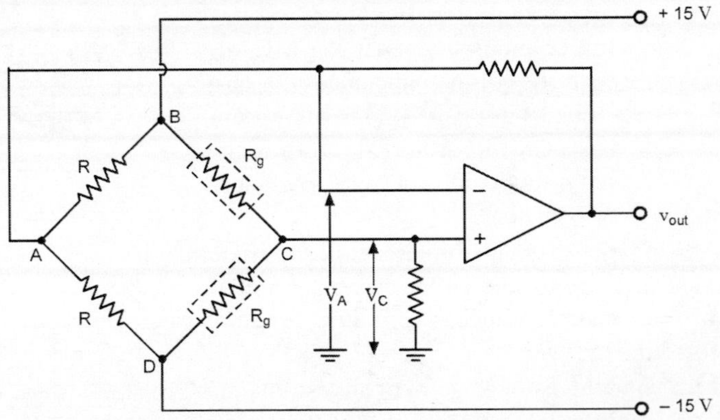

Fig. 3.22. ***Differential Amplifier Connected to a Strain Gauge Wheatstone Bridge Network***

In this case the difference will be considered as source unbalance resistance, and there will be an error in output voltage v_{out} due to the common-mode signals at the input terminals of the amplifier. The higher the value of voltage V_{BD}, the larger will be the error. Thus, for situations where very-low level voltages present along with high common-mode voltages have to be amplified, the instrumentation amplifier is used.

For applications that call for very high-gain common-mode voltages or isolation impedances with galvanic isolation, isolation amplifiers with transformer or optical coupling are recommended for use.

3.6.1. Instrumentation Amplifier With Linear Gain Control

The arrangement given in Fig. 3.21 can be modified to have an instrumentation amplifier with continuously adjustable gain, as illustrated in Fig. 3.23. The advantage of this modified circuit is that gain is linearly proportional to the adjustable resistance R_4. The set position of R_4 gives the direct indication of the gain.

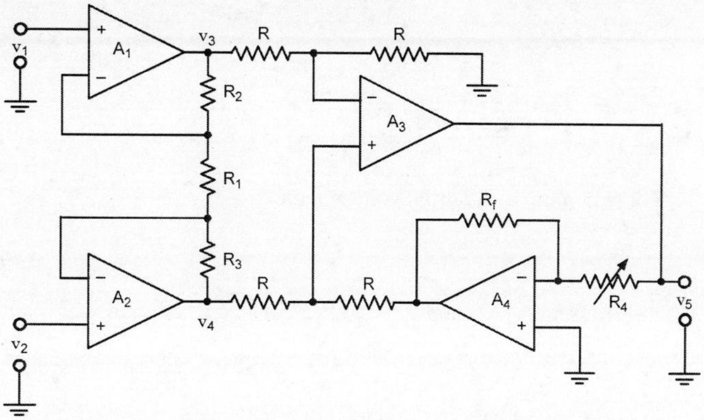

Fig. 3.23. *Instrumentation Amplifier With Linear Gain Control*

The gain of the amplifier circuit depicted in Fig. 3.21 is

$$\frac{v_5}{v_1 - v_2} = 1 + \frac{2R_2}{R_1} \quad ...(3.17)$$

where $R_3 = R_2$

Though the gain can be varied by R_1 alone, its dependence on R_1 is nonlinear. From the modified circuit shown in Fig. 3.23, the gain is given as

$$\frac{v_5}{v_1 - v_2} = \left[1 + \frac{2R_2}{R_1}\right]\frac{R_x}{R_f} \quad ...(3.18)$$

where R_x is the set value of R_4. The gain is thus linearly proportional to R_x. The ratio of $\frac{R_2}{R_1}$ in Fig. 3.23 can be so chosen as to provide nominal gain in the first stage so that additional drift errors due to the following stages are non-significant.

Example 3.8. Determine the output voltage for the circuit shown in Fig. 3.24. It is given that $R_1 = R_2 = R_3 = R_4 = R_5 = R_6 = 10\ k\Omega$ and $R_P = 1\ k\Omega$.

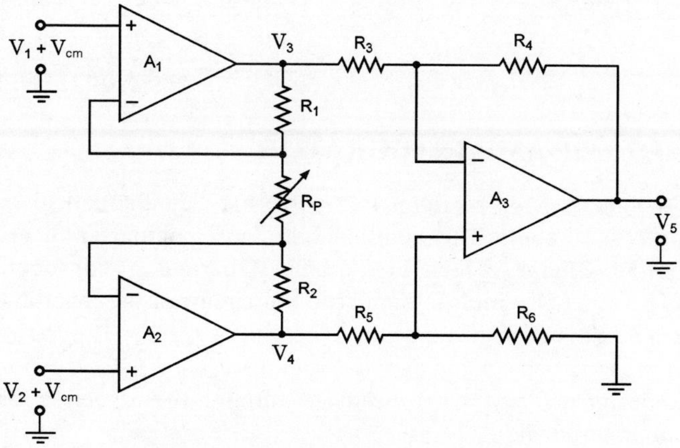

Fig. 3.24. *Instrumentation Amplifier*

Solution : Resistance, $R = 10\ k\Omega$

Potentiometer resistance, $R_P = 1\ k\Omega$

Substituting above values in Eq. (3.16) we have

$$\text{Output voltage, } V_5 = \left(1 + \frac{2R}{R_P}\right)(V_2 - V_1)$$

$$= \left(1 + \frac{2 \times 10\ k\Omega}{1\ k\Omega}\right)(V_2 - V_1)$$

$$= 21(V_2 - V_1) \quad \textbf{Ans.}$$

Example 3.9. Design an instrumentation amplifier to have a variable differential gain in the range 5–200. Use a 50 kΩ potentiometer.

Solution : R_1 is a potentiometer and it is a 50 kΩ potentiometer.

Let us choose $R_3 = R_4 = 15$ kΩ.

The differential gain of the instrumentation amplifier depicted in Fig. 3.25 comes out to be

$$A_d = 1 + \frac{2R_2}{R_1} \qquad \ldots(i)$$

∵ From Eq. (3.17),

$$A_d = \frac{\left(1 + \frac{2R_2}{R_1}\right)(V_2 - V_1)}{(V_2 - V_1)} = 1 + \frac{2R_2}{R_1}$$

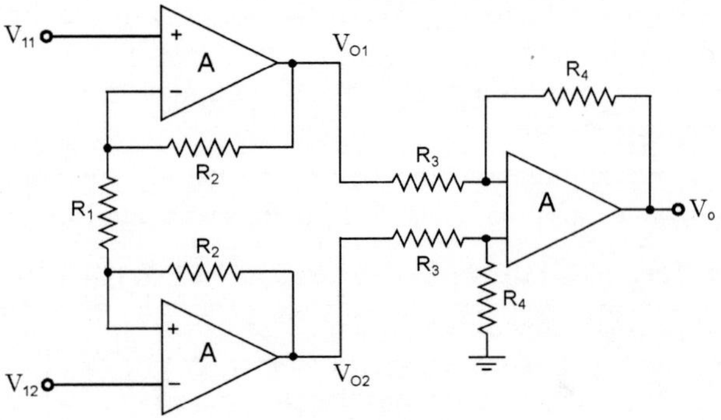

Fig. 3.25

For Minimum differential voltage gain

$$A_d = A_{d\,min} = 5$$

$$R_1 = R_{1\,max} = 50 \text{ k}\Omega \qquad ∵ A_d \text{ will be minimum only when } R_1 \text{ will be maximum}$$

Substituting $A_d = 5$ and $R_1 = 50$ kΩ in Eq. (*i*), we have

$$5 = 1 + \frac{2R_2}{50 \text{ k}\Omega}$$

$$\text{or } R_2 = 100 \text{ k}\Omega$$

For Maximum differential voltage gain

$$A_d = A_{d\,max} = 200$$

Substituting $A_d = 200$, $R_2 = 100$ kΩ in Eq. (*i*), we have

$$200 = 1 + \frac{2 \times 100 \text{ k}\Omega}{R_1}$$

$$\text{or } R_1 = \frac{200 \text{ k}\Omega}{199} \cong 1 \text{ k}\Omega$$

For maximum value of A_d, R_1 will have minimum value, therefore,

$$R_{1\,min} = 1 \text{ k}\Omega$$

Thus

$$R_1 = 1\text{–}50 \text{ k}\Omega \text{ potentiometer}$$

$$R_2 = 100 \text{ k}\Omega$$

$$R_3 = R_4 = \text{Say, } 15 \text{ k}\Omega \text{ each}$$ **Ans.**

3.7. OP-AMP INTEGRATOR

An integrator is a circuit that performs a mathematical operation called *integration*. Integration is a process of continuous addition. The most popular application of an integrator is to produce a *ramp* of output voltage, which is a linearly increasing or decreasing voltage. The circuit is shown in Fig. 3.26 (*a*) which is similar to the circuit of the inverting amplifier shown in Fig. 2.42 except that the feedback is through capacitor C instead of resistor R_f. The input need not be sinusoidal, and is, therefore, represented by a lower case symbol $v = v(t)$. Correspondingly, the current as a function of time is designated by $i = i(t)$. Feedback through the capacitor forces a virtual ground to exist at the inverting input terminal.

The virtual ground equivalent circuit shows that an expression between input and output voltages can be derived from the current i, which flows from input to output. Recall that virtual ground means that the voltage at the junction point of resistor R and capacitor C can be considered to be at ground (since $V_{in} \cong 0$ V) but no current passes into the ground at that point.

Hence $i(t) = \frac{v(t)}{R}$ and output voltage $v_{out(t)} = -\frac{1}{C}\int i(t)\,dt$

Combining the above two equations we have

$$v_{out}(t) = -\frac{1}{C}\int \frac{v(t)}{R}\,dt = \frac{-1}{RC}\int v(t)\,dt + A \quad ...(3.19)$$

where A is an integration constant and is proportional to the value of the output voltage v_{out} at time $t = 0$ second.

Above Eq. (3.19) shows that the output voltage is the integral of the input voltage, with an inversion and scale factor of $\frac{1}{RC}$. The ability to integrate a given signal provides the analog computer with the ability to solve differential equations and, therefore, permits setup of a large variety of electrical circuit analogs of the physical-system operations.

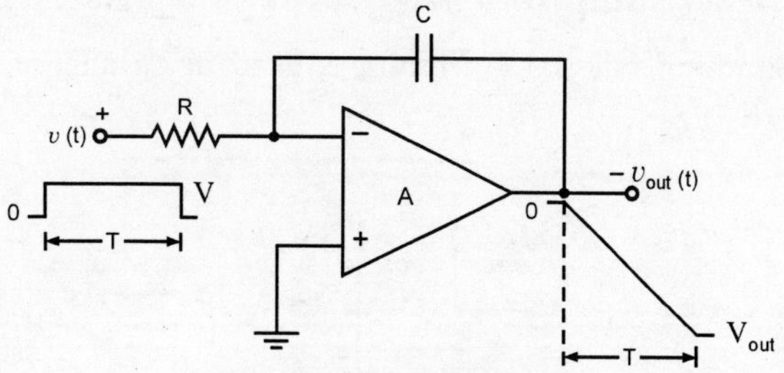

(a) Op-Amp Integrator

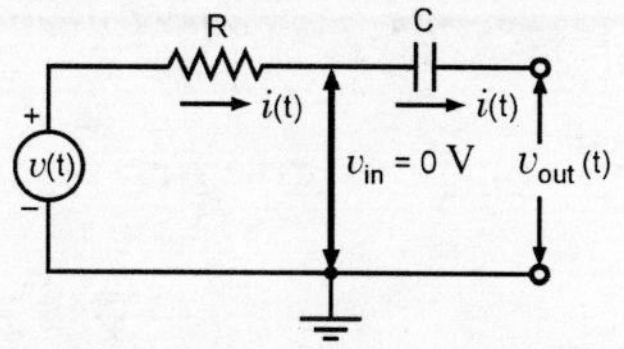

(b) Equivalent Circuit of Op-Amp Integrator

Fig. 3.26

If the input voltage is a step voltage, then the output voltage will be a ramp or linearly changing voltage, as discussed above; if the input voltage is a square wave, the output voltage will be a triangular wave as depicted in Fig. 3.27; and if the input voltage is a sinusoidal one, the output voltage will be cosine wave as shown in Fig. 3.28.

Integrators are widely used in ramp or sweep generators, filters, analog computers etc.

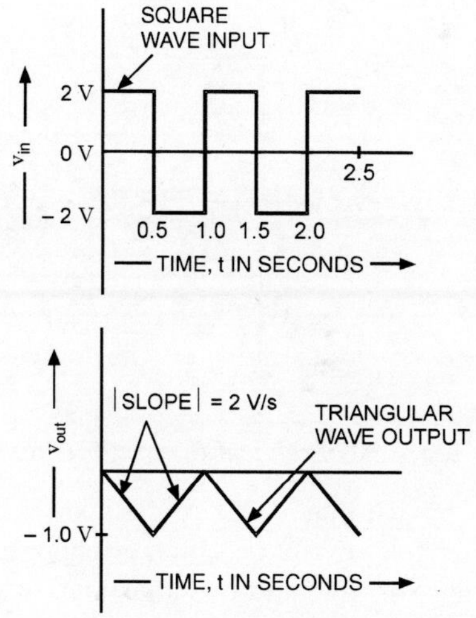

Fig. 3.27. ***Input and Ideal Output Waveforms***

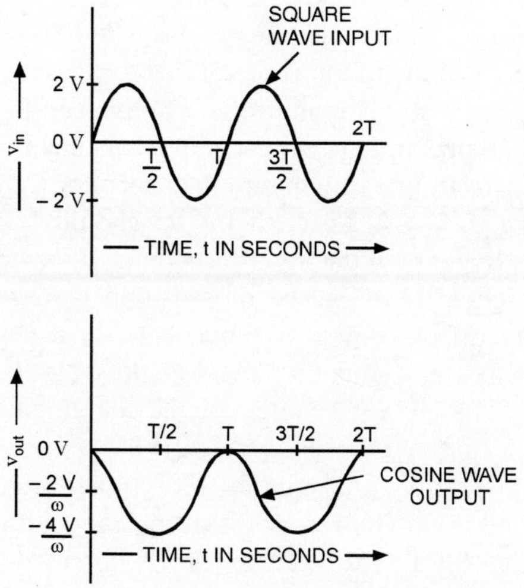

Fig. 3.28. ***Input and Ideal Output Waveforms***

3.7.1. Summing Integrator

Summing integrator circuit, as used in an analog computer, is illustrated in Fig. 3.29 (*a*).

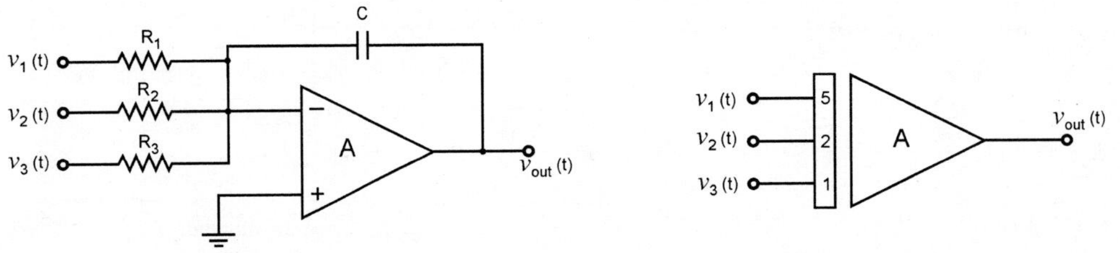

(*a*) *Summing Integrator Circuit* (*b*) *Analog-Computer Integrator-Circuit Representation*

Fig. 3.29

Resulting voltage is given as

$$v_{out}(t) = -\left[\frac{1}{R_1C}\int v_1(t)\,dt + \frac{1}{R_2C}\int v_2(t)\,dt + \frac{1}{R_3C}\int v_3(t)\,dt\right] \quad ...(3.20)$$

Analog computer integrator circuit representation is given in Fig. 3.29 (*b*). The analog computer representation only indicates the scale factor for each input for R_1 = 200 kΩ, R_2 = 500 kΩ, R_3 = 1 MΩ and C = 1 μF.

3.7.2. DC Stabilization

The integrator circuit shown in Fig. 3.26 (*a*) produces output voltage proportional to the time integral of the input signal provided that the input signal frequency is not too low. The circuit shown in Fig. 3.26 (*a*) needs a slight modification to make it practical. Because a capacitor is open to dc signals, there is no negative feedback at zero frequency. In absence of negative feedback, the circuit treats any input offset voltage as a valid input voltage. The result is that the capacitor charges and the output goes into positive or negative saturation, where it stays indefinitely. Thus the process of integration is destabilized.

The effect of input offset voltage can be reduced by reducing the voltage gain at zero frequency by introducing a resistor R′ in parallel with the capacitor C, as illustrated in Fig. 3.30 (*a*). The resistor R′ should be at least 10 times larger than the input resistor R. If the added resistor R′ is equal to 10 R, the closed-loop gain is 10 and the output offset voltage is reduced to an acceptable level. When a valid input voltage is present, the additional resistor has almost no effect on the charging of a capacitor, so that the output voltage is still almost a perfect ramp.

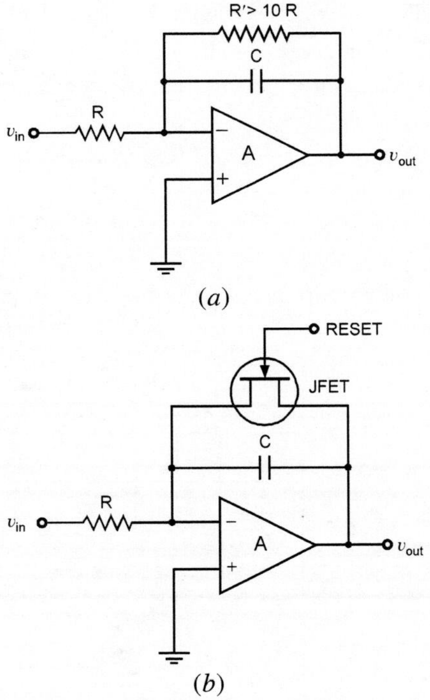

Fig. 3.30. ***Practical Integrator Circuits***

The effect of input offset voltage can also be reduced by using a JFET switch as shown in Fig. 3.30 (*b*). The reset voltage on the gate of the JFET is either 0 V or – V_{CC} which is enough to cut off the JFET. Thus JFET can be set to a low resistance when the integrator is idle and to a high resistance when the integrator is active.

The JFET discharges the capacitor in preparation for the next input pulse. Just before the beginning of the next input pulse, the reset voltage is made equal to zero volt. This discharges the capacitor. At the instant the next pulse begins, the reset voltage becomes $-V_{CC}$ which cuts off the JFET. The integrator then produces an output voltage ramp.

Example 3.10. A 10 mV, 2 kHz sinusoidal signal is applied to the inverting input terminal of an op-amp integrator for which R = 50 kΩ and C = 2 μF. Determine the output voltage.

Solution : The equation for the sinusoidal signal of rms value of 10 mV and frequency 2 kHz is given as

$$v_{in}(t) = \sqrt{2}\, V_{rms} \sin 2\pi f t$$

$$= \sqrt{2} \times 10 \sin 4{,}000\pi t \text{ mV} \qquad \text{assuming } v_{in}(t) = 0 \text{ at } t = 0$$

$$\text{Scale factor} = -\frac{1}{CR} = -\frac{1}{2 \times 10^{-6} \times 50 \times 10^{3}} = -10$$

So output voltage,

$$v_{out}(t) = \frac{-1}{RC}\int v_{in}(t)\, dt = -10 \int_{o}^{t} \sqrt{2} \times 10 \sin 4{,}000\,\pi t = -141.42\left[\frac{-\cos 4{,}000 t}{4{,}000\pi}\right]_0^t$$

$$= 0.0113 (\cos 4{,}000\, t - 1) \text{ mV} \quad \textbf{Ans.}$$

Example 3.11. When will the output get saturated in Fig. 3.31.

Solution : The output equation is given as

$$v_{out} = -\frac{1}{RC}\int_0^t v(t)\, dt + A = \frac{-1}{500 \times 10^{3} \times 10 \times 10^{-6}}\int_0^t -0.5\, dt$$

Constant of integration A being constant $= [0.1t]_0^t$

Output will get saturated when it attains the value of supply voltage *i.e.,* of 12 V.

Thus, $v_{out} = 0.1t = 12$ V

or Time duration required for saturation of output voltage.

$$t = \frac{12}{0.1} = 120 \text{ seconds} \quad \textbf{Ans.}$$

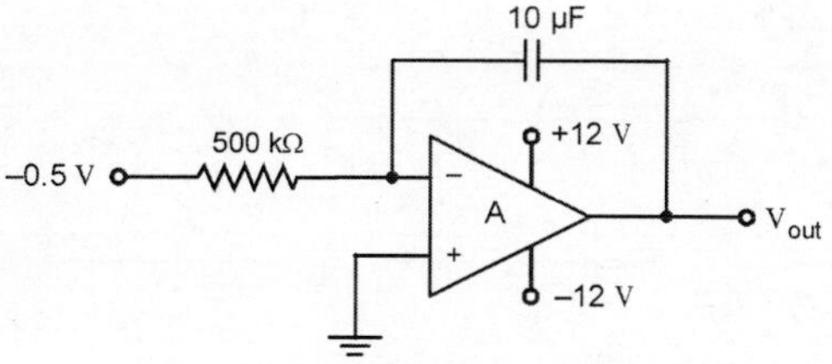

Fig. 3.31

Example 3.12. Draw the circuit of an inverting integrator using ideal op-amp. Calculate the output voltage after (*i*) 0.5 ms (*ii*) 5 ms of the application of an step input voltage of 50 mV. Take R = 50 Ω, C = 0.1 μF and the supply voltage of the op-amp is ±15 V.

Solution : The output equation is given as

$$v_{out} = -\frac{1}{RC}\int_0^t v(t)\, dt = -\frac{1}{50 \times 0.1 \times 10^{-6}}\int_0^t 50 \times 10^{-3}\, dt = [-10{,}000t]_0^t = -10{,}000t$$

(*i*) When $t = 0.5 \text{ ms} = 5 \times 10^{-4}$ second

Output voltage, $v_{out} = -10{,}000 \times 5 \times 10^{-4} = -5$ V **Ans.**

(*ii*) When $t = 5 \text{ ms} = 5 \times 10^{-3}$ second

Output voltage, $v_{out} = -10{,}000 \times 5 \times 10^{-3} = -50$ V

But output voltage will get saturated when it attains the value of supply voltage *i.e.,* of –15 V.

So output voltage will be –15 V. **Ans.**

Example 3.13. Determine the capacitor voltage at the end of the pulse in Fig. 3.32. Also determine the closed-loop time constant if the op-amp has a differential voltage gain of 10^5.

Solution : Because of virtual ground on the inverting input, the input current,

$$I = \frac{V_{in}}{R} = \frac{10}{2.2\text{ k}\Omega} = 4.545 \text{ mA}$$

The output voltage at the end of pulse is given by equation

$$V = \frac{IT}{C} \quad \because \quad V = \frac{Q}{C} \text{ or } \frac{V}{T} = \frac{Q/T}{C} \text{ or } \frac{V}{T} = \frac{I}{C},$$

charging current being constant

$$= \frac{(4.545 \text{ mA})(1 \text{ ms})}{1 \, \mu\text{F}} = 4.545 \text{ V} \quad \textbf{Ans.}$$

The output voltage starts at 0 V and decreases linearly to –4.545 V, as shown in Fig. 3.26 (*a*).

Because of Miller effect, the closed-loop time constant is

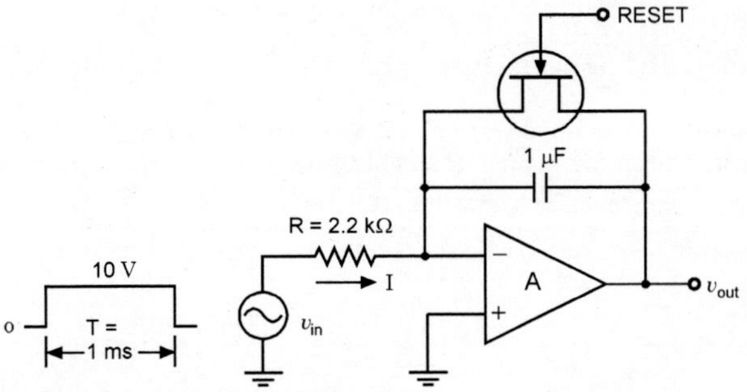

Fig. 3.32

$$RC' = (2.2 \text{ k}\Omega) \times (1 \, \mu\text{F}) \times (10^5) = \textbf{220 s} \quad \textbf{Ans.}$$

Since the closed-loop time constant (220s) is much greater than the pulse width of 1 ms, only the very earliest part of the exponential charge is involved. Because of this, the output voltage is almost a perfect ramp.

Example 3.14. Find R_1 and R_f in the lossy integrator so that peak gain is 20 dB and the gain is 3 dB down from its peak when ω = 10,000 rad/second. Use capacitance of 0.01 µF.

Solution : Practical integrating circuit is also called a **lossy integrator** because the parallel combination of R′ and C behaves as a practical capacitor and thus dissipates power unlike an ideal capacitor.

Circuit diagram is shown in Fig. 3.33.

Applying KCL at point A we have

$$\frac{V_1}{R_1} + \frac{V_{out}}{1/sC} + \frac{V_{out}}{R_f} = 0$$

$$\text{or} \quad \left|\frac{V_{out}}{V_1}\right| = -\left|\frac{R_f/R_1}{(1+sCR_f)}\right|$$

Substituting $s = j\omega$ we have

$$\frac{V_{out}}{V_1} = \frac{R_f/R_1}{\sqrt{(\omega C R_f)^2 + 1}}$$

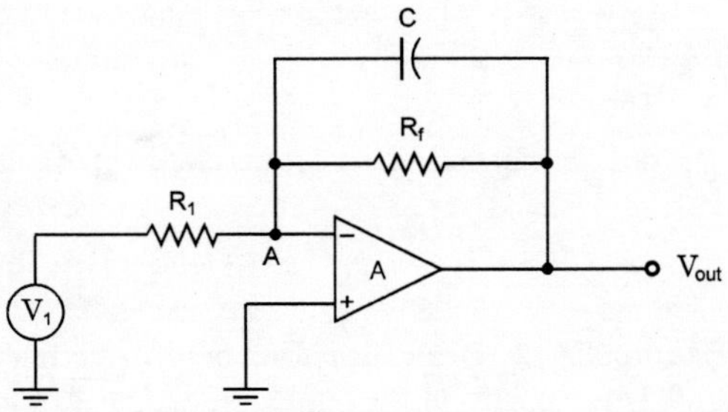

Fig. 3.33

At ω = 0 (peak gain)

$$\frac{V_{out}}{V_1} = \frac{R_f}{R_1}$$

$$\because \quad 20 \log_{10} \frac{R_f}{R_1} = 20$$

$$\therefore \quad \frac{R_f}{R_1} = 10$$

$$\omega_{3 \text{ dB}} = \frac{1}{CR_f}$$

$$\text{or} \quad 10{,}000 = \frac{1}{0.01 \times 10^{-6} \times R_f} \quad \because \quad C = 0.01 \, \mu\text{F (given)}$$

$$\text{or} \quad R_f = 10 \text{ k}\Omega$$

$$\text{and} \quad R_1 = \frac{R_f}{10} = 1 \text{ k}\Omega \quad \textbf{Ans.}$$

Example 3.15. For the following op-amp circuit sketch v_{out} for 50 ms after the switch S is opened. Assume ideal op-amp. [*U.P. Technical Univ. Analog Integrated Circuits 2004-05*]

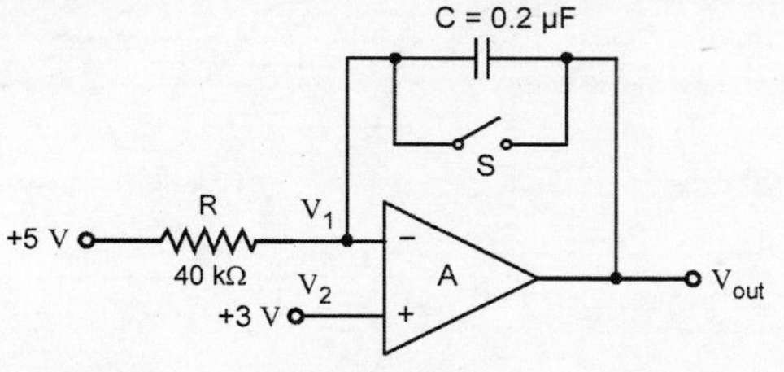

Fig. 3.34

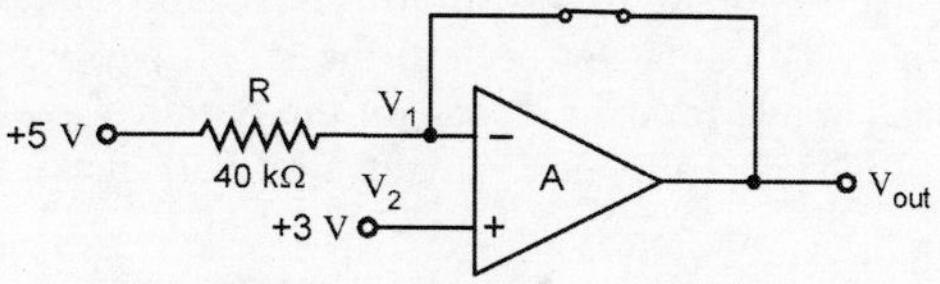

Fig. 3.35

Solution : When switch S is closed, the capacitor remains short circuited, and the circuit becomes as shown in Fig. 3.35.

Since op-amp is assumed ideal,

$$V_2 = V_1$$

and, therefore, output is

$$V_{out} = V_1 = V_2 = +3\text{ V}$$

When the switch S is opened the circuit acts like an integrator and output equation is given as

$$v_{out} = -\frac{1}{RC}\int_0^t (V_{in} - V_1)\, dt + V_{out}\Big|_{t=0}$$

$$= \frac{-1}{40\times10^3 \times 0.2\times10^{-6}}\int_0^{50\,\text{ms}} (5-3)\, dt + 3$$

$$= \frac{-1\times2\times50\times10^{-3}}{8\times10^{-3}} + 3 = -12.5 + 3 = -9.5\text{ V}$$

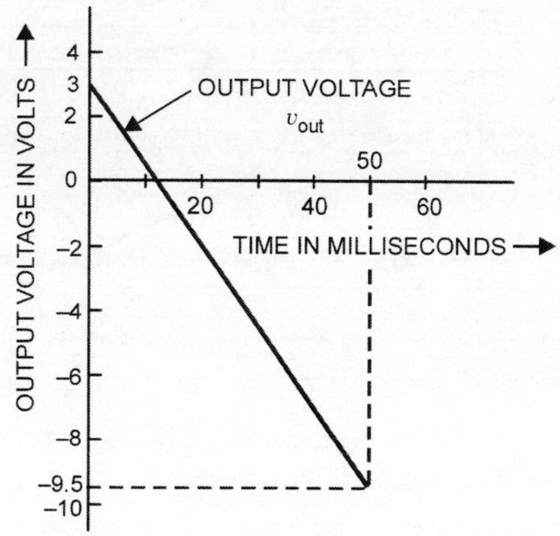

Fig. 3.36

The v_{out} graph is shown in Fig. 3.36.

Example 3.16. The time constant (RC) of an op-amp integrator circuit is 0.1 ms. Determine the output response for the following input signals :

(*i*) 1 V peak sine wave at 5 kHz.

(*ii*) Step voltage $V_i = 1$ V for $0 \le t \le 0.5$ ms

(*iii*) 2 V P-P square wave of 5 kHz.

Draw the input and output waveforms. [*G.G.S.I.P. Univ. Analog Electronics-II May 2010*]

Solution : (*i*) For an input of 1 V peak sine wave at 5 kHz, the output is

$$v_{out}(t) = -\frac{1}{RC}\int v_{in}(t)\, dt = \frac{-1}{0.1\times10^{-3}}\int_0^t 1 \sin 2\pi\times5{,}000t\, dt$$

$$= -10{,}000\int_0^t \sin 10{,}000\pi t\, dt = \frac{-10{,}000}{10{,}000\pi}\left[-\cos 10{,}000\pi t\right]_0^t$$

$$= 0.318 \cos 10{,}000\pi t$$

Thus the output is a cosine wave with a peak amplitude of 0.318 V only as shown in Fig. 3.37 (*a*).

(*ii*) When the input is a step voltage $V_{in} = 1$ V for $0 \le t \le 0.5$ ms, then the output voltage at $t = 0.5$ ms is

$$v_{out} = -\frac{1}{RC}\int_0^{0.5\,\text{ms}} 1\, dt = \frac{-1}{0.1\times10^{-3}}[t]_0^{0.5\,\text{ms}} = -10^4 \times 0.5\times10^{-3} = -5\text{ V}$$

The output voltage is a ramp function with a slope of 10^4 V per second or 10 V/ms and is shown in Fig. 3.37 (*b*).

(*iii*) The output waveform for an input of 2 V peak-to-peak square wave of 5 kHz is shown in Fig. 3.37 (*c*). It is to be noted that the input is of constant amplitude of 1 V from 0 to 0.1 ms and –1 V from 0.1 ms to 0.2 ms. The output for each of these half periods will be ramps as mentioned above for step input. So, the output waveform will be a triangular wave of peak value, for the first half cycle, given as

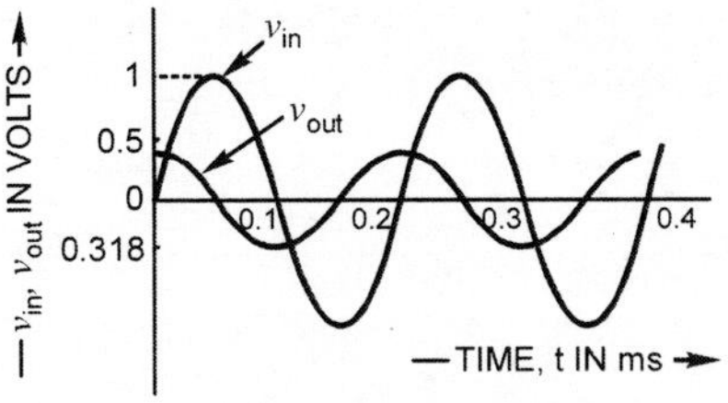

Fig. 3.37 (*a*)

$$v_{out} = \frac{-1}{CR}\int_0^t V_{max}\, dt = \frac{-1}{0.1 \times 10^{-3}}\int_0^{0.1\,ms} 1\, dt$$

$$= -10^{-4} \times 0.1 \times 10^{-3} = -1 \text{ V}$$

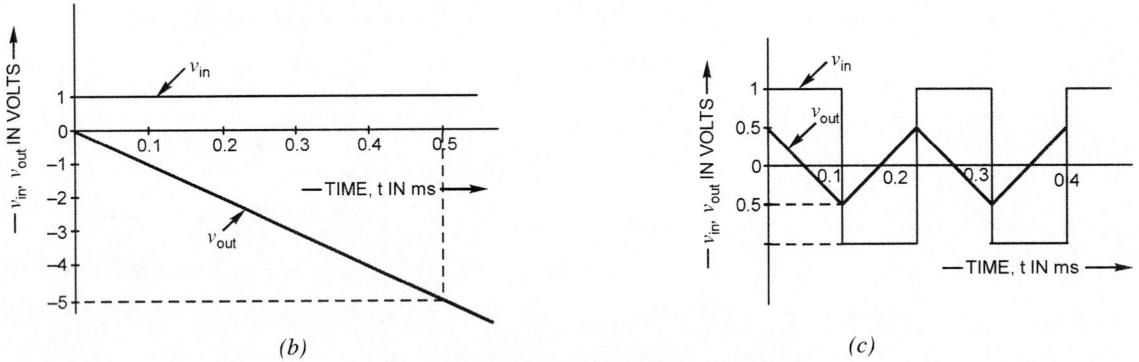

Fig. 3.37. ***Input and Output Waveforms***

This gives the total change in the output voltage over the first half cycle from 0 to 0.1 ms. Similarly, integration over the next half cycle develops a positive change of 1 V.

Example 3.17. For the circuit shown in Fig. 3.38, R_1 = 10 kΩ, R_f = 100 kΩ, C_f = 10 nF. Derive and plot the response for the following input signals:

(*i*) Sine wave of 1 V peak amplitude and 5 kHz

(*ii*) Square wave of 1 V peak amplitude and 5 kHz.

[*G.G.S.I.P. Univ. Analog Electronics-II May 2011*]

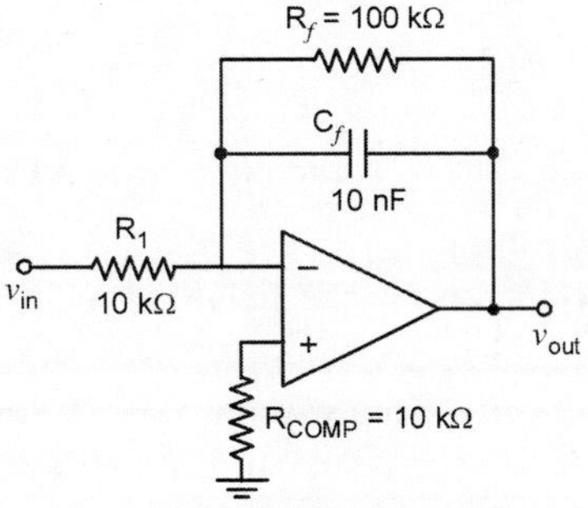

Fig. 3.38

Solution : The time constant (RC) of the given integrator circuit is

$$R_1C_f = 10 \times 1{,}000 \times 10 \times 10^{-9}$$

$$= 1 \times 10^{-4} \text{ second} = 0.1 \text{ ms}$$

The output response of the circuit as shown in Fig. 3.37 (*a*) and (*c*). For details refer to Solution of Example 3.16 (*i*) and (*iii*).

3.8. OP-AMP DIFFERENTIATOR

Its function is to provide an output voltage proportional to the rate of change of the input voltage. It is an inverse mathematical operation to that of an integrator.

Differentiator circuit can be obtained by interchanging the resistor and capacitor of the integrator circuit, as shown in Fig. 3.39.

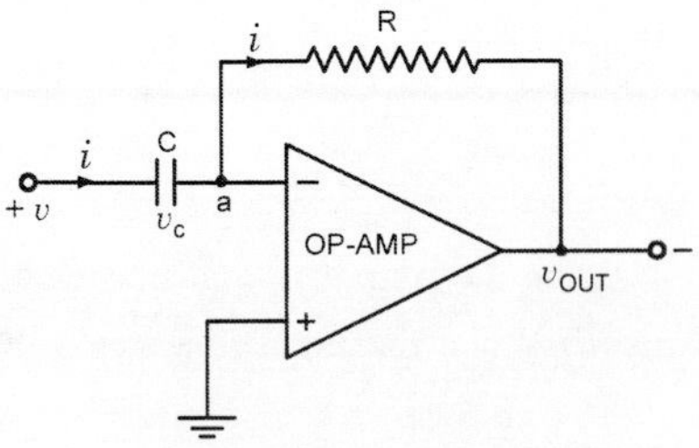

Fig. 3.39. ***Op-Amp Differentiator***

Let i be the rate of change of charge *i.e.*, dq/dt

Now charge, $q = Cv_c$

$$\text{So } i = \frac{dq}{dt} = \frac{d}{dt}Cv_c = C\frac{dv_c}{dt}$$

Taking inverting input terminal as virtual ground

$$\text{Output voltage } v_{out} = -iR = -\left[C\frac{dv_c}{dt}\right]R$$

$$= -CR\frac{dv_c}{dt} \qquad ...(3.21)$$

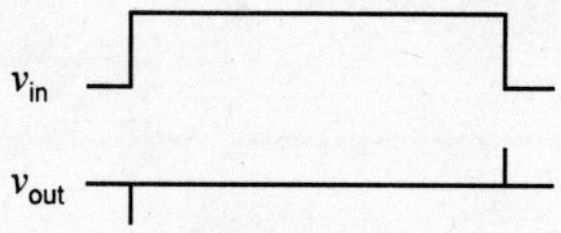

Fig. 3.40

i.e., output voltage is proportional to the derivative of the input voltage, the constant of proportionality being −CR.

One common application of the op-amp differentiator is to produce very narrow spikes, as shown in Fig. 3.40. The advantage of an op-amp differentiator over a simple R-C differentiator is that the spikes produced come from a low-impedance source, which makes driving typical load resistance easier.

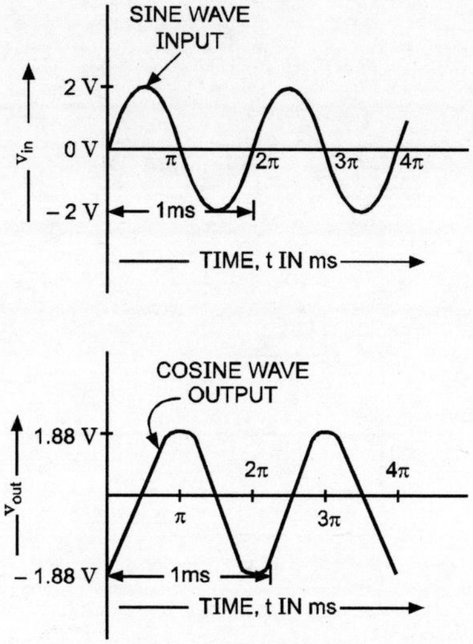

Fig. 3.41. ***Input and Output Waveforms***

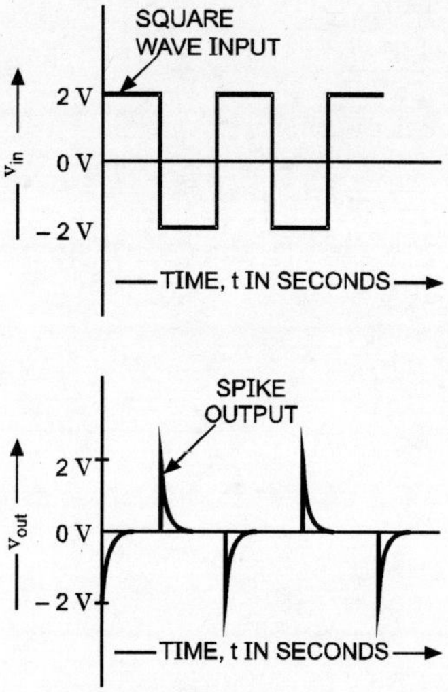

Fig. 3.42. ***Input and Output Waveforms***

As the differentiator performs an inverse operation to that of an integrator, a cosine wave input will generate a sine wave output, or a triangular input will produce a square wave output. But the circuit shown in Fig. 3.39 will not be able to do so because of some practical problems like unstability and susceptibility to high-frequency noise.

A differentiator is rarely used in analog computers as it tends to amplify (the output voltage magnitude being proportional to the frequency of the input signal) noise drift and other unwanted disturbances in the system.

3.8.1. Practical Op-Amp Differentiator

Both of the problems of unstability and susceptibility to high-frequency noise can be overcome by modifying the circuit given in Fig. 3.39 to the circuit shown in Fig. 3.43.

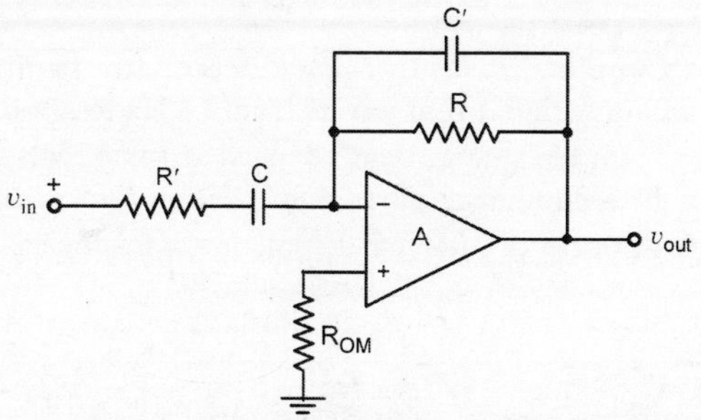

Fig. 3.43. ***Practical Differentiator***

Example 3.18. The input to an op-amp differentiator circuit shown in Fig. 3.39 is a sinusoidal voltage of peak value of 10 μV and frequency 2 kHz. Determine the output voltage if R = 50 kΩ and C = 2 μF.

Solution : The equation for the sinusoidal signal of peak value of 10 μV and frequency 2 kHz is given as

$$v_{in} = V_{max} \sin 2\pi ft = 10 \sin 2 \times 2{,}000\pi t = 10 \sin 4{,}000\pi t \ \mu V$$

$$\text{Scale factor} = CR = 2 \times 10^{-6} \times 50 \times 10^{3} = 0.1$$

So output voltage, $v_{out} = -CR \dfrac{dv_c}{dt} = -0.1 \dfrac{d}{dt}(10 \sin 4{,}000\pi t) \ \mu V$

$$= -0.1 \times 10 \frac{d}{dt} (\sin 4{,}000\pi t) \ \mu V = 4{,}000\pi \cos 4{,}000\pi t \ \mu V \text{ or } 12.56 \cos 4{,}000\pi t \text{ mV} \quad \textbf{Ans.}$$

Example 3.19. Design a differentiator to differentiate an input signal that varies in frequency from 10 Hz to 1 kHz. The input is a sine wave peak-to-peak amplitude 3 V at 200 Hz. Sketch the output voltage waveform.

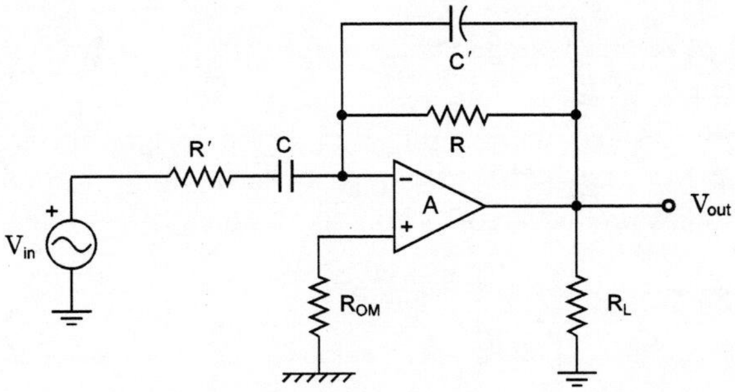

Fig. 3.44

Solution : The circuit is shown is Fig. 3.44

$$f_a = 1 \text{ kHz}$$

$$\text{Also } f_a = \frac{1}{2\pi RC}$$

Let C = 0.1 μF then

$$R = \frac{1}{2\pi f_a C} = \frac{1}{2\pi \times 1 \times 10^3 \times 0.1 \times 10^{-6}}$$

$$= 1.59 \text{ k}\Omega = 1.5 \text{ k}\Omega \text{ (say)}$$

$$f_b = 20 f_a = 20 \times 1 = 20 \text{ kHz}$$

$$\text{Also, } f_b = \frac{1}{2\pi R'C}$$

$$\therefore R' = \frac{1}{2\pi f_b C} = \frac{1}{2\pi \times 20 \times 0.1 \times 10^{-6}}$$

$$= 79.5 \ \Omega$$

Since $R'C = RC'$

Let R' be of 82 Ω

$\therefore C' = 0.0055$ μF

$R_{OM} = R = 1.5$ kΩ. **Ans.**

$V_p = 1.5$ V, $f = 200$ Hz ...(given)

$$V_{in} = V_p \sin \omega t = 1.5 \sin (2\pi \times 200)\, t$$

$$= 1.5 \sin 400\pi t$$

$$v_{out} = -RC \frac{dv_{in}}{dt}$$

$$= -1.5 \times 10^3 \times 0.1 \times 10^{-6} \frac{d}{dt} [1.5 \sin 400\pi t]$$

$$= -0.2827 \cos 400\pi t \quad \textbf{Ans.}$$

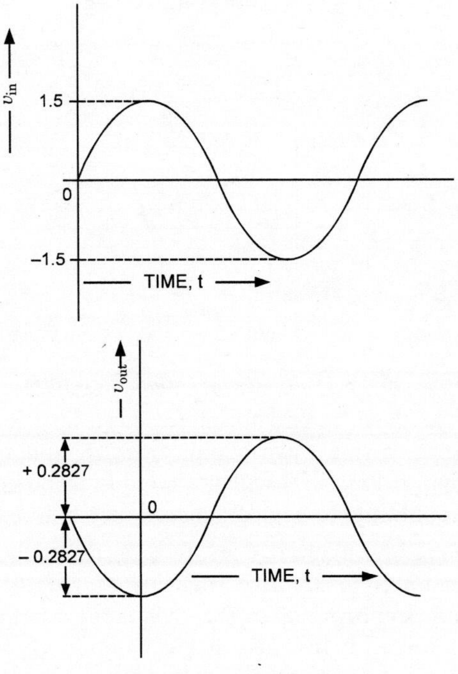

Fig. 3.45. ***Waveforms***

Example 3.20. (*a*) Design a differentiator to differentiate an input signal that varies from 10 Hz to about 1 kHz.

(*b*) If a sine wave of 1 V peak at 1,000 Hz is applied to a differentiator of part (*a*), draw its output waveform.

Solution : (*a*) Refer to the solution of Example 3.19

(*b*) $V_P = 1$ V, $f = 1{,}000$ Hz

$$v_{in} = V_P \sin \omega t = 1 \sin 2\pi ft = 1 \sin (2\pi \times 1{,}000)\, t = 1 \sin 2000\, \omega t$$

and $$v_{out} = -RC \frac{dv_{in}}{dt} = -1.5 \times 10^3 \times 0.1 \times 10^{-6} \frac{d}{dt}[\sin 2000\pi t] = -1.5 \times 10^3 \times 0.1 \times 10^{-6} \times 2000\pi \cos 2000\pi t$$

$$= -0.942 \cos 2000\pi t$$

The input and output waveforms are as shown in Fig. 3.41.

Example 3.21. Explain the working of differentiator. It is excited by the input shown in Fig. 3.46. Find the output voltage if $R_f = 2\ k\Omega$, $C_i = 0.1\ \mu F$.

[G.G.S.I.P. Univ. Delhi Analog Electronics-II May 2011]

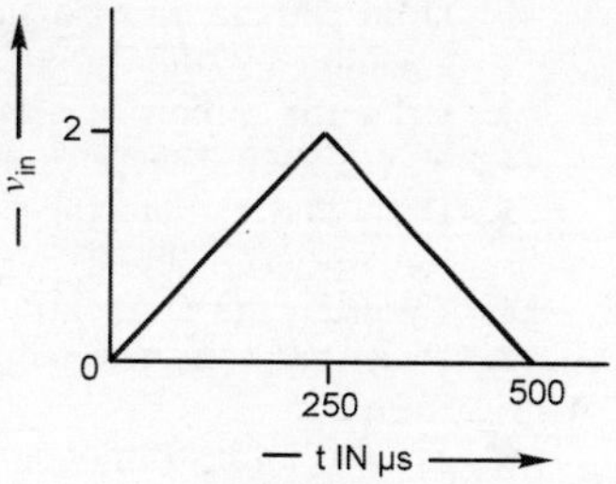

Fig. 3.46

Solution : Output voltage,

$$V_{out} = -R_f C_i \frac{d}{dt}(V_{in})$$

$$= -R_f C_i \frac{\Delta V_{in}}{\Delta t}$$

$$= -2 \times 10^3 \times 0.1 \times 10^{-6} \times \frac{2-0}{250\ \mu s}$$

$\because$ Voltage changes from 0 to 2 V in 250 μs

$$= 2 \times 10^3 \times 0.1 \times 10^{-6} \times \frac{2}{250 \times 10^{-6}}$$

$= -1.6$ V **Ans.**

Similarly for the input when changes from 2 V to 0 V in 250 μs

$$V_{out} = -2 \times 10^3 \times 0.1 \times 10^{-6} \times \frac{0-2}{250 \times 10^{-6}} = 1.6 \text{ V} \quad \textbf{Ans.}$$

The output waveform is shown in Fig. 3.47.

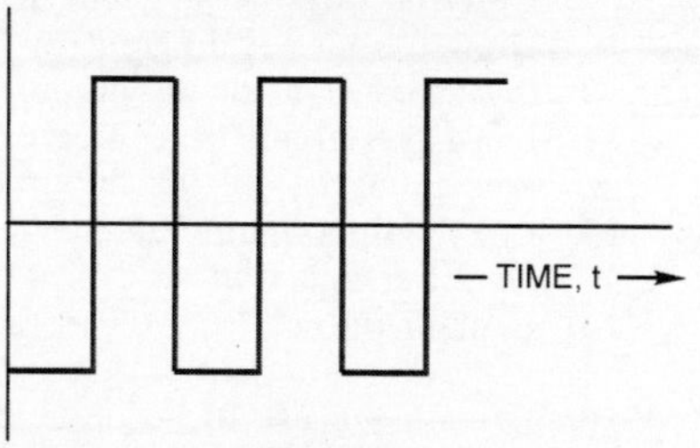

Fig. 3.47

HIGHLIGHTS

1. In linear circuits (such as adders, subtractors, integrators, differentiators, instrumentation amplifiers, voltage-to-current converters, current-to-voltage converters), the output signal varies directly with the input signal.
2. Summing, scaling and average amplifiers can be constructed by using inverting, noninverting and differential configurations.
   If a number of signals are connected to the inverting input, the output is proportional to the sum of the input signals, making possible the solution of linear algebraic and differential equations on an analog computer.
3. Since differential amplifier amplifies the difference of two input signals applied to the inverting and noninverting terminals of an op-amp, it can be used as subtractor circuit.
4. A circuit that produces a current proportional to input signal voltage is called a voltage-to-current converter. It is an ideal circuit for low-voltage dc and ac voltmeters, LED and zener diode tester.
   A device that produces a voltage proportional to input signal current is called a current-to-voltage converter. One of the most common uses of the current-to-voltage converter is in sensing current from photodetectors and in digital-to-analog converter applications.
5. An input voltage can be used to control an output voltage or current, or an input current can be used to control an output voltage or current. Such controlled sources are suitable for use in different instrumentation circuits.
6. The low-level signal outputs of electrical transducers often need to be amplified before further processing, and this is done with instrumentation amplifiers.
   Obviously, all kinds of amplifiers are used in instrumentation; however, a particular class and configuration conventionally is given the name *instrumentation amplifier*, mainly to distinguish from the simpler op-amp circuits.
   In practice, an instrumentation amplifier is a specific combination of dc op-amps wired up with feedback, and needs very little design effort from the user.
7. If a capacitor is included as a feedback element, the output voltage is the integral of the input voltage. If a capacitor is included as an input element, the output voltage is the derivative of the input voltage.
   An integrator is a circuit that performs a mathematical operation called the integration. Integrators are widely used in ramp or sweep generators, filters, analog computers etc.
   The function of an op-amp differentiator is to provide an output voltage proportional to the rate of change of the input voltage. One common application of the op-amp differentiator is to produce very narrow spikes.

EXERCISES

1. Realize op-amp as inverting adder, averaging and scaling amplifier.

2. Draw the circuit of a summing operational amplifier using inverting amplifier configuration. Write an equation for the output voltage of this circuit.
3. Draw the circuit of a subtractor using op-amp and explain its working.
4. How do you use an op-amp as voltage-to-current converter ? Draw the circuit and explain its operation.
5. Draw and explain the operation of a voltage-to-current converter when load is grounded. Is there any limitation on the size of the load when grounded ? **[G.G.S.I.P. Univ. Analog Electronics-II May 2011]**
6. Describe the application of op-amp as current-to-voltage converter.
7. Describe the instrumentation amplifier. What is the most desired characteristic of it and maximum limit to that ?
8. What is an instrumentation amplifier ? How does it differ from an ordinary op-amp ? Draw the schematic diagram of an instrumentation amplifier and explain it.
9. Describe differential mode instrumentation amplifier. Find its output.
10. Draw the circuit diagram for differential instrumentation amplifier using transducer bridge and derive the expression for the output voltage at a given input voltage.
11. Draw the circuit of a three op-amp instrumentation amplifier, and derive an expression for the output voltage. What are the important features ? **[G.G.S.I.P. Univ. Analog Electronics Dec. 2009]**
12. List the important features of an instrumentation amplifier. Draw the circuit of a practical instrumentation amplifier and find an expression for its output voltage. **[G.G.S.I.P. Univ. Analog Electronics May 2010]**
13. Explain the working of a practical instrumentation amplifier. **[G.G.S.I.P. Univ. Analog Electronics-II May-June 2009]**
14. Explain the use of an op-amp as integrator. Draw the circuit and obtain the expression for its output voltage.
15. Draw the circuit of an op-amp integrator and explain the working. Indicate the input and output waveforms.
16. Explain how an op-amp can be used as integrator. Also derive expression for the output.
17. Draw op-amp differentiator and integrator circuits. **[G.G.S.I.P. Univ. Analog Electronics-II May-June 2009]**
18. Explain the working of an ideal op-amp differentiator. What are its disadvantages and how to overcome these ? **[G.G.S.I.P. Univ. Analog Electronics-II May 2011]**
19. Explain op-amp as integrator and differentiator.
20. Explain the difference between the integrator and differentiator and give one application of each.
21. Describe the working of practical differentiator circuit. Derive the expression for output voltage.
22. Write short notes on the following :
    (*i*) Summing, scaling and averaging amplifier (*ii*) Op-amp subtractor
    (*iii*) Instrumentation amplifier (*iv*) Op-amp as integrator.

SHORT ANSWER TYPE QUESTIONS WITH ANSWERS ....

Q. 1. Give the examples of linear circuits.

Ans. Adder, subtractor, differentiator, integrator, voltage-to-current converter, current-to-voltage converter, instrumentation amplifier fall under the category of linear circuits.

Q. 2. What is an adder or summing amplifier ?

Ans. Adder or summing amplifier is a circuit that provides an output voltage proportional to or equal to the algebraic sum of two or more input voltages multiplied by a constant gain factor.

Q. 3. What is scaling amplifier ?

Ans. If the input voltage is amplified by a different factor *i.e.,* weighted differently at the output, the circuit becomes a scaling or weighted amplifier. The circuit shown in Fig. 3.1(*a*) can be used as scaling or weighted amplifier. The output voltage is given by Eq. $V_{out} = -R_f\left[\frac{V_1}{R_1}+\frac{V_2}{R_2}+\frac{V_3}{R_3}\right]$ where $R_1 \# R_2 \# R_3$.

Q. 4. What is another name for voltage-to-current converter ?

Ans. The another name for voltage-to-current converter is transconductance amplifier.

Q. 5. What is current-to-voltage converter ?

Ans. A device that produces a voltage proportional to input signal current is called a current-to-voltage converter.

Q. 6. What is difference between positive scalar and voltage-to-current converter ?

Ans. The circuit of voltage-to-current converter is similar to that of a positive scalar except that feedback resistor R_f is replaced by load resistor R_L.

Q. 7. Name any two instrumentation amplifiers ?

Ans. The two instrumentation amplifiers may be

1. Instrumentation amplifier using transducer bridge.
2. Instrumentation amplifier using temperature controller.

Q. 8. Why a potentiometer is provided in an instrumentation amplifier ?

Ans. A potentiometer is provided, in an instrumentation amplifier, to permit adjustment of the scale factor of the circuit.

Q. 9. What is an integrator ?

Ans. An integrator is a circuit that performs a mathematical operation called integration.

Q. 10. What are the applications of integrators ?

Ans. Integrators are widely used in ramp or sweep generators, filters, analog computers etc.

Q. 11. Why integrators are preferred over differentiators ?

Ans. Op-amp is mostly used as an integrator rather as a differentiator, because in differentiator at high frequency, gain is high and so high frequency noise is also amplified which absolutely abstract the differentiated signal.

Q. 12. What will be the output shape of an op-amp integrator with a square-wave input ?

Ans. The output of an op-amp integrator with a square-wave input will be a triangular wave, as shown in Fig. 3.27.

Q. 13. What will be the output shape of an op-amp integrator with a sinusoidal input ?

Ans. The output of an op-amp integrator with a sinusoidal input will be a cosine wave, as shown in Fig. 3.28.

Q. 14. What do you mean by differentiator ?

Ans. A differentiator or differentiation amplifier is a circuit that performs the mathematical operation of differentiation *i.e.,* the output voltage is proportional to the rate of change of the input voltage. It is an inverse mathematical operation to that of integrator.

A differentiator is rarely used in analog computers as it tends to amplify (the output voltage magnitude being proportional to the frequency of the input signal) noise drift and other unwanted disturbances in the system.

Q. 15. What will be the output shape of an op-amp differentiator with a sinusoidal input ?

Ans. The output of an op-amp differentiator with a sinusoidal input will be a cosine wave as shown in Fig. 3.41.

Q. 16. What will be the output shape of an op-amp differentiator with a square-wave input ?

Ans. The output of an op-amp differentiator with a square-wave input will be a spike output, as shown in Fig. 3.42.

PROBLEMS

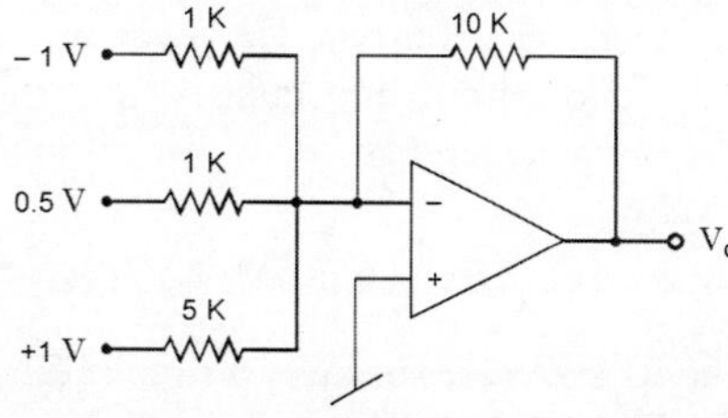

Fig. 1

1. Find out the output voltage of the following circuit. [**Ans.** 3 V]
2. Calculate the output voltage of an op-amp summing amplifier for following sets of voltage and resistors. Use $R_f = 1\ M\Omega$ in all cases.
   (*a*) $V_1 = 1\ V, V_2 = +2\ V, V_3 = +3\ V, R_1 = 500\ k\Omega, R_2 = 1\ M\Omega, R_3 = 1\ M\Omega$
   (*b*) $V_1 = -2\ V, V_2 = +3\ V, V_3 = +1\ V, R_1 = 200\ k\Omega, R_2 = 500\ k\Omega, R_3 = 1\ M\Omega$ [**Ans.** (*a*) –7 V (*b*) +3 V]
3. Design an adder circuit using an op-amp to get the output expressions as
   $$V_{out} = -(V_1 + 5\ V_2 + 25\ V_3)$$
   where V_1, V_2 and V_3 are the inputs. Given that $R_f = 50\ k\Omega$. [**Ans** $R_1 = 50\ k\Omega$; $R_2 = 10\ k\Omega$; $R_3 = 2\ k\Omega$]
4. An op-amp has feedback resistor $R_f = 10\ k\Omega$ and the resistances on the input sides are $R_1 = 10\ k\Omega$, $R_2 = 2\ k\Omega$ and $R_3 = 1\ k\Omega$. The corresponding inputs are + 9, – 5 and – 2 V respectively. Noninverting terminal is grounded. Determine the output voltage.[**Ans.** –9 + 25 + 20 = 36 V]
5. Draw the circuit diagram of an op-amp adder circuit to provide $V_{out} = -V_1 + 3\ V_2 - 5\ V_3$.
6. Design a differential amplifier using an op-amp to provide an output voltage $V_{out} = 5\ (V_2 - V_1)$. Use minimum value of resistance as 20 kΩ. [**Ans.** R = 20 kΩ; $R_f = 100\ k\Omega$]
7. A 10 mV, 5 kHz sinusoidal signal is applied to the inverting input terminal of an op-amp integrator for which R = 100 Ω and C = 1 μF. Determine output voltage. [**Ans.** 4.5 cos (10,000 π – 1) μV]
8. Design a practical integrator circuit to properly process input sinusoidal waveform up to 1 kHz. The input amplitude of signal is 10 mV.

❑❑❑

CHAPTER 4

Operational Transconductance Amplifier (OTA)

INSIDE THIS CHAPTER

4.1. INTRODUCTION

The operational amplifiers considered up to this point have been voltage amplifiers. The input signal is a voltage and the output signal is a voltage. Another type of op-amp is an *operational transconductance amplifier (OTA)* which produces an output current proportional to an input voltage. The constant of proportionality is the transconductance of the amplifier. Thus, the OTA is voltage controlled current source (VCCS). There is usually an additional input for a current to control the amplifier's transconductance. Ideally, the OTA has infinite input and output impedances, and the transconductance is assumed to be constant. However, real OTAs have noninfinite input and output impedances, and the transconductance is frequency dependent.

4.2. IC OPERATIONAL TRANSCONDUCTANCE AMPLIFIER

The IC OTAs are specially designed single chip transconductance amplifiers in which the input is voltage and the output is current such that

$$I_{out} = g_m V_d = g_m (V_1 - V_2) \quad ...(4.1)$$

where g_m is the transconductance or gain of the OTA. The unique feature of an OTA is that g_m can be varied over a wide range by means of an external control current I_c. Usually, g_m is proportional to I_c so that

$$g_m = kI_c \quad ...(4.2)$$

where $k = \dfrac{e}{2KT}$ and $\dfrac{KT}{e}$ = 25 mV at 25°C. The typical range of I_c is 0.1 μA to 1 mA.

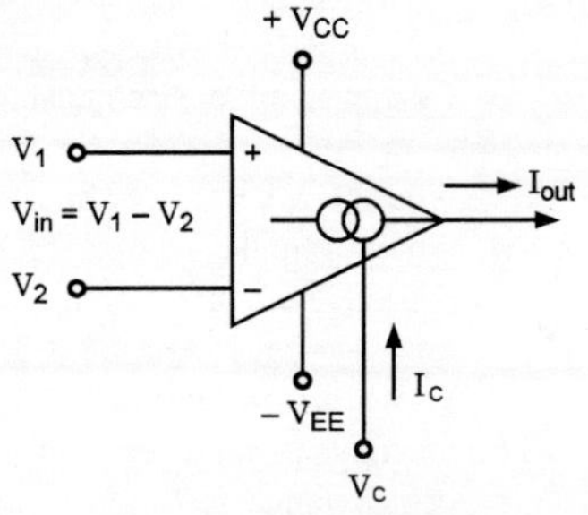

Fig. 4.1. *OTA Symbol*

Normally, OTAs are operated without feedback in linear applications. This is possible because the magnitude of the resistance attached to its output controls its output voltage. The major drawback is that if the differential input signal is not small, it may lead to distortion. Typically input signal remains below 100 mV peak-to-peak.

OTA finds applications in implementing programmable amplifiers and integrators in audio- processing and electronic music synthesis. They are also employed as current switches in sample and hold applications. One of the important use of OTAs using VLSI technique is in neural networks. Popular OTAs are LM 3080, LMI 3600/3700 (National Semiconductor) CA 3080 (RCA) and the NE 5517 (Signetics).

The simplified internal circuit diagram of an OTA is shown in Fig. 4.2.

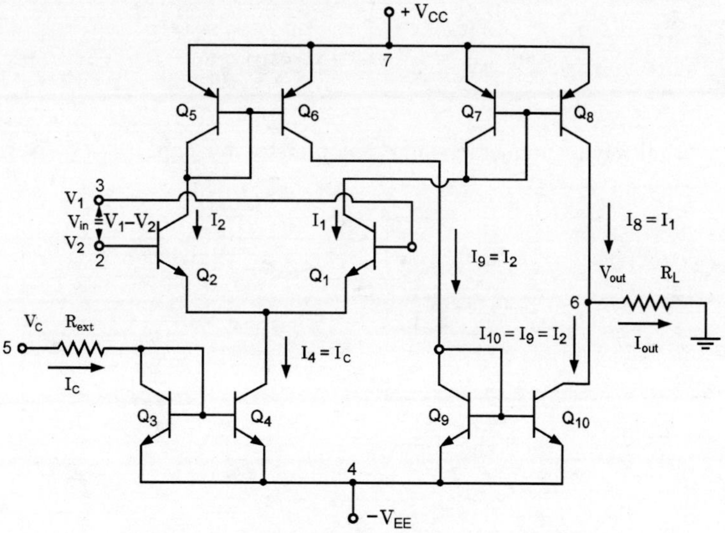

Fig. 4.2. *Simplified Internal Circuit of OTA*

Transistor pairs $Q_1 - Q_2$ form a differential amplifier. Transistors $Q_3 - Q_4$ accept the controlling current I_c whose value is being controlled by an external resistor R_{ext} and controlling voltage V_c. Since transistors $Q_3 - Q_4$ form current mirror,

$$I_4 = I_c \qquad \text{...[4.3 (a)]}$$

Also, transistors $Q_5 - Q_6$ form current mirror, so that the current coming from collector of Q_6 to Q_9 duplicates I_2.

So that, $I_9 = I_2$ ...[4.3 (*b*)]

Again transistors $Q_9 - Q_{10}$ form current mirror pair so that

$$I_{10} = I_9 = I_2 \qquad \text{...(4.4)}$$

The transistors $Q_7 - Q_8$ form current mirror pair and collector current of Q_8 duplicates I_1 so that

$$I_8 = I_1 \qquad \text{...(4.5)}$$

Now, at the output, applying Kirchoff's current law, we have,

$$I_8 = I_{out} + I_{10} \qquad \text{...(4.6)}$$

$$\text{or } I_{out} = I_8 - I_{10} = I_1 - I_2 \qquad \text{...(4.7)}$$

∴ The voltage gain of the amplifier can be expressed as

$$A = \frac{V_{out}}{V_{in}} = \frac{I_{out}\, R_L}{V_{in}}$$

$$= g_m R_L \qquad \text{...(4.8)}$$

∴ The transconductance g_m can be calculated as [behaving like diode]

$$I_1 = I_s\, [\exp\,(V_1/V_T) - 1]$$

$$\text{and } \cong I_s \exp\,(V_1/V_T) \qquad \text{...(4.9)}$$

$$I_2 = I_s\, [\exp\,(V_2/V_T) - 1]$$

$$\cong I_s \exp\,(V_2/V_T) \qquad \text{...(4.10)}$$

where I_s = Reverse saturation current of transistor Q_1 or transistor Q_2 (assuming same characteristics)

and V_T – Volt equivalent of temperature

$$= \frac{kT}{e} = \frac{T}{11{,}600} \qquad \text{(where T is in K)}$$

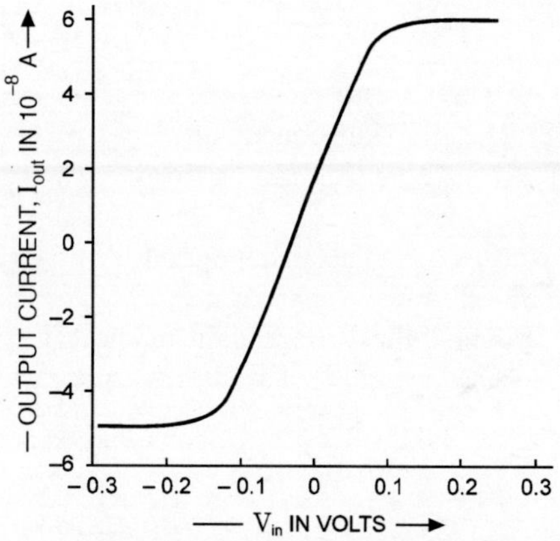

Fig. 4.3. *OTA Transfer Characteristic*

Now the controlling current is given as

$$I_c = I_1 + I_2 = I_s\left[\exp\left(\frac{V_1}{V_T}\right) + \exp\left(\frac{V_2}{V_T}\right)\right]$$

$$\text{or } I_s = \frac{I_c}{\exp\left(\frac{V_1}{V_T}\right) + \exp\left(\frac{V_2}{V_T}\right)}$$

$$\text{and } I_{out} = I_1 - I_2 = I_s\left[\exp\left(\frac{V_1}{V_T}\right) - \exp\left(\frac{V_2}{V_T}\right)\right] = \frac{I_c}{[e^{V_1/V_T} + e^{V_2/V_T}]}[e^{V_1/V_T} - e^{V_2/V_T}]$$

Multiplying numerator and denominator by $\exp\left(-\frac{V_1+V_2}{2V_T}\right)$ we have,

$$I_{out} = I_c \frac{\left[e^{\left(\frac{V_1}{V_T}-\frac{V_1+V_2}{2V_T}\right)} - e^{\left(\frac{V_2}{V_T}-\frac{V_1+V_2}{2V_T}\right)}\right]}{\left[e^{\left(\frac{V_1}{V_T}-\frac{V_1+V_2}{2V_T}\right)} + e^{\left(\frac{V_2}{V_T}-\frac{V_1+V_2}{2V_T}\right)}\right]} = I_c \frac{\left[e^{\left(\frac{V_1-V_2}{2V_T}\right)} - e^{-\left(\frac{V_1-V_2}{2V_T}\right)}\right]}{\left[e^{\left(\frac{V_1-V_2}{2V_T}\right)} + e^{-\left(\frac{V_1-V_2}{2V_T}\right)}\right]}$$

$$I_{out} = I_c \tanh\left(\frac{V_1-V_2}{2V_T}\right) \qquad \text{...(4.11)}$$

The above equation depicts that output current I_{out} is a function of differential voltage V_{in} *i.e.*, $(V_1 - V_2)$. Thus a transconductance amplifier normally computes a tan-hyperbolic. For smaller range of inputs the output current behaves linearly and smoothly shifts to saturation. Transfer characteristic of OTA is given in Fig. 4.3.

Equation (4.11) can be assumed as

$$I_{out} \simeq I_c \frac{V_1 - V_2}{2V_T} \qquad \because \text{ [for small values of } \theta, \tanh\theta \simeq \theta]$$

$$\text{or } \frac{I_{out}}{V_{in}} \simeq \frac{I_c}{2V_T}$$

$$\therefore \quad \text{Transconductance, } g_m = \left|\frac{\partial I_{out}}{\partial V_{in}}\right| = \frac{I_c}{2V_T} \qquad \text{...(4.12)}$$

$$\therefore \quad \text{The voltage gain, } A = g_m R_L = \frac{I_c R_L}{2V_T} \qquad \text{...(4.13)}$$

Thus, the voltage gain of the OTA is controlled externally by a controlling current I_c.

$$\text{At } T = 27\,°C = 300\text{ K}$$

$$V_T = 25\text{ mV}$$

$$\therefore \quad g_m = \frac{I_c}{2V_T} = \frac{I_c}{2 \times 25 \times 10^{-3}\text{ V}} = 20\, I_c \text{ per volt} \qquad \text{...(4.14)}$$

$$\therefore \quad I_{out} = (20I_c)\, V_{in} \qquad ...(4.15)$$

The OTA IC CA 3080 has few limitations as follows:

(*i*) It behaves linearly only when $I_{out} < 400\ \mu A$ and $V_{in} < 20$ mV. Thus input voltage is limited to 20 mV only. (See Eq. 4.15).

(*ii*) The inputs are fed directly to bases of transistors (Q_2 and Q_1) whose impedances are of the order of 10 kΩ to 30 kΩ. Therefore, loading effects may also occur when input signals are applied directly. Hence, op-amp voltage followers should be used to avoid loading effects.

Principal Differences From Standard Op-Amps. There are two main differences between the OTA and the conventional op-amp:

1. An OTA is a voltage controlled current source whereas the conventional op-amp is a voltage controlled voltage source (VCVS). OTA, being a current source, has a very high output impedance, in contrast to the op-amp's very low output impedance. It is very well known that the low output impedance is a desirable feature in general amplifiers used to drive resistive loads.
2. Another difference between OTAs and op-amps is that it is possible to design circuits using the OTA without using negative feedback. Thus, instead of using feedback for reducing the sensitivity of a circuits performance to device parameter, the transconductance g_m is treated as a design parameter in place of resistors and capacitors in op-amps based circuits.

4.3. OTA VOLTAGE AMPLIFIER

A basic OTA voltage amplifier circuit is shown in Fig. 4.4.

At room temprature (T = 300 K), the output current,

$$I_{out} = (20\ I_c)\, V_{in}$$

Now, if the whole output current I_{out} flows into the R_L. This will occur only if $[I_B^+ << I_{out}]$, thus

$$V_{out} = I_{out}\, R_L$$
$$= (20\ I_c)\, V_{in}\, R_L \qquad ...(4.16)$$

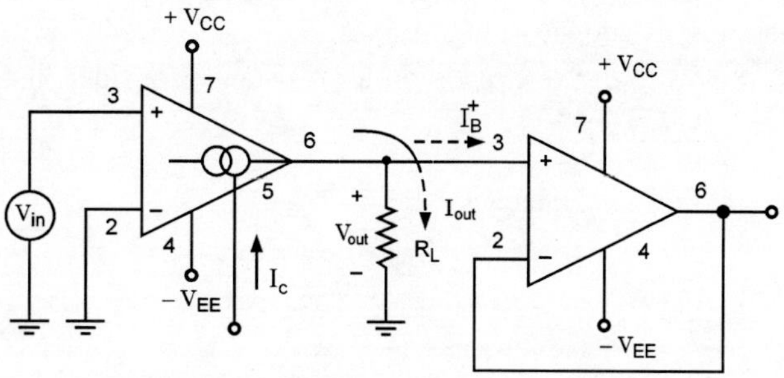

Fig. 4.4. ***Basic Voltage Amplifier Circuit of an OTA***

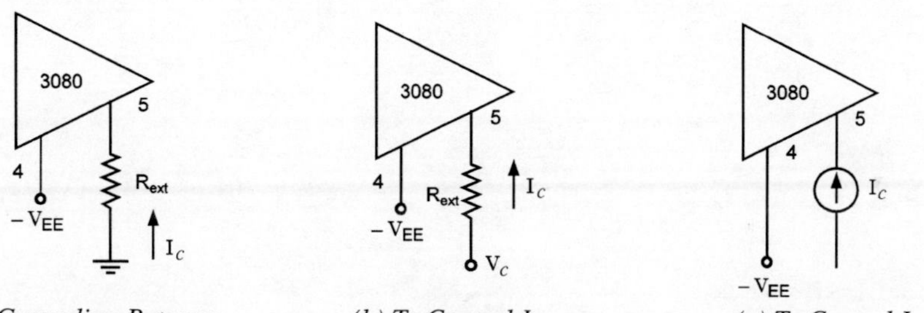

(*a*) *Grounding Between R_{ext} and Pin 5*

(*b*) *To Control I_c Control Voltage V_c is Used*

(*c*) *To Control I_c Current Source is Used*

Fig. 4.5. ***Circuit Arrangements For Setting Value of Control Current I_c***

Thus, the R_L also contributes to voltage gain. And to have $I_B^+ << I_{out}$, we select an op-amp voltage follower to avoid loading effects.

The controlling current I_c also alters the output, so we connect an external resistance R_{ext} (between pair 5 and ground, as shown in Fig. 4.5 (*a*) so that

$$I_c = \frac{|-V_{EE}| - V_{BE}(Q_3)}{R_{ext}} \quad \text{...(4.17)}$$

One approach, as shown in Fig. 4.5 (*b*), to set controlling current I_c is using a control voltage V_c, so that

$$I_c = \frac{V_c + |-V_{EE}| - V_{BE(Q_3)}}{R_{ext}} \quad \text{...(4.18)}$$

$V_{BE(Q_3)} = 0.6$ V = Forward voltage drop of the diode connected transistor Q_3 in Fig. 4.2.

Another approach for setting controlling current I_c is by using a current source, as shown in Fig. 4.5 (*c*). (One may employ FET, BJT, an op-amp, an I_c current chip or another 3080 as current source).

4.4. OTA AS A PROGRAMMABLE RESISTOR

We may employ I_c OTA as a programmable resistor whose resistance is controlled by controlling current I_c.

We can define resistance, using Eq. (4.16)

$$R = \frac{V_{in}}{I_{out}} = \frac{1}{20 I_C} \quad \text{...(4.19)}$$

Two OTAs can be employed to behave as programmable resistor, as shown in Fig. 4.6.

Now to derive expression for resistance consider first the OTA IC OT_2.

$$\text{Here, } I_{c_2} = \frac{|-V_{EE}| - 0.6\text{V}}{R_{ext}}$$

$$\simeq \frac{|-V_{EE}|}{R_{ext}} \quad \text{...(4.20)}$$

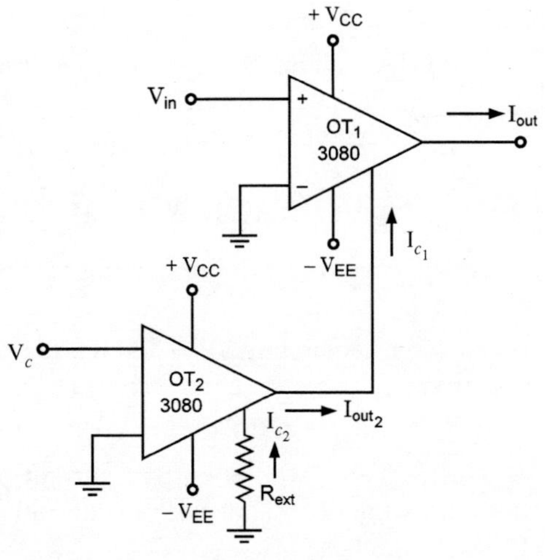

Fig. 4.6

$$\therefore \quad I_{out_2} = (20)\, I_{c_2} V_c = 20 \cdot \frac{|-V_{EE}|}{R_{ext}} \cdot V_c \quad \text{...(4.21)}$$

For the other IC 3080 OT_1 say,

$$I_{c_1} = I_{out_2}$$

$$\text{and } I_{out} = 20 I_{c_1} V_{in} = (20)^2 \left[\frac{|-V_{EE}|}{R_{ext}} \cdot V_c\right] V_{in} = \left[(20)^2 \cdot \frac{|-V_{EE}|}{R_{ext}}\right] V_c \cdot V_{in} \quad \text{...(4.22)}$$

$$\therefore \quad R = \frac{V_{in}}{I_{out}} = \frac{R_{ext}}{(400)\,|-V_{EE}|\,V_c} \quad \text{...(4.23)}$$

Here V_{in} must be less than 20 mV and $I_0 < 400$ μA.

4.5. OTA CIRCUIT FOR REALIZATION OF A GROUNDED RESISTOR

Figure 4.7 shows an OTA circuit that realizes a grounded resistor with a value that can be controlled by current I_c. The input impedance is

$$\frac{V_{in}}{I_{in}} = Z_{in} = \frac{1}{g_m + Y_{in} + Y_{out}} \qquad ...(4.24)$$

$$\text{where } Y_{in} = j\omega C_{in} + \frac{1}{R_{in}}$$

$$\text{and } Y_{out} = j\omega C_{out} + \frac{R_{in}}{R_{out}}$$

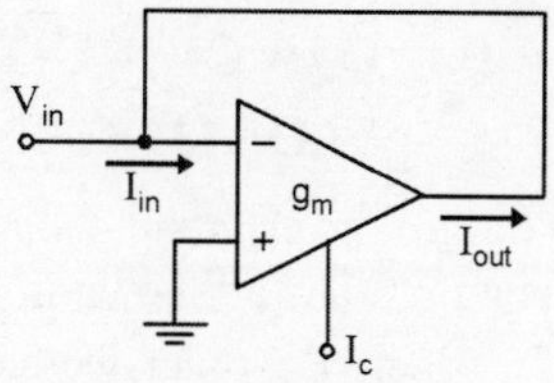

Fig. 4.7. *OTA Realization of Grounded Resistor*

For frequencies low enough and parasitic resistances large enough, $Z_{IN} \simeq \frac{1}{g_m}$.

If the input leads of the OTA are interchanged, the circuit yields a short circuit stable negative resistor with $Z_{IN} \simeq -\frac{1}{g_m}$.

4.6. OTA CIRCUIT FOR REALIZATION OF A FLOATING RESISTOR

Figure 4.8 depicts a circuit that can realize a floating resistor. Using the ideal model and solving for I_1 and I_2, we have

$$I_1 = (V_1 - V_2) \times g_{m_1}$$

$$\text{If } g_{m_1} = g_{m_2} \text{ then } I_2 = -I_1$$

and the circuit realizes a floating resistor of value $\frac{1}{g_m}$.

The two circuits shown in Figs. 4.7 and 4.8 permits the construction of active filters without the use of resistors. Thus, transconductance C filters can be realized and such filters are attractive for monolithic I_c realizations.

For active filters, adders (summers) and integrators are also requried which are discussed in the forthcoming sections.

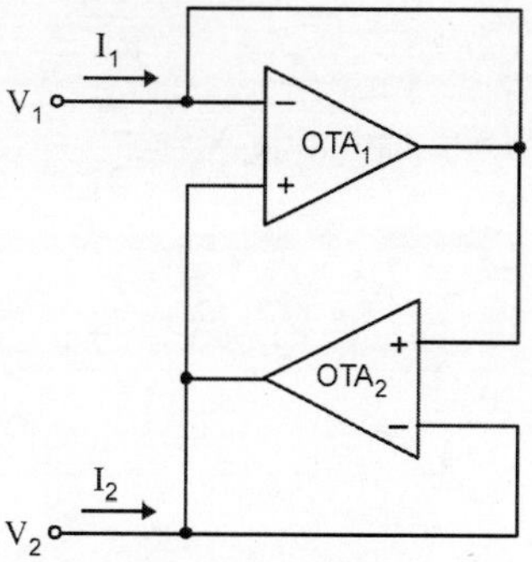

Fig. 4.8. *OTA Realization of Floating Resistor*

4.7. OTA ADDER (SUMMER)

OTA adder is shown in Fig. 4.9 for which the output voltage is given as

$$V_{out} = \frac{V_1 g_{m_1}}{G} + \frac{V_2 g_{m_2}}{G}$$

where G = 1/R. This circuit is versatile because the resistor R can be replaced by OTA ground resistor circuit shown in Fig. 4.7 to obtain an all-OTA circuit.

More input voltages can be accommodated simply by adding more OTAs at the input, and interchanging the input leads of one of the OTAs at the input side of the adder changes the sign of the corresponding term in the output.

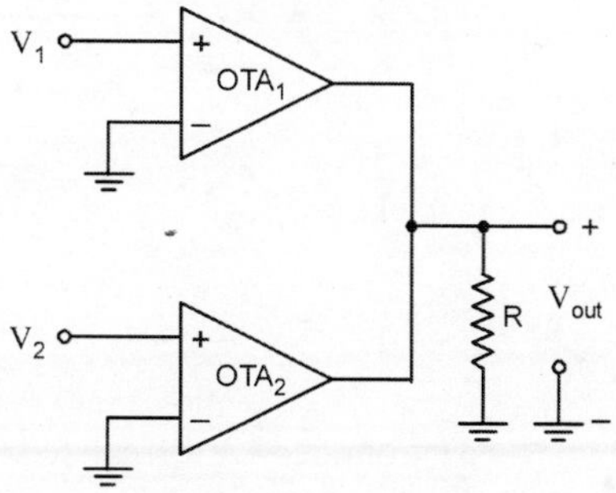

Fig. 4.9. *OTA Adder*

4.8. OTA LOSSY INTEGRATOR

A lossy differential input OTA integrator is shown in Fig. 4.10 whose output voltage is given by the expression

$$V_{out} = (V_1 - V_2) \times \frac{g_m R}{1 + j\omega CR} \qquad ...(4.25)$$

and, again, the resistor R can be replaced by the circuit shown in Fig. 4.7. Removal of resistor results in a lossless integrator.

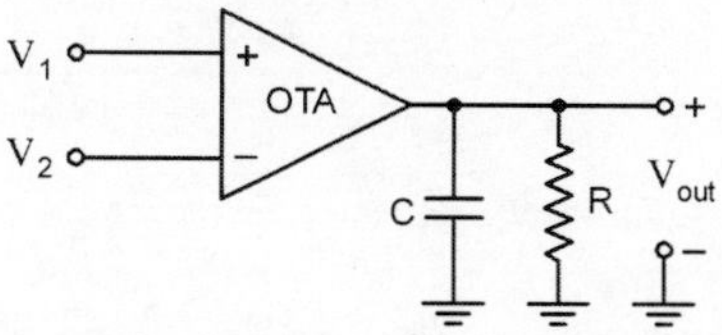

Fig. 4.10. *OTA Lossy Integrator*

4.9. OTA CIRCUITS FOR REALIZATION OF GROUNDED INDUCTOR AND FLOATING INDUCTOR

The grounded inductor can be realized by using 2 OTAs as shown in Fig. 4.11.

Using Eq. (4.1), we have

$$I_{out_1} = g_{m_1} V_1 \quad ...(4.26)$$

and $$I_{out_2} = -g_{m_2} V_c \quad ...(4.27)$$

Also we have

$$I_1 = -I_{out_2} = g_{m_2} V_c \quad ...(4.28)$$

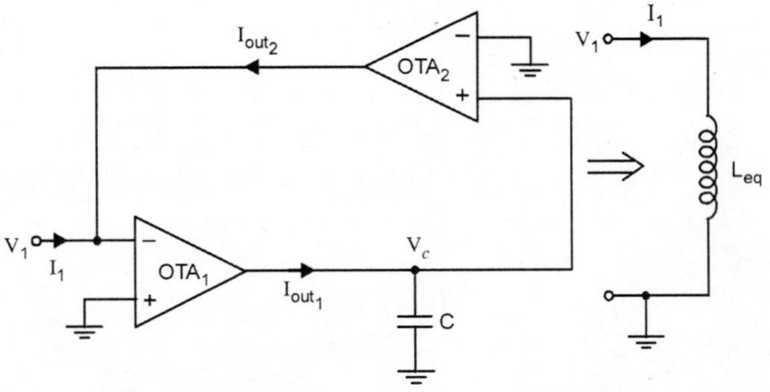

Fig. 4.11. ***Realization of Grounded Inductor***

where V_c is the voltage across capacitor and is given as

$$V_c = \frac{I_{out_1}}{j\omega C}$$

Substituting $V_c = \dfrac{I_1}{g_{m_2}}$ from Eq. (4.28), we have

$$I_1 = \frac{g_{m_2} I_{out_1}}{j\omega C}$$

or $$I_{out_1} = \frac{j\omega C I_1}{g_{m_2}} \quad ...(4.29)$$

From Eqs. (4.26) and (4.29), we have

$$g_{m_1} V_1 = I_{out_1} = \frac{j\omega C I_1}{g_{m_2}}$$

or $$\frac{V_1}{I_1} = Z_{IN} = \frac{j\omega C}{g_{m_1} g_{m_2}} = j\omega L_{eq}$$

or $$L_{eq} = \frac{C}{g_{m_1} g_{m_2}} \quad ...(4.30)$$

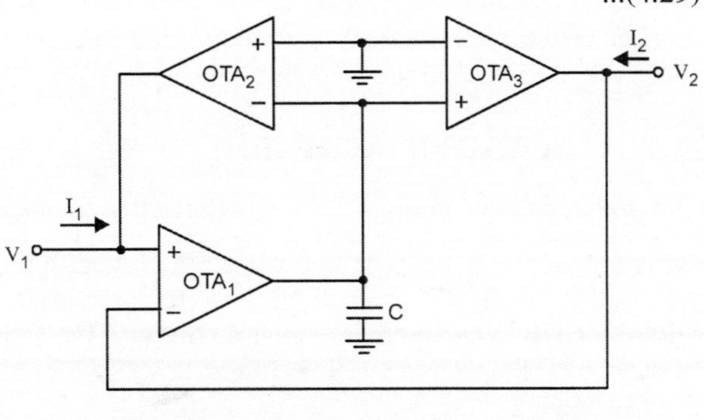

Fig. 4.12. ***Realization of Floating Inductor***

Hence grounded L_{eq} is realized.

The floating inductor can be realized by using 3 OTAs as shown in Fig. 4.12.

From circuit shown in Fig. 4.12.

$$I_1 = \frac{g_{m_1} g_{m_2} (V_2 - V_1)}{j\omega C} \quad ...(4.31)$$

and $$I_2 = \frac{-g_{m_1} g_{m_3} (V_1 - V_2)}{j\omega C} \quad ...(4.32)$$

if $g_{m_2} = g_{m_3} = g_m$, then

$I_2 = -I_1$, and the floating inductor realized ideally is

$$L = \frac{C}{g_m g_{m_1}} \qquad ...(4.33)$$

However, the parasitic resistors and capacitors can seriously influence the floating inductor realized. Unwanted resistance and capacitance will exist across both ports, the inductor itself will be lossy, the value of inductor will be changed from its ideal value, and a current source will be included in the overall equivalent circuit of the inductor. Whether these departures from ideal behaviour will prevent the OTA realization from being utilized depends on the particular application.

4.10. OTA IN SAMPLE AND HOLD CIRCUITS

The OTA can be used in sample and hold circuits, as shown in Fig. 4.13.

To sample and hold the input signal the control terminal is biased on (= + V_{CC}) and off (= – V_{EE}), so that the holding capacitor C_H can hold the signal.

When the circuit is sampling the input signal, the control voltage V_c is high (= + V_{CC}) and the OTA output charges the holding capacitor C_H up to value V_{in}. When the sampled signal is required to hold (*i.e.*, V_c = – V_{EE}) then I_c reduces to zero, and the OTA output virtually behaves as an open circuit so that the holding capacitor holds the sampled input voltage.

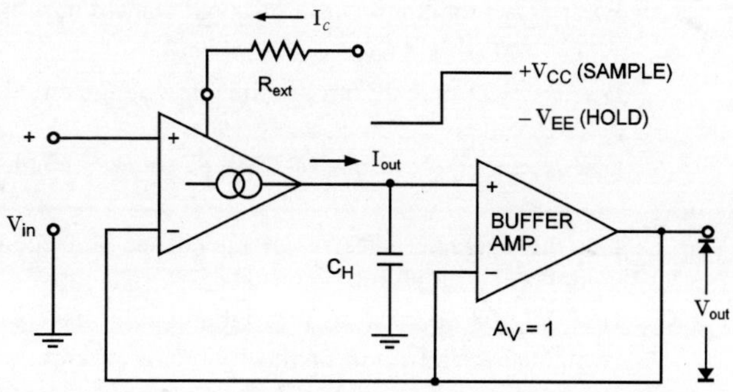

Fig. 4.13. ***OTA Applications in Sample and Hold Circuit***

How long the holding capacitor stores (holds) the voltage depends upon the output impedance of OTA and input impedance of buffer amplifier.

4.11. LIMITATIONS OF OTA

1. The restriction in behaving linearly when $I_{out} < 400\ \mu A$ and $V_{in} < 20$ mV.
2. Controlling current I_c is always required to be varied to adjust gain or resistance and this also affects the other parameters of the amplifier such as offset voltage, input bias current and slew rate etc. To reduce such effects, the input and output need to be bufferred, and frequency compensating techniques should be used.
3. The OTA also shows temperature dependency [Eq. (4.11)] has a volt equivalent term

$$V_T = \frac{kT}{e} = \frac{T}{11{,}600} = 25 \text{ mV at } T = 300 \text{ K.}$$

∴ Proportionality constant 20 per volt is valid only at room temperature Thus, the performance of OTA is inversely proportional to temperature.

4. Effect of loading of Q_2 and Q_1 has already been discussed.

HIGHLIGHTS ....

1. An OTA is a voltage controlled current source (VCCS), more specifically the term "operational" comes from the fact that it takes the difference of two voltages as the input for current conversion. The ideal transfer characteristic is

$$I_{out} = g_m(V_1 - V_2) = g_m V_d$$

With the ideally constant transconductance g_m as the proportionality factor between the two. In reality the transcondutance is also a function of the input differential voltage and dependent on temperature.

Popular OTAs and LM3080, LMI 3600/3700 (National Semiconductor, CA 3080 (RCA) and the NE 5517 (Signetics).

OTA finds applications in implementing programmable amplifiers and integrators in audio-processing and electronic music synthesis. They are also used as current switches in sample and hold applications.

EXERCISES ....

1. Draw and explain the internal diagram of an OTA and explain its main features.
[G.G.S.I.P. Univ. Analog Electronics-II May 2011]
2. Explain the difference between OTA and conventional type op-amp.
3. What is a programmable transconductance amplifier ? Explain its basic principle.
4. What are OTA's ? Give one application.
5. Discuss electronically programmable voltage amplifier using OTA. How controlling current controls the output voltage ?
6. Discuss the applications of OTA as programmable voltage amplifiers.
[G.G.S.I.P. Univ. Analog Electronics-II, May 2011]
7. List the characteristics of an operational transconductance amplifier (OTA). Draw the inverting and noninverting amplifiers using OTA.
8. How IC 3080 is used as a programmable resistor ?
9. Explain the realization of the following circuits using OTAs :
   (*i*) Grounded resistor. (*ii*) Floating resistor.
   (*iii*) Grounded inductor. (*iv*) Floating inductor.
10. Explain with circuit diagram : (*i*) OTA adder, (*ii*) OTA integrator.
11. Explain using circuit diagram how an OTA can be used in sample and hold applications.
12. Draw and explain operation of sample and hold circuit using OTA.
13. Discuss the limitations of OTA.
14. Write short notes on the following:
   (*i*) OTA and its applications.
   (*ii*) Limitations of OTA.

❑❑❑

Unit-II

WAVEFORM GENERATION

Contents of This Unit

WAVEFORM GENERATION : Sinewave Generator (Phase Shift, Wein Bridge, Hartley and Colpitts), Ramp and Sawtooth Generators; Linearity of Waveform, Astable Multivibrator, OTA-C Oscillators, Crystal Oscillator, Voltage Controlled Oscillators.

CHAPTER 5

Transistor Oscillators (Sinusoidal)

5.1. INTRODUCTION

INSIDE THIS CHAPTER

An oscillator is the basic element of all ac signal sources and generates sinusoidal signals of known frequency and amplitude. It is one of the basic and useful instruments used in electrical and electronic measurements. For example, an oscillator finds wide applications in electronic communication equipment. In AM (amplitude modulation) and FM (frequency modulation) superheterodyne receivers, "local" oscillators are used to assist in the reduction of the incoming radio frequency (RF) to a lower intermediate frequency (IF). Oscillator circuits are also employed in the "exciter" section of a transmitter to generate the RF carrier. Other applications include their use as "clocks" in digital systems such as microcomputers, in the sweep circuits found in TV sets and oscilloscopes.

Since sinusoidal waveforms are encountered so frequently in electronic measurement work, the oscillator (sinewave generator) represents the largest single category of signal generators. This device covers the frequency range from a few Hz to many GHz.

Although we speak of an oscillator as "generating" a sinusoidal signal, it is to be noted that it does not create energy, but merely acts as an *energy converter*. It simply converts unidirectional current drawn from a dc source of supply into alternating current of desired frequency. The function of an oscillator is reverse of that of a rectifier and, therefore, sometimes called *inverter*. However, we generally think of oscillator circuits as providing an ac *voltage signal*.

It is noteworthy here that although an alternator (ac generator) generates sinusoidal ac power of 50 Hz, it cannot be called an oscillator. An alternator is a mechanical device that has rotating parts, converts mechanical energy into ac energy but cannot produce ac energy of high frequency* whereas an

* An alternator cannot be employed for generating frequencies exceeding 1,000 Hz because for generating higher frequencies, either the speed of rotation of the armature has to be made extremely high or the number of poles has to be made large and both options are impracticable.

oscillator is a non-rotating electronic device that converts dc energy into ac energy of frequency ranging from a few Hz to many GHz.

Though alternator generates large amount of ac power but for several applications such as radio transmitters and receivers, radars etc., an oscillator is preferred owing to its numerous advantages over alternators for such applications. These advantages are :

(*i*) Portable and cheap in cost.

(*ii*) An oscillator is a non-rotating device. Consequently, there is no wear and tear and hence longer life.

(*iii*) Frequency of oscillation may be conveniently varied.

(*iv*) Voltage or currents of any frequency (20 Hz to extremely high frequencies) adjustable over a wide range can be generated.

(*v*) Frequency once set remains constant for a considerable period of time.

(*vi*) Voltages free from harmonics as well as rich with harmonics can be generated by sinusoidal oscillators and relaxation oscillators respectively.

(*vii*) Operation of an oscillator is silent, as there is no moving part in it.

(*viii*) High operation efficiency—due to absence of moving part, there is no wastage of energy owing to friction.

For having a sinusoidal oscillator, we require an amplifier with positive feedback.* The idea is to use the feedback signal in place of an input signal. If the loop gain and phase are correct, there will be an output signal even though there is no external input signal. In other words, an oscillator is an amplifier that is modified by positive feedback to supply its own input signal.

The difference between an amplifier and an oscillator can be explained with the help of Fig. 5.1.

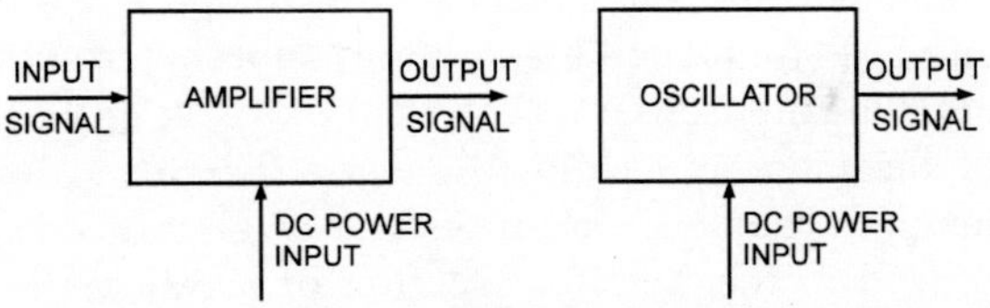

Fig. 5.1

As we know that, an amplifier strengthens the input signal without any change in its waveform and frequency. The additional power required comes from the external dc source. Thus an amplifier is essentially an *energy converter* that draws energy from a dc supply and converts it into ac energy at signal frequency, the energy conversion process being controlled by the input signal. On the other hand an oscillator does not require any external signal either to start or maintain the process of energy conversion and the energy conversion process is controlled by the oscillator circuit itself, as illustrated in Fig. 5.1. In an oscillator, the output signal frequency depends on the passive components employed in the circuit and can be varied as per needs. Oscillator may provide fixed or variable frequency.

When employed as an instrument, an oscillator may be called with different names such as *test oscillator* (or *signal generator*), standard signal generator (or function generator) depending on the function performed by the device, type of signal produced by it, order of sophistication etc.

Oscillators may be broadly divided into two categories viz. *harmonic* oscillators and *relaxation* oscillators. Both types can include active devices such as BJTs, FETs and op-amps, and passive components such as resistors, inductors and capacitors. In harmonic oscillators, the energy always flows in one direction—from the active to passive components. However, in relaxation oscillators, the energy is exchanged between the

* If the amplifier output signal increases in magnitude after applying the feedback, then it is called the positive feedback. The positive feedback is cumulative. The amplified signal when fed back and passed through the amplifier becomes even stronger. Theoretically signal approaches infinity but in practice a condition is reached when the amplifier gives an output signal without any need of input signal.

active and passive components. In harmonic oscillators, the frequency of oscillations is determined by the feedback path. However, in relaxation oscillators, the frequency is determined by time constants–specifically, the charge and discharge time constants during the exchange of energy. Harmonic oscillators can develop low-distortion sinusoidal output waveforms, but relaxation oscillators can only generate nonsinusoidal waveforms such as sawtooth, square or triangular. Harmonic oscillators will be discussed here.

Oscillators can also be categorised on the basis of design principle used, and the frequency range over which they are employed.

The oscillators, according to the design principle used are feedback oscillators and negative resistance oscillators. *Feedback oscillators* are widely employed.

The oscillators, according to operating frequency, may be classified as follows :

Type of Oscillators	*Approximate Range*
1. Audio frequency (AF) oscillators	20 Hz–20 kHz
2. Radio frequency (RF) oscillators	20 kHz–30 MHz
3. Very low frequency (VLF) oscillators	15–100 kHz
4. Low frequency (LF) oscillators	100–500 kHz
5. Broadcast oscillators	500 kHz–1.5 MHz
6. Video frequency oscillators	0–5 MHz
7. High frequency (HF) oscillators	1.5–30 MHz
8. Very high frequency (VHF) oscillators	30–300 MHz
9. Ultra high frequency (UHF) oscillators	300–3,000 MHz
10. Microwave oscillators	Beyond 3 GHz (3,000 MHz)

5.2. OPERATION OF OSCILLATOR

The use of positive feedback that results in a feedback amplifier having closed-loop gain $|A_f|$ exceeding unity and satisfies the phase conditions results in operation as an oscillator circuit. An oscillator circuit then provides a constantly varying output signal. If the output signal varies sinusoidally, the circuit is referred to as a *sinusoidal oscillator* and on the other hand if the output voltage rises quickly to one voltage level and later drops quickly to another voltage level, the circuit is usually referred to as a pulse or *square-wave generator*.

To understand how an oscillator produces an output signal without an external input signal, let us consider the feedback circuit shown in Fig. 5.2 (*a*), where V_{in} is the voltage of ac input driving the input terminals *bc* of an amplifier having voltage gain A. The amplified output voltage is

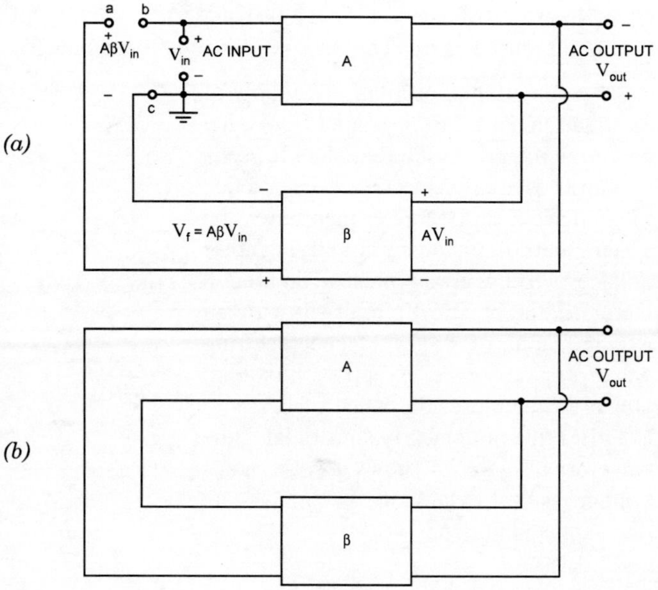

Fig. 5.2

$$V_{out} = AV_{in} \quad ...(5.1)$$

This voltage drives a feedback circuit that is usually a *resonant circuit*, as we get maximum feedback at one frequency. The feedback voltage returning to point *a* is given by equation

$$V_f = A\beta V_{in} \quad ...(5.2)$$

where β is the gain of feedback network.

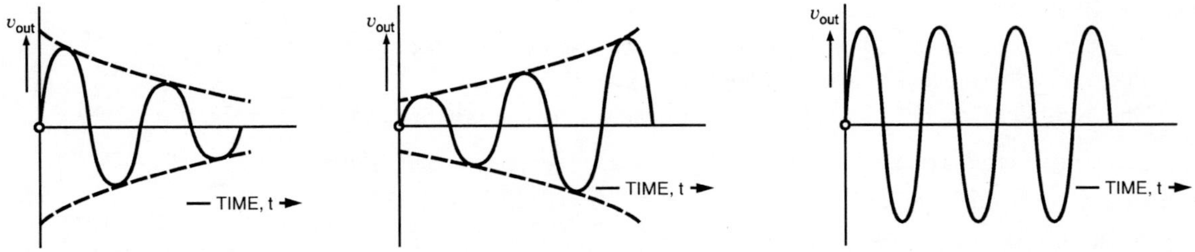

(a) Oscillations Die out *(b) Oscillations Buildup* *(c) Oscillations of Constant Amplitude*

Fig. 5.3

If the phase shift through the amplifier and feedback circuit is zero, then $A\beta V_{in}$ is in phase with the input signal V_{in} that drives the input terminals of the amplifier.

Now we connect point '*a*' to point '*b*' and simultaneously remove voltage source V_{in}, then feedback voltage $A\beta V_{in}$ drives the input terminals *bc* of the amplifier, as shown in Fig. 5.2 (*b*). In case Aβ is less than unity, $A\beta V_{in}$ is less than V_{in} and the output signal will die out, as illustrated in Fig. 5.3 (*a*). On the other hand if Aβ is greater than unity, the output signal will build up, as illustrated in Fig. 5.3 (*b*). If Aβ is equal to unity, $A\beta V_{in}$ equals V_{in} and the output signal is a steady sinewave, as illustrated in Fig. 5.3 (*c*). In this case the circuit supplies its own input signal and produces a sinusoidal output.

Certain conditions are required to be fulfilled for sustained oscillations and these conditions are that (*i*) the loop gain of the circuit must be equal to (or slightly greater than) unity and (*ii*) the phase shift around the circuit must be zero. These two conditions for sustained oscillations are called *Barkhausen criteria.*

For initiation of oscillations, supply of an input signal is not essential. Only the condition βA = 1 must be satisfied for self-sustained oscillations to result. In practice βA is made slightly greater than unity, the system starts oscillating by amplifying noise voltage* which is always present. Saturation factors in the practical circuits *provide* an average value of βA of 1. The resulting waveforms are never exactly sinusoidal. However, the closer the value of βA is to exactly 1, the more nearly sinusoidal is the waveform. Figure 5.4 shows how the noise voltage results in a buildup of a steady-state oscillation condition.

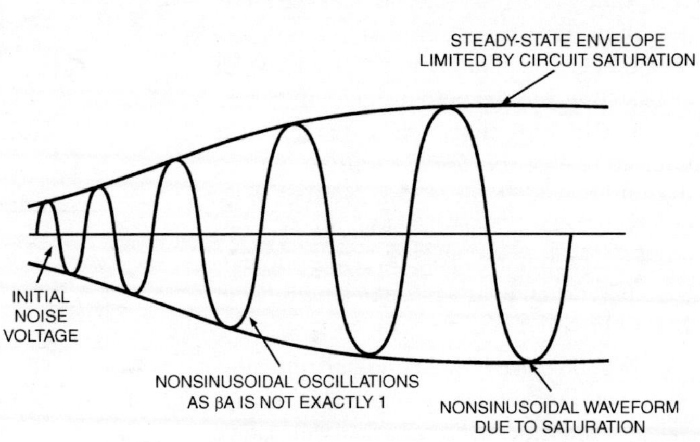

Fig. 5.4. ***Buildup of Steady-state Oscillations***

* Because of the ambient temperature, the free electrons contained by resistors move randomly in different directions and generate a *noise voltage* across the resistor. The motion is so random that it contains frequencies to over 1,000 GHz. Each resistor may, thus be thought of as a small ac voltage source producing all frequencies.

Another way of seeing how the feedback circuit provides operation as an oscillator is obtained by noting the denominator in the basic feedback equation, $A_f = \dfrac{A}{1 + \beta A}$. When $\beta A = -1$ or magnitude 1 at a phase angle of 180°, the denominator becomes zero and the gain with feedback, A_f, becomes infinite. Thus, an infinitesimal signal (noise voltage) can provide a measurable output voltage, and the circuit acts as an oscillator even without an input signal.

By deliberate design the phase shift around the loop is made 0° at the resonant frequency. Above and below the resonant frequency, the phase shift is different from 0°. Thus, oscillations are obtained at only one frequency, the resonant frequency of the feedback circuit.

To understand and apply the Barkhausen criterion, we must consider both the gain and phase shift of $A\beta$ as a function of frequency. Reactive elements, capacitance in particular, contained in the amplifier and/or feedback causes the gain magnitude and phase shift to vary with the frequency. In general, there will be only one frequency at which the gain magnitude is unity and at which, simultaneously, the total phase shift is equivalent to 0° (in phase—a multiple of 360°). The *system will oscillate at the frequency that satisfies those conditions.* Designing an oscillator amounts to selection of reactive components and their incorporation into circuit in such a way that the conditions will be satisfied at a desired frequency.

To show the dependence of the loop gain $A\beta$ on frequency, we express $A\beta(j\omega)$, a complex phasor that can be expressed in both polar and rectangular form :

$$A\beta(j\omega) = |A\beta|\angle\theta = |A\beta|\cos\theta + j|A\beta|\sin\theta \qquad ...(5.3)$$

where $|A\beta|$ is the magnitude of the loop gain, a function of frequency, and θ is the phase shift, also a function of frequency. For satisfaction of Barkhausen criterion, we must have

$$|A\beta| = 1 \qquad ...(5.4)$$

$$\text{and } \theta = \pm 360^\circ\, n \qquad ...(5.5)$$

where n is any integer, including 0. In polar and rectangular forms, the **Barkhausen criterion** is expressed as

$$A\beta(j\omega) = 1 \angle \pm 360^\circ n = 1 + j0 \qquad ...(5.6)$$

5.3. ESSENTIALS OF AN OSCILLATOR CIRCUIT

From the above discussion it can be inferred that an oscillator circuit must have the following three elements :

1. Oscillatory circuit or frequency determining element.
2. Amplifier.
3. Positive feedback network.

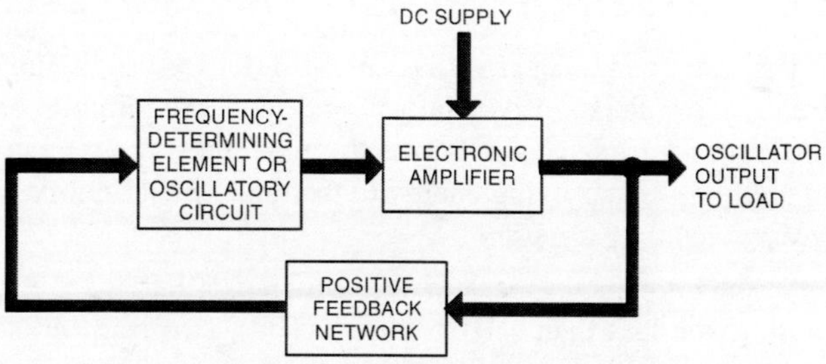

Fig. 5.5. ***Block Diagram of An Oscillator***

The oscillatory circuit or a frequency-determining element can be an inductance-capacitance network (L-C tank), resistance-capacitance (R-C network), or a quartz crystal, depending on the frequency and waveshape desired.

The electronic amplifier receives dc power from the battery or dc power supply and converts it into ac power for supply to the tank circuit. The oscillations occurring in the tank circuit are applied to the input of the electronic amplifier. Because of the amplifying properties of the amplifier, we get increased output of these oscillations. This amplified output of oscillations is because of dc power supplied from the external source (a battery or power supply). The output of the amplifier can be supplied to the tank circuit to meet the losses.

The *feedback network* supplies a part of the output power to the frequency-determining element in correct phase to aid the oscillations. In other words, feedback circuit must be a *positive feedback network.* The block diagram of an oscillator is shown in Fig. 5.5.

5.4. FREQUENCY STABILITY OF OSCILLATOR

An oscillator having initially been set at a particular frequency does not maintain its initial frequency, but instead drift and wander about in frequency, sometimes uniformly in one direction, sometimes quite erratically. The *frequency stability* of an oscillator is a measure of its ability to maintain as nearly a fixed frequency as possible over as long a time interval as possible. These deviations in frequency are caused due to variations in the values of circuit features (circuit components, transistor parameters, supply voltage, stray capacitances, output load etc.) that determine the oscillator frequency.

As the circuit operates, it warms up. Consequently the values of resistors, inductors and capacitors, which are frequency determining factors in such circuits, are changed with temperature. Variations in temperature also cause variations in transistor parameters. However, the variation or drift in oscillator frequency is slow because of slow variations in R, L, C and transistor parameters with time.

The other major factor responsible for deviation in frequency is *variations in power supply* (operating voltage applied to the active device). However, this problem can be overcome by using regulated power supply.

Any variation in load coupled to the tank circuit may cause a change in effective resistance of the circuit by transformer action which, in turn, causes the drift in frequency.

Other factors responsible for drift in oscillator frequency are :

Inter-element capacitances, operating point of active device and mechanical vibrations. The effects of *variations in inter-element capacitances* can be neutralised by introducing a swamping capacitor across the offending elements (the introduced capacitor becomes the part of the tank circuit). The *operating point Q of the active device* is so chosen as to restrict the operation of the device on the linear portion of its characteristic because operation on nonlinear portion causes the variations in transistor parameters which, in turn, affect the oscillator frequency stability. The instability in oscillator frequency caused due to *mechanical vibrations* can be avoided by isolating the oscillator circuit from the source of mechanical vibrations.

5.5. TYPES OF TRANSISTOR OSCILLATORS

A transistor can be operated as an oscillator for producing continuous undamped oscillations of any desired frequency if tank (or oscillatory) and feedback circuits are properly connected to it. All oscillators under different names have similar function *i.e.,* they generate continuous undamped output. However, they differ in methods of supplying energy to the tank or oscillatory circuit to meet the losses and the frequency ranges over which they are used.

The frequency spectrum over which oscillators are employed to produce sinusoidal signals is extremely wide (from less than 1 Hz to many GHz). However, no single oscillator design is practical for generating signals over this entire range. Instead, a variety of designs are employed, each of which generates sinusoidal outputs most advantageously over various portions of the frequency spectrum.

Oscillators, which use *inductance-capacitance* (L-C) circuits as their tank or oscillatory circuits, are very popular for generating high frequency (*e.g.,* 10 kHz to 100 MHz) outputs. The most widely used LC oscillators are the *Hartley* and *Colpitt's oscillators.* Although they slightly differ from one another in their electronic circuitry but they have virtually identical frequency ranges and frequency-stability characteristics. However, such oscillators are not suitable for generating low-frequency sinusoidal outputs. This is due to the fact that the components required in construction of low-frequency L-C resonant circuits are too bulky and heavy. So *resistor-capacitor* (R-C) *oscillators* are generally employed for generating low-frequency (from 1 Hz to about

1 MHz) sinusoidal signals. Two most common R-C oscillators are the *Wien bridge* and *phase shift types.*

Other less frequently used oscillators are the *crystal oscillators* and the *negative resistance oscillators.* The operating frequency ranges of various types of most commonly used oscillators are given below :

Type of Oscillator	*Approximate Frequency Ranges*
Wien bridge oscillator	1 Hz — 1 MHz
Phase shift oscillator	1 Hz — 10 MHz
Hartley oscillator	10 kHz — 100 MHz
Colpitt's oscillator	10 kHz — 100 MHz
Negative resistance oscillator	> 100 MHz
Crystal oscillator	Fixed frequency

The oscillators of various types will be discussed in the following sections.

5.6. L-C OSCILLATORS

Oscillators, which use inductance-capacitance (L-C) circuits as their tank or oscillatory circuits are called L-C *oscillators.* L-C oscillators are very popular for generating high-frequency outputs (*e.g.,* 10 kHz to 100 MHz). There is a large variety of L-C oscillators such as tuned collector oscillators, tuned-base oscillators, Colpitt's oscillators, Hartley oscillators, Clapp oscillators, crystal oscillators etc.

5.7. GENERAL FORM OF AN L-C OSCILLATOR

In the general form of an oscillator depicted in Fig. 5.6 (*a*), any of the active devices such as vacuum tube, BJT, FET or op-amp may be used in the amplifier section, Z_1, Z_2 and Z_3 are the reactive elements constituting the feedback tank circuit which determines the frequency of oscillation. Here, Z_1 and Z_2 serve as an ac voltage divider for the output voltage and feedback signal. Thus, the voltage across Z_2 is the feedback signal. The equivalent circuit is drawn in Fig. 5.6 (*b*) with the following two assumptions :

(*i*) h_{re} of transistor is negligibly small and, therefore, the feedback source $h_{re}V_{out}$ is negligible.

(*ii*) h_{oe} of the transistor is very small *i.e.,* the output resistance $\frac{1}{h_{oe}}$ is very large and, therefore, $\frac{1}{h_{oe}}$ is omitted from the equivalent circuit.

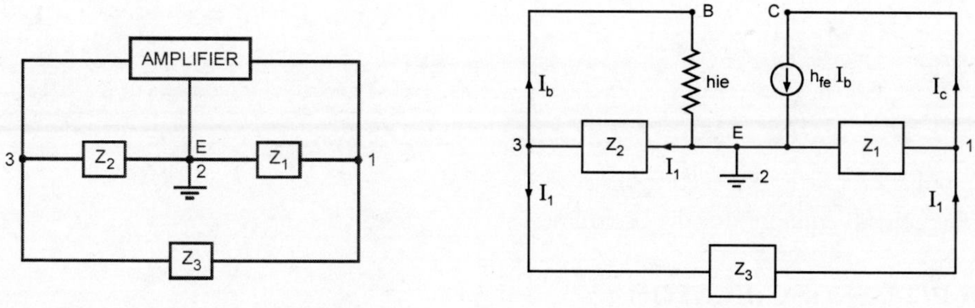

(*a*) *General Form of an Oscillator* (*b*) *Equivalent Circuit*

Fig. 5.6

Let us determine the load impedance between output terminals 1 and 2. Here Z_2 and h_{ie} are in parallel and their resultant impedance is in series with impedance Z_3.

The equivalent impedance is in parallel with impedance Z_1.

Thus load impedance between output terminals is given as

$$\mathbf{Z}_L = \mathbf{Z}_1 \parallel [\mathbf{Z}_3 + (\mathbf{Z}_2 \parallel h_{ie})] = \mathbf{Z}_1 \parallel \left[\mathbf{Z}_3 + \frac{\mathbf{Z}_2\, h_{ie}}{\mathbf{Z}_2 + h_{ie}}\right]$$

$$= \mathbf{Z}_1 \parallel \left[\frac{\mathbf{Z}_3 (\mathbf{Z}_2 + h_{ie}) + \mathbf{Z}_2\, h_{ie}}{\mathbf{Z}_2 + h_{ie}}\right] = \mathbf{Z}_1 \parallel \left[\frac{h_{ie} (\mathbf{Z}_2 + \mathbf{Z}_3) + \mathbf{Z}_2\, \mathbf{Z}_3}{\mathbf{Z}_2 + h_{ie}}\right]$$

$$\text{or } \frac{1}{\mathbf{Z}_L} = \frac{1}{\mathbf{Z}_1} + \frac{\mathbf{Z}_2 + h_{ie}}{h_{ie} (\mathbf{Z}_2 + \mathbf{Z}_3) + \mathbf{Z}_2\, \mathbf{Z}_3} = \frac{h_{ie} (\mathbf{Z}_1 + \mathbf{Z}_2 + \mathbf{Z}_3) + \mathbf{Z}_1\, \mathbf{Z}_2 + \mathbf{Z}_2\, \mathbf{Z}_3}{\mathbf{Z}_1 \left[h_{ie} (\mathbf{Z}_2 + \mathbf{Z}_3) + \mathbf{Z}_2\, \mathbf{Z}_3\right]}$$

$$\text{or } \mathbf{Z}_L = \frac{\mathbf{Z}_1 \left[h_{ie} (\mathbf{Z}_2 + \mathbf{Z}_3) + \mathbf{Z}_2\, \mathbf{Z}_3\right]}{h_{ie} (\mathbf{Z}_1 + \mathbf{Z}_2 + \mathbf{Z}_3) + \mathbf{Z}_1\, \mathbf{Z}_2 + \mathbf{Z}_2\, \mathbf{Z}_3} \qquad \text{...(5.7)}$$

The voltage gain of a CE amplifier without feedback is given as

$$A = \frac{-h_{fe}\mathbf{Z}_L}{h_{ie}} \qquad \text{...(5.8)}$$

The output voltage between terminals 1 and 2 is given as

$$V_{out} = \left[\mathbf{Z}_3 + \frac{\mathbf{Z}_2 h_{ie}}{\mathbf{Z}_2 + h_{ie}}\right] I_1 = \left[\frac{h_{ie} (\mathbf{Z}_2 + \mathbf{Z}_3) + \mathbf{Z}_2\, \mathbf{Z}_3}{\mathbf{Z}_2 + h_{ie}}\right] I_1$$

The voltage feedback to the input terminals 2 and 3 is given as

$$V_f = \frac{\mathbf{Z}_2 h_{ie}}{\mathbf{Z}_2 + h_{ie}}\, I_1$$

$$\text{So feedback fraction, } \beta = \frac{V_f}{V_{out}} = \frac{\mathbf{Z}_2 h_{ie}}{h_{ie} (\mathbf{Z}_2 + \mathbf{Z}_3) + \mathbf{Z}_2\mathbf{Z}_3} \qquad \text{...(5.9)}$$

Applying the criterion of oscillation *i.e.*, $A\beta = 1$, we have

$$\frac{-h_{fe}\mathbf{Z}_L}{h_{ie}} \cdot \frac{\mathbf{Z}_2 h_{ie}}{h_{ie} (\mathbf{Z}_2 + \mathbf{Z}_3) + \mathbf{Z}_2\mathbf{Z}_3} = 1$$

$$\text{or } \frac{h_{fe}\mathbf{Z}_1[h_{ie} (\mathbf{Z}_2 + \mathbf{Z}_3) + \mathbf{Z}_2\mathbf{Z}_3]}{h_{ie} (\mathbf{Z}_1 + \mathbf{Z}_2 + \mathbf{Z}_3) + \mathbf{Z}_1\mathbf{Z}_2 + \mathbf{Z}_2\mathbf{Z}_3} \cdot \left[\frac{\mathbf{Z}_2}{h_{ie} (\mathbf{Z}_2 + \mathbf{Z}_3) + \mathbf{Z}_2\mathbf{Z}_3}\right] = -1$$

$$\text{or } \frac{h_{fe}\mathbf{Z}_1\mathbf{Z}_2}{h_{ie} (\mathbf{Z}_1 + \mathbf{Z}_2 + \mathbf{Z}_3) + \mathbf{Z}_1\mathbf{Z}_2 + \mathbf{Z}_2\mathbf{Z}_3} = -1$$

$$\text{or } h_{ie}(\mathbf{Z}_1 + \mathbf{Z}_2 + \mathbf{Z}_3) + \mathbf{Z}_1\mathbf{Z}_2 + \mathbf{Z}_2\mathbf{Z}_3 = -h_{fe}\, \mathbf{Z}_1\, \mathbf{Z}_2$$

$$\text{or } h_{ie} (\mathbf{Z}_1 + \mathbf{Z}_2 + \mathbf{Z}_3) + \mathbf{Z}_1\, \mathbf{Z}_2 (1 + h_{fe}) + \mathbf{Z}_2\, \mathbf{Z}_3 = 0 \qquad \text{...(5.10)}$$

This is the general equation for the oscillator.

5.8. COLPITT'S OSCILLATOR

The Colpitt's oscillator circuit is a superb circuit and is widely used in commercial signal generators up to 100 MHz. The basic circuit of a transistor Colpitt's Oscillator is shown in Fig. 5.7. It basically consists of a single stage inverting amplifier and an L-C phase shift network, as obvious from the circuit diagram shown. The two series capacitors C_1 and C_2 form the potential divider used for providing the feedback voltage—the voltage developed across capacitor C_2 provides the regenerative feedback required for sustained oscillations. Parallel combination of R_E and C_E along with resistors R_1 and R_2 provides the stabilized self bias. The collector supply voltage V_{CC} is applied to the collector through a radio frequency choke (RFC) which permits an easy flow of direct current but at the same time it offers very high impedance to the high frequency currents. The presence of coupling capacitor C_{C_2} in the output circuit does not permit the dc currents to go to the tank circuit (the flow of dc current in a tank circuit reduces its Q). The radio frequency energy developed across RFC is capacitively coupled to the tank circuit through the capacitor C_{C_2}. The output of the phase shift L-C network is coupled from the junction of L and C_2 to the amplifier input at base through coupling capacitor C_{C_1}, which blocks dc but provides path to ac. Transistor itself produces a phase shift of 180º and another phase shift of 180º is provided by the capacitive feedback. Thus a total phase shift of 360º is obtained which is an essential condition for developing oscillations. The output voltage is derived from a secondary winding L' coupled to the inductance L. The frequency is determined by the tank circuit and is varied by gang-tuning the two capacitors C_1 and C_2. It is to be noted that capacitors C_1 and C_2 are ganged. As the tuning is varied, values of both capacitors vary simultaneously, the ratio of the two capacitances remaining the same.

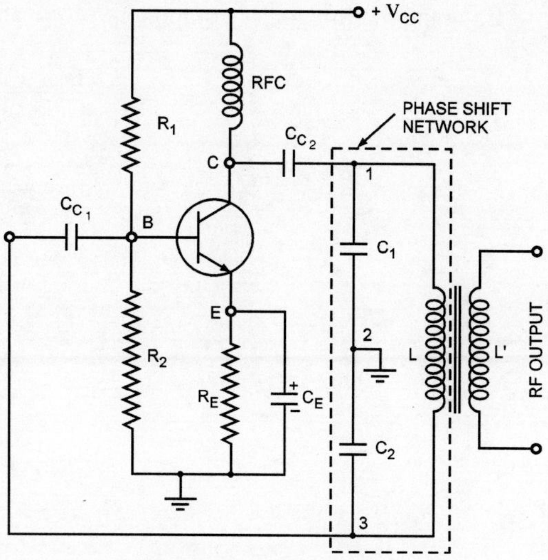

Fig. 5.7. ***Basic Circuit For Transistor Colpitt's Oscillator***

Working. When the collector supply voltage V_{CC} is switched on, the capacitors C_1 and C_2 are charged. These capacitors C_1 and C_2 discharge through the coil L, setting up oscillations of frequency $f = \frac{1}{2\pi}\sqrt{\frac{1}{LC_1}+\frac{1}{LC_2}}$.

The oscillations across capacitor C_2 are applied to the base-emitter junction and appear in the amplified form in the collector circuit. Of course, the amplified output in the collector circuit is of the same frequency as that of the oscillatory circuit. This amplified output in the collector circuit is supplied to the tank circuit in order to meet the losses. Thus the tank circuit is getting energy continuously from the collector circuit to make up for the losses occurring in it and, therefore, ensures undamped oscillations. The energy supplied to the tank circuit is of correct phase, as already explained, and if Aβ exceeds unity, oscillations are sustained in the circuit.

Frequency of Oscillation :

Here referring to Fig. 5.6

$$\mathbf{Z}_1 = \frac{1}{j\omega C_1} = -\frac{j}{\omega C_1};\ \mathbf{Z}_2 = \frac{1}{j\omega C_2} = \frac{-j}{\omega C_2} \text{ and } \mathbf{Z}_3 = j\omega L$$

Substituting these values in Eq. (5.10), we have

$$h_{ie}\left[\frac{-j}{\omega C_1} - \frac{j}{\omega C_2} + j\omega L\right] + \left[\frac{-j}{\omega C_1}\cdot\frac{-j}{\omega C_2}\right](1+h_{fe}) + \left[\frac{-j}{\omega C_2}\right]j\omega L = 0$$

$$\text{or } -jh_{ie}\left[\frac{1}{\omega C_1} + \frac{1}{\omega C_2} - \omega L\right] - \frac{1+h_{fe}}{\omega^2 C_1 C_2} + \frac{L}{C_2} = 0 \qquad ...(5.11)$$

Equating the imaginary component of the above Eq. (5.11) to zero we have

$$h_{ie}\left[\frac{1}{\omega C_1}+\frac{1}{\omega C_2}-\omega L\right]=0$$

$$\text{or}\quad \frac{1}{\omega C_1}+\frac{1}{\omega C_2}=\omega L \qquad \because h_{ie}\neq 0$$

$$\text{or}\quad \frac{C_1+C_2}{\omega C_1C_2}=\omega L$$

$$\text{or}\quad \omega^2=\frac{C_1+C_2}{LC_1C_2}$$

$$\text{or}\quad \omega=\sqrt{\frac{C_1+C_2}{LC_1C_2}}=\sqrt{\frac{1}{LC_1}+\frac{1}{LC_2}}$$

$$\text{or}\quad f=\frac{1}{2\pi}\sqrt{\frac{1}{LC_1}+\frac{1}{LC_2}} \qquad \text{...(5.12)}$$

The above Eq. (5.12) gives the frequency of oscillation. Equating the real component of Eq. (5.11) to zero we have

$$\frac{1+h_{fe}}{\omega^2C_1C_2}=\frac{L}{C_2}$$

$$\text{or}\quad 1+h_{fe}=\omega^2LC_1$$

$$\text{or}\quad 1+h_{fe}=\frac{C_1+C_2}{LC_1C_2}\times LC_1=\frac{C_1+C_2}{C_2}=1+\frac{C_1}{C_2}$$

$$\text{or}\quad h_{fe}=\frac{C_1}{C_2} \qquad \text{...(5.13)}$$

As for other oscillator circuits, the loop gain must be greater than unity to ensure that the circuit oscillates.

$$\text{So } A\beta \geq 1 \text{ or } A\geq\frac{1}{\beta}\geq\frac{C_2}{C_1} \qquad \text{...(5.14)}$$

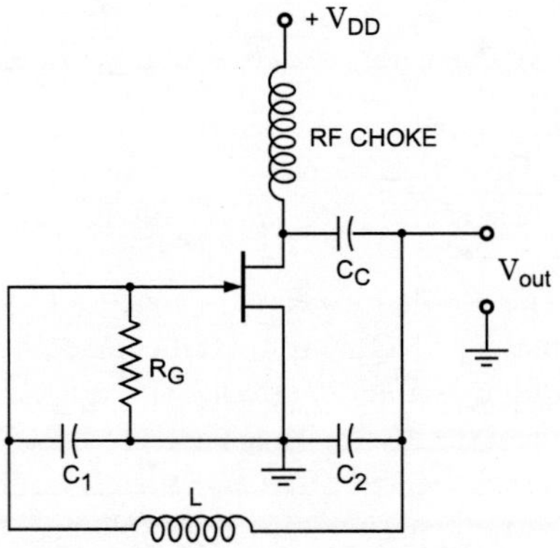

Fig. 5.8. ***FET Colpitt's Oscillator***

FET Colpitt's Oscillator. A practical version of an FET Colpitt's oscillator is shown in Fig. 5.8. Here, R_G is the gate biasing resistor. The RFC choke performs two functions. (*i*) It keeps ac current out of the dc drain supply and (*ii*) it provides drain load. As seen, C_1 is in the input circuit whereas C_2 is in the output circuit. The oscillator frequency can be determined by the same relation as in case of transistor Colpitt's oscillator *i.e.*, by relation

$$f_0=\frac{1}{2\pi}\sqrt{\frac{1}{LC_1}+\frac{1}{LC_2}} \qquad \text{...(5.15)}$$

Example 5.1. In a transistor Colpitt's oscillator shown in Fig. 5.7, L = 100 μH, L_{RFC} = 0.6 mH, C_1 = 0.001 μF, C_2 = 0.01 μF and C_C = 10 μF. Determine (*i*) operating frequency (*ii*) feedback fraction (*iii*) minimum gain to sustain oscillations.

Solution : Inductance, $L = 100\ \mu H = 100 \times 10^{-6}$ or 1×10^{-4} H

Capacitance, $C_1 = 0.001\ \mu F = 0.001 \times 10^{-6}$ F

Capacitance, $C_2 = 0.01\ \mu F = 0.01 \times 10^{-6}$ F

$$(i)\ \text{Operating frequency, } f = \frac{1}{2\pi}\sqrt{\frac{1}{LC_1}+\frac{1}{LC_2}} = \frac{1}{2\pi}\sqrt{\frac{1}{10^{-4}\times 0.001\times 10^{-6}}+\frac{1}{10^{-4}\times 0.01\times 10^{-6}}} = 528 \text{ kHz}$$

$$(ii)\ \text{Feedback fraction, } \beta = h_{fe} = \frac{C_1}{C_2} = \frac{0.001\times 10^{-6}}{0.01\times 10^{-6}} = 0.1 \textbf{ Ans.}$$

$$(iii)\ \text{Minimum gain to sustain oscillations, } A_{min} = \frac{1}{\beta} = \frac{1}{0.1} = 10 \textbf{ Ans.}$$

Example 5.2. A Colpitt's oscillator is designed with $C_1 = 100$ pF and $C_2 = 7{,}500$ pF. The inductance is variable. Determine the range of inductance values, if the frequency of oscillations is to vary between 950 kHz and 2,050 kHz.

Solution : Resonant frequency is given as

$$f_0 = \frac{1}{2\pi}\sqrt{\frac{1}{LC_1}+\frac{1}{LC_2}} \qquad \text{...Refer to Eq. (5.12)}$$

$$\text{or} \quad L = \frac{1}{4\pi^2 f_0^2}\left(\frac{1}{C_1}+\frac{1}{C_2}\right)$$

For frequency of oscillations of 950 kHz, $C_1 = 100$ pF and $C_2 = 7{,}500$ pF.

$$L_1 = \frac{1}{4\pi^2(950\times 10^3)^2}\left(\frac{1}{100\times 10^{-12}}+\frac{1}{7{,}500\times 10^{-12}}\right)$$

$$= \frac{1}{4\pi^2(950\times 10^3)^2}\times\frac{76\times 10^{12}}{7{,}500} = 0.284 \text{ mH}$$

For frequency of oscillations of 2,050 kHz, $C_1 = 100$ pF and $C_2 = 7{,}500$ pF

$$L_2 = \frac{1}{4\pi^2(2{,}050\times 10^3)^2}\times\left(\frac{1}{100\times 10^{-12}}+\frac{1}{7{,}500\times 10^{12}}\right)$$

$$= \frac{1}{4\pi^2(2{,}050)^2\times 10^6}\times\frac{76\times 10^{12}}{7{,}500} = 0.061 \text{ mH}$$

Thus, the range of required inductance is 0.061 mH to 0.284 mH **Ans.**

Example 5.3. Determine the range of variable inductance, if frequency of oscillation is to vary between 950 kHz to 2,050 kHz. Assume $C_1 = 7{,}500$ pF and $C_2 = 100$ pF. [*G.G.S.I.P. Univ. Analog Electronics-II May 2011*]

Solution : Refer to Solution of Example 5.2.

5.9. HARTLEY OSCILLATOR

The transistor Hartley oscillator is as popular as Colpitt's oscillator and is widely used as a local oscillator in radio receivers. The circuit arrangement is shown in Fig. 5.9. Hartley oscillator circuit is similar to Colpitt's oscillator circuit, except that phase shift network consists of two inductors L_1 and L_2 and a capacitor C instead of two capacitors and one inductor. The output of the amplifier is applied across inductor L_1 and the voltage across inductor L_2 forms the feedback voltage. The coil L_1 is inductively coupled to coil L_2, the combination functions as an auto-transformer. However, because of direct connection, the junction of L_1 and L_2 cannot be directly grounded. Instead, another capacitor C_L is used. The operation of the circuit is similar to that of the Colpitt's oscillator circuit.

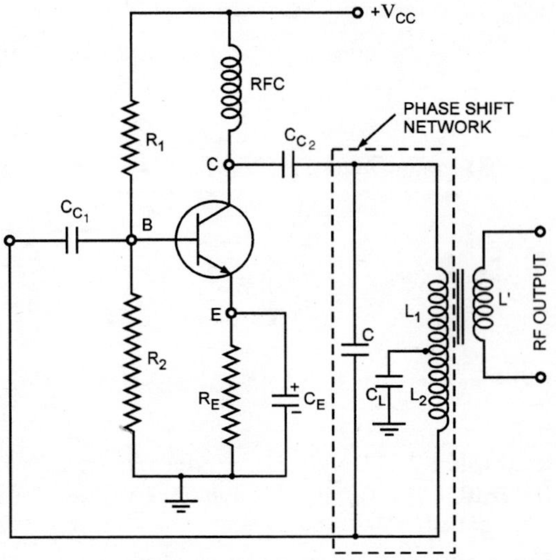

Fig. 5.9. *Basic Circuit For Hartley Oscillator*

Considering the fact that there exists mutual inductance between coils L_1 and L_2 because the coils are wound on the same core, their net effective inductance is increased by mutual inductance M. So in this case effective inductance is given by the equation

$$L = L_1 + L_2 + 2M$$

and resonant or oscillation frequency is given by the equation

$$f = \frac{1}{2\pi\sqrt{[C(L_1 + L_2 + 2M)]}}$$, as derived below :

Frequency of Oscillation. The general Eq. (5.10) derived in Art. 5.7 is reproduced here

$$h_{ie}(\mathbf{Z}_1 + \mathbf{Z}_2 + \mathbf{Z}_3) + \mathbf{Z}_1\,\mathbf{Z}_2\,(1 + h_{fe}) + \mathbf{Z}_2\,\mathbf{Z}_3 = 0$$

Here $\mathbf{Z}_1 = j\omega L_1 + j\omega M$; $\mathbf{Z}_2 = j\omega L_2 + j\omega M$ and $\mathbf{Z}_3 = \frac{1}{j\omega C} = -j/\omega C$

Substituting these values in general Eq. (5.10) we get

$$h_{ie}\left[(j\omega L_1 + j\omega M) + (j\omega L_2 + j\omega M) - \frac{j}{\omega C}\right] + (j\omega L_1 + j\omega M)(j\omega L_2 + j\omega M)(1 + h_{fe}) + (j\omega L_2 + j\omega M)\left(\frac{-j}{\omega C}\right) = 0$$

$$\text{or}\quad j\omega h_{ie}\left[L_1 + L_2 + 2M - \frac{1}{\omega^2 C}\right] - \omega^2 (L_2 + M)\left[(L_1 + M)(1 + h_{fe}) - \frac{1}{\omega^2 C}\right] = 0 \quad ...(5.16)$$

Equating the imaginary part of above Eq. (5.16) to zero we get

$$\omega h_{ie}\left[L_1 + L_2 + 2M - \frac{1}{\omega^2 C}\right] = 0$$

$$\text{or}\quad L_1 + L_2 + 2M - \frac{1}{\omega^2 C} = 0 \qquad \because \omega\, h_{ie} \neq 0$$

$$\text{or}\ \omega^2 C = \frac{1}{L_1 + L_2 + 2M}$$

$$\text{or}\ f = \frac{\omega}{2\pi} = \frac{1}{2\pi\sqrt{C(L_1 + L_2 + 2M)}} \quad ...(5.17)$$

The above Eq. (5.17) gives the frequency of oscillation.

Equating the real component of Eq. (5.16) to zero we get

$$(L_1 + M)(1 + h_{fe}) - \frac{1}{\omega^2 C} = 0$$

$$\text{or } (1 + h_{fe}) = \frac{1}{\omega^2 C(L_1 + M)} = \frac{L_1 + L_2 + 2M}{L_1 + M} = 1 + \frac{L_2 + M}{L_1 + M}$$

$$\text{or } h_{fe} = \frac{L_2 + M}{L_1 + M} \quad ...(5.18)$$

As for other oscillator circuits, the loop gain must be greater than 1 to ensure that circuit oscillates.

$$\text{So } A\beta \geq 1$$

$$\text{or } A \geq \frac{1}{\beta}$$

$$\geq \frac{L_1 + M}{L_2 + M} \quad ...(5.19)$$

FET Hartley Oscillator. An FET Hartley oscillator circuit is shown in Fig. 5.10. The noteworthy point is that inductors L_1 and L_2 have a mutual coupling M, which is to be taken into account in determination of the equivalent inductance for the resonant circuit. This circuit operates at a frequency given by Eq. (5.17).

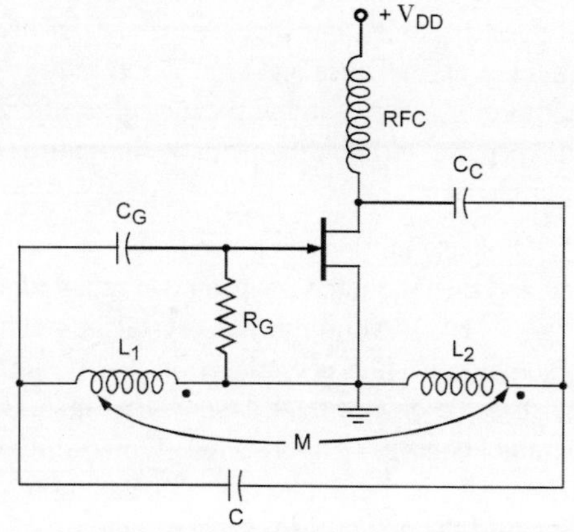

Fig. 5.10. ***FET Hartley Oscillator***

Example 5.4. In a transistorized Hartley oscillator, the tank circuit has the capacitance of 100 pF. The value of inductance between the collector and tapping point is 30 mH and the value of inductance between the tapping point and the transistor base is 1 × 10⁻⁸ H. Determine the frequency of oscillations. Neglect the mutual inductance.

Solution : Effective inductance, $L = L_1 + L_2 + 2M = 30 \times 10^{-3} + 1 \times 10^{-8} + 0 \simeq 0.03$ H

Capacitance, $C = 100 \text{ pF} = 100 \times 10^{-12} \text{ F} = 1 \times 10^{-10}$ F

$$\text{Frequency of oscillation, } f_o = \frac{1}{2\pi\sqrt{LC}} = \frac{1}{2\pi \times \sqrt{0.03 \times 10^{-10}}} = 91.9 \text{ kHz} \quad \textbf{Ans.}$$

Example 5.5. Determine the oscillation frequency of a transistor Hartley oscillator shown in Fig. 5.9 with circuit values L_1 = 1 mH, L_2 = 100 μH, M = 50 μH and C = 100 pF.

Solution : Effective inductance, $L = L_1 + L_2 + 2M = 1 \times 10^{-3} + 100 \times 10^{-6} + 2 \times 50 \times 10^{-6} = 1{,}200$ μH

Capacitance, $C = 100 \text{ pF} = 10^{-10}$ F

$$\text{Frequency of oscillation, } f_o = \frac{1}{2\pi\sqrt{LC}} = \frac{1}{2\pi\sqrt{1{,}200 \times 10^{-6} \times 10^{-10}}} = 459 \text{ kHz} \quad \textbf{Ans.}$$

5.10. CRYSTAL OSCILLATORS

In crystal oscillators, the usual electrical resonant circuit is replaced by a mechanically vibrating crystal. The crystal (usually quartz) has a high degree of stability in holding constant at whatever frequency the crystal is originally cut to operate. The crystal oscillators are, therefore, used whenever great stability is needed, for example, in communication transmitters, and receivers, digital clocks etc.

A quartz crystal exhibits a very important property known as piezoelectric effect. When a mechanical pressure is applied across the faces of the crystal, a voltage proportional to the applied mechanical pressure appears across the crystal. Conversely, when a voltage is applied across the crystal surfaces, the crystal is distorted by an amount proportional to the applied voltage. An alternating voltage applied to a crystal causes it to vibrate at its natural frequency.

Besides quartz, the other substances that exhibit the piezoelectric effect are Rochelle salt and tourmaline. Rochelle salt exhibits the greatest piezoelectric effect, but its applications are limited to manufacture of microphones, headsets and loudspeakers. It is because the Rochelle salt is mechanically the weakest and strongly affected by moisture and heat. Tourmaline is most rugged but shows the least piezoelectric effect. Quartz is a compromise between the piezoelectric effect of Rochelle salt and the mechanical strength of tourmaline. It is inexpensive and readily available in nature. It is mainly the quartz crystal that is used in radio frequency (RF) oscillators.

For use in electronic oscillators, the crystal is suitably cut and then mounted between two metal plates, as shown in Fig. 5.11 (*a*). Although the crystal has electromechanical resonance but the crystal action can be represented by an electrical resonant circuit, as shown in Fig. 5.11 (*b*). The crystal actually behaves as a series R-L-C circuit in parallel with C_M where C_M is the capacitance of the mounting electrodes. Because the crystal losses, represented by R, are small, the equivalent crystal Q is high—typically 20,000. Values of Q up to 10^6 can be obtained by making use of crystals.

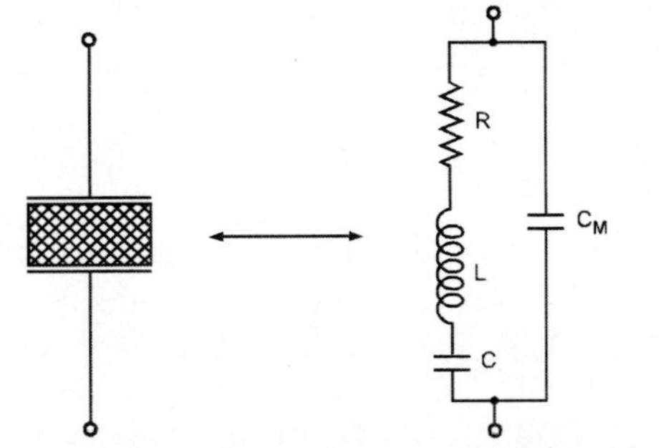

(a) Crystal Mounting *(b) Electrical Equivalent Circuit of a Crystal*

Fig. 5.11

Because of presence of C_M, the crystal has two resonant frequencies. One of these is the series resonant frequency f_s at which $2\pi f L = \dfrac{1}{2\pi f C}$ and in this case the crystal impedance is very low. The other is parallel resonance frequency f_p which is due to parallel resonance of capacitance C_M and the reactance of the series circuit. In this case crystal impedance is very high. The impedance versus frequency curve of the crystal is shown in Fig. 5.12. In order to use the crystal properly it must be connected in a circuit so that its low impedance in the series-resonant operating mode or high impedance in the antiresonant or parallel resonant operating mode is selected.

Two resonant frequencies are given by the equations

$$\text{Series resonant frequency, } f_s = \frac{1}{2\pi\sqrt{LC}} \quad \text{...(5.20)}$$

$$\text{Parallel resonant frequency, } f_p = \frac{1}{2\pi}\sqrt{\frac{1 + C/C_M}{LC}} \quad \text{...(5.21)}$$

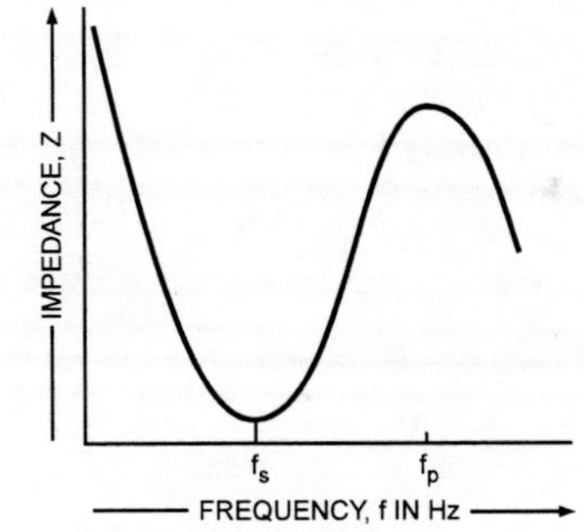

Fig. 5.12. ***Crystal Impedance Versus Frequency Curve***

It appears that f_p is higher than f_s but the two frequencies are very close to each other. It is due to the fact that the ratio $\dfrac{C}{C_M}$ is very small.

To stabilize the frequency of an oscillator, a crystal may be operated at either its series or parallel resonant frequency.

To excite a crystal for operation in the series-resonant mode it may be connected as a series element in a feedback path, as shown in Fig. 5.13 (*a*). In this mode of operation the crystal impedance is the smallest and the amount of positive feedback is the largest. Resistors R_1, R_2 and R_E provide a voltage-divider stabilized dc bias circuit, the capacitor C_E provides ac bypass of the emitter resistor R_E and the radio frequency coil (RFC) provides for dc bias while decoupling any ac signal on the power lines from affecting the output signal. The voltage feedback signal from the collector to the base is maximum when the crystal impedance is minimum (*i.e.*, in series-resonant mode). The coupling capacitor C_C has negligible impedance at the circuit operating frequency but blocks any dc between collector and base. The circuit shown in Fig. 5.13 (*a*) is generally called the *Pierce crystal.* The resulting circuit frequency of oscillations is set by the series resonant frequency of the crystal. Variations in supply voltage, transistor parameters, etc. have no effect on the circuit operating frequency which is held stabilized by the crystal. The circuit frequency stability is set by the crystal frequency stability, which is good.

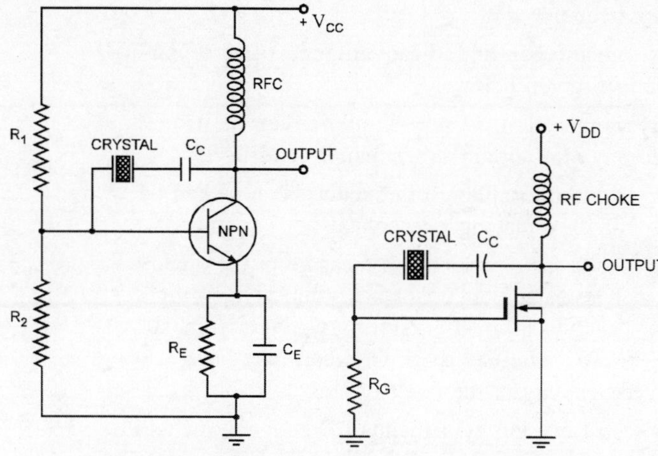

(*a*) *Oscillator With Crystal Operating in Series Resonance*

(*b*) *FET Circuit*

Fig. 5.13

FET circuit with crystal operating in series resonance is shown in Fig. 5.13 (*b*).

Use of FET is desirable because its high input impedance results in light loading of the crystal which (*i*) results in good stability and (*ii*) does not lower the Q value.

Oscillator circuit with crystal operating in parallel resonance (a modified Colpitt's oscillator circuit) is illustrated in Fig. 5.14. Since the parallel-resonant impedance of a crystal is of a maximum value, it is connected in parallel. C_1 and C_2 form a capacitive voltage divider which returns a portion of the output voltage to the transistor emitter. Transistor NPN combined with R_1, R_2, RFC, and R_E, constitutes a common base circuit. Capacitor C_3 provides an ac short circuit across R_2 to ensure that the transistor base remains at a fixed voltage level. As the output voltage increases positively, the emitter voltage also increases, and since the base voltage is fixed, the base-emitter voltage is reduced. The reduction in V_{BE} causes collector current I_C to diminish, and this in turn causes the collector voltage V_C to increase positively. Thus, the circuit is applying its own input, and a state of oscillation exists. The crystal in parallel with C_1 and C_2 permits maximum voltage feedback from the collector to emitter when its impedance is maximum, *i.e.*, at its parallel resonant frequency. At other frequencies, the crystal impedance is low, and so the resultant feedback voltage is too small to sustain oscillations. The oscillation frequency is stabilized at the parallel resonant frequency of the crystal.

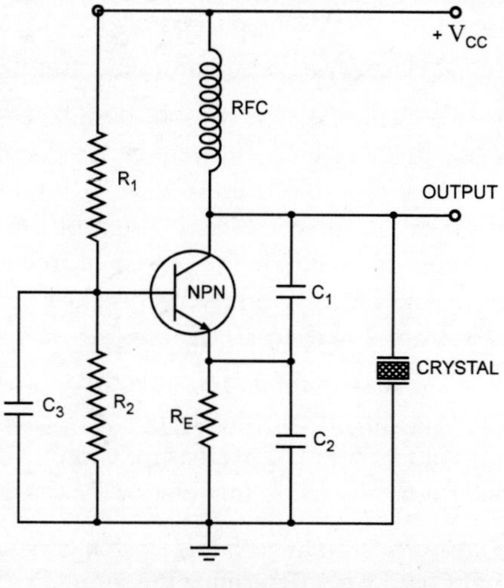

Fig. 5.14. ***Oscillator With Crystal Operating in Parallel Resonance***

A Miller crystal controlled oscillator circuit is shown in Fig. 5.15. A tuned L-C circuit in the drain section is adjusted near the crystal parallel-resonant frequency. The maximum gate-source signal occurs at the crystal antiresonant frequency, controlling the circuit operating frequency.

The advantages and disadvantages of a crystal oscillator are given below :

Advantages. 1. It is very simple circuit as it does not need any tank circuit other than crystal itself.

2. Different oscillation frequencies can be had by simply replacing one crystal with another.

3. The Q factor, which is a measure of the quality of resonance circuit of a crystal, is very high. The Q factor of a crystal may range from 10^4 to 10^6 whereas the L-C circuit may have a Q factor only of the order of 100.

4. Most crystals will maintain frequency drift to within a few cycles at 25°C. For greater frequency stability, the crystal is often contained in an insulated enclosure termed as crystal oven in which the temperature is thermostatically controlled. In this way it is possible to have frequency drifts less than 1 part in 10^{10}.

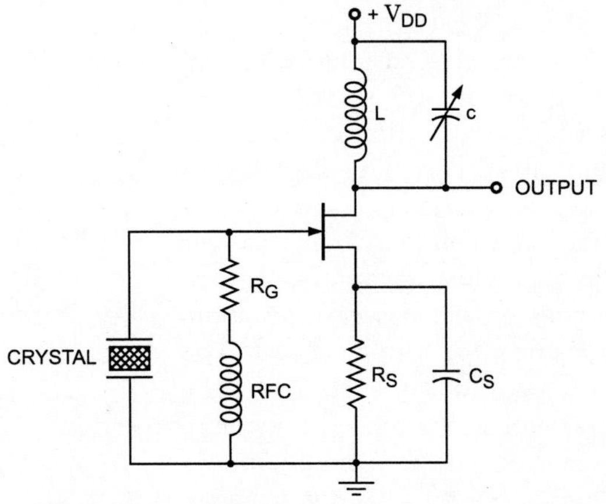

Fig. 5.15. ***Miller Crystal Controlled Oscillator***

Disadvantages. 1. The crystal oscillators have a very limited tuning range (or not at all). They are used for frequencies exceeding 100 kHz.

2. The crystal oscillators are fragile and, therefore, can only be used in low power circuits.

Crystal oscillators must be designed to provide a load capacitance on the crystal as per specifications listed by the manufacturer. This requirement is essential for obtaining oscillations at the specified frequency. It is also important from the point of view of limiting the power supplied to the crystal to the specified maximum. Too much crystal power causes distortion in the oscillator waveform. It also causes overheating of the crystal, consequently rendering the resonant frequency unstable. More important is that the thin plated electrodes may be melted off an overdriven crystal, destroying the device. Typical maximum drive levels for plated crystals vary from 2 mW to 10 mW.

The maximum permissible drive power limits the ac voltages that may be applied across the crystal and consequently affects the design of oscillator circuits. Crystal manufacturers usually specify the resistance of individual crystal, as well as maximum drive power. From these two, the maximum crystal ac voltage may be determined by using the relation $P = V^2/R$.

Example 5.6. The parameters of a crystal oscillator equivalent circuit are L_s = 0.8 H; C_s = 0.08 pF, R_s = 5 kΩ and C_p = 1.0 pF. Determine the resonance frequencies f_s and f_p.

Solution : Series resistance, $R_s = 5\text{ k}\Omega = 5{,}000\ \Omega$

Series inductance, $L_s = 0.8$ H

Series capacitance, $C_s = 0.08\text{ pF} = 8 \times 10^{-14}$ F

Parallel capacitance, C_p or $C_M = 1.0\text{ pF} = 1 \times 10^{-12}$ F

$$\text{Series resonant frequency, } f_s = \frac{1}{2\pi\sqrt{L_s C_s}} = \frac{1}{2\pi\sqrt{0.8 \times 8 \times 10^{-14}}} = 629\text{ kHz } \textbf{Ans.}$$

$$\text{Parallel resonant frequency, } f_p = \frac{1}{2\pi}\sqrt{\frac{1 + C_s/C_p}{L_s C_s}} = \frac{1}{2\pi}\sqrt{\frac{1 + 8 \times 10^{-14}/1 \times 10^{-12}}{0.8 \times 8 \times 10^{-14}}} = 654\text{ kHz } \textbf{Ans.}$$

Example 5.7. Calculate the value of inductance (L) offered by the crystal at oscillation frequency f_s = 1 MHz.

Solution : Series resonant frequency, $f_s = 1 \text{ MHz} = 1 \times 10^6$ Hz

Assume $C_s = 0.1$ pF

$$\text{Since } f_s = \frac{1}{2\pi\sqrt{L_s C_s}}$$

$$\text{So } L_s = \frac{1}{4\pi^2 C_s f_s^2} = \frac{1}{4\pi^2 \times 0.1 \times 10^{-12} \times 1 \times 10^{12}} = 0.2533 \text{ H} \quad \textbf{Ans.}$$

Example 5.8. Prove that ratio of the parallel to series resonant frequencies of crystal is given approximately by $\frac{f_p}{f_s} \cong \left(1 + \frac{1}{2}\frac{C}{C'}\right)$, where C′ = mounting capacitance. If C = 0.04 pF and C′ = 2.0 pF, by what percentage is the parallel resonant frequency greater than the series resonant frequency.

Solution :

$$f_p = \frac{1}{2\pi}\sqrt{\frac{1 + \frac{C}{C'}}{LC}} \quad \text{...}(i)$$

$$f_s = \frac{1}{2\pi\sqrt{LC}} \quad \text{...}(ii)$$

Dividing Eq. (*i*) by Eq. (*ii*), we have

$$\frac{f_p}{f_s} = \left(1 + \frac{C}{C'}\right)^{1/2} \cong 1 + \frac{1}{2}\frac{C}{C'} \quad \textbf{Proved.}$$

Parallel resonant frequency is greater than the series resonant frequency by percentage

$$= \frac{1}{2}\frac{C}{C'} \times 100 = \frac{1}{2} \times \frac{0.04}{2.0} \times 100 = 1\% \quad \textbf{Ans.} \quad \text{by approximate expression}$$

$$= \left[\left(1 + \frac{C}{C'}\right)^{1/2} - 1\right] \times 100 = \left(\sqrt{1 + \frac{0.04}{2.0}} - 1\right) \times 100 = 0.995\% \quad \textbf{Ans.}$$

5.11. AUDIO OSCILLATORS

So far, we have considered the oscillators which use L-C tuned circuit that causes a phase shift of 180° due to inductive or capacitive coupling in addition to a 180° phase shift produced by the transistor itself. The oscillators employing L-C elements, called the *L-C oscillators,* are very popular for generating high frequency oscillations but they cannot be employed for generation of low frequency oscillations as they become too bulky and expensive. *R-C oscillators* are commonly used for generating audio frequencies as they provide good frequency stability and proper waveform. Also, with the advent of IC technology, R-C network is the only feasible solution, as it is very difficult to make a too high value inductance in an integrated circuit. Two commonly used R-C oscillators are (*i*) R-C phase shift oscillator and (*ii*) Wien bridge oscillator. So these two audio oscillators will be discussed here.

5.11.1. Basic Principles of R-C Oscillators

For producing oscillations in an oscillator circuit we need positive feedback which means that the voltage signal feedback should be in phase with the input signal. For providing a positive feedback at one particular frequency, an inverting amplifier may be used with a feedback network that causes a phase shift of 180° at the desired frequency of oscillation, as shown in Fig. 5.16 (*a*). The 180° phase shift in the feedback signal can be obtained by a suitable network consisting of three R-C sections, as shown in Fig. 5.16 (*b*).

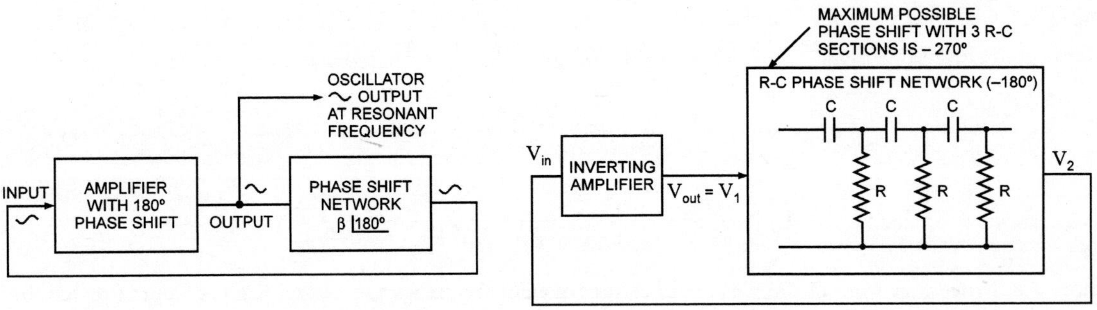

(*a*) (*b*)

Fig. 5.16

When a phase shift network such as that indicated in Fig. 5.16 is used in a phase shift oscillator, the R's and C's must be selected so as to produce a phase shift of 180° at the desired frequency of oscillation. The output of the voltage amplifier is fed to the input to the phase shift network. Thus $V_1 = V_{out}$. The output resistance of the amplifier is designed to be very small in comparison to the input impedance of the phase shift network. The output voltage of the phase shift network, V_2 is fed into the input of the amplifier *i.e.,* $V_2 = V_{in}$. The amplifier's input impedance must be much larger than the output impedance of the phase shift network.

Alternatively, a positive feedback can be obtained by using two stages of amplifiers each giving a phase shift of 180°. A part of this output is fed back to the input through a feedback network without causing any further phase shift. Wien bridge oscillator operates on this principle.

5.12. PHASE SHIFT OSCILLATOR

We select the so *called phase shift oscillator* (Fig. 5.17) as a first example as it exemplifies very simply the principles set forth above. The circuit is drawn to show clearly the amplifier and feedback network. The circuit consists of a common source FET amplifier followed by a three-section R-C phase shift network. The amplifier stage is self-biased with a capacitor bypassed source resistor R_S and a drain bias resistance R_D. The output of the last section is supplied back to the gate. If the loading of the phase shift network on the amplifier can be assumed to be negligible, a phase shift of 180° between the amplified output voltage V_{out} and the input voltage V_{in} at the gate is produced by the amplifier itself. The three-section R-C phase shift network produces an additional phase shift, which is a function of frequency and equals 180° at some frequency of operation. At this frequency the total phase shift from the gate around the circuit and back to gate will be exactly zero. This particular frequency will be the one at which the circuit will oscillate provided that the magnitude of the amplification is sufficiently large. In a FET phase shift oscillator voltage series feedback [*i.e.,* feedback voltage proportional to the output voltage V_{out} and supplied in series with the input signal at the gate] is used.

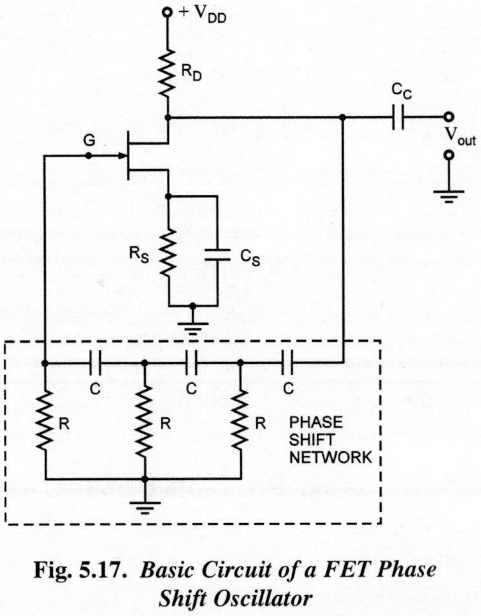

Fig. 5.17. ***Basic Circuit of a FET Phase Shift Oscillator***

The frequency of the oscillator output depends upon the values of capacitors C and resistors R used in the phase shift network. Using basic RC circuit analysis technique, it can be shown that the network phase shift is 180° when

$$X_C = \sqrt{6}R \text{ or } \frac{1}{2\pi fC} = \sqrt{6}R \text{ or } f = \frac{1}{2\pi RC\sqrt{6}} \quad ...(5.22)$$

The frequency can be adjusted over a wide range if variable capacitors are used.

As well as phase shifting, the R-C network attenuates the amplifier output. Network analysis shows that when the necessary phase shift of 180° is obtained, this network attenuates the output voltage by a factor of $\frac{1}{29}$. This means that the amplifier must have a voltage gain of 29 or more.

When the amplifier voltage gain is 29 and feedback factor of R-C network, $\beta = \frac{1}{29}$ then the loop gain is $\beta A = \frac{1}{29} \times 29 = 1$, the amplifier phase shift of $-180°$ combined with the network phase shift of $+180°$ gives a loop phase shift of zero. Both of these conditions are necessary to satisfy the Barkhausen criteria.

If the amplifier gain is much greater than 29, the oscillator output waveform is likely to be distorted. When the gain is slightly greater than 29, the output is usually a reasonably pure sinusoidal.

The advantages and disadvantages of phase shift oscillators are given below :

Advantages. 1. It is cheap and simple circuit as it contains resistors and capacitors (not bulky and expensive high value inductors).

2. It provides good frequency stability.
3. The phase shift oscillator circuit is much simpler than the Wien bridge oscillator circuit because it does not need negative feedback and the stabilization arrangements.
4. The output is sinusoidal that is quite distortion free.
5. They have a wide frequency range (from a few Hz to several hundred kHz).
6. They are particularly suitable for low frequencies, say of the order of 1 Hz, as these frequencies can be easily obtained by using R and C of large values.

Disadvantages. 1. The output is small. It is due to smaller feedback.

2. It is difficult for the circuit to start oscillations as the feedback is usually small.
3. The frequency stability is not as good as that of Wien bridge oscillator.
4. It needs high voltage (12 V) battery so as to develop sufficiently large feedback voltage.

Applications. FET phase shift oscillator is used for generating signals over a wide frequency range. The frequency may be varied from a few Hz to 200 Hz by employing one set of resistors with three capacitors ganged together to vary over a capacitance range in the 1 : 10 ratio. Similarly the frequency ranges of 200 Hz to 2 kHz, 2 kHz to 20 kHz and 20 kHz to 200 kHz can be obtained by using other sets of resistors.

Transistor Phase Shift Oscillator. The circuit arrangement of a phase shift oscillator using N-P-N transistor in CE configuration is shown in Fig. 5.18. As usual, the voltage divider $R_1 - R_2$ provides dc emitter base bias, R_E and C_E combination provides temperature stability and prevent ac signal degeneration and collector resistor R_C controls the collector voltage. The oscillator output voltage is capacitively coupled to the load by C_C.

In case of a transistor phase shift oscillator, the output of the feedback network is loaded appreciably by the relatively small input resistance (h_{ie}) of the transistor. Hence, instead of employing voltage-series feedback (as used in case of FET phase shift oscillator), voltage-shunt feedback is used for a transistor phase shift oscillator, as shown in Fig. 5.18. In this circuit, the feedback signal is coupled through the feedback resistor R' in series with the amplifier stage input resistance h_{ie}. The value of R' should be such that when added with amplifier stage input resistance h_{ie}, it is equal to R *i.e.*, $R' + h_{ie} = R$.

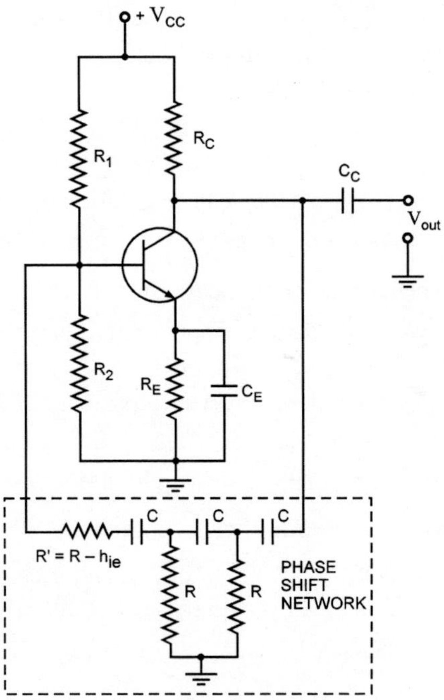

Fig. 5.18. ***Basic Circuit of a Transistor Phase Shift Oscillator***

Operation. The circuit is set into oscillations by any random or variation caused in the base current, that may be either due to noise inherent in the transistor or minor variation in voltage of dc power supply. This variation in base current is amplified in collector circuit. The output of the amplifier is supplied to an R-C feedback network. The R-C network produces a phase shift of 180° between output and input voltages. Since CE amplifier produces a phase reversal of the input signal, total phase shift becomes 360° or 0° which is essential for *regeneration* or for sustained oscillations. The output of this network is thus in the same phase as the originally assumed input to the amplifier and is applied to the base terminal of the transistor. Thus, sustained variation in collector current between saturation and cutoff values are obtained. R-C phase shift network is the frequency determining network, as already explained in case of FET phase shift oscillator.

Frequency of Oscillation. The equivalent circuit for the analysis of a transistor phase shift oscillator (circuit shown in Fig. 5.18) is shown in Fig. 5.19. The equivalent circuit shown in Fig. 5.19 is simplified if the following assumptions are made.

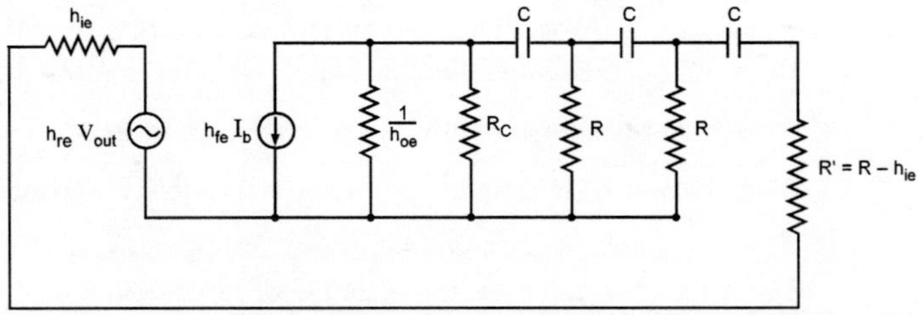

Fig. 5.19. ***Equivalent Circuit of Transistor Phase-Shift Oscillator Circuit Shown in Fig. 5.18***

(*i*) h_{re} of the transistor is usually negligibly small and, therefore, $h_{re}V_{out}$ is omitted from the circuit.

(*ii*) h_{oe} of the transistor is very small *i.e.*, $\frac{1}{h_{oe}}$ is much larger than R_C. Thus the effect of h_{oe} can be neglected.

Making above assumptions and replacing current source by equivalent voltage source, the simplified equivalent circuit is shown in Fig. 5.20.

Applying Kirchhoff's voltage law to the three loops shown in Fig. 5.20 we have

$$\left(R + R_C + \frac{1}{j\omega C}\right) I_1 - RI_2 + h_{fe} I_b R_C = 0 \quad ...(5.23)$$

$$-RI_1 + \left(2R + \frac{1}{j\omega C}\right) I_2 - RI_b = 0 \quad ...(5.24)$$

$$0 - RI_2 + \left(2R + \frac{1}{j\omega C}\right) I_b = 0 \quad ...(5.25)$$

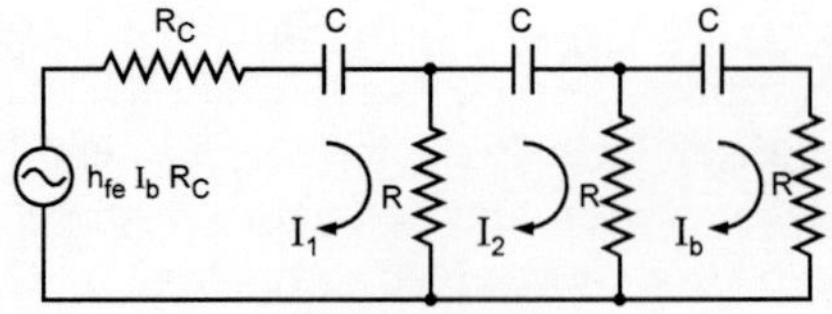

Fig. 5.20

As the currents I_1, I_2 and I_b are non-vanishing, the determinant of the coefficients of I_1, I_2 and I_b must be zero. Substituting $\frac{1}{\omega C} = X_C$ we have

$$\begin{vmatrix} (R + R_C - jX_C) & (-R) & (h_{fe}R_C) \\ (-R) & (2R - jX_C) & (-R) \\ 0 & (-R) & (2R - jX_C) \end{vmatrix} = 0 \qquad ...(5.26)$$

$$\text{or } (R + R_C - jX_C)\,[(2R - jX_C)^2 - R^2] + R\,[(-R)(2R - jX_C) - h_{fe}R_C(-R)] = 0$$

$$\text{or } (R + R_C - jX_C)(3R^2 - X_C^2 - j4RX_C) - R[2R^2 - jRX_C - h_{fe}R_CR] = 0$$

$$\text{or } R^3 + R^2R_C(3 + h_{fe}) - 5RX_C^2 - R_CX_C^2 - 6jR^2X_C - j4RR_CX_C + jX_C^3 = 0 \qquad ...(5.27)$$

Equating the imaginary component of the above equation to zero we have

$$6R^2X_C + 4RR_CX_C - X_C^3 = 0$$

$$\text{or } X_C = \sqrt{6R^2 + 4RR_C}$$

$$\text{or } 2\pi fC = \frac{1}{\sqrt{6R^2 + 4RR_C}} \quad \because X_C = \frac{1}{\omega C} = \frac{1}{2\pi fC}$$

$$\text{or } f = \frac{1}{2\pi RC\sqrt{6 + \frac{4R_C}{R}}} \qquad ...(5.28)$$

If $R = R_C$, then

$$f = \frac{1}{2\pi RC\sqrt{10}} \qquad ...(5.29)$$

The above equation gives frequency of oscillation.

Equating the real component of Eq. (5.27) to zero we have

$$R^3 + R^2R_C(3 + h_{fe}) - X_C^2(5R + R_C) = 0$$

$$\text{or } R^3 + R^2R_C(3 + h_{fe}) - (6R^2 + 4RR_C)(5R + R_C) = 0$$

$$\text{or } -29R^3 - 23R^2R_C + h_{fe}R^2R_C - 4RR_C^2 = 0$$

$$\text{or } \frac{-29R}{R_C} - 23 + h_{fe} - 4\frac{R_C}{R} = 0$$

$$\text{or } h_{fe} = 23 + 29\frac{R}{R_C} + 4\frac{R_C}{R} \qquad ...(5.30)$$

For the loop gain to be greater than unity, the requirement of the current gain of the transistor is found to be

$$h_{fe} > 23 + 29\frac{R}{R_C} + 4\frac{R_C}{R} \qquad ...(5.31)$$

If $R = R_C$, then

$$h_{fe} > (23 + 29 + 4) \textit{ i.e.,} > 56 \qquad ...(5.32)$$

Applications. The phase shift oscillator is well suited to the range of frequencies from several hertz to several hundred kilohertz (20 Hz to 200 kHz), and so includes the audio frequency range (up to 20 kHz). For generating different audio frequencies, variable air capacitors are employed as circuit elements in the phase shift network. It is possible to vary the frequency in the range of about 1 : 10 because the range of capacitors can be varied in the ratio of 10 : 1 (typically from 40 pF to 450 pF). For variations of frequency over a large range, the three capacitors are usually ganged so as to vary the capacitance of the three capacitors simultaneously. Such a variation keeps the input impedance to the phase shift network constant and also keeps constant the magnitudes of β and $A\beta$. Thus the amplitude of oscillations will remain unaffected as the frequency is adjusted. The phase shift oscillator is operated in class A so as to keep distortion to the minimum. Frequency range from 20 Hz to 200 Hz, 200 Hz to 2 kHz, 2 kHz to 20 kHz and 20 kHz to 200 kHz can be obtained by using different set of resistors.

Phase shift oscillators are not suitable for higher frequency operation because at higher frequency, the internal phase shift of the transistor and reduction in h_{fe} cause difficulties in designing the circuit. The frequency of the oscillator cannot be changed easily.

Example 5.9. In a phase shift oscillator shown in Fig. 5.17, $R_1 = R_2 = R_3 = 800$ kΩ and $C_1 = C_2 = C_3 = 100$ pF. Determine the frequency of oscillation.

Solution :

$$R_1 = R_2 = R_3 = R = 800 \text{ k}\Omega = 8 \times 10^5 \ \Omega$$
$$C_1 = C_2 = C_3 = C = 100 \text{ pF} = 10^{-10} \text{ F}$$

$$\text{Frequency of oscillation, } f_o = \frac{1}{2\pi RC\sqrt{6}} = \frac{1}{2\pi \times 8 \times 10^5 \times 10^{-10} \times \sqrt{6}} = 812 \text{ Hz } \textbf{Ans.}$$

Example 5.10. An R-C phase shift oscillator has to be designed to provide a sine wave of frequency 1 kHz. If the value of capacitance is 0.016 μF, calculate the value of the resistances.

Solution : Let the phase shift oscillator be BJT one.

$$C_1 = C_2 = C_3 = C = 0.016 \times 10^{-6} \text{ F}$$
$$f_0 = 1 \text{ kHz} = 1{,}000 \text{ Hz}$$

$$\text{Since } f_0 = \frac{1}{2\pi RC\sqrt{10}} \qquad \text{Refer to Eq. (5.29)}$$

$$R = \frac{1}{2\pi f_0 C\sqrt{10}} = \frac{1}{2\pi \times 1{,}000 \times 0.016 \times 10^{-6}\sqrt{10}}$$

$$= 3{,}146 \ \Omega = 3.3 \text{ k}\Omega \text{ (standard value) } \textbf{Ans.}$$

5.13. WIEN BRIDGE OSCILLATOR

It is one of the most popular type of oscillators used in audio and sub-audio frequency ranges (20 – 20 kHz). This type of oscillator is simple in design, compact in size, and remarkably stable in its frequency output. Furthermore, its output is relatively free from distortion and its frequency can be varied easily. However, the maximum frequency output of a typical Wien bridge oscillator is only about 1 MHz.

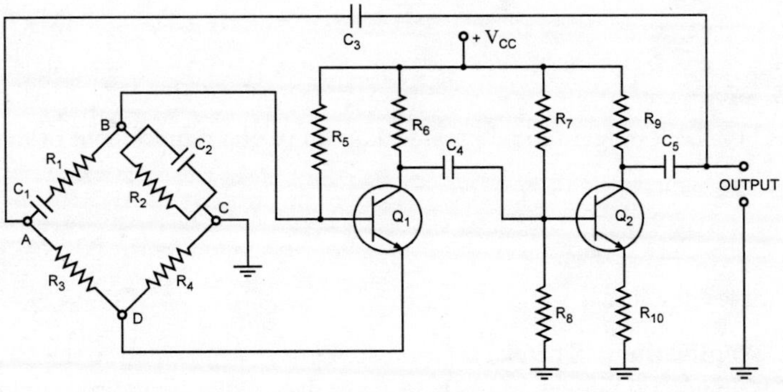

Fig. 5.21. *Wien Bridge Oscillator Circuit*

This is also, in fact, a phase shift oscillator. It employs two transistors, each producing a phase shift of 180°, and thus producing a total phase shift of 360° or 0°.

The circuit diagram of Wien bridge oscillator is shown in Fig. 5.21. It is essentially a two-stage amplifier with an R-C bridge circuit. R-C bridge circuit (Wien bridge) is a lead-lag network (R_1-C_1 and R_2-C_2). The phase shift across the network lags with increasing frequency and leads with decreasing frequency. By adding Wien bridge feedback network, the oscillator becomes sensitive to a signal of only one particular frequency. This particular frequency is that at which Wien bridge is balanced and for which the phase shift is 0°. If the Wien bridge feedback network is not employed and output of transistor Q_2 is fed back to transistor Q_1 for providing regeneration required for producing oscillations, the transistor Q_1 will amplify signals over a wide range of frequencies and thus direct coupling would result in poor frequency stability. Thus, by employing Wien bridge feedback network frequency stability is increased.

In the bridge circuit R_1 in series with C_1, R_3, R_4 and R_2 in parallel with C_2 form the four arms.

From the analysis of the bridge circuit it is obvious that the bridge will be balanced only when

$$R_3\left[\frac{R_2}{1+j\omega C_2R_2}\right] = R_4\left(R_1 - \frac{j}{\omega C_1}\right)$$

$$\text{or } R_2R_3 = R_4(1+j\omega C_2R_2)(R_1 - j/\omega C_1)$$

$$\text{or } R_2R_3 - R_4R_1 - \frac{C_2}{C_1}R_2R_4 + \frac{jR_4}{\omega C_1} - j\omega C_2R_2R_1R_4 = 0$$

Separating real and imaginary terms we have

$$R_2R_3 - R_4R_1 - \frac{C_2}{C_1}R_2R_4 = 0$$

$$\text{or } \frac{C_2}{C_1} = \frac{R_3}{R_4} - \frac{R_1}{R_2} \quad ...(5.33)$$

$$\text{and } \frac{R_4}{\omega C_1} - \omega C_2R_2R_1R_4 = 0$$

$$\text{or } \omega^2 = \frac{1}{C_1C_2R_1R_2}$$

$$\text{or } \omega = \frac{1}{\sqrt{C_1C_2R_1R_2}}$$

$$\text{and } f = \frac{1}{2\pi\sqrt{R_1C_1R_2C_2}} \quad ...(5.34)$$

$$\text{if } C_1 = C_2 = C \text{ and } R_1 = R_2 = R, \text{ then } f = \frac{1}{2\pi CR} \quad ...(5.35)$$

$$\text{and } R_3 = 2R_4 \quad ...(5.36)$$

Thus, we see that in a bridge circuit the output will be in phase with the input only when the bridge is balanced *i.e.,* at resonant frequency given by Eq. (5.35). At all other frequencies the bridge is off-balance *i.e.,* the voltage fed back and output voltage do not have the correct phase relationship for sustained oscillations.

So this bridge circuit can be used as feedback network for an oscillator, provided that the phase shift through the amplifier is zero. This requisite condition is achieved by using a two-stage amplifier, as illustrated in Fig. 5.21. In this arrangement the output of the second stage is supplied back to the feedback network and the voltage across the parallel combination C_2R_2 is fed to the input of the first stage. Transistor Q_1 serves as an oscillator and amplifier whereas the transistor Q_2 as an invertor to cause a phase shift of 180°. The circuit uses positive and negative feedbacks. The positive feedback is through R_1, C_1, R_2, C_2 to transistor Q_1 and negative feedback is through the voltage divider $R_3 - R_4$ to the emitter of transistor Q_1. Resistors R_3 and R_4 are used to stabilize the amplitude of the output.

The two transistors Q_1 and Q_2 thus cause a total phase shift of 360° and ensure proper positive feedback. The negative feedback is provided in the circuit to ensure constant output over a range of frequencies. This is achieved by taking resistor R_4 in the form of a temperature sensitive lamp, whose resistance increases with the increase in current. In case the amplitude of the output tends to increase, more current would provide more negative feedback. Thus the output would regain its original value. A reverse action would take place in case the output tends to fall.

The amplifier voltage gain, $A = \frac{R_3 + R_4}{R_4} = \frac{R_3}{R_4} + 1 = 3$ ...(5.37) $\because$ from Eq. (5.36) $R_3 = 2R_4$

The above corresponds with the feedback network attentuation of 1/3. Thus, in this case, voltage gain A must be equal to or greater than 3, to sustain oscillations.

To have a voltage gain of 3 is not difficult. On the other hand, to have a gain as low as 3 may be difficult. For this reason also negative feedback is essential.

Operation. The circuit is set in oscillation by any random change in base current of transistor Q_1, that may be due to noise inherent in the transistor or variation in voltage of dc supply. This variation in base current is amplified in collector circuit of transistor Q_1 but with a phase shift of 180°. The output of transistor Q_1 is fed to the base of second transistor Q_2 through capacitor C_4. Now a still further amplified and twice phase-reversed signal appears at the collector of the transistor Q_2. Having been inverted twice, the output signal will be in phase with the signal input to the base of transistor Q_1. A part of the output signal at transistor Q_2 is fed back to the input points of the bridge circuit (points A-C). A part of this feedback signal is applied to emitter resistor R_4 where it produces degenerative effect (or negative feedback). Similarly, a part of the feedback signal is applied across the base-bias resistor R_2 where it produces regenerative effect (or positive feedback). At the rated frequency, effect of regeneration is made slightly more than that of degeneration so as to obtain sustained oscillations.

The continuous frequency variation in this oscillator can be had by varying the two capacitors C_1 and C_2 simultaneously. These capacitors are variable air-gang capacitors. We can change the frequency range of the oscillator by switching into the circuit different values of resistors R_1 and R_2.

FET Wien bridge oscillator is shown in Fig. 5.22.

The advantages and disadvantages of Wien bridge oscillators are given below :

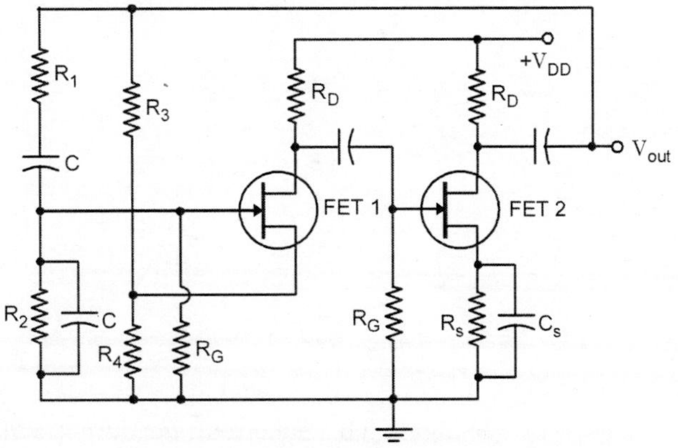

Fig. 5.22. ***FET Wien Bridge Oscillator***

Advantages. 1. It provides a stable low distortion sinusoidal output over a wide range of frequency.

2. The frequency range can be selected simply by using decade resistance boxes.
3. The frequency of oscillation can be easily varied by varying capacitances C_1 and C_2 simultaneously.
4. The overall gain is high because of two transistors.

Disadvantages. 1. The circuit needs two transistors and a large number of other components.

2. The maximum frequency output is limited because of amplitude and the phase shift characteristics of the amplifier.

Example 5.11. Design the R-C elements of a Wien bridge oscillator for operation at f_0 = 10 kHz.

Solution : Using values of R_1 equal to R_2 and C_1 equal to C_2 and taking each of R_1 and R_2 of 200 kΩ,

$$\text{Required value of } C = \frac{1}{2\pi f R} = \frac{1}{2\pi \times 10 \times 10^3 \times 200 \times 10^3} = 79.6 \text{ pF } \textbf{Ans.} \qquad \text{Refer to Eq. (5.35)}$$

We can use $R_3 = 400\ \text{k}\Omega$ and $R_4 = 200\ \text{k}\Omega$ **Ans.**

Example 5.12. Design the R-C elements of a Wien bridge oscillator shown in Fig. 5.21 for operation at *f* = 15 kHz.

Solution : Using values of R_1 equal to R_2 and C_1 equal to C_2 and taking each of R_1 and R_2 of 200 kΩ,

$$\text{Required value of } C = \frac{1}{2\pi f R} = \frac{1}{2\pi \times 15 \times 10^3 \times 200 \times 10^3} = 53 \text{ pF } \textbf{Ans.} \qquad \text{Refer to Eq. (5.35)}$$

We can use $R_3 = 400\ \text{k}\Omega$ and $R_4 = 200\ \text{k}\Omega$ **Ans.**

Example 5.13. Find the frequency of oscillations of a Wien bridge oscillator with R = 20 kΩ and C = 1,000 pF.

Solution :

$$R = R_1 = R_2 = 20\ \text{k}\Omega = 20{,}000\ \Omega$$

$$C = C_1 = C_2 = 1{,}000 \text{ pF} = 1 \times 10^{-9} \text{ F}$$

$$\text{Frequency of oscillation, } f = \frac{1}{2\pi RC} = \frac{1}{2\pi \times 20{,}000 \times 1 \times 10^{-9}} = 7.96 \text{ kHz } \textbf{Ans.}$$

Example 5.14. If resistors and capacitors in the circuit have the values of 8.2 kΩ and 0.05 μF, determine the frequency of oscillation. [*G.G.S.I.P. Univ. Analog Electronics-II May 2009*]

Solution :

$$R = R_1 = R_2 = 8.2\ \text{k}\Omega = 8{,}200\ \Omega$$

$$C = C_1 = C_2 = 0.05\ \mu\text{F} = 0.05 \times 10^{-6} \text{ F}$$

$$\text{Frequency of oscillation, } f = \frac{1}{2\pi RC} = \frac{1}{2\pi \times 8{,}200 \times 0.05 \times 10^{-6}} = 388.2 \text{ Hz } \textbf{Ans.}$$

5.14. SELECTION OF OSCILLATOR

While selecting an oscillator for a particular application, the following particulars are to be considered.

1. ***Frequency Range.*** The oscillator selected for a particular application should be capable of supplying an output signal whose upper and lower frequency limits exceed those required by the application.
2. ***Power and/or Voltage Requirements.*** The oscillator selected for a particular application should be capable of generating the pertinent quantity with a magnitude large enough to meet the requirement.
3. ***Accuracy and Dial Resolution.*** The accuracy of an oscillator specifies how closely the output frequency corresponds to the frequency indicated on the dial of the instrument. Dial resolution indicates to what percentage of the output frequency value the dial setting can be read.
4. ***Amplitude and Frequency Stability.*** The amplitude stability is a measure of an oscillator's ability of maintaining a constant voltage amplitude with variations in the output signal frequency.

   Frequency stability determines how closely the oscillator maintains a constant frequency over a given time period. Sometimes the frequency stability is included in the accuracy specifications of the oscillator.
5. ***Waveform Distortion.*** This quantity is a measure of how closely the output waveform of the oscillator resembles a pure sinusoidal signal. Sometimes the oscillator is employed as a source in a test used

for measuring the tendency of a circuit to distort a sinusoidal signal. In such tests, the distortion caused by the oscillator should be much less than the anticipated distortion because of the circuit under test.

6. ***Output Impedance.*** The output impedance of an oscillator specifies the impedance value of the load which must be connected to it for maximum power transfer. It is very important that the output impedance of the oscillator be equal to the characteristic impedance of the system to which it is to be connected.

HIGHLIGHTS

1. An oscillator is a device that generates a periodic, ac output signal without any form of input signal required. Since sinusoidal waveforms are encountered so frequently in electronic measurement work, sinusoidal oscillators represent the largest single category of signal generators. These devices cover the frequency range from a few Hz to many GHz.
2. The active devices used in oscillators may be BJTs, FETs or op-amps, and passive components used may be resistors, inductors and capacitors.
3. The introduction of positive feedback in an amplifier causes an increase in voltage gain. If the loop gain $A\beta \geq 1$, the Barkhausen criterion for oscillation is satisfied, and an output remains even without the input.

   For linear oscillation, in which the output is sinusoidal with little distortion, the loop gain, $A\beta$, should be only slightly greater than unity.

   The other condition that is required to be fulfilled for sustained oscillations is that the phase shift around the circuit must be zero.
4. The three elements that an oscillator must have are: 1. Oscillatory circuit or frequency determining element, 2. Amplifier, 3. Positive feedback network.
5. The oscillators may be categorised broadly into two classes viz. audio frequency oscillators and radio frequency oscillators. Phase shift and Wien bridge oscillators fall under the category of audio oscillators and are employed for generating low frequency sinusoidal signals. Colpitt's and Hartley oscillators fall under the category of radio frequency oscillators and are very popular for generating high frequency (*e.g.,* 10 kHz to 100 MHz) outputs.

   RF sinusoidal oscillators, also called the L-C oscillators, make use of L-C circuits as their tank or oscillatory circuits. L-C oscillators cannot be used for generation of AF signals because the components required in construction of low frequency L-C resonant circuits are too bulky and heavy.
6. Oscillators, which use inductance-capacitance (L-C) circuits as their tank or oscillatory circuits are called *L-C oscillators*. L-C oscillators are very popular for generating high frequency outputs (*e.g.,* 10 kHz to 100 MHz). There is a large variety of L-C oscillators such as tuned collector oscillators, tuned base oscillators, Colpitt's oscillators, Hartley oscillators, Clapp oscillators, crystal oscillators etc.
7. The Colpitt's oscillator is an example of a resonant circuit oscillator and is characterised by a split capacitor in its tank circuit: The frequency of oscillation is given by

$$f = \frac{1}{2\pi}\sqrt{\frac{1}{LC_1} + \frac{1}{LC_2}}$$

8. The Hartley oscillator has a tapped coil in the tank circuit. Like the Colpitt's, the Hartley oscillator is used for generation of radio frequency (RF) sinusoidal signals. The frequency of oscillation is given by

$$f = \frac{1}{2\pi\sqrt{C(L_1 + L_2 + 2M)}}$$

9. In crystal oscillators, the usual electrical resonant circuit is replaced by a mechanically vibrating crystal. The crystal (usually quartz) has a high degree of stability in holding constant at whatever frequency the crystal is originally cut to operate. The crystal oscillators are, therefore, used whenever great stability is needed, for example, in communication transmitters, and receivers, digital clocks etc.
10. A phase shift oscillator relies upon an R-C phase shift network to provide the necessary phase relationship between output and input to an amplifier. This network attenuates the output voltage by a factor β =

$\frac{1}{29}$. So the amplifier must have a voltage gain of 29 or more. The frequency of oscillation is given by

$f = \frac{1}{2\pi RC\sqrt{6}}$ where R is in ohms and C in farads.

11. A Wien bridge oscillator makes use of a frequency-selective bridge circuit to provide low distortion sinusoidal oscillations at a frequency given by

$$f = \frac{1}{2\pi\sqrt{R_1R_2C_1C_2}} = \frac{1}{2\pi RC}$$

if $R_1 = R_2 = R$ and $C_1 = C_2 = C$.

12. The (*i*) frequency range, (*ii*) power and/or voltage requirements, (*iii*) accuracy and dial resolution, (*iv*) amplitude and frequency stability, (*v*) waveform distortion and (*vi*) output impedance of an oscillator determines its suitability for a particular application.

EXERCISES ....

1. Explain oscillator principles and frequency stability. **[G.G.S.I.P. Univ. Analog Electronics-II May-June 2009]**
2. Describe the conditions for sustained oscillation and frequency stability in positive feedback amplifier. Classify the oscillators giving frequency range and the type of waveform generated. **[G.G.S.I.P. Univ. Analog Electronics-II May-June 2009]**
3. Explain the two conditions for oscillation in feedback amplifier. **[G.G.S.I.P. Univ. Analog Electronics-II May 2011]**
4. Give the two Barkhausen conditions required for sinusoidal oscillations to be sustained with the help of neat diagrams.
5. Sketch the topology for a generalized resonant-circuit oscillator using impedances Z_1, Z_2, Z_3. At what frequency will the circuit oscillate ?
6. Draw the circuit diagram of a Colpitt's oscillator using transistor. Derive an expression for its frequency of oscillations. Deduce the starting condition for this oscillator.
7. Draw a neat circuit diagram of a Colpitt's oscillator using NPN transistor. Give its equivalent circuit. Derive expressions for the following : (*i*) The frequency of oscillations (*ii*) The maximum gain for sustained oscillations.
8. State Barkhausen criteria and explain the working of Hartley oscillator.
9. Explain briefly, how the oscillations are maintained in Hartley oscillator. Also derive the expression of frequency of oscillation.
10. What is crystal oscillator ? How it works ? Draw its equivalent circuit and derive results for impedance, resonant frequency.
11. What is a piezoelectric crystal ? Sketch the circuit symbol, an electrical equivalent model and the variation of the reactance of the crystal with frequency. Explain how this crystal can be incorporated in an electronic circuit to form a stable oscillator.
12. What are the advantages of using crystal oscillators ? Mention its applications.
13. Explain the condition for self sustained sinusoidal oscillations. Name an oscillator which gives highly stable fixed frequency oscillations. Explain its working. **[G.G.S.I.P. Univ. Analog Electronics-II May 2011]**
14. Compare R-C phase shift and crystal oscillators.
15. What are the requirements of an oscillator circuit ? Draw a neat diagram of a phase shift oscillator using BJT. What advantage has the phase shift oscillator in the audio frequency.
16. For an R-C phase shift oscillator, derive the expression for the frequency of oscillation and condition for sustained oscillations. **[G.G.S.I.P. Univ. Analog Electronics-II May 2010]**
17. Draw a neat circuit diagram of the R-C Wein bridge oscillator and derive the expression for its condition of oscillation and frequency of oscillation. **[G.G.S.I.P. Univ. Analog Electronics-II February 2009]**
18. Explain the working of Wein bridge oscillator and derive the expression for the frequency of oscillation. **[G.G.S.I.P. Univ. Analog Electronics-II May 2011]**
19. Explain the working of Wein bridge oscillator. Derive the frequency of oscillation and condition for oscillation. **[G.G.S.I.P. Univ. Analog Electronics-II, May 2011]**

20. Sketch the circuit for a Wien bridge oscillator. What determines the frequency of oscillation ? Will oscillations take place if the bridge is balanced ?

21. Sketch the circuit schematic of a Wien bridge oscillator and explain its operation. Derive an expression for the frequency of oscillation and suggest a suitable scheme for amplitude stabilization of the oscillator.

22. Write the important factors in choosing a certain type of oscillator.

Why do we use R-C oscillators for audio frequencies and L-C oscillators for RF frequencies ? Why does not an L-C tank circuit produce sustained oscillations ? Explain how the low frequency parasitics can be suppressed in an oscillator.

23. Write short notes on any *four* of the following :

(*a*) Basic principles of sinusoidal oscillator

(*b*) Crystal oscillator

(*c*) Clap oscillator

(*d*) Hartley oscillator

(*e*) Phase shift oscillator

24. Write short notes on the following :

(*i*) Stability criteria for an oscillator.

(*ii*) Hartley and Colpitt's oscillators.

(*iii*) Hartley oscillator.

(*iv*) Crystal oscillator.

(*v*) R-C phase shift oscillator.

(*vi*) Wien-bridge oscillator.

SHORT ANSWER TYPE QUESTIONS WITH ANSWERS

Q. 1. What is an oscillator ?

Ans. Oscillator may be defined as a circuit which generates an ac output signal of very high frequency without requiring any externally applied input signal or it may be defined as an electronic source of alternating current (or voltage) having sinusoidal or non-sinusoidal (square, sawtooth or pulse) waveshape. It can also be defined as a circuit that converts dc energy into very high frequency ac energy.

Q. 2. What is an oscillator ? How does it differ from an amplifier ?

Ans. Refer to Answer of Q. 1.

The difference between an amplifier and an oscillator can be explained with the help of Fig. 1. As we know that, an amplifier strengthens the input signal without any change in its waveform and frequency. The additional power required comes from the external dc source. Thus an amplifier is essentially an *energy converter* that draws energy from a dc supply and converts it into ac energy at signal frequency, the energy conversion process being controlled by the input signal. On the other hand an oscillator does not require any external signal either to start or maintain the process of energy conversion and the energy conversion process is controlled by the oscillator circuit itself, as illustrated in Fig. 1. In an oscillator, the output signal frequency depends on the passive components employed in the circuit and can be varied as per needs. Oscillator may provide fixed or variable frequency.

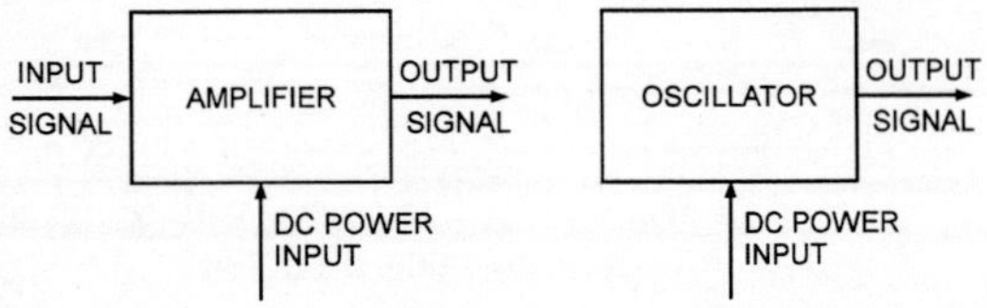

Fig. 1

Q. 3. What is Barkhausen criterion for the feedback oscillator.

Ans. Barkhausen conditions for sustained oscillations are : (*i*) the loop gain of the circuit must be ≥ 1 and (*ii*) the phase shift around the circuit must be zero.

Q. 4. Explain why positive feedback and not negative feedback is necessary to produce oscillations.

Ans. With negative feedback, feedback factor $A\beta$ becomes less than unity, $A\beta V_{in}$ less than V_{in} and therefore output signal will die out while with positive feedback $A\beta$ exceeds unity, $A\beta V_{in}$ exceeds V_{in} and oscillations are produced. That is why positive feedback is necessary to produce oscillations.

Q. 5. Every electronic oscillator can be considered to be an amplifier with infinite gain. Explain where you get the starting signal voltage from.

Ans. Due to the thermal energy, electrons are at random vibrations. These vibrations give rise to small potentials. All oscillators use positive feedback. The small voltage due to movement of electrons due to thermal energy is repeatedly amplified in the oscillator circuit and potential buildup occurs. This results in oscillations. To make these oscillations, dc biasing voltage is essential.

Q. 6. Why are L-C resonant circuits impractical at audio frequencies ?

Ans. L-C resonant circuits are impractical at audio frequencies because components required in their construction for low frequency operation are too bulky and heavy and expensive.

Q. 7. How does Hartley oscillator differ from Colpitt's oscillator in construction ?

Ans. Hartley oscillator circuit is similar to Colpitt's oscillator except that phase shift network consists of two inductors and one capacitor instead of two capacitors and one inductor.

Q. 8. Why is crystal oscillator used in radio transmitter ?

Ans. In crystal oscillators, the usual electrical resonant circuit is replaced by a mechanically vibrating crystal. The crystal (usually quartz) has a high degree of stability in holding constant at whatever frequency the crystal is originally cut to operate. The crystal oscillators are, therefore, used whenever great stability is needed, for example, in communication transmitters, and receivers, digital clocks etc.

Q. 9. Why do you use three R-C sections in R-C phase shift oscillator ?

Ans. In a phase shift oscillator, if the loading of the phase-shift network on the amplifier can be assumed to be negligible, a phase shift of 180° between the amplified output voltage v_{out} and the input voltage v_{in} is produced by the amplifier. The three section R-C phase-shift network is provided to produce an additional phase shift, which is a function of frequency and equals 180° at some frequency of operation. Thus, at the frequency of operation, the total shift from the gate around the circuit and back to the gate will be exactly zero.

Q. 10. Why we use two transistors in a Wien bridge oscillator ?

Ans. Wien bridge oscillator employs two transistors so as to obtain 360° or 0° phase shift between output and input voltages.

Q. 11. Sometimes a lamp is used in one of the resistance arms of a Wien bridge oscillator. Why ?

Ans. A lamp is used in one of the resistance arms of a Wien bridge oscillator so as to provide a negative feedback, in proportion to the output current, in the circuit and thus ensure constant output over a range of frequencies.

Q. 12. Why is –ve feedback provided in Wien bridge oscillator ?

Ans. Negative feedback is provided in Wien bridge oscillator to stabilize the amplitude of the output.

PROBLEMS

1. Determine the operating frequency of a Colpitt's oscillator shown in Fig. 5.7 if L = 50 μH, L_{RFC} = 0.8 mH, C_1 = 0.01 μF, C_2 = 0.02 μF, C_C = 20 μF. [**Ans.** 276 kHz]
2. Determine the oscillation frequency of a transistor Hartley oscillator shown in Fig. 5.9 with circuit values L_1 = 150 μH, L_2 = 1.5 mH, M = 75 μH and C = 150 pF. [**Ans.** 306 kHz]
3. In a phase shift oscillator shown in Fig. 5.17 if $R_1 = R_2 = R_3$ = 400 kΩ and $C_1 = C_2 = C_3$ = 50 pF. Determine the frequency of oscillation. [**Ans.** 3.25 kHz]
4. Design the phase shifting network for oscillating frequency of 1 kHz. [**Ans.** $R_1 = R_2 = R_4$ = 200 kΩ, R_3 = 400 kΩ, C = 796 pF]
5. In a Wien bridge oscillator shown in Fig. 5.21 if $R_1 = R_2$ = 110 kΩ and $C_1 = C_2$ = 1,600 pF. Determine the frequency of oscillation. [**Ans.** 904 Hz]

❑❑❑

CHAPTER

6

IC Waveform Generators

6.1. INTRODUCTION

Basically, the function of an oscillator (waveform generator) is to generate alternating current or voltage waveforms. There are two distinctly different types of waveform generators : sinusoidal oscillators, which utilise some form of resonance and non-sinusoidal oscillators or function generators, which employ switching mechanism implemented with a multivibrator circuit.

INSIDE THIS CHAPTER

6.2. SINUSOIDAL WAVEFORM GENERATORS

A sinusoidal oscillator can be realized by placing a frequency-selective network in the feedback path of an amplifier (a transistor or an op-amp). The circuit will oscillate at the frequency at which the total phase shift around the loop is zero, provided that the magnitude of the loop gain at this frequency is equal to, or slightly greater than unity.

Sinusoidal oscillators using discrete transistors have already been discussed in Chapter 5. Here we will discuss only the sinusoidal oscillators using op-amp as an active device. The theory discussed in case of discrete transistor oscillators will equally apply with IC oscillators.

6.3. PHASE SHIFT OSCILLATOR

The R-C phase shift oscillator [Fig. 6.1 (*a*)] which generates waveform at audio frequency consists of an op-amp which operates as an amplifier and provides 180° phase shift as it is used as an inverting amplifier. The extra 180° phase shift is provided by a R-C cascaded network which is connected as the feedback circuit. Thus, the total phase shift becomes 360° (or 0°).

The R-C cascaded network provides 180° phase shift at some specific frequency say, f_0 (called frequency of oscillation) and the amplifier gain is sufficiently large to produce oscillations.

We can easily calculate the transfer function for the feedback network. The feedback network has been redrawn, as, shown in Fig. 6.1 (*b*).

Writing KVL equation for each loop, we have

$$V_f(s) = I_1(s)\left(\frac{1}{sC} + R\right) - R I_2(s) \qquad ...[6.1(a)]$$

$$0 = -RI_1(s) + \left(2R + \frac{1}{sC}\right) I_2(s) - RI_3(s) \qquad ...[6.1(b)]$$

$$-V_{out}(s) = -RI_2(s) + \left(R + \frac{1}{sC}\right) I_3(s) \qquad ...[6.1(c)]$$

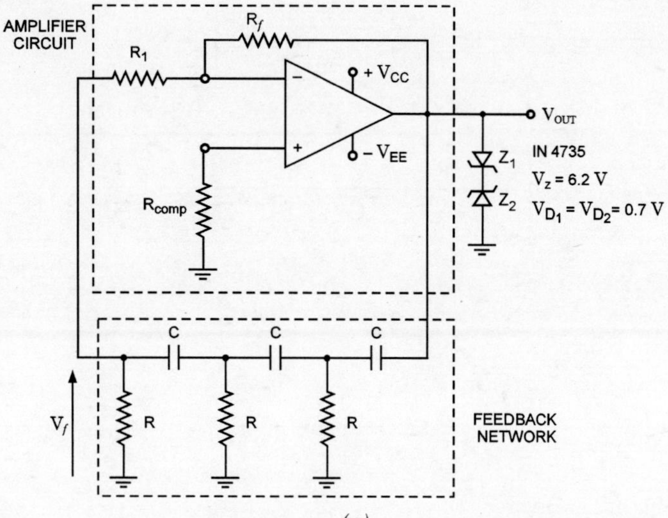

(*a*)

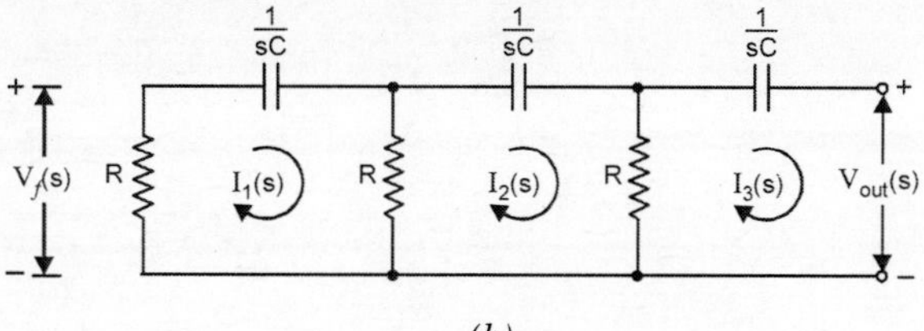

(*b*)

Fig. 6.1

In matrix form :

$$\begin{bmatrix} V_f(s) \\ 0 \\ -V_{out}(s) \end{bmatrix} = \begin{bmatrix} R + \frac{1}{sC} & -R & 0 \\ -R & 2R + \frac{1}{sC} & -R \\ 0 & -R & R + \frac{1}{sC} \end{bmatrix} \begin{bmatrix} I_1(s) \\ I_2(s) \\ I_3(s) \end{bmatrix} \qquad ...(6.2)$$

$$I_1(s) = \frac{\begin{bmatrix} V_f(s) & -R & 0 \\ 0 & 2R + \frac{1}{sC} & -R \\ -V_{out}(s) & -R & R + \frac{1}{sC} \end{bmatrix}}{\Delta} = \frac{V_f(s)\left(R^2 + \frac{3R}{sC} + \frac{1}{s^2C^2}\right) - V_{out}(s) \cdot R^2}{\Delta} \qquad ...(6.3)$$

$$\text{where } \Delta = \begin{bmatrix} R + \frac{1}{sC} & -R & 0 \\ -R & 2R + \frac{1}{sC} & -R \\ 0 & -R & R + \frac{1}{sC} \end{bmatrix} = \frac{3R^2}{sC} + \frac{4R}{s^2C^2} + \frac{1}{s^3C^3} \qquad ...(6.4)$$

Now from Fig. 6.1(*b*)

$$-V_f(s) = I_1(s)R \qquad ...(6.5)$$

$$\text{or } -V_f(s) = \frac{\left[V_f(s)\left(R^2 + \frac{3R}{sC} + \frac{1}{s^2C^2}\right) - V_{out}(s) R^2\right] R}{\Delta}$$

$$\text{or } -V_f(s)\Delta = V_f(s)\left(R^3 + \frac{3R^2}{sC} + \frac{R}{s^2C^2}\right) - V_{out}(s)R^3$$

$$\text{or}\quad V_f(s)\left[R^3 + \frac{3R^2}{sC} + \frac{R}{s^2C^2} + \Delta\right] = V_{out}(s)\,R^3$$

Substituting the value of Δ from Eq. (6.4) in above equation, we have

$$\text{or}\quad V_f(s)\left[R^3 + \frac{3R^2}{sC} + \frac{R}{s^2C^2} + \frac{3R^2}{sC} + \frac{4R}{s^2C^2} + \frac{1}{s^3C^3}\right] = V_{out}(s)\,R^3$$

$$\text{or}\quad \frac{V_f(s)}{V_{out}(s)} = \frac{R^3}{R^3 + \dfrac{6R^2}{sC} + \dfrac{5R}{s^2C^2} + \dfrac{1}{s^3C^3}} \qquad ...(6.6)$$

$$\text{or feedback gain, } \beta = \frac{V_f(s)}{V_{out}(s)} = \frac{1}{1 + \dfrac{6}{sCR} + \dfrac{5}{s^2C^2R^2} + \dfrac{1}{s^3C^3R^3}} \qquad ...(6.7)$$

Substituting $s = j\omega$ in Eq. (6.7), we have

$$\beta = \frac{1}{1 + \dfrac{6}{j\omega CR} + \dfrac{5}{(j\omega CR)^2} + \dfrac{1}{(j\omega CR)^3}}$$

$$= \frac{1}{1 - \dfrac{5}{(\omega CR)^2} - j\left[\dfrac{6}{\omega CR} - \dfrac{1}{(\omega CR)^3}\right]} \qquad ...(6.8)$$

To satisfy frequency of oscillation criteria, equate the imaginary components of Eq. (6.8) to zero (so that $A\beta = 1$ for which β must be real quantity).

$$\text{Hence}\quad \frac{6}{\omega CR} - \frac{1}{(\omega CR)^3} = 0$$

$$\text{or}\quad \frac{1}{\omega CR}\left(6 - \frac{1}{\omega^2C^2R^2}\right) = 0$$

$$\text{or}\quad \frac{1}{\omega CR} = \sqrt{6}$$

$$\text{or } f = \frac{1}{2\pi CR\sqrt{6}} \qquad ...(6.9)$$

This is frequency of oscillation.

Also, the loop gain, $\beta A = -1$ (Barkhausen criterion)

or $\beta A = 1\angle 180^\circ$

$$\text{or}\quad \frac{1}{\left[1 - \dfrac{5}{(2\pi f CR)^2}\right]}A = 1\angle 180^\circ$$

$$\text{or}\quad \frac{1}{[1 - 5(\sqrt{6})^2]}A = 1\ 1\angle 180^\circ$$

$$\text{or}\quad A \geq -29 \qquad ...(6.10)$$

i.e., the inverting amplifier should have gain of at least 29.

or say $\left|\frac{R_f}{R_1}\right| = 29$ ...[6.11(*a*)]

i.e., $R_f = 29\ R_1$. ...[6.11(*b*)]

Thus, the circuit will produce a sinusoidal waveform of frequency *f* if the gain is at least 29 and total phase shift of the circuit is 360°.

The gain is kept greater than 29 so that any variation in circuit parameters cannot make $|A\beta| < 1$. Otherwise oscillations will die out.

The op-amp 741 can be used only at low frequencies (<1 kHz). For high frequencies LM 318 or LF 351 are employed.

Example 6.1. Design a phase shift oscillator using FET or op-amp to work on 2 kHz frequency. Assume that input resistance without feedback is very high.

Solution : Frequency of oscillation, $f_0 = 2\text{ kHz} = 2{,}000\text{ Hz}$

Assume $C = 0.1\ \mu F = 1 \times 10^{-7}\ F$

$$f_0 = \frac{1}{2\pi\sqrt{6}\,RC}$$

$$\text{or } R = \frac{1}{2\pi\sqrt{6} f_0 C} = \frac{1}{2\pi \times \sqrt{6} \times 2{,}000 \times 1 \times 10^{-7}} = 0.325\text{ k}\Omega \textbf{ Ans.}$$

Now to provide required gain to oscillator circuit and to avoid loading of the amplifier due to R-C cascaded network

$$R_1 \geq 10R = \geq 10 \times 0.325 \geq 3.25\text{ k}\Omega \text{ say, } 3.3\text{ k}\Omega$$

$$\text{and } R_F = 29R_1 = 29 \times 3.3 = 95.7\text{ k}\Omega \simeq 100\text{ k}\Omega \textbf{ Ans.}$$

Example 6.2. Design a phase-shift oscillator for an output frequency of 1 kHz. 741 IC op-amp with $V_{CC} \pm 10$ V may be used for it.

Solution : For 741 IC op-amp, $I_{B\,(max)} = 500\text{ nA}$

Let current through resistor R_1, $I_1 = 100 \times I_{B\,(max)} = 100 \times 500 \times 10^{-9} = 50\ \mu A$

Output voltage, $V_{out} = \pm (V_{CC} - 1\text{ V}) = \pm (10\text{ V} - 1\text{ V}) = \pm 9\text{ V}$

$$V_{in} = \frac{9\text{ V}}{29}$$

$$R_1 = \frac{V_{in}}{I_1} = \frac{9/29}{50 \times 10^{-6}} = 6.2\text{ k}\Omega$$

5.6 kΩ resistor may be used for R_1, being standard value resistor.

$$R_f = A \times R_1 = 29 \times 5.6 = 162.4\text{ k}\Omega$$

180 kΩ resistor may be used to provide A > 29.

$$R_{comp} = R_f = 180\text{ k}\Omega$$
$$R = R_1 = 5.6\text{ k}\Omega$$

$$\text{Capacitance of capacitor, } C = \frac{1}{2\pi R f\sqrt{6}} = \frac{1}{2\pi \times 5.6 \times 1{,}000 \times 1{,}000\sqrt{6}} = 0.0116\ \mu F$$

0.01 μF capacitor may be used.

$$\text{Thus } R_{comp} = R_f = 180\text{ k}\Omega$$
$$R = R_1 = 5.6\text{ k}\Omega$$
$$C = 0.01\ \mu F.$$

6.4. WIEN BRIDGE OSCILLATOR

It is another commonly used oscillator in audio and sub-audio frequency range (20–20 kHz). It is simple to design, compact and provides good frequency stability. The output waveform is relatively free from distortion and its frequency can be varied easily. The maximum frequency output of a typical Wien bridge oscillator is limited to about 1 MHz.

As shown in Fig. 6.2 (*a*) the Wien bridge, which is a lead-lag network of R-C, is introduced in between the amplifier input and the output.

The bridge has a R-C network in series connection and a parallel RC network in adjacent arm. The phase shift across the network lags with increasing frequency and leads with decreasing frequency. By introducing Wien bridge feedback network the oscillator becomes sensitive to a signal of only one particular frequency. This particular frequency is that at which Wien bridge is balanced and for which the phase shift is 0°. If the Wien bridge feedback network is not employed the output of amplifier is fed back to op-amp inputs for providing regeneration required for producing oscillations, the op-amp will amplify signals over a wide range of frequencies and thus direct coupling would result in poor frequency stability.

Thus by employing Wien bridge feedback network frequency stability is increased.

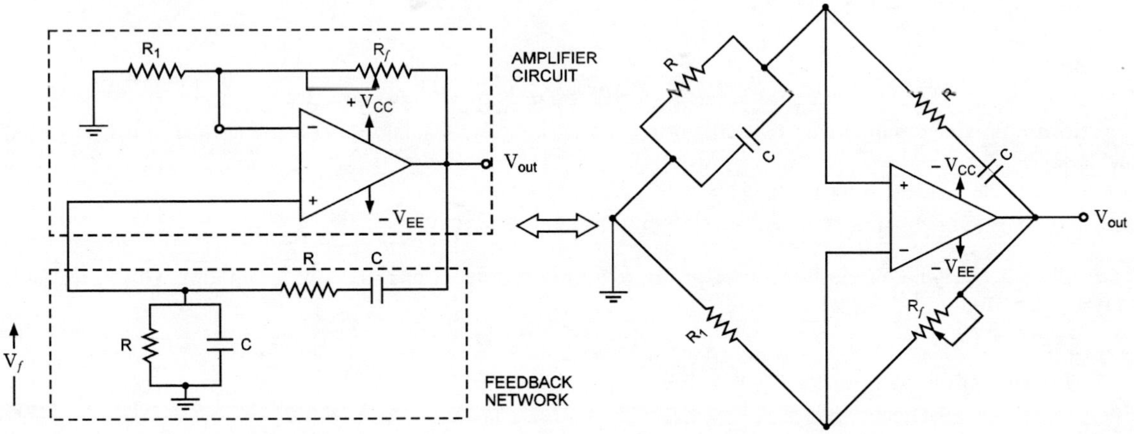

(*a*) *Wien Bridge Oscillator* (*b*) *Equivalent Circuit in Bridge Form*

Fig. 6.2

Now the feedback circuit can be redrawn to find $\beta = \dfrac{V_f}{V_{out}}$, as shown in Fig. 6.3.

$$\text{where } Z_1(s) = R \parallel \frac{1}{sC} = \frac{R}{1+sCR}$$

$$Z_2(s) = R + \frac{1}{sC} = \frac{1+sCR}{sC}$$

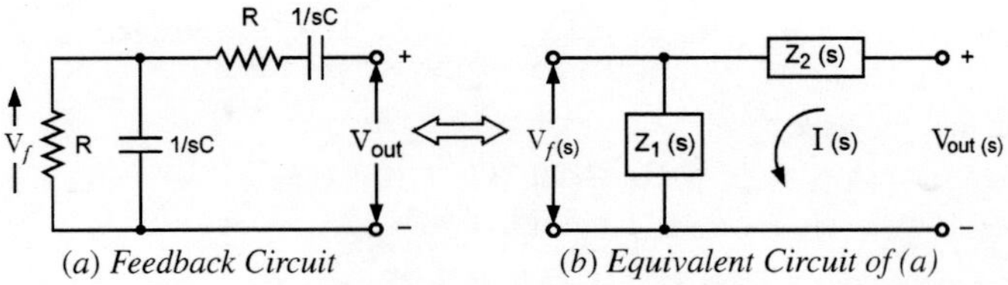

(*a*) *Feedback Circuit* (*b*) *Equivalent Circuit of* (*a*)

Fig. 6.3

Now Fig. 6.3(*b*) provides

$$\frac{V_f(s)}{V_{out}(s)} = \frac{I(s)\cdot Z_1(s)}{I(s)\cdot[Z_1(s)+Z_2(s)]} = \frac{R/(1+sCR)}{\dfrac{R}{1+sCR}+\dfrac{1+sCR}{sC}}$$

and feedback gain

$$\beta = \frac{V_f(s)}{V_{out}(s)} = \frac{sCR}{1+s^2C^2R^2+3sCR} = \frac{R}{\left(\dfrac{1}{sC}+sCR^2+3R\right)} \quad ...(6.12)$$

Substituting $s = j\omega$

$$\beta = \frac{R}{3R+j(\omega R^2C-1/\omega C)} \quad ...(6.13)$$

For unity feedback gain $A\beta = 1$

So β must be a real quantity.

Equating imaginary term to zero, we have

$$\omega R^2C - \frac{1}{\omega C} = 0$$

$$\text{or } \omega = \frac{1}{RC}$$

$$\therefore \text{ Frequency of oscillation, } f = \frac{1}{2\pi CR} \quad ...(6.14)$$

$$\text{Also loop gain } \beta A = 1$$

$$\text{or } \frac{1}{3}A = 1 \quad \because \beta = \frac{1}{3} \quad ...(6.15)$$

$$\text{or } A = 3$$

$$1 + \frac{R_f}{R_1} = 3$$

$$\text{or } R_f = 2R_1 \quad ...(6.16)$$

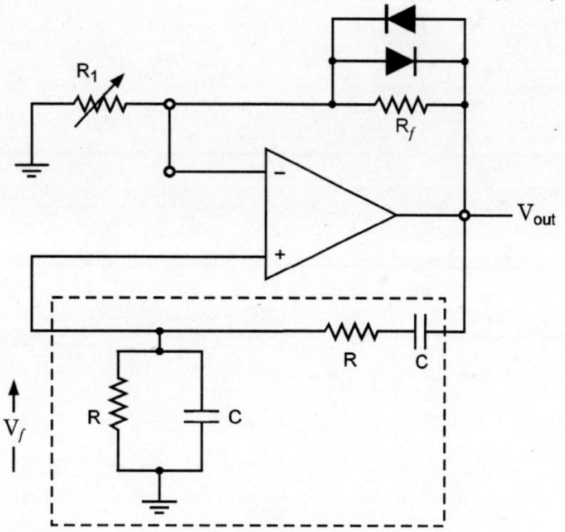

Fig. 6.4. ***Adjusted Negative Feedback Circuits For Amplifier***

Thus for sustained oscillations Eqs. (6.14) and (6.16) must be simultaneously satisfied.

Practically the gain of op-amp, A > 3 and if A < 3, the oscillations will die out or fail to start when power is on. For A > 3, oscillations will be growing.

Amplifier circuit is slightly improved by connecting variable resistor R_1 so that initially gain is adjustable to start oscillations. R_f is connected across two diodes, as shown in Fig. 6.4. Now the output signal grows until the drop across R_f = Cutin voltage of diode (0.7 V for silicon diode).

Example 6.3. Design a Wien bridge oscillator for an output frequency of 10 kHz. 741 IC op-amp with $V_C = \pm 10$ V may be used.

Solution : The amplifier maximum input current,

$$I_{B\,max} = 500 \text{ nA}$$

Let current through resistor R_1 be equal to 100 times $I_{B\,max}$

$$\text{So } I_1 = 100 \times 500 \times 10^{-9} = 50\ \mu A$$

$$\text{Output voltage, } V_{out} = \pm(V_{CC} - 1\text{ V}) = \pm(10\text{ V} - 1\text{ V}) = \pm 9\text{ V}$$

$$R_f + R_1 = \frac{9\text{ V}}{50\ \mu A} = \frac{9}{50\times10^{-6}} = 180\text{ k}\Omega$$

From Eq. (6.16) $\quad R_f = 2\,R_1$

$$\text{So } R_1 = \frac{180}{3} = 60\text{ k}\Omega \text{ (use 56 k}\Omega\text{ standard value resistor)}$$

$$R_f = 2R_1 = 2 \times 56 = 112\text{ k}\Omega \text{ (use 120 k}\Omega\text{ standard value resistor)}$$

This will make $R_f > 2R_1$ and $A > 3$

$$\text{Let } R = R_1 = 56\text{ k}\Omega$$

From Eq. (6.14)
$$C = \frac{1}{2\pi f R} = \frac{1}{2\pi \times 10 \times 10^3 \times 56 \times 10^3} = 284\text{ pF.}$$

300 pF standard capacitor may be used.

Thus $R_1 = 56\text{ k}\Omega$; $R_f = 120\text{ k}\Omega$, $R = 56\text{ k}\Omega$; and $C = 300$ pF **Ans.**

Example 6.4. Determine the frequency of oscillation of the circuit shown in Fig. 6.5. Assume op-amp to be ideal.

Solution : For an ideal op-amp

$$V_1 = V_2 = V_a$$

At Node 2,
$$\frac{V_a}{R_2} + \frac{V_a - V_{out}}{R_1} = 0$$

$$\text{or } V_{out} = V_a\left(\frac{1}{R_2} + \frac{1}{R_1}\right) \times R_1 = V_a\left(1 + \frac{R_1}{R_2}\right)$$

At Node 1,
$$\frac{V_a}{R} + j\omega CV_a + \frac{V_a - V_{out}}{R + \dfrac{1}{j\omega C}} = 0$$

$$\text{or } V_a\left(\frac{1}{R} + j\omega C\right)\left(R + \frac{1}{j\omega C}\right) + V_a - V_{out} = 0$$

$$\text{or } V_a \frac{(1 + j\omega CR)^2}{j\omega CR} + V_a - V_{out} = 0$$

$$\text{or } V_a (1 + 3j\omega CR + j^2\omega^2C^2R^2) = j\omega CRV_{out}$$

$$\text{or } 1 + 3j\omega CR - \omega^2C^2R^2 = j\omega CR\left(1 + \frac{R_1}{R_2}\right)$$

This is characteristic equation.

Substituting value of R_2 and R_1, we get

$$1 + 3j\omega CR - \omega^2C^2R^2 = 3j\omega CR$$

$$\text{or } \omega^2C^2R^2 = 1$$

$$\text{or } \omega = \frac{1}{RC} = \frac{1}{1 \times 10^3 \times 4.7 \times 10^{-6}} = 212.76 \text{ radians/second}$$

$$\text{or } f = \frac{\omega}{2\pi} = \frac{212.76}{2\pi} = 33.86 \text{ Hz } \textbf{Ans.}$$

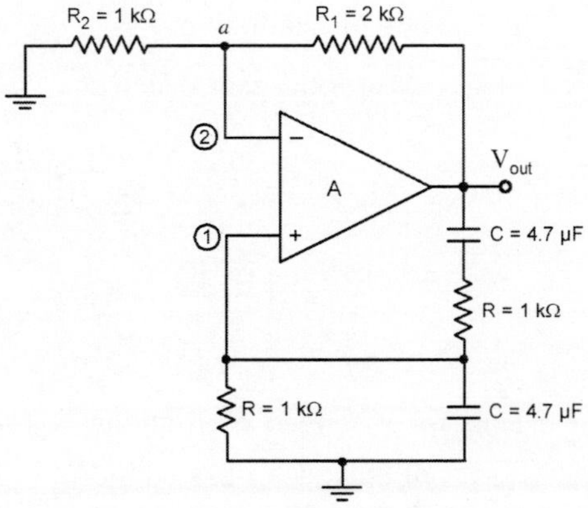

Fig. 6.5

6.5. COLPITT'S OSCILLATOR

The Colpitt's oscillator circuit is a superb circuit and is widely used in commercial signal generators up to 100 MHz. Colpitt's oscillator, using an inverting amplifier and a phase shift network consisting of an inductor and two capacitors is shown in Fig. 6.6. In this circuit an L-C network (L, C_1 and C_2) provides the required phase shift between amplifier output voltage and feedback voltage and acts as a filter to pass the desired oscillating frequency and block all other frequencies.

The filter circuit resonates at the desired oscillating frequency. For resonance.

$$X_L = X_C$$

where X_C is the reactance of the equivalent capacitance in parallel with the inductance. This provides the resonance or oscillating frequency,

$$f_0 = \frac{1}{2\pi\sqrt{LC}} \quad \text{...(6.17)}$$

$$\text{where } C = \frac{1}{\frac{1}{C_1} + \frac{1}{C_2}} = \frac{C_1C_2}{C_1 + C_2}$$

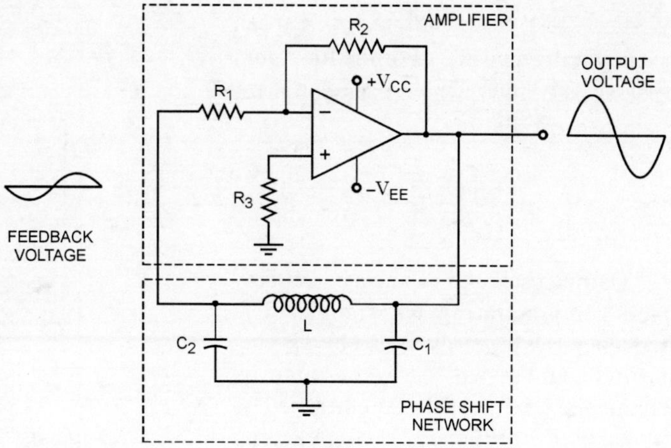

Fig. 6.6

Consideration of L-C network shows that its attenuation (from the amplifier output to input) is because of potential divider effect of L and C_2. This gives

$$\beta = \frac{X_{C_2}}{X_L - X_{C_2}}$$

It can be shown that for 180° phase shift

$$X_L - X_{C_2} = X_{C_1}$$

$$\text{So, } \beta = \frac{X_{C_2}}{X_{C_1}} = \frac{C_1}{C_2} \quad \text{...(6.18)}$$

As for oscillator circuits, the loop gain must be greater than unity to ensure that the circuit oscillates. So

$$A\beta \geq 1$$

$$\text{or } A \geq \frac{1}{\beta} \geq \frac{C_2}{C_1} \quad \text{...(6.19)}$$

In derivation of above equation, it has been assumed that the resistance of the inductor is negligibly small in comparison to the inductor impedance *i.e.*, Q factor $\left(\frac{\omega L}{R}\right)$ of the inductor is very large.

This is to be taken into account while selecting an inductor L. It has also been assumed that operational amplifier input impedance is very much larger than the impedance of capacitor C_2. Because of the negligible inductor resistance and amplifier input impedance, and because of stray capacitance effects where a high-frequency oscillator is involved, the amplifier gain usually has to be substantially larger than C_2/C_1.

The capacitors C_1 and C_2 are ganged. As the tuning is varied, the values of both capacitors increase or decrease simultaneously, but the ratio of the two capacitances remains the same.

6.6. HARTLEY OSCILLATOR

Hartley oscillator, using an inverting amplifier and a phase shift network made up of two inductors and a capacitor is shown in Fig. 6.7.

The attenuation of the phase shift network is

$$\beta = \frac{X_{L_2}}{X_{L_2} - X_C}$$

For a 180° phase shift, $(X_{L_2} - X_C)$ can be shown to equal X_{L_1}. For the loop gain to be at least unity, which is essential for sustained oscillations

$$A \geq \frac{1}{\beta} \geq \frac{X_{L_1}}{X_{L_2}} \geq \frac{L_1}{L_2} \qquad ...(6.20)$$

Hartley oscillator can also be suitably used for generating RF signals. The frequency can be easily varied by varying the inductances which can be done by making the core movable. Another method of varying frequency is of varying capacitance. Hartley oscillator is not suitable for low-frequency work because at low frequency, the value of inductance required becomes large.

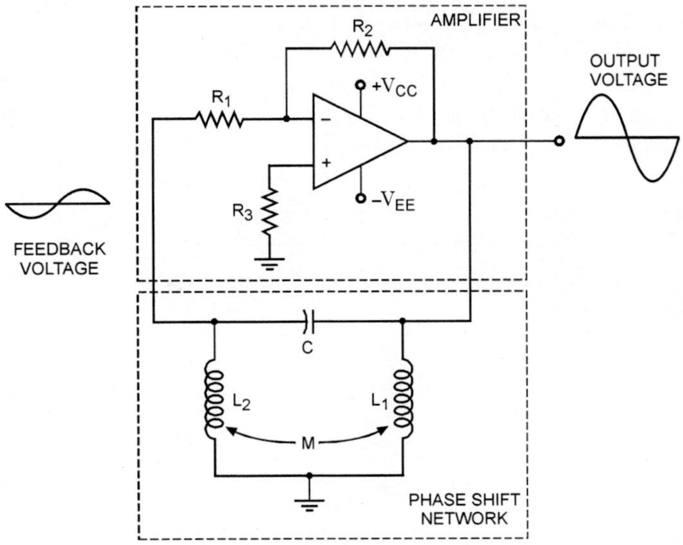

Fig. 6.7

6.7. CRYSTAL OSCILLATOR

Circuit diagram of a crystal oscillator using op-amp is shown in Fig. 6.8.

Equivalent circuit of a crystal may be given by Fig. 6.9.

The resonant frequency of the circuit is determined by the series resonance of the circuit made up of C_1, C_2, C_s and L_s; C_1 and C_2 are much larger than C_s. So the resonant frequency is almost entirely dependent on the value of C_s. Resonant frequency is given by

$$f = \frac{1}{2\pi\sqrt{L_s C_s}} \qquad ...(6.21)$$

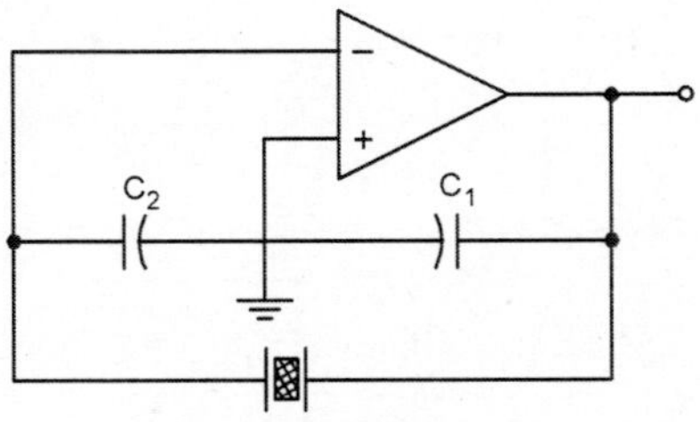

Fig. 6.8

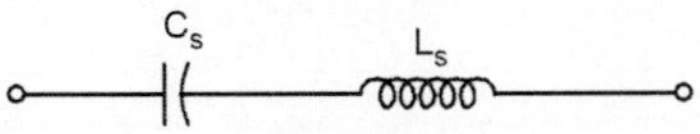

Fig. 6.9

6.8. QUADRATURE OSCILLATOR

The quadrature oscillator generates two signals (sine and cosine) that are in quadrature, that is, out of phase by 90°. Figure 6.10 shows the circuit arrangement of a quadrature oscillator. Output of op-amp A_1 is labelled as sine and output of A_2 as cosine, although the actual location of sine and cosine is arbitrary. The circuit requires two op-amps and three RC arrangements.

The op-amp A_1 works as a noninverting integrator. The second op-amp A_2 is working as a pure integrator. The feedback circuit consists of a resistance R_3 and a capacitor C_3 which is connected at the output of A_2 as a voltage divider section. A_1 and A_2 constitute the amplifier stage.

The total phase shift of 360° is achieved in two parts. The op-amp A_2, as a pure integrator, provides –270° or 90° phase shift and remaining –90° or 270° is achieved by the voltage divider R_3 C_3 section and op-amp A_1. The total phase shift of 360° (or 0°) is achieved only at one frequency f_0, called frequency of oscillation.

The oscillation frequency is given by

$$f_0 = \frac{1}{2\pi CR} = \frac{0.159}{RC} \quad ...(6.22)$$

where $R_1C_1 = R_2C_2 = R_3C_3 = RC$

At frequency f_0,

$A\beta = 1$

$$A = \frac{1}{\beta} = 1.414$$

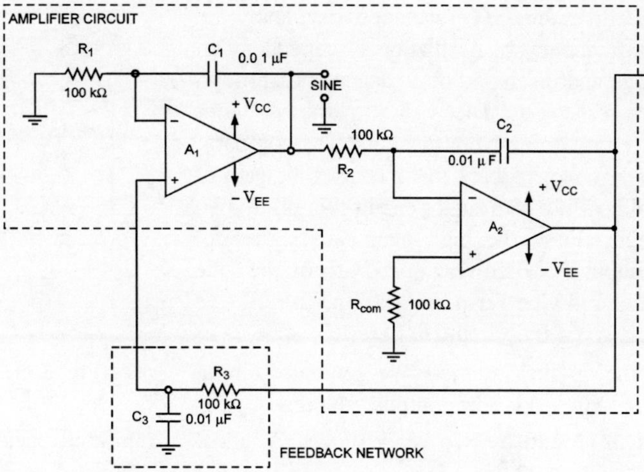

Fig. 6.10. *Quadrature Oscillator, A_1 and A_2 Dual Op-amp 1458/353*

Example 6.5. Design a quadrature oscillator of frequency f_0 = 318 Hz using an op-amp 1458/772.

Solution : Taking C = 0.015 μF

$$R = \frac{0.159}{f_0 C} = \frac{0.159}{318 \times 1.5 \times 10^{-8}}$$

$= 33.33$ kΩ (say, 33 kΩ)

Thus $C_1 = C_2 = C_3 = 0.015$ μF

and $R_1 = R_2 = R_3 = 33$ kΩ

Example 6.6. What are the two requirements for oscillations ? Discuss quadrature oscillator and design it to operate at a frequency of 1.5 kHz.

Solution : Let C= 0.01 μF

$$R = \frac{0.159}{f_0 C} = \frac{0.159}{1.5 \times 10^3 \times 0.01 \times 10^{-6}} = 10.6 \text{ k}\Omega \text{ (say, 10 k}\Omega)$$

Thus $C_1 = C_2 = C_3 = 0.01$ μF and $R_1 = R_2 = R_3 = 10$ kΩ

6.9. NON-SINUSOIDAL WAVEFORM GENERATORS

Most digital systems require some kind of a timing waveform, for instance, a source of trigger pulses is required for all clocked sequential systems. In digital systems, a rectangular waveform is most desirable (unlike analog systems where sinusoidal signals are often used). The generators of rectangular waveforms are referred to as *multivibrators*. Multivibrators are of three types viz. astable (or free-running) multivibrators, monostable multivibrators (or one-shot) and bistable multivibrators (or flip-flops).

Until a few years ago, multivibrators used to be designed using discrete devices, such as vacuum triodes, bipolar junction transistors (BJTs), field-effect transistors (FETs) etc., which have become obsolete now because of the availability of various integrated circuits (ICs). The ICs used in multivibrators are (*i*) op-amps (*ii*) timers (*iii*) logic gates and (*iv*) monostable multi-vibrators (MMVs).

Here we will be dealing with use of op-amps in non-sinusoidal waveform generators or relaxation oscillators.

6.10. OP-AMP SQUARE-WAVE GENERATOR (FREE-RUNNING OR ASTABLE MULTIVIBRATOR)

The non-sinusoidal waveform generators are also called *relaxation oscillators.* The op-amp relaxation oscillator shown in Fig. 6.11 is a *square-wave generator.* In general, square waves are relatively easy to

* Refer to Art. 11.2.

produce. Like the UJT relaxation oscillator, the circuit's frequency of oscillation is dependent on the charge and discharge of a capacitor C through feedback resistor R_f. The "heart" of the oscillator is an inverting op-amp comparator.* The comparator uses *positive* feedback that increases the gain of the amplifier. In a comparator circuit this offers two advantages. First, the high gain causes the op-amp's output to switch very quickly from one state to another and vice versa. Second, the use of + ve feedback gives the circuit *hysteresis*.

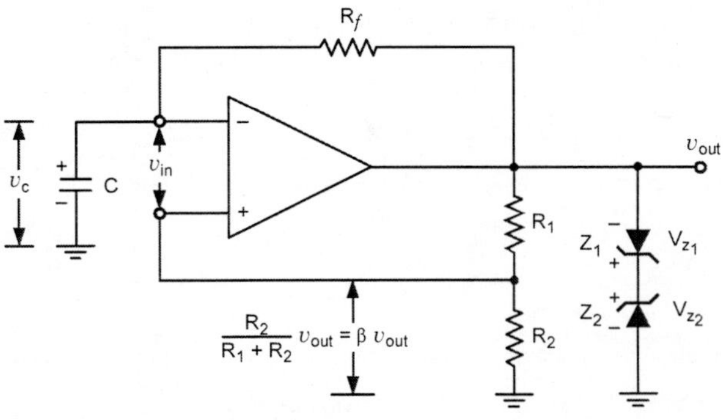

Fig. 6.11. ***Op-Amp Square-Wave Generator***

In the op-amp square-wave generator circuit given in Fig. 6.11, the output voltage v_{out} is shunted to ground by two zener diodes Z_1 and Z_2 connected back to back and is limited to either V_{z_2} or $-V_{z_1}$. A fraction of the output is fed back to the noninverting (+) input terminal. Combination of R_f and C acting as a low-pass R-C circuit is used to integrate the output voltage v_{out} and the capacitor voltage v_c is applied to the inverting input terminal in place of external signal.

The differential input voltage is given as

$$v_{in} = v_c - \beta\, v_{out} \qquad \text{where } \beta = \frac{R_2}{R_1 + R_2}$$

When v_{in} is positive, $v_{out} = -V_{z_1}$ and

When v_{in} is negative $v_{out} = +V_{z_2}$

Consider an instant of time when $v_{in} < 0$. At this instant $v_{out} = +V_{z_2}$, and the voltage at the noninverting (+) input terminal is βV_{z_2}, the capacitor C charges exponentially towards V_{z_2} with a time constant $R_f C$. The output voltage remains constant at V_{z_2} until v_c equals βV_{z_2}. When it happens, comparator output reverses to $-V_{z_1}$. Now v_c changes exponentially towards $-V_{z_1}$ with the same time constant and again the output makes a transition from $-V_{z_1}$ to $+V_{z_2}$ when v_c equals $-\beta V_{z_1}$.

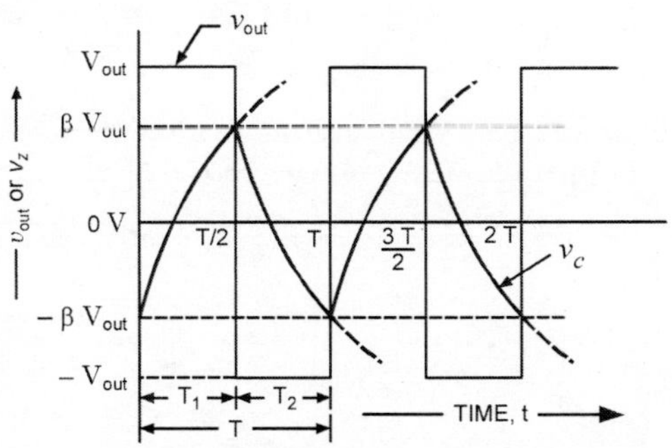

Fig. 6.12. ***Output and Capacitor Voltage Waveforms***

Let $V_{z_1} = V_{z_2}$

The time period, T, of the output square wave is determined using the charging and discharging phenomena of the capacitor C. The voltage across the capacitor, v_c when it is charging from $-\beta V_z$ to $+V_z$ is given by

$$v_c = V_z\left[1 - (1+\beta)e^{-t/R_f C}\right] = V_z\left[1 - (1+\beta)e^{-t/\tau}\right] \qquad \text{...(6.23)} \quad \text{where } \tau = R_f C$$

The waveforms of the capacitor voltage v_c and output voltage v_{out} (or v_z) are shown in Fig. 6.12.

$$\text{When } t = \frac{T}{2}, \quad v_c = +\beta V_z \text{ or } +\beta V_{out}$$

$$\therefore \quad \beta V_z = V_z\left[1 - (1+\beta)e^{-T/2\tau}\right]$$

$$\text{or } e^{-T/2\tau} = \frac{1-\beta}{1+\beta}$$

$$\text{or } T = 2\tau \log_e \frac{1+\beta}{1-\beta} = 2R_f C \log_e \left[\frac{1 + \dfrac{R_2}{R_1 + R_2}}{1 - \dfrac{R_2}{R_1 + R_2}} \right]$$

$$= 2R_f C \log_e \left(1 + \frac{2R_2}{R_1}\right) \qquad \text{...(6.24)}$$

The frequency, $f = \frac{1}{T}$, of the square wave is independent of output voltage V_{out}. This circuit is also known as *free-running* or *astable* multivibrator because it has two quasistable states. The output remains in one state for time T_1 and then makes an abrupt transition to the second state and remains in that state for time T_2. The cycle repeats itself after time $T = (T_1 + T_2)$ where T is the time period of the square wave.

The op-amp square wave generator is useful in the frequency range of about 10 Hz–10 kHz. At higher frequencies, the op-amp's slew rate limits the slope of the output square wave. The symmetry of the output waveform depends on the matching of two zener diodes Z_1 and Z_2. The unsymmetrical square wave, called rectangular wave, $(T_1 \neq T_2)$ can be had by using different constants for charging the capacitor C to $+V_{out}$ and $-V_{out}$.

6.11. PULSE GENERATOR (MONOSTABLE MULTIVIBRATOR)

A monostable multivibrator (MMV) has one stable state and one quasistable state. The circuit remains in its stable state till an external triggering pulse causes a transition to the quasistable state. The circuit comes back to its stable state after a time period T. Thus it generates a single output pulse in response to an input pulse and is referred to as a one-shot or single shot.

Monostable multivibrator circuit illustrated in Fig. 6.13 (*a*) is obtained by modifying the astable multivibrator circuit shown in Fig. 6.11 by connecting a diode D_1 across capacitor C so as to clamp v_c at v_d during positive excursion.

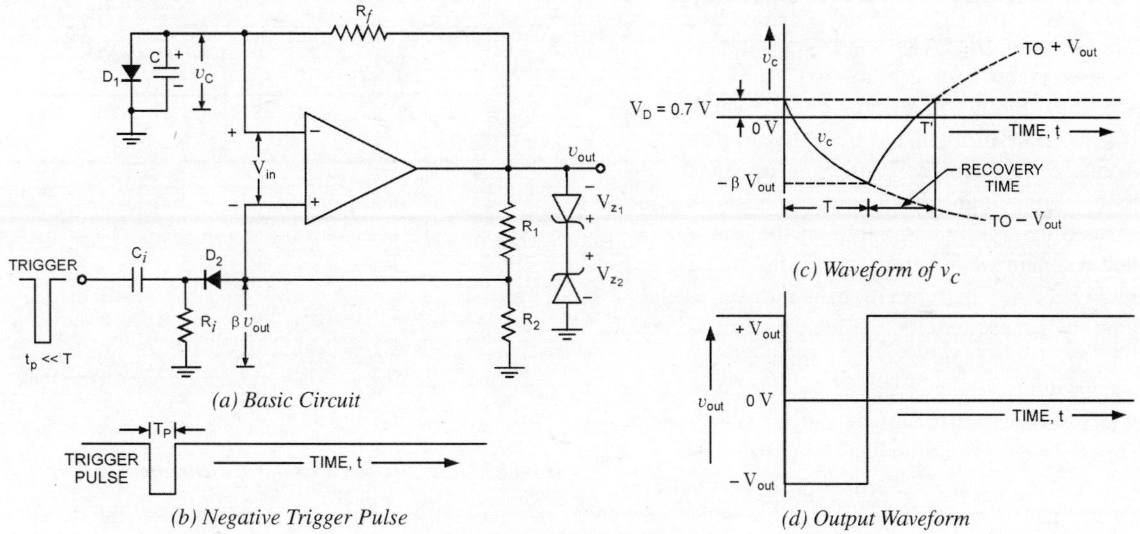

Fig. 6.13. ***Pulse Generator***

Under steady-state condition, this circuit will remain in its stable state with the output $v_{out} = +V_{out}$ or $+V_z$ and the capacitor C is clamped at the voltage V_D (*on*-voltage of diode $V_D = 0.7$ V). The voltage V_D must be less than βV_{out} for $v_{in} < 0$. The circuit can be switched to the other state by applying a negative pulse with amplitude greater than $\beta V_{out} - V_D$ to the noninverting (+) input terminal.

When a trigger pulse with amplitude greater than $\beta V_{out} - V_D$ is applied, v_{in} goes positive causing a transition in the state of the circuit to $-V_{out}$. The capacitor C now charges exponentially with a time constant $\tau = R_f C$ toward $-V_{out}$ (diode D_1 being reverse biased). When capacitor voltage v_c becomes more negative than $-\beta v_{out}$, v_{in} becomes negative and, therefore, output swings back to $+V_{out}$ (steady-state output). The capacitor now charges towards $+V_{out}$ till v_c attains V_D and capacitor C becomes clamped at V_D. The trigger pulse, capacitor voltage waveform and output voltage waveform are shown in Figs. 6.13 (*b*), 6.13 (*c*) and 6.13 (*d*) respectively.

The width of the trigger pulse T_p must be much smaller than the duration of the output pulse generated *i.e.*, $T_p << T$. For reliable operation the circuit should not be triggered again before T.

During the quasistable state, the capacitor voltage is given as

$$v_c = -V_{out} + (V_{out} + V_D)\, e^{-t/\tau} \qquad \text{...(6.25) where } \tau = R_f C$$

At instant $t = T$,

$$v_c = -\beta V_{out}$$

$$\text{So} \quad -\beta V_{out} = -V_{out} + (V_{out} + V_D)\, e^{-T/\tau}$$

$$\text{or} \quad e^{-T/\tau} = \frac{V_{out}(1-\beta)}{V_{out} + V_D}$$

$$\text{or} \quad T = \tau \log_e \frac{V_{out} + V_D}{V_{out}(1-\beta)} = R_f C \log_e \frac{1 + V_D / V_{out}}{1-\beta} \qquad \text{...(6.26)}$$

Usually $V_D << V_{out}$ and if $R_1 = R_2$

$$\text{so that} \quad \beta = \frac{R_2}{R_1 + R_2} = \frac{1}{2}, \text{ then}$$

$$T = R_f C \log_e 2 = 0.693\, R_f\, C. \qquad \text{...(6.27)}$$

6.12. TRIANGULAR-WAVE GENERATOR

The op-amp triangular-wave generator is another example of a relaxation oscillator. We know that the integrator output waveform will be triangular if the input to it is a square wave (refer to Art. 3.7). It means that a triangular-wave generator can be formed by simply cascading an integrator and a square wave generator, as illustrated in Fig. 6.14. This circuit needs a dual op-amp, two capacitors, and at least five resistors. The rectangular-wave output of the square-wave generator drives the integrator which produces a triangular output waveform. The rectangular-wave swings between $+V_{sat}$ and $-V_{sat}$ with a time period determined from Eq. (6.24). The triangular waveform has the same period and frequency as the square-waveform. Peak-to-peak value of output triangular-waveform can be obtained from the following equation.

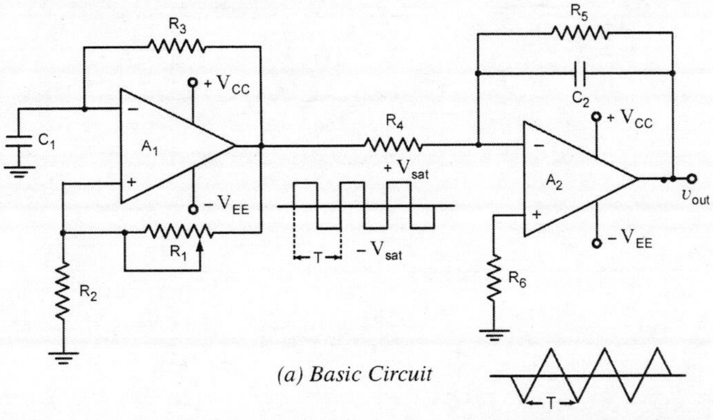

(a) Basic Circuit

(b) Output Waveform

Fig. 6.14. ***Triangular Waveform Generator***

$$V_{out\ (peak\text{-}to\text{-}peak)} = \frac{V_{in}}{4fR_5C_2} \quad ...(6.28)$$

The input of integrator A_2 is a square wave and its output is a triangular waveform. But the output of integrator will be triangular wave only when $5R_4C_2 > T/2$ where T is the time period of square wave. As a general rule, R_4C_2 should be equal to T. It may also be necessary to shunt the capacitor C_2 with resistance $R_5 = 10R_4$ and connect an offset voltage compensating network at the noninverting (+) input terminal of op-amp A_2 so as to obtain a stable triangular wave. Since the frequency of the triangular-wave generator like any other oscillator, is limited by the op-amp slew rate, a high slew-rate op-amp, like LM 301, should be used for the generation of relatively higher frequency waveforms.

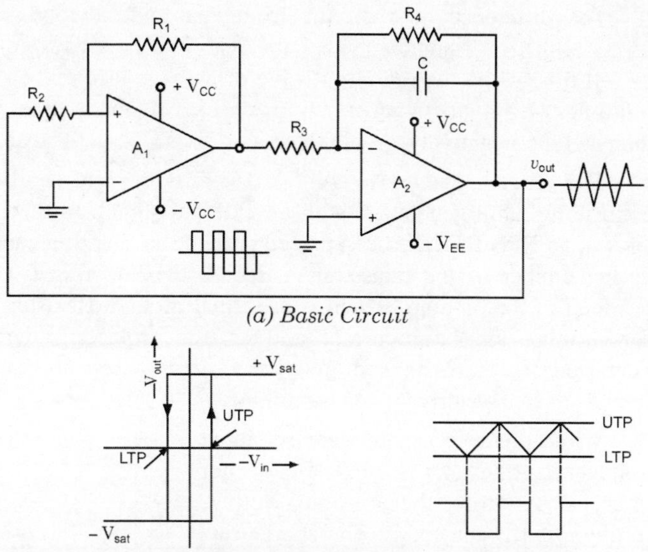

(a) Basic Circuit

(b) Transfer Characteristic of Schmitt Trigger (c)

Fig. 6.15. ***Feedback Circuit With Schmitt Trigger and Integrator Producing Triangular Output Waveform***

Another triangular-waveform generator, that needs fewer components is shown in Fig. 6.15 (*a*). The arrangement consists of a noninverting Schmitt trigger* A_1 and an integrator A_2. The output of a Schmitt trigger is a rectangular wave that drives an integrator. The output of the integrator is a triangular wave, which is fed back and used to drive the Schmitt trigger. Thus first stage drives the second, and the second drives the first. But the question arises how does the circuit get started in the first place? This is explained below.

When the Schmitt trigger is connected to power supplies for the first time, the output of the Schmitt trigger must be either *low* or *high*. When the Schmitt trigger output is *low*, the output of the integrator will be a rising ramp while for Schmitt trigger *high* output, the integrator will produce falling ramp. Either way, the triangular waveform has started, and the positive feedback will keep it going.

The transfer characteristic of the Schmitt trigger is shown in Fig. 6.15 (*b*). When the output is *low*, the input must increase to the UTP to switch the output to *high*. Likewise, when the output is *high*, the input must fall to the LTP to switch the output to *low*.

The triangular wave produced by the integrator is capable of driving the Schmitt trigger. When the output of Schmitt trigger is *low*, the integrator develops a rising ramp which increases till it reaches UTP, as illustrated in Fig. 6.15 (*c*). At this point the output of the Schmitt trigger switches to the *high* state and forces the triangular wave to reverse in direction. The negative or falling ramp produced by the integrator now falls till it reaches LTP, where another Schmitt output change occurs.

6.13. SAWTOOTH-WAVE GENERATOR

Sometimes it is felt necessary to provide a relatively slow linear ramp with a rapid fall (or rise in the case of a negative ramp) at its end. This is a sawtooth wave. Also, in applications such as time-base generators and power control circuits, the sawtooth must be triggered by (or be synchronised with) some control signal.

* Refer to Art. 6.15.

The difference between the triangular and sawtooth waveforms is that in triangular waves the rise time is always equal to its fall time while the sawtooth waveforms have different rise and fall times *i.e.,* sawtooth wave may rise positively many times faster than it falls negatively or vice versa.

The circuit shown in Fig. 6.16 provides the ability of controlling ramp generation with an external signal. In the circuit shown, an NPN BJT has been placed around the charging capacitor C and emitter of the transistor is tied to the inverting (–) input terminal of the op-amp, which is at virtual ground. Resistor R_B is for limiting the base current and so for protecting the BJT. However, R_B is to be kept relatively small to ensure that the transistor can be driven into saturation.

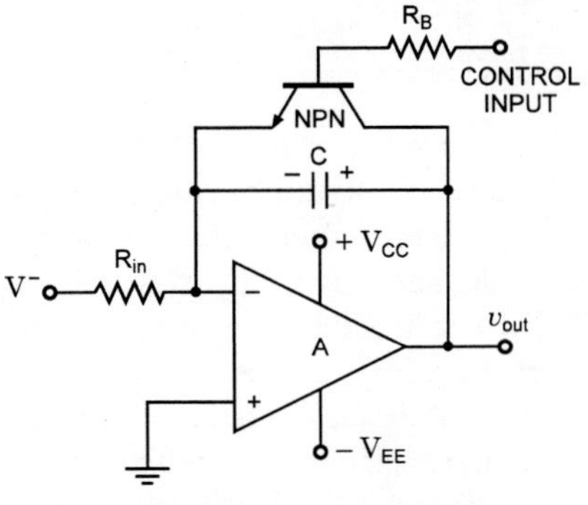

Fig. 6.16. ***BJT Controlled Ramp Generator***

With a zero or negative control input voltage, the transistor is *off*. The capacitor charges up from the op-amp output, through C, R_{in} and to V^-. The charge rate is given as

$$\text{Rate} = \frac{V^-}{R_{in} \times C}$$

If the control voltage is not changed, the capacitor C will eventually charge up, and hold the output at $+V_{sat}$.

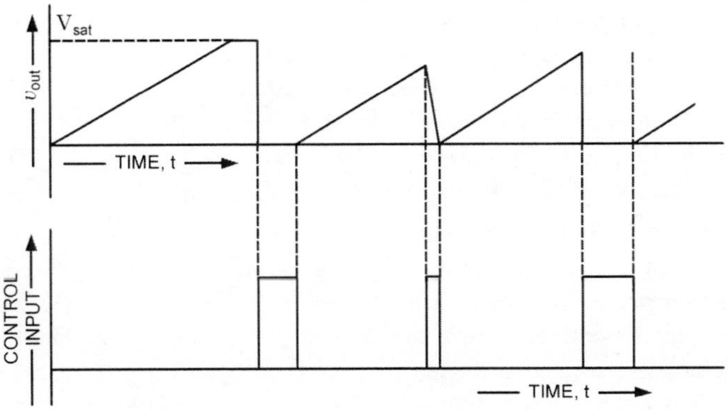

Fig. 6.17. ***Control Input Signal's Effect on The Ramp Generator's Output***

However, when a positive control input is applied, the transistor gets turned *on*. If this voltage is large enough to force transistor into saturation the capacitor is effectively short circuited. The capacitor C rapidly discharges.

The output voltage falls to zero (actually about 0.2 V) and stays there as long as positive control voltage keeps the transistor saturated. The expected obtainable waveform is given in Fig. 6.17.

For control of negative going ramps, the circuit shown in Fig. 6.16 will require several minor changes.

First, the charging voltage, connected to R_{in}, polarity will have to be reversed to V^+. This reverses the direction of charging current. It means capacitor will also have to be reversed, if it is an electrolytic one. The emitter of the transistor must be connected to virtual ground (the inverting input terminal of op-amp). To allow the capacitor to discharge from left to right, NPN transistor would have to be replaced by PNP transistor. In this case, a zero or positive control input would keep the PNP transistor *off*, while a negative control input would be required to turn the transistor *on*. This is opposite to that shown in Figs. 6.16 and 6.17. It means that control signal must be inverted, making sure that it goes negative to turn the PNP transistor *on*.

A programmable unijunction transistor (PUT) can also be used in place of BJT across the capacitor in the circuit shown in Fig. 6.16.

A PUT controlled sawtooth generator circuit is shown in Fig. 6.18. When power is first applied, the programmable unijunction transistor (PUT) is *off*. The capacitor C begins to charge up and the output voltage rises. This continues until the output voltage (which is also the PUT anode voltage) is about 0.7 V above the control input (the gate voltage). The PUT gets switched *on*. The capacitor C is shorted out through PUT and, therefore, capacitor gets immediately discharged through the PUT.

The output voltage, which is equal to the voltage across the capacitor, falls. When the current through the PUT falls below its holding current I_H, it goes *off* and the cycle repeats. When the PUT turns *off*, approximately 1 V is usually left on the capacitor. The output waveform is shown in Fig. 6.19.

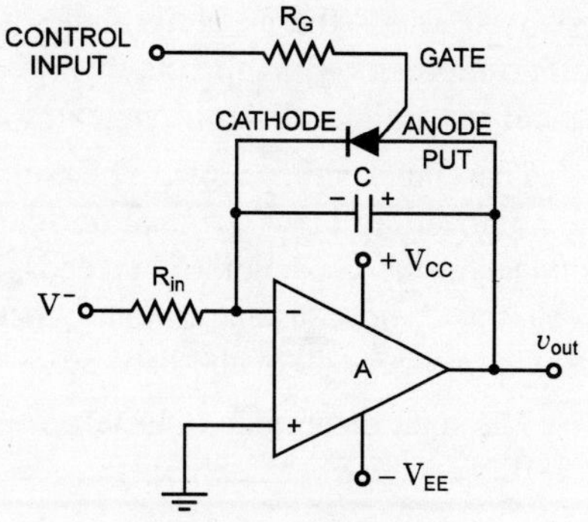

Fig. 6.18. ***PUT Controlled Sawtooth Generator***

The time period of the PUT controlled sawtooth generator depends on the charge rate (V/RC) and the control voltage $V_{control}$. This is obvious from Fig. 6.19.

$$\text{Time period, } T = \frac{\text{Distance}}{\text{Rate}}$$

$$= \frac{(V_{control} + 0.7\text{ V}) - 1\text{ V}}{\frac{|-V|}{RC}} \cong \frac{V_{control}\ RC}{|-V|}$$

and frequency, $$f = \frac{1}{T} = \frac{|-V|}{V_{control}\ RC} \qquad \text{...(6.29)}$$

The PUT controlled sawtooth generator can be used as a voltage-to-frequency converter.

Precautions : 1. The cathode of the PUT must be tied to ground or virtual ground and current flows only from anode to cathode. So PUT cannot be used to control a negative ramp generator.

2. To *turn-off* the current through the PUT must drop below its holding current I_H (specified by the manufacturer). When the PUT is *on*, a current equal to that used to charge the capacitor, in addition to capacitor discharge current, flows through the PUT. This current flows through R to V^- must be below I_H.

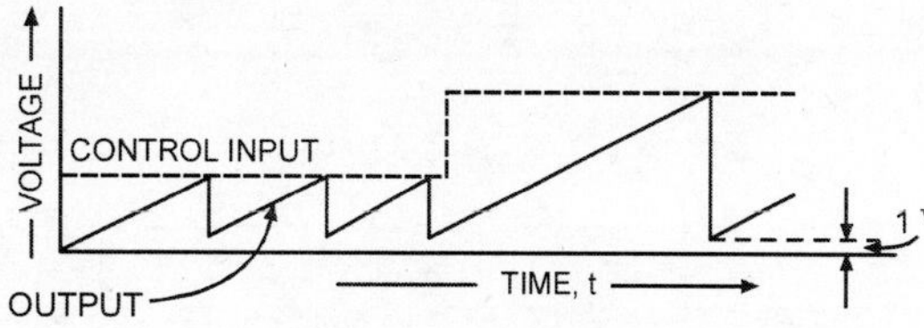

Fig. 6.19. ***Control Input and Output From a PUT Controlled Sawtooth Generator***

i.e., $$I = \frac{V'}{R} < I_H$$

Failing which, once the PUT goes *on*, it will be held *on* by this charge current, even when the capacitor has fully discharged. This charge current can be lowered by increasing R or reducing negative voltage V^-. However, both of these factors affect the charge rate and, therefore, the frequency. For keeping frequency unaffected the changes in either R or V^- will have to be balanced with appropriate changes in C.

6.14. VOLTAGE-CONTROLLED OSCILLATOR (VCO)

In most cases, the frequency of an oscillator is determined by the time constant RC. However, in cases or applications such as *frequency modulation (FM)*, *tone generators*, and *frequency-shift keying* (FSK), the frequency is to be controlled by means of an input voltage, called the *control voltage*. This can be achieved in a *voltage-controlled oscillator* (VCO). A VCO is a circuit that provides an oscillating output signal (typically of square wave or triangular waveform) whose frequency can be adjusted over a range by a dc voltage. An example of a VCO is the 566 IC unit, that provides simultaneously the square wave and triangular wave outputs as a function of input voltage. The frequency of oscillation is set by an external resistor R_1 and a capacitor C_1 and the voltage V_C applied to the control terminal 5. Figure 6.20 (*a*) shows that the 566 IC unit contains current sources to charge and discharge an external capacitor C_1, at a rate set by an external resistor R_1 and the modulating dc input voltage. A Schmitt trigger circuit is employed to switch the current sources between charging and discharging the capacitor, and the triangular voltage produced across the capacitor and square wave from the Schmitt trigger are provided as outputs through buffer amplifiers. Both the output waveforms are buffered so that the output impedance of each is 50 Ω. The typical magnitude of the triangular wave and the square wave are 2.4 $V_{peak\text{-}to\text{-}peak}$ and 5.4 $V_{peak\text{-}to\text{-}peak}$.

The output voltage swing of the Schmitt trigger (ST) is designed to V^+ and 0.5 V^+ and the voltage at the noninverting input terminal of Schmitt trigger swings from 0.5 V^+ to 0.25 V^+. In Fig. 6.20 (*c*), when the voltage on the capacitor C_1 exceeds 0.5 V^+ during charging, the output of ST goes LOW (0.5 V^+). The capacitor C_1 now discharges and when it is at 0.25 V^+, the ST output goes HIGH (V^+). The charging and discharging periods of the capacitors are the same.

The total change in capacitor voltage is from 0.25 V^+ to 0.5 V^+ *i.e.*, Δv = 0.25 V^+.

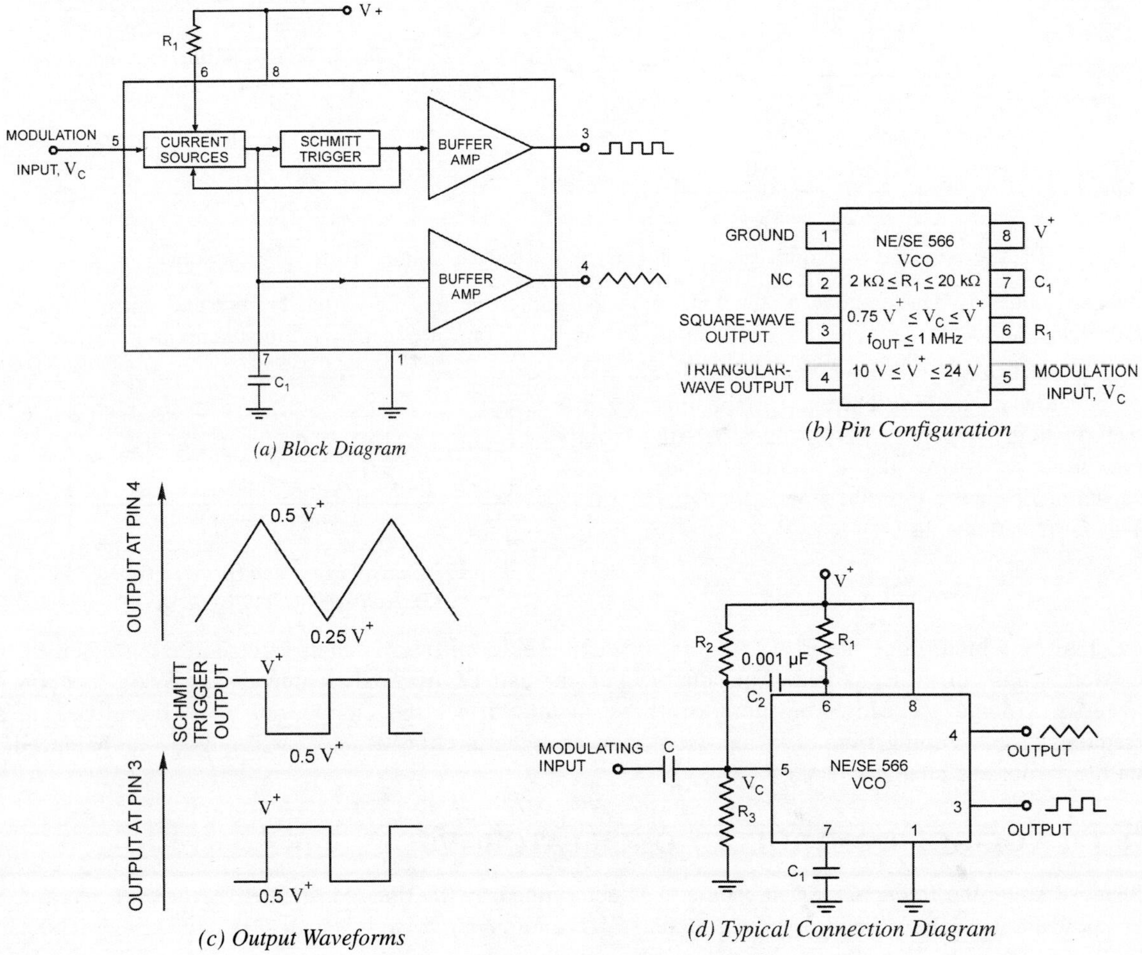

(a) Block Diagram

(b) Pin Configuration

(c) Output Waveforms

(d) Typical Connection Diagram

Fig. 6.20. ***Voltage Controlled Oscillator***

The capacitor charges with a constant current source, so

$$\frac{\Delta v}{\Delta t} = \frac{i}{C_1}$$

$$\text{or} \quad \frac{0.25\,V^+}{\Delta t} = \frac{i}{C_1}$$

$$\text{or} \quad \Delta t = 0.25\,V^+ \times \frac{C_1}{i}$$

The time period of the triangular waveform is $2\Delta t$.

$$\text{The frequency of oscillator, } f_0 = \frac{1}{T} = \frac{1}{2\,\Delta t} = \frac{i}{0.5\,V^+C_1}$$

But $i = \dfrac{V^+ - V_c}{R_1}$ where V_c is the voltage at pin 5. Thus,

$$f_0 = \frac{2(V^+ - V_c)}{R_1C_1V^+} \qquad ...(6.30)$$

The output frequency of the VCO can be changed varying either (*i*) R_1 (*ii*) C_1 or (*iii*) the voltage V_c at the modulation input terminal 5. The voltage V_c can be varied by connecting R_2R_3 circuit as shown in Fig. 6.20 (*d*).

Figure 6.20 (*d*) shows the pin connection of the 566 unit. The VCO can be programmed over a 10-to-1 frequency range by proper selection of an external resistor and capacitor, and then modulated over a 10-to-1 frequency range by a control voltage, V_c.

The voltage controlled oscillators (VCOs) are commonly used in converting low-frequency signals such as EEG (electroencephalograms) or ECG (electrocardiograms) into an audio-frequency (AF range).

6.15. SCHMITT TRIGGER OR REGENERATIVE COMPARATOR

A Schmitt trigger circuit is a fast-operating voltage level detector. When the input voltage arrives at the upper or lower trigger levels, the output changes rapidly. The circuit operates with almost any type of input waveform, and it gives a pulse-type output.

The circuit of an op-amp Schmitt trigger circuit is shown in Fig. 6.21 (*a*). The input voltage v_{in} is applied to the inverting input terminal and the feedback voltage goes to the noninverting terminal. This means the circuit uses *positive voltage feedback* instead of negative feedback *i.e.*, in this circuit feedback voltage aids the input voltage rather than opposing it. For instance, assume the inverting input voltage to be slightly positive. This will produce a negative output voltage. The voltage divider feeds back a negative voltage to the noninverting input, which results in a larger negative output voltage. This feeds back more negative voltage until the circuit is driven into negative saturation. If the input voltage were slightly negative instead of positive, the circuit would be driven into the positive saturation. This is the reason the circuit is also referred to as *regenerative comparator*.

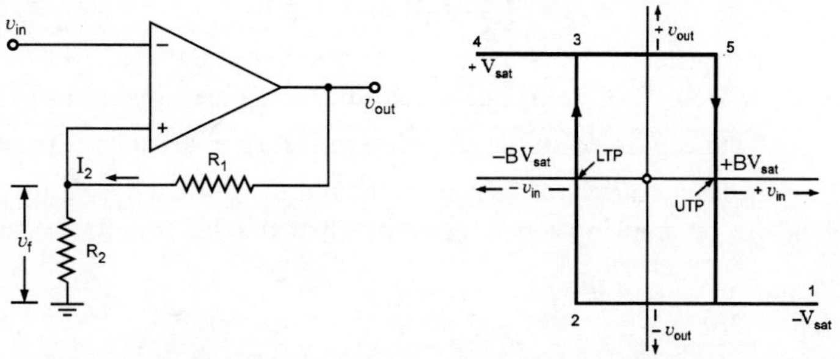

(*a*) Circuit Diagram (*b*) Input-Output Characteristics

Fig. 6.21

When the circuit is positively saturated, a positive voltage is fed back to the noninverting input. This positive input holds the output in the *high* state. Similarly, when the output voltage is negatively saturated, a negative voltage is fed back to the noninverting input, holding the output in the *low* state. In either case, the positive feedback reinforces the existing output state.

$$\text{The feedback fraction, } B = \frac{R_2}{R_1 + R_2} \qquad ...(6.31)$$

When the output is positively saturated, the reference voltage applied to the noninverting input is

$$v_{ref} = +BV_{sat} \qquad ...(6.32)$$

When the output is negatively saturated, the reference voltage is

$$v_{ref} = -BV_{sat} \qquad ...(6.33)$$

The output voltage will remain in a given state until the input voltage exceeds the reference voltage for that state. For instance, if the output is positively saturated, the reference voltage is $+BV_{sat}$. The input voltage v_{in} must be increased slightly above $+BV_{sat}$ to switch the output voltage from positive to negative, as shown in Fig. 6.21 (*b*). Once the output is in the negative state, it will remain there indefinitely until the input voltage becomes more negative than $-BV_{sat}$. Then the output switches from negative to positive. This can be explained from the input-output characteristics of the Schmitt trigger shown in Fig. 6.21 (*b*).

Assume that input voltage v_{in} is greater than the $+BV_{sat}$, and output voltage v_{out} is at its negative extreme (point 1). The voltage across R_2 in Fig. 6.21 (*a*) is a negative quantity. $-\frac{v_{out} \times R_2}{R_1 + R_2}$ or $-BV_{sat}$. As a result, v_{in} must be reduced to this negative voltage level (point 2 on the characteristics) before the output switches positively (point 3). If the input voltage is made more negative than the $-BV_{sat}$, the output remains at $+v_{out}$ (points 3 to 4). For the output to go negative once again, v_{in} must be increased to the $+BV_{sat}$ level (point 5 on the characteristics).

In Fig. 6.21 (*b*), the trip points are defined as the two input voltages where the output changes states. The upper *trip point* (abbreviated UTP) has a value

$$UTP = BV_{sat}$$

and the lower trip point has a value

$$LTP = -BV_{sat}$$

The difference between the trip points is the hysteresis H and is given as

$$H = +BV_{sat} - (-BV_{sat}) = 2BV_{sat} \qquad ...(6.34)$$

The hysteresis is caused due to positive feedback. If there were no positive feedback, B would be equal to zero and the hysteresis would disappear, because the trip points would be both equal to zero.

Hysteresis is desirable in a Schmitt trigger because it prevents noise from causing false triggering.

To design a Schmitt trigger, potential divider current I_2 is once again selected to be very much larger than the op-amp input bias current. Then the resistor R_2 is calculated from equation

$$R_2 = \frac{UTP}{I_2} \qquad ...(6.35)$$

and R_1 is determined from $R_1 = \frac{V_{out} - UTP}{I_2}$ ...(6.36)

Example 6.7. An inverting voltage comparator uses an op-amp with saturation voltage of ±12 V, $R_1 = 2$ kΩ, $R_F = 390$ kΩ. The reference voltage is –3 V. Determine two voltages at which comparator will switch.

Solution : Feedback fraction, $B = \frac{R_1}{R_1 + R_F} = \frac{2\ k\Omega}{2\ k\Omega + 390\ k\Omega} = 0.0051$

Refer to Fig. 6.21(*a*)

$UTP = \beta V_{sat} = 0.0051 \times 12\ V = 0.0612\ V$ or 6.12 mV **Ans.**

$LTP = -\beta V_{sat} = -0.0051 \times 12\ V = -0.0612\ V$ or -61.2 mV **Ans.**

Example 6.8. The operational amplifier circuitry is given in Fig. 6.22. Determine its gain and indicate its applicability.

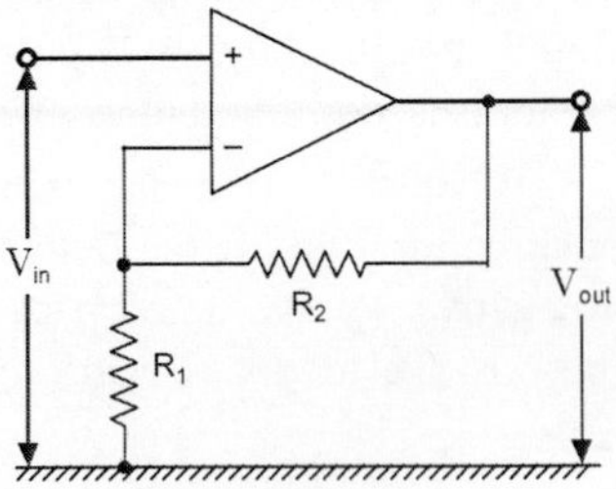

Fig. 6.22

Solution :

$$V_{in} = V_{out} \times \frac{R_1}{R_1 + R_2}$$

$$\text{Gain} = \frac{V_{out}}{V_{in}} = \frac{R_1 + R_2}{R_1} \quad \textbf{Ans.}$$

The given amplifier circuitry is referred to as a Schmitt trigger or regenerative comparator. Its applications are :

(*i*) Use as comparator.

(*ii*) As a zero-crossing detector.

(*iii*) As a square-wave generator.

(*iv*) Used in phase meter, pulse-time modulation and amplitude distribution analyzer.

Example 6.9. Design a Schmitt Trigger with V_{UT} = 5 V and V_{LT} = – 5 V. What is hysteresis voltage ?

Solution : Assume $V_{sat} = \pm 10$ V

$$V_{UT} = \left(\frac{R_1}{R_1 + R_2}\right) V_{sat} = 5$$

$$V_{LT} = \left(\frac{-R_1}{R_1 + R_2}\right) V_{sat} = -5$$

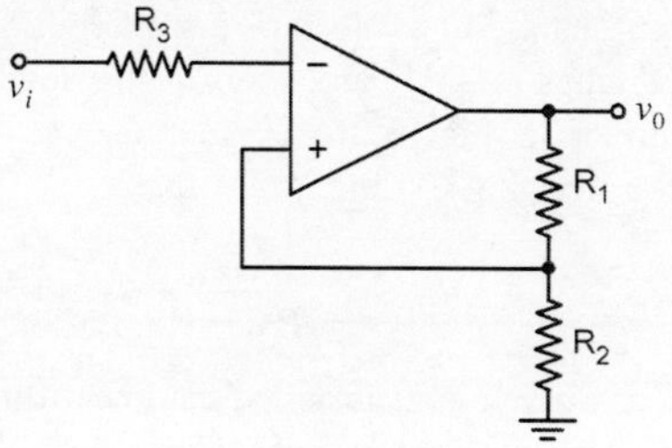

Fig. 6.23

Now designing requirements are such that

$$\frac{10 R_1}{R_1 + R_2} = 5$$

$$\text{Hysteresis voltage} = V_{hy} = V_{UT} - V_{LT} = 5 - (-5) = 10 \text{ V} \quad \textbf{Ans.}$$

Example 6.10. Explain the working of the circuit shown in Fig. 6.23. Draw the output waveform for a sine wave applied at the input. [*G.G.S.I.P. Univ. Analog Electronics-II May 2011*]

Solution : Given circuit is the circuit of Schmitt trigger. For the working of circuit refer to Art. 6.15. Input and output waveforms are shown in Fig. 6.24.

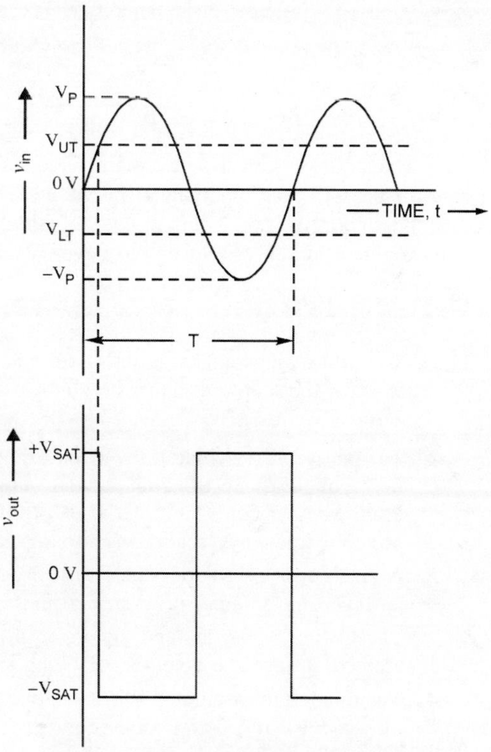

Fig. 6.24. ***Input and Output Waveforms***

Example 6.11. A comparator circuit using positive feedback is shown in Fig. 6.25. A sine wave input is applied. Derive the output response. Explain hysteresis in this circuit.

[*G.G.S.I.P. Univ. Analog Electronics-II May 2011*]

Solution : The circuit shown in Fig. 6.25 is of Schmitt trigger. The input voltage v_{in} is applied to the inverting input terminal and feedback voltage to the noninverting input terminal. The input voltage v_{in} triggers the output voltage v_{out} every time it exceeds certain voltage levels, known as upper threshold voltage V_{UT} and lower threshold voltage V_{LT}. The hysteresis width is the

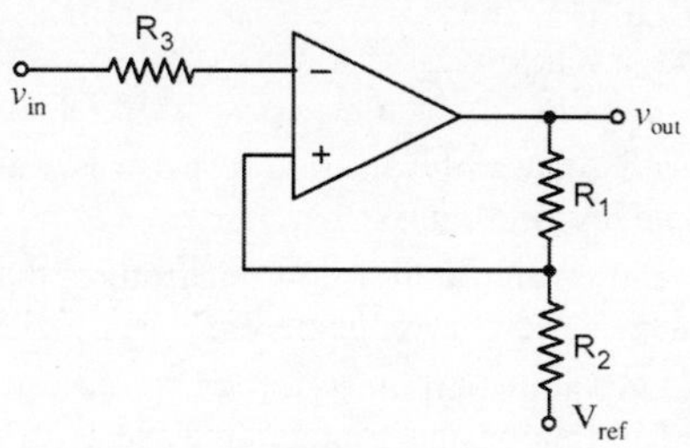

Fig. 6.25

difference of these two threshold voltages.

i.e., $V_H = V_{UT} - V_{LT}$

These two threshold voltages are determined as follows:

When the output is positively saturated, *i.e.* $v_{out} = +V_{sat}$, the voltage at the noninverting terminal can be had by using superposition theorem

$$V_{UT} = \frac{R_1}{R_1 + R_2} V_{ref} + \frac{R_2}{R_1 + R_2} V_{sat} \quad ...(i)$$

This voltage is referred to as upper threshold voltage V_{UT}. As long as v_{in} is less than V_{UT}, the output voltage remains constant at $+V_{sat}$. When v_{in} just exceeds V_{UT}, the output regeratively switches to $-V_{sat}$ and stays at this level as long as $v_{in} > V_{UT}$ as shown in Fig. 6.26 (*a*). For $v_{out} = -V_{sat}$, the voltage at the noninverting (+) input terminal is

$$V_{LT} = \frac{R_1}{R_1 + R_2} V_{ref} - \frac{R_2}{R_1 + R_2} V_{sat} \quad ...(ii)$$

This voltage is called the lower threshold voltage. The input voltage v_{in} must become lesser than V_{LT} so as to make output voltage v_{out} to switch from $-V_{sat}$ to $+V_{sat}$. A regenerative transition occurs as depicted in Fig. 6.26 (*b*) and the output voltage returns from $-V_{sat}$ to $+V_{sat}$, almost instantaneously. The complete transfer characteristics are shown in Fig. 6.26 (*c*).

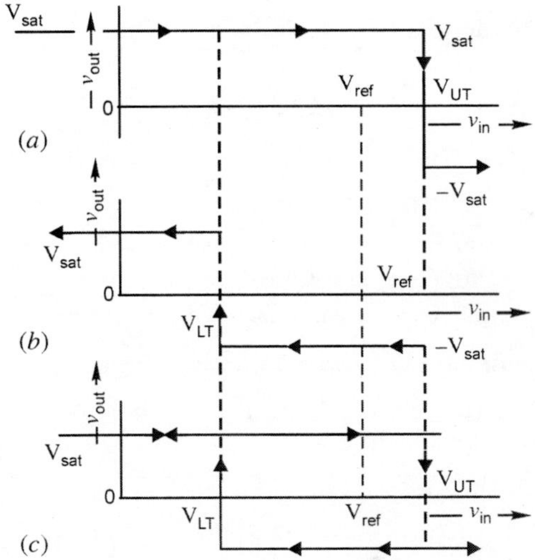

(a) Transfer Characteristic for increasing v_{in}
(b) Transfer Characteristic for decreasing v_{in}
(c) Composite input-output curve

Fig. 6.26

As already mentioned above, hysteresis width,

$$V_H = V_{UT} - V_{LT} = \frac{2R_2}{R_1 + R_2} V_{sat} \quad ...(iii)$$

Because of the hysteresis, the circuit triggers at a higher voltage for increasing signals than for decreasing ones. In case the peak-to-peak input signal v_{in} were small than V_H then the Schmitt trigger circuit, having responded at a threshold voltage by a transition in one direction would never reset itself, *i.e.*, once the output has jumped to, say, V_{sat}, it would remain at this level and never return to $-V_{sat}$.

HIGHLIGHTS ....

1. To build a sinusoidal oscillator, we need to use an amplifier with positive feedback. For the oscillator to start, the loop gain must be slightly greater than unity and the phase shift around the loop must be zero.
2. The phase shift oscillator, the Wien bridge oscillator, the quadrature oscillator are popular configurations for frequencies up to about 1 MHz. For higher frequencies L-C tuned or crystal tuned oscillators are used. Popular configurations are Colpitt's and Hartley circuits. Crystal oscillators provide the highest possible frequency accuracy and stability.
3. A phase shift oscillator consists of three R-C networks, each providing a phase shift of 60 degrees, and an inverting op-amp, providing a phase shift of 180 degrees, for a total phase shift of 360 degrees.

   The Colpitt's, Hartley and crystal oscillator circuits use preferably discrete transistors rather than op-amps, but have the potential of being very high frequency oscillators.
4. The quadrature oscillator generates two signals (sine and cosine) that are in quadrature, that is, and out of phase by 90°. The circuit requires 2 op-amps and three R-C arrangements.
5. A square-wave output waveform is generated if the op-amp output is forced to swing respectively between positive and negative saturation. One way of obtaining the triangular wave is to integrate the square wave. Thus, the triangular-wave generator can be designed by simply cascading an integrator and a square-wave generator.

The sawtooth waveform has a rise time many times longer than fall time or vice versa. The triangular-wave generator using a comparator and an integrator can be converted to provide a sawtooth waveform.

6. In most cases, the frequency of an oscillator is determined by the time constant RC. However, in cases or applications such as FM, tone generators, and frequency-shift keying (FSK), the frequency is to be controlled by means of an input voltage, called the *control voltage*. This can be achieved in a *voltage-controlled oscillator* (VCO). A VCO is a circuit that provides an oscillating output signal (typically of square wave or triangular waveform) whose frequency can be adjusted over a range by a dc voltage.
7. A Schmitt trigger uses a comparator with a positive feedback and converts an irregular waveform to a square or pulse waveform. In the Schmitt trigger, the input voltage triggers the output every time, it exceeds certain voltage levels known as upper threshold and lower threshold voltages.

EXERCISES

1. For an R-C phase shift oscillator, derive the expression for the frequency of oscillation and the condition for sustained oscillations. **[G.G.S.I.P. Univ. Analog Electronics May 2010]**
2. Explain how an op-amp can be used as R-C phase shift oscillator.
3. Show that the circuit given in Fig. 1 will work as an oscillator at $f = \frac{1}{2\pi RC}$, if $R_1 = 2R_2$.
4. What is Wien bridge ? What are its uses ?
   Show how variable frequency oscillator can be built using an operational amplifier with a bridge. Derive an expression for the frequency of oscillation of the circuit.
5. Explain op-amp Wien bridge oscillator and state conditions for sustained oscillations. **[G.G.S.I.P. Univ. Analog Electronics II May 2010]**

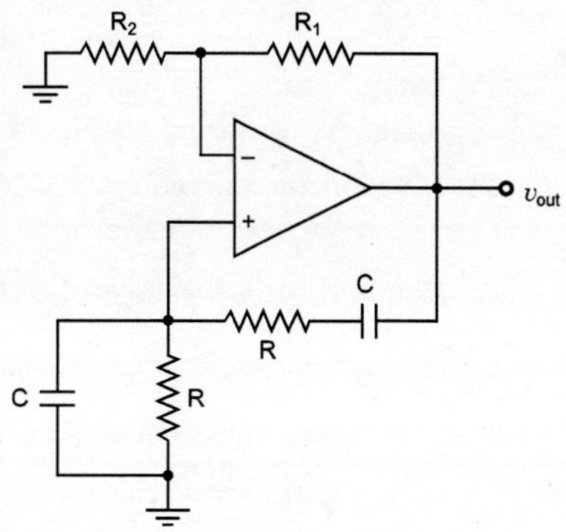

Fig. 1

6. Draw a neat circuit diagram of the R-C Wein bridge oscillator and derive the expressions for its condition of oscillation and frequency of oscillation. **[G.G.S.I.P. Univ. Analog Electronics II February 2009]**
7. Explain the working of a Wien-bridge oscillator. Derive the frequency of oscillation and the condition for oscillation. **[G.G.S.I.P. Univ. Analog Electronics-II May 2011]**
8. Draw the circuit of an op-amp oscillator to generate radio-frequency signals. Determine the expression for the frequency of oscillation and the condition for oscillation. **[G.G.S.I.P. Univ. Analog Electronics-II May 2010]**
9. Explain the basic principle of operation of crystal oscillator using op-amp with neat block diagram.
10. What are the two requirements for oscillations ? Discuss quadrature oscillator.
11. Explain quadrature oscillator. **[G.G.S.I.P. Univ. Analog Electronics-II May-June 2009]**
12. How is a square wave signal generated ? Draw circuit diagram and derive equations for the frequency of oscillations. Does it have high frequency limitation ? Discuss. **[G.G.S.I.P. Univ. Analog Electronics-II May-June 2009]**
13. Explain the working of a square waveform generator and derive an expression for the frequency of oscillation. Suggest a method for obtaining asymmetrical square wave. **[G.G.S.I.P. Univ. Analog Electronics-II May 2010]**
14. Draw and explain the working of an op-amp square-wave generator. Derive an expression for the frequency of oscillation. **[G.G.S.I.P. Univ. Analog Electronics December 2009]**
15. Explain the operation of an astable multivibrator and derive the expression for frequency of oscillations. **[G.G.S.I.P. Univ. Analog Electronics-II May 2011]**
16. Explain the working of an op-amp square-wave generator. Derive an expression for the frequency of oscillations. What is the maximum amplitude of the square wave and how it can be varied. **[G.G.S.I.P. Univ. Analog Electronics-II May 2011]**

17. Draw the circuit of a multivibrator which can generate a single output pulse of duration T. Explain the working and derive the expression for the pulse width T.

[G.G.S.I.P. Univ. Analog Electronics-II May 2010]

18. How does a sawtooth wave differ from a triangular wave ? How the ramp generation can be controlled by using (*i*) bipolar junction transistor (BJT) and (*ii*) programmable unijunction transistor (PUT) ? Discuss the problems associated with PUT-controlled sawtooth wave generator.

19. Draw and explain the working of a triangular waveform generator using minimum components. Derive an expression for the frequency of oscillation. **[G.G.S.I.P. Univ. Analog Electronics-II May 2011]**

20. Explain how the triangular wave generator can be converted into a ramp generator ?

[G.G.S.I.P. Univ. Analog Electronics-II May 2011]

21. Discuss the operation of voltage-controlled oscillator with neat diagrams.

22. Describe the circuit of a square-wave generator which generates square, triangular and sine waveshapes.

[M.D. Univ. Electronic Measurement and Instrumentation Dec. 2007; May 2008]

23. Draw a block schematic of voltage controlled oscillator (VCO). Explain its working and derive an expression for its frequency. **[G.G.S.I.P. Univ. Analog Electronics-II May 2010]**

24. Explain the operation of VCO with the help of block diagram.

[G.G.S.I.P. Univ. Analog Electronics-II May 2011]

25. Explain the advantage of Schmitt trigger over basic comparator. How will it convert sine wave into square wave ? **[G.G.S.I.P. Univ. Analog Electronics-II May 2011]**

26. Write short notes on the following :
   (*i*) Phase shift and Wien bridge oscillators
   (*ii*) Colpitt's and Hartley oscillators.
   (*iii*) Quadrature oscillator.
   (*iv*) Square-wave generator.
   (*v*) Sawtooth wave generator.
   (*vi*) Triangular-wave generator
   (*vii*) Schmitt trigger

SHORT ANSWER TYPE QUESTIONS WITH ANSWERS

Q. 1. What is an oscillator ?

Ans. Oscillator may be defined as a circuit which generates an ac output signal of very high frequency without requiring any externally applied input signal or it may be defined as an electronic source of alternating current (or voltage) having sinusoidal or non-sinusoidal (square, sawtooth or pulse) waveshape. It can also be defined as a circuit that converts dc energy into very high-frequency ac energy.

Q. 2. What are the conditions which are to be satisfied for the successful operation of phase-shift oscillator ?

Ans. The phase-shift oscillator will generate sinusoidal waveform only if the gain is ≥ 29 and total phase-shift of the circuit is 360°.

Q. 3. Why is square-wave generator also called the astable multivibrator ?

Ans. Square-wave generator is also called the free-running or astable multivibrator because it has two quasi stable states.

Q. 4. Give the frequency range of square-wave generator.

Ans. 10 Hz to 10 kHz.

Q. 5. Write an application of astable multivibrator.

Ans. Astable multivibrator or free running multivibrator has no stable state *i.e.*, it keeps oscillating between high and low states. Its applications include (*i*) free-running ramp generator and (*ii*) square-wave generator.

Q. 6. Write an application of monostable multivibrator.

Ans. A monostable multivibrator (MMV) has one stable state and one quasi stable state. The circuit remains in its stable state till an external triggering pulse causes a transition to the quasi stable state. The circuit comes back to its stable state after a time period T. Thus it generates a single output pulse in response to an input pulse and is referred to as a one-shot or single shot.

The applications of MMV include (*i*) frequency divider; (*ii*) pulse stretcher.

Q. 7. Why are crystal oscillators used in communication transmitters and receivers ?

Ans. Crystal oscillators are used in communication transmitters and receivers because of their greater frequency stability.

Q. 8. What is difference between a triangular wave and a sawtooth wave ?

Ans The difference between a triangular wave and a sawtooth wave is that in a triangular wave the rise time is always equal to its fall time while the sawtooth waveform has different rise time and fall time.

Q. 9. What is a VCO ? Give two applications that require VCO.

Ans. The oscillator whose output frequency is controlled by varying the magnitude of the input voltage is called a voltage-controlled oscillator (VCO). Applications requiring VCO are frequency modulation (FM) and frequency shift keying (FSK).

Q. 10. What is Schmitt trigger ?

Ans. A Schmitt trigger uses a comparator with a positive feedback and converts an irregular waveform to a square or pulse waveform. In the Schmitt trigger, the input voltage triggers the output every time, it exceeds certain voltage levels known as upper threshold and lower threshold voltages.

Q. 11. Why hysteresis is desirable in a Schmitt trigger ?

Ans. Hysteresis is desirable in a Schmitt trigger because it prevents noise from causing false triggering.

Q. 12. Give the applications of Schmitt trigger.

Ans. Applications of Schmitt trigger are :

(*i*) Use as a comparator.

(*ii*) As a zero-crossing detector.

(*iii*) As a square-wave generator.

(*iv*) Used in phase meter, pulse-time modulation and amplitude distortion analyzer.

PROBLEMS

1. Design a Wien bridge oscillator that will oscillate at 2 kHz.
2. Design a phase shift oscillator of frequency 1 kHz.
3. Design a quadrature oscillator to operate at frequency of 2.5 kHz.
4. Consider the Fig. 2 given an ideal op-amp.

   (*i*) Compute condition of oscillation. (*ii*) Find the frequency of oscillation.

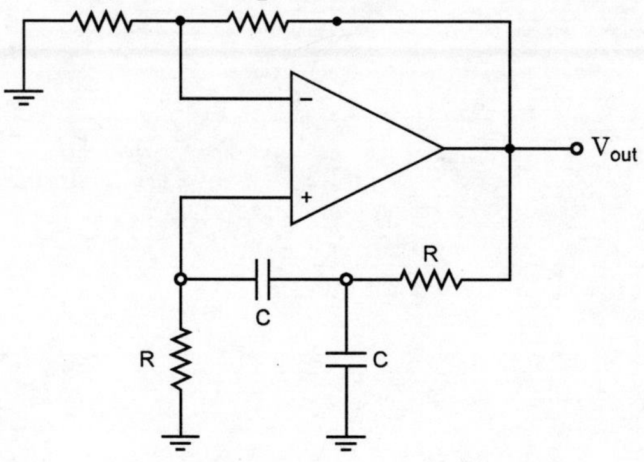

Fig. 2

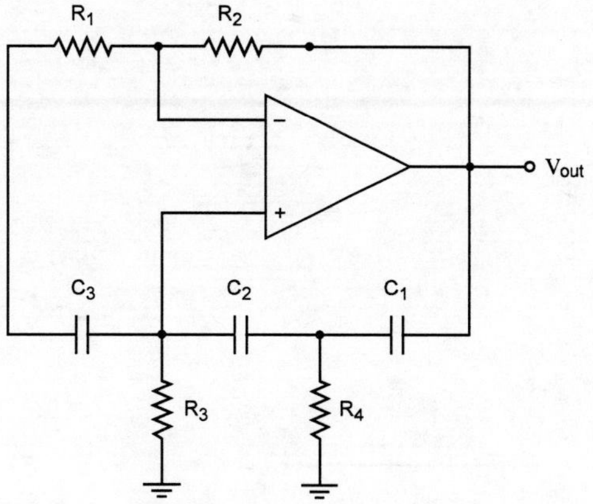

Fig. 3

5. Given the circuit of Fig. 3. Calculate frequency of oscillation if R_1 = 200 kΩ; R_2 = 1.6 MΩ; R_3 = 20 kΩ; R_4 = 2 kΩ, C_1 = 30 μF, C_2 = 3 μF; C_3 = 0.3 μF.

6. For the circuit shown in Fig. 4, calculate frequency of oscillation if R_1 = R_2 = 22 kΩ; R_3 = 33 kΩ; C = 3.3 μF.

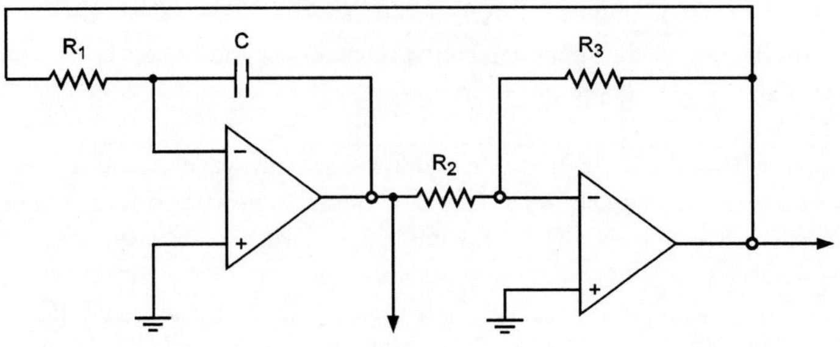

Fig. 4

7. Determine the trip points and hysteresis in the circuit given in Fig. 5 if V_{sat} = 13.5 V.

[**Ans.** +0.281 V, –0.281 V, 0.562 V]

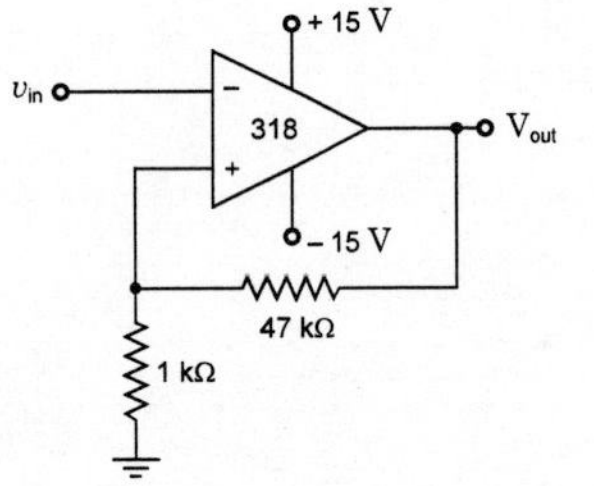

Fig. 5

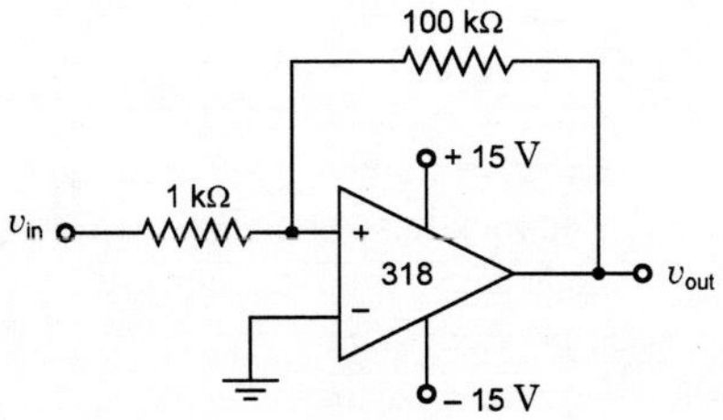

Fig. 6

8. Determine the trip points and hysteresis in the noninverting Schmitt trigger given in Fig. 6. V_{sat} = 12 V.

[**Ans.** +0.12 V, –0.12 V, 0.24 V]

9. Design a Schmitt trigger circuit, as shown in Fig. 6.21 (*a*). The circuit trigger points are to be approximately ±5 V and the supply voltage is ± 20 V. Use a 741 op-amp. [**Ans.** R_1 = 270 kΩ, R_2 = 100 kΩ]

10. Design a Schmitt trigger circuit to have trigger voltages of ±2 V. The supply voltage is ±18 V. Use a 741 op-amp. [**Ans.** 330 kΩ, 39 kΩ]

❑❑❑

Unit-III

POWER AMPLIFIERS AND WAVESHAPING

CONTENTS IN THIS UNIT

POWER AMPLIFIERS : Power Dissipation in Transistors; Harmonic Distortion; Amplifier Classification (Class-A, Class-B, Class-C, Class A-B), Efficiency, Push-Pull and Complementary Push-Pull Amplifiers, Tuned Amplifiers.

LINEAR AND NONLINEAR WAVESHAPING : Clipping and Clamping Circuits; Comparators, Log/Antilog Circuits Using Op-Amps, Precision Rectifiers.

CHAPTER

7

Power Amplifiers

INSIDE THIS CHAPTER

7.1. INTRODUCTION

An amplifier system consists of signal pickup transducer, followed by small signal amplifier(s), a large signal amplifier and an output transducer. A *transducer* is used to convert one form of energy into another type. For example, a microphone is used to convert acoustical energy into electrical energy. Conversely, a loudspeaker is used to convert electrical energy into acoustical energy.

The input transducer produces small electrical (typically voltage) signal, that needs sufficient amplification to operate some output device such as a loudspeaker, a servomotor, a solenoid or relay. The factors of prime interest in small signal voltage amplifiers are usually linearity and gain. Since the signal voltage and current from the input transducer is usually very small, the amount of power handling capacity and power efficiency are of slight concern. The functions of voltage amplifiers are to present a high resistance to the input transducer to minimize loading effects and to provide a large enough voltage signal to the large signal amplifier stages to operate such output devices (loudspeaker, servomotor etc.). A large signal amplifier must operate efficiently and be capable of handling large amounts of power–typically, a few watts to hundred of watts. Large signal amplifier that drives the output transducer demands even more consideration than the small signal voltage amplifiers that we have focused so far. The factors of greatest concern to the large signal power amplifiers are the power efficiency of the circuit, the maximum amount of power that the circuit is capable of handling, and impedance matching to the output device.

Power amplifier is meant to raise the power level of the input signal. In order to get large power at the output, it is necessary that the input-signal voltage is large. That is why, in an electronic system, a voltage amplifier always precedes the power amplifier, as illustrated in the block diagram (Fig. 7.1). Also, that is why power amplifiers are called large signal amplifiers.

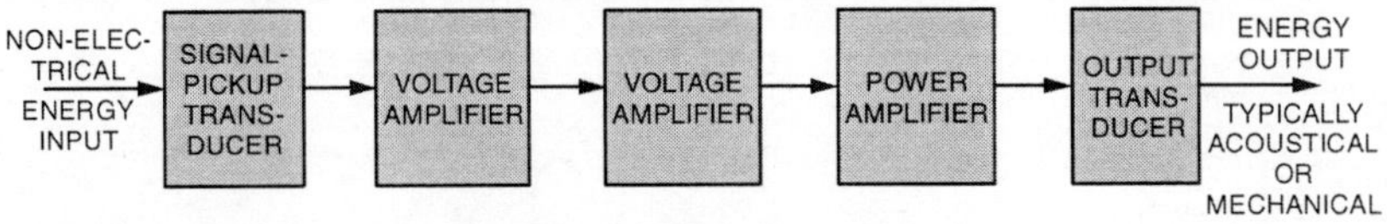

Fig. 7.1. ***Amplifier System***

In fact, power amplifier does not amplify power.* What a power amplifier actually does is that it draws power from dc supply connected to the output circuit and converts it into useful ac signal power. The type of ac power available at the output terminals of the power amplifier is controlled by the input signal. Thus a *power amplifier may be defined as a device that converts dc power into ac power and whose action is controlled by the input signal.*

The transistors employed in power amplifiers are called **power transistors**. They differ from other transistors in the following respects :

(*i*) The base is made thicker to handle large currents *i.e.,* in power amplifiers, transistors with comparatively smaller β are used.

(*ii*) The area of collector region of a power transistor is made considerably larger in order to dissipate the heat developed in the transistor during operation. Moreover, heat sinks are used for improving the heat dissipation.

(*iii*) The emitter and base layers are heavily doped. The contact area between the base layers and base leads is in ring like form so that the area is increased. By doing so ohmic resistance between emitter and base is reduced and due to low resistance, small power is required at input.

7.2. DIFFERENCE BETWEEN VOLTAGE AND POWER AMPLIFIERS

The primary function of the *voltage amplifier* is to raise the voltage level of the signal. It is designed to achieve the largest possible gain. Only very little power can be drawn from its output.

The voltage gain of an amplifier is given by the expression $A_v = \beta R_C/R_{in}$. So in order to achieve high voltage amplification (*i*) transistors having thin base *i.e.,* the transistors with high β (exceeding 100) are used (*ii*) input resistance R_{in} is kept quite low in comparison to collector load R_C and (*iii*) to permit high collector load, which is necessary to give high voltage gain, the voltage amplifiers are always operated at low collector current (of the order of 1 mA). R-C coupling is preferred for coupling various stages of voltage amplifiers because it is smaller in size, light in weight, cheaper in cost and it occupies less space.

A power amplifier is required to deliver a large amount of power and as such it has to handle large currents. So in order to achieve high power amplification, (*i*) large sized power transistors are used in order to dissipate the heat produced in transistor during operation, (*ii*) transistor having thick base (*i.e.,* transistors with comparatively smaller β) are used in order to handle large currents (*iii*) transformer coupling is used for impedance matching and (*iv*) collector resistance is made low. If its value is high, power losses in the collector resistor will be increased. Moreover, voltage drop across it will increase and, therefore, operating voltage will be reduced. This is the reason that low collector load resistance is used.

The noteworthy point is that we cannot obtain a power amplifier by simply replacing a transistor of low power dissipation in a voltage amplifier circuit by another transistor with high power dissipation.

* No device can amplify power because the idea of amplification of power will be against the basic principles of physics.

Let us consider a voltage divider biased common emitter voltage amplifier circuit, with replaced transistor, shown in Fig. 7.2. Let I_{CQ} and V_{CEQ} be the collector current and collector-emitter voltage respectively at the operating point.

Now collector–emitter voltage,

$$V_{CEQ} = V_{CC} - I_{CQ}(R_C + R_E) \quad ...(7.1)$$

Total dc power supplied to the amplifier from collector supply

$$= V_{CC} I_{CQ} \quad ...(7.2)$$

DC power input to transistor $= V_{CEQ} I_{CQ}$ ...(7.3)

Out of total dc power of $V_{CC} I_{CQ}$ drawn from the dc supply, $V_{CEQ} I_{CQ}$ is only the power that is converted into useful ac power and rest is wasted.

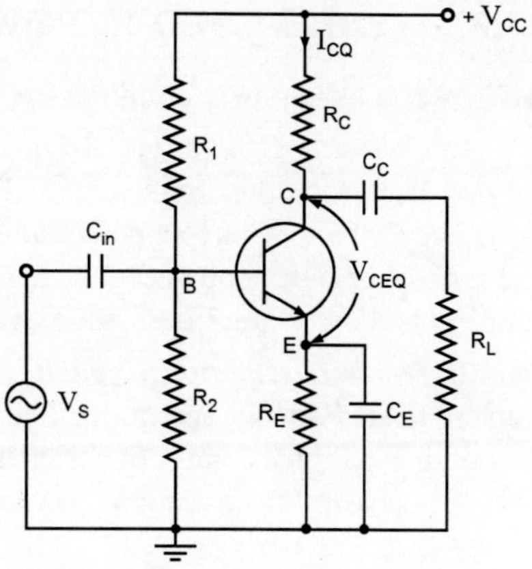

Fig. 7.2. ***Simple Amplifier Circuit***

i.e., Power wasted $= V_{CC} I_{CQ} - V_{CEQ} I_{CQ}$

$$= I_{CQ}[V_{CC} - V_{CEQ}]$$

$$= I_{CQ}[I_{CQ}(R_C + R_E)] \quad \because \text{from Eq. (7.1) } V_{CC} - V_{CEQ} = I_{CQ}(R_C + R_E)$$

$$= I_{CQ}^2 (R_C + R_E) \quad ...(7.4)$$

The $I^2_{CQ}(R_C + R_E)$ is the power that gets wasted unnecessarily in heating of resistors. This is the reason that a voltage amplifier cannot be used as a power amplifier.

Power wastage can be reduced by short circuiting collector and emitter resistors, R_C and R_E, but it is not possible. Emitter resistor R_E cannot be eliminated because it is part of biasing network. In the absence of R_E, the stabilization of the operating point is adversely affected. Collector resistor R_C also cannot be short circuited because in that case load resistor R_L will be short circuited (from ac point of view) and there will be no transfer of power to the load, R_L.

However, collector resistor R_C can be replaced by an inductor whose dc resistance is zero (acts as a short circuit for dc) and ac impedance is high. By doing so, two things are achieved. Power loss in the amplifier circuit is reduced because dc power loss in the choke is almost zero. In addition we can afford to use lower voltage supply V_{CC} for the same amplifier because no dc voltage drop occurs across the choke. This is the reason that transformer coupling is preferred over R-C coupling in power amplifiers.

Still more improvements can be made in the above circuit so as to make it more efficient in operation and will be discussed in this unit.

The comparison between voltage and power amplifiers is tabulated as below :

S. No.	Type of Amplifier	Voltage Amplifier	Power Amplifier
1.	Current gain, β	High, exceeding 100	Low, 20 – 50
2.	Collector load, R_C	High, about 10 kΩ	Low, 5 – 20 Ω
3.	Input voltage	Low, a few mV	High, 2 – 4 volts
4.	Collector current, I_C	Low, about 1 mA	High, exceeding 100 mA
5.	Power output	Low	High, a few watts to hundred of watts
6.	Power dissipation capacity	Less than 0.5 W	More than 0.5 W
7.	Output impedance	High, about 10 kΩ	Low, about 200 Ω
8.	Coupling	Usually R-C coupling	Invariably transformer coupling or tuned circuit

7.3. TERMS USED IN POWER AMPLIFIERS

The criteria for a power amplifier are collector efficiency, power dissipation capacity and distortion.

1. *Collector Efficiency*. As mentioned earlier, power amplifier converts dc power into ac power at output. Therefore, the effectiveness of a power amplifier is measured by its capability of converting dc power from supply battery into ac power at the output. This is known as *collector efficiency* and may be defined as the ratio of ac output power to the zero signal power (or dc power drawn from the supply battery of a power amplifier). Power amplifiers are designed to provide maximum collector efficiency.

2. *Power Dissipation Capacity*. As mentioned earlier, a power transistor handles large currents and gets up heated during operation. This may cause the junction temperature to rise to a dangerous level. The power dissipation capacity of a transistor can be increased by making a suitable provision for rapid conduction of heat away from the junction and this is achieved by attaching a metal case, called the *heat sink*, to the power transistor case.

Power dissipation capacity of a power transistor is defined as its ability to dissipate the heat developed in it.

3. *Distortion* is defined as the change of output waveshape from the input waveshape of the amplifier. Since power amplifier handles large signals and for large input signals, the transistor operation becomes nonlinear, the problem of distortion arises. Lesser the distortion, better is the amplifier.

7.4. CLASSIFICATION OF POWER AMPLIFIERS

The power amplifiers are primarily divided into two categories viz. *audio power amplifiers* and *radio power amplifiers*.

Audio power amplifiers, also called the *small signal power amplifiers*, raise the power levels of signals that have audio frequency range (20 Hz–20 kHz).

Radio power amplifiers, also called *large signal power amplifiers,* raise the power level of signals that have radio frequency range. They amplify a specific frequency or narrowband of frequencies while rejecting all other frequencies.

Classification According To Mode of Operation. Transistor power amplifiers handle large signals. Many of them are driven so hard by the input large signal that collector current is either cutoff or is in saturation region during a large portion of the input cycle. So such amplifiers are generally classified according to their mode of operation. This classification is based on the amount of transistor bias and amplitude of the input signal. It takes into account the portion of the cycle for which the transistor conducts. They are classified as below :

1. *Class A Power Amplifiers*. In this case, transistor is so biased that the output current flows for the entire cycle of the input signal. Thus the operating point is so selected that the transistor operates only over the linear region of its load line. So such an amplifier can amplify input signal of small amplitude. As the transistor operates over the linear portion of load line, the output waveform is exactly similar to input waveform. So class A amplifiers are characterised by a high fidelity of the output. Such amplifiers are used where freedom from distortion is prime aim. Operation is restricted only over a small central region of the load line so such amplifiers can be used for amplifying signals of small amplitude. Also ac power output per transistor is small. The maximum possible overall efficiency with resistive load is 25%. The maximum possible collector efficiency with resistive load is 50%. In case an output transformer is used, both of these efficiencies are 50%.

2. *Class B Power Amplifiers*. In this case, the transistor bias and signal amplitude are such that output current flows only during positive half cycle of the input signal. At zero signal, the collector current is zero and no biasing system is required in class B amplifiers. The operating point is selected at collector cutoff voltage. Because of total absence of negative half cycle from the output the signal distortion is high. Zero signal input represents the best condition for class B amplifiers because of zero collector current. The transistor dissipates more power with increase in signal strength. In comparison to class A amplifiers average current is less, power dissipation is less. So overall efficiency is increased. The theoretical efficiency in class B operation is about 78.5% while it is only 50% in class A operation.

3. Class AB Power Amplifiers. An amplifier may be biased at a dc level above the zero base current level of class B power amplifiers and above one-half the supply voltage level of class A; this bias condition is class AB. Class AB operation still needs a push-pull connection to achieve a full output cycle, but the dc bias level is usually closer to zero base current level for better power efficiency. For class AB operation the output signal swing occurs between 180° and 360° and is neither class A nor class B operation.

4. Class C Power Amplifiers. A class C power amplifier is biased for operation for less than 180° of the input signal cycle and will operate only with a tuned or resonant circuit which provides a full cycle of operation for the tuned or resonant frequency. Such power amplifiers are, therefore, employed in special areas of tuned circuits, such as radio or communications.

5. Class D Power Amplifiers. Class D power amplifiers are designed to operate with digital or pulse type signals. Using digital techniques makes it possible to have a signal that varies over the entire cycle (using sample-and-hold circuitry) to recreate the output from many pieces of input signal. The main advantage of class D power amplifiers is that it is *on* (using power) only for short intervals and the overall efficiency can practically be very high (above 90%).

Classification Based on Deriving Output. Transistor amplifiers, according to the method of deriving output, are classified as single-ended, double-ended (or push-pull) and complementary symmetry push-pull power amplifiers.

Single-ended power amplifier uses single transistor and derives output power w.r.t. one end permanently grounded.

Double-ended or push-pull amplifier uses two transistors in a single stage. It consists of two loops in which the transistor collector currents flow in opposite directions but add in the load.

Complementary symmetry push-pull power amplifier uses two transistors having complementary symmetry (one N-P-N and another P-N-P). The term complementary arises from the fact that one transistor is the N-P-N type and the other is P-N-P type. They have symmetry as they are made with the same material and technology and are of same maximum ratings.

Classification Based on Use. The amplifiers, according to use, may be classified as voltage, power, current or general purpose amplifiers. In general, the load of an amplifier is impedance. The two most important special cases are the idealized resistive load and the tuned circuit operating near its resonant frequency. Class AB and class B operation are used with untuned power amplifiers whereas class C operation is employed with tuned radio frequency amplifiers. Many important waveshaping functions may be accomplished by class B or class C overdriven amplifiers.

7.5. PRACTICAL POWER AMPLIFIER STAGES

Power amplifier is meant to amplify a weak signal until sufficient power is available to operate an output device such as a loudspeaker, a solenoid or a relay. Power amplifier, to provide the desired power amplification, has generally 3 stages (voltage amplification stage, driver stage and output stage), as shown by a block diagram in Fig. 7.3.

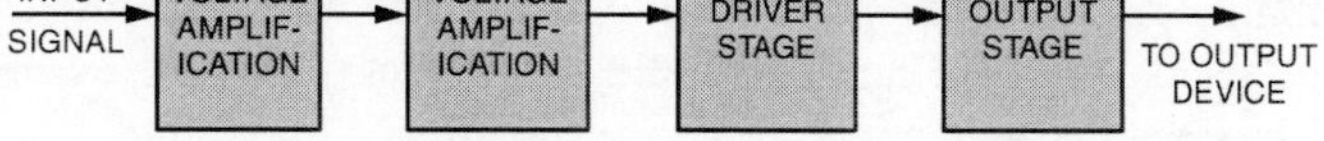

Fig. 7.3. ***Stages in a Practical Power Amplifier***

1. *Voltage Amplification Stage.* As already mentioned in Art. 7.1, the signal developed by the input transducer is very small (of the order of few millivolts) and needs sufficient amplification so as to operate the output device. Therefore, for raising the level of the weak input signal, it is amplified in two or more stages, R-C coupling is usually employed.

2. *Driver Stage.* The stage that immediately precedes the output stage is called the *driver stage*. The output from the last voltage amplification stage is fed to the driver stage and output from the driver stage is fed to the output stage. The driver stage renders power amplification as usual. Here, concentrated effort is made to provide maximum power gain and so transformer coupled class A power amplifier is employed in this stage. The driver transformer is usually a step-down transformer and facilitates impedance matching.

3. *Output Stage.* The output stage essentially consists of a power amplifier and is meant for transferring maximum power to the output device. In order to transfer maximum power at high efficiency, push-pull arrangement is employed in the output stage (if a single transistor is used as a class A amplifier in the output stage to provide faithful amplification, the operation efficiency will be quite low). In this arrangement two transistors are used in class B operation and are fed from the centre-tapped secondary of the transformer whose primary forms the collector load for the driver stage.

7.6. CLASS A POWER AMPLIFIERS (DIRECT COUPLED WITH RESISTIVE LOAD)

A class A power amplifier is defined as a power amplifier in which output current flows for the entire cycle (360°) of the input signal, as illustrated in Fig. 7.4. In other words, the transistor remains forward biased throughout the input cycle.

A schematic circuit of a series fed class A large signal amplifier using resistive load R_C is shown in Fig. 7.5. The term "series fed" is derived from the fact that the load R_C is connected in series with the transistor output. The only difference between this circuit and the small signal amplifier circuits considered previously is that the signals handled by the large signal circuit are in the range of volts and the transistor used is a power transistor capable of operating in the range of few watts. This circuit is seldom used for power amplification because of its poor collector efficiency but will give clear understanding of class A operation to the readers. The output characteristics with operating point Q are shown in Fig. 7.6. I_{CQ} and V_{CEQ} represent no signal collector current and collector-emitter voltage respectively. When an input ac signal is applied to the amplifier given in Fig. 7.5, the output will vary from its dc bias operating voltage and current. A small input signal, as indicated in Fig. 7.6, will cause the base current to vary above and below the dc bias point, which will then cause the collector current (output) I_c to vary from the dc bias point set as well as the collector-emitter voltage V_{ce} to vary around its dc bias value. With the strengthening of the input signal, the output will vary further around the established dc bias point until either the current or voltage attains a limiting condition. For the current this limiting condition is either zero current at the lower end or $\frac{V_{CC}}{R_C}$ at the

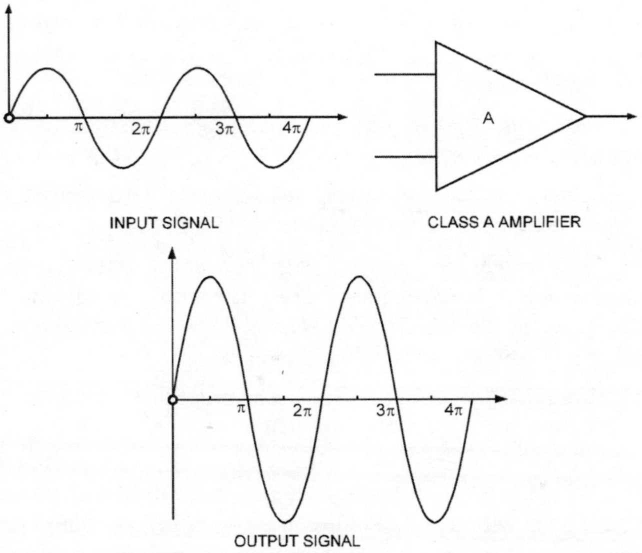

Fig. 7.4

upper end. For the collector-emitter voltage this limit is either 0 V or the supply voltage V_{CC}. Let the output current vary between limits $I_{c\,min}$ and $I_{c\,max}$ and similarly the collector-emitter voltage between limits $V_{ce\,min}$ and $V_{ce\,max}$.

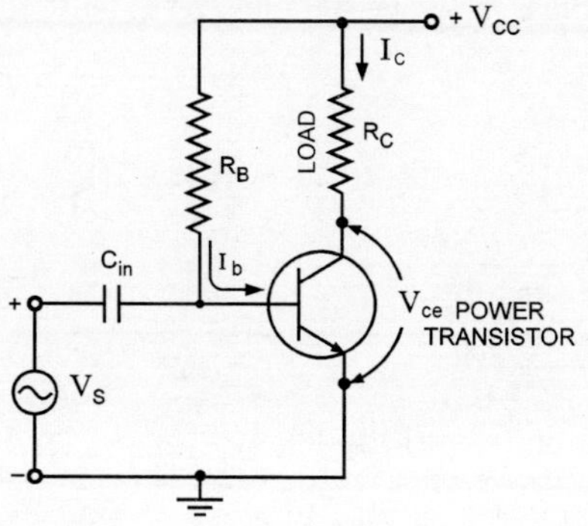

Fig. 7.5. *Series Fed Class A Large Signal Amplifier*

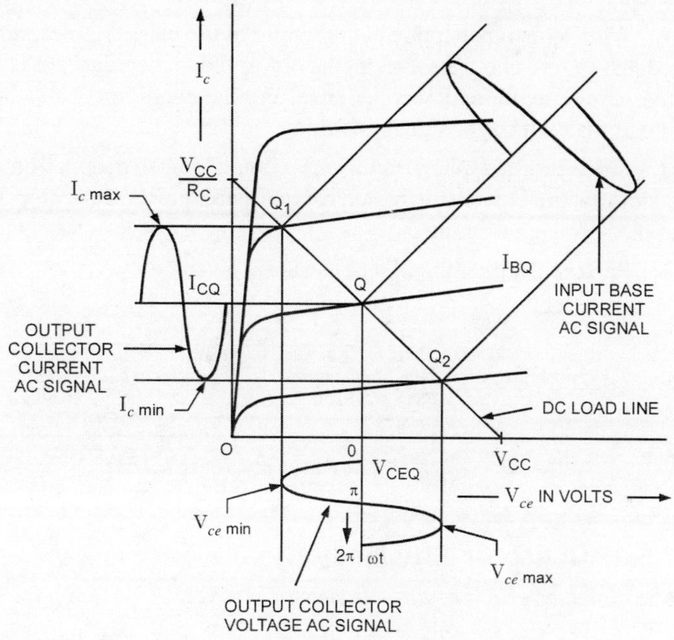

Fig. 7.6. *Transistor Output Characteristics Showing DC Load Line and AC Signals*

7.6.1. Power Distribution

Input power from the collector supply V_{CC},

$$P_{in\,(dc)} = V_{CC}\, I_{CQ} \quad \text{...(7.5)}$$

The power drawn from the collector supply is used in the following two components

(*i*) Power dissipated in collector load as heat,

$$P_{RC\,(dc)} = I^2_{CQ}\, R_C \quad \text{...(7.6)}$$

(*ii*) Power supplied to the transistor,

$$P_{tr\,(dc)} = P_{in\,(dc)} - P_{RC\,(dc)} = V_{CC}\, I_{CQ} - I^2_{CQ}\, R_C \quad \text{...(7.7)}$$

Power supplied to the transistor, $P_{tr\,(dc)}$ is further subdivided into

(*a*) ac power developed across the load resistor constituting ac power output and is given as

$P_{out\,(ac)} = I_c^2 R_C = I_c V_{ce}$ ...(7.8 *a*) where I_c and V_{ce} are the rms values of collector current and collector-emitter voltage

$$= \frac{1}{2}\frac{I_{c\,(peak\text{-}to\text{-}peak)}}{\sqrt{2}} \times \frac{1}{2}\frac{V_{ce\,(peak\text{-}to\text{-}peak)}}{\sqrt{2}} \quad \text{...(7.8 } b\text{)}$$

$$= \frac{I_{c\,(peak\text{-}to\text{-}peak)} \times V_{ce\,(peak\text{-}to\text{-}peak)}}{8} \quad \text{...(7.8 } c\text{)}$$

(*b*) Power dissipated, in the form of heat, by the transistor itself. The cause of power dissipation in transistor is explained below :

Consider an N-P-N transistor. The potential difference across the depletion layer formed near the collector junction is called the barrier potential. This potential gives the P-region (base) slightly more energy than N-region (collector). Thus when electrons emitted from emitter cross the base junction and enter the collector region, they give up energy in the form of heat and it is this energy that the transistor has to dissipate to the surrounding.

With zero signal applied at the input of the class A power amplifier, ac power developed across the load reduces to zero and therefore, all the power fed to the transistor is wasted in the form of heat. Thus, a *transistor dissipates* maximum power under zero signal condition. Thus the device is cooler when delivering power to a load than with zero signal condition.

Since in class A operation, maximum power dissipation in the transistor occurs under zero signal condition, the power dissipation capacity of a power transistor, for class A operation, must be at least equal to the zero signal rating.

Power flow diagram of a transistor is given below :

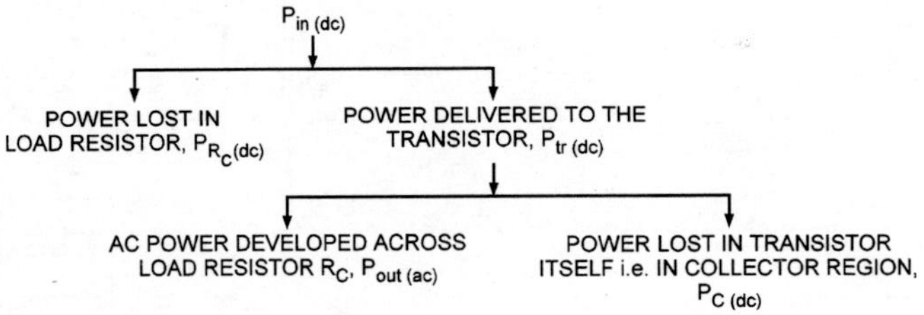

7.6.2. Collector Efficiency

The collector efficiency of a transistor is given as

$$\eta_{collector} = \frac{\text{Average ac power output, } P_{out\,(ac)}}{\text{Average dc power input to the transistor, } P_{tr\,(dc)}}$$

7.6.3. Power Efficiency

A measure of the ability of an active device to convert the dc power of supply into the ac (signal) power delivered to the load is called the *power* or *conversion* or *theoretical efficiency*. By definition the efficiency is

$$\eta = \frac{\text{AC power delivered to the load, } P_{out\,(ac)}}{\text{Total power drawn from dc supply, } P_{in\,(dc)}}$$

Now ac power delivered to the load,

$$P_{out\,(ac)} = \frac{I_{c\,(peak\text{-}to\text{-}peak)} \times V_{ce\,(peak\text{-}to\text{-}peak)}}{8} \quad \text{...Refer to Eq. (7.8 } c\text{)}$$

$$= \frac{(I_{c\,max} - I_{c\,min}) \times (V_{ce\,max} - V_{ce\,min})}{8}$$

and from Eq. (7.5), total power drawn from dc supply,

$$P_{in\,(dc)} = V_{CC}\,I_{CQ}$$

$$\text{So power efficiency, } \eta = \frac{P_{out\,(ac)}}{P_{in\,(dc)}} = \frac{(I_{c\,max} - I_{c\,min}) \times (V_{ce\,max} - V_{ce\,min})}{8\,V_{CC}\,I_{CQ}} \quad \text{...(7.9)}$$

Note : RMS value $= \frac{1}{2}\left[\frac{\text{peak-to-peak}}{\sqrt{2}}\right] = \frac{1}{2\sqrt{2}}$ peak-to-peak.

7.6.4. Maximum Power and Efficiency

If the operating point Q is set at the midpoint of the maximum signal swing, the resulting maximum power condition may be achieved. For the circuit shown in Fig. 7.5, this would be determined using

$$\text{Maximum } V_{ce\,(\text{peak-to-peak})} = V_{CC}$$

$$\text{Maximum } I_{c\,(\text{peak-to-peak})} = \frac{V_{CC}}{R_C}$$

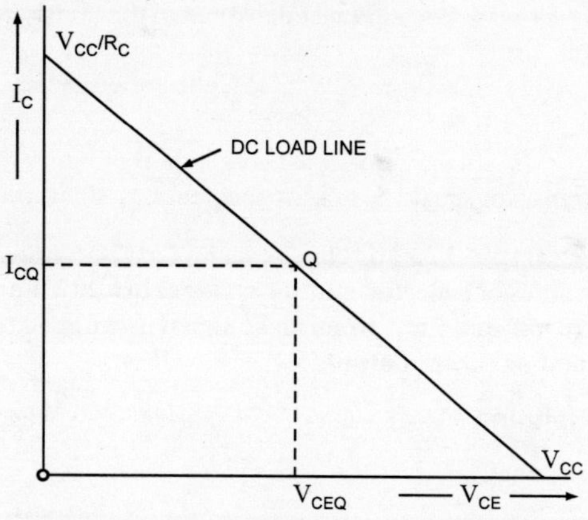

Fig. 7.7

From Eq. (7.8 *c*) we have maximum ac power developed across the load resistor,

$$P_{out\,(\text{ac})\,\max} = \frac{1}{8}\cdot\frac{V_{CC}}{R_C}\times V_{CC} = \frac{V_{CC}^2}{8\,R_C} \qquad \text{...(7.10)}$$

For the quiescent point Q,

$$I_{CQ} = \frac{V_{CC}/R_C}{2} \qquad \text{...(7.11)}$$

and dc power drawn from dc supply,

$$P_{in\,(\text{dc})\,\max} = V_{CC}I_{CQ} = \frac{V_{CC}\times\dfrac{V_{CC}}{R_C}}{2} = \frac{V_{CC}^2}{2\,R_C} \qquad \text{...(7.12)}$$

Power supplied to the transistor,

$$P_{tr}(\text{dc}) = V_{CC}I_{CQ} - I_{CQ}^2R_C \qquad \text{...Refer to Eq. (7.7)}$$

$$= V_{CC}\cdot\frac{V_{CC}}{2\,R_C} - \left(\frac{V_{CC}}{2\,R_C}\right)^2 R_C = \frac{V_{CC}^2}{2R_C} - \frac{V_{CC}^2}{4R_C} = \frac{V_{CC}^2}{4R_C} \qquad \text{...(7.13)}$$

So maximum efficiency of an amplifier (class A power) is given as

$$\eta_{\max} = \frac{P_{out\,(\text{ac})\,\max}}{P_{in\,(\text{dc})\,\max}} = \frac{V_{CC}^2/8R_C}{V_{CC}^2/2R_C} = \frac{2R_C}{8R_C} = 0.25 \text{ or } 25\% \qquad \text{...(7.14)}$$

This is the maximum percent efficiency for a series-fed class A power amplifier. Since this maximum efficiency will occur only under ideal conditions and for the maximum ac signal swings, most series-fed class A power amplifiers have power efficiencies much less than 25%.

From Eqs. (7.10) and (7.13), we have

$$\frac{P_{\text{out (ac) max}}}{P_{\text{tr (dc)}}} = \frac{V_{CC}^2/8R_C}{V_{CC}^2/4R_C} = \frac{1}{2} \qquad \text{...(7.15)}$$

Thus, we see that a class A power amplifier with a direct coupled resistive load can supply only one-half the amount of ac power to the load that the transistor can dissipate and the dc power input is four times the amount of ac signal power in the load.

Class A power amplifiers have got advantages of simple construction and distortionless output voltage.

The drawbacks of class A power amplifiers are very low efficiency and large power dissipation in the power transistors.

Example 7.1. A certain power transistor meant for Class A operation has a zero signal power dissipation of 20 watts. If the ac output power is 5 W, find (*a*) Collector efficiency and (*b*) power rating of transistor.

Solution : AC power output, $P_{out\,(ac)} = 5$ W

Power delivered to the transistor, $P_{tr\,(dc)}$ = Zero-signal power dissipation = 20 W

(*a*) Collector efficiency, $\eta_{collector} = \dfrac{P_{out\,(ac)}}{P_{tr\,(dc)}} \times 100 = \dfrac{5}{20} \times 100 = 25\%$ **Ans.**

(*b*) Since zero-signal condition represents worst condition for the transistor, the power dissipation capacity of the transistor should be at least equal to zero-signal power dissipation *i.e.,* 20 W **Ans.**

Example 7.2. For class A series fed CE large signal amplifier using resistance load, the maximum and minimum values of collector-emitter voltages are 20 V and 10 V and maximum and minimum values of collector current are 10 mA and 5 mA when an ac signal is supplied to it. Determine rms values of collector voltage and collector current and ac power output.

Solution : RMS value of collector-emitter voltage,

$$V_{ce\,rms} = \frac{V_{ce\,(peak\text{-}to\text{-}peak)}}{2\sqrt{2}} = \frac{V_{ce\,max} - V_{ce\,min}}{2\sqrt{2}} = \frac{20-10}{2\sqrt{2}} = 3.54 \text{ V } \textbf{Ans.}$$

Similarly rms value of collector current, $I_{c\,rms} = \dfrac{I_{c\,max} - I_{c\,min}}{2\sqrt{2}} = \dfrac{10-5}{2\sqrt{2}} = 1.77$ mA **Ans.**

Hence ac power output, $P_{out\,(ac)} = V_{ce\,rms} \times I_{c\,rms} = 3.54 \times 1.77 \times 10^{-3} = 6.25$ mW **Ans.**

Example 7.3. Calculate the input power, output power delivered to the transistor, power lost in transistor, collector efficiency and overall efficiency of the amplifier shown in Fig. 7.8 for an input voltage resulting in base current of 10 mA peak. Take β = 25 and V_{BE} = 0.7 V.

Solution : Saturation current, $I_{C\,(sat)} = \dfrac{V_{CC}}{R_C} = \dfrac{20}{20} = 1$ A or 1,000 mA

$$V_{CE} = V_{CC} = 20 \text{ V}$$

Now dc load line is drawn joining points (20 V, 0) and (0, 1000 mA), as shown in Fig. 7.9.

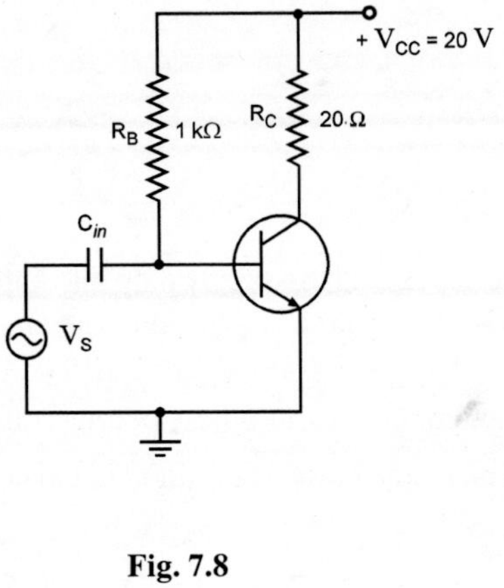

Fig. 7.8

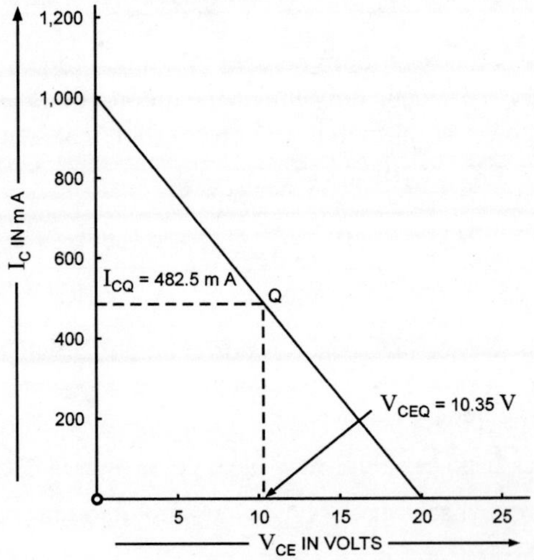

Fig. 7.9

For operating point Q

$$\text{Base current, } I_B = \frac{V_{CC} - V_{BE}}{R_B} = \frac{20 - 0.7}{1\,k\Omega} = 19.3 \text{ mA}$$

$$\text{Collector current, } I_{CQ} = \beta\, I_B = 25 \times 19.3 = 482.5 \text{ mA}$$

$$\text{Collector-emitter voltage, } V_{CEQ} = V_{CC} - I_{CQ}\, R_C = 20 - 482.5 \times 10^{-3} \times 20 = 10.35 \text{ V}$$

The operating point Q (10.35 V, 482.5 mA) is marked on the dc load line.

$$\text{Peak value of collector current, } I_{c\,peak} = \beta\, I_{b\,peak} = 25 \times 10 \text{ mA peak} = 250 \text{ mA}$$

$$\text{Input power, } P_{in\,(dc)} = V_{CC}\, I_{CQ} = 20 \times 482.5 \times 10^{-3} = 9.65 \text{ W } \textbf{Ans.}$$

$$\text{Output power, } P_{out\,(ac)} = \left(\frac{I_{c\,peak}}{\sqrt{2}}\right)^2 R_C = \left(\frac{250 \times 10^{-3}}{\sqrt{2}}\right)^2 \times 20 = 0.625 \text{ W } \textbf{Ans.}$$

$$\text{Power delivered to the transistor, } P_{tr\,(dc)} = V_{CC}\, I_{CQ} - I_{CQ}^2 R_C$$

$$= 20 \times 482.5 \times 10^{-3} - (482.5 \times 10^{-3})^2 \times 20 = 9.65 - 4.66 = 4.99 \text{ W } \textbf{Ans.}$$

$$\text{Power lost in transistor} = P_{tr\,(dc)} - P_{out\,(ac)} = 4.99 - 0.625 = 4.365 \text{ W } \textbf{Ans.}$$

$$\text{Collector efficiency, } \eta_{collector} = \frac{P_{out\,(ac)}}{P_{tr\,(dc)}} \times 100 = \frac{0.625}{4.99} \times 100 = 12.53\% \textbf{ Ans.}$$

$$\text{Overall efficiency, } \eta_{overall} = \frac{P_{out\,(ac)}}{P_{in\,(dc)}} = \frac{0.625}{9.65} \times 100 = 6.48\% \textbf{ Ans.}$$

$$\text{Power rating of transistor} = \text{Zero-signal power dissipation} = V_{CEQ}\, I_{CQ} = 10.35 \times 482.5 \times 10^{-3} = 5 \text{ W } \textbf{Ans.}$$

7.7. CLASS A POWER AMPLIFIERS (TRANSFORMER COUPLED WITH RESISTIVE LOAD)

This is also sometimes referred to as *single ended power amplifier*. The term "single ended" (denoting only one transistor) is used to distinguish it from the push-pull amplifier using two transistors.

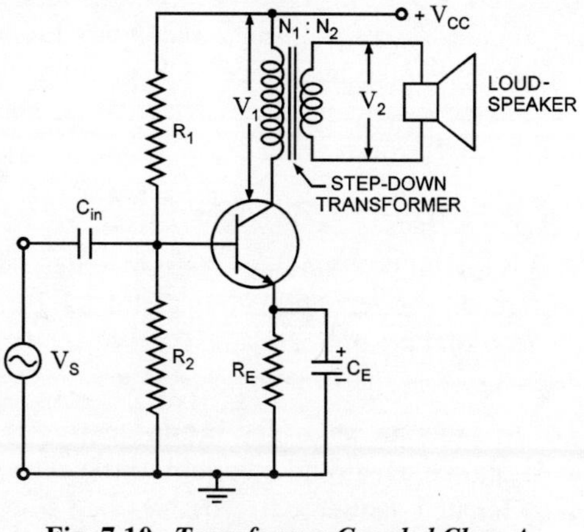

Fig. 7.10. ***Transformer Coupled Class A Power Amplifier***

In case of a direct coupled class A power amplifier shown in Fig. 7.5, the quiescent current flows through the collector resistive load and causes large wastage of dc power in it. This dc power dissipated in the load resistor does not contribute to the useful ac output power. Furthermore, it is generally inadvisable to pass the dc through the output device such as in a voice coil of a loudspeaker. For these reasons an arrangement using a suitable transformer for coupling the load (say, a loudspeaker) to the amplifier is usually employed, as shown in Fig. 7.10. This arrangement enhances the conversion efficiency by a factor of 2 by eliminating the dc power dissipation in the load. Since the primary of the transformer has negligible dc resistance, there is negligible dc power lost at this point. The ac power is, however, coupled magnetically across the transformer into the load R_L. This coupling method also prevents a large dc current from flowing through the load, which otherwise could be harmful if the load were a loudspeaker, since it would cause saturation of the magnetic circuit and impair the reproduction of the audio signal.

This arrangement also permits impedance matching.

In a power amplifier circuit shown in Fig. 7.10, R_1 and R_2 provide potential divider biasing and emitter resistor R_E is meant for bias stabilization. The emitter bypass capacitor C_E is meant for preventing negative feedback in the emitter circuit. The input capacitor C_{in} couples ac signal voltage to the base of the transistor but blocks any dc from the previous stage. A step-down transformer of suitable turn ratio $\left(a = \frac{N_1}{N_2}\right)$ is provided to couple the high impedance collector circuit to low impedance load.

7.7.1. Impedance Matching

The power transferred from the power amplifier to the load (say a loudspeaker) will be maximum only if the amplifier output impedance equals the load impedance R_L. This is in accordance with the maximum power transfer theorem. If we were not able to achieve the above condition, lesser power will be transferred to the load R_L, though the amplifier is capable of delivering more power, and rest of power developed would be lost in the active device. Hence for transfer of maximum power from amplifier to the output device matching of amplifier output impedance with the impedance of output device is necessary. This is accomplished by using a step-down transformer of suitable turn ratio. The transformer impedance matching circuit is shown separately in Fig. 7.11, where R_L' is the resistance looking into the primary of the transformer and is given as

$$\frac{R'_L}{R_L} = \frac{V_1/I_1}{V_2/I_2} = \frac{V_1}{V_2} \cdot \frac{I_2}{I_1} = \frac{N_1}{N_2} \cdot \frac{N_1}{N_2} = \left(\frac{N_1}{N_2}\right)^2 \quad \text{where } \frac{V_1}{V_2} = \frac{N_1}{N_2} \text{ and } \frac{I_2}{I_1} = \frac{N_1}{N_2}$$

Thus the ratio of the transformer input and output resistances varies directly as the square of the transformer turn ratio :

$$\frac{R'_L}{R_L} = \left(\frac{N_1}{N_2}\right)^2 = a^2$$

$$\text{or} \quad R'_L = a^2 R_L \qquad \text{...(7.16)}$$

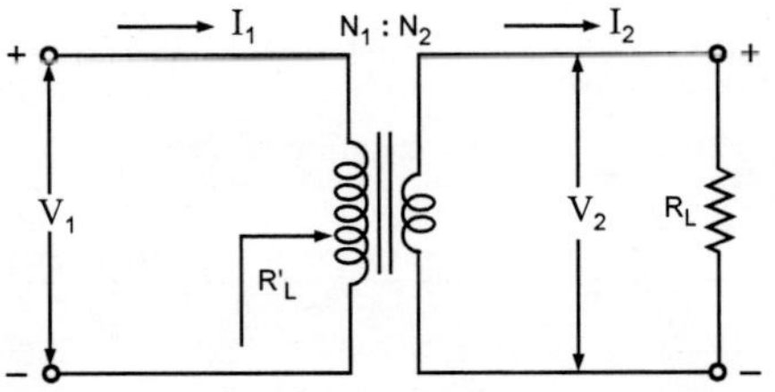

Fig. 7.11

where a is ratio of primary to secondary turns of step-down transformer, R_L is the resistance of load connected across the transformer secondary and R'_L is effective resistance looking into the transformer primary.

7.7.2. Circuit Operation

In this circuit dc (winding) resistance determines the dc load line. Typically, this resistance is quite small (assumed to be zero) providing dc load to be a vertical line rising from V_{CC}, as shown in Fig. 7.12. When an ac signal is applied to the base of the transistor the collector current will vary around the operating point Q.

In order to have maximum ac power output, the peak value of collector current due to input ac signal alone should be equal to the zero signal collector current. To achieve this, the operating point Q is located at the centre of the load line. This is achieved by adjusting the biasing circuit (R_1, R_2 and R_E).

When ac signal is applied, collector current fluctuates from maximum to minimum (zero), and operating point Q moves up and down the load line. At the peak of the positive half cycle of the input signal, the total collector current $I_{c\,max} = 2\,I_{CQ}$ and collector-emitter voltage $V_{ce\,min} = 0$ while at the peak of the negative half cycle of the input signal, the collector current $I_{c\,min} = 0$ and collector-emitter voltage $V_{ce\,max} = 2\,V_{CC}$. Thus collector-emitter voltage varies in opposite phase to the collector current. The variation of collector voltage appears across primary of the transformer. Now ac voltage is induced in the transformer secondary which in turn develops ac power and supplies to the load.

7.7.3. Circuit Analysis

In an ideal transformer, there is no voltage drop in primary so $V_{CEQ} = V_{CC}$ and power input to the transistor becomes equal to the dc power drawn from the collector supply V_{CC} as power lost in the primary is negligibly small.

Thus power input to the transistor, P_{tr} = Power drawn from collector supply, $P_{in\,(dc)} = V_{CC} I_{CQ}$ and overall efficiency becomes equal to collector efficiency and

$$= \frac{P_{out\,(ac)}}{V_{CC}\, I_{CQ}}$$

Under condition of development of maximum ac power, voltage swings from $V_{ce\,max}$ to zero and collector current from $I_{c\,max}$ to zero. So

$$V_{rms} = \frac{1}{\sqrt{2}}\left[\frac{V_{ce\,max} - V_{ce\,min}}{2}\right] = \frac{V_{ce\,max}}{2\sqrt{2}} = \frac{2\,V_{CC}}{2\sqrt{2}} = \frac{V_{CC}}{\sqrt{2}}$$

$$\text{and } I_{rms} = \frac{1}{\sqrt{2}}\left[\frac{I_{c\,max} - I_{c\,min}}{2}\right] = \frac{I_{c\,max}}{2\sqrt{2}} = \frac{2I_{CQ}}{2\sqrt{2}} = \frac{I_{CQ}}{\sqrt{2}}$$

AC power developed across the load, $P_{out\,(ac)} = V_{rms}\, I_{rms} = \frac{V_{CC}}{\sqrt{2}} \cdot \frac{I_{CQ}}{\sqrt{2}} = \frac{V_{CC}\, I_{CQ}}{2}$

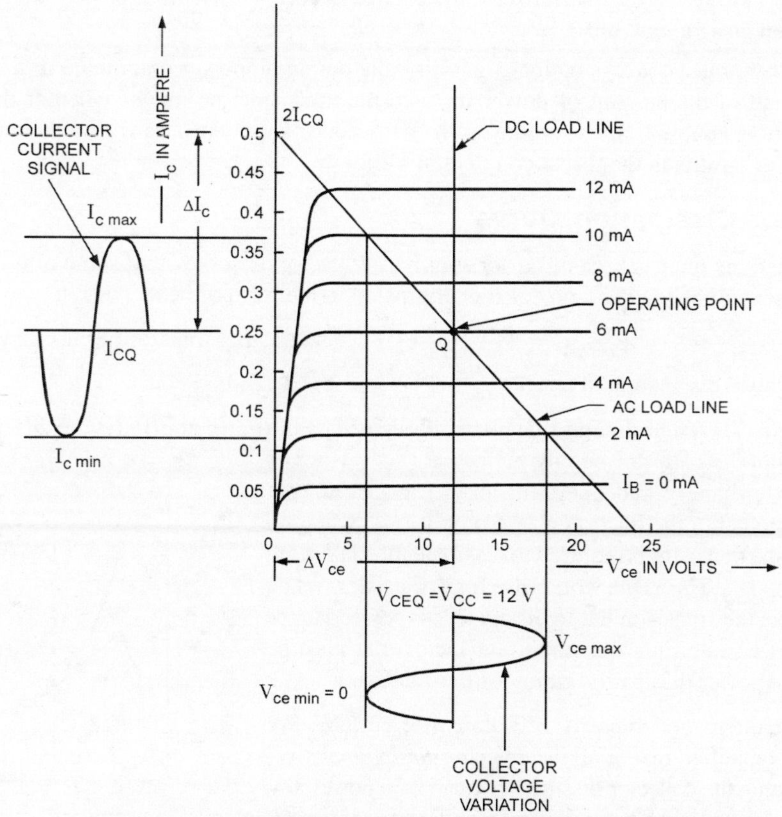

Fig. 7.12. ***Graphical Representation of Transformer Coupled Class A Audio Power Amplifier***

$$\text{Collector efficiency, } \eta_{collector} = \frac{P_{out\,(ac)}}{V_{CC}\, I_{CQ}} = \frac{V_{CC}\, I_{CQ}/2}{V_{CC}\, I_{CQ}} = 0.5 \text{ or } 50\% \qquad ...(7.17)$$

Thus for a transformer coupled class A power amplifier the maximum theoretical efficiency is 50%. In practice, the efficiency of such an amplifier is somewhat less than 50%. It is about 30%.

The efficiency of a transformer coupled class A power amplifier can be given as

$$\%\eta = 50 \times \left(\frac{V_{ce\,max} - V_{ce\,min}}{V_{ce\,max} + V_{ce\,min}} \right)^2 \% \qquad ...(7.18)$$

The larger the value of $V_{ce\,max}$ and smaller the value of $V_{ce\,min}$ the closer the efficiency approaches the theoretical limit of 50%. Well-designed circuits can approach the limit of 50%. The larger the amount of power handled by the amplifier, the more critical the efficiency becomes. For a few watts of power a simpler, cheaper circuit with less than maximum efficiency is acceptable (and sometimes desirable). For power levels in the tens to hundreds of watts efficiency as close as possible to the theoretical maximum would be desired.

The maximum power dissipated in the transistor is under zero signal condition *i.e.,* $V_{CC}\, I_{CQ}$ where I_{CQ} is the collector current under zero signal condition.

Since collector current flows at all times throughout the whole cycle, no part of the signal is cutoff and thus the output waveform obtained in this case is exact replica of the input waveform. Thus such amplifiers have least distortion. However, they suffer from the drawbacks of (*i*) low output power and (*ii*) low collector efficiency (less than 50%; about 30%).

Thus, for a given transistor dissipation capability, the output ac power obtainable in a load is the same (*i.e.,* ideally one half of the amount of power input to the transistor) no matter whether the load is direct coupled or transformer coupled. However, the drain on the power supply in a transformer coupled amplifier is ideally one half of what it is in the direct coupled amplifier.

7.7.4. Collector Dissipation Curve

In a power amplifier, the main source of power supply is collector dc supply and some of it is converted into ac power and appears in the load and rest of it is dissipated in the form of heat. Thus in a power amplifier

$$P_{in\,(dc)} = P_{out\,(ac)} + P_d, \qquad \text{neglecting transformer losses being very small}$$

$$\text{or Power dissipation, } P_d = P_{in\,(dc)} - P_{out\,(ac)} = V_{CC}\, I_{CQ} - P_{out\,(ac)} \qquad ...(7.19)$$

From above Eq. (7.19) it is obvious that power dissipated by the transistor (P_d) is the difference of average dc power input from the collector supply battery, $P_{in\,(dc)}$ (which is constant for a given battery and operating point) and ac output power, $P_{out\,(ac)}$ supplied to the load. For zero output (*i.e.,* when no signal is applied), the transistor is required to dissipate the maximum power *i.e.,* $P_d = V_{CC}\, I_{CQ}$. This is the worst case for a transistor. When the signal is applied, the transistor has to dissipate less power. Hence in designing a power amplifier, the important factor that is to be kept in mind is the power dissipation rating of the transistor.

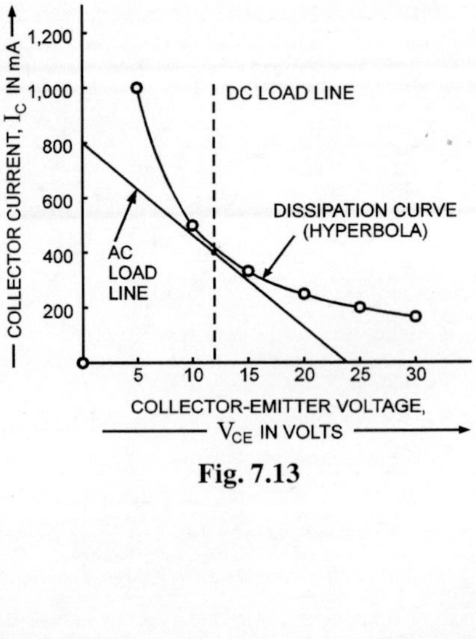

Fig. 7.13

Let the power dissipation capacity of a transistor be 5 W. For designing a power amplifier, first of all its *collector dissipation curve* is plotted. For plotting the collector dissipation curve of a power amplifier collector current I_C is determined corresponding to some arbitrary values of collector-emitter voltage V_{CE} from the relation

$P_d = V_{CE} I_C = 5$ W, the power dissipation capacity of the transistor. Some sample calculations are given below in tabular form. The curve obtained from these values is a hyperbola, as shown in Fig. 7.13.

S.No.	*V_{CE} in volts*	*I_C in mA*	*P_d in watts*
1	5	1,000	5
2	10	500	5
3	15	333	5
4	20	250	5
5	25	200	5
6	30	167	5

For using this transistor in a power amplifier, its operating point must be below the dissipation curve, otherwise transistor will get damaged. This is the importance of dissipation curve.

7.7.5. Drawbacks of Single Ended Transformer Coupled Amplifier

A single ended transformer coupled amplifier has the following drawbacks :

1. The total harmonic distortion is unacceptably high for most audio applications except in cheaper amplifiers. The second harmonic distortion is usually the greatest offender.
2. Maximum output power is delivered at a load resistance that is different from the value for minimum harmonic distortion. Consequently, the amplifier is designed to operate into a load that provides slightly less than maximum output power, but a significantly lower total distortion.
3. The output transformer is subject to saturation problems due to flow of direct current in transformer primary.

7.7.6. Output Transformer Saturation

For the proper operation of a transformer, the necessary requirement is that the flux in the transformer core varies in proportion with the variation in primary current. The transformer core is said to be saturated if the core flux does not change in proportion with the change in primary current. The effects of transformer core saturation are as follows:

1. With the increase in primary current of transformer, it is possible that there may not be a correct corresponding increase in magnetic flux to induce the proper voltage in the secondary.
2. The primary current may increase to a very high value due to sudden reduction in the primary impedance because the saturation of the core will reduce the primary inductance to a very low value.

In a single ended transformer coupled amplifier, the relatively large dc collector current flowing under zero signal condition may cause a large magnetic flux density in the transformer core, that may result in saturation of transformer core. Then, when the collector current increases due to increase in input signal, it is possible that magnetic flux may not increase in proportion to the collector current and so proper voltage may not be induced in the secondary. On the negative swing, however, the collector current would be able to fall and reduce the flux by the correct amount, inducing in the secondary a voltage larger than was obtained on the positive half-cycle, as shown in Fig. 7.14. Thus saturation of output transformer core may result in distorted output.

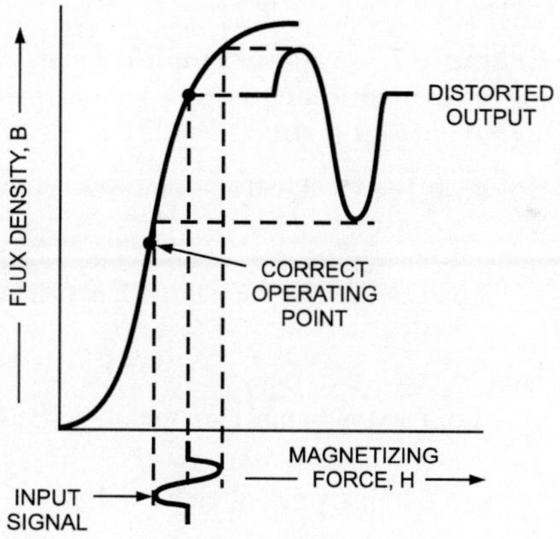

Fig. 7.14

Example 7.4. A transformer coupled class A large signal amplifier has maximum and minimum values of collector-emitter voltage of 25 V and 2.5 V. Determine its collector efficiency.

Solution :

$$V_{ce\,max} = 25\text{ V}$$
$$V_{ce\,min} = 2.5\text{ V}$$

From Eq. (7.18)

$$\text{Collector efficiency, } \eta_{collector} = 50 \times \left(\frac{V_{ce\,max} - V_{ce\,min}}{V_{ce\,max} + V_{ce\,min}}\right)^2 \% = 50 \times \left(\frac{25 - 2.5}{25 + 2.5}\right)^2 = 33.5\% \textbf{ Ans.}$$

Example 7.5. A class A transformer-coupled amplifier uses 25:1 transformer to derive 4 Ω load. Calculate the effective load (seen by the transistor connected to the large turns side of the transformer).

Solution : Load resistance, $R_L = 4\ \Omega$

Turn ratio, $a = 25$

Load as seen by transistor connected to large turns side of the transformer,

$$R'_L = a^2 \times R_L = 25^2 \times 4 = 2{,}500\ \Omega \text{ or } 2.5\text{ k}\Omega \textbf{ Ans.}$$

Example 7.6. A power amplifier working in class A operation has a transformer as a load. If the transformer has a turn ratio of 10 and the secondary load is 100 Ω, find maximum ac power output given that zero-signal collector current is 0.14 A.

Solution : Load resistance, $R_L = 100\ \Omega$

Turn ratio, $a = 10$

Load, as seen by the transformer primary,

$$R'_L = a^2 R_L = 10^2 \times 100 = 10{,}000\ \Omega$$

For maximum power output, $I_{max} = 2I_{CQ}$ and $I_{min} = 0$

$$\text{So, rms value of current, } I_{rms} = \frac{1}{\sqrt{2}}\left[\frac{I_{max} - I_{min}}{2}\right] = \frac{1}{2\sqrt{2}} \times 2I_{CQ} = \frac{I_{CQ}}{\sqrt{2}}$$

$$\text{Maximum power output, } P_{out\,(ac)} = I_{rms}^2 R'_L = \frac{I_{CQ}^2}{2} R'_L = \frac{(0.14)^2}{2} \times 10{,}000 = 98\text{ W} \textbf{ Ans.}$$

Example 7.7. A power amplifier working in class A operation has a transformer as load. If the transformer has a turn ratio of 10 and secondary load is 100 Ω. Find the maximum ac power output. Given that zero signal current is 0.1 A.

Solution : Load resistance, $R_L = 100\ \Omega$

Turn ratio, $a = 10$

Load, as seen by the transformer primary,

$$R'_L = a^2 R_L = 10^2 \times 100 = 10{,}000\ \Omega$$

For maximum power output, $I_{max} = 2I_{CQ}$ and $I_{min} = 0$

$$\text{So, rms value of current, } I_{rms} = \frac{1}{\sqrt{2}}\left[\frac{I_{max} - I_{min}}{2}\right] = \frac{1}{2\sqrt{2}} \times 2I_{CQ} = \frac{I_{CQ}}{\sqrt{2}}$$

$$\text{Maximum power output, } P_{out\,(ac)} = I^2_{rms} R'_L = \frac{1}{2} I_{CQ}^2 R'_L = \frac{1}{2} \times (0.1)^2 \times 10{,}000 \text{ watts} \qquad \because I_{CQ} = 0.1\text{ A}$$

$$= 50 \text{ watts } \textbf{Ans.}$$

Example 7.8. A transformer coupled class A power amplifier draws a current of 200 mA from a collector supply of 10 V, when no signal is applied to it. Determine (*i*) maximum output power (*ii*) maximum collector efficiency and (*iii*) power rating of the transistor.

If the load connected across the transformer secondary is of 2 Ω and transformer turn ratio is 5 : 1, comment on the impedance matching.

Solution : Collector supply voltage, $V_{CC} = 10$ V

Zero-signal collector current, $I_{CQ} = 200 \text{ mA} = 0.2$ A

(*i*) Maximum output power, $P_{out\,(ac)} = \dfrac{V_{CC}\, I_{CQ}}{2} = \dfrac{10 \times 0.2}{2} = 1$ W **Ans.**

DC power input, $P_{in\,(dc)} = V_{CC} I_{CQ} = 10 \times 0.2 = 2$ W

(*ii*) Maximum collector efficiency, $\eta_{max} = \dfrac{P_{out\,max}}{P_{in\,(dc)}} \times 100 = \dfrac{1}{2} \times 100 = 50\%$ **Ans.**

(*iii*) Power rating of the transistor = Zero-signal power dissipated $= V_{CC}\, I_{CQ} = 10 \times 0.2 = 2$ W **Ans.**

Load, $R_L = 2\ \Omega$

Load, as seen by the transformer primary, $R'_L = a^2 R_L = (5)^2 \times 2 = 50\ \Omega$, which is also the ratio of $\dfrac{V_{CC}}{I_{CQ}}\left(\dfrac{10}{0.2}\right)$. Thus the load is properly matched.

Example 7.9. A power transistor is to be used as a Class A transformer coupled amplifier and is to deliver a maximum of 5 W to a 4 Ω load. Operating point is adjusted for symmetrical clipping with collector supply voltage of 20 V. Assume ideal characteristics with V_{min} = 0 V. Calculate :

(*i*) **Transformer turns ratio.**

(*ii*) **Peak collector current.**

(*iii*) **Operating point values of I_{CEQ} and V_{CEQ}.**

(*iv*) **Power dissipation rating of transistor.**

(*v*) **Collector circuit efficiency.**

Solution : (*iii*) $V_{CEQ} = V_{CC} = 20$ V **Ans.**

$$I_{CQ} = \frac{2 \times P_{out\,(ac)}}{V_{CC}} = \frac{2 \times 5}{20} = 0.5 \text{ A } \textbf{Ans.}$$

(*ii*) Peak collector current, $I_{CQ\,max} = 2I_{CQ} = 2 \times 0.5 = 1$ A

$$R'_L = \frac{2 \times P_{out\,(ac)}}{I_{CQ}^2} = \frac{2 \times 5}{0.5^2} = 40\ \Omega$$

(*i*) Transformer turn ratio, $a = \sqrt{\dfrac{R'_L}{R_L}} = \sqrt{\dfrac{40}{4}} = 3.16$ **Ans.**

(*iv*) Power dissipation rating of transistor $= V_{CC} I_{CQ} = 20 \times 0.5 = 10$ W **Ans.**

(*v*) Maximum collector efficiency, $\eta = \dfrac{P_{out\,max}}{P_{in\,(dc)}} = \dfrac{5}{20 \times 0.5} = 50\%$ **Ans.**

Example 7.10. A transistor rated for a maximum collector dissipation of 100 W operates a single ended class A output stage from a 10 V supply. Calculate the approximate values of (*i*) maximum undistorted ac power output, (*ii*) the quiescent current and (*iii*) turns ratio of the output transformer.

Assume load resistance of 16 Ω and overall collector efficiency as 0.5.

Solution : Maximum collector dissipation rating of transistor,

$$P_{in} = V_{CC}I_{CQ} = 100 \text{ W}$$

Overall collector efficiency of, $\eta = 0.5$

(*i*) Maximum undistorted ac power output,

$$P_{out(ac)} = V_{CC}I_{CQ} \times \text{overall collector efficiency} = 100 \times 0.5 = 50 \text{ W} \quad \textbf{Ans.}$$

(*ii*) Quiescent current, $I_{CQ} = \dfrac{P_{in\,(dc)}}{V_{CC}} = \dfrac{P_{in}}{V_{CC}} = \dfrac{100}{10} = 10 \text{ A}$ **Ans.**

Load seen by the transformer primary,

$$R'_L = \frac{V_{CC}}{I_{CQ}} = \frac{10}{10} = 1\,\Omega$$

Load resistance, $R_L = 16\,\Omega$

(*iii*) Transformer turns-ratio (primary to secondary)

$$a = \sqrt{\frac{R'_L}{R_L}} = \sqrt{\frac{1}{16}} = \frac{1}{4} \quad \textbf{Ans.}$$

Example 7.11. A transformer coupled class A audio power amplifier driving an 8 Ω speaker is shown in Fig. 7.15. If the circuit component values result in a dc base current of 6 mA and the input signal V_{in} results in a peak base swing of 4 mA, determine (*i*) $V_{ce\,max}$ (*ii*) $V_{ce\,min}$ (*iii*) $I_{c\,max}$ (*iv*) $I_{c\,min}$ (*v*) rms values of load current and voltage (*vi*) ac power developed across the load (*vii*) power dissipated by the transistor and (*viii*) the efficiency of the amplifier circuit. Transistor output characteristics are shown in Fig. 7.16.

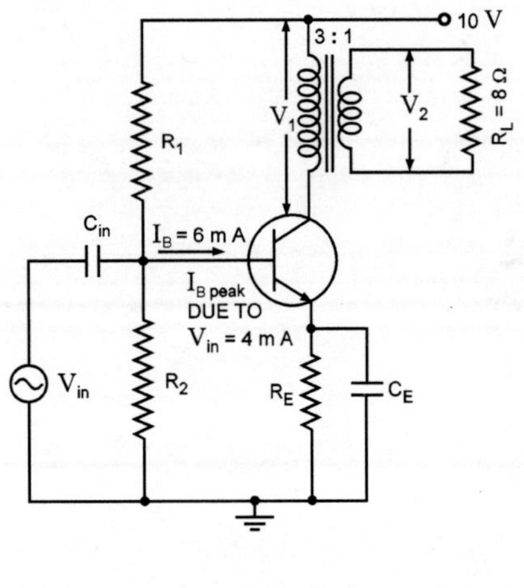

Fig. 7.15

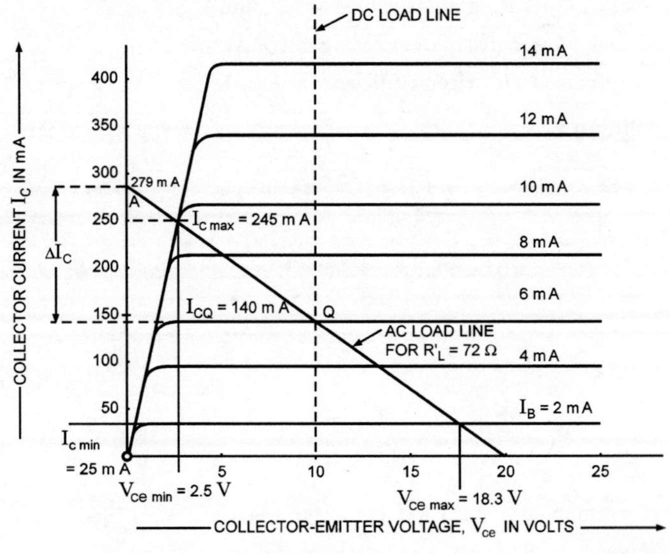

Fig. 7.16

Solution : $V_{CEQ} = V_{CC} = 10$ V, so dc load line drawn vertically from the voltage point $V_{CEQ} = 10$ V.

For base current $I_B = 6$ mA, and operating point Q in Fig. 7.16 gives

$$V_{CEQ} = 10 \text{ V and } I_{CQ} = 140 \text{ mA}$$

$$\text{Effective ac resistance } R'_L = a^2 R_L = (3)^2 \times 8 = 72\ \Omega$$

$$\text{Current swing, } \Delta I_C = \frac{\Delta V_{CEQ}}{R'_L} = \frac{10 \text{ V}}{72\ \Omega} = 139 \text{ mA}$$

Now point A is marked corresponding to collector current of 140 + 139 = 279 mA on the current axis. AC load line is now drawn joining points A and Q.

For the given peak base current swing of 4 mA, from Fig. 7.16

(*i*) $V_{ce\,max} = 18.3$ V **Ans.**

(*ii*) $V_{ce\,min} = 2.5$ V **Ans.**

(*iii*) $I_{c\,max} = 245$ mA **Ans.**

(*iv*) $I_{c\,min} = 25$ mA **Ans.**

(*vi*) AC power developed across transformer primary,

$$P_{out\,(ac)} = \frac{(V_{ce\,max} - V_{ce\,min})(I_{c\,max} - I_{c\,min})}{8} = \frac{(18.3 - 2.5)(245 - 25) \times 10^{-3}}{8}$$

$$= 0.434 \text{ W } \textbf{Ans.}$$

(*v*) RMS value of voltage across primary,

$$V_{1\,rms} = \frac{V_{ce\,max} - V_{ce\,min}}{2\sqrt{2}} = \frac{18.3 - 2.5}{2\sqrt{2}} = 5.59 \text{ V}$$

RMS value of voltage across load,

$$V_{L\,rms} = V_{2\,rms} = \left(\frac{1}{a}\right) V_{1\,rms} = \left(\frac{1}{3}\right) \times 5.59 = 1.86 \text{ V } \textbf{Ans.}$$

$$\text{RMS value of load current, } I_{rms} = \frac{V_{2\,rms}}{R_L} = \frac{1.86}{8} = 232.5 \text{ mA } \textbf{Ans.}$$

$$\text{Power input, } P_{in\,(dc)} = V_{CC} \times I_{CQ} = 10 \times 140 \times 10^{-3} = 1.4 \text{ W}$$

(*vii*) Power dissipated in the transistor,

$$P_d = P_{in\,(dc)} - P_{out\,(ac)} = 1.4 - 0.434 \text{ W} = 0.966 \text{ W } \textbf{Ans.}$$

$$\text{(viii) Amplifier circuit efficiency, } \eta = \frac{P_{out\,(ac)}}{P_{in\,(dc)}} \times 100 = \frac{0.434}{1.4} \times 100 = 31\% \textbf{ Ans.}$$

7.8. CLASS B POWER AMPLIFIERS

In class B operation the transistor is so biased that zero signal collector current is zero. Hence class B operation does not need any biasing system. The operating point is set at cutoff, as illustrated in Fig. 7.17. It remains forward biased for only half cycle of the input signal *i.e.,* its conduction angle is 180º.

As illustrated in Fig. 7.17, during the positive half cycle of the input ac signal, the circuit is forward biased and, therefore, collector current flows. On the other hand, during negative half cycle of the input ac signal, the circuit is reverse biased and no collector current flows.

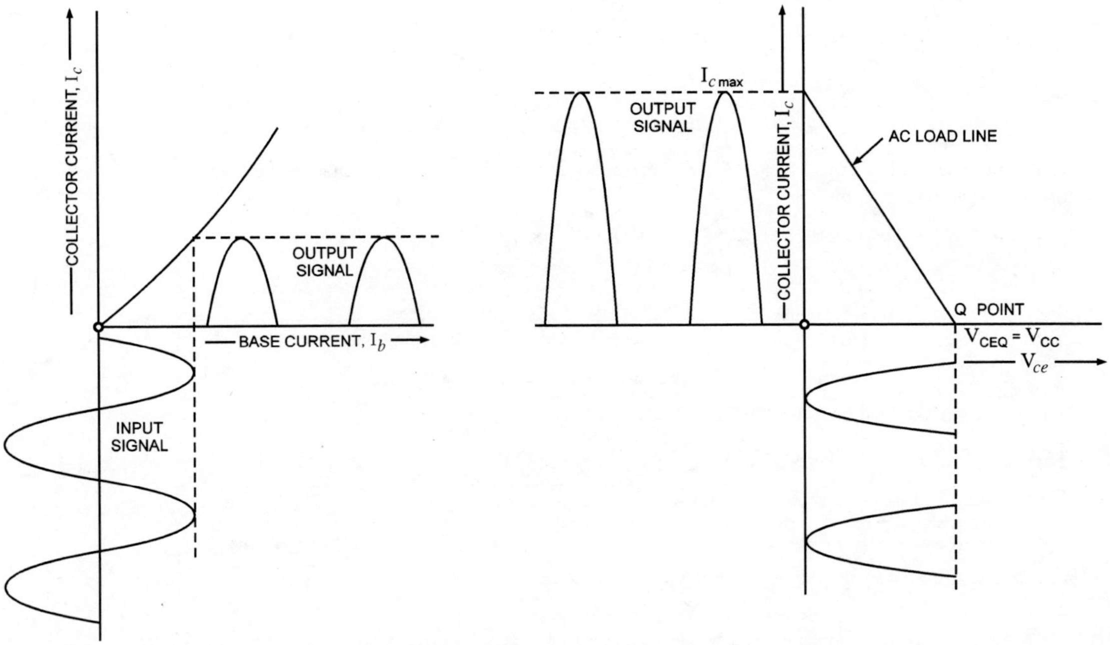

Fig. 7.17. *Class B Operation*

7.8.1. Power and Efficiency Calculations

Input dc power, $P_{in\,(dc)} = V_{CC}\, I_{dc}$ where I_{dc} is the average or direct current taken from the collector supply.

If $I_{c\,max}$ is the maximum or peak value of collector or output current, then

$$I_{dc} = \frac{1}{2\pi}\int_0^{\pi} I_{c\,max} \sin\theta \, d\theta$$

$$= \frac{I_{c\,max}}{2\pi}\left[-\cos\theta\right]_0^{\pi} = \frac{I_{c\,max}}{\pi} \qquad \because i_c = I_{c\,max}\sin\theta \text{ for } 0 < \theta < \pi$$

$$\therefore P_{in\,(dc)} = \frac{V_{CC}\, I_{c\,max}}{\pi} \qquad ...(7.20)$$

$$\text{RMS value of collector current, } I_{c\,rms} = \frac{I_{c\,max}}{\sqrt{2}}$$

$$\text{RMS value of output voltage, } V_{rms} = \frac{V_{CC}}{\sqrt{2}}$$

Hence output power during half cycle,

$$P_{out\,(ac)} = \frac{1}{2}\frac{I_{c\,max}}{\sqrt{2}} \times \frac{V_{CC}}{\sqrt{2}} = \frac{I_{c\,max}\, V_{CC}}{4} \qquad ...(7.21)$$

In above equation factor $\frac{1}{2}$ is used because power is developed during one half cycle only.

$$\text{DC power loss in load, } P_{R_{C(dc)}} = I_{dc}^2 R_C = \left(\frac{I_{c\,max}}{\pi}\right)^2 R_C \quad ...(7.22)$$

DC power loss in collector region or transistor

$$= P_{in\,(dc)} - P_{R_C\,(dc)} - P_{out\,(ac)} \quad ...(7.23)$$

$$\text{Overall efficiency, } \eta_{overall} = \frac{P_{out\,(ac)}}{P_{in\,(dc)}} = \frac{I_{c\,max}\,V_{CC}/4}{I_{c\,max}\,V_{CC}/\pi} = \frac{\pi}{4} = 0.785 \text{ or } 78.5\% \quad ...(7.24)$$

Conclusions. 1. Severe distortion occurs because of total absence of negative half cycle from the output.

2. As average current in class B operation $\frac{(I_{c\,max})}{\pi}$ is less than that in class A operation, so power dissipation is less and consequently overall efficiency is high [78.5% when peak input signal makes $V_{ce\,min} = 0$]. In general, overall efficiency is given as

$$\eta_{overall} = 78.5\left[1 - \frac{V_{ce\,min}}{V_{CC}}\right]\% \quad ...(7.25)$$

3. Zero signal input represents the best condition for class B operation as collector current is zero. This is quite contrary to class A operation. In class B operation the transistor dissipation increases with increased input signals.

4. Half sine-wave type of collector current contains very pronounced even harmonics (particularly the second one). Such impulses can be made useful either by using push-pull arrangement or by employing tuned amplifiers.

Example 7.12. Determine the overall efficiency of class B power amplifier when V_{CC} = 20 V and $V_{ce\,min}$ = 2.5 V.

Solution : From Eq. (7.25)

$$\text{Overall efficiency, } \eta_{overall} = 78.5\left[1 - \frac{V_{ce\,min}}{V_{CC}}\right]\% = 78.5\left(1 - \frac{2.5}{20}\right) = 68.69\% \textbf{ Ans.}$$

Example 7.13. For a class B amplifier providing a 20 V peak signal to 16 Ω load (speaker) and a power supply of V_{CC} = 30 V, determine the input power, output power and circuit efficiency.

Solution : Maximum value of output signal, $V_{ce\,max} = 20$ V

Maximum (or peak) value of collector or output current,

$$I_{c\,max} = \frac{V_{ce\,max}}{R_L} = \frac{20}{16} = 1.25 \text{ A}$$

$$\text{Input power, } P_{in(dc)} = \frac{V_{CC}\,I_{c\,max}}{\pi} = \frac{30 \times 1.25}{\pi} = 11.94 \text{ W} \quad \textbf{Ans.}$$

$$\text{Output power, } P_{out(ac)} = \frac{V_{CC}\,I_{c\,max}}{4} = \frac{30 \times 1.25}{4} = 9.375 \text{ W} \quad \textbf{Ans.}$$

$$\text{Circuit efficiency, } \eta = \frac{P_{out(ac)}}{P_{in(dc)}} \times 100 = \frac{9.375}{11.94} \times 100 = 78.5\% \quad \textbf{Ans.}$$

7.9. HARMONIC DISTORTION IN POWER AMPLIFIERS

Output signal variations of less than 360° of the signal cycle are considered to have distortion. This means that the output signal is no longer just an amplified version of the input signal but in some way a distorted or changed from that of the input. The poor quality of music coming from a radio or hi-fi system with the music or voice no longer sounding like that which was originally recorded or transmitted is the result of distortion.

A good amplifier should not only give an enlarged version of the input signal at the output, but should also provide a faithful reproduction of the input waveforms. One type of distortion that is common to most of the amplifiers is called the *amplitude* or *harmonic distortion*, which is caused due to nonlinearity of the active device employed for amplification. The active device (semiconductor device or tube) used for amplification may not increase equally all portions of the input signal over its positive and negative excursions. Harmonic distortion increases as we go from class A operation to class C operation. Even in class A operation some distortion occurs.

When nonlinear distortion is present, the output waveform contains components of frequencies which are harmonics (integer multiples) of the input signal frequency. There may be present second, third and higher order harmonic components, but the most important in terms of the amount of distortion for the classes of operation we will consider is the second harmonic. For a signal occurring in class AB or class B, the distortion may be mainly even harmonics of which the second harmonic component is the greatest.

An instrument such as a spectrum analyzer permits the measurement of the harmonics present in the signal by providing a display of the fundamental component of signal and a number of harmonics on a CRT screen. Similarly, a wave analyzer instrument allows a more precise measurement of the harmonic components of a distorted signal by filtering out each of these components and providing a reading of these components, one at a time. In any case the technique of considering any distorted signal as containing a fundamental component and harmonic components is practical and useful.

7.9.1. Determination of Second Harmonic Distortion

A collector current waveform is shown in Fig. 7.18, with the quiescent, minimum and maximum, signal levels, and the time they occur marked on the waveform. An equation which approximately describes the distorted signal waveform is

$$i_C \simeq I_{CQ} + I_0 + I_1 \cos(\omega t) + I_2 \cos(2\omega t) \quad ...(7.26)$$

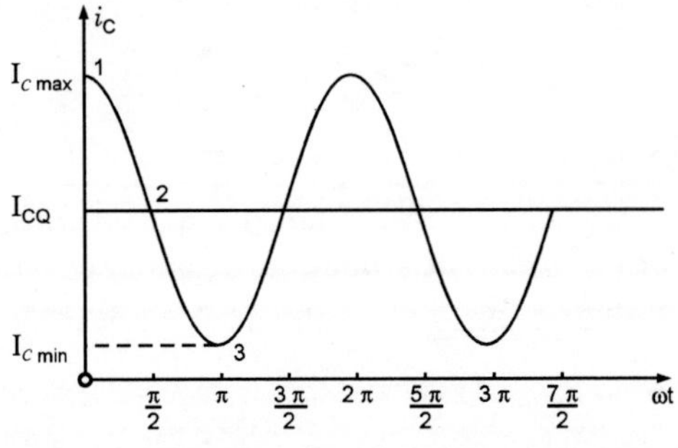

Fig. 7.18. ***Waveform For Obtaining Second Haramonic Distortion***

The given current waveform contains the original quiescent current I_{CQ}, that occurs with zero input signal, an additional dc current I_0, due to the nonzero average of the distorted signal; the fundamental component of the distorted ac signal I_1; and a second harmonic component I_2, at twice the fundamental frequency. Although other harmonics are present, only the second is considered here. Equating the resultant current from Eq. (7.26) at a few points in the cycle to that shown on the current waveform we get

At point 1 *i.e.*, when $\omega t = 0$

$$i_C = I_{c\,max} = I_{CQ} + I_0 + I_1 \cos(0) + I_2 \cos(0) = I_{CQ} + I_0 + I_1 + I_2 \quad ...(7.27)$$

At point 2 *i.e.*, when $\omega t = \frac{\pi}{2}$

$$i_C = I_{CQ} = I_{CQ} + I_0 + I_1 \cos\left(\frac{\pi}{2}\right) + I_2 \cos\left(\frac{2\pi}{2}\right) = I_{CQ} + I_0 - I_2 \quad ...(7.28)$$

$$\text{or} \quad I_0 = I_2 \qquad ...(7.29)$$

At point 3 *i.e.*, when $\omega t = \pi$

$$I_c = I_{c\,min} = I_{CQ} + I_0 + I_1 \cos(\pi) + I_2 \cos(2\pi) = I_{CQ} + I_0 - I_1 + I_2 \qquad ...(7.30)$$

Solving Eqs. (7.27), (7.29) and (7.30), we get

$$I_0 = I_2 = \frac{I_{c\,max} + I_{c\,min} - 2I_{CQ}}{4} \qquad ...(7.31)$$

$$\text{and} \quad I_1 = \frac{I_{c\,max} - I_{c\,min}}{2} \qquad ...(7.32)$$

By definition the per cent of the second harmonic distortion is given as

$$D_2 = \frac{I_2}{I_1} \times 100\% = \frac{\frac{1}{2}(I_{c\,max} + I_{c\,min}) - I_{CQ}}{I_{c\,max} - I_{c\,min}} \times 100 \qquad ...(7.33)$$

The second harmonic distortion is the per cent of the second harmonic component present in the output current waveform with respect to the amount of the fundamental component. Obviously, 0% distortion is the ideal condition of no distortion.

In a similar way, the per cent of second harmonic distortion present in the output voltage waveform w.r.t. the amount of the fundamental component can be determined and is given as

$$D_2 = \frac{V_2}{V_1} \times 100\,\% = \frac{\frac{1}{2}(V_{ce\,max} + V_{ce\,min}) - V_{CEQ}}{V_{ce\,max} - V_{ce\,min}} \times 100 \qquad ...(7.34)$$

7.9.2. Five Point Method of Determination of Harmonic Distortion in Power Amplifiers

The method of determination of second harmonic distortion is called the *3-point method* as it involved equating the assumed waveform of the output current/voltage to the measured voltage at three points in the signal cycle. Using an assumed output signal equation containing more harmonic terms, alongwith choosing more points in the waveform, results in obtaining relations for the magnitude of the harmonic components at higher harmonic frequencies. Using a 5-point method provides the dc component, first harmonic (fundamental), second harmonic, third harmonic, and fourth harmonic components. The harmonic distortion for each of these components is then defined as

$$D_2 = \frac{I_2}{I_1};\ D_3 = \frac{I_3}{I_1};\ D_4 = \frac{I_4}{I_1} \text{ and so on.} \qquad ...(7.35)$$

Total distortion may be defined, in general, using the individual distortion components

$$D = \sqrt{D_2^2 + D_3^2 + D_4^2 + \cdots} \qquad ...(7.36)$$

When distortion occurs, the output power computed on the basis of non-distortion is no longer correct. For example, the power computed from Eq. (7.8*c*) gives output power only for the non-distorted case. In case of a distorted output, the output power due to the fundamental component of the distorted signal is given as

$$P_1 = \frac{I_1^2 R_C}{2}$$

$$\text{Similarly } P_2 = \frac{I_2^2 R_C}{2};\ P_3 = \frac{I_3^2 R_C}{2} \text{ and so on.}$$

Total power due to all harmonic components of the distorted signal is given as

$$P = (I_1^2 + I_2^2 + I_3^2 + \ldots)\frac{R_C}{2}$$

$$= \left[1 + \left(\frac{I_2}{I_1}\right)^2 + \left(\frac{I_3}{I_1}\right)^2 + \ldots\right]\frac{I_1^2 R_C}{2}$$

$$= [1 + D_2^2 + D_3^2 + \ldots]\,P_1 = (1 + D^2)\,P_1 \quad \ldots(7.37)$$

If the total distortion is 10% *i.e.*, 0.1, then the total power is $[1 + (0.1)^2]\,P_1 = 1.01\,P_1$

$$\text{Percentage increase in power} = \frac{1.01P_1 - P_1}{P_1} \times 100 = 1.0\%$$

i.e., a 10% distortion represents a power increase of only 1% of the fundamental. Thus, only a small error is introduced in using only the fundamental term P_1 for determination of the output power.

Example 7.14. An output waveform displayed on an oscilloscope provided the following measured values.

(*a*) $V_{CE\,min} = 1.2$ V, $V_{CE\,max} = 22$ V, $V_{CEQ} = 10$ V

(*b*) $V_{CE\,min} = 2$ V, $V_{CE\,max} = 18$ V, $V_{CEQ} = 10$ V

Determine the percent second harmonic distortion in each case.

Solution : (*a*) According to Eq. (7.34)

$$D_2 = \frac{\frac{1}{2}(V_{CE\,max} + V_{CE\,min}) - V_{CE\,Q}}{V_{CE\,max} - V_{CE\,min}} \times 100 = \frac{\frac{1}{2}(22 + 1.2) - 10}{22 - 1.2} \times 100 = 7.7\% \text{ Ans.}$$

$$(b)\ D_2 = \frac{\frac{1}{2}(18 + 2) - 10}{18 - 2} \times 100 = 0\% \text{ i.e., no distortion}$$

Example 7.15. A sinusoidal signal $V_s = 1.95 \sin 400t$ is applied to a power amplifier. The resulting current is

$$i_0 = 12 \sin 400\,t + 1.2 \sin 800\,t + 0.9 \sin 1{,}200\,t + 0.4 \sin 1{,}600\,t$$

Calculate (*i*) the total harmonic distortion (*ii*) the percentage increase in power because of distortion.

Solution : Harmonic distortion of each component is given as

$$D_2 = \frac{V_2}{V_1} = \frac{1.2}{12} = 0.1$$

$$D_3 = \frac{V_3}{V_1} = \frac{0.9}{12} = 0.075$$

$$D_4 = \frac{V_4}{V_1} = \frac{0.4}{12} = 0.0333$$

$$\text{Total distortion, } D = \sqrt{D_2^2 + D_3^2 + D_4^2} = \sqrt{(0.1)^2 + (0.075)^2 + (0.0333)^2}$$

$$= 0.1294 \text{ or } 12.94\% \text{ Ans.}$$

$$\text{Total power, } P = (1 + D^2)\,P_1 = [1 + (0.1294)^2]P_1 = 1.01674P_1$$

Percentage increase in power because of distortion

$$= \frac{P - P_1}{P_1} \times 100 = \frac{(1.01674 - 1)\,P_1}{P_1} \times 100 = 1.674\% \textbf{ Ans.}$$

Example 7.16. For a given input current $i_i = 50 \times 10^{-3} \sin 2\omega t$, the output current given by a class A amplifier is $i_0 = 2 \times 10^{-2} + 50 \times 10^{-2} \sin 200t + 10^{-4} \sin 400t + 3 \times 10^{-6} \cos 600t$. Determine the distortion factor for it.

Solution : Considering the magnitude of fundamental component, second and third harmonic components as

$I_0 = 2 \times 10^{-2}$ dc component

$I_1 = 50 \times 10^{-2}$ fundamental component

$I_2 = 10^{-4}$ second harmonic component

$I_3 = 3 \times 10^{-6}$ third harmonic component

Percentage harmonic distortion of various components is given by

$$D_2 = \frac{I_2}{I_1} \times 100 = \frac{10^{-4}}{50 \times 10^{-2}} \times 100 = 0.02\%$$

$$D_3 = \frac{I_3}{I_1} \times 100 = \frac{3 \times 10^{-6}}{50 \times 10^{-2}} \times 100 = 0.0006\%$$

$$\text{Distortion factor, D} = \sqrt{D_2^2 + D_3^2} = \sqrt{(0.02)^2 + (0.0006)^2} = 0.02\% \textbf{ Ans.}$$

7.10. VARIATION OF OUTPUT POWER IN TRANSFORMER COUPLED POWER AMPLIFIER

The variations of output power and the distortion with the reflected load R'_L, for a typical transistor amplifier, are shown in Fig. 7.19.

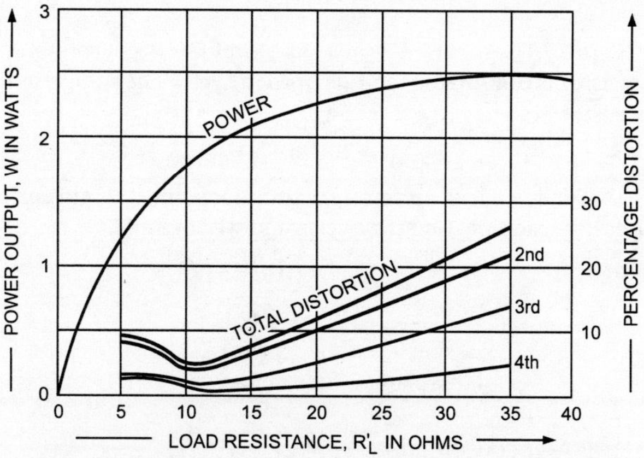

Fig. 7.19. ***Power Output and Distortion For a Typical Transistor Amplifier as a Function of Load Resistance, R'_L***

The curves shown in Fig. 7.19 reveal that the load that provides minimum total harmonic distortion is different from the load that gives maximum power output. Fortunately, the power curve is fairly flat in the region of maximum power so that a well-designed amplifier would work into a load that provides power slightly less than the maximum output power, but a significantly lower distortion.

7.11. CLASS AB POWER AMPLIFIERS

In class AB power amplifiers, the biasing circuit is so adjusted that the operating point Q lies near the cutoff voltage. During a small portion of negative half cycle and for complete positive half cycle of the signal, the input circuit remains forward biased and hence collector current flows. But during a small portion (less than half cycle) of the negative cycle, the input circuit is reverse biased and, therefore, no collector current flows during this period. Class AB operation needs a push-pull connection to achieve a full output cycle.

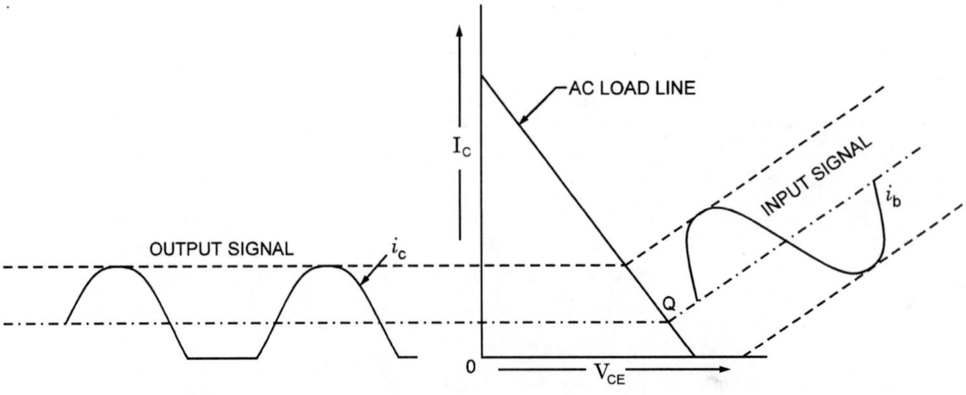

Fig. 7.20. *Class AB Operation*

7.12. CLASS C POWER AMPLIFIERS

A class C power amplifier is biased to operate for less than 180° of the input signal cycle, as shown in Fig. 7.21. The tuned circuit in the output, however, will provide a full cycle of output signal for the fundamental or resonant frequency of the tuned circuit (L and C tank circuit) of the output. The use of such amplifiers is, therefore, limited for a fixed frequency, as occurs in communication circuits. Operation of a class C circuit is not intended primarily for large signal or power amplifiers.

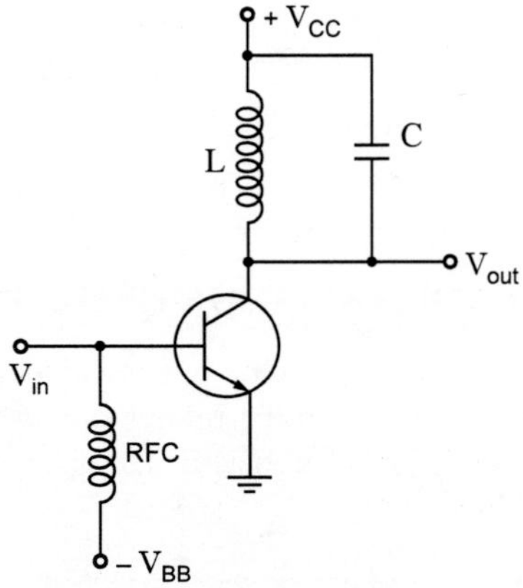

Fig. 7.21. *Class C Amplifier Circuit*

Example 7.17. A class C transistor amplifier is operating at 150 kHz. The transistor is conducting for 1 μs in each cycle. The saturation values for the transistor are :

$$I_{C(SAT)} = 100 \text{ mA}; \quad V_{CE\,(SAT)} = 0.2 \text{ V}$$

Assuming ideal pulse approximations and the output swinging over the entire load line, find the average power dissipation.

Solution :

$$I_{C\,(SAT)} = 100 \text{ mA}$$

$$V_{CE\,(SAT)} = 0.2 \text{ V}$$

$$\text{Time period, T} = \frac{1}{\text{Frequency, } f} = \frac{1}{150 \times 10^3} = \frac{20}{3} \ \mu\text{s}$$

$$\text{Power consumption, P} = I_{C\,(SAT)} \times V_{CE\,(SAT)} = 100 \times 10^{-3} \times 0.2 = 20 \text{ mW}$$

Average power dissipation,

$$P_{av} = \frac{1}{T}\int_0^t P\,dt$$

$$= \frac{P}{\frac{20}{3}\ \mu\text{s}}\int_0^{1\,\mu\text{s}} dt$$

$$= \frac{20 \times 3}{20\ \mu\text{s}}\left[t\right]_0^{1\,\mu\text{s}}$$

$$= \frac{3 \times 1\ \mu\text{s}}{\mu\text{s}} = 3 \text{ mW} \ \textbf{Ans.}$$

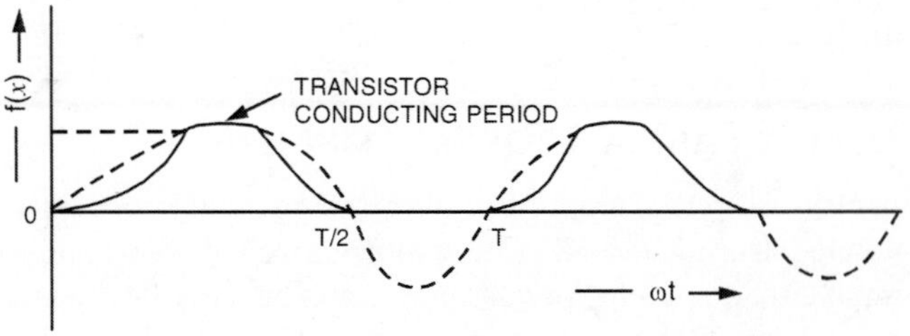

Fig. 7.22

7.13. PUSH-PULL AMPLIFIERS

As already mentioned in Art. 7.4, *double-ended or push-pull* amplifier makes use of two identical transistors in a single stage. It consists of two loops in which the transistor collector currents flow in opposite directions but add in the load. Push-pull amplifiers eliminate all of the three drawbacks listed in Art. 7.7.5.

The push-pull amplifier is a power amplifier and is frequently used in the output stages of electronic circuits. It is employed whenever high output at high efficiency is required. The audio power amplifiers used in transistor receivers, tape recorders, record-players, public address systems etc. make use of such circuits. These systems are usually operated by batteries or cells and, therefore, amplifier efficiency is of prime importance.

7.13.1. Basic Principle of Working of Push-Pull Amplifier

The basic principle on which a push-pull amplifier operates is that the input signal is converted before amplification, into two separate signals, which are identical except for a 180° phase difference. Each of these signals is applied as the input to one of the transistors of the push-pull amplifier. Since the inputs to the two transistors are 180° out of phase, so the output of the transistors. The output transformer is connected in the collector circuits of the transistors in such a way that the collector currents of the two transistors combine to provide an overall signal having the same waveform as the original input signal. The magnitude of this combined output signal is larger than the input signal. Push-pull amplifiers can be operated in class A, class AB or class B mode.

7.14. CLASS A PUSH-PULL AMPLIFIER

A class A push-pull amplifier circuit is shown in Fig. 7.23. By class A push-pull amplifier means that current flows in the output of the active device (each transistor) for the entire of the input cycle.

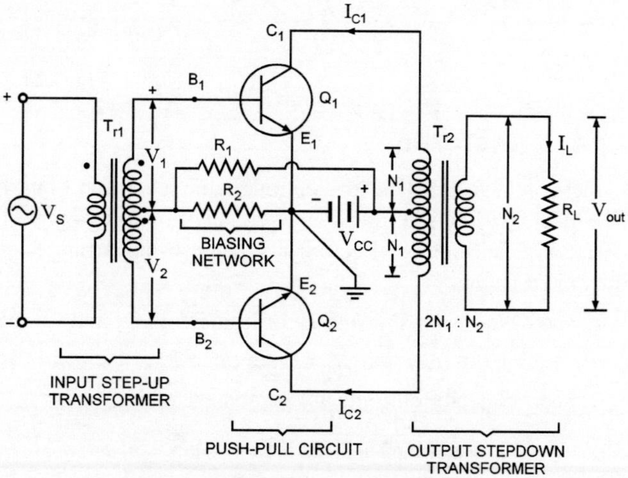

Fig. 7.23. ***Push-Pull Amplifier Circuit***

7.14.1. Circuit Arrangement

The circuitry of a typical push-pull amplifier is shown in Fig. 7.23. As already mentioned, push-pull amplifier uses two identical transistors, say Q_1 and Q_2. The emitter terminals of the two transistors Q_1 and Q_2 connected together. The input signal is applied to the inputs of two transistors through centre tapped step-up transformer T_{r_1} which provides opposite polarity signals to the two transistor inputs. The collector terminals of both the transistors are connected to end terminals of centre tapped primary of output transformer T_{r_2}. The power supply V_{CC} is connected between the emitter terminals and the centre tap of primary of output transformer. Resistors R_1 and R_2 are used to provide the proper bias for the circuit. The load R_L is connected across the secondary of the output transformer T_{r_2}. The turn ratio ($2N_1 : N_2$) of the output transformer is chosen so as to match the load with the output impedance of the amplifier and, therefore, transfer maximum power. The quiescent currents of the two transistors, which are equal in magnitude, flow in opposite directions through each half of primary of the output transformer T_{r_2}, so no saturation of the magnetic core occurs.

7.14.2. Circuit Operation

When the base current of one transistor is being driven positive with respect to the quiescent point Q, the collector current increases, thus causing a decrease in collector potential relative to ground. At the same time, however, a reverse action takes place in the base circuit of the second transistor, *i.e.,* base current decreases causing a drop in the collector current with a consequent rise in collector potential w.r.t. ground. This means that the ac current flowing through the transformer primary winding is in the same direction. As i_{C_1} increases (*i.e.,* pulls), the current i_{C_2} decreases (*i.e.,* pushes). Hence the name *push-pull amplifier.* The overall operation results in ac voltage induced in the secondary of the output transformer and thus ac power is delivered to the load. The difference of two collector currents is illustrated in Fig. 7.24.

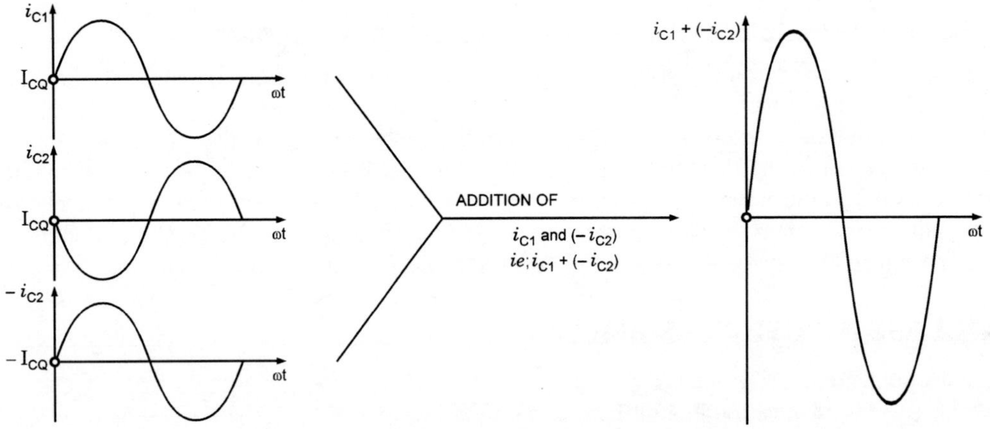

Fig. 7.24

7.14.3. Distortion

In view of the fact that the load current in the secondary of the output transformer is proportional to the difference of collector currents, it follows that the harmonic content will be less than it is in either of the collector currents because of cancellation effects. Thus the push-pull arrangement yields much less distortion in the output. This is illustrated below.

The base currents i_{b_1} and i_{b_2} of transistors Q_1 and Q_2 respectively are expressed as

$i_{b_1} = I_b \sin(\omega t)$ and $i_{b_2} = I_b \sin(\omega t + \pi)$

Their collector currents are expressed as

$$i_{C_1} = I_C + I_1 \sin \omega t + I_2 \sin 2\,\omega t + I_3 \sin 3\,\omega t + \ldots$$

$$i_{C_2} = I_C + I_1 \sin(\omega t + \pi) + I_2 \sin(2\omega t + 2\pi) + I_3 \sin(3\omega t + 3\pi) = I_C - I_1 \sin \omega t + I_2 \sin 2\omega t - I_3 \sin 3\omega t + \ldots$$

and $i_{C_1} - i_{C_2} = 2\,I_1 \sin \omega t + 2\,I_3 \sin 3\omega t + \ldots$

The output voltage induced in the output transformer secondary, being proportional to $(i_{C_1} - i_{C_2})$,

$$v_{out} = K\,(i_{C_1} - i_{C_2}) = 2K\,[I_1 \sin \omega t + I_3 \sin 3\omega t + I_5 \sin 5\omega t + \ldots]$$

From the above equation for output voltage v_{out} it is obvious that all even harmonics are eliminated. This conclusion was reached on the assumption that the two transistors are identical. If their characteristics differ appreciably, the appearance of even harmonics is not ruled out.

7.14.4. Advantages

1. Because of absence of even harmonics in the output of the push-pull amplifier, such a circuit gives more output per active device for a given amount of distortion or less distortion for a given power output per transistor.
2. As already mentioned earlier, the dc components of the collector current oppose each other magnetically in the transformer core. This eliminates any tendency toward core saturation and consequent nonlinear distortion that may arise from the curvature of the transformer magnetization curve.
3. Another advantage of this system is that the effects of ripple voltage that may be contained in the power supply because of inadequate filtering will be balanced out. This cancellation results because the currents produced by the ripple voltage are in opposite directions in the transformer winding, and so will not appear in the load. Of course, the power supply hum will also act on the voltage amplifier stages, and so will be part of the input to the power stage. This hum will not be eliminated by the push-pull circuit.

7.14.5. Disadvantages

The drawbacks of push-pull amplifiers are (*i*) requirement of two identical* or matched pair transistors (*ii*) need of use of driver stage to furnish two equal and opposite voltages at the input and (*iii*) need of bulky and expensive transformers.

7.15. CLASS B PUSH-PULL AMPLIFIER

The circuitry for the class B push-pull operation is the same as that for the class A operation except that the devices are biased at cutoff. The transistor circuit of Fig. 7.23 operates as class B if $R_2 = 0$ because a transistor is essentially at cutoff if the base is shorted to the emitter. The advantages of class B operation over class A operation are given below :

1. It is possible to obtain greater power output—class B push-pull amplifier provides practically 4-times the power supplied by a single ended amplifier provided load resistance remains the same.
2. It gives higher operating efficiency (theoretical 78.5%). It is primarily due to the fact that no power is drawn by the circuit under zero signal condition.

Because of above advantages, a push-pull class B transistor circuit is preferred in systems where the power supply is limited, such as those operating from solar cells or a battery. The automatic cancellation of all even-order harmonics from the output makes class B push-pull amplifiers highly desirable for communication sound equipment.

The drawbacks of class B push-pull amplifiers are that the harmonic distortion is higher, self-bias cannot be used, and the supply voltages must have good regulation.

Class B push-pull amplifiers are widely employed for audio work in portable record-players, as stereo amplifiers and in high-fidelity radio receivers.

7.15.1. Circuit Arrangement

Circuitry for a class B push-pull amplifier is shown in Fig. 7.25. As illustrated in the figure, the bases of the two CE-connected identical transistors Q_1 and Q_2 have been connected to the opposite ends of the secondary of the input transformer T_{r_1}. Load resistance R_L is coupled to the collector terminals of the two transistors through output transformer T_{r_2}. The emitter terminals of the transistors are connected to the centre tap of the

* If the parameters of the two transistors used in the circuit are not identical, there will be unequal amplification of the two halves of input signal.

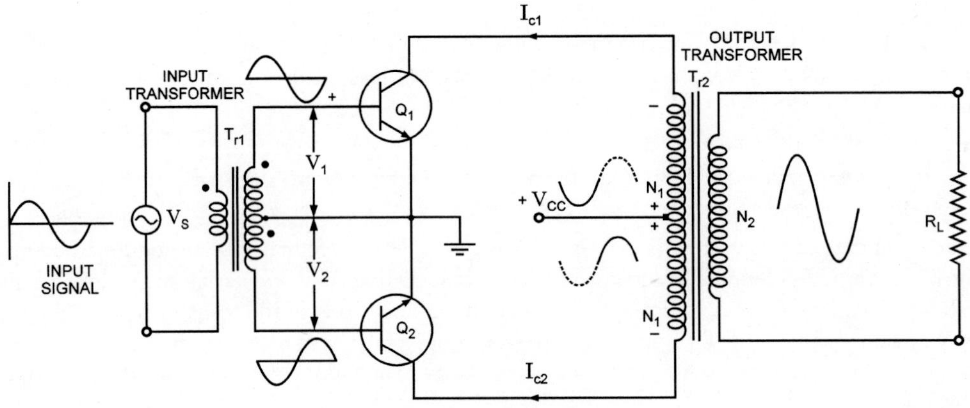

Fig. 7.25. ***Class B Push-Pull Amplifier***

secondary of the input transformer T_{r_1} while the supply V_{CC} is connected to the centre tap on the primary of output transformer. This is done in order to have balanced circuits. Since zero bias is required for cutoff, the two bases of the two transistors have been grounded.

7.15.2. Circuit Operation

When the input signal is applied, the centre tapped secondary of the input transformer develops two signals which are identical but in phase opposition. The transistors, Q_1 and Q_2 are driven by these two signals. Thus when V_1 is going positive, V_2 is going negative, so that transistor Q_2 is being biased further off when transistor Q_1 is being biased on. As the collector current in Q_1 increases from zero, it produces a half sine wave of voltage across the upper half of the primary of the output transformer, as shown. When the positive half cycle of the input signal to Q_1 begins to go negative, the signal at Q_2 base is commencing to go positive. Thus, as Q_1 becomes biased off again, Q_2 is biased on and a half cycle of voltage waveform is generated across the lower primary winding of output transformer.

The effect of the two half cycles in separate halves of the primary of output transformer is to produce a magnetic flux in the transformer core, which flows first in one direction and then in the opposite direction. This flux links with the secondary of the output transformer and generates a complete sine wave output, which is passed on to the load. In the class B circuit, the two output transistors are said to be operating in push-pull.

As explained above, a full sine wave is developed at the output of the transformer. When no signal is present, both transistors remain off, and thus no current is drawn from dc supply source V_{CC}. Hence there is no dissipation of power with zero signal. Some power is dissipated in each transistor only while the transistor is conducting.

The load resistance (usually a loudspeaker) is connected across the secondary of the output transformer. The turn-ratio $2N_1 : N_2$ of the transformer is chosen so that load resistance R_L is matched with the output impedance of the transistor. Under matched conditions, maximum power is delivered to the load by the amplifier.

7.15.3. Conversion Efficiency of Class B Amplifier

The power provided to the load R_L coupled to the amplifier is drawn from the dc power supply V_{CC} and is considered an input dc power. The input dc power is given as

$P_{in\,(dc)} = V_{CC}\, I_{dc}$ where I_{dc} is the average or direct current taken from the dc supply V_{CC}.

In class B operation the current drawn from a single power supply is a full-wave rectified signal, thus

$$I_{dc} = \frac{1}{\pi}\int_0^{\pi} I_{c\,max} \sin\theta\, d\theta = \frac{2I_{c\,max}}{\pi}$$

$$\text{Thus } P_{in\,(dc)} = \frac{2}{\pi} V_{CC} I_{c\,max} \quad ...(7.38)$$

$$\text{AC power output, } P_{out\,(ac)} = V_{rms} I_{rms} = \frac{V_{CC}}{\sqrt{2}} \cdot \frac{I_{c\,max}}{\sqrt{2}} = \frac{V_{CC} I_{c\,max}}{2} \quad ...(7.39)$$

Total collector power dissipation for the two transistors,

$$= P_{in\,(dc)} - P_{out\,(ac)} = \frac{2}{\pi} V_{CC} I_{c\,max} - \frac{V_{CC} I_{c\,max}}{2} = 2I_{c\,max}\left(\frac{V_{CC}}{\pi} - \frac{V_{CC}}{4}\right)$$

$$\text{Overall efficiency, } \eta_{overall} = \frac{P_{out\,(ac)}}{P_{in\,(dc)}} = \frac{V_{CC}\dfrac{I_{c\,max}}{2}}{2\,V_{CC}\dfrac{I_{c\,max}}{\pi}} = \frac{\pi}{4} = 0.785 \text{ or } 78.5\% \quad ...(7.40)$$

Example 7.18. Determine the input power, output power and efficiency resulting in a class-B push-pull amplifier providing a signal of 20 V peak to a 16 Ω load, using a single supply of V_{CC} = 30 V.

Solution : Maximum value of collector current,

$$I_{c\,max} = \frac{V_{CC}}{R'_L} = \frac{30}{16} = 1.875 \text{ A}$$

$$\text{Input power, } P_{in(dc)} = \frac{2}{\pi} V_{CC} I_{cmax} = \frac{2}{\pi} \times 30 \times 1.875 = 35.8 \text{ watts } \textbf{Ans.}$$ ...Refer to Eq. (7.38)

$$\text{AC power output, } P_{out\,(ac)} = \frac{V_{CC} I_{cmax}}{2} = \frac{30 \times 1.875}{2} = 28.125 \text{ watts } \textbf{Ans.}$$ ...Refer to Eq. (7.39)

$$\text{Overall efficiency, } \eta = \frac{P_{out\,(ac)}}{P_{in\,(dc)}} \times 100 = \frac{28.125}{35.8} \times 100 = 78.5\% \textbf{ Ans.}$$

Example 7.19. A class B push-pull amplifier must deliver 10 W of audio power to the output load. (*i*) If the output transformer is 80% efficient, what is the minimum power drain on the power supply under optimum conditions and (*ii*) what is the minimum average dissipation rating required for each transistor ?

Solution : Power delivered to load, P_L = 10 W

$$\text{AC power output of amplifier, } P_{out\,(ac)} = \frac{P_L}{\text{Output transformer efficiency}} = \frac{10}{0.8} = 12.5 \text{ W}$$

Efficiency of a class B push-pull amplifier under optimum conditions, η = 78.5% = 0.785

(*i*) Minimum power drain on the power supply

$$= P_{in\,(dc)} = \frac{P_{out\,(ac)}}{\eta} = \frac{12.5}{0.785} = 15.9 \text{ W } \textbf{Ans.}$$

(*ii*) Minimum average power dissipation, $P_d = P_{in\,(dc)} - P_{out\,(ac)} = 15.9 - 12.5 = 3.4$ W

$$\text{Minimum average dissipation rating required for each transistor} = \frac{P_d}{2} = \frac{3.4}{2} = 1.7 \text{ W } \textbf{Ans.}$$

Example 7.20. A class B push-pull power amplifier is supplied with V_{CC} = 50 V. The signal swings the collector voltage down to V_{min} = 5 V. The total dissipation in both transistors is 40 W. Find the total power and conversion efficiency.

Solution :

DC supply voltage, $V_{CC} = 50\text{ V}$

$V_{min} = 5\text{ V}$

Total power dissipation, $P_d = 40\text{ W}$

Also total power dissipation, $P_d = P_{in\,(dc)} - P_{out\,(ac)}$

$$\therefore 40 = \frac{2}{\pi} V_{CC} I_{c\,max} - \frac{I_{c\,max}}{2} (V_{CC} - V_{c\,min})$$

$$= I_{c\,max}\left[\frac{2V_{CC}}{\pi} - \frac{V_{CC} - V_{c\,min}}{2}\right] = I_{c\,max}\left[\frac{2 \times 50}{\pi} - \frac{50-5}{2}\right] = 9.33 I_{c\,max}$$

$$\text{or } I_{c\,max} = \frac{40}{9.33} = 4.287\text{ A}$$

Total power input, $P_{in\,(dc)} = \frac{2}{\pi} V_{CC} I_{c\,max} = \frac{2}{\pi} \times 50 \times 4.287 = 136.45\text{ W}$ **Ans.**

AC power output, $P_{out}\,(ac) = \frac{I_{c\,max}}{2} (V_{CC} - V_{c\,min}) = \frac{4.287}{2} \times (50 - 5) = 96.46\text{ W}$

Conversion efficiency, $\eta = \frac{P_{out\,(ac)}}{P_{in\,(dc)}} \times 100 = \frac{96.46}{136.45} \times 100 = 70.7\%$ **Ans.**

Example 7.21. Consider the idealised push-pull class B power amplifier shown in Fig. 7.25. Given that $R_2 = 0$, $V_{CC} = 20$ V, $N_2 = 2N_1$, $R_L = 20\ \Omega$ and the transistors have $h_{FE} = 20$. The input is sinusoid. For the maximum output signal at $V_{max} = V_{CC}$ determine (*i*) the output signal power, (*ii*) the collector dissipation in each transistor.

Solution : Load seen by the transformer primary,

$$R'_L = R_L \times \left(\frac{N_1}{N_2}\right)^2 = R_L \left(\frac{N_1}{2N_1}\right)^2 = \frac{20}{4}\Omega = 5\ \Omega$$

Peak value of ac output current, $I_{max} = \frac{V_{max}}{R'_L} = \frac{V_{CC}}{R'_L}$

(*i*) Thus ac power output, $P_{out(ac)} = I^2_{rms} R'_L = \frac{I^2_{max}}{2} R'_L = \frac{1}{2}\frac{V^2_{CC}}{R'_L} = \frac{V^2_{CC}}{2R'_L} = \frac{(20)^2}{2 \times 5} = 40\text{ W}$ **Ans.**

Power input, $P_{in\,(dc)} = \frac{2}{\pi} V_{CC} I_{c\,max} = \frac{2}{\pi} V_{CC} \times \frac{V_{CC}}{R'_L} = \frac{2V^2_{CC}}{\pi R'_L} = \frac{2 \times 20^2}{\pi \times 5} = 50.93\text{ W}.$

Total collector power dissipation for the two transistors

$$= P_{in(dc)} - P_{out(ac)} = 50.93 - 40 = 10.93\text{ W}$$

(*ii*) Collector dissipation in each resistor $= \frac{10.93}{2} = 5.465\text{ W}.$ **Ans.**

7.16. CROSSOVER DISTORTION

In addition to the distortion introduced due to the nonlinearity of the collector characteristics and due to non-matching of the two transistors, there is one more source of distortion, that is caused by nonlinearity of the input characteristic.

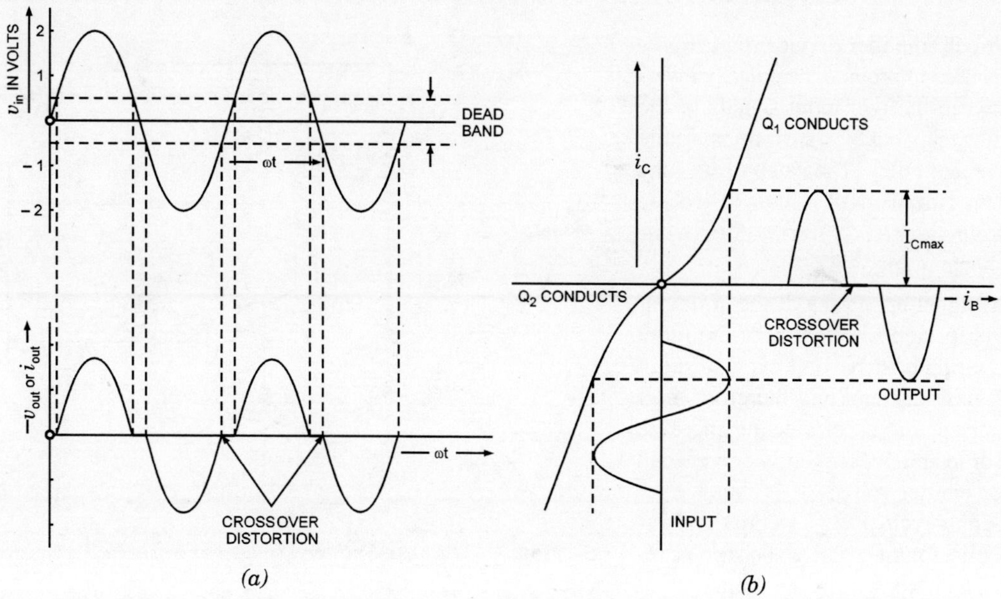

Fig. 7.26. ***Crossover Distortion***

Recall that silicon transistors must have at least 0.5 V to 0.6 V of forward base-emitter bias before they will go into conduction. Since in class B push-pull amplifiers the forward bias is produced by the input signal, both of the transistors will be non-conducting (or off), when the input signal is approximately ± 0.5 V. This forms a "dead band" in the input and produces *crossover distortion* in the output. In simple words, crossover distortion occurs as a result of one transistor cutting off before the other begins conducting, as illustrated in Fig. 7.26 (*a*). The distortion introduced is called the crossover distortion because it occurs during the time operation crossover from one transistor to the other in the push-pull amplifier. The same is shown in Fig. 7.26 (*b*) by using transfer characteristics of the two transistors.

To eliminate crossover distortion, it is necessary to add a small amount of forward bias to take the transistors to the average of conduction or slightly beyond. This does slightly lowers the efficiency of the circuit and there is a waste of standby power, but it alleviates the crossover distortion problem. Technically the operation of transistors lies between class B and class A mode. Therefore, the circuit operation is often referred to as being class AB operation (refer to Art. 7.17).

7.17. CLASS AB PUSH-PULL AMPLIFIERS

The basic circuit of class AB push-pull amplifier is the same as that of class A push-pull amplifier shown in Fig. 7.25 except that the voltage drop across resistor R_2 is so adjusted that it is approximately equal to cutin voltage (0.5 V for S*i* and 0.1 V for G*e*). Now operation becomes AB type *i.e.,* the collector current flows for more than half the cycle of the input signal but less than the complete cycle. The distortion introduced in class AB push-pull amplifier is less than that in class B push-pull amplifier but more than in class A push-pull amplifier. The drawbacks of class AB push-pull amplifiers in comparison to class B push-pull amplifiers are lower conversion efficiency and wastage of standby power.

7.18. PHASE-SPLITTER CIRCUIT FOR PUSH-PULL AMPLIFIER [OR PHASE INVERTER]

In a push-pull amplifier circuit an input transformer is used to provide the polarity inversion between the two input signals to the push-pull circuit, but the transformer is quite expensive and bulky. The same result may be achieved by using a circuit called the *phase-splitter* or *phase invertor*. Such a circuit is shown in Fig. 7.27.

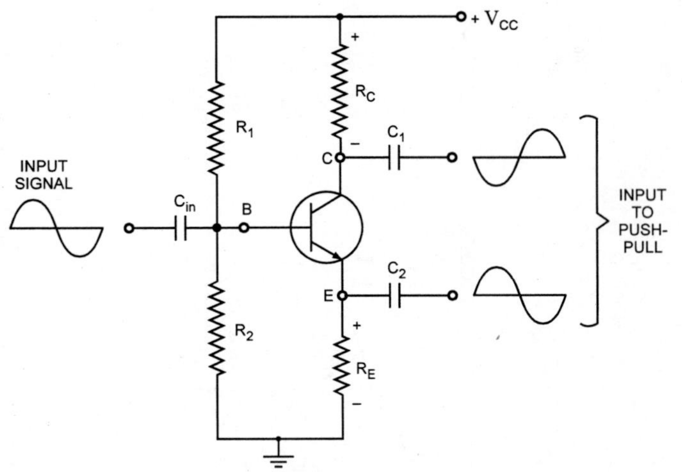

Fig. 7.27. ***Phase-Splitter Circuit For Push-Pull Amplifier***

The signal appearing at the collector is of polarity opposite to that of the input signal. During positive half cycle of input voltage, base current, and therefore, collector current increases causing increase in I_cR_C drop and consequently decrease in collector-emitter voltage V_{ce}. Thus we get negative half cycle of V_{ce} *i.e.,* phase of signal at collector of a CE transistor is reversed. The output from the emitter is of the same polarity as the input. Thus the two output signals obtained from this circuit are of opposite polarity. Values of resistors R_C, R_E and β are chosen so as to make the voltage gain for the output signal at collector to be unity. The gain for signal at emitter is 1. Thus the circuit gives two signals of equal magnitude but of opposite polarities. The use of input transformer can thus be eliminated.

However, the circuit explained above has three shortcomings as follows :

1. Since this circuit is of the emitter-follower type, there is no voltage gain.
2. Since the emitter and collector currents are not equal, the output voltages will have some imbalance.
3. This phase inverter could not be used to drive a class B push-pull amplifier because output voltages are only equal if driving a resistive load over the entire cycle.

This circuit can be used on a class A push-pull circuit, where the load is constant on each output.

7.19. COMPLEMENTARY SYMMETRY PUSH-PULL AMPLIFIERS

The use of transformers, at input as well as at output ends, in the push-pull amplifier shown in Fig. 7.23 makes it bulky and expensive especially in this age of the integrated circuits. Another drawback of the circuit given in Fig. 7.23 is that it needs two out-of-phase signals which necessitates an input tapped transformer or phase inverter, and thus makes the driver circuitry quite complicated.

The above two drawbacks of an ordinary push-pull amplifier have been overcome in the complementary symmetry push-pull amplifier.

This arrangement uses two transistors having complementary symmetry (one N-P-N and another P-N-P). The term complementary arises from the fact that one transistor is the N-P-N type and the other is P-N-P type. They have symmetry as they are made with the same material and technology and are of same maximum rating.

The circuit of a complementary push-pull class AB power amplifier is shown in Fig. 7.28. The resistors R_1 and R_2 provide the voltage divider bias to forward bias the emitter-base junction of transistor Q_1 and similarly resistors R_3 and R_4 provide the voltage divider bias for emitter junction of transistor Q_2. The resistors are so selected that under zero signal condition, the operating point is at cutoff and so no collector current flows.

The signal applied at the input goes to the base of both the transistors. The two transistors conduct in opposite half cycles of the input. For example during the positive half cycle of the input signal, the NPN transistor Q_1 is forward biased and conducts while the PNP transistor Q_2 is reverse biased and so does not conduct. This results in a half cycle of output voltage across the load, resistor R_L. Similarly during the negative half cycle only the PNP transistor Q_2 is forward biased and conducts and develops second half cycle of the output voltage across the load resistor R_L. Transistor Q_1 being reverse biased does not conduct during the negative half cycle of the input signal. Thus, during a complete cycle of input, a complete cycle of the output signal is developed.

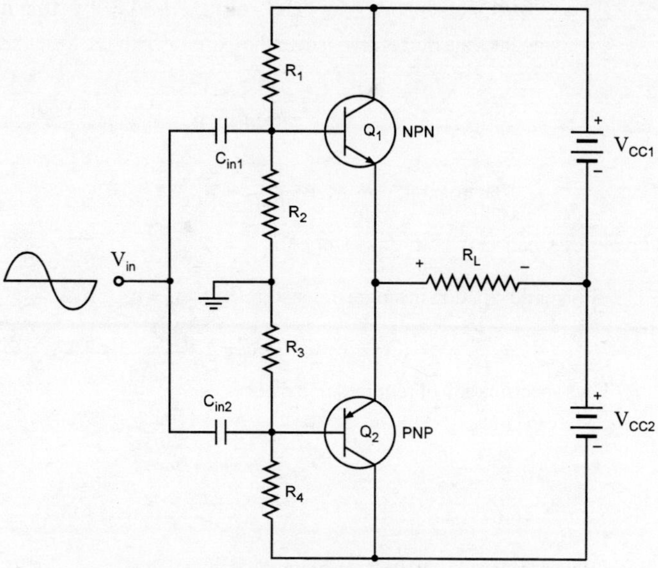

Fig. 7.28. ***Complementary Symmetry Push-Pull Class B Power Amplifier***

Since collector current from each transistor flows through the load during alternate half cycles of the input signal, no centre tapped output transformer is required. However, for impedance matching an output transformer (without centre tapping) can be used.

One obvious drawback of this circuit arrangement is the requirement for two supply voltages. Another drawback of this circuit is difficulty of obtaining matched complementary transistors. If there is an unbalance in the characteristics of the two transistors, even harmonics will no longer be cancelled and considerable distortion will be introduced. Very often negative feedback is used in power amplifiers to reduce nonlinear distortion.

Example 7.22. Design a push-pull amplifier to deliver 200 mW to a load of 6 Ω. Assume transformer efficiency to be 70% and V_{CC} = 12 V.

Solution : Load delivered to the primary of the output transformer,

$$P'_{out\,(ac)} = \frac{P_{out\,(ac)}}{\text{Transformer efficiency, } \eta} = \frac{200 \text{ mW}}{0.70} = 285 \text{ mW}$$

$$\text{Reflected load resistance, } R'_L = \frac{V_{CC}^2}{2P'_{out\,(ac)}} = \frac{12^2}{2 \times 285 \times 10^{-3}} = 252\ \Omega$$

As primary is centred tapped, collector-to-collector resistance,

$$R_{C\text{-}C} = 4R'_L = 4 \times 252 = 1{,}008\ \Omega$$

$$\text{Transformer turn ratio, } \frac{N_1}{N_2} = \sqrt{\frac{R_{C\text{-}C}}{R_L}} = \sqrt{\frac{1{,}008}{6}} = 13$$

Secondary voltage of output transformer,

$$V_{rms} = \sqrt{200 \times 10^{-3} \times 6} = 1.1 \text{ V}$$

Primary voltage of output transformer = Turn ratio × output voltage = 13 × 1.1 = 14.3 V ≃ 14.5 V (say)

Thus output transformer will be of rating 1.1/14.5 V with a minimum power rating of 300 mW (> 285 mW)

Assuming the transistor is driven over the extreme of the ac load line, the maximum collector current in each transistor,

$$I_{c\,max} = \frac{V_{CC}}{R'_L} = \frac{12}{252} = 48 \text{ mA}$$

The dc input power, $P_{in\,(dc)} = V_{CC} \times I_{dc} = V_{CC} \times \frac{2I_{c\,max}}{\pi} = 12 \times \frac{2}{\pi} \times 48 \text{ mA} = 367 \text{ mW}$

Power dissipation in both transistors, $2P_C = P_{in\,(dc)} - P_{out\,(ac)} = 367 - 200 = 167 \text{ mW}$

Power dissipation in single transistor $= \frac{167}{2} = 83.5 \text{ mW}$

$$V_{ce\,max} = 2\,V_{CC} = 2 \times 12 \text{ V} = 24 \text{ V}$$

Thus specification of required transistors

$$P_{C\,max} > 83.5 \text{ mW}$$

$$V_{ce\,max} = 24 \text{ V}$$

$$I_{c\,max} = \frac{48 \text{ mA}}{2} = 24 \text{ mA}$$

Select a transistor 2N109, from data sheet, which has $V_{ce\,max}$ of 25 V and power dissipation capacity of 165 mW. **Ans.**

7.20. QUASI-COMPLEMENTARY PUSH-PULL AMPLIFIER

In practical power amplifier circuits, it is preferable to employ NPN transistors for both high-current output devices. As the push-pull connection needs complementary devices, a PNP high power transistor is required. A practical means of having complementary operation while using the same, matched NPN transistors for the output is provided by a quasi-complementary circuit, as illustrated in Fig. 7.29.

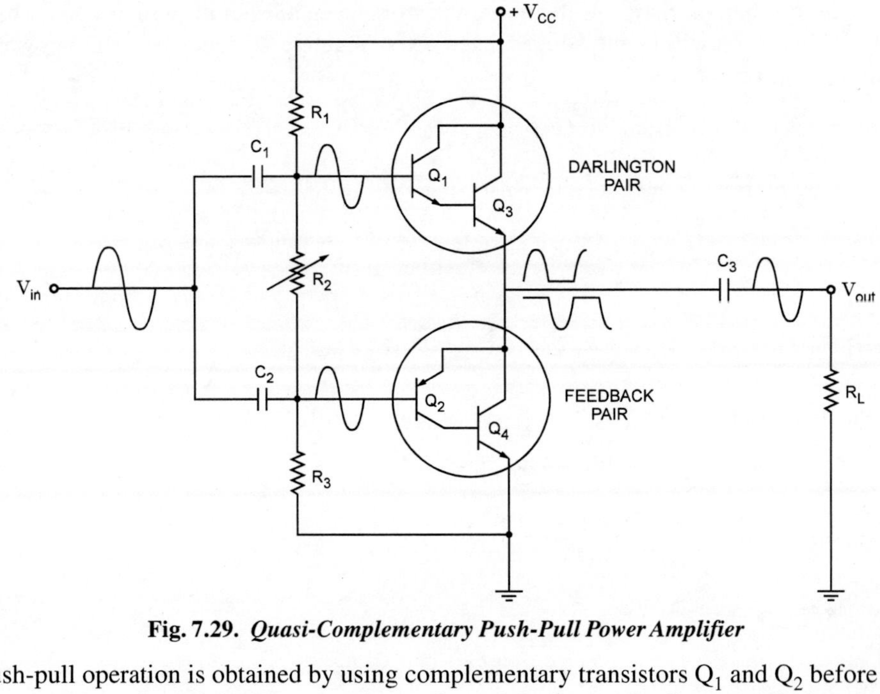

Fig. 7.29. ***Quasi-Complementary Push-Pull Power Amplifier***

The push-pull operation is obtained by using complementary transistors Q_1 and Q_2 before the matched NPN output transistors Q_3 and Q_4. Here transistors Q_1 and Q_3 form a Darlington pair that provides output from a low-impedance emitter follower. Transistors Q_2 and Q_4 form a feedback pair that provides a low-impedance drive to the load. Variable resistor R_2 is provided for adjusting the dc bias condition so as to minimize crossover distortion. The single input signal applied to the push-pull stage provides full cycle output to the load R_L.

The quasi-complementary push-pull amplifier is considered to be the most popular type of power amplifier used nowadays.

HIGHLIGHTS

1. Power amplifier is meant to raise the power level of the input signal. In order to get large power at the output, it is necessary that the input signal voltage is large. That is why, in an electronic system, a voltage amplifier always precedes the power amplifier, as illustrated in the block diagram (Fig. 7.1). Also, that is why power amplifiers are called large signal amplifiers.
2. A power amplifier is required to deliver a large amount of power and as such it has to handle large current. So in order to achieve high power amplification, (*i*) large sized power transistors are used in order to dissipate the heat produced in transistor during operation, (*ii*) transistor having thick base (*i.e.*, transistors with comparatively smaller β) are used in order to handle large currents (*iii*) transformer coupling is used for impedance matching and (*iv*) collector resistance is made low. If its value is high, power losses in the collector resistor will be increased. Moreover, voltage drop across it will increase and, therefore, operating voltage will be reduced. This is the reason that low collector load resistance is used.
3. The power amplifiers are primarily divided into two categories viz. *audio power amplifiers* and *radio power amplifiers*.

   Audio power amplifiers, also called the *small signal power amplifiers*, raise the power levels of signals that have audio frequency range (20 Hz – 20 kHz).

   Radio power amplifiers, also called *large signal power amplifiers* raise the power level of signals that have radio frequency range. They amplify a specific frequency or narrowband of frequencies while rejecting all other frequencies.
4. Important features of large signal amplifiers are : Necessity of impedance matching; Use of power transistors; Presence of harmonic distortion in the output; Capability of handling a large power; Bulky in size.
5. A class A power amplifier is defined as a power amplifier in which output current flows for the full cycle duration of the input ac signal.
6. In a class A amplifier utilising a direct coupled resistive load, the maximum conversion efficiency is given as

$$\eta_{max} = \frac{P_{out\,(ac)\,max}}{P_{in\,(dc)}} \times 100 = 25\%$$

7. In a class A power amplifier, maximum power dissipation occurs in the transistor under zero signal condition. The power dissipation capacity of a power transistor, for class A operation, must be at least equal to the zero signal rating.
8. When a signal is applied to class A power amplifier, the power dissipated in the transistor reduces by an amount equal to the ac power delivered to the load.
9. A class A power amplifier with direct coupled resistive load can supply only one-half the amount of ac power to the load that the transistor can dissipate and the dc power input is four times the amount of ac signal power in the load.
10. Class A power amplifiers have the advantages of simple construction and distortionless output voltage and disadvantages of very low efficiency and large power dissipation in power transistors.
11. Transformer coupled class A power amplifier is sometimes referred to a single ended power amplifier. The term "single ended" (denoting only one transistor) is used to distinguish it from the push-pull amplifier using two transistors.
12. For transfer of maximum power from amplifier to the output device, matching of amplifier output impedance with the impedance of output device is necessary. This is accomplished by using step-down transformer of suitable turn ratio.

The ratio of the transformer input and output resistances varies directly as the square of the transformer turn ratio:

$$\frac{R'_L}{R_L} = \left(\frac{N_1}{N_2}\right)^2 = a^2,$$

where R_L is the resistance of load connected across the transformer secondary and R'_L is the effective resistance looking into the transformer primary.

13. A transformer coupled class A power amplifier has a maximum theoretical efficiency of 50%, but its maximum ac power delivering capability is still one-half of its maximum dissipation capability.

14. The operating point of a transformer coupled power amplifier is loacated at,

$$I_{CQ} = \frac{2P'_{out\,(ac)}}{V_{CC}} \text{ and } V_{CEQ} = V_{CC}$$

ignoring collector and emitter resistances

where $P'_{out(ac)}$ is the power delivered to the primary of the output transformer.

15. A graph of power output and total harmonic distortion against load resistance indicates occurrence of maximum output power at a higher value of load resistance than that which provides minimum distortion.

16. A single ended transformer coupled amplifier has two additional disadvantages. The total harmonic distortion is unacceptably high with the second harmonic contributing the most; dc current flow in the primary of the transformer causes saturation problem.

17. In class B operation, no biasing system is required; the operating point is set at cutoff. Conduction angle is 180°. Less power dissipation resulting in higher overall efficiency; Severe distortion; Increase of transistor power dissipation with increase in input signal.

18. In class AB operation, the biasing circuit is so adjusted that the operating point Q lies near the cutoff voltage. During a small portion (less than half a cycle) of the negative cycle, no collector current flows in class AB operation. Class AB operation needs a push-pull arrangement to achieve a full output cycle.

19. A class A push-pull amplifier consists of two active devices (transistors), driven by two signals having a phase difference of 180°, and a centre-tapped output transformer.

20. The operation of a push-pull amplifier is such that all the even harmonics are eliminated from the output without changing the magnitudes of the fundamental and odd harmonics.

21. A pure class B push-pull amplifier has both of its active devices biased at cutoff, with centre tapped transformers used both at the input and output. Thus, no power is drawn under zero signal condition.

22. Crossover distortion occurs as a result of one transistor cutting off before the other begins conducting in class B push-pull amplifier. The distortion introduced is called the crossover distortion because it occurs during the time of operation crossover from one transistor to the other in the push-pull amplifier.

23. Class AB operation eliminates the crossover distortion by slightly forward biasing both transistors to permit a small idling current to flow at all times.

24. The maximum conversion efficiency in case of a class B push-pull amplifier is 78.5% while in case of class AB operation it is 70%.

25. The relation between maximum output power $P'_{out\,(ac)}$ and reflected load resistance R'_L is given by

$$R'_L = \frac{V_{CC}^2}{2\,P'_{out\,(ac)}}$$

26. The transistor dissipation in a class B amplifier is zero under quiescent or zero signal condition but goes through a maximum power dissipation $2\,P_{(ac)\,max}$ twice for each cycle of the input signal, having an average power dissipation of $P_{(ac)\,av}$ over a complete cycle.

27. The current flowing from the power supply in a class B push-pull amplifier flows in full-wave rectifier pulses, so the power supply must have good regulation.

28. Use of high power transistors makes possible to dispense with the output transformer by placing the load so that each collector current pulse passes directly through the load, but in opposite directions, thus providing the advantage of push-pull.

29. Complementary symmetry amplifiers do not depend upon either an input or an output transformer. They make use of opposite types of transistors, NPN and PNP, so that no phase-splitting device is required.

30. The output transformer is selected on the basis of the collector-to-collector resistance, $R_{CC} = 4\, R'_L$.

EXERCISES

1. Explain the difference between a voltage amplifier and a power amplifier by giving an example. **[G.G.S.I.P. Univ. Analog Electronics-II May 2011]**
2. Differentiate between voltage and power amplifiers, Give the classification of power amplifiers and the use of each class.
3. Distinguish amongst class A, class B and class C operation of amplifiers.
4. Explain how the position of Q-point varies in case of classes A, B and C operation.
5. Draw the circuit of a simple class A amplifier which can provide power amplification and explain its working.
6. Explain how ac power is developed across a loudspeaker in a single-ended power amplifier. Explain why a power amplifier is also known as the large-signal amplifier ?
7. Prove that a class A power amplifier is cooler in the presence of signal than in the absence of signal.
8. Define conversion efficiency. Determine the maximum value of conversion efficiency for a series fed class A power amplifier. **[G.G.S.I.P. Univ. Analog Electronics-II May 2011]**
9. Discuss a transistor class A power amplifier with output transformer as load. Discuss its working and find an expression for its efficiency.
10. For class A CE amplifier show that $P_{cc} = \dfrac{V_{CC}^2}{R_L}$ which is constant and independent of signal current as long as distortion is negligible. Draw the required circuit diagram and deduce the relation to prove the above equation. Show that locus of all operating points at which collector dissipation is exactly $P_{C\,max}$ lies on a hyperbola.
11. Show that the maximum efficiency of class B operation will not exceed 78.5 per cent.
12. Compare the efficiency and power dissipated by each transistor if a transformer coupled amplifier effective load is driven by (*a*) class-A amplifier, (*b*) class-B amplifier.
13. Discuss the operation of class B power amplifier. Hence derive expression for maximum undistorted output power, maximum transistor power dissipation and stage efficiency. What types of distortion occurs in class B power amplifier output. How to minimize this distortion. Derive the expression for maximum efficiency of modified circuit.
14. What are the various types of distortion present in power amplifier ? How do you graphically determine the second harmonic distortion ?
15. Distinguish between power amplifiers and voltage amplifiers. Discuss a method of determining the total harmonic distortion of a power amplifier.
16. Discuss the harmonic distortion as present in power amplifiers. Give its analysis.
17. Discuss five point method of computing harmonic distortion.
18. How are power amplifiers classified ? Explain class A, B and C power amplifiers. Derive the general expression for the collector efficiency of a power amplifier. Why does the output stage employ push-pull arrangement ? Why is the push-pull circuit called so.
19. Why class B and class C amplifiers have more efficiency than class A amplifier ? **[G.G.S.I.P. Univ. Analog Electronics-II May 2011]**
20. State classification of power amplifiers. Draw circuit diagrams and state their characteristics. **[G.G.S.I.P. Univ. Analog Electronics-II May-June 2009]**
21. Explain transformer coupled class B push-pull amplifier. **[G.G.S.I.P. Univ. Analog Electronics-II May-June 2009]**
22. Draw the circuit diagram of a class B push-pull amplifier and explain its operation. Derive an expression for its maximum conversion efficiency.

23. Draw the circuit diagram of class A push-pull amplifier and explain its operation with proper waveforms. Explain why even harmonics are not present in a push-pull amplifier ?
24. What is the difference between a voltage amplifier and a power amplifier ?
Derive the conversion efficiency of a class B power amplifier and clearly explain the reasons. Why are most of the power amplifiers used in practice designed to operate in class AB stage ?
25. Explain the working of a complementary push-pull amplifier and derive an expression for its converion efficiency. **[G.G.S.I.P. Univ. Analog Electronics-II May 2011]**
26. Draw the circuit of a class B push-pull amplifier and explain its operation. What is crossover distortion ? How is it minimised ? Explain why even harmonics are not present in a push-pull amplifier.
27. A class B push-pull amplifier gives crossover distortion. Explain and suggest a circuit to eliminate it. **[G.G.S.I.P. Univ. Analog Electronics-II May 2011]**
28. Explain how push-pull configuration reduce the harmonic distortion in an amplifier ? **[G.G.S.I.P. Univ. Analog Electronics-II May 2011]**
29. What type of bias is used in a true class B push-pull amplifier ? Explain how it is obtained in a transistor amplifier.
30. What is crossover distortion ? What are the main factors that lead to crossover distortion and how they can be removed ?
31. What is the basic advantage of push-pull operation of transistors ? Draw the circuit diagram of transformer-coupled push-pull power amplifier using NPN transistors and explain its operation.
32. Explain crossover distortion, its causes and remedy.
What penalty is paid for biasing an amplifier into class AB operation ?
33. Explain how two complementary transistors in a class B push-pull amplifier act simultaneously as phase inverters and an output push-pull pair.
34. Draw and explain the working of a complementary symmetry push-pull amplifier. Derive an expression for its efficiency. **[G.G.S.I.P. Univ. Analog Electronics-II May 2010]**
35. Write short notes on the following :
   (*i*) Classification of power amplifiers.
   (*ii*) Harmonic distortion.
   (*iii*) Class A push-pull amplifier.
   (*iv*) Crossover distortion.

SHORT ANSWER TYPE QUESTIONS WITH ANSWERS ....

Q. 1. What is meant by power amplifier ?

Ans. Power amplifier is meant to raise the power level of the input signal and may be defined as a device that converts dc power into ac power and whose action is controlled by the input signal.

Q. 2. How do you classify amplifiers ? What are small signal amplifiers ? How are they different from large signal amplificr ?

Ans. When the input signal is so weak as to produce small fluctuations in the collector current compared to its quiescent value, the amplifier is known as *small signal amplifier.* On the other hand, when the fluctuations in collector current are large *i.e.,* beyond the linear portion of its characteristics, the amplifier is called the large signal amplifier. In case of a small signal amplifier, the transistor can be represented by a linear circuit model, and a simple analytical method can be used for determination of various performance characteristics of the amplifier. But in case of a large signal power amplifier, the equivalent linear circuit model of the transistor and analytical method cannot be applied and graphical method is preferred because of involvement of certain nonlinear operation in it. Voltage amplifiers and audio power amplifiers are the small signal power amplifiers while radio power amplifiers are the large signal power amplifiers.

Q. 3. Why a power amplifier is always preceded by a voltage amplifier.

Ans. Signal voltage at the input is quite small but a large voltage input signal is necessitated so as to provide large power at the output. So voltage amplifier is required for raising the voltage level of the weak input signal before it is fed to the power amplifier.

Q. 4. Why is a power amplifier called a large signal amplifier ? Explain.

Ans. Power amplifiers are required to handle large voltage signals so as to deliver large power at the output. That is why power amplifiers are called large signal amplifiers.

Q. 5. What is the collector junction temperature for *Ge* and *Si* in case of transistor power amplifier ?

Ans. The collector junction temperature for *Ge* lies in the range of 150°C to 225°C while in case of silicon it ranges 60°C to 100°C.

Q. 6. What is meant by power dissipation capacity of a power transistor ?

Ans. Power dissipation capacity of a power transistor is its capability to dissipate the heat developed in it.

Q. 7. Why a class A power amplifier is cooler in the presence of signal than in the absence of signal ?

Ans. With the zero signal applied at the input of the class A power amplifier, ac power developed across the load reduces to zero and therefore, all the power drawn from collector supply V_{CC} is wasted in the form of heat. Thus, a power transistor dissipates maximum power under zero-signal condition. Thus the device is cooler in the presence of signal than in the absence of signal.

Q. 8. What is the maximum collector circuit efficiency of transformer-coupled class A power amplifier?

Ans. 50 per cent.

Q. 9. Why do we use transformer in the output stage of an amplifier ?

Ans. For transfer of maximum power from amplifier to the output device matching of amplifier output impedance with the impedance of the output device is necessary and this is accomplished by using a step-down transformer of suitable turn ratio at the output.

Q. 10. What principal factor contributes to the doubling of the conversion efficiency in a transformer coupled load amplifier compared with an amplifier with a direct coupled resistive load ?

Ans. Negligible power loss in the transformer primary (in the collector load) is the major factor that contributes to the doubling of the conversion efficiency in a transformer coupled load amplifier compared with an amplifier with a direct coupled resistive load.

Q. 11. What advantages does the transformer provide in a transformer coupled power amplifier ?

Ans. Transformer provide the following advantages in a transformer coupled power amplifier.

1. Power loss in the primary of the transformer, used for coupling the load to the amplifier, is negligible and, therefore, conversion efficiency is doubled in a transformer coupled-load amplifier compared with a direct coupled resistive load.
2. Output transformer makes the transfer of maximum power to the load through impedance matching.
3. Output transformer prevents a large dc current from flowing in the load, which otherwise could be harmful if the load were a loudspeaker.

Q. 12. Under what conditions would the dc load line in a transformer coupled amplifier not be drawn vertically on the output characteristics ? Which way would it slope ?

Ans. If the resistance of the transformer primary is significant and biasing stabilizing resistance is included, the dc load line will slope slightly to the left.

Q. 13. What equations specify the operating point in terms of the output power delivered to the transformer primary ?

Ans. The operating point of a transformer coupled amplifier is specified by the following equations:

$$I_{CQ} = \frac{2P'_{out\,(ac)}}{V_{CC}}$$

where $P'_{out(ac)}$ is the power delivered to the transformer primary and $V_{CEQ} = V_{CC}$
where V_{CC} is the collector supply voltage in volts.

Q. 14. What should be the effective ac resistance reflected into the primary of the output transformer in a transformer coupled power amplifier ?

Ans. Effective ac resistance reflected into the primary of the output transformer in a transformer coupled power amplifier is given by

$$R'_L = \frac{V_{CC}^2}{2P'_{out\,(ac)}}$$

where V_{CC} is the collector supply voltage in volts and $P'_{out\,(ac)}$ is the power delivered to the transformer primary, in watts.

Q. 15. What two equations may be used for determination of the output power of an amplifier ?

Ans. The two equations used for determination of output power of an amplifier are:

$$1.\ P'_{out\,(ac)} = \frac{I_1^2 R'_L}{2}$$

$$\text{and } 2.\ P'_{out\,(ac)} = \frac{I_{C\max} - I_{C\min}}{2\sqrt{2}} \times \frac{V_{CE\max} - V_{CE\min}}{2\sqrt{2}}$$

Q. 16. Why may an amplifier be operated with a load resistance lower than the value that would provide maximum output power ?

Ans. From the power output and distortion load resistance curves, it is seen that the load providing minimum total harmonic distortion is lower than the load that provides maximum power output. That is why, an amplifier is operated with a load resistance lower than the value that would provide maximum output power so that output with least distortion is obtained.

Q. 17. Comment on the maximum efficiency of class B operation.

Ans. The maximum possible value of efficiency for class B operation is 78.5%. The efficiency of class B operation exceeds that of class A operation because with zero excitation there is no current in class B operation whereas power is drawn from supply source in class A operation.

Q. 18. Although a class-B single ended power amplifier has high efficiency, yet it is less used in practical circuits. Explain why ?

Ans. Severe distortion occurs, because of total absence of negative half cycles from the output in a class-B single ended power amplifier so it is less used in practical circuits.

Q. 19. What is the use of class C amplifiers ?

Ans. The use of class C amplifiers is limited for a fixed frequency, as occurs in communication circuits.

Q. 20. What is push-pull amplifier ?

Ans. Push-pull amplifier uses two active devices (transistors), needs two input signals having a phase difference of 180°, and usually requires a centre-tapped output transformer.

Q. 21. What is class A push-pull amplifier ?

Ans. Class A push-pull amplifier is that in which current flows in the output of each transistor for the whole of the input cycle.

Q. 22. Why does the output stage employ push-pull arrangement ?

Ans. Push-pull arrangement is used in the output stage in order to have high output power at high efficiency with less distortion.

Q. 23. Draw the circuit of push-pull amplifier. Explain why is it called push-pull ?

Ans. Refer to Fig. 7.23. The circuit is called a push-pull arrangement because during one-half cycle one transistor is in conducting (pushing) state and the other is in non-conducting (pulling) state and in another half cycle, the states of the two transistors are reversed. Moreover, by doing so, the transistors also amplify the signal, hence, the name push-pull amplifier.

Q. 24. What is crossover distortion ? How can it be minimized ?

Ans. Crossover distortion occurs in class B push-pull amplifiers as a result of one transistor cutting off before the other begins conducting. The distortion introduced is called the crossover distortion because it occurs during the time operation crossover from one transistor to the other in the push-pull amplifier.

For elimination of crossover distortion, it is necessary to add a small amount of forward bias to take the transistors to the average of conduction or slightly beyond. Technically the operation of transistors will lie between class B and class A mode (*i.e.*, class AB operation).

Q. 25. The complementary symmetry push-pull amplifiers are called so. Why ?

Ans. The complementary symmetry push-pull amplifier uses two transistors having complementary symmetry (one N-P-N and another P-N-P). The term complementary arises from the fact that one transistor is the N-P-N and the other is P-N-P. They have the symmetry as they are made with the same material and technology and are of same maximum ratings.

Q. 26. What is the maximum collector circuit efficiency of a complementary push-pull class B power amplifier ?

Ans. Maximum collector circuit efficiency of a class B complementary push-pull power amplifier is 78.5%.

Q. 27. What are the drawbacks of complementary symmetry push-pull class B power amplifier ?

Ans. One obvious drawback of this circuit arrangement is the requirement for two supply voltages. Another drawback of this circuit is difficulty of obtaining matched complementary transistors. If there is an unbalance in the characteristics of the two transistors, even harmonics will no longer be cancelled and considerable distortion will be introduced. Very often negative feedback is used in power amplifiers to reduce nonlinear distortion.

PROBLEMS

1. A power transistor working in class A operation has zero signal power dessipation of 10 watts. If the ac output power is 4 watts, find (*i*) collector efficiency (*ii*) power rating of the transistor. [**Ans.** 40%; 10 W]

2. For a class A power amplifier, the maximum and minimum values of collector-emitter voltage are 20 V and 4 V respectively. V_{CC} = 25 V. Determine overall efficiency of the amplifier. [**Ans.** 16%]

3. For class A, CE large signal amplifier shown in Fig. 7.5, V_{CC} = 25 V, V_{CEQ} = 12.5 V, I_{CQ} = 0.5 A, R_C = 25 Ω. If the output current varies by ± 0.25 A when an input signal is applied to the base, determine (*i*) input power (*ii*) output power (*iii*) power delivered to the transistor (*iv*) power lost in transistor, (*v*) collector efficiency and (*vi*) overall efficiency. [**Ans.** (*i*) 12.5 W (*ii*) 0.78125 W (*iii*) 6.25 W (*iv*) 5.47 W (*v*) 12.5% (*vi*) 6.25%]

4. A class A amplifier shown in Fig. 7.10 operates from V_{CC} = 20 V, draws a no-signal current of 5 A and feeds a load of 40 Ω through a step-up transformer of N_2/N_1 = 3.16. Find (*a*) whether the amplifier is properly matched for maximum power transfer (*b*) maximum ac signal power output (*c*) maximum dc power input (*d*) conversion efficiency at maximum signal input. [**Ans.** (*a*) properly matched (*b*) 50 W (*c*) 100 W (*d*) 50%]

5. A transformer coupled class A transistor power amplifier uses a transistor with β = 100 and supplies to load of 8 Ω connected across the secondary of the transformer. If the maximum and minimum values of collector-emitter voltage and currents are 20 V and 2.4 V and 225 mA and 25 mA respectively. Determine (*i*) appropriate value of zero signal collector current (*ii*) zero signal base current (*iii*) dc power input (*iv*) ac power output (*v*) transformer turn ratio (*vi*) collector efficiency and (*vii*) transistor power rating.

[**Ans.** (*i*) 125 mA (*ii*) 1.25 mA (*iii*) 1.4 W (*iv*) 0.44 W (*v*) 3.317 : 1 (*vi*) 31.43% (*vii*) 1.4 W]

6. An output signal displayed on an oscilloscope provided the maximum, minimum and quiescent values of collector current to be 240 mA, 24 mA and 125 mA respectively. Determine the second harmonic distortion. [**Ans.** 3.24%]

7. Using a 5-point method of determination of harmonic components gives the following results :

D_2 = 0.1, D_3 = 0.03 and D_4 = 0.015, with I_1 = 5 A and R_C = 10 Ω. Determine the total distortion, fundamental power component, total power and percentage increase in power because of distortion.

[**Ans.** 10.55%; 125 W; 126.4 W; 1.12%]

8. A class B push-pull amplifier is supplied with V_{CC} = 40 V. The signal swings the collector voltage down to V_{min} = 8 V. The dissipation in both transistors total 38 W. Determine the (*i*) total power input (*ii*) total power developed across the load (*iii*) power rating of each transistor and (*iv*) overall efficiency.

[**Ans.** (*i*) 101.86 W (*ii*) 64 W (*iii*) 19 W (*iv*) 62.83%]

9. Determine the input power, output power and efficiency resulting in a class B push-pull amplifier providing a signal of 20 V peak to a 16 Ω load, using a single supply of V_{CC} = 30 V. [**Ans.** 35.81 W; 28.125 W; 78.5%]

❑❑❑

CHAPTER

8

Tuned Amplifiers

8.1. INTRODUCTION

INSIDE THIS CHAPTER

Amplifiers, on the basis of frequency range of operation, may be classified as dc amplifiers (from zero to about 10 Hz), audio-frequency amplifiers or simply audio amplifiers (20 Hz to 20 kHz) and radio-frequency amplifiers (a few kHz to hundred of MHz). Though audio amplifiers can be operated at radio frequencies but they suffer from two major drawbacks. Firstly their efficiency drops at radio frequency and secondly, their gain is independent of signal frequency over a large bandwidth because of resistive loads. Thus the audio amplifiers do not permit the selection of a particular desired frequency while rejecting all other unwanted frequencies. However, sometimes selection of a desired frequency or a narrowband of frequencies for amplification is our need, such as in radio/television the receiver is required to pick up and amplify the desired radio frequency while rejecting all others. This is achieved in radio receivers with the help of tuned amplifiers.

Amplifier amplifying a signal of specific frequency or narrowband of frequencies is known as *tuned amplifier*. Tuned amplifiers are mostly employed for amplification of high or radio frequency signals, because radio frequencies are generally single and tuned circuits permit their selection. Such amplifiers cannot be used for amplification of audio-frequency signals because of the following reasons :

(*i*) A tuned amplifier selects and amplifies a single frequency while audio frequencies are mixtures of frequencies in the frequency range of 20 Hz – 20 kHz and are not single. It is desired that all these frequencies should be equally amplified for proper reproduction of the signal. So tuned amplifiers cannot be used for this purpose.

(*ii*) For amplification of low frequency signals large valued* inductors and capacitors (L and C) are required and this makes the tuned circuit bulky and expensive.

* The resonant frequency of a parallel tuned circuit is given as $f_r = \frac{1}{2\pi}\sqrt{1/LC}$.

It is noteworthy here that R-C coupled and transformer coupled amplifiers, which are comparatively cheap, can be conveniently employed for amplification of low frequency signals.

Such amplifiers are widely used in radio and TV circuits where radio frequency signals are to be handled.

A tuned amplifier employes a tuned circuit as shown in Fig. 8.1. Because of the well-known phenomenon called *resonance*, the tuned circuit is capable of selecting a particular or relatively narrowband of frequencies. The centre of this frequency band is the *resonant frequency* of the tuned circuit.

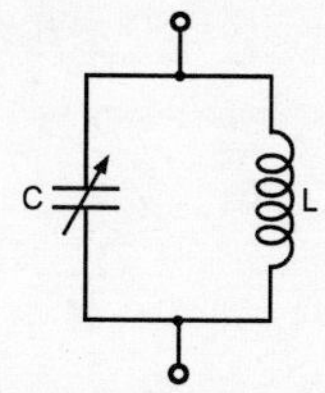

Fig. 8.1. ***Tuned Circuit***

Resonant L-C circuit as a load provides high impedance and so the tuned transistor amplifiers can provide high gain because gain of a transistor amplifier depends directly on the value of its load impedance.

The frequency response curve of the amplifier assumes the same shape as the resonance curve of the tuned circuit. Obviously, only a narrowband of frequencies around the resonant frequency f_r would be amplified well while other frequencies would be discriminated against.

The difference between the tuned amplifier and other amplifiers is that tuned amplifiers are designed for specific, usually narrow, bandwidth.

8.2. CLASSIFICATION OF TUNED AMPLIFIERS

Tuned amplifiers, like audio amplifiers, may also be divided into two categories viz *small signal tuned amplifiers* and *large signal tuned amplifiers*. Small signal tuned amplifiers are used for amplification of small signals at radio frequencies and operated under class A operation as power involved is small. In class A operation, as already known, distortion is negligibly small. Large signal tuned amplifiers are employed for amplification of large signals, at radio frequencies. These amplifiers are operated under class B or class C operation because of involvement of large power. But in tuned circuits nonlinear distortion is eliminated because of high selectivity of the load impedance, and therefore, output is nearly sinusoidal. With the removal of distortion high amplifier efficiency can be achieved by operating the transistor in its nonlinear region.

Small signal tuned amplifiers may further be classified into single tuned amplifiers and double tuned amplifiers. In single tuned amplifier, one parallel tuned circuit is employed while in double tuned amplifier, two inductively coupled tuned circuits are employed per stage and both tuned circuits being tuned to the same frequency.

8.3. SINGLE TUNED AMPLIFIERS

The circuit arrangement of a single tuned amplifier is given in Fig. 8.2. The output may be taken either with the help of a capacitive coupling, as shown in Fig. 8.2 (*a*) or by an inductive coupling as shown in Fig. 8.2 (*b*), R_1, R_2 and R_E form the biasing and stabilization circuits. C_E is the bypass capacitor. A parallel tuned L-C circuit connected in the collector circuit, the impedance of which depends upon the frequency, acts as a collector load. Capacitor C is generally variable so that the resonant frequency of the circuit may be varied. Inductor L can also be made variable. If the input signal has the same frequency as the resonant frequency of L-C circuit, large amplification will be obtained because of high impedance of L-C circuit at resonant frequency.

Circuit Operation. The high frequency signal to be amplified is applied between base and emitter. The resonant frequency of parallel L-C circuit is made equal to the frequency of input signal by varying C (or L). Now the tuned circuit will offer very high impedance to the signal frequency and thus large output will appear across it. In case the input signal is complex consisting of many frequencies, then only the frequency corresponding to the resonant frequency of tuned circuit will be amplified and all other frequencies will be rejected by the tuned circuit. Thus the desired frequency signal is selected and amplified by tuned circuit.

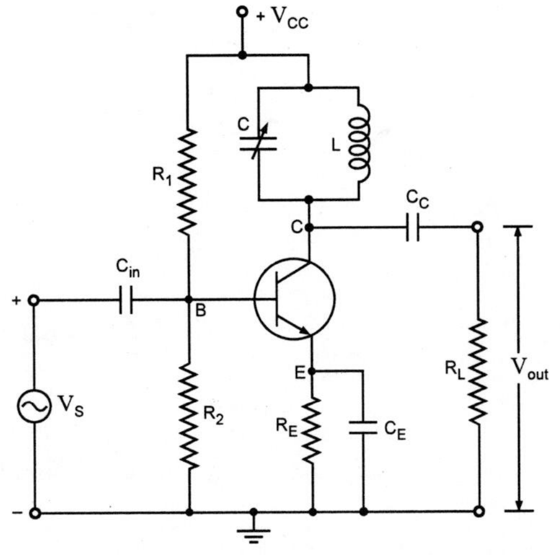

(a) Tuned Amplifier With Capacitive Coupling

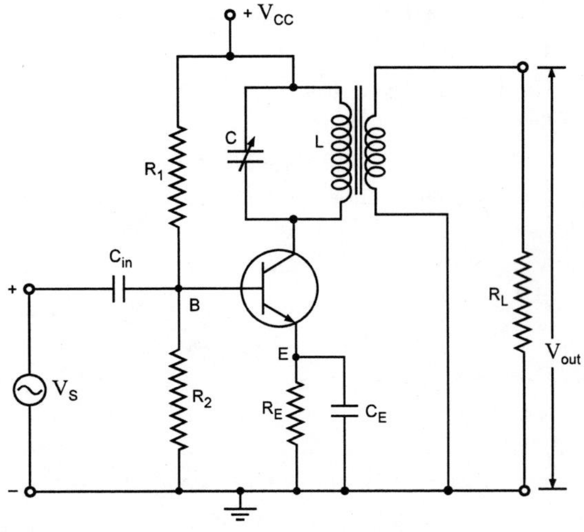

(b) Tuned Amplifier With Inductive Coupling

Fig. 8.2. ***Tuned Amplifiers***

Voltage Gain and Frequency Response Curve. The voltage gain of an amplifier is given as

$$A_v = \frac{\beta R_{ac}}{r_{in}} = \frac{\beta L/CR}{r_{in}} = \frac{\beta L}{CRr_{in}} \quad \text{...(8.1)} \quad \because R_{ac} = \text{Tuned circuit impedance, } Z = L/CR$$

where r_{in} is the input resistance.

At the resonant frequency, the impedance of the parallel resonant circuit is very high and is purely resistive. Therefore, when the circuit is tuned to resonant frequency, the voltage across R_L is maximum. In other words, the voltage gain is maximum at resonant frequency, f_r. On both sides of the resonant frequency, the impedance of the tuned circuit decreases, so the voltage. The frequency response curve of a tuned amplifier is shown in Fig. 8.3 which is similar to impedance frequency curve for a parallel resonant circuit. The bandwidth of the tuned amplifier is given as

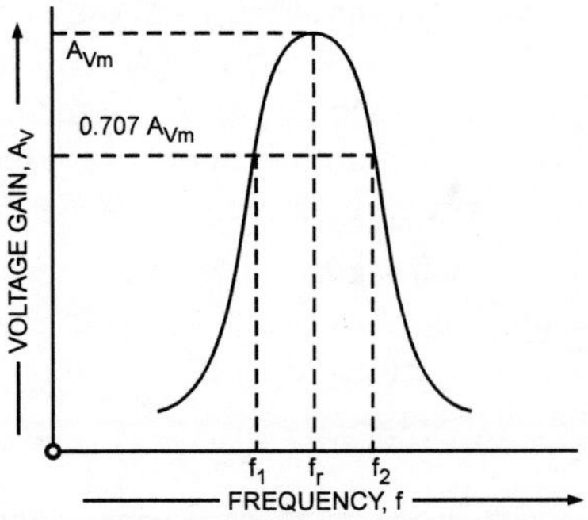

Fig. 8.3. ***Frequency Response Curve***

$$BW = f_2 - f_1 = \frac{f_r}{Q} \quad \text{...(8.2)}$$

The amplifier will amplify any frequency well within this range.

Limitations. Tuned voltage amplifiers are usually employed in radio frequency stage of wireless communication systems, where such circuits are assigned the work of selecting the desired carrier frequency and of amplifying the permitted passband around the selected carrier frequency. In other words, tuned amplifiers are required to be highly selective. But high selectivity requires a tuned circuit with a high Q factor. A high Q circuit will give a high voltage gain, but at the same time, it will give much reduced bandwidth because the bandwidth is inversely proportional to the Q factor. It means that a tuned amplifier with reduced bandwidth may not be able to amplify equally the complete band of signals and result in poor reproduction. This is called the potential *instability* in tuned amplifiers.

Example 8.1. The tuned amplifier circuit shown in Fig. 8.2 (*a*) consists of tuned circuit having R = 10 Ω, L = 20 mH and C = 0.05 μF. Determine the (*i*) resonant frequency (*ii*) Q factor of the tank circuit and (*iii*) bandwidth of the amplifier.

Solution : (*i*) Resonant frequency, $f_r = \frac{1}{2\pi}\sqrt{\frac{1}{LC} - \frac{R^2}{L^2}} = \frac{1}{2\pi}\sqrt{\frac{1}{20 \times 10^{-3} \times 0.05 \times 10^{-6}} - \frac{10^2}{(20 \times 10^{-3})^2}}$

$= 5$ kHz **Ans.**

(*ii*) Q factor of tank circuit, $Q = \frac{2\pi f_r L}{R} = \frac{2\pi \times 5 \times 10^3 \times 20 \times 10^{-3}}{10} = 62.8$ **Ans.**

(*iii*) Bandwidth, $BW = \frac{f_r}{Q} = \frac{5 \times 1{,}000}{62.8} = 79.6$ Hz **Ans.**

8.4. EFFECT OF CASCADING SINGLE TUNED AMPLIFIERS ON BANDWIDTH

In order to have a high overall gain, several identical stages of tuned amplifiers can be cascaded. The overall voltage gain is the product of the voltage gains of individual stages. Simultaneously, the high voltage gain is accompanied by a narrower bandwidth than for a single stage.

Consider n stages of single tuned direct coupled amplifiers connected in cascade. The relative gain of a single tuned amplifier with respect to the gain at resonant frequency f_r is given by equation

$$\left|\frac{A}{A_{resonance}}\right| = \frac{1}{\sqrt{1 + (2\delta Q_e)^2}}$$

where Q_e is effective circuit magnification factor or quality factor and δ is the fractional frequency variation.

Now, the gain of n stage cascaded amplifier becomes

$$\left|\frac{A}{A_{resonance}}\right|^n = \left[\frac{1}{\sqrt{1+(2\delta Q_e)^2}}\right]^n = \frac{1}{[1+(2\delta Q_e)^2]^{n/2}} \quad ...(8.3)$$

The 3-dB frequencies for the n stage cascaded amplifier can be determined by equating $\left|\frac{A}{A_{resonance}}\right|^n$ to $\frac{1}{\sqrt{2}}$

$$\left|\frac{A}{A_{resonance}}\right|^n = \frac{1}{[\sqrt{1 + (2\delta Q_e)^2}\,]^n} = \frac{1}{\sqrt{2}}$$

$$\text{or } 1 + (2\delta Q_e)^2 = 2^{1/n}$$

$$\text{or } 2\delta Q_e = \pm\sqrt{2^{1/n} - 1}$$

Substituting $\delta = \frac{\omega - \omega_r}{\omega_r} = \frac{f - f_r}{f_r}$, we have

$$2\left(\frac{f - f_r}{f_r}\right)Q_e = \pm\sqrt{2^{1/n} - 1}$$

$$\text{or } 2(f - f_r)Q_e = \pm f_r\sqrt{2^{1/n} - 1}$$

Now $$f_2 - f_r = +\frac{f_r}{2Q_e}\sqrt{2^{1/n} - 1}$$

Similarly $$f_r - f_1 = +\frac{f_r}{2Q_e}\sqrt{2^{1/n} - 1}$$

Thus, the bandwidth of n stage identical amplifier is

$$B_{1n} = f_2 - f_1 = (f_2 - f_r) + (f_r - f_1)$$

$$= \frac{f_r}{2Q_e}\sqrt{2^{1/n}-1} + \frac{f_r}{2Q_e}\sqrt{2^{1/n}-1}$$

$$= \frac{f_r}{Q_e}\sqrt{2^{1/n}-1} \qquad \text{...(8.4)}$$

$$= B_1\sqrt{2^{1/n}-1} \qquad \text{...(8.5)}$$

where B_{1n} is the bandwidth of n stages of the cascade amplifier and B_1 is the bandwidth for single stage. Bandwidth of n stages B_{1n} is equal to B_1 multiplied by a factor of $2^{1/n} - 1$

When $n = 2$; $\sqrt{2^{1/n}-1} = 0.643$

$n = 3$; $\sqrt{2^{1/n}-1} = 0.51$

Thus, the bandwidth is reduced to 64.3% for two stages and 51% for 3 stages of cascade amplifier. In order to maintain a prescribed 3-dB bandwidth, Q of the tuned circuit should be reduced.

8.5. FET TUNED AMPLIFIER

Figure 8.4 (*a*) depicts the circuit of an FET tuned amplifier. It uses a first-order JFET model for the 2N 5484 [I_{DSS} = 2.41 mA, V_P = –1.2 V]. Thus,

$$\text{Drain current, } I_D = I_{DSS}\left(1 - \frac{V_{GS}}{V_P}\right)^2 = 2.41\left(1 + \frac{V_{GS}}{1.2}\right)^2$$

$$\text{and } I_D = 4.29 - 3.91 I_D \text{ when gate voltage is 4.29 V}$$

Solving for the quiescent bias values I_D = 1.19 mA and V_{GS} = – 0.35 V.

A SPICE computation provides V_{GS} = – 0.4 V and I_D = 1.24 mA. The noteworthy point is that the first order does not allow for *channel length modulation*.

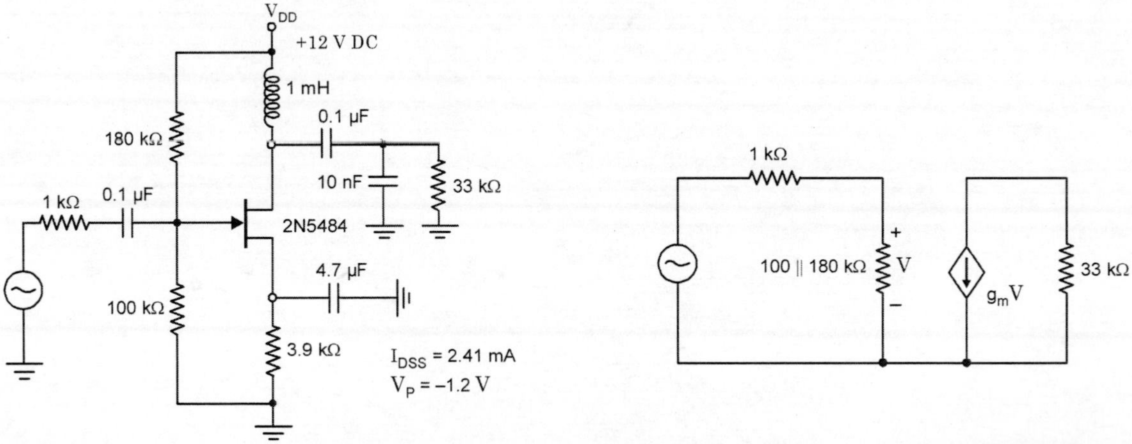

(a) Circuit Diagram *(b) Small Signal Model For Circuit Shown in Fig. 8.4 (a)*

Fig. 8.4. ***FET Tuned Amplifier***

Small signal model for operation at resonance is shown in Fig. 8.4 (*b*). Using the computed bias current and voltage values the transconductance g_m is 2.84 mA/V. The voltage gain at resonance then is 92.3 (39.3 dB).

8.6. DOUBLE TUNED AMPLIFIERS

The problem of potential instability with a single tuned amplifier is overcome in a double tuned amplifier which consists of inductively coupled two tuned circuits—one (L_1, C_1) in the collector circuit and the other (L_2, C_2) in the output circuit, as shown in Fig. 8.5. A change in the coupling of the two tuned circuits results in change in the shape of the frequency response curve. By proper adjustment of the coupling between the two coils of the two tuned circuits, the required results (high selectivity, high voltage gain and required bandwidth) may be obtained.

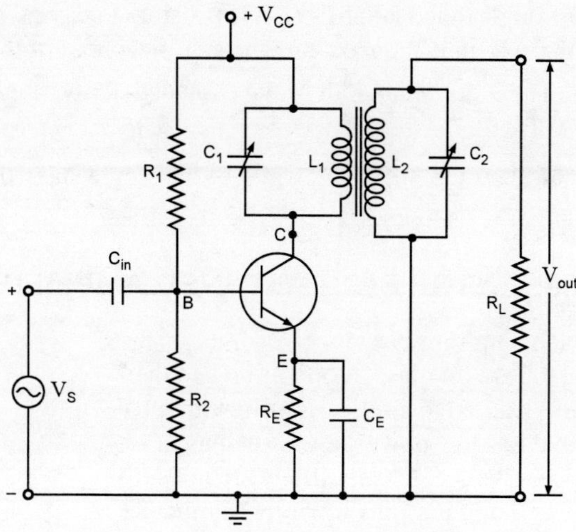

Fig. 8.5. *Double Tuned Amplifier*

Circuit Operation. The high frequency signal to be amplified is applied to the input terminal of the amplifier. The resonant frequency of tuned circuit connected in the collector circuit is made equal to signal frequency by varying the value of C_1. Now the tuned circuit L_1-C_1 offers very high impedance to the signal frequency and therefore, large output is developed across it. The output from this tuned circuit (L_1-C_1) is transferred to the second tuned circuit (L_2-C_2) through mutual induction. The frequency response of a double tuned circuit depends upon the magnetic coupling of L_1 and L_2.

Frequency Response Curve. The frequency response curves of a double tuned amplifier for different coefficients of coupling are shown in Fig. 8.6. Because of coupling of coils L_1 and L_2, a portion of load resistance is coupled into the tank circuit L_1-C_1 and affects it in exactly the same manner as though a resistor had been inserted in series with the coil L_1.

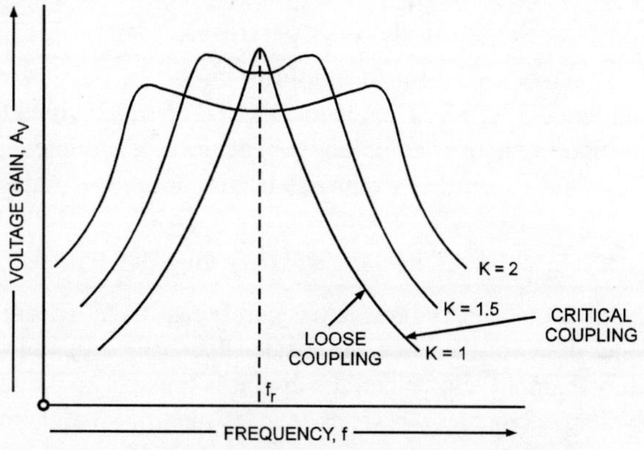

Fig. 8.6. *Frequency Response Curves of a Double Tuned Amplifier*

When the coils L_1 and L_2 are spaced apart, flux created by coil L_1 will not link the secondary coil, and these coils will be said to be loosely coupled and under such conditions the resistance reflected from the load (or tuned circuit L_2-C_2) will be small. Thus the resonance curve will be sharp with high value of Q, as shown in Fig. 8.6. When the two coils L_1 and L_2 are placed very close together, they are said to be tightly coupled. Under such conditions, the reflected resistance will be larger and consequently Q will be smaller.

From Fig. 8.6 it is obvious that bandwidth increases with the increase in degree of coupling. Naturally, the determining factor in a double tuned circuit is the coupling, not the Q factor. For a given frequency, the tighter the coupling, the greater is the bandwidth.

It is to be noted that the most suitable response curve is when optimum coefficient of coupling exists between the two tuned circuits. The circuit is then highly selective and also provides sufficient amount of gain for a particular band of frequencies.

8.7. EFFECT OF CASCADING DOUBLE TUNED AMPLIFIERS ON BANDWIDTH

Like single tuned amplifiers, it is possible to cascade double tuned amplifiers so that the overall bandwidth of the system is reduced, and the side steepness of the response is increased as illustrated in Fig. 8.3.

The 3-dB bandwidth of the cascaded double tuned amplifier is given by equation

$$B_{2n} = B_2[2^{1/n} - 1]^{1/4} \quad ...(8.6)$$

where B_2 is the 3-dB bandwidth of the single stage double tuned amplifier and n is the number of identical stages cascaded.

8.8. LARGE SIGNAL TUNED AMPLIFIERS

Tuned amplifier circuits given in Figs. 8.2 and 8.5 are for class A operation and their use is limited to applications in which RF signal has low power levels such as in radio receivers. Such amplifiers are not suitable for applications involving larger RF power, because of lower efficiency of class A operation, such as for excitation of transmitting antenna. For such applications, large signal tuned amplifiers are employed because they are operated in class C operation that has a very high efficiency and is capable of delivering more power in comparison to that of class A operation.

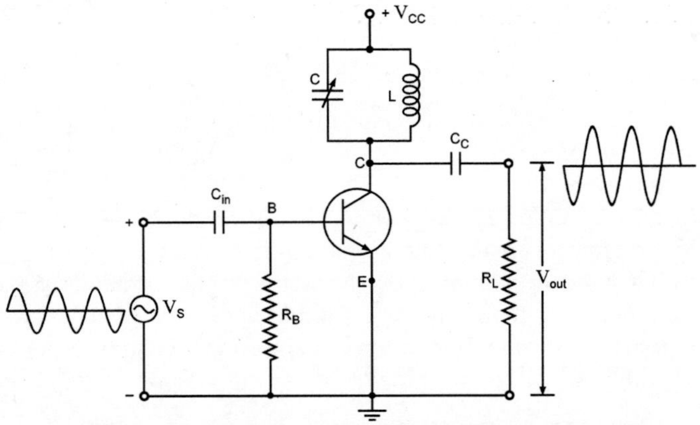

Fig. 8.7. ***Tuned Class C Amplifier***

Large signal tuned amplifier circuit is shown in Fig. 8.7. The resonant tuned circuit is tuned to the frequency of the input signal. When the circuit has a high Q factor (*i.e.,* inductive reactance of inductor is much larger than its resistance), parallel resonance occurs at approximately frequency given by $f_r = \frac{1}{2\pi}\sqrt{\frac{1}{LC}}$. At the resonant frequency, the impedance of the parallel resonant circuit is very large and is purely resistive. (This approximation assumes that the Q of the circuit is greater than 10,* a condition usually satisfied in tuned amplifiers). When the circuit is tuned** to the resonant frequency, the voltage across load resistance R_L is maximum and sinusoidal.

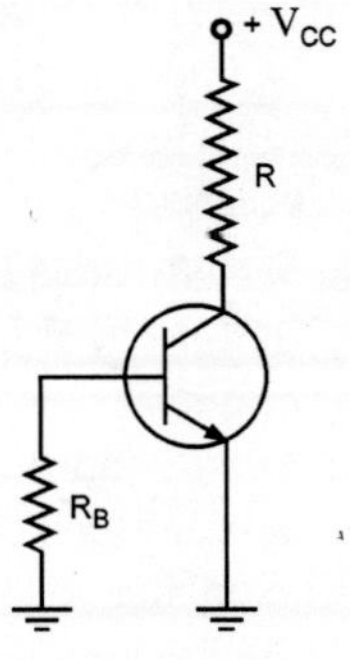

Fig. 8.8. ***DC Equivalent Circuit of a Tuned Class C Amplifier***

DC equivalent circuit is shown in Fig. 8.8. It is to be noted that there is *no bias* applied to the transistor. Therefore, its Q point is at cutoff on the dc load line. Since there is no dc bias, so V_{BE} is zero. This means that no collector current can flow until the input signal attain a value greater than

* In practical class C amplifiers, the Q factor of the coil is typically 50 or more, and the Q factor of the circuit is 10 or more.

** One way of tuning a resonant circuit to the input frequency is to look for a dip in the direct current drawn from the collector supply V_{CC}. For this a dc ammeter may be connected in series with the V_{CC} supply. When the tank is adjusted to resonance, the ammeter will give minimum reading because of maximum impedance at resonance. This indicates that the circuit is resonant to the input frequency.

approximately 0.7 V. DC collector resistance R is the dc resistance of the RF inductor L, typically a few ohms.

Load Lines. The dc load line appears to be almost vertical, as shown in Fig. 8.9. This is because the dc collector resistance R is very small. There is no danger of thermal runaway because the transistor has no current aside from leakage. The Q point is located at cutoff with no risk of thermal runaway.

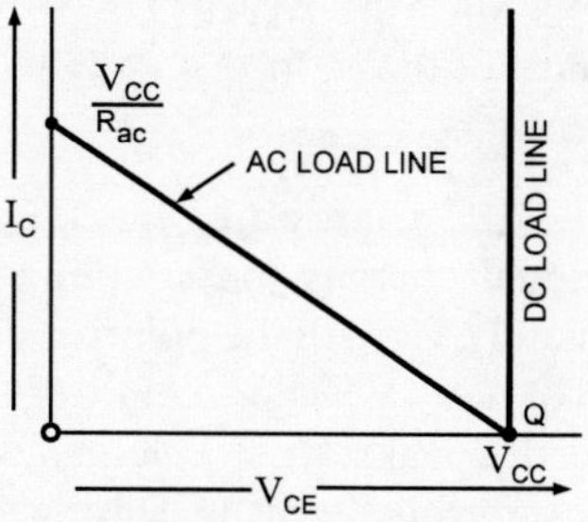

Fig. 8.9. *Load Lines*

For a CE amplifier, $I_{C\,(sat)} = I_{CQ} + \frac{V_{CEQ}}{R_{ac}}$ and $V_{CE\,(off)} = V_{CEQ} + I_{CQ}R_{ac}$

and for a class C amplifier shown in Fig. 8.7 $I_{CQ} = 0$ and $V_{CEQ} = V_{CC}$.

Therefore, the foregoing equations reduce to $I_{C\,(sat)} = \frac{V_{CC}}{R_{ac}}$ and $V_{CE\,(cut)} = V_{CC}$.

AC load line is shown in Fig. 8.9. When the transistor is conducting, its operating point swings upward along the ac load line. As before, R_{ac} is the ac resistance seen by the collector. Therefore ac saturation current in a class C amplifier is V_{CC} / R_{ac}, and the maximum voltage swing is V_{CC}.

AC Equivalent Circuit. *Approximate ac equivalent circuit,* assuming Q of the resonant circuit to be greater than 10, is shown in Fig. 8.10. In this equivalent circuit, the series resistance of the inductor is lumped into the collector resistance. In a class C amplifier, the input capacitor is part of a negative dc clamper. On the input side of a class C amplifier, therefore, the signal is negatively clamped. On the output side, the collector current source drives a parallel resonant tank circuit. At resonance, the peak-to-peak load voltage attains a maximum.

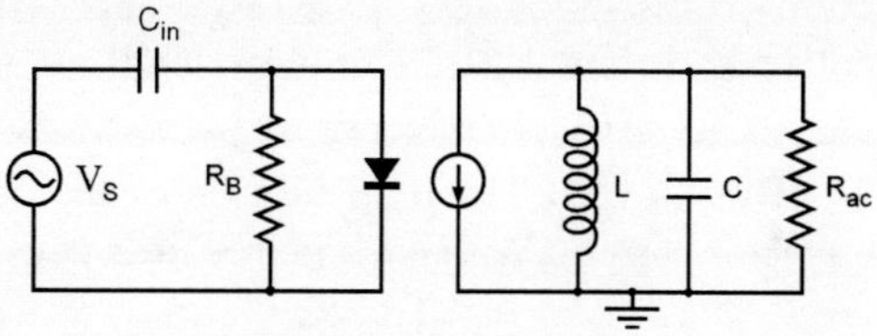

Fig. 8.10. *AC Equivalent Circuit of a Tuned Class C Amplifier*

AC Collector Resistance. Any coil or inductor has some series resistance R. Although in schematic diagrams, it is never shown as a separate component. The Q factor of the inductor is given as $Q_L = \frac{X_L}{R}$ where X_L and R are the inductive reactance and resistance of the inductive coil and Q_L is quality factor of the coil. The quality factor Q for the whole circuit is less than that for the coil because it includes the effect of load resistance as well as the coil resistance. The series resistance of the inductor, however, can be replaced by a parallel resistance R_P, as illustrated in Fig. 8.11 (*b*). This equivalent resistance is given by an approximate expression $R_P = Q_L X_L$. For $X_L >> 10$, this relation gives an error of less than 1%.

In Fig. 8.11 (*b*), it is to be noted that all the losses in the coil are now represented by the parallel resistance R_P, series resistance R no longer exists in the equivalent circuit. Because of this, X_L neutralize X_C at resonance, leaving only R_P in parallel with R_L. Thus ac resistance seen by the collector at resonance is given as

$$R_{ac} = R_P \,||\, R_L$$

and the quality factor Q of the circuit is given as

$$Q = \frac{R_{ac}}{X_L} = \frac{R_P \,||\, R_L}{X_L} \quad \text{...(8.7)}$$

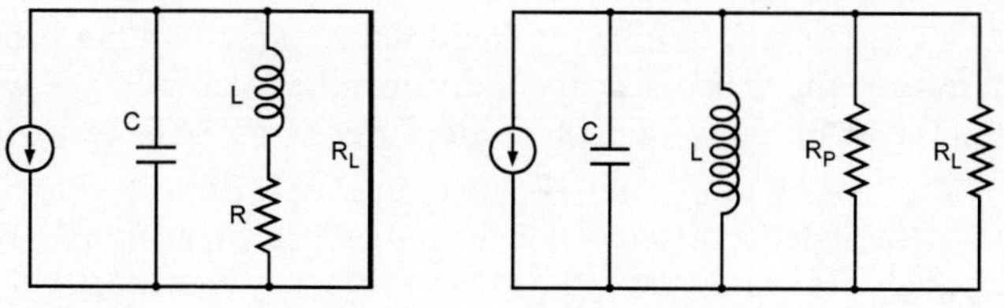

(*a*) *R Representing Coil Losses* (*b*) *R_P Representing Coil Losses*

Fig. 8.11

Since the overall Q is 10 or more, the operation is narrowband. Also, since the Q of the coil is 50 or more, most of the ac load is delivered to the load resistor and only a small amount of power is wasted in the coil resistance.

DC Clamping. In the Fig. 8.12 (*a*), the input signal charges the coupling capacitor to approximately V_P with the polarity shown. During the positive half cycles, the emitter diode conducts briefly at the peaks; this replenishes any capacitor charge lost during the cycle. During the negative half cycles, the only discharge path is through R_B. As long as the period T of the input signal is much smaller than the time constant R_BC, the capacitor loses only a small amount of its charge. For replenishment of the lost capacitor charge, the base voltage must swing slightly above 0.7 V to turn on the emitter diode briefly at each positive peak, as shown in Fig. 8.12 (*a*). The conduction angle of the base and collector current, therefore, is much less than 180°. This is why the collector current is a train of narrow pulses, as shown in Fig. 8.12 (*b*).

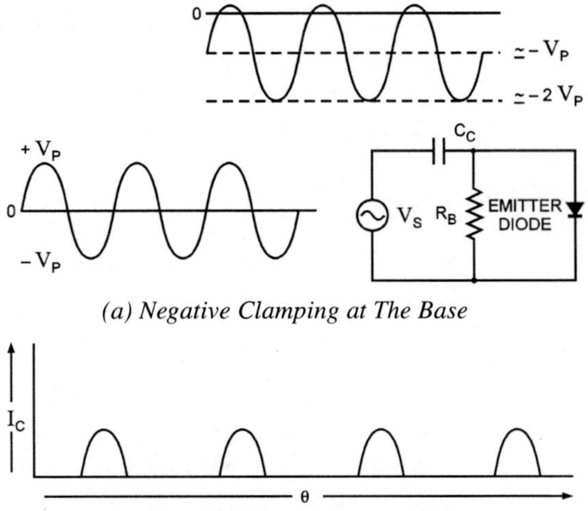

(*a*) *Negative Clamping at The Base*

(*b*) *Narrow Pulses of Collector Current*

Fig. 8.12

Duty Cycle. The brief turn-on of the emitter diode at each positive peak produces narrow pulses of collector current. With such pulses, it is convenient to use the duty cycle, defined as

$$D = \frac{W}{T} \qquad \text{...(8.8)}$$

where D is duty cycle, W is width of pulse and T is period of pulses.

For instance, the pulse width shown on an oscilloscope is of 0.25 μs and a period shown is 1.75 μs, the duty cycle is

$$D = \frac{0.25}{1.75} = 0.143 \text{ } i.e., \text{ } 14.3\%.$$

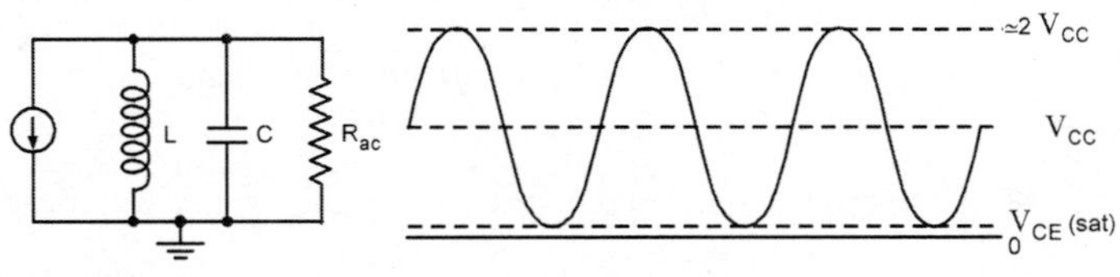

(*a*) *AC Equivalent Circuit* (*b*) *Collector Voltage*

Fig. 8.13

Filtering of Harmonics. As discussed earlier, nonsinusoidal wave consists of fundamental frequency *f* and its harmonics *i.e.*, 2*f*, 3*f*, 4*f* and so on. In Fig. 8.13 (*a*), the collector current source drives the resonant tank circuit with the nonsinusoidal current shown in Fig. 8.12 (*b*). If the tank circuit is resonant to the fundamental frequency, then all the harmonics are filtered out, and the load voltage is sinusoidal of fundamental frequency, as shown in Fig. 8.13 (*b*).

As discussed earlier, the maximum voltage swing along the ac load line is approximately V_{CC}. Therefore, under full-signal conditions, the load voltage swings from approximately $V_{CE\,(sat)}$ to $2\,V_{CC}$. Thus the ac output of a class C amplifier is $\cong 2\,V_{CC}$ because $V_{CE\,(sat)}$ is about equal to zero.

The class C amplifier is rather unusual. First it clamps the input signal negatively to get highly distorted current pulses. Then it makes use of a high-Q resonant circuit to restore the fundamental frequency. This improves the stage efficiency. Furthermore, because of the narrow current pulses, the transistor power dissipation is lower than with class A or class B. The overall effect is less current drain, which means a higher stage efficiency. The efficiency of class C approaches 100 per cent.

Frequency Response and Bandwidth. The frequency response curve of a class C tuned amplifier is shown in Fig. 8.14. Voltage gain attains maximum value at the resonant frequency. On both sides of the resonant frequency, the voltage gain decreases, as illustrated in Fig. 8.14. The higher the Q of the circuit, the faster the gain drops off on either side of the resonant frequency.

The bandwidth of the tuned amplifier is given as

$$BW = f_2 - f_1 = \frac{f_r}{Q} \quad ...(8.9)$$

From above relation it is obvious that with a large Q, a small bandwidth, equivalent to sharp tuning, is obtained. Class C amplifiers always have a circuit Q >> 10. This means that the bandwidth is less than 10% of the resonant frequency. For this reason, class C amplifiers are also sometimes called the *narrowband amplifiers*. The output of a narrowband amplifier is a large sinusoidal voltage at resonance with a rapid fall above and below resonance.

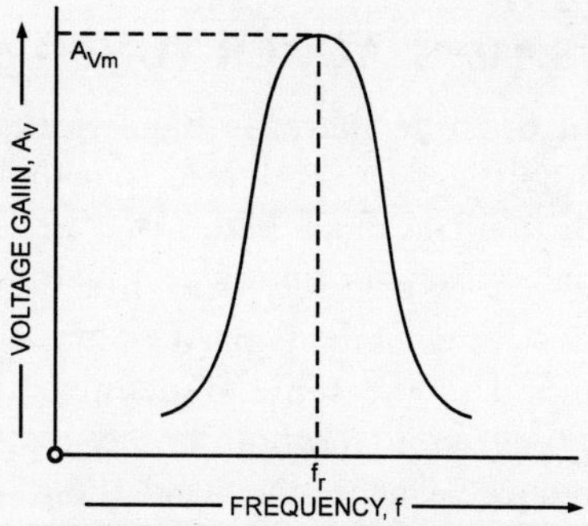

Fig. 8.14. ***Frequency Response***

Example 8.2. In a class C tuned amplifier circuit shown in Fig. 8.7, C = 600 pF and the coil has L = 25 μH and R = 20 Ω and R_L = 2 MΩ. Determine resonant frequency, dc load, ac load and quality factor of the circuit.

Solution :

Resonant frequency, $f_r = \frac{1}{2\pi\sqrt{LC}} = \frac{1}{2\pi\sqrt{25 \times 10^{-6} \times 600 \times 10^{-12}}} = 1.3$ MHz **Ans.**

DC load, R_{dc} = Resistance of coil = 20 Ω **Ans.**

Inductive coil reactance, $X_L = 2\pi f_r L = 2\pi \times 1.3 \times 10^6 \times 25 \times 10^{-6} = 204\ \Omega$

Quality factor of coil, $Q_L = \frac{X_L}{R} = \frac{204}{20} = 10.2$

Equivalent parallel resistance of inductor, $R_P = Q_L X_L = 10.2 \times 204 = 2.082\ k\Omega$

AC load, $R_{ac} = R_P \parallel R_L = 2.082\ k\Omega \parallel 2\ M\Omega = 2.08\ k\Omega$ **Ans.**

Quality factor of the circuit, $Q = \frac{R_{ac}}{X_L} = \frac{2.08 \times 10^3}{204} = 10.2$ **Ans.**

8.9. OSCILLATIONS IN A TUNED AMPLIFIER

A tuned amplifier produces some significant spurious output voltage V_{out} due to transient disturbance even when there is no applied input voltage. A portion of this voltage $(-\beta V_{out})$ could be fed back to the input side through interelectrode capacitance, lead inductances, stray wiring, variation in power supply, ageing of transistor or ground coupling etc. The exact path of this feedback is often complex. But the signal fedback is called *positive feedback signal*. This positive feedback signal when coupled to the input reinforces the input signal *i.e.,* it gets added in phase. The combined signal gets amplified and provides increased output voltage βAV_{out}. When this is equal to V_{out} *i.e.,* $\beta A = 1$, the spurious output has generated itself. That means the amplifier oscillations. These parasitic oscillations are frequent in tuned amplifiers.

Parasitic oscillations may be eliminated by the following means :

1. By changing the circuit parameters.
2. By rearranging the wiring and circuit components.
3. By the use of additional bypassing or shielding to reduce spurious pickup.
4. By changing the transistor.
5. By eliminating the feedback by using neutralization.

8.10. STAGGER TUNED AMPLIFIERS

In order to increase bandwidth, double tuned amplifiers are preferred, but their alignment is difficult. A much better overall response can be had by *stagger tuning,* as illustrated in Fig. 8.15.

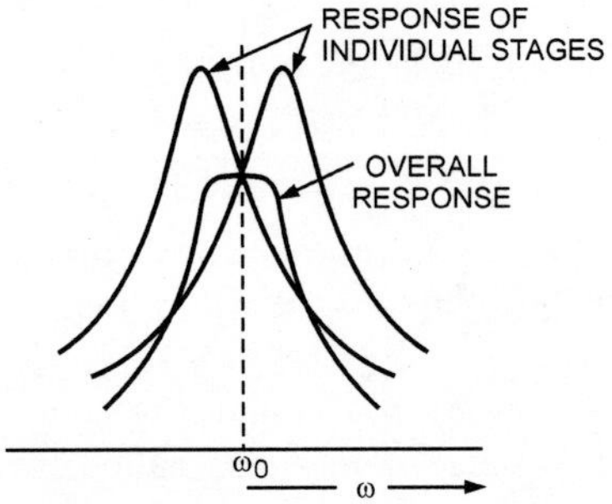

Fig. 8.15. ***Stagger Tuning The Individual Resonant Circuits Resulting in an Overall Response With a Passband Flatter Than That With Synchronous Tuning***

If two or more tuned circuits are cascaded and tuned to the same frequency, it is called the *synchronous tuning.* The overall bandwidth is reduced. On the other hand, if the tuned circuits are cascaded and they are tuned to different frequencies, it is possible to have increased bandwidth with more desirable characteristics (*i.e.,* flat passband with steeper sides). This technique is called the *stagger tuning.*

A two-stage stagger tuned voltage amplifier is shown in Fig. 8.16. In stagger tuned circuits, two single tuned cascaded amplifiers having a certain bandwidth are taken and the resonant frequencies of the two tuned circuits are so adjusted that they are separated by an amount equal to the bandwidth of each stage. Stagger tuned amplifiers are usually designed so that the overall response exhibits maximum flatness around the center frequency f_0. It needs a number of tuned circuits operating in union. If more number of stages are employed flatter will be the passband and steeper will be the gain fall of outside the passband. Because of stagger tuning there is a loss of voltage gain. The overall frequency response of a stagger tuned amplifier is obtained by adding individual response together. Since the resonant frequencies of different tuned circuits are displaced or staggered, they are referred to as *stagger tuned amplifiers.*

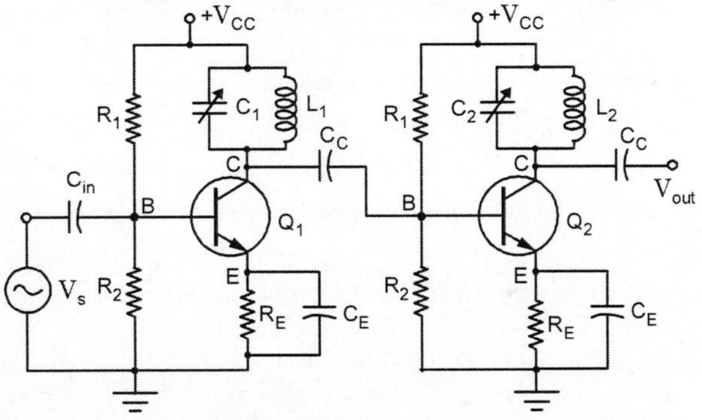

Fig. 8.16. ***A Two-Stage Stagger Tuned Voltage Amplifier***

The main advantage of stagger tuned amplifiers is increased bandwidth. Its drawbacks are reduced selectivity and critical tuning of many tank circuits. They are used in RF amplifier stage in radio receivers.

The resultant staggered pair will have a bandwidth $\sqrt{2}$ times that of each individual single tuned circuits. The overall selectivity function will be identical in form with that of a single stage double tuned amplifier. The gain of the single tuned amplifier is given as

$$\frac{A}{A_{\text{resonance}}} = \frac{1}{1 + j2\delta Q_e} = \frac{1}{1 + jx} \quad \text{where } x = 2Q_e\delta$$

As bandwidth is given as $B = \frac{f_0}{Q_e}$ and under 3-dB frequency condition $\delta = \frac{1}{2Q_e}$, the equation for bandwidth can be written as

$$B = 2\delta f_0$$

Since one stage is tuned to a frequency $\delta_0 f_0$ below f_0 and the other stage is tuned to a frequency $\delta_0 f_0$ above f_0, the corresponding selectivity function of the circuits are

$$\left[\frac{A}{A_{\text{resonance}}}\right]_1 = \frac{1}{1 + j(x - 1)}$$

and $$\left[\frac{A}{A_{\text{resonance}}}\right]_2 = \frac{1}{1 + j(x+1)}$$

Multiplying the relative gains of the two amplifiers, we have the overall gain function as

$$\left[\frac{A}{A_{\text{resonance}}}\right]_{\text{pair}} = \left[\frac{A}{A_{\text{resonance}}}\right]_1 \times \left[\frac{A}{A_{\text{resonance}}}\right]_2 = \frac{1}{1 + j(x-1)} \times \frac{1}{1 + j(x+1)}$$

$$= \frac{1}{1 - (x^2 - 1) + j(x+1) + j(x-1)}$$

$$= \frac{1}{2 - x^2 + 2jx}$$

The magnitude of the resultant function is

$$\left| \left[\frac{A}{A_{\text{resonance}}}\right]_1 \times \left[\frac{A}{A_{\text{resonance}}}\right]_2 \right| = \frac{1}{\sqrt{(2 - x^2)^2 + (2x)^2}} = \frac{1}{\sqrt{4 + x^4}} = \frac{1}{\sqrt{4 + (2\delta_0 Q)^4}} = \frac{1}{2}\frac{1}{\sqrt{1 + 4\delta_0^4 Q^4}} \quad \text{...(8.10)}$$

where δ_0 is the value of δ referred to new frequency f_0 and Q is the value of Q_e for each circuit referred to ω_0. The above Eq. (8.10), if compared with that of double tuned amplifier, we find that the nature of variation is identical.

HIGHLIGHTS

1. Sometimes selection of a desired frequency or a narrowband of frequencies for amplification is our need, such as in radio/television the receiver is required to pick up and amplify the desired radio frequency while rejecting all others. This is achieved in radio receivers with the help of tuned amplifier.
2. Tuned amplifiers, like audio amplifiers, may also be divided into two categories viz. *small signal tuned amplifiers* and *large signal tuned amplifiers*. Small signal tuned amplifiers are used for amplification of small signals at radio frequencies and operated under class A operation as power involved is small. In class A operation, as already known, distortion is negligibly small. Large signal tuned amplifiers are employed for amplification of large signals, at radio frequencies. These amplifiers are operated under class B or class C operation because of involvement of large power.
3. Single tuned amplifiers use one parallel resonant circuit as the load impedance in each stage and all the tuned circuits are tuned to the same frequency. Single tuned amplifiers have the limitation of the potential instability.
4. The problem of potential instability with a single tuned amplifier is overcome in a double tuned amplifier, which use two inductively coupled tuned circuits per stage, both the tuned circuits being tuned to the same frequency.
5. Stagger tuned amplifiers use a number of single tuned stages in cascade, the successive tuned circuits being tuned to slightly different frequencies. The main advantage of stagger tuned amplifiers is increased bandwidth. Its drawbacks are reduced selectivity and critical tuning of many tank circuits.

EXERCISES

1. Explain in brief the function of a tank circuit in a tuned voltage amplifier.
2. Differentiate between a tuned voltage amplifier and a basic voltage amplifier. Draw circuit of a tuned voltage amplifier and explain its working.
3. Draw the circuit diagram of a tuned amplifier and explain its working.
4. Draw the circuit diagram of a collector tuned amplifier and derive expressions for the voltage gain at the tuned frequency and bandwidth.
5. Draw the circuit diagrams of single and double tuned amplifiers and their frequency response.

6. Explain in brief the advantages of using double tuned circuit over a single tuned circuit.
7. Explain the reasons for potential instability in tuned amplifiers.
8. Mention the reasons of oscillations in a tuned amplifier. Briefly explain the methods used to stabilise the tuned amplifiers against oscillations.
9. What is stagger tuned amplifier ? Explain its working.
10. Write short notes on the following :
    (*i*) Tuned amplifiers.
    (*ii*) Single tuned amplifiers.
    (*iii*) Double tuned amplifiers.
    (*iv*) Large signal tuned amplifiers.
    (*v*) Oscillations in a tuned amplifier.
    (*vi*) Stagger tuned amplifier.

SHORT ANSWER TYPE QUESTIONS WITH ANSWERS

Q. 1. What is a tuned amplifier ?

Ans. The amplifier used for amplifying a signal of specific frequency or narrowband of frequencies is known as a tuned amplifier.

Q. 2. What is the fundamental difference between audio amplifiers and tuned amplifiers ?

Ans. The fundamental difference between audio amplifiers and tuned amplifiers is the bandwidth they are expected to amplify. Audio amplifiers amplify a major portion of audio-frequency spectrum (20 Hz – 20 kHz) equally well throughout while tuned amplifiers amplify a relatively narrow portion of RF spectrum, rejecting all other frequencies.

Q. 3. Give the typical value of Q of a coil used in class C tuned amplifiers.

Ans. 50 or more.

Q. 4. What is the typical value of Q for the circuit of a class C tuned amplifier ?

Ans. 10 or more.

Q. 5. Why the quality factor of the whole circuit is less than that of the inductor ?

Ans. The quality factor Q for the whole circuit is less than that of the inductor because it includes the effect of load resistance as well as the coil resistance.

Q. 6. Explain in brief the effect of changing Q of the coil used in tank circuit, on its bandwidth.

Ans. In case of series resonance, higher value of Q factor means not only higher voltage magnification, but also a higher selectivity of the tuning coil and that is why it is necessary that the coil be of high inductance and low resistance. In fact, Q factor of a series resonant circuit may be defined as the ratio of resonant frequency to bandwidth

$$i.e., \; Q_r = \frac{f_r}{\Delta f} = \frac{\dfrac{1}{2\pi\sqrt{LC}}}{R/2\pi L} = \frac{1}{R}\sqrt{\frac{L}{C}}$$

The quality factor Q of a parallel tuned circuit is of utmost importance as it determines the sharpness of the resonance curve and hence the selectivity of the circuit. The higher the value of Q, the more selective is the tuned circuit.

Q. 7. Tuned-class C amplifiers are used for applications involving larger RF power. Why ?

Ans. Tuned amplifiers operating in class A operation are not suitable for applications involving large RF power, because of lower efficiency of class A operation. For applications involving larger RF power, like for excitation of transmitting antenna, tuned class C amplifiers are employed because class C operation has a very high efficiency and is capable of delivering more power in comparison to that of class A operation.

Q. 8. Define stagger tunning.

OR

What is stagger tunning ? Explain in brief.

Ans. A much better overall response can be had by stagger-tuning the individual stages, as illustrated in Fig. 8.15. Stagger tuned amplifiers are usually designed so that the overall response exhibits maximal flatness around the centre frequency f_o. It needs a number of tuned circuits operating in union. The overall frequency response of a stagger tuned amplifier is obtained by adding the individual response together. Since the resonant frequencies of different tuned circuits are displaced or staggered, they are referred to as *stagger tuned amplifier.* The main advantage of stagger tuned amplifier is increased bandwidth. Its drawbacks are reduced selectivity and critical tuning of many tank circuits. They are used in RF amplifier stage in radio receivers.

A two-stage stagger tuned voltage amplifier is shown in Fig. 8.16. The stagger tuning in the circuit, can be achieved by resonating the tuned circuits L_1C_1 and L_2C_2 to slightly different frequencies.

Q. 9. Justify the name of stagger tuned amplifiers.

Ans. Since the resonant frequencies of different tuned circuits used in stagger tuned amplifiers are displaced or staggered, they are referred to as stagger tuned amplifiers.

Q. 10. What is the advantage of stagger tuned amplifier ?

Ans. Stagger tuning results in increased bandwidth.

Q. 11. What are the drawbacks of stagger tuned amplifiers ?

Ans. The drawbacks of stagger tuned amplifiers are:

(*i*) Reduced selectivity.

(*ii*) Tuning of many tank (L-C) circuits precisely at different frequencies is quite cumbersome.

Q. 12. Give the application of stagger tuned amplifiers.

Ans. Stagger tuned amplifiers are used in RF amplifier stage in radio receivers.

❑❑❑

CHAPTER

9

Linear Waveshaping

9.1. INTRODUCTION

INSIDE THIS CHAPTER

Waveshaping may be defined as a process of generating new waveforms from older waveforms by employing certain physical systems. Depending upon the physical system involved, the methods of waveshaping may be classified into (*i*) linear waveshaping and (*ii*) nonlinear waveshaping.

Linear waveshaping involves passage of signal through linear systems (such as R-C, R-L and R-L-C circuits) and the operations involved are linear such as integration, differentiation, summation, filtering etc.

9.2. HIGH-PASS R-C CIRCUIT

The circuit is shown in Fig. 9.1. The reactance of the capacitor is given as

$$X_C = \frac{1}{2\pi f C} \quad \text{or} \quad X_C \propto \frac{1}{f}$$

i.e., Reactance X_C of the circuit capacitor decreases with the increase in frequency. Thus at low frequencies the capacitor C offers considerable reactance and so blocks them, but at higher frequencies it offers little reactance and allows them to pass through it. The output voltage is developed across resistor R.

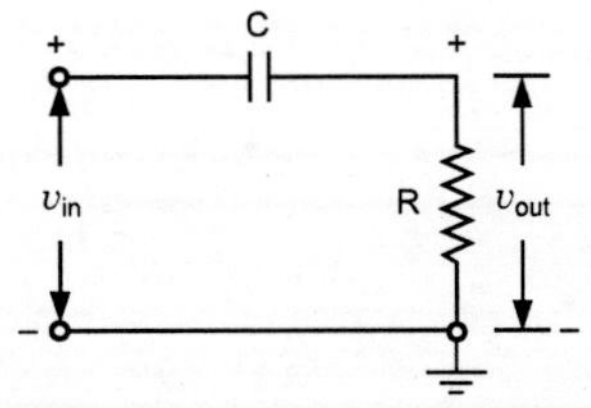

Fig. 9.1. *High-Pass R-C Circuit*

With the increase in frequency, the reactance of the capacitor decreases and, therefore, the output and gain increase. At very high frequencies, the capacitive reactance becomes very small, so the output becomes almost equal to the input and gain becomes equal to unity. Since this circuit attenuates the low-frequency signals and allows transmission of high-frequency signals with little or no attenuation, it is called a high-pass circuit.

Because of its blocking property for dc or low frequency input signals the capacitor acts like an open circuit and blocks the signal. This is the reason that the capacitor in the circuit is called the *blocking capacitor*. This circuit is widely employed as a *coupling circuit* to provide isolation between input and output.

Now response of the high-pass filter circuit for different types of inputs will be found.

9.2.1. Sinusoidal Input

For a sinusoidal input the waveshape of the output will remain the same; the output will differ only in amplitude and phase angle depending on the frequency of the input signal. Here we are concerned only with amplification of the signal which is expressed as

$$\text{Gain A} = \frac{V_{out}}{V_{in}}$$

The current through the circuit is given as

$$I = \frac{V_{in}}{R - jX_C}$$

and output voltage, V_{out} = Voltage drop across resistor $R = I\,R = \frac{R}{R - j\,X_C}\,V_{in}$

$$\text{So gain, A} = \frac{V_{out}}{V_{in}} = \frac{R}{R - j\,X_C} = \frac{1}{1 - \frac{j\,X_C}{R}} = \frac{1}{1 - \frac{j}{2\pi f RC}} \quad ...(9.1) \because X_C = \frac{1}{\omega C} = \frac{1}{2\pi f C}$$

From above equation it is obvious that for a given circuit, the gain A depends upon the frequency of the input signal.

The frequency response curve for the circuit is given in Fig. 9.2. From frequency response curve it is obvious that all frequencies above cutoff frequency f_1 are passed while those below are attenuated. At cutoff frequency $R = X_C$, the phase angle is 45° and output power is half of the input power. The output voltage is 0.707 times the maximum value of voltage V_{max}. The cutoff frequency is given as

$$f_1 = \frac{1}{2\pi RC} \quad ...(9.2)$$

Substituting $\frac{1}{2\pi RC} = f_1$ in Eq. (9.1), we have

$$\text{Gain A} = \frac{1}{1 - j\,f_1/f} \quad ...(9.3)$$

Now magnitude of A and phase angle θ are given as

$$|A| = \frac{1}{\sqrt{1 + \left(\frac{f_1}{f}\right)^2}} \quad ...(9.4)$$

and phase angle $\theta = \text{Tan}^{-1}\frac{f_1}{f}$ ...(9.5)

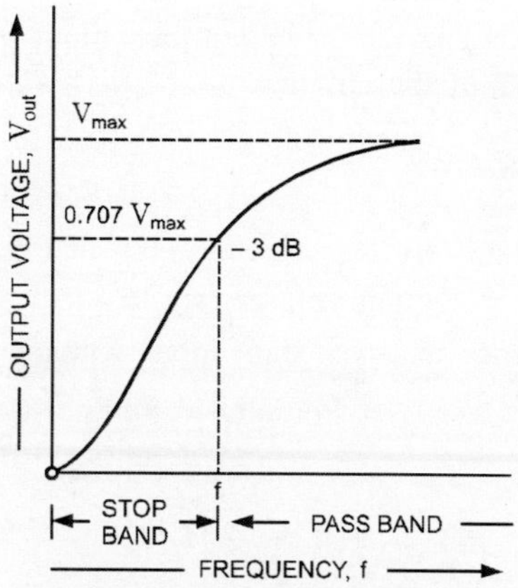

Fig. 9.2. ***Frequency Response Curve For High-Pass R-C Circuit***

As $f \to \infty$; $\frac{f_1}{f} = 0$ and $|A| \to 1$

When $f = f_1$; $|A| = \frac{1}{\sqrt{2}} = 0.707$

f_1 is also known as low 3-dB frequency.

$f = f_1$ also implies that $f = \frac{1}{2\pi RC}$ or $R = \frac{1}{2\pi f C} = X_C$.

So f_1 is the frequency at which capacitive reactance is equal to the resistance of the circuit.

The maximum possible value of gain (unity) is approached asymptotically at high frequencies.

The lower cutoff frequency of a high-pass circuit, f_1 produces a tilt.

For a 10 per cent change in capacitor voltage, the time or pulse width involved is

$$t = 0.1 \text{ RC} = \text{Pulse Width (PW)}$$

$$\text{or} \quad \frac{\text{PW}}{\text{RC}} = 0.1 = \text{Fractional tilt}$$

$$\text{or Fractional tilt} = \frac{\text{PW}}{\text{RC}} = 2\pi f_1 \text{PW} \qquad \text{...(9.6)} \quad \because \; f_1 = \frac{1}{2\pi\text{RC}}$$

The above equation applies when the tilt is 10 per cent or less. When the tilt exceeds 10%, the output voltage should be treated as exponential instead of linear and is given by an expression

$$\text{V}_{out} = \text{V}_{final} - (\text{V}_{final} - \text{V}_{initial})\, e^{-t/\tau} \qquad \text{...(9.7)}$$

where τ = RC, is called the time constant of the circuit.

9.2.2. Step Input

A step voltage is zero for all times when $t < 0$ and maintains a voltage level, say V, for all times when $t \geq 0$. The transition from voltage level 0 to V occurs at $t = 0$ and is instantaneous. Instant of time just before the transition occurs is denoted by $t = 0^-$ and instant of time just after the transition has taken place is denoted by $t = 0^+$.

At instant $t = 0^-$, since the input is zero, the output is also zero. At instant $t = 0$, the input suddenly becomes V. Since the voltage across a capacitor cannot change instantaneously, the output voltage remains V. At $t = 0^+$ the input voltage remains unchanged at V, the capacitor starts charging exponentially with time constant RC. As the capacitor C charges exponentially to V, the output voltage V_{out} falls exponentially to zero. When the capacitor gets charged to V volts, the output voltage will be obviously zero, as illustrated in Fig. 9.3.

The rate at which the output voltage drops to zero exponentially depends upon the time constant RC of the circuit.

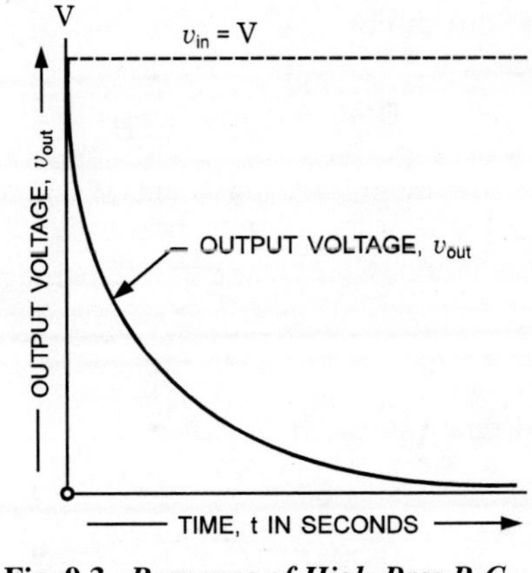

Fig. 9.3. ***Response of High-Pass R-C Circuit To Step Input***

$$\text{Input voltage, } v_{in} = 0 \quad \text{for } t < 0$$
$$= \text{V} \quad \text{for } t \geq 0$$

From the circuit shown in Fig. 9.1.

$$\text{Output voltage, } v_{out} = i\,\text{R}$$

Since at any instant input voltage,

$$v_{in} = \text{Potential difference across resistance R} + \text{potential difference across capacitor.}$$

$$v_{in} = i\text{R} + \frac{1}{\text{C}}\int i\,dt$$

Taking Laplace of this

$$v_{in}(s) = \text{R}(s) + \frac{i(s)}{\text{C}s}$$

Assuming zero charge on capacitor at time instant $t = 0$

$$i(s) = \frac{v_{in}(s)}{\text{R} + \dfrac{1}{\text{C}s}}$$

and $v_{out}(s) = Ri(s) = \dfrac{v_{in}(s)}{R + \dfrac{1}{Cs}} \times R = \dfrac{v_{in}(s)}{1 + \dfrac{1}{RCs}} = \dfrac{V}{s + \dfrac{1}{RC}}$ $\quad \because v_{in}(s) = \dfrac{V}{s}$

Taking inverse Laplace we have

$$v_{out}(t) = V \cdot e^{-t/RC} = V e^{-t/\tau} \quad ...(9.8) \quad \text{where } \tau = \text{Time constant of the circuit} = RC$$

Note that the output voltage is equal to input voltage or initial value at $t = 0$; 0.607 times of initial value at $t = 0.5$ T; 0.368 times of initial value at $t =$ T; 0.135 times of initial value at $t = 2$T and 0 at $t = \infty$. The output has completed more than 95 per cent of its total change after 3T and more than 99 per cent of its swing when $t > 5$T. Hence, although the steady-state is approached asymptotically, we may assume for most applications that the final value has been reached after 5T.

The voltage across a capacitor can change instantaneously only when an infinite current passes through it.

9.2.3. Pulse Input

An ideal pulse voltage waveform is given in Fig. 9.4. Here V is amplitude of pulse and t_p is pulse duration or pulse width.

Input voltage, $v_{in} = 0$ for $t < 0$

$= V$ for $t \geq 0 \leq t_p$ *i.e.*, $0 \leq t \leq t_p$

$= 0$ for $t > t_p$.

A pulse can be considered as a sum of two step voltages (*i.e.*, a step of amplitude + V starting at $t = 0$ and a step voltage of – V appearing at $t = t_p$, as illustrated in Fig. 9.5 (*a*) and 9.5 (*b*) respectively. The sum of these two steps is the pulse shown in Fig. 9.4.

For $t < 0$, the input is zero and the output is also zero. At $t = 0$, there is a positive step of V at the input. Since voltage across the capacitor cannot change instantaneously, the changes in input appear at the output in totality. Hence the output at $t = 0$ is V. For time t, $0 \leq t \leq t_p$ input is constant at V, the capacitor charges to V and the output falls exponentially to zero.

$$v_{out} = V e^{-t/CR}.$$

The above expression was derived in the previous subsection

At $t = t_p$

$$v_{out} = V e^{-t_p/CR}$$

At the input at, $t = t_p$, there is a negative step of V and again this step will appear at the output

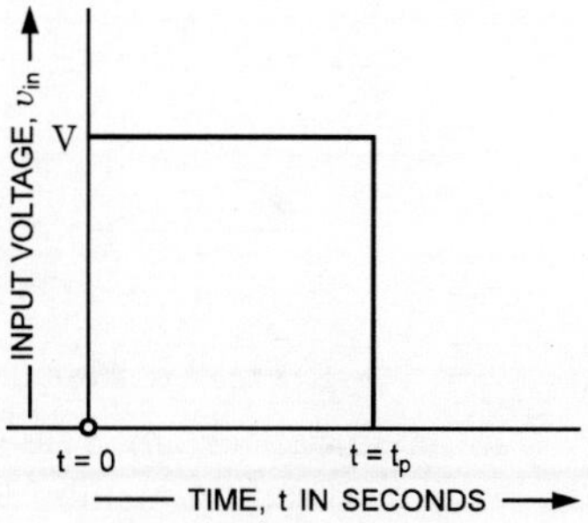

Fig. 9.4. ***Pulse***

At $t = t_p$

$$v_{out} = V e^{-t_p/CR} - V = V(e^{-t_p/CR} - 1)$$

Since $V > V e^{-t_p/CR}$ at $t = t_p$, v_{out} is negative, as illustrated in Fig. 9.6. Now, input is zero, the capacitor discharges to zero exponentially and the output voltage rises to 0 exponentially. For $t > t_p$, output voltage is given as

$$v_{out} = V(e^{-t_p/CR} - 1)\, e^{(t - t_p)/CR} \quad ...(9.9)$$

The rate of exponential fall or rise depends upon the time constant RC. The response for RC >> t_p and for RC << t_p is shown in Fig. 9.7 (*a*) and (*b*). If the time constant is very small, we get positive and negative spikes. The process of converting pulses into spikes is known as *peaking*.

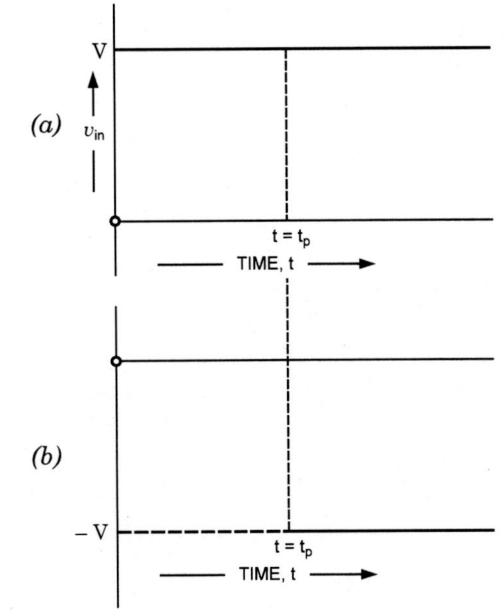

Fig. 9.5

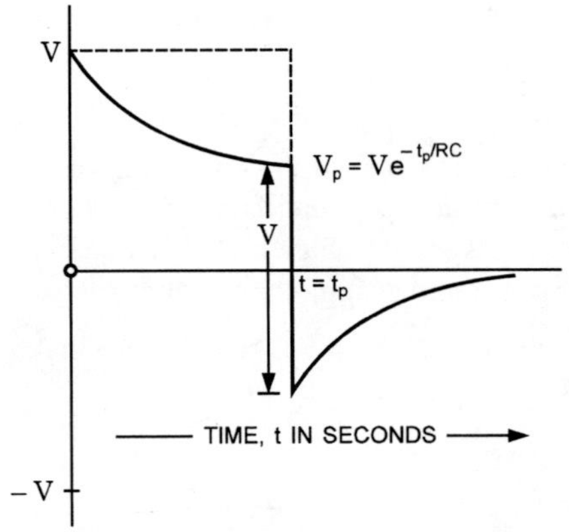

Fig. 9.6

The areas A_1 and A_2 above and below the zero axis are equal which means dc component of output voltage is always zero.

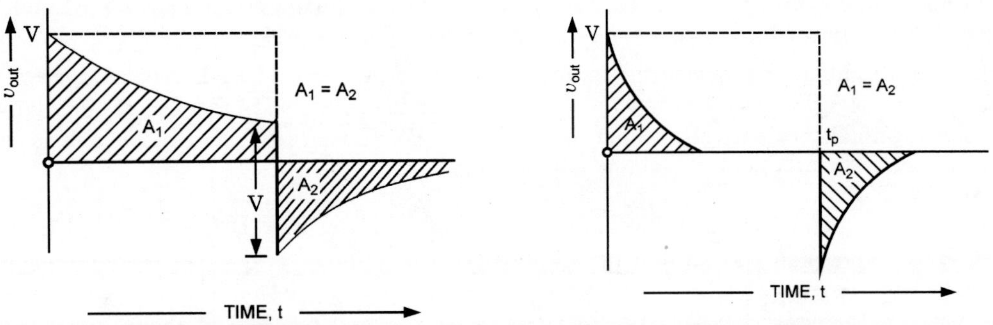

(a) For RC >> i_p *(b) For RC << i_p*

Fig. 9.7. *Response of R-C Circuit To Pulse Input For Different Time Constants*

In any case, for any value of RC, there must always be a tilt and an undershoot. The undershoot can be minimised by making RC very large in comparison to pulse width t_p. However, the area below the zero axis will always be equal to the area above the axis $A_1 = A_2$. That is why output below zero axis rises to zero very slowly for RC >> t_p.

9.2.4. Square-Wave Input

If a number of pulses of alternative positive and negative amplitude occur respectively, a square-wave is formed. Also a square-wave consists of a number of positive and negative unit step functions having their discontinuity at certain intervals of time.

A square-wave may alternatively be defined as a waveform which maintains itself at one constant level V_1 for a time duration of T_1 and another constant level V_2 for time duration of T_2 and which repeats itself at time $T = T_1 + T_2$.

For any periodic input waveform, the output is also a periodic waveform of same period and the average level of the steady state output signal is always zero and is independent of the dc level of the input. This is due to blocking property of the capacitor. This can be shown as below.

Since at any instant, input voltage v_{in} is equal to potential difference across resistor plus potential difference across the capacitor,

$$v_{in} = \frac{1}{C}\int i\,dt + v_{out}$$

Differentiating above equation w.r.t. time. we have

$$\frac{dv_{in}}{dt} = \frac{i}{C} + \frac{dv_{out}}{dt} = \frac{v_{out}}{RC} + \frac{dv_{out}}{dt} \quad \because i = \frac{v_{out}}{R}$$

Multiplying by dt and then integrating over time period T, we have

$$\int_0^T dv_{in} = v_{in}(T) - v_{in}(0) = \frac{1}{RC}\int_0^T v_{out}\,dt + v_{out}(T) - v_{out}(0)$$

Under steady-state conditions, the output waveform is repetitive with a period T so that

$$v_{out}(T) = v_{out}(0) \text{ and } v_{in}(T) = v_{in}(0)$$

$$\text{So } \int_0^T v_{out}\,dt = 0$$

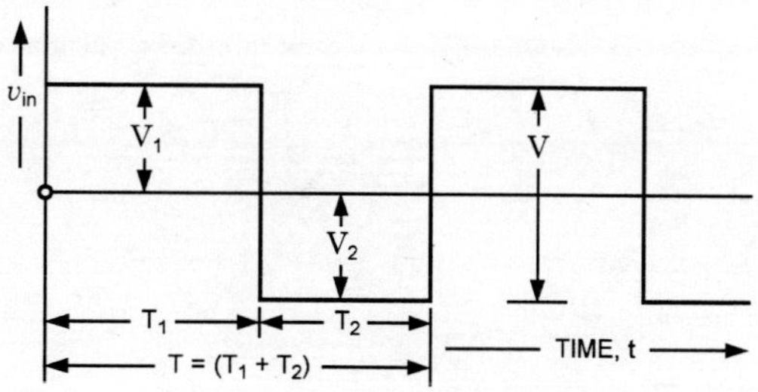

Fig. 9.8. ***Square-Wave Input***

The above integral represents the area under the output waveform over one cycle. It implies that the average level of a steady-state output signal is always zero. It shows that output must extend in both the positive and negative directions of zero axis, so that the area above the axis must be equal to the area below the axis making resultant area equal to zero.

For square-wave, output is same as for pulse input only the output in this case is repetitive with alternate positive and negative amplitude.

The shape of the output waveform in this case is going to depend upon the time constant RC. If $\frac{RC}{T_1}$ and $\frac{RC}{T_2}$ are both arbitrarily large in comparison with unity, the output will be identical to the input waveform. On the other hand, if $\frac{RC}{T_1}$ and $\frac{RC}{T_2}$ are both very small as compared to unity, the output will consist of alternate positive and negative peaks with having peak-to-peak amplitude twice that of the input, as illustrated in Fig. 9.9 (*b*).

Analysis. For a constant voltage applied to the circuit shown in Fig. 9.1, output at any instant is given as

$$v_{out} = V e^{-t/CR}$$

When a square-wave signal shown in Fig. 9.8 is applied to the circuit shown in Fig. 9.1, the output is as follows.

From Fig. 9.9 (*c*)

At time $t = 0$, $\quad V_{out} = V_1$

At time $t = T_1$, $\quad V_1' = V_1 e^{-T_1/RC}$; $\quad V_1' - V_2 = V$

At time $t = T_1 + T_2$, $\quad V_2' = V_2 e^{-T_2/RC}$; $\quad V_1 - V_2' = V$

For a symmetrical square-wave

$$T_1 = T_2 = T/2$$
$$V_1 = -V_2$$
$$\text{and } V_1' = -V_2'$$

Substituting these values in above equations, we have

$$V_1' = V_1\, e^{-T_1/RC}$$
$$V_1' + V_1 = V$$

Eliminating V_1', we have

$$V_1 = V - V_1\, e^{-T_1/RC}$$
$$\text{or } V_1\,[1 + e^{-T_1/RC}] = V$$
$$\text{or } V_1 = \frac{V}{1 + e^{-T/2RC}} \quad \because T_1 = T/2$$
$$\text{and } V_1' = V_1\, e^{-T_1/RC}$$
$$= \frac{V}{1 + e^{-T/2RC}} \cdot e^{-T/2RC}$$
$$= \frac{V}{1 + e^{T/2RC}}$$

For $\frac{T}{2RC} << 1$ these reduce to

$$V_1 = \frac{V}{1 + 1 - \frac{T}{2RC}}$$

by taking the exponential series and ignoring the higher terms

$$V_1 = \frac{V}{2 - \frac{T}{2RC}} = \frac{V}{2}\left[1 - \frac{T}{4RC}\right]^{-1}$$

$$= \frac{V}{2}\left[1 + \frac{T}{4RC}\right]$$ by taking binomial and ignoring the higher terms.

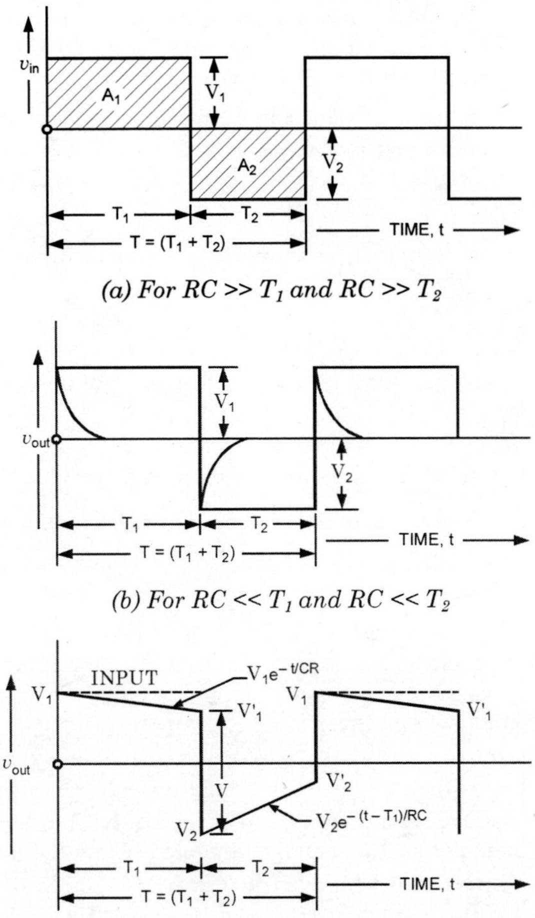

(a) For $RC >> T_1$ and $RC >> T_2$

(b) For $RC << T_1$ and $RC << T_2$

(c)

Fig. 9.9. ***Response To a Square Wave***

Similarly $V_1' = \frac{V}{2}\left[1 - \frac{T}{4RC}\right]$

The percentage tilt P is defined as

$$P = \frac{V_1 - V_1'}{V/2} \times 100 = \frac{T}{2RC} \times 100$$

Since the lower cutoff frequency – 3-dB point is given by $f_1 = \frac{1}{2\pi RC}$.

$$\therefore P = \pi \frac{f_1}{f} \times 100 \text{ per cent} \qquad ...(9.10)$$

where $f = \frac{1}{T}$ is the frequency of the square-wave applied to the R-C circuit.

9.2.5. Ramp Input

A waveform which is zero for $t < 0$ and which increases linearly with time for $t > 0$, $v = \alpha t$, is known as a *ramp* or *sweep voltage*. Such a waveform is shown in Fig. 9.10 (*a*). When the high-pass R-C circuit shown in Fig. 9.1 is excited by a ramp input $v_{in}(t) = \alpha t$ where α is the slope of the ramp.

Since at any instant input voltage v_{in} is equal to *pd* across resistor plus *pd* across the capacitor,

$$v_{in} = \frac{1}{C}\int i\,dt + v_{out}$$

Differentiating above equation w.r.t. time, we have

$$\frac{dv_{in}}{dt} = \frac{i}{C} + \frac{dv_{out}}{dt} = \frac{v_{out}}{RC} + \frac{dv_{out}}{dt} \quad ...(9.11) \because i = \frac{v_{out}}{R}$$

The above equation has the solution, for $v_{out} = 0$ at $t = 0$,

$$v_{out} = \alpha\, RC(1 - e^{-t/\tau}) = \alpha\, RC(1 - e^{-t/RC})$$

For times t which are very small in comparison with RC, we have

$$v_{out} = \alpha RC\left[1 - \left\{1 + \left(\frac{-t}{RC}\right) + \left(\frac{-t}{RC}\right)^2 \cdot \frac{1}{2!} + \left(\frac{-t}{RC}\right)^3 \cdot \frac{1}{3!} + ...\right\}\right]$$

$$= \alpha RC\left[\frac{t}{RC} - \frac{t^2}{2(RC)^2}\right] = \alpha t - \frac{\alpha t^2}{2RC} = \alpha t\left(1 - \frac{t}{2RC}\right) \quad ...(9.12)$$

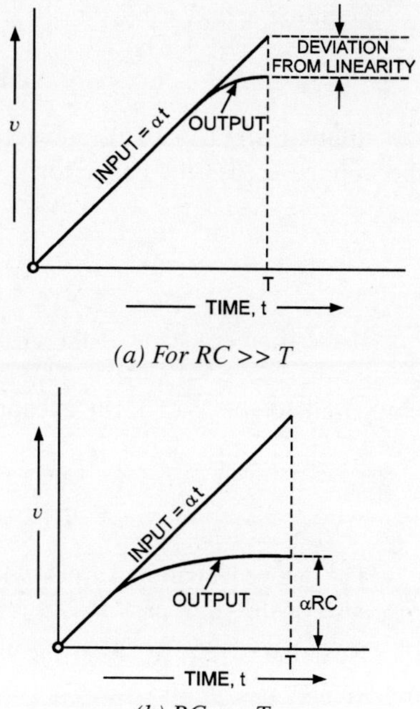

(a) For RC >> T

(b) RC << T

Fig. 9.10. ***Response of a High-Pass R-C Circuit To a Ramp Voltage***

Figure 9.10 depicts the response of the high-pass circuit for a ramp input when (*a*) RC >> T and (*b*) RC << T where T is the ramp duration. The output signal falls away slightly from the input, as depicted in Fig. 9.10 (*a*). As a measure of the departure from linearity, let us define the transmission error e_t as the difference between input and output divided by the input. The error at time t = T is then

$$e_t = \frac{v_{in} - v_{out}}{v_{in}} = \frac{\alpha t - \alpha t\left[1 - \frac{t}{2RC}\right]}{\alpha t} \simeq \frac{T}{2RC} = \pi f_1 T \quad ...(9.13)$$

where $f_1 = \frac{1}{2\pi RC}$ is again the lower 3-dB frequency of the high-pass RC circuit.

For large values of t in comparison with RC, the output approaches the constant value α RC as depicted in Fig. 9.10 (*b*).

9.2.6. Exponential Input

When an exponential input

$$v_{in} = V(1 - e^{-t/T}) \quad ...(9.14)$$

shown in Fig. 9.11 is applied to a high-pass R-C circuit shown in Fig. 9.1, Eq. (9.11) becomes

$$\frac{V}{\tau}e^{-t/\tau} = \frac{v_{out}}{RC} + \frac{dv_{out}}{dt} \quad ...(9.15)$$

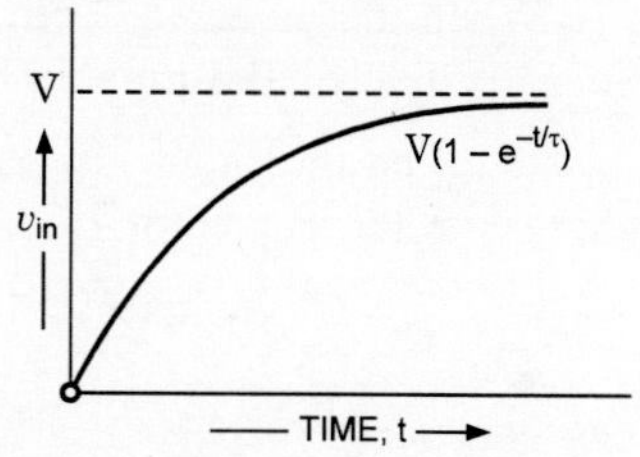

Fig. 9.11. ***Exponential Input***

Defining x and n by

$$x \equiv \frac{t}{\tau} \quad \text{and} \quad n \equiv \frac{RC}{\tau}$$

the solution of Eq. (9.15), subjected to the condition that initially the voltage across the capacitor is zero, is given by the expression

$$v_{out} = \frac{Vn}{n-1}(e^{-x/n} - e^{-x}) \quad \text{if } n \neq 1 \qquad ...(9.16)$$

and

$$v_{out} = Vxe^{-x} \quad \text{if } n = 1 \qquad ...(9.17)$$

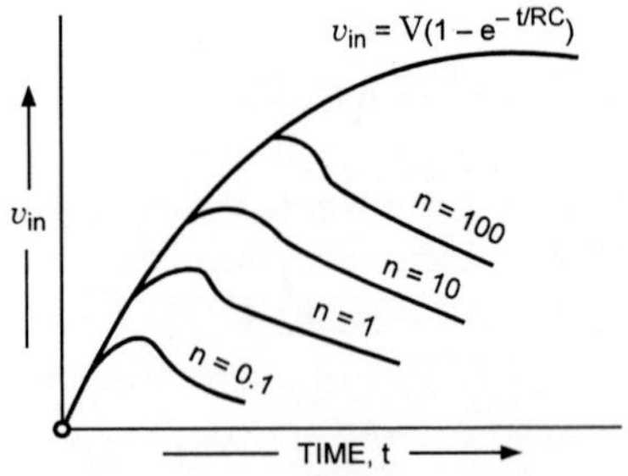

Fig. 9.12. ***Response of a High-Pass R-C Circuit To An Exponential Input***

If the time constant of the circuit is very high *i.e.*, RC is much greater than $\tau(n >> 1)$, the second term of the Eq. (9.16) is negligible compared to the first term except for very small values of time. Then

$$v_{out} \simeq \frac{Vn}{n-1}e^{-x/n} \simeq Ve^{-t/RC} \qquad ...(9.18)$$

The above equation agrees with the way the circuit should behave for an ideal step voltage. The response of the high-pass R-C circuit for different values of n is depicted in Fig. 9.12.

Near the origin of time, the output follows the input. Also, the smaller the circuit time constant, the smaller will be the output peak and narrower will be the pulse. The larger the circuit time constant, the larger will be the peak output but also the wider will be the pulse. A value of RC is chosen to provide the best compromise between these two conflicting characteristics for the particular application at hand. The choice is seldom critical.

9.2.7. R-C Differentiator

A circuit that gives an output voltage proportional to the derivative of its input, is known as a *differentiating circuit.*

$$i.e., \text{ Output } \alpha \frac{d}{dt} \text{ (input)}$$

Figure 9.1 shows a typical differentiating circuit. The output voltage across R will be the derivative of the input voltage. It is important to note that merely an R-C series circuit will not provide the voltage across R as a derivative of the input. It is also necessary to set the proper circuit values. The conditions to be fulfilled are:

(*i*) The time constant RC of the circuit should be much smaller than the time period of the input signal *i.e.*, RC << T.

(*ii*) The value of X_C $\left(i.e., \frac{1}{2\pi fC}\right)$ should not be smaller than 10 times of R *i.e.*, $X_C \geq 10R$.

When the given R-C circuit fulfills the above conditions then the voltage appearing across R will be derivative of the input. This is shown below.

If v_{in} be the input alternating voltage, i is the resulting alternating current then the instantaneous charge on the capacitor is given as

$q = Cv_C$ where v_C is the instantaneous voltage across the capacitor C.

$$\text{Current, } i = \frac{dq}{dt} = \frac{d}{dt}(q) = \frac{d}{dt}Cv_C = C\frac{d}{dt}v_C \qquad ...(9.19)$$

As the capacitive reactance is much larger than R ($X_C \geq 10R$), the voltage across the capacitor may be considered equal to the input voltage *i.e.*, $v_C = v_{in}$

$$i.e., \; i = C\frac{d}{dt}v_{in}$$

$$\text{Output voltage, } v_{out} = iR = RC\frac{d}{dt}v_{in}$$

$$\text{or } v_{out} \;\alpha\; \frac{d}{dt}v_{in} \qquad ...(9.20)$$

$$\text{Hence output voltage } \alpha \; \frac{d}{dt}\text{ (input)}$$

If the input voltage is the square-wave, output wave should be impulses of infinite amplitude and alternating polarity, occurring precisely at discontinuities of input. But in practice, the output impulses never exceed V maximum input amplitude, due to voltage drop across R, which is actually not negligible. Thus if the R-C circuit is to be employed as a differentiator, input voltage V_{in} should appear across the capacitor C in totality at every instant. To realise this condition both R and C must be very small and $X_C >> R$.

9.3. APPLICATIONS OF HIGH-PASS R-C CIRCUITS

R-C circuit with RC >> T is employed in R-C coupling of amplifiers where distortion and differentiation of waveform is to be avoided. R-C circuit with RC << T is employed to generate pips for triggering electronic circuits such as multivibrators, flip-flops etc.

Example 9.1. An oscilloscope has a coupling capacitor of 0.4 μF and an input resistance of 2.5 MΩ. Determine the minimum square-wave frequency that can be displayed. The tilt is not to exceed 0.25 per cent.

Solution : Coupling capacitor of the oscilloscope,

$$C_C = 0.4\ \mu F = 0.4 \times 10^{-6}\ F$$

$$\text{Input resistance of the oscilloscope, } R_{in} = 2.5\ M\Omega = 2.5 \times 10^{6}\ \Omega$$

$$\text{Fractional tilt} = 0.25\% = \frac{0.25}{100} = 0.0025$$

The circuit is a high-pass R-C filter

$$\text{Pulse-width, PW} = \text{Fractional tilt} \times R_{in} \times C_C = 0.0025 \times 0.4 \times 10^{-6} \times 2.5 \times 10^{6}$$
$$= 0.0025 \text{ second}$$

$$\text{Square-wave frequency, } f = \frac{1}{T} = \frac{1}{2\,PW} = \frac{1}{2 \times 0.0025} = 200 \text{ Hz} \quad \textbf{Ans.}$$

Example 9.2. An amplifier gives an output of 1,000 Hz with a rise time of 350 ns and a tilt of 5 per cent. Find upper and lower 3-dB frequencies.

Solution :

$$\text{Rise time } t_r = 350 \text{ ns} = 350 \times 10^{-9} \text{ s}$$

$$\text{Output frequency, } f = 1{,}000 \text{ Hz}$$

$$\text{Upper 3-dB frequency, } f_2 = \frac{0.35}{t_r} = \frac{0.35}{350 \times 10^{-9}} = 1 \text{ MHz} \quad \textbf{Ans.}$$

$$\text{Lower 3-dB frequency, } f_1 = \frac{P \times f}{\pi \times 100} = \frac{5 \times 1{,}000}{\pi \times 100} = 15.92 \text{ Hz} \quad \textbf{Ans.}$$

$$\because \text{ Percentage tilt, } P = \frac{\pi f_1}{f} \times 100 \simeq 5\%$$

Example 9.3. The limited ramp shown in Fig. 9.13 by dotted line is applied to an R-C differentiator. Draw the output waveforms when T is 25%, 100% and 400% of RC.

Solution : For ramp input, Slope, $\alpha = \frac{V}{T}$

where T is the duration of the ramp and V is the amplitude of ramp at t = T

Output for ramp input is given as

$$v_{out}(t) = \alpha RC\,(1 - e^{-t/RC})$$

(*i*) For T = 25% of RC

The output at t = T is

$$= \frac{V}{T} \times 4T\,(1 - e^{-T/4T}) = 0.885 \text{ V volts}$$

(*ii*) For T = 100% of RC

The output at t = T is

$$= \frac{V}{T} \times T(1 - e^{-T/T}) = 0.632 \text{ V volts}$$

(*iii*) For T = 400% of RC

The output at t = T is

$$= \frac{V}{T} \times \frac{T}{4}(1 - e^{-T/0.25T}) = \frac{V}{4}(1 - e^{-4}) = 0.245 \text{ V volts}$$

The outputs for different constants are shown in Fig. 9.13.

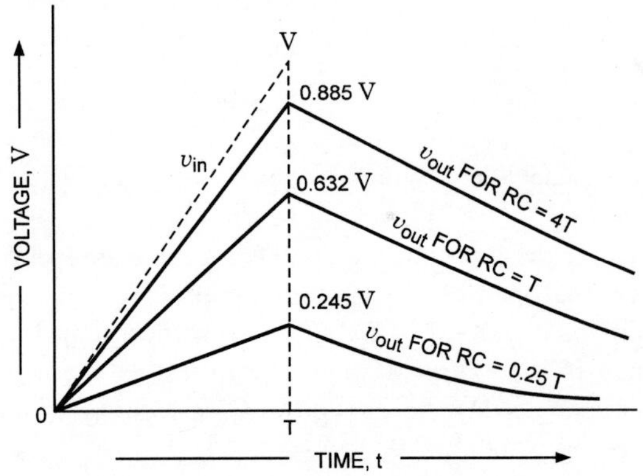

Fig. 9.13. ***Input and Output Waveforms of an R-C Differentiator***

9.4. LOW-PASS R-C CIRCUIT

The low-pass R-C circuit is shown in Fig. 9.14. In a low-pass R-C circuit, the output voltage v_{out} is taken across the capacitor. Resistance offers fixed opposition. Since the reactance offered by the capacitor C falls with the increase in frequency, low frequency signal develops across the capacitor but signal of frequency above cutoff frequency f_2 develops negligible voltage across capacitor C. At zero frequency, the capacitor acts as an open circuit and output is same as input. However, with the increase in frequency, the capacitive reactance decreases and so the output voltage. At infinite frequency, the capacitive reactance of the circuit will be zero, and therefore, output voltage will also be zero. Since it passes low frequency signals and blocks the high frequency signals, it is called the *low-pass circuit*.

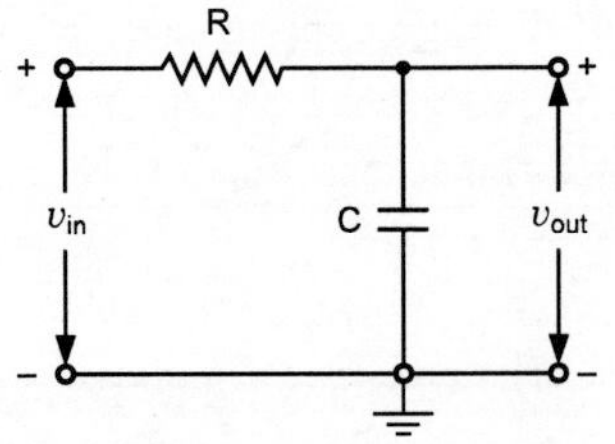

Fig. 9.14. ***Low-Pass R-C Circuit***

9.4.1. Sinusoidal Input

For a sinusoidal input the waveshape of the output will remain the same; the output will differ only in amplitude and phase angle depending on frequency of the input signal. So only gain of the circuit needs consideration.

$$\text{Gain A} = \frac{V_{out}}{V_{in}}$$

The current through the circuit is given as

$$I = \frac{V_{in}}{R - j\,X_C}$$

$$\text{Output voltage, } V_{out} = -j I X_C = \frac{-j X_C}{R - j X_C} V_{in}$$

$$\text{So gain } A = \frac{V_{out}}{V_{in}} = \frac{-jX_C}{R - jX_C} = \frac{1}{1 - \dfrac{R}{jX_C}}$$

$$= \frac{1}{1 + j2\pi fCR} \quad \text{...(9.21)}$$

From the above equation it is obvious that for a circuit, the gain A depends upon the frequency of the input signal.

The frequency response curve for the circuit is given in Fig. 9.15. From frequency response curve it is obvious that all frequencies below cutoff frequency f_2 are passed while those above are attenuated. At cutoff frequency $R = X_C$, the phase angle is 45° and output power is half of the input power. The output voltage is 0.707 times the maximum value of voltage V_{max}. The cutoff frequency f_2 is given as

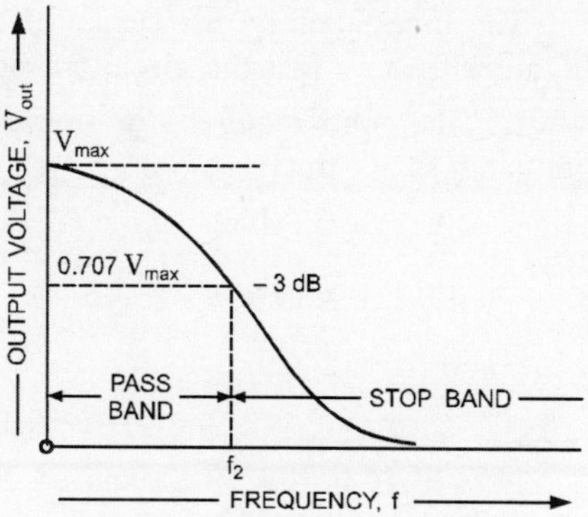

Fig. 9.15. ***Frequency Response Curve For Low-Pass R-C Circuit***

$$f_2 = \frac{1}{2\pi RC} \quad \text{...(9.22)}$$

Substituting $\dfrac{1}{2\pi RC} = f_2$ in Eq. (9.21), we have

$$\text{Gain } A = \frac{1}{1 + j\dfrac{f}{f_2}} \quad \text{...(9.23)}$$

Now magnitude of A and phase angle θ are given as

$$|A| = \frac{1}{\sqrt{1 + \left(\dfrac{f}{f_2}\right)^2}} \quad \text{...(9.24)}$$

$$\text{and phase angle } \theta = \text{Tan}^{-1} \frac{f}{f_2}$$

As $f \to 0$; $\quad \dfrac{f}{f_2} = 0$ and $|A| \to 1$

When $f = f_2$; $|A| = \dfrac{1}{\sqrt{2}} = 0.707$

f_2 is also known as high 3-dB frequency

$$f = f_2 \text{ also implies that } f = \frac{1}{2\pi RC} \text{ or } R = \frac{1}{2\pi fC} = X_C$$

So f_2 is the frequency at which capacitive reactance X_C is equal to the resistance of the circuit.

9.4.2. Step Input

At instant $t = 0^-$, since the input is zero, the output is also zero. At instant $t = 0$, the input suddenly becomes V. Since the voltage across a capacitor cannot change instantaneously, the output voltage remains zero. At $t = 0^+$, the input voltage remains unchanged at V, the capacitor starts charging exponentially to voltage V and the output is the same as the capacitor voltage. The response of the low-pass R-C circuit to a step input is shown in Fig. 9.16. The output voltage is given as

$$v_{out} = V(1 - e^{-t/CR}) \quad \text{...(9.25)}$$

The time taken by the capacitor to rise from 0.1 to 0.9 of its final value is called the *rise time*. It indicates how fast the circuit responds to a discontinuity in voltage.

Let the time required for output voltage to attain one-tenth of its final value be $t = t_1$. So at $t = t_1$

$$v_{out} = 0.1 \text{ V}$$

$$\text{and } 0.1 \text{ V} = \text{V}(1 - e^{-t_1/\text{CR}})$$

$$\text{or } e^{-t_1/\text{CR}} = 0.9$$

$$\text{or } t_1 = -\text{CR} \log_e 0.9 \simeq 0.1 \text{ CR}$$

Now if t_2 is the time required by the capacitor to attain 0.9 times its final value, then

$$0.9 \text{ V} = \text{V}(1 - e^{-t_2/\text{CR}})$$

$$\text{or } e^{-t_2/\text{CR}} = 0.1$$

$$\text{or } t_2 = -\text{CR} \log_e 0.1 = 2.3 \text{ CR}$$

So rise time, $t_r = t_2 - t_1 = 2.3 \text{ CR} - 0.1 \text{ CR}$

$$= 2.2 \text{ CR} = \frac{2.2}{2\pi f_2} = \frac{0.35}{f_2} \quad \text{...(9.26)} \quad \because f_2 = \frac{1}{2\pi \text{RC}}$$

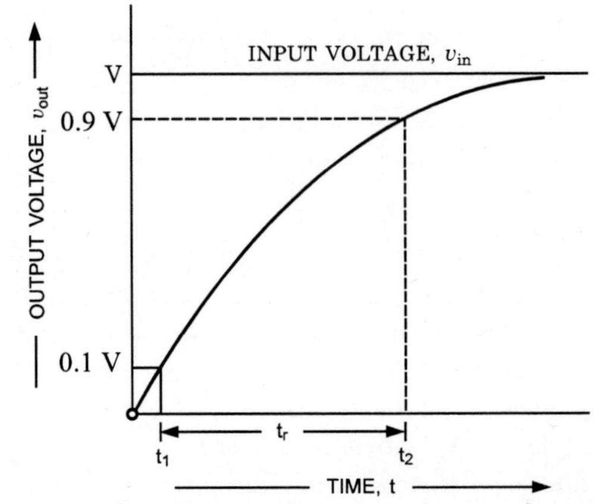

Fig. 9.16. ***Response of Low-Pass R-C Circuit To a Step Input***

Since bandwidth of a circuit, $\Delta f = f_2 - f_1$.

As the lower cutoff frequency f_1 for the circuit is zero, therefore, Bandwidth $\Delta f = f_2$.

Thus rise time is proportional to the time constant τ and inversely proportional to the bandwidth or upper 3-dB frequency.

9.4.3. Pulse Input

The response to a pulse, for times less than the pulse width t_p is the same as that for a step input because pulse signal is same as step input for $t < t_p$. However, at the end of the pulse, as the input becomes zero, the output also drops exponentially to zero. The reason is that capacitor voltage falls exponentially to zero, as the input becomes zero.

Thus output for pulse input is given as

$$v_{out} = \text{V}(1 - e^{-t/\text{CR}}) \text{ for } t < t_p.$$

At $t = t_p$, $\quad v_{out} = \text{V}(1 - e^{-t_p/\text{CR}}) = \text{V}_p \text{ (say)}$

At $t = t_p$, input voltage becomes zero but since the voltage across a capacitor cannot change instantaneously, output remains the same as it is at $t = t_p$. After that capacitor starts getting discharged through resistance R and voltage across it drops exponentially to zero. Hence output voltage for $t > t_p$.

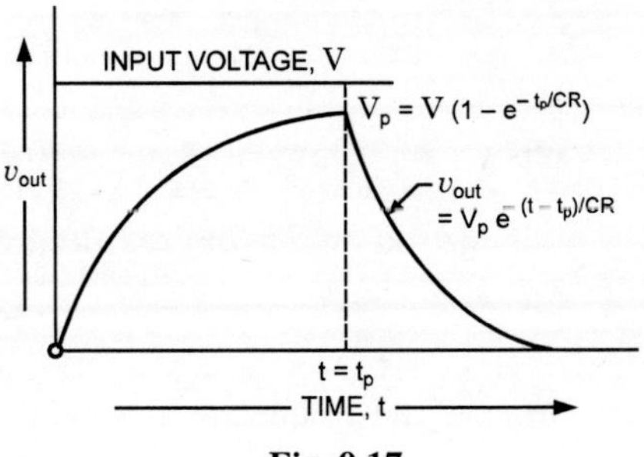

Fig. 9.17

$$v_{out} = \text{V}_p \, e^{-(t - t_p)/\text{CR}}$$

The above equation is discharging equation of capacitor delayed by time t_p. Hence output voltage must be decreasing towards zero for $t > t_p$. This is illustrated in Fig. 9.17.

It shows that output voltage will always extend beyond pulse width t_p. This is because the charge stored on capacitor during pulse cannot leak off instantaneously.

It is also observed from Fig. 9.17 that output shape is different from that of input *i.e.,* there is distortion. Distortion may be reduced by making rise time much smaller than pulse width t_p. For a circuit having f_2 equal to $\frac{1}{t_p}$, $t_r = 0.35\, t_p$. For this value of t_r output will be of the shape shown in Fig. 9.18, which is not exact but approximately reproduction of the input.

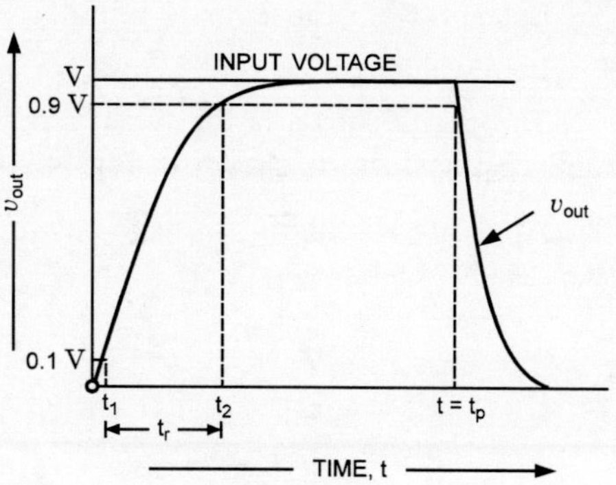

Fig. 9.18

As a thumb rule it can be said :

A pulse shape will be preserved if the 3-dB frequency is approximately equal to the reciprocal of the pulse width (PW).

Thus to pass a 0.2 μs pulse reasonably well needs a circuit with an upper cutoff frequency of the order of 5 MHz.

9.4.4. Square-Wave Input

Consider a periodic waveform whose instantaneous value is constant at V′ w.r.t. ground for a time T_1 and then changes abruptly to V″ for a time T_2. A reasonable reproduction of the input is obtained if the rise time t_r is small compared with the pulse width. The steady-state response is drawn in Fig. 9.19 (*b*). If the time constant RC is comparable with the period of input square-wave, the output will have the appearance, as illustrated in Fig. 9.19 (*c*). If the time constant is very large compared with the input wave period, the output will consist of exponential sections which are essentially linear as illustrated in Fig. 9.19 (*d*).

The input appears pulses of opposite polarity and added alternately. Like pulse output for time $0 < t < T_1$, for square-wave input also output is $v_{out} = V'(1 - e^{-t/RC})$ and at $t = T_1$ it is $V'(1 - e^{-T_1/RC})$ if capacitor is initially uncharged.

Since for succeeding cycles capacitor may remain charged to some value at the begining of the new cycle, so taking a generalized case, and then analysing the circuit shown in Fig. 9.14 for $0 < t < T_1$.

Initially charged capacitor is equivalent to an uncharged capacitor in series with a voltage source of magnitude and polarity equal to that initial charge. This is illustrated in Fig. 9.20.

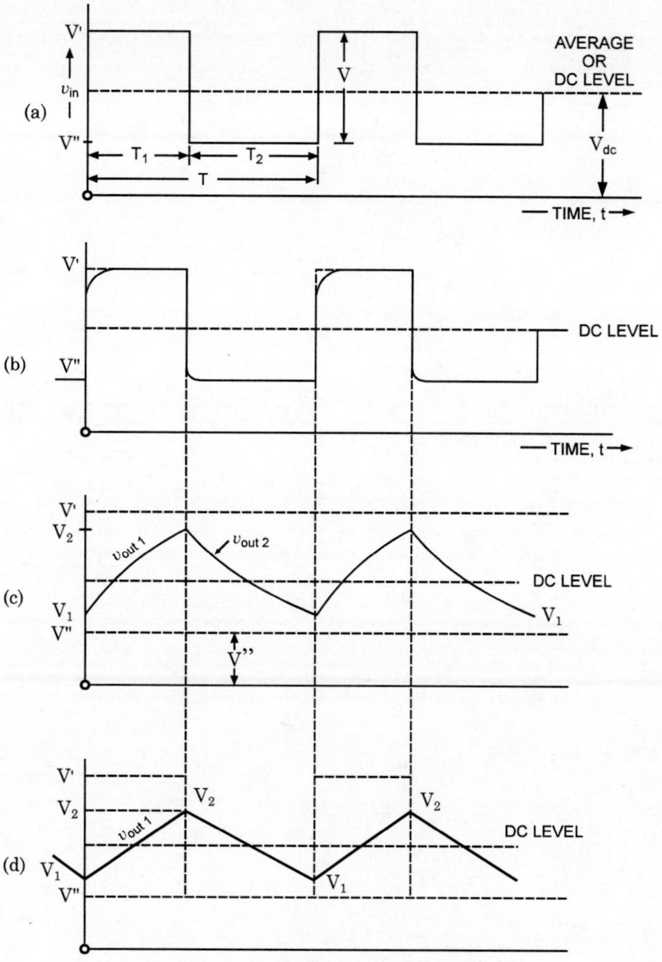

Fig. 9.19. ***Response To Square Wave Input***

Let at $t = 0^+$ capacitor is initially charged to a voltage V_1 and for $0 < t < T_1$

Input voltage, $v_{in} = V'$

The circuit shown in Fig. 9.14 is redrawn for time $0 < t < T_1$ and is shown in Fig. 9.20 (*a*) and its Laplace transform is shown in Fig. 9.20 (*b*).

Applying Kirchhoff's voltage law to the circuit, we have

$$\frac{V'}{s} - R\,i(s) - \frac{i(s)}{Cs} - \frac{V_1}{s} = 0$$

$$\text{or} \quad i(s) = \frac{V' - V_1}{s\left(R + \dfrac{1}{Cs}\right)}$$

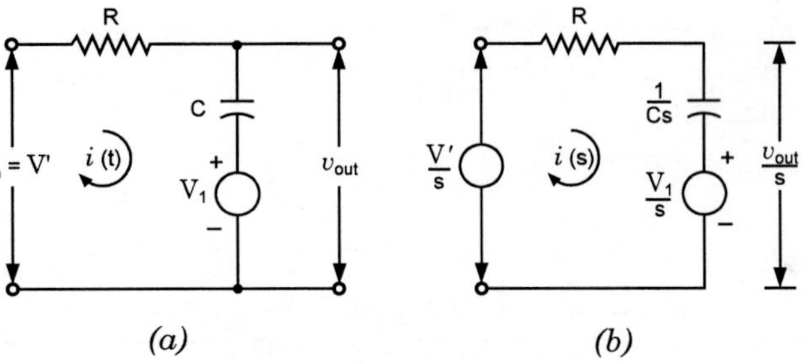

Fig. 9.20

$$v_{out}(s) = i(s)/Cs + \frac{V_1}{s}$$

$$= \frac{V' - V_1}{Cs^2\left(R + \dfrac{1}{Cs}\right)} + \frac{V_1}{s} = \frac{V' - V_1 + RCs\,V_1 + V_1}{s\,(RCs + 1)}$$

$$\text{or} \quad v_{out}(s) = \frac{V' + RCs\,V_1}{s\,(RCs + 1)} = \frac{V'}{s} + \frac{(V_1 - V')RC}{RCs + 1} = \frac{V'}{s} + \frac{V_1 - V'}{s + \dfrac{1}{RC}}$$

Taking inverse Laplace, we have

$$v_{out}(t) = V' + (V_1 - V')\,e^{-t/RC}$$

So output voltage for $0 < t < T_1$

$$v_{out\,1}(t) = V' + (V_1 - V')\,e^{-t/RC}$$

Similarly for $T_1 \le t \le T_2$, if initial voltage across capacitor is V_2 and input voltage is constant at V'' and output voltage = $V_{out\,2}$ (say)

$$v_{out\,2} = V'' + [(V_2 - V'')e^{-(t - T_1)/RC}]$$

Since at $t = T_1$. $v_{out\,1} = V_2$

and $t = T_1 + T_2$, $v_{out\,2} = V_1$

Then at $t = T_1$, $\quad v_{out1} = V_2 = V' + (V_1 - V')e^{-T_1/RC}$

$$\text{or } V_2 = V'(1 - e^{-T_1/RC}) + V_1 e^{-T_1/RC}$$

and at $t = T_1 + T_2$, $\quad v_{out\,2} = V_1 = V'' + [(V_2 - V'')\,e^{-(T_1 + T_2 - T_1)/RC}]$

$$\text{or } V_1 = V'' + (V_2 - V'')\,e^{(-T_2)/RC}$$

For symmetrical wave

$$V_1 = -V_2;\; V' = -V'' \text{ and } T_1 = T_2 = T/2$$

Therefore, $\quad V_1 = -V' + (-V_1 + V')e^{-T/2\,RC}$

$$= -V' - V_1 e^{-T/2\,RC} + V'\,e^{-T/2\,RC}$$

$$\text{or } V_1 (1 + e^{-T/2\,RC}) = V' (e^{-T/2\,RC} - 1)$$

$$\text{or } V_1 = \frac{V' (e^{-T/2\,RC} - 1)}{e^{-T/2\,RC} + 1}$$

If peak-to-peak voltage of input square-wave is V, then V' = V/2

$$\text{and } V_1 = \frac{V}{2} \frac{(e^{-T/2\,RC} - 1)}{(e^{-T/2\,RC} + 1)} \qquad \text{...(9.27)}$$

$$\text{and } V_2 = -\frac{V}{2} \frac{(e^{-T/2\,RC} - 1)}{(e^{-T/2\,RC} + 1)} = \frac{V}{2} \frac{(1 - e^{-T/2\,RC})}{(1 + e^{-T/2\,RC})} \qquad \text{...(9.28)}$$

9.4.5. Ramp Input

For an input of the form $v_{in} = \alpha t$, the voltage across the resistor is given by equation

$$v_R = \alpha\, RC(1 - e^{-t/RC}) \qquad \text{...Refer to Art. 9.2.5}$$

The voltage across the capacitor is

$$v_{out} = v_{in} - v_R = \alpha t - \alpha\, RC(1 - e^{-t/RC})$$

$$= \alpha\,(t - RC) + \alpha RC e^{-t/RC} \qquad \text{...(9.29)}$$

When the time constant τ = RC is very small in comparison to the total ramp time T, the ramp will be transmitted with minimum distortion. It is seen from Fig. 9.21 (*a*) that the output follows the input but is delayed by one time constant RC from the input (except near the origin) where there is a distortion. If the time constant is large in comparison to the sweep duration *i.e.,* RC >> T, the output will be highly distorted, as depicted in Fig. 9.21 (*b*).

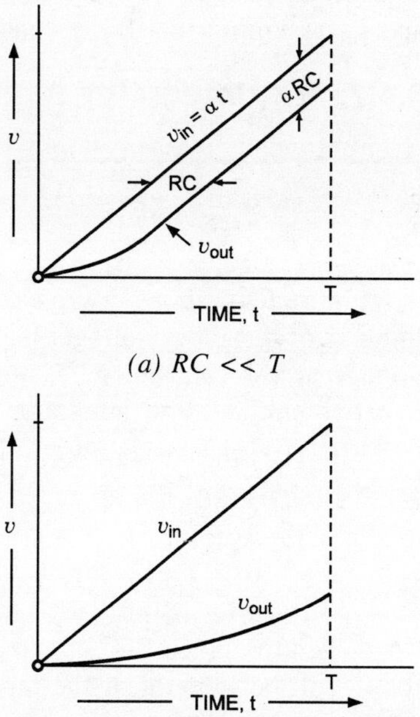

(*a*) *RC << T*

(*b*) *RC >> T*

Fig. 9.21. ***Response of a Low-Pass R-C Circuit For a Ramp Input***

The transmission error is defined as the difference between input and output divided by the input at t = T. For RC << T, we find transmission error

$$e_t \simeq \frac{RC}{T} = \frac{1}{2\pi f_2 T} \qquad \text{...(9.30)}$$

where f_2 is the upper 3-dB frequency.

By expanding the exponential term in Eq. (9.29), we find

$$v_{out} = \frac{\alpha t^2}{2\,RC} \qquad \text{...(9.31)}$$

This shows that a quadratic response is obtained for a linear input and thus the circuit acts as an integrator for RC >> T.

9.4.6. Exponential Input

For an input of the form $v_{in} = V\,(1 - e^{-t/\tau})$ the voltage across the resistor is given by Eq. (9.16) for $n \neq 1$. Hence the voltage across the capacitor is equal to the difference of input voltage less voltage across the resistor. Thus

$$v_{out} = v_c = v_{in} - v_R = V(1 - e^{-t/\tau}) - \frac{Vn}{n-1}(e^{-x/n} - e^{-x}) \quad \text{for } n \neq 1$$

$$\text{or } \frac{v_{out}}{V} = 1+\frac{1}{n-1}e^{-x} - \frac{n}{n-1}e^{-x/n} \quad ...(9.32)$$

$$\text{and } \frac{v_{out}}{V} = 1 - (1 + x)\, e^{-x} \quad ...(9.33) \text{ for } n = 1$$

The parameters x and n are defined by $x \equiv \frac{t}{\tau}$ and $n \equiv \frac{RC}{T}$.

Above Eqs. (9.32) and (9.33) give the voltage across the capacitor of a low-pass R-C circuit when excited by an exponential input of rise time $t_{r_1} = 2.2\tau$.

If an exponential of rise time t_{r_1} is passed through a low-pass R-C circuit with rise time $t_{r_2} = 2.2RC$, the rise time of the output waveform t_r will be given by an empirical relation, $t_r = 1.05\sqrt{t_{r_1}^2 + t_{r_2}^2}$. This is same as the rise time obtained when a step is applied to a cascade of two circuits of rise times t_{r_1} and t_{r_2} assuming that the second circuit does not load the first one.

9.4.7. R-C Integrator

A circuit that gives an output voltage directly proportional to the integral of its input is known as *integrating circuit.*

$$\textit{i.e.,}\ \text{Output} \propto \int \text{Input}$$

Figure 9.14 shows a typical integrating circuit. The output voltage across capacitor C will be the integral of the input voltage. It is important to note that merely an R-C series circuit will not provide voltage across capacitor C as an integral of the input. The conditions to be fulfilled for making an R-C series circuit as an integrator are given below.

(*i*) The time constant RC of the circuit should be very large as compared to the time period of the input signal *i.e.,* RC >> T.

(*ii*) The value of R should not be smaller than 10 times the capacitive reactance *i.e.,* $R \geq 10\, X_C$ or $\geq \frac{10}{2\pi f C}$.

When the given R-C circuit fulfills the above conditions, then the voltage appearing across capacitor C will be integral of the input. This is shown below.

Let v_{in} be the input alternating voltage and i be the resulting alternating current. Since R is very large in comparison of capacitive reactance X_C, the voltage drop across resistance R may be considered equal to input voltage *i.e.,*

$$v_{in} = v_R$$

$$\text{and } i = \frac{v_R}{R} = \frac{v_{in}}{R}$$

The charge on the capacitor at any instant

$$q = \int i\, dt = \frac{1}{R}\int v_{in}\, dt$$

$$\text{and output voltage, } v_{out} = \frac{q}{C} = \frac{1}{CR}\int v_{in}\, dt \quad ...(9.34)$$

$$\text{or } v_{out}\ \alpha \int v_{in}\, dt$$

$$\text{Hence output voltage } \alpha \int \text{Input}$$

The output waveform from an integrating circuit depends upon the time constant of the circuit and shape of the input wave. When the input wave is a square wave, output will be a triangular wave. This is because integral of a constant is a linear ramp. For a rectangular input wave, the output will be a saw-tooth wave.

Integrators find application in generation of triangular and ramp waveforms, indiscriminating between pulses of different lengths as in separation of horizontal and vertical synchronising pulses in television, in analog computers as integrator blocks.

9.5. APPLICATIONS OF LOW-PASS R-C CIRCUIT

R-C circuits with time constant larger than the time period of the input wave (*i.e.*, RC >> T) are used as bypass capacitors. Low-pass filters are also employed in generation of triangular and ramp waveforms. It also finds application in discrimination between pulses of different lengths.

Integrators are almost invariably preferred over differentiators in analog computer applications. The reasons are given below :

1. Since the gain of an integrator falls with the increase in frequency whereas the gain of the differentiator increases nominally linearly with frequency, it is easier to stabilize the former than the latter with respect to spurious oscillations.
2. An integrator is less sensitive to noise voltages than a differentiator because of its limited bandwidth.
3. The amplifier of a differentiator may get overloaded in case the input waveform changes very rapidly.
4. It is more convenient to introduce initial conditions in an integrator.

Example 9.4. A pulse generator has an output resistance of 1 kΩ and is connected to an oscilloscope of input capacitance of 25 pF. Find the fastest rise time which the oscilloscope can display.

Solution : Resistance, $R = 1\text{ k}\Omega = 1{,}000\ \Omega$

Capacitance, $C = 25\text{ pF} = 25 \times 10^{-12}\text{ F}$

Time constant, $\tau = RC = 1{,}000 \times 25 \times 10^{-12} = 25\text{ ns}$

Fastest rise time of oscilloscope $= 2.2\,\tau = 2.2 \times 25\text{ ns} = 55\text{ ns}$ **Ans.**

Example 9.5. A step generator of 100 Ω impedance applies a 10 V step of 2.2 ns rise time to an R-C series circuit. Voltage across the resistor is found to be a pulse of 1 V amplitude. The resistance R = 100 Ω. Determine the capacitance C.

Solution : Maximum value of output voltage,

$$V_{out\ max} = 1\text{ V}$$

$$R = 100\ \Omega$$

Rise time of step input, $t_r = 2.2\text{ ns}$

$$\text{Time constant of the exponential input, } \tau = \frac{t_r}{2.2} = \frac{2.2}{2.2} = 1\text{ ns}$$

$$V_{out\ max} = Vn^{1/1-n} = 1\text{ volt}$$

$$\text{or}\quad 10n^{1/1-n} = 1$$

$$\text{or}\quad n^{1/1-n} = 0.1$$

By trial and error n is found to be 0.14

$$\text{Capacitance, } C = \frac{n \times \tau}{R} = \frac{0.14 \times 1 \times 10^{-9}}{100} = 1.4 \times 10^{-12}\text{ F}$$

$$= 1.4\text{ pF} \quad \textbf{Ans.} \qquad \because\ n = \frac{RC}{\tau}$$

Example 9.6. In the circuit diagram shown in Fig. 9.22, determine (*a*) rise time of capacitor voltage (*b*) time for capacitor to charge to 63.2% of its maximum voltage and (*c*) time for capacitor to be fully charged.

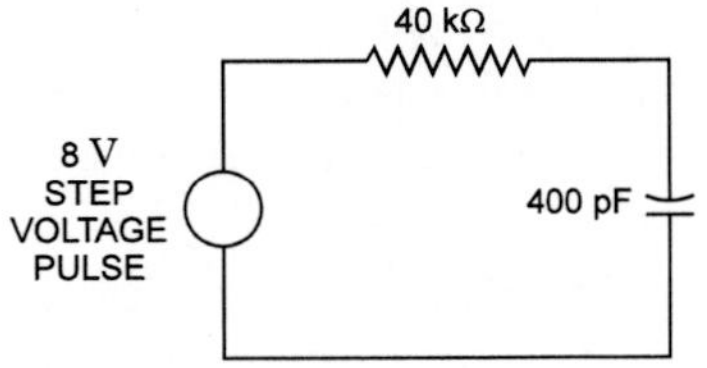

Fig. 9.22

Solution : The given circuit operates as a low-pass filter.

$$R = 40\ k\Omega = 40 \times 10^3\ \Omega$$

$$C = 400\ pF = 400 \times 10^{-12}\ F$$

(*a*) Rise time of capacitor voltage,

$$t_r = 2.2 \times R \times C = 2.2 \times 40 \times 10^3 \times 400 \times 10^{-12} = 35.2\ \mu s \quad \textbf{Ans.}$$

(*b*) Time for capacitor to charge to 63.2% of maximum voltage is τ and

$$\tau = RC = 40 \times 10^3 \times 400 \times 10^{-12} = 16\ \mu s \quad \textbf{Ans.}$$

(*c*) Time for capacitor to be fully charged (99% of maximum value)

$$= 5\tau = 5 \times 16\ \mu s = 80\ \mu s \quad \textbf{Ans.}$$

Example 9.7. A symmetrical wave whose average value is zero is applied to an R-C integrator. The time constant of the circuit is equal to half the period of the square wave. Determine the peak-to-peak value of the output wave if the peak-to-peak value of the input wave is 3 V.

Solution : The time constant (RC) of the circuit is comparable to the period of the input waveform which is symmetrical, the output must also be symmetrical with respect to zero level *i.e.*, the positive and negative peaks of the output must be equal in magnitude but opposite in polarity.

$$v_{out} = \frac{V_{in}}{2}\left[\frac{1 - e^{\frac{-T}{2RC}}}{1 + e^{\frac{-T}{2RC}}}\right] = \frac{3}{2} \times \frac{1 - e^{-1}}{1 + e^{-1}} = 0.693\ V$$

It means peak-to-peak value of the output wave is from –0.693 V to +0.693 V **Ans.**

The input step waveform is shown in dotted lines whereas the output waveform is shown in solid lines in Fig. 9.23.

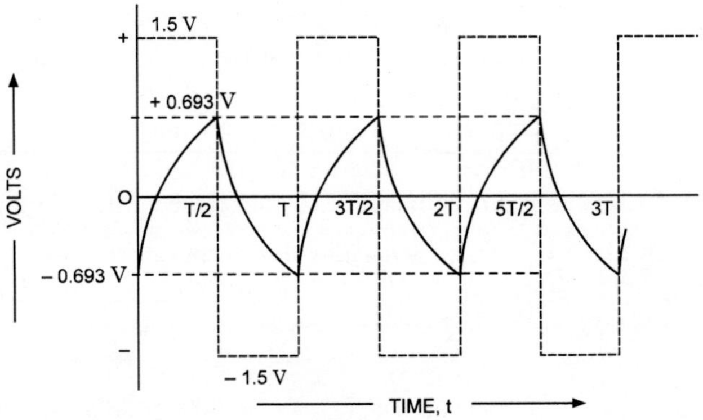

Fig. 9.23

HIGHLIGHTS

1. Linear waveshaping involves passage of signal through linear systems such as R-C, R-L, R-L-C circuits and the operations involved are linear operations such as integration, differentiation, summation, filtering etc. On the other hand, nonlinear waveshaping involves passage of signal through a nonlinear system. Nonlinear waveshaping permits transformation of analog signal to digital signal and vice versa. The clipping operation and rejection of negative peaks by rectification are nonlinear operations.
2. A high-pass circuit blocks and attenuates all low-frequency signals but allows high-frequency signals to pass through it. The lower cutoff frequency of a high-pass circuit is finite and is given by $f_1 = \frac{1}{2\pi RC}$ but its upper cutoff frequency is ∞, therefore, its bandwidth is infinite. The high-pass RC circuit is referred to as capacitive coupling circuit. The capacitor in the high-pass R-C circuit blocks the dc component of the input from going to output and is, therefore, known as the blocking capacitor.
3. The output of a high-pass R-C circuit excited by a square input exhibits a tilt when the time constant of the circuit is high.

   Fractional tilt is defined as the ratio of pulse width to the time constant RC.

4. A high-pass R-C circuit acts as a differentiator if (*i*) the time constant of the circuit is much smaller than the time period of the input signal *i.e.*, $\tau << T$, and (*ii*) the value of $X_C \left(i.e. \dfrac{1}{2\pi fC} \right)$ is much larger than 10R *i.e.*, $X_C >> 10R$.

5. A low-pass circuit passes low-frequency signals through it but attenuates the high-frequency signals. The lower cutoff frequency of a low-pass circuit is zero and its upper cutoff frequency is equal to its bandwidth and is given by $f_2 = \dfrac{1}{2\pi RC}$.

6. A low-pass R-C circuit acts as an integrator if (*i*) the time constant of the circuit is very large as compared to the time period of the input signal *i.e.*, $\tau >> T$. (*ii*) the value of R is not smaller than 10 times the capacitive reactance *i.e.*, $R \geq 10X_C$ or $\geq \dfrac{10}{2\pi fC}$.

7. R-C circuits with time constant larger than time period of the input signal are used as bypass capacitors. Low-pass filter circuits are also used in generation of triangular and ramp waveforms. It also finds application in discrimination between pulses of different lengths.

8. Integrators are almost invariably preferred over differentiators in analog computer applications.

EXERCISES

1. What is meant by linear waveshaping ? Give some examples of linear waveshaping circuits.
2. What do you mean by a high-pass-circuit ? Define lower cutoff frequency of a high-pass circuit. Derive an expression for the lower cutoff frequency of a high-pass circuit.
3. Describe with circuit diagram response of a high-pass R-C circuit to a sinusoidal voltage input.
4. Describe with circuit diagram response of a high-pass R-C circuit to a step voltage input.
5. Obtain the response of a high-pass circuit to ramp and square wave input. Discuss the effect of RC on the response.
6. If RC time constant of a high-pass circuit filter is made increasingly smaller in comparison of duration of the input waveform, is width of the output pulse increased or decreased ? Explain your answer.
7. What is differentiator ? How does it operate ? Show that the output of a differentiator circuit is derivative of the input. What assumptions are made in the derivation ?
8. What is meant by a low-pass circuit ? What is the upper cutoff frequency of a low-pass R-C circuit. Derive an expression for the upper cutoff frequency of a low-pass circuit.
9. Consider an R-C circuit shown in Fig. 1. A symmetrical square wave shown in Fig. 2 is applied as input to this circuit. Obtain an expression for peak-to-peak value of the output voltage. What is peak-to-peak voltage for τ = RC ?
10. What is low-pass R-C circuit ? Derive an expression of output voltage for square wave input and draw input-output characteristics of this circuit.
11. Under what conditions does an R-C circuit act as a differentiator or as an integrator. Draw and explain the circuit diagram of an R-C integrator.
12. Integrators are almost invariably preferred over differentiators in analog computer applications. Explain.

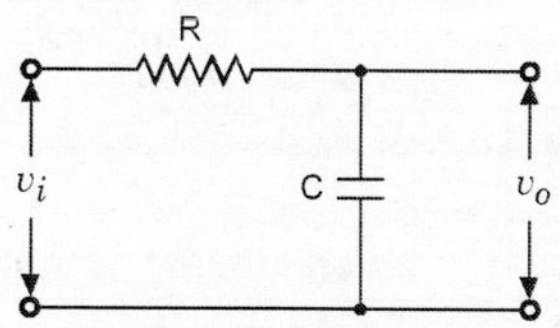

Fig. 1. *An R-C Circuit*

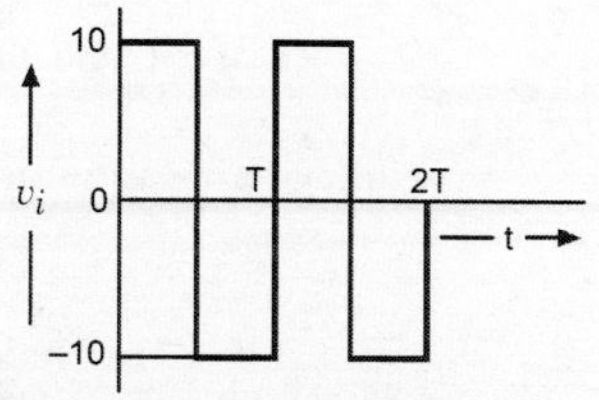

Fig. 2. *A Symmetrical Square Wave of Voltage v_i*

SHORT ANSWER TYPE QUESTIONS WITH ANSWERS

Q. 1. What is waveforming ?

Ans. Waveforming is a process of generating new waveshape from another waveform using some network.

Q. 2. Define linear waveshaping and give examples.

Ans. Linear waveshaping involves passage of signal through linear systems such as R-C, R-L, R-L-C circuits and the operations involved are linear operations such as integration, differentiation, summation, filtering etc.

Q. 3. Give some examples of linear and nonlinear waveshaping circuits.

Ans. Linear waveshaping circuits — R-C, R-L, R-L-C circuits.

Nonlinear waveshaping circuits — clipping and clamping circuits.

Q. 4. What is difference between linear and nonlinear waveshaping circuits ?

Ans. Linear waveshaping involves passage of signal through linear systems such as R-C, R-L, R-L-C circuits and the operations involved are linear operations such as integration, differentiation, summation, filtering etc. On the other hand, nonlinear waveshaping involves passage of signal through a nonlinear system. Nonlinear waveshaping permits transformation of analog signal to digital signal and vice versa. The clipping operation and rejection of negative peaks by rectification are nonlinear operations.

Q. 5. What is meant by high-pass circuit ?

Ans. The circuit that blocks and attenuates low frequencies but allow high-frequency signals to pass through it, is called the high-pass circuit.

Q. 6. In a high-pass R-C circuit why capacitor is named as blocking capacitor ?

Ans. Because of its blocking property for dc or low frequency input signals the capacitor acts like an open circuit and blocks the signal and so the capacitor is called the blocking capacitor.

Q. 7. What is meant by lower 3-dB frequency of a high-pass circuit ?

Ans. Lower 3-dB frequency of a high-pass circuit is the frequency at which capacitive reactance is equal to resistance of the circuit or the frequency at which the output voltage is 0.707 times the maximum value of output voltage.

Q. 8. What is meant by a fractional tilt ?

Ans. Fractional tilt is defined as the ratio of pulse width (PW) to the time constant τ(RC).

Q. 9. What is main reason of applying a step voltage input in a high-pass R-C circuit ? **[P.T.U. December 2008]**

Ans. The main reason of applying a step voltage input in a high-pass R-C circuit is that for times less than the pulse width t_p the output voltage is the same as that for a step input because pulse signal is same as step input for $t < t_p$. However, at the end of pulse, as the input becomes zero, the output also drops exponentially to zero. The reason is that capacitor voltage falls exponentially to zero, as the input becomes zero.

Q. 10. What is meant by a differentiating circuit ?

Ans. A circuit that gives an output voltage proportional to the derivative of its input, is known as a differentiating circuit.

Q. 11. Why does a high-pass filter act as a differentiator ?

Ans. If the time constant of R-C circuit (Fig. 9.1) is much smaller than the time period of the input signal *i.e.,* RC << T and the value of $X_C \left(i.e. \dfrac{1}{2\pi fC} \right)$ is not smaller than 10 times of R *i.e.,* $X_C \geq 10R$, the voltage appearing across R will be derivative of the input

$$i.e., \quad \text{Output voltage} \;\alpha\; \frac{d}{dt}(\text{input})$$

So a high-pass filter with RC << T and $X_C \geq 10R$ acts as a differentiator.

Q. 12. What are the applications of high-pass R-C circuits ?

Ans. R-C circuit with RC >> T is employed in R-C coupling of amplifiers where distortion and differentiation of waveform is to be avoided. R-C circuit with RC << T is employed to generate pips for triggering electronic circuits such as multivibrators, flip-flops etc.

Q. 13. What is meant by low-pass circuit ?

Ans. Low-pass circuit is that which passes low-frequency signals through it but blocks the high-frequency signals.

Q. 14. What is integrator ?

Ans. An integrator is that circuit which gives an output voltage directly proportional to the integral of its input.

Q. 15. What are the applications of a low-pass R-C circuit ?

Ans. R-C circuits with time constant larger than time period of the input signal are used as bypass capacitors. Low-pass filter circuits are also used in generation of triangular and ramp waveforms. It also finds application in discrimination between pulses of different lengths.

Q. 16. Why does a low-pass filter act as an integrator ?

Ans. If the time constant of the R-C series circuit (Fig. 9.14) is much larger than the time period of the input signal *i.e.,* RC >> T and the value of R is not smaller than 10 times of capacitive reactance *i.e.,* $R \geq 10X_C$ or $\geq \frac{10}{2\pi fC}$, the output voltage across capacitor C will be the integral of the input voltage *i.e.,* output voltage $\propto \int \text{Input}$

So a low-pass filter with RC >> T and $R \geq 10X_C$ acts as an integrator.

PROBLEMS

1. A step generator of 50 Ω impedance applies a 5 V step of 3.3 ns rise time to an R-C series circuit. Voltage across the resistor is found to be a pulse of 0.5 V amplitude. Determine the capacitance C if R = 50 Ω. [**Ans.** 4.2 pF]
2. A pulse generator has an output resistance of 2 kΩ and is connected to an oscilloscope of input capacitance of 50 pF. Determine the fastest rise time which the oscilloscope can display. [**Ans.** 220 ns]
3. Design a differentiating network to have 8 V trigger pulse with a square-wave input of 10 V amplitude, 50 per cent duty cycle and 100 Hz repetition rate. Source and load resistances may be taken equal to 50 Ω and 1,000 Ω respectively. [**Ans.** τ = 0.1 ms; R′ = 250 Ω; C = 0.5 μF]
4. Rectangular pulses of duty cycle δ and repetition period 10 ms are impressed on a series low-pass R-C circuit having R = 10 kΩ and C = 100 μF. Determine (*i*) minimum pulse duration of pulses which the circuit can generate without appreciable distortion and (*ii*) 3-dB bandwidth. [**Ans.** (*i*) 6.2 μs, (*ii*) 160 kHz]
5. A 1 kHz square-wave output from an amplifier has rise time of 250 ns and tilt 10 per cent. Determine upper and lower cutoff frequencies. [**Ans.** 1.4 MHz; 31.84 Hz]

❑❑❑

CHAPTER

10

Nonlinear Waveshaping

10.1. INTRODUCTION

INSIDE THIS CHAPTER

As already explained in chapter 9, waveshaping methods can be classified into two categories viz., linear and nonlinear waveshaping. Linear waveshaping makes use of resistors, reactive elements like capacitors and inductors and operational amplifiers (op-amps) operating in linear region. A typical example of linear active waveshaping is an integrator using an op-amp. Nonlinear waveshaping may use not only linear passive elements such as resistors, inductors and capacitors, but also nonlinear elements such as diode, zener diode and transistors operating in nonlinear region.

The circuits used for nonlinear waveshaping are of two types viz. clipping circuits and clamping circuits. The output of a clipping circuit appears as if a portion of the input signal were clipped off but a clamper circuit simply clamps (*i.e.,* lifts up or down) the input signal to a different dc level.

10.2. CLIPPING CIRCUITS

A waveshaping circuit which controls the shape of the output waveform by removing or clipping a portion of the applied wave is known as *clipping circuit*. A half-wave rectifier is an example of the simplest form of clipper as it passes only the positive (or negative) half cycle of an alternating waveform and clips off the other half cycle. Clipping circuits find extensive use in radars, digital computers and other electronic systems for removing unwanted portions of the input signal voltages above or below a specified level. Another application is in radio receivers for communication circuits where noise pulses, that rise well above the signal amplitude, are clipped down to the desired level. Clipping circuits are also referred to as *voltage limiters*, *amplitude selectors* or *slicers*.

Clipping circuit consists of nonlinear and linear devices. The nonlinear devices generally used for clipping are diodes and transistors. According to nonlinear devices used, clippers may be classified as diode clippers and transistor clippers.

According to biasing, the clippers may be classified as unbiased clippers and biased clippers.

According to configuration used the clippers may be (*i*) series diode clippers (*ii*) parallel or shunt diode clippers (*iii*) a series combination of diode, resistor and reference supply (*iv*) Multidiode clippers consisting of several diodes, resistors and reference voltages (*v*) two emitter-coupled transistors operating as an overdriven difference amplifier.

According to level of clipping the clippers may be (*i*) positive clippers (*ii*) negative clippers (*iii*) biased clippers and (*iv*) combination clippers.

10.3. DIODE CLIPPERS

For a clipping circuit at least two components—an ideal diode and resistor are required and sometimes a dc battery is also employed for fixing the clipping level. The diode acts as a closed switch when forward biased and an open switch when reverse biased. The input waveform can be clipped at different levels by simply changing the voltage of the battery and by interchanging the positions of the various elements.

Depending on the orientation of the diode, the positive or negative region of the input signal is "clipped" off and accordingly the diode clippers may be positive or negative clippers.

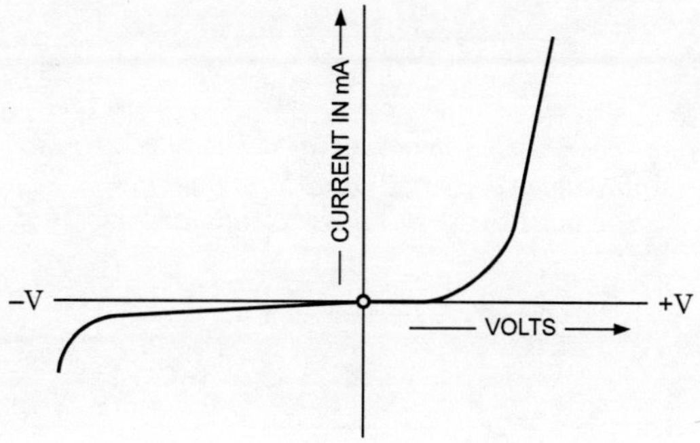

Fig. 10.1 ***V-I Characteristics of a Practical Diode***

There are two general categories of clippers: series and parallel (or shunt). The series configuration is defined as one where diode is in series with the load, while the shunt clipper has the diode in a branch parallel to the load.

V-I characteristics of a practical diode are depicted in Fig. 10.1. V-I characteristics of an idealized diode (approximated by a curve which is piecewise linear and continuous) are shown in Figs. 10.2 (*a*), (*b*) and (*c*) respectively. The idealized diode characteristic exhibits a discontinuity in slope at the voltage V_γ, and this point of slope discontinuity is called a *break point*. The break point occurs at $V_\gamma = 0.2$ V for germanium and $V_\gamma = 0.6$ V for silicon. For a backward diode, V_γ is almost zero.

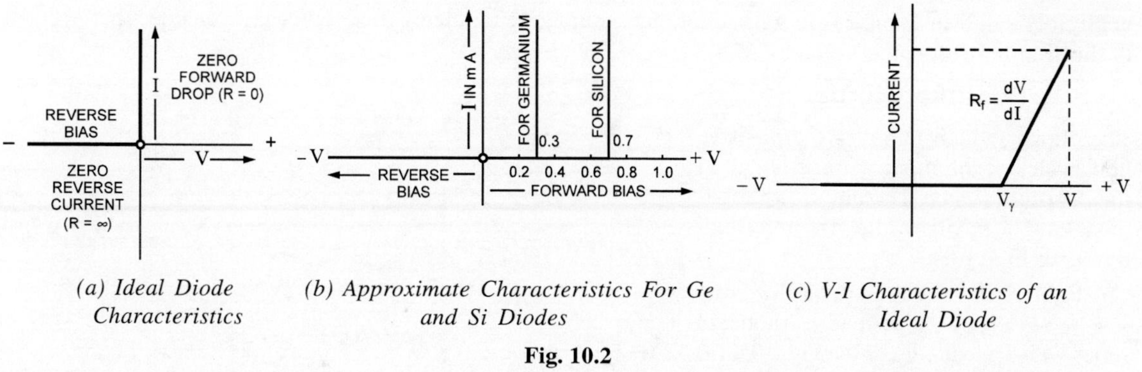

(a) Ideal Diode Characteristics *(b) Approximate Characteristics For Ge and Si Diodes* *(c) V-I Characteristics of an Ideal Diode*

Fig. 10.2

Using the piecewise linear model of a diode, the clipping circuit given in Fig. 10.3 (*b*) has the transmission characteristic depicted in Fig. 10.3 (*c*). The transmission characteristic — a plot of the output voltage v_{out} against the input voltage v_{in} also exhibits piecewise linear discontinuity. A break point occurs at the voltage $V + V_\gamma$. To the left of the break point (for $v_{in} < V + V_\gamma$), the diode is reverse biased (off). In this region, the input signal v_{in} may be transmitted directly to the output, as there is no load across the output to cause a drop across the series resistor R. To the right of the break point (for $v_{in} > V + V_\gamma$), the diode is forward biased (on) and increments Δv_{in} in the input are attenuated and appear at the output as increment $\Delta v_{out} = \Delta v_{in} \times \dfrac{R_f}{R_f + R}$, in which R_f is the diode forward resistance.

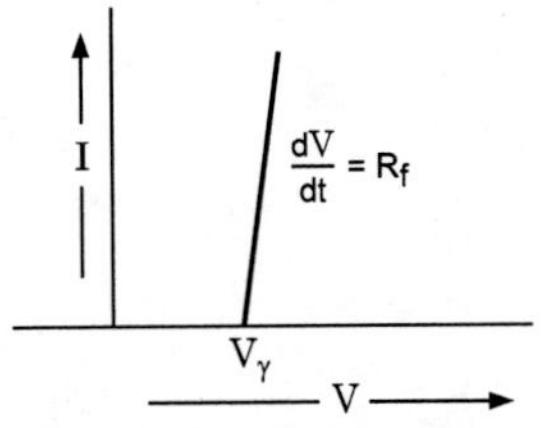

(a) V-I Characteristic of a Diode Approximated By a Broken Straight Line Having Piecewise Linear Continuity

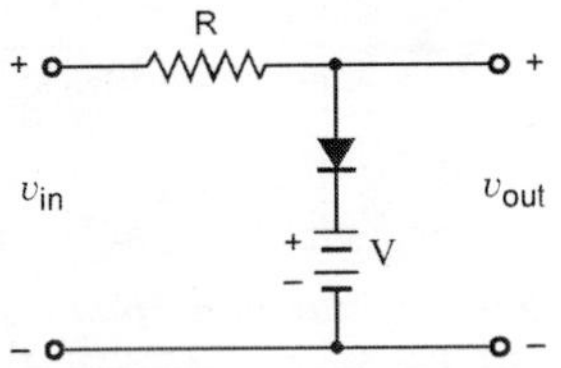

(b) Diode Clipping Circuit

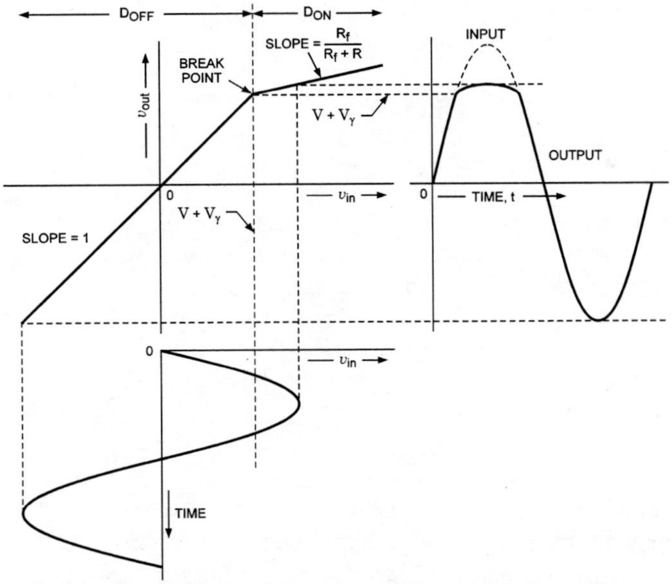

(c) Piecewise Linear Transmission Characteristic of The Circuit—A Sinusoidal Input and Clipped Output

Fig. 10.3. ***Diode Clipping***

Figure 10.3 (*c*) depicts a sinusoidal input signal of amplitude large enough so that the signal makes excursions past the break point. The corresponding output exhibits a suppression of positive peak of the signal. In case forward resistance R_f is much smaller than series resistance $R (R_f << R)$, this suppression will be very pronounced, and the positive output excursion will be sharply limited at the voltage $V + V_\gamma$. The output will appear as if the positive peak had been 'clipped off' or 'sliced off'. Usually V_γ is negligibly small in comparison with reference voltage V. In such a case one may consider that V itself is the limiting reference voltage.

10.3.1. Positive Clipper

The clipper which removes the positive half cycles of the input voltage is called the *positive clipper*. The circuit arrangements for a positive clippers are illustrated in Fig. 10.4.

Figure 10.4 (*a*) illustrates the *positive series clipper circuit* (*i.e.,* diode in series with the load). From the Fig. 10.4 (*a*) it is seen that while the input is positive, diode D is reverse biased and so the output remains at zero *i.e.,* positive half cycle is clipped off. During the negative half cycle of the input, the diode is forward biased and so the negative half cycle appears across the output.

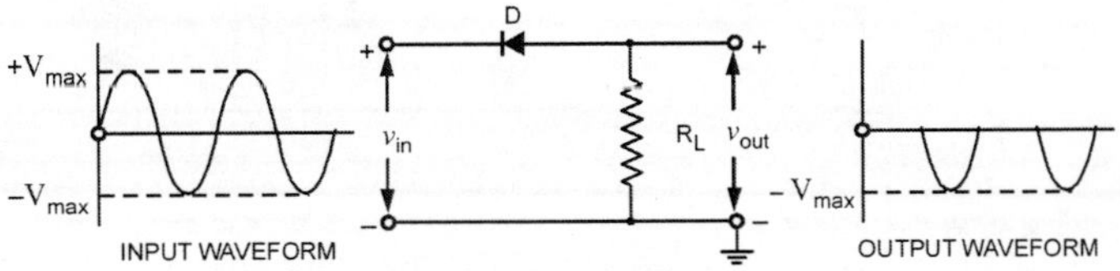

(a) Positive Series Clipper

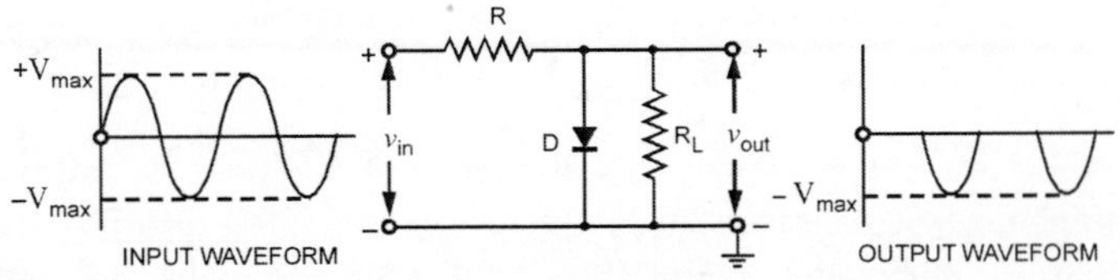

(b) Positive Shunt Clipper

Fig. 10.4

Figure 10.4 (*b*) illustrates the *positive shunt clipper circuit* (*i.e.,* diode in parallel with the load). From the Fig. 10.4 (*b*) it is seen that while input side is positive, the diode D is forward biased and conducts heavily (*i.e.,* diode acts as a closed switch). So the voltage drop across the diode or across the load resistance R_L is zero. Thus output voltage during the positive half cycles is zero, as shown in the output waveform. During the negative half cycles of the input signal voltage, the diode D is reverse biased and behaves as an open switch. Consequently the entire input voltage appears across the diode or across the load resistance R_L if R is much smaller than R_L.

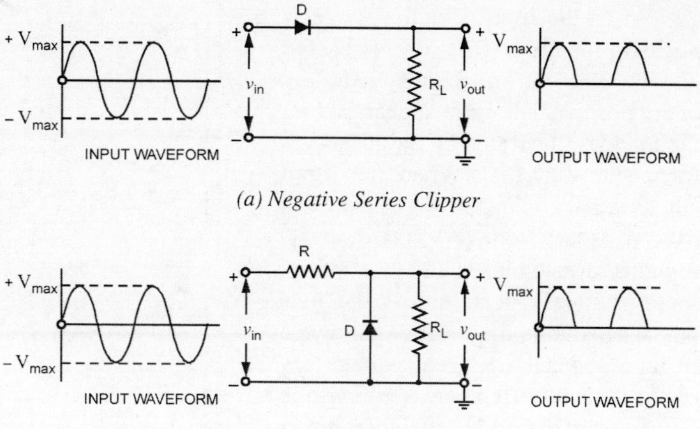

(*a*) *Negative Series Clipper*

(*b*) *Negative Shunt Clipper*

Fig. 10.5

Actually the circuit behaves as a voltage divider with an output voltage of $\frac{-R_L}{R+R_L} V_{max} \simeq -V_{max}$ when $R_L >> R$.

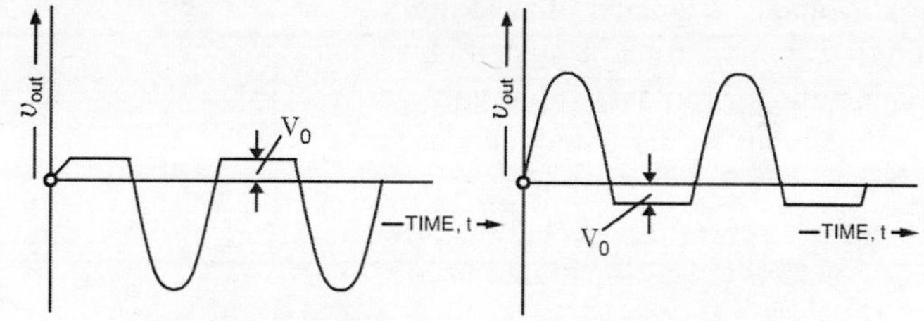

(*a*) *Output Waveform For Positive Clipper*

(*b*) *Output Waveform For Negative Clipper*

Fig. 10.6

Note : If the diode in Figs. 10.4 (*a*) and 10.4 (*b*) is reconnected with reversed polarity, the circuits will become for a negative series clipper and negative shunt clipper respectively.

The negative series and negative shunt clippers are shown in Figs. 10.5 (*a*) and 10.5 (*b*) respectively. The negative clipper is one which removes all the negative half cycles of the input voltage.

In the above discussion, the diode is considered to be ideal one. If second approximation for diode is considered the barrier potential, V_0 (0.7 V for silicon and 0.3 V for Ge) of diode, will be taken into account. Then the output waveforms for positive and negative clippers will be of the shape shown in Fig. 10.6.

10.3.2. Biased Clipper

Sometimes it is desired to remove a small portion of positive or negative half cycles of the signal voltage. Biased clippers are employed for this purpose. The circuit diagram for a *biased negative clipper* (*i.e.,* for removing a small portion of each negative half cycle) is illustrated in Fig. 10.7 (*a*). The action of the circuit is explained below.

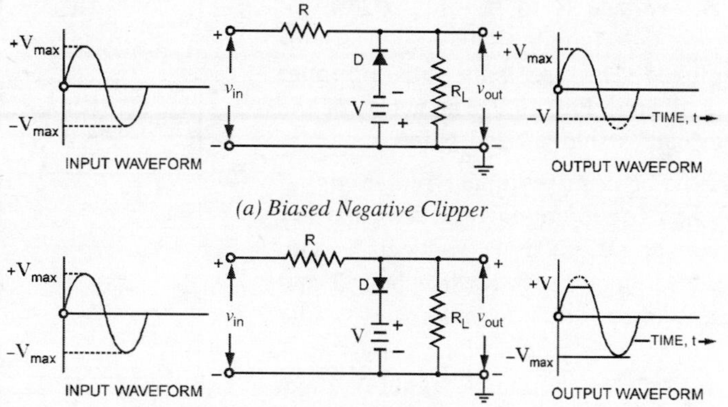

(*a*) *Biased Negative Clipper*

(*b*) *Biased Positive Clipper*

Fig. 10.7

When the input signal voltage is *positive*, the diode D is reverse biased and behaves as an open switch, the entire positive half cycle appears across the load, as illustrated by output waveform [Fig. 10.7 (*a*)]. When the input signal voltage is negative but does not exceed battery voltage V, the diode D remains reverse biased and most of the input voltage appears across the output. When during the negative half cycle of input signal, the signal voltage exceeds the battery voltage V, the diode D is forward biased *i.e.,* conducts heavily. The output voltage is equal to –V and stays at –V as long as the input signal voltage is greater than battery voltage V in magnitude. Thus a biased negative clipper removes input voltage when the input signal voltage exceeds the battery voltage. Clipping can be changed by reversing the battery and diode connections, as illustrated in Fig. 10.7 (*b*).

Some of other biased clipper circuits are given in Fig. 10.8. While drawing the waveshape of the output basic principles discussed above are followed. The diode has been considered as an ideal one.

10.3.3. Combination Clipper

When a portion of both positive and negative of each half cycle of the input voltage is to be clipped (or removed), combination clipper is employed. The circuit for such a clipper is given in Fig. 10.9. The action of the circuit is summarized below.

For positive input voltage signal when input voltage exceeds battery voltage $+V_1$, diode D_1 conducts heavily while diode D_2 is reverse biased and so voltage $+V_1$ appears across the output. This output voltage $+V_1$ stays as long as the input signal voltage exceeds $+V_1$. On the other hand for the negative input voltage signal, the diode D_1 remains reverse biased and

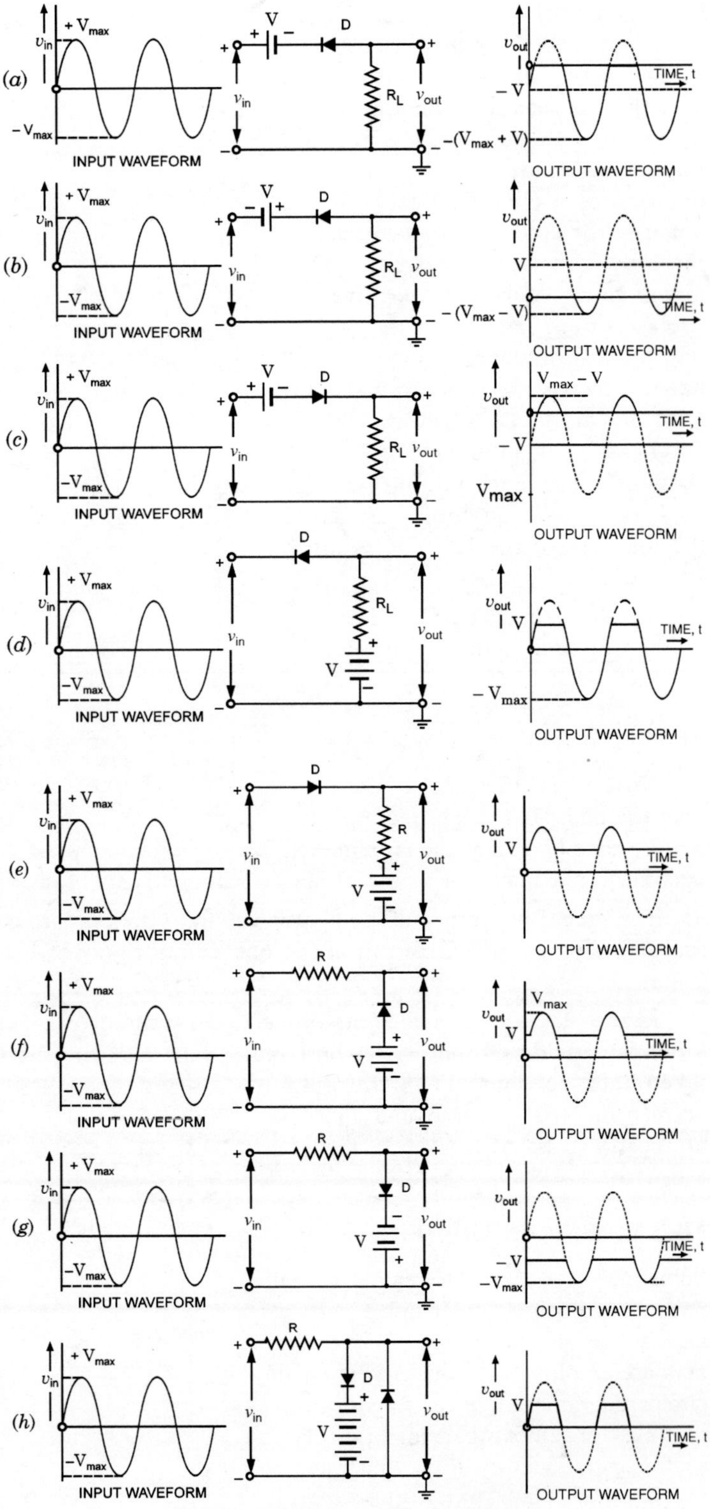

Fig. 10.8. ***Different Clipping Circuits***

diode D_2 conducts heavily only when input voltage exceeds battery voltage V_2 in magnitude.

Thus during the negative half cycle the output stays at $-V_2$ so long as the input signal voltage is greater than $-V_2$.

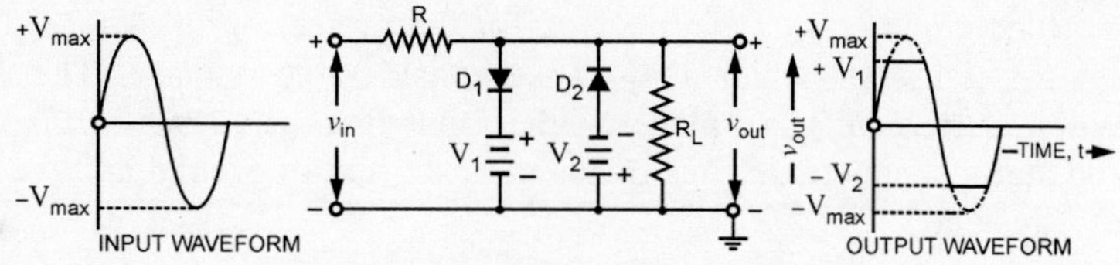

Fig. 10.9. ***Combination Clipper***

10.3.4. Zener Diode as a Peak Clipper

Zener diodes can be used for clipping off the input waveform in waveshaping circuits, as illustrated in Fig. 10.10.

Two similar zener diodes D_1 and D_2 are shunted across the input sinusoidal voltage of peak value of V_{max} volts. Let the zener voltage of each of zener diodes be V_Z volts (V_Z being lesser than V_{max}).

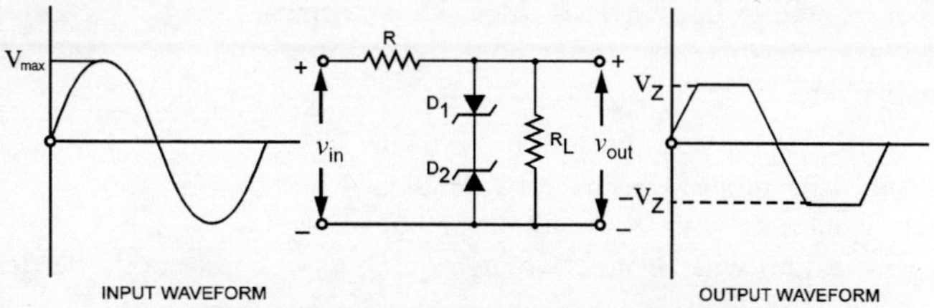

Fig. 10.10. ***Zener Diode as Peak Clipper***

We have seen that a zener diode acts like a 'short' (or very low resistance) when forward biased and 'open' (or very high resistance) when reverse biased till it goes into breakdown at V_Z. The action of the circuit is explained below.

For positive input half cycle diode D_1 is forward biased and acts like a 'short' while diode D_2 acts like an open (being reverse biased) up to V_Z. Thereafter diode D_2 goes into breakdown and holds the output voltage constant at V_Z till input voltage falls below V_Z in the later part of the positive half cycle. This is because during this duration of time both of the diodes act as 'short'. When the supply voltage falls below V_Z, diode D_2 comes out of the breakdown and again acts as an open across the load circuit. During the negative input half cycle, the actions of diodes D_1 and D_2 are reversed. The result is that the output voltage waveform is clipped off on both peaks, as illustrated in Fig. 10.10.

10.3.5. Drawbacks of Series and Shunt Diode Clippers

In series clippers, when diode is in 'off' position, there should be no transmission of input signal to output. But in case of high-frequency signals transmission occurs through diode capacitance which is undesirable. This is the drawback of using diode as a series element in such clippers.

In shunt clippers, when diode is in the 'off' condition, transmission of input signal should take place to output. But in case of high-frequency input signals, diode capacitance affects the circuit operation adversely and the signal gets attenuated (*i.e.*, it passes through diode capacitance to ground).

10.3.6. Applications of Diode Clippers

There are numerous applications and it is beyond the scope of the book to discuss or enumerate all of them. They find wide application in television receivers for separating synchronising signals from composite picture signals and also in television transmitters at the time of processing the picture signals. They are also employed for generation of different waveforms, such as trapezoidal, square or rectangular waves. Series clippers are employed as noise limiters in FM transmitters by clipping excessive noise peaks above a specified level.

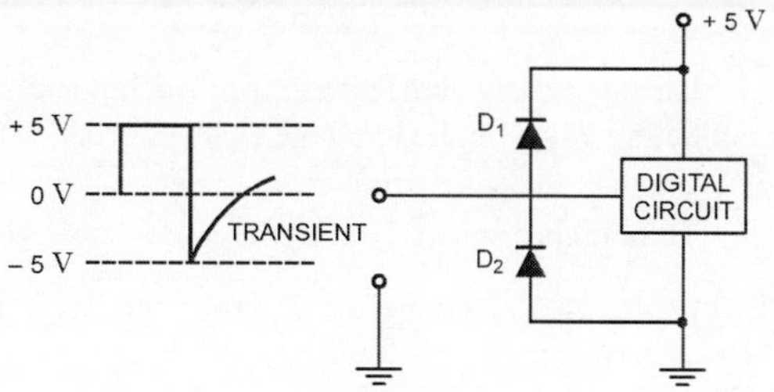

Fig. 10.11

The diode clipper can be used for the protection of different types of circuits *e.g.* a digital circuit against transients which may cause considerable damage. The arrangement is shown in Fig. 10.11. When a transient takes place on the input line, as shown in Fig. 10.11, diode D_2 gets forward biased and starts conducting. Thus transient is shorted to ground and the circuit is protected from damage due to transient.

Example 10.1. Determine the peak output voltage for a positive series clipper circuit shown in Fig. 10.12. The input signal is sinusoidal of peak value 15 V and the barrier voltage for silicon diode is of 0.7 V.

Solution : During the positive half cycle of the input signal, the diode D remains reverse biased and so does not conduct. Thus output voltage remains zero throughout the positive half cycle.

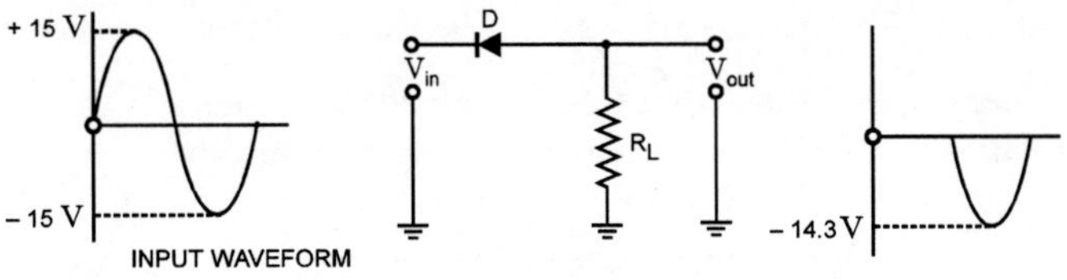

Fig. 10.12

During the negative half cycle of the input signal, the diode remains forward biased and so conducts heavily. As a result the output voltage (peak value) will be input voltage (peak value) less diode barrier voltage. *i.e.,*

$$V_{out\,(peak)} = V_{in\,(peak)} - V_0 = 15 - 0.7 = 14.3 \text{ V} \quad \textbf{Ans.}$$

Example 10.2. Determine the peak output voltage for a negative shunt clipper circuit shown in Fig. 10.13. The input signal is sinusoidal of peak value 12 V, the barrier voltage for silicon diode is 0.7 V, series resistance is of 500 Ω and load resistance is of 2.5 kΩ.

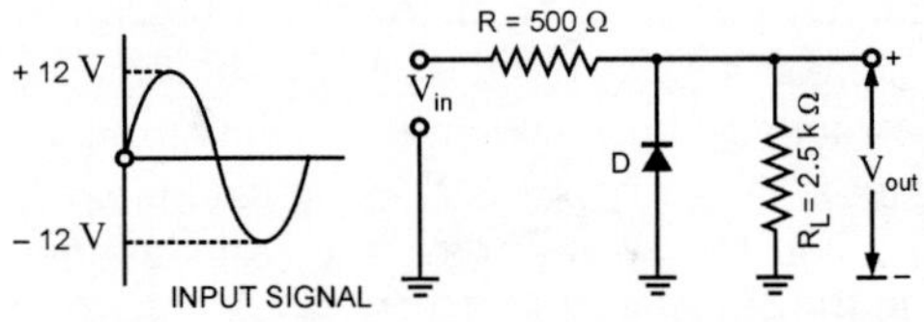

Fig. 10.13

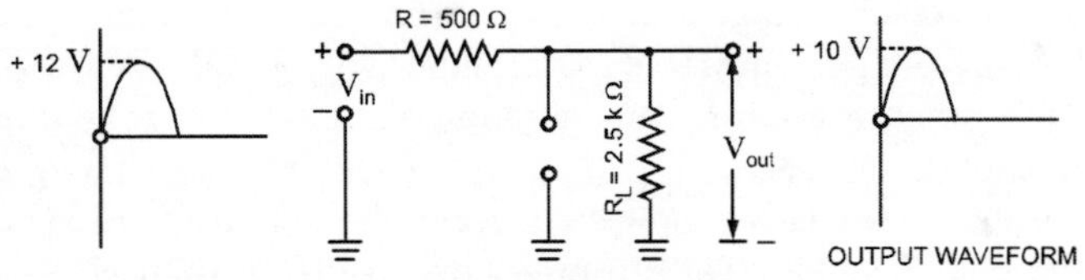

Fig. 10.14. ***For Positive Half Cycle***

Solution : During the positive half cycle of input signal voltage, the diode remains reverse biased and behaves as an open switch, as illustrated in Fig. 10.14. Under this condition

$$V_{out\,(peak)} = \text{Peak voltage across } R_L$$

$$= V_{in\,(peak)} \times \frac{R_L}{R + R_L}$$

$$= 12 \times \frac{2{,}500}{500 + 2{,}500} = 10 \text{ V}$$

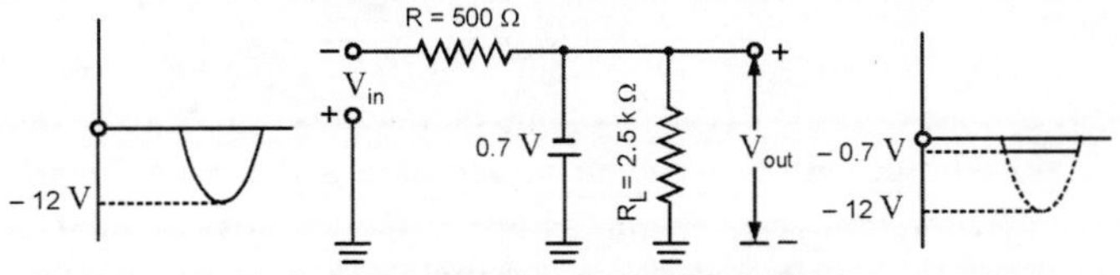

Fig. 10.15. ***For Negative Half Cycle***

During the negative half cycle of the input sinusoidal signal, the diode remains forward biased. The diode starts conducting when applied voltage exceeds – 0.7 V. Since diode D and load resistor R_L are in parallel, voltage across them cannot exceed – 0.7 V. Thus negative half cycle beyond – 0.7 V is clipped.

Thus output voltage during negative half cycle is given as $V_{out} = -0.7$ V.

Example 10.3. Determine V_o and required PIV rating of each of the diodes of the following circuit (Fig. 10.16).

Solution : Rectified output, $V_{dc} = \dfrac{2V_{s\,max}}{\pi} = \dfrac{2 \times 100}{\pi} = 63.66$ V **Ans.**

PIV rating of each diode $= V_{s\,max} = 100$ V **Ans.**

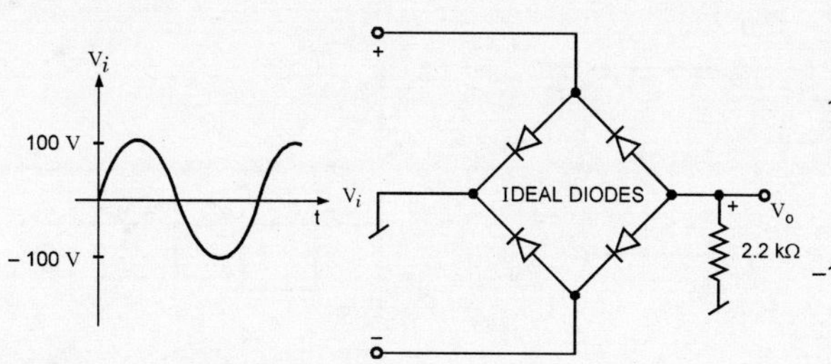

Fig. 10.16

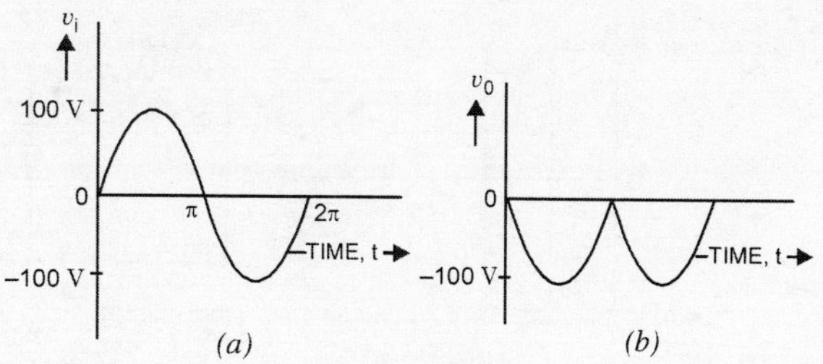

Fig. 10.17

Example 10.4. Determine V_{out} for the network given in Fig. 10.18.

Solution : From the direction of the diode and the polarity of the dc supply of 2 V, it can be observed that the diode will be in the 'ON' state for a part of the +ve region of the input signal.

The diode will be ON for those inputs satisfying the condition $V_{in} < 2$ V.

Thus, the transition state of the diode is $V_{in} = 2$ V.

When $V_{in} < 2$ V, the diode is ON and the output will appear as shown in Fig. 10.21. Hence $V_{out} = 2$ V.

When $V_{in} > 2$ V the diode is OFF and the output will appear as shown in Fig. 10.21 where $V_{out} = V_{in}$ **Ans.**

During negative half cycle diode will be ON and, therefore, output voltage $V_{out} = 2$ V, as shown in the Fig. 10.21.

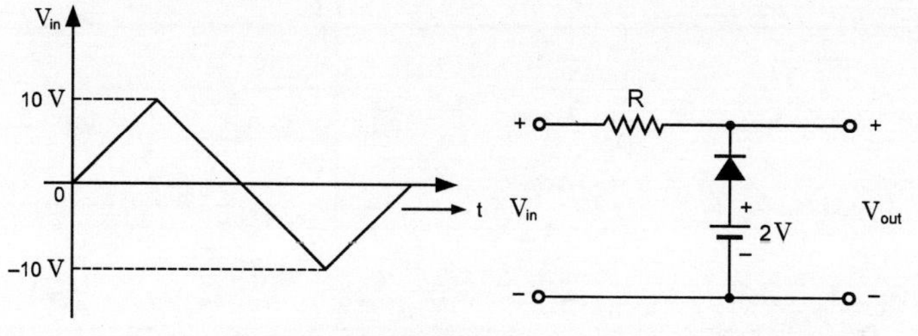

Fig. 10.18

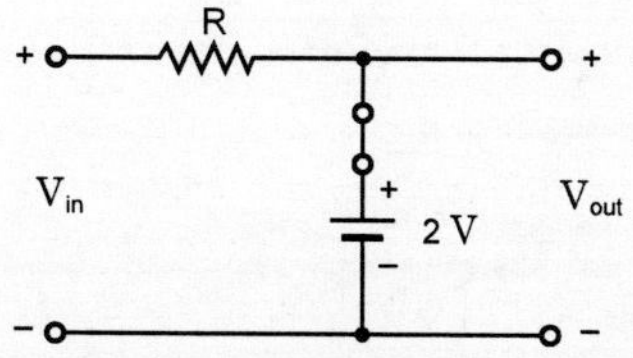

Fig. 10.19

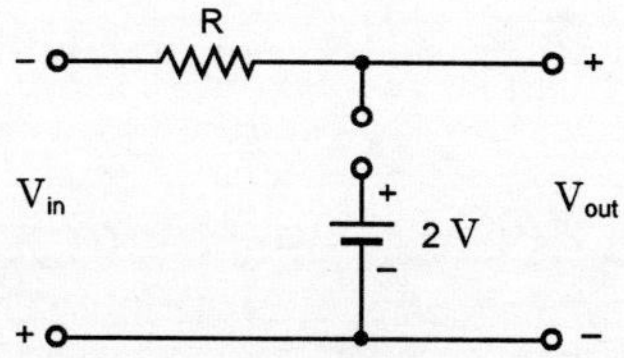

Fig. 10.20

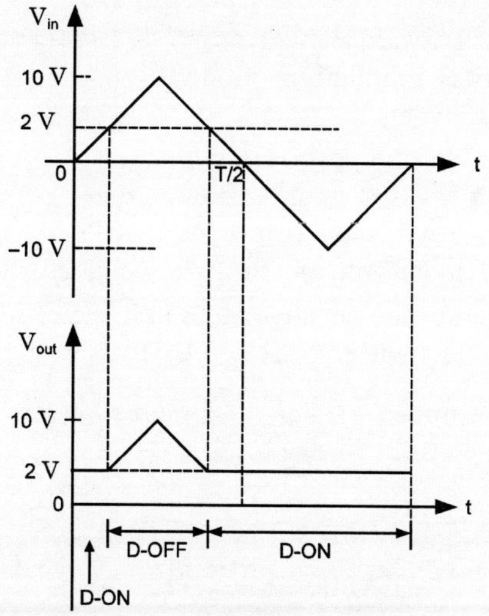

Fig. 10.21

Example 10.5. Sketch the output voltage V_{out} for the circuit shown in Fig. 10.22. Assume diodes D_1 and D_2 to be ideal diodes.

Solution : (*a*) For positive half cycle of input wave,

(1) D_1 will be forward biased,

(2) D_2 will be reverse biased *i.e.,* not conducting, and therefore, output will be zero.

(*b*) For negative cycle of input wave

(1) D_1 will be reverse biased,

(2) D_2 will be forward biased and, therefore, conducting.

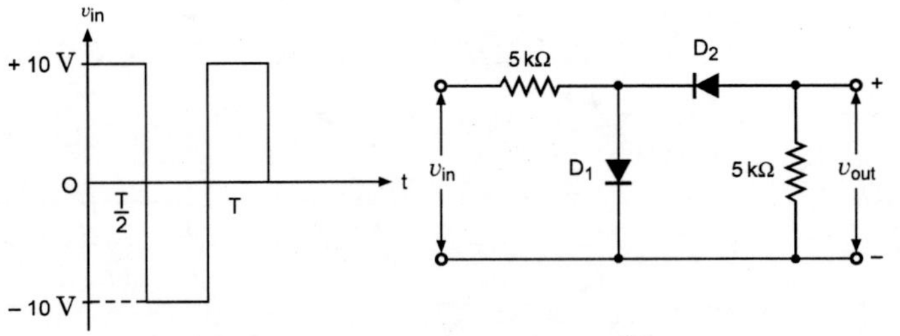

(a) *(b)*

Fig. 10.22

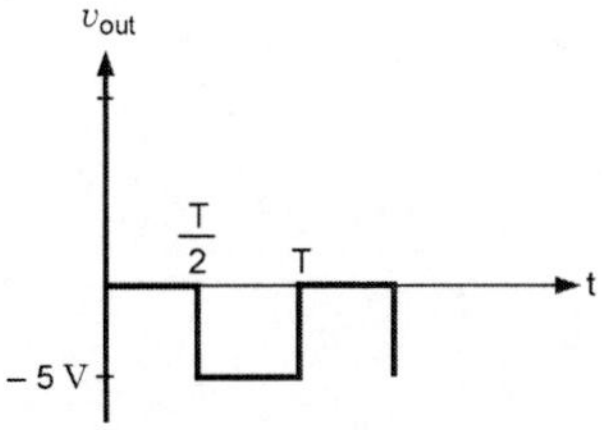

Fig. 10.23

The current flowing through the load resistance,

$$I_L = \frac{v_{in}}{(5+5)\times 10^3}$$

and output voltage, $v_{out} = I_L \times R_L = I_L \times 5 \times 10^3 = \frac{v_{in}}{10 \times 10^3} \times 5 \times 10^3 = \frac{v_{in}}{2}$

Output waveform is shown in Fig. 10.23.

Example 10.6. Draw the output waveform for the circuit shown in Fig. 10.24.

Solution : Forward voltage drop across silicon diode = 0.7 V.

Forward voltage drop across germanium diode = 0.3 V.

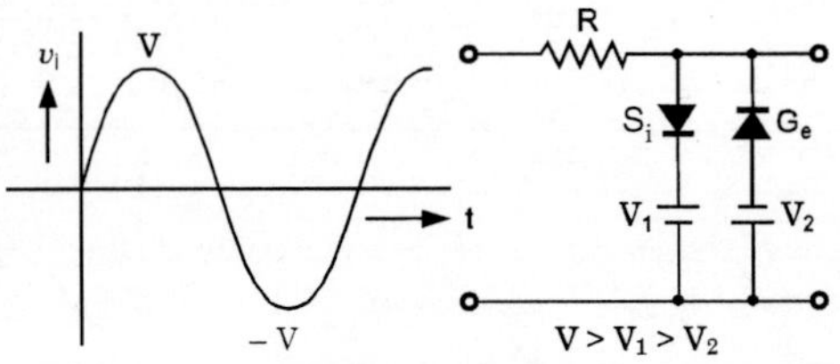

Fig. 10.24

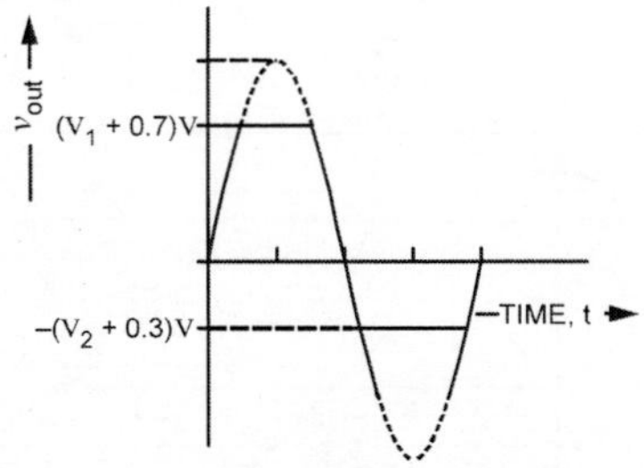

Fig. 10.25

For the positive input half cycle, germanium diode will remain reverse biased and silicon diode will conduct when $v_{in} \geq (V_1 + 0.7)$ V.

For the negative input half cycle, silicon diode will remain reverse biased and germanium diode will conduct when $v_{in} \geq (V_2 + 0.3)$ V.

Example 10.7. For the circuit shown in Fig. 10.26, sketch and explain the output waveform. Assume the diode to be ideal.

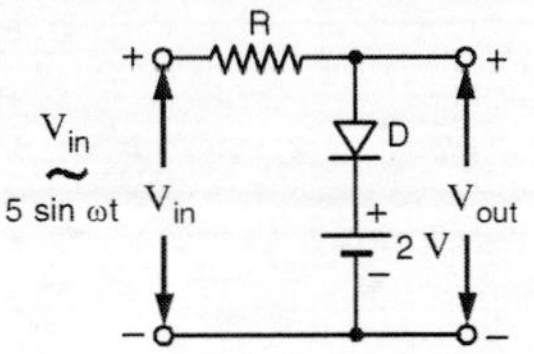

Fig. 10.26

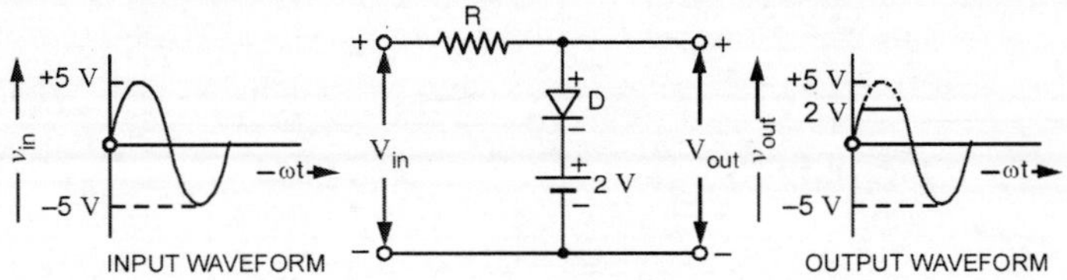

Fig. 10.27

Solution : During positive half cycles, the diode remains reverse biased for $0 < V_{in} < 2$ V, and the output voltage $v_{out} = v_{in}$.

When $V_{in} \geq 2$V diode D is forward biased and, therefore, conducts and output voltage $V_{out} = 2$ V

During negative half cycle the diode D remains reverse biased and therefore $v_{out} = v_{in}$

Output waveform is shown in Fig. 10.27.

Example 10.8. Design a clipping circuit to clip a sinusoidal wave of 2 V peak voltage at 100 Hz above +1 volt. Draw input and output waveforms. [*G.G.S.I.P. Univ. Analog Electronics-II May-June 2009*]

Solution : Biased positive clipper circuit shown in Fig. 10.28 will clip the sinusoidal wave of 2 V peak volt at 100 Hz above +1 V, as detailed below :

During positive half cycle, the diode remains reverse biased for $0 < v_{in} < 1$ V, and the output voltage $v_{out} = v_{in}$ as shown in Fig. 10.28.

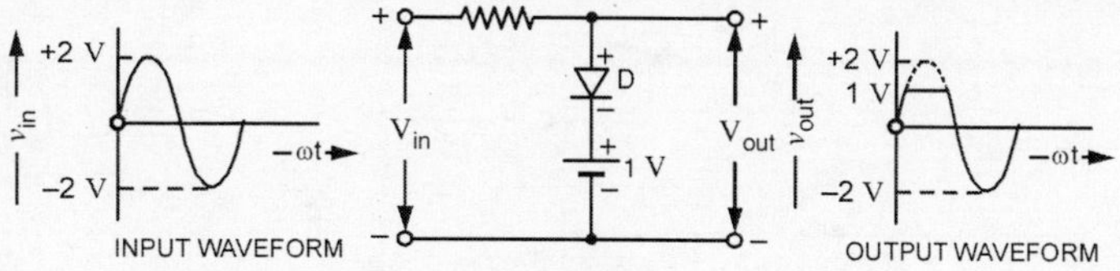

Fig. 10.28

When $v_{in} \geq 1$ V diode D is forward biased and, therefore, output voltage $v_{out} = 1$ V.

During negative half cycle the diode remains reverse biased and therefore, $v_{out} = v_{in}$ as shown in Fig. 10.28.

Example 10.9. The two similar zener diodes Z_1 and Z_2 have been joined back-to-back across the input sinusoidal voltage of peak value ± 25 V. Both zener diodes have zener voltage of 20 V. Find the output waveform.

If the peak value of input signal is increased to ± 60 V and zener diodes are replaced by zener diodes each having zener voltage of 10 V.

Solution : The given circuit will appear as shown in Fig. 10.29. We know that a zener diode acts like a 'short' (or very low resistance) when forward biased and 'open' (or very high resistance) when reverse biased till it goes into breakdown at V_Z.

During positive input half cycle, zener diode Z_1 is forward biased and so acts like a short but zener diode Z_2 acts like an open up to 20 V. Thereafter Z_2 goes into breakdown and holds the output voltage constant at 20 V till input voltage falls below 20 V in the later part of the positive half cycle. When the supply voltage falls below 20 V, the diode Z_2 comes out of the breakdown and again acts an 'open' across the load circuit. During negative input half cycle, the action of zener diodes Z_1 and Z_2 are reversed. The result is that the output voltage waveform is clipped off on both peaks, as illustrated in Fig. 10.29 (*a*).

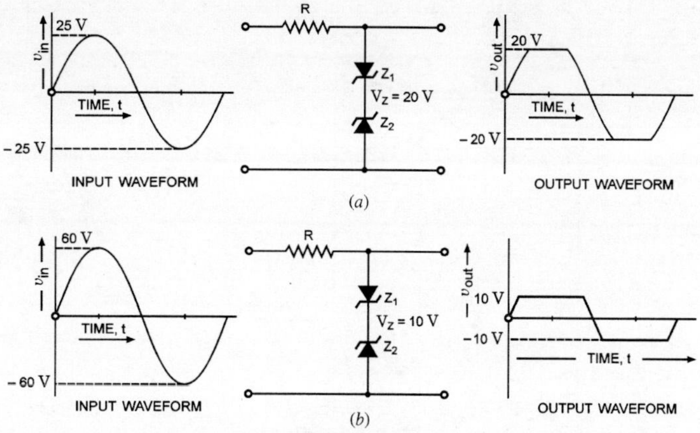

Fig. 10.29

When the peak value of the input signal is increased to 60 V and zener diodes of lesser zener voltage (10 V) are used, the circuit appears as shown in Fig. 10.29 (*b*) and the output waveform obtained is almost of square shape.

10.4. TRANSISTOR AS A CLIPPER

The transistor has two pronounced nonlinearities—one occurs when the transistor passes from cutin region to the active region and the second occurs when it passes from active region to the saturation region. If any input signal makes excursion which takes the transistor across the boundary between cutin region and active region, or across the boundary between the active region and saturation region, a portion of the input signal waveform will be clipped off. Portion of the input waveform which keeps the transistor in the active region shall appear at the output without any distortion. In such a case, it is the input current rather than the input voltage that should have the waveform of the signal of interest. Obvious reason is that over a large signal excursion in the active region, the transistor output current responds linearly to the input current but is related quite nonlinearly to the input voltage. Therefore, a current drive is used in a transistor clipper, as illustrated in Fig. 10.30. The resistor R_B must be large compared to the input resistance of the transistor in the active region. The input base current then has the waveform of input voltage and

$$i_B = \frac{v_{in} - \text{base-to-emitter cutin voltage}}{R_B}.$$

Waveforms for the transistor clipper for ramp input are shown in Fig. 10.31.

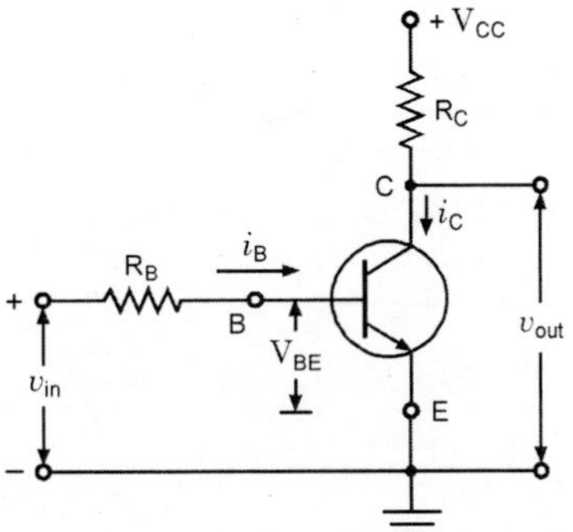

Fig. 10.30. ***Transistor as a Clipper***

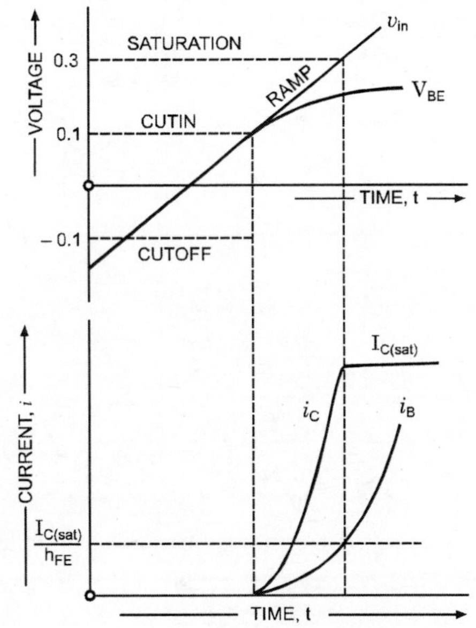

Fig. 10.31. ***Transistor Clipper Waveforms For Ramp Input***

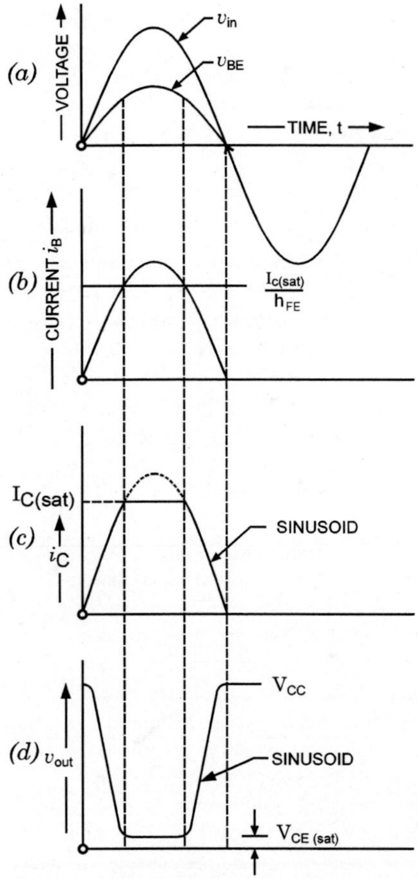

Fig. 10.32. ***Transistor Clipper Waveforms For Sinusoidal Input***

The voltages are taken for a germanium transistor. At –0.1 V, the transistor is in the cutoff region. At 0.1 V, it starts conducting and goes into the active region. At 0.3 V, the transistor goes into the saturation region and voltage V_{BE} is limited to 0.3 V. The input base current i_B increases as the transistor passes from cutoff region to active region and eventually into saturation. In the active region, the output current (collector current i_C) will be of the same form as the input base current. In saturation region, however, collector current will become constant and becomes $I_{C\,(sat)}$.

Waveforms for the transistor clipper for sinusoidal input are shown in Fig. 10.32.

10.5. EMITTER COUPLED CLIPPER

An emitter-coupled clipper using two transistors is shown in Fig. 10.33. It is a two level clipper in which the input is applied to the base of one of the transistors and the base of the other transistor is kept at a fixed voltage.

Consider initially that input voltage v_{in} is negative enough to ensure that transistor Q_1 is in cutoff. In this condition only transistor Q_2 carries current. Let us also consider that V_{BB_2} has been adjusted so that transistor Q_2 is in its active region and current in the emitter resistance R_E remains constant. This will be valid only if

$$|V_{BE_2}| < |V_{BB_2} + V_{EE}|$$

As the input voltage rises, transistor Q_1 will eventually come out of cutoff, both transistors will be carrying current, but current in transistor Q_2 decreases while current in transistor Q_1 increases, the sum of currents in the two transistors remaining unchanged and equal to emitter resistance current I. The input signal appears at the output, amplified but not inverted.

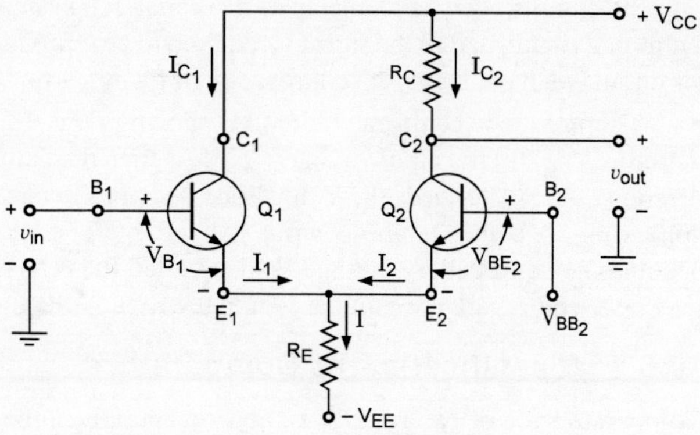

Fig. 10.33. ***Emitter Coupled Clipping Circuit***

As input voltage v_{in} continues its excursion in positive direction, the common emitter follows the base of transistor Q_1. As the base of transistor Q_2 is fixed, a point will be reached when the rising emitter voltage cuts off transistor Q_2. In summary, the input signal is amplified but limited twice, once by the cutoff of transistor Q_1 and once by the onset of cutoff in transistor Q_2.

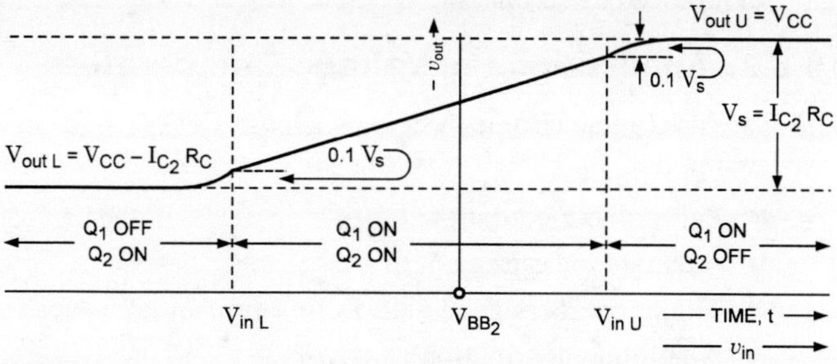

Fig. 10.34. ***Transfer Characteristic of an Emitter-Coupled Clipper***

The total range Δv_{out} over which the output can follow the input is IR_E and is, therefore, adjustable through the adjustment of current I.

The absolute voltage of the portion of the input waveform selected for transmission may be selected through an adjustment of a biasing voltage on which v_{in} is superimposed or through an adjustment of V_{BB2}.

The total range of input voltage Δv_{in} between clipping limits is $\dfrac{\Delta v_{out}}{A}$, where A is the gain of the amplifier stage. The transfer characteristic of an emitter-coupled clipper is given in Fig. 10.34.

10.6. COMPARATORS

A circuit used to mark the instant when an arbitrary waveform attains some particular reference level is called a *comparator.* The clipping circuits can also be used to perform the operation of comparison. In this

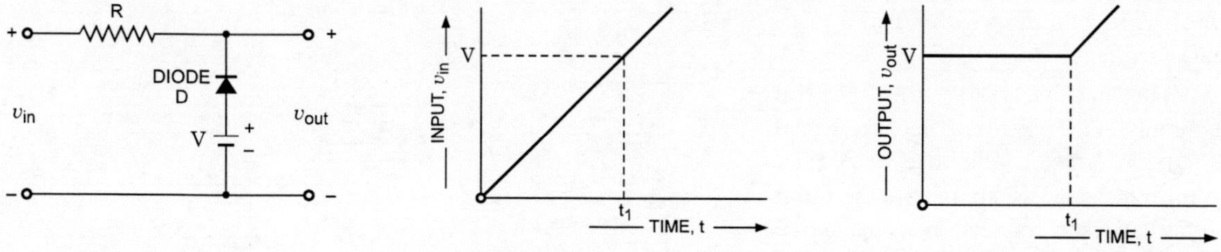

(a) Circuit Diagram *(b) Comparison With a Ramp Input Signal and Corresponding Output Waveform*

Fig. 10.35. ***Diode Comparator***

case the circuits become elements of a comparison system. The only difference between the two is that, in a clipping circuit, part of the signal is required to be reproduced without any distortion, whereas in a comparator circuit we have no interest in reproducing any part of the signal waveform.

A simple diode comparator circuit is shown in Fig. 10.35 (*a*). The diode D remains forward biased and, therefore, '*on*' till the input voltage v_{in} is less than the battery voltage V and the output remains constant at V. As soon as the v_{in} exceeds the V the diode becomes reverse biased and, therefore, switches *off* and the output voltage v_{out} becomes equal to input voltage v_{in}. The break point is that point when the input voltage v_{in} becomes equal to battery voltage V, also called the *reference voltage,* at time instant $t = t_1$. Thus the circuit may be used to mark the point at which the input voltage attains the reference level V.

10.6.1. Classification of Comparators

Comparators are of two types viz., non-regenerative type and regenerative type.

Clipping circuits are examples of non-regenerative comparators. Blocking oscillators, Schmitt trigger circuits are the examples of regenerative comparators in which a positive feedback is employed to have an infinite forward gain (unity loop gain). The Schmitt trigger comparator generates a step input whereas the blocking oscillator comparator generates a pulse output waveform. Operational amplifiers and tunnel diodes may also be used as comparators.

10.6.2. Applications of Voltage Comparators

These are used in electronic systems such as:

- Phase meters.
- Pulse time modulation.
- Accurate measurement of time.
- Timing markers generated from a sinusoidal wave.
- Amplitude distribution analyzers.
- ADC (analog-to-digital converters).
- Conversion of sinusoidal waveform into square waveform.

10.7. DIODE CLAMPING CIRCUITS

A circuit that places either the +ve or –ve peak of a signal at a desired level is known as a *clamping circuit.* It simply adds or subtracts a dc component to the input signal. In steady state, the output waveform is an exact replica of the input waveform. The clamper is also referred to as a *dc restorer* and a (signal) *level shifter.*

The necessity to establish the extremity of the positive or negative signal excursion at some reference level (*i.e.,* to introduce a dc component to the input signal) arises in connection with a signal that has passed through a capacitive coupling network and lost its dc component, as in case of television

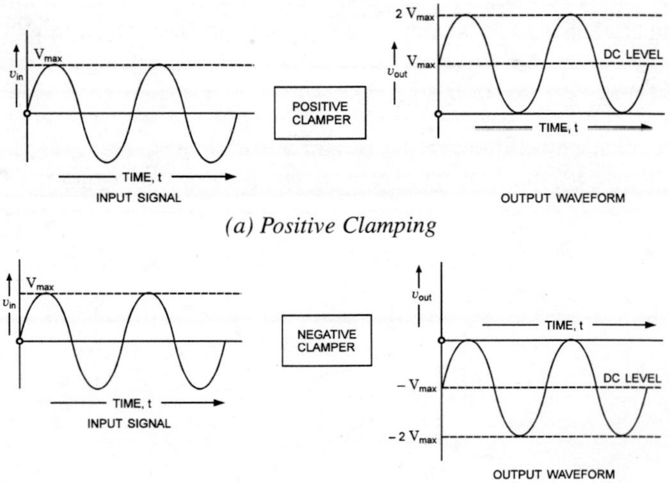

(*a*) *Positive Clamping*

(*b*) *Negative Clamping*

Fig. 10.36

receivers. However, normally the dc component introduced is not identical with the dc component lost in transmission.

In fact, a clamp circuit adds dc component (positive or negative) to the input signal so as to push it either on the positive side, as illustrated in Fig. 10.36 (*a*) or on the negative side, as illustrated in Fig. 10.36 (*b*). When the circuit pushes the signal on the positive side or upward, the negative peak of the signal coincides with the zero* level and the circuit is called the *positive clamper*. On the other hand, when the signal is pushed on the negative side or downward, as illustrated in Fig. 10.36 (*b*), the positive peak of the input signal coincides with the zero level and the circuit is called the *negative clamper*.

For a clamping circuit at least three components—a diode, a capacitor and a resistor are required. Sometimes an independent dc supply is also required to cause an additional shift. The important points regarding clamping circuits are :

(*i*) Waveform remains the same but its level is shifted either upward or downward.

(*ii*) The clamping circuit does not change the peak-to-peak or rms value of the waveform. Thus referring to Fig. 10.36, the input waveform and output waveform have the same peak-to-peak value *i.e.*, 2 V_{max}. If the input voltage and clamped output are measured by an ac voltmeter, the readings will be the same.

(*iii*) The clamping circuit, of course, alters the peak and average values of the waveform. In Fig. 10.36 the input waveform has a peak value of V_{max} and average value over a complete cycle is zero. The clamped output varies from 2 V_{max} and 0 (or 0 and – 2 V_{max}). Thus the peak value of the clamped output is 2 V_{max} and average value is V_{max}.

(*iv*) Both resistors R and C affect the waveform.

(*v*) The values of R and C should be taken such that the time constant of the circuit τ = RC is large enough to ensure that the voltage across the capacitor C does not change significantly during the interval the diode is non-conducting. For good clamping action, the circuit time constant τ = RC should be at least ten times the time period of the input signal voltage.

It is advantageous to first consider the condition under which the diode becomes forward biased.

Clamping circuits are often employed in television receivers as dc restorers. The incoming composite video signal is normally processed through capacitively coupled amplifiers which eliminate the dc component thereby losing the black and white reference levels and the blanking level. These reference levels have to be restored before applying the video signal to the picture tube. This is accomplished by employing clamping circuits. Clamping circuits are also used in storage counter, analog frequency meter, capacitance meter, divider and staircase waveform generator.

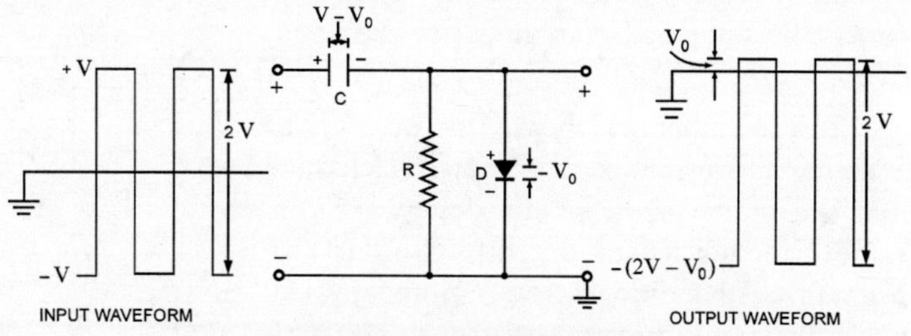

Fig. 10.37. *Negative Clamper*

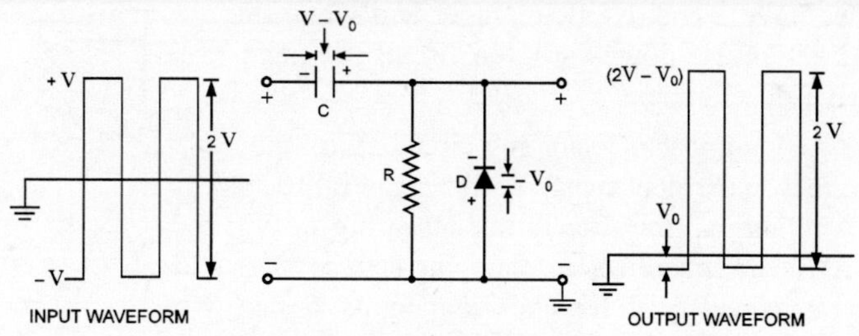

Fig. 10.38. *Positive Clamper*

Consider a *negative clamping circuit*, a circuit that shifts the original signal in a vertical downward direction, shown in Fig. 10.37. With the positive input, the diode D is forward biased and capacitor C is

* Assuming diode to be an ideal one.

charged with the polarity shown. During the positive half cycle of input, the output voltage is equal to barrier potential of the diode, V_o and the capacitor is charged to $(V - V_o)$. When the input goes negative, the diode is reverse biased and acts as an open circuit and thus has no effect on the capacitor voltage. R being of very high value cannot discharge C by very much during the –ve portion of the input waveform. Thus during negative input, the output voltage being the sum of the input voltage and the capacitor voltage is equal to $-V - (V - V_o)$ or $-(2V - V_o)$. Thus peak-to-peak output being the difference of the negative and positive peak voltage levels is equal to $V_o - [-(2V - V_o)]$ or 2 V.

If the diode in Fig. 10.37 is reconnected with reversed polarity, the circuit becomes a positive clamping circuit. The positive clamping circuit shifts the original signal in a vertical upward direction. A positive clamping circuit is shown in Fig. 10.38. It contains a diode D and a capacitor C as are contained in a negative clamper. The only difference in the circuit is that the polarity of the diode is reversed. Because of this reason the circuit acts as a positive clamper. The remaining explanation regarding the working of the circuit is the same as is explained for the negative clamper.

To remember which way the dc level of a signal moves, look at Fig. 10.37. Notice that the diode arrow points downward, the same direction as the dc shift. Similarly in Fig. 10.38, the diode arrow points upward, again the same direction as the dc shifts. It means that, when the diode points upward, we have a positive dc clamper and when the diode points downward, the circuit is a negative dc clamper.

A number of clamping circuits with their effect on the input signal are shown in Fig. 10.39. Although all the waveforms appearing in Fig. 10.39 are square waves, clamping networks operate equally well for sinusoidal inputs. In fact, one approach to the analysis of clamping networks with sinusoidal inputs is to replace the sinusoidal wave signal by a square wave of

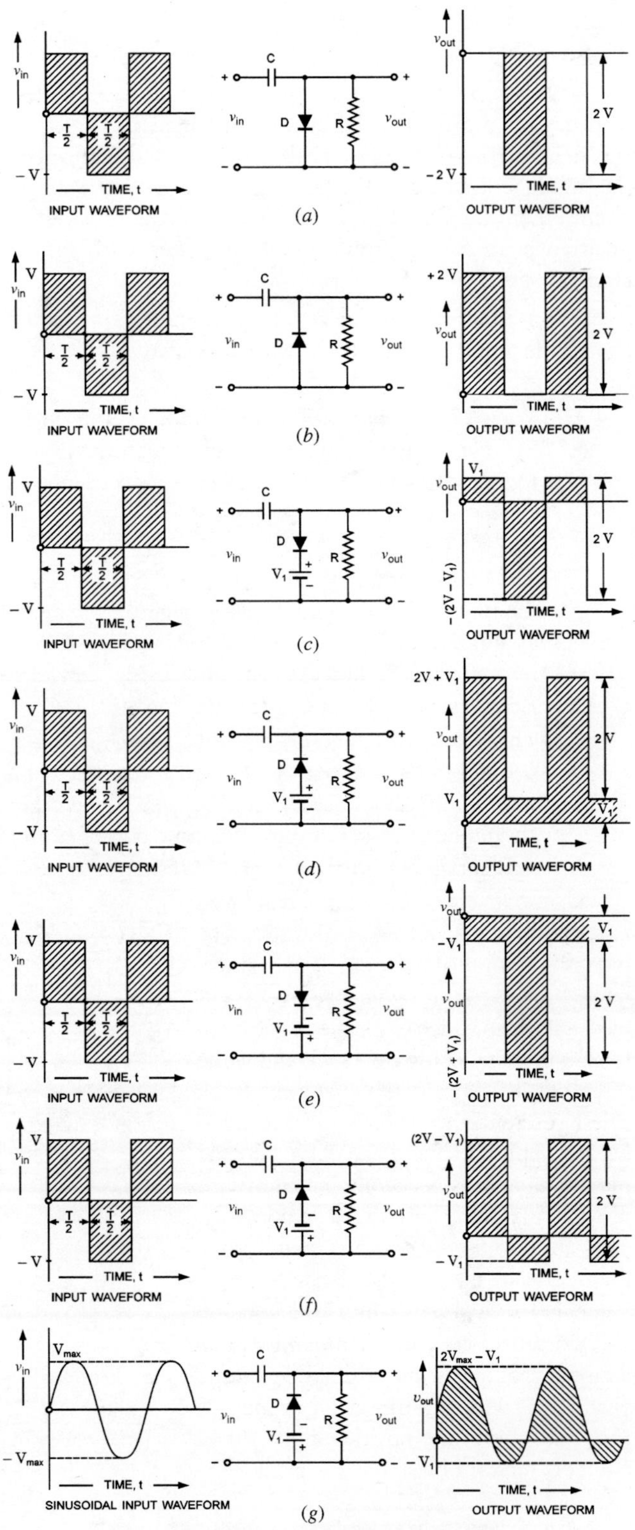

Fig. 10.39. *Different Clamping Circuits*

the same peak values. The resulting output will then form an envelope for the sinusoidal response, as illustrated in Fig. 10.39 (*g*) for a network appearing in Fig. 10.39 (*f*). The diodes have been assumed to be ideal and 5 RC >> T/2 in drawing the output waveforms.

Example 10.10. Give proper justification and draw the output voltage across diode (assume ideal) for the input waveform shown below (Fig. 10.40).

Solution : For the circuit shown in Fig. 10.40 (*a*), when the input level is –10 V, the diode is reverse biased.

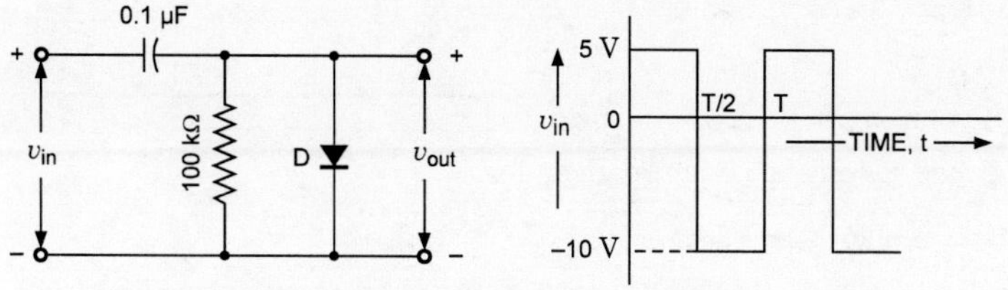

(*a*) *Circuit Diagram* (*b*) *Input Waveform*

Fig. 10.40

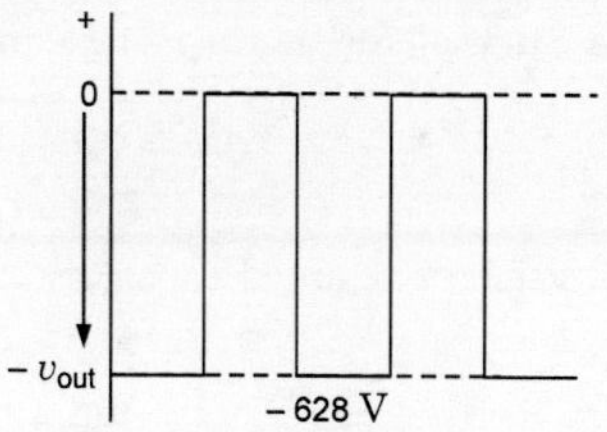

Fig. 10.41

$$\text{Assume frequency, } f = 1.000 \text{ Hz}$$

$$\text{Reactance of capacitor at 1,000 Hz} = \frac{1}{2\pi fC} = \frac{1}{2\pi \times 1{,}000 \times 0.1 \times 10^{-6}} = 1{,}592\ \Omega$$

$$\text{Current at } -10 \text{ V} = \frac{-10}{1{,}592} = -6.28 \text{ mA}$$

$$\text{Output voltage, } v_{out} = \text{Voltage across diode} = -6.28 \times 10^{-3} \times 100 \times 1{,}000 = -628 \text{ V}$$

When the input v_{in} goes to + 5 V, the capacitor cannot charge as its time constant is very large as compared to the duration of the pulse.

$$\text{RC} = 100 \times 10^3 \times 0.1 \times 10^{-6} = 10 \text{ ms}$$

$$\text{Pulse width of input} = \frac{T}{2} = \frac{1}{2f} = \frac{1}{2000} = 0.5 \text{ ms}$$

The diode D cannot conduct as it is an ideal diode.

Example 10.11. For the clamping circuit shown in Fig. 10.42 sketch for V_o.

Solution : Given circuit is of positive clamper.

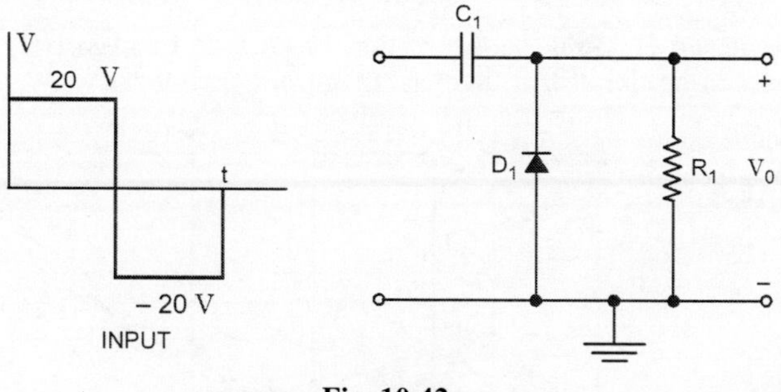

Fig. 10.42

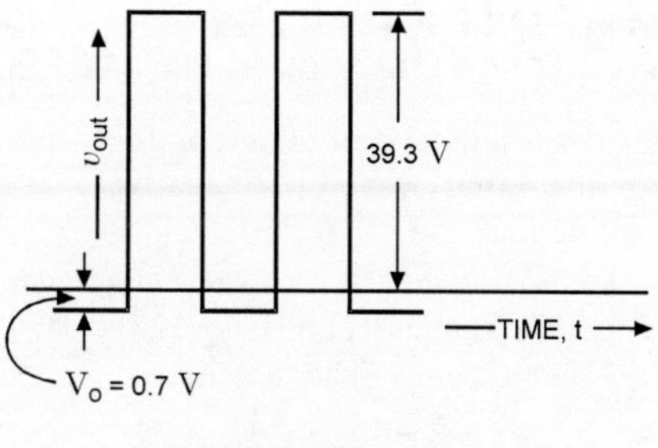

Fig. 10.43

In positive half cycle capacitor will shift the voltage to positive direction and output voltage will be

$$V_{out} = 2V_{in} - V_F = (2 \times 20 - 0.7) \text{ V} = 39.3 \text{ V} \text{ assuming silicon diode.}$$

During negative half cycle, the diode will be forward biased and output voltage will be equal to the forward voltage drop of diode *i.e.*, 0.7 V.

The output waveform is shown in Fig. 10.43.

Example 10.12. Design a diode clamper circuit to restore the bottom peaks of the input signal to zero level. Use a silicon diode with forward resistance, R_f = 50 Ω and reverse resistance, R_r = 400 kΩ. The frequency of the input voltage signal is 5 kHz. Assume RC = 20 T.

Solution : Since negative peaks are to be clamped to zero level, the required diode clamping circuit will be an unbiased positive clamping circuit as given in Fig. 10.44.

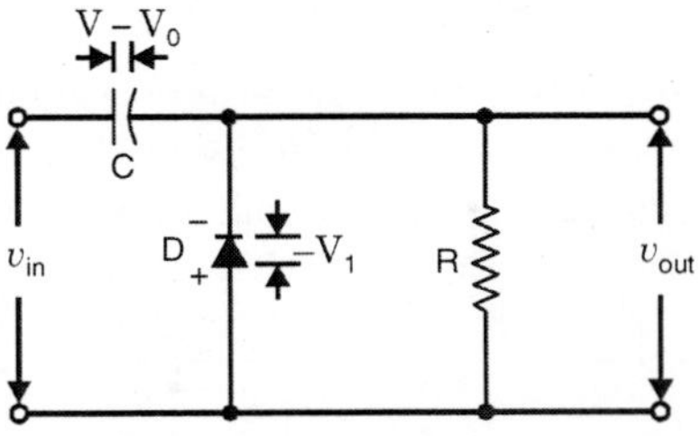

Fig. 10.44

$$RC = 20T$$

The external resistance to be connected is given as

$$R = \sqrt{R_f \times R_r} = \sqrt{50 \times 400 \times 1{,}000}$$

$$= 4{,}472\ \Omega \text{ or } 4.472\ k\Omega \quad \textbf{Ans.}$$

The value of required capacitance to be connected in the circuit is given by

$$C = \frac{20T}{R} = \frac{20}{R} \times \frac{1}{f} = \frac{20}{4{,}472 \times 5 \times 10^3} = 0.895\ \mu F \quad \textbf{Ans.}$$

Example 10.13. Determine V_o for the following circuit (Fig. 10.45). Also name the configuration.

Solution : The given circuit is of positive clamper. In positive half cycle, capacitor will shift the voltage in positive direction, but battery will be in negative direction.

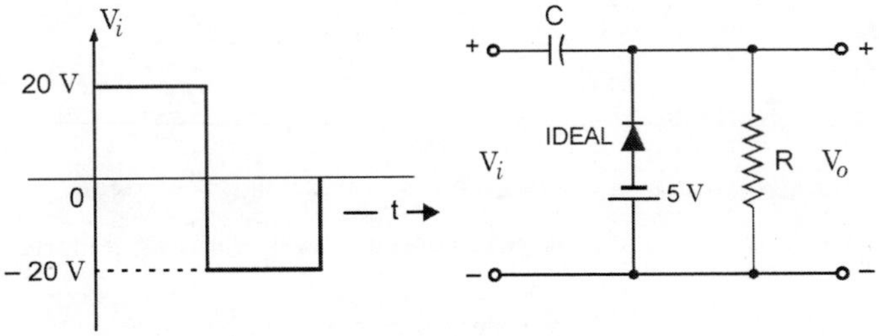

Fig. 10.45

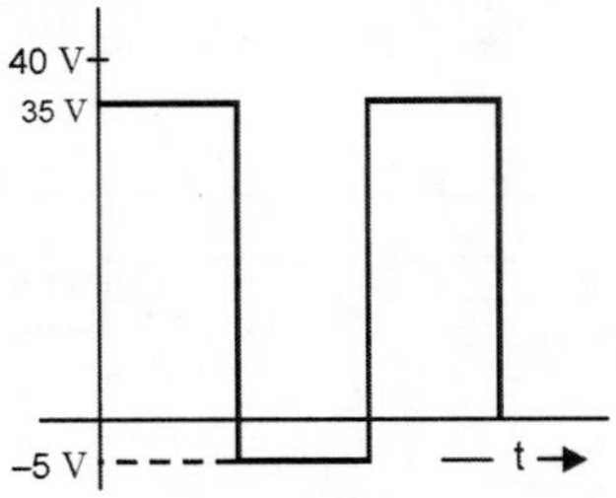

Fig. 10.46. *Output Waveform*

So output voltage, $V_o = 2V_i - V_1 = 2 \times 20 - 5 = 35$ V

In negative half cycle, diode will be forward biased and output voltage V_o will be equal to battery voltage

i.e., $V_o = -V_1 = -5$ V **Ans.**

Output waveform is shown in Fig. 10.46.

Example 10.14. A 125 V peak square-wave with a period of 25 ms, shown in Fig. 10.47, is to be negatively clamped at 50 V. Draw the circuit diagram required to accomplish it. Sketch the output waveform.

Solution : The circuit diagram for a clamper required to clamp the input waveform negatively at 50 V is given in Fig. 10.48 (*a*).

The diode conducts when the input swings to + 125 V and the capacitor C gets charged to 125 – 50 *i.e.*, 75 V in a few cycles. Under steady-state condition

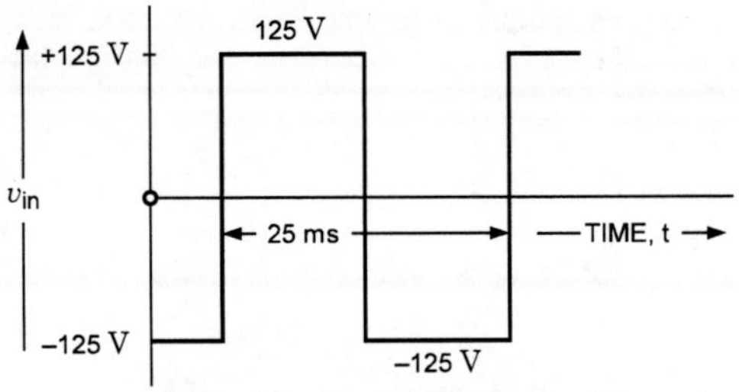

Fig. 10.47. *Input Waveform*

$$v_{out} = v_{in} - 75$$

When input voltage, v_{in} = + 125 V

Output voltage, $v_{out} = v_{in} - 75 = 125 - 75 = 50$ V

When input voltage, v_{in} = – 125 V

Output voltage, $v_{out} = -125 - 75 = -200$ V

Thus output swings between + 50 V and – 200 V as shown in Fig. 10.48(*b*).

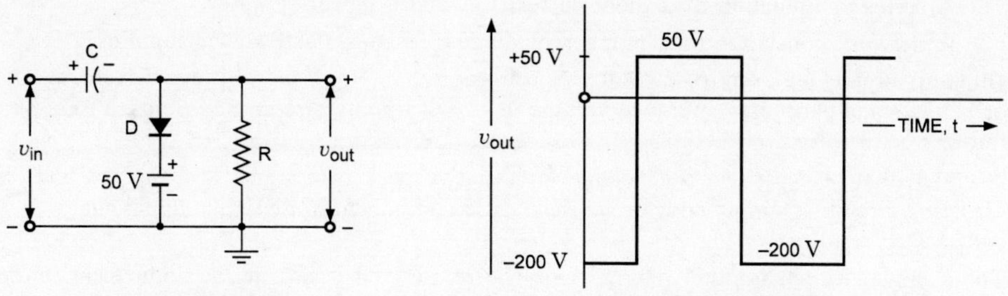

(a) Circuit Diagram (b) Output Waveform

Fig. 10.48

Example 10.15. Explain the circuit given below (Fig. 10.49) with the help of suitable input.

Solution : Let the input be a square wave of maximum value 20 V.

The diode conducts when the input swings goes to –20 V and the capacitor charges to –(20 + 5) *i.e.*, –25 V in a few cycles. Under steady-state conditions

$$v_{out} = v_{in} - (-25) = (v_{in} + 25) \text{ V}$$

When v_{in} = +20 V, v_{out} = (20 + 25) = 45 V

When v_{in} = –20 V, v_{out} = (–20 + 25) = 5 V

Output waveform is shown in Fig. 10.50.

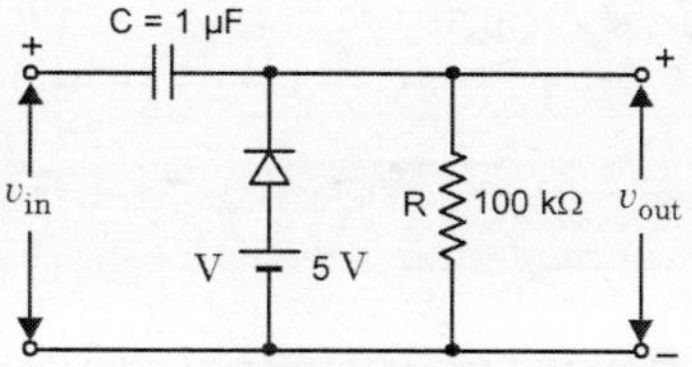

Fig. 10.49

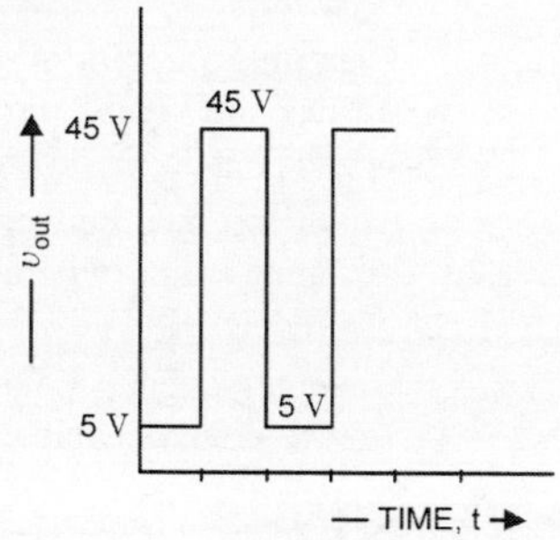

Fig. 10.50

Example 10.16. Design a diode clamper circuit to restore the +ve peaks of 1 kHz input signal to a voltage level to 5 V. Assume voltage drop across diode as 0.7 V. Use silicon diode with forward resistance R_f = 200 Ω and reverse resistance R_r = 500 kΩ. Assume RC = 20T.

Solution : Voltage drop across diode,

$$V_r = 0.7 \text{ V}$$

$$\text{Clamping level} = V_R + V_r$$

$$\text{or} \quad V_R = \text{Clamping level} - V_r = 5 - 0.7 = 4.3 \text{ V}$$

$$\text{Forward resistance}, R_f = 200\ \Omega$$

$$\text{Reverse resistance}, R_r = 500\ \text{k}\Omega$$

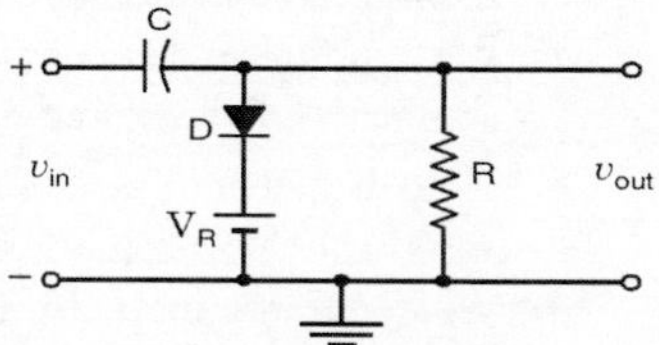

Fig. 10.51

External resistance required, $R = \sqrt{200 \times 500 \times 1{,}000} = 10\ \text{k}\Omega$ **Ans.**

Capacitance of capacitor required, $C = \dfrac{20T}{R} = \dfrac{20}{f \times R} = \dfrac{20}{1{,}000 \times 10^4} = 2\ \mu\text{F}$ **Ans.**

HIGHLIGHTS ....

1. When any wave is applied at the input of a nonlinear device (such as a diode or transistor), the shape of the output wave varies nonlinearly with input wave. This process is called the nonlinear waveshaping.
2. Clipping is the process of clipping or removing a part of the input waveform.
3. A waveshaping circuit which controls the shape of the output waveform by removing or clipping a portion of the input wave is known as the clipping circuit.

   Clipping circuits are also referred to as voltage (or current) limiters, amplitude selectors, or slicers.

   Clipping circuits do not require energy storage elements and are available in two configurations.

1. A series combination of a diode, a resistor, and a power supply.
2. A network consisting of a number of diodes, resistors, and power supplies.

Clipping circuits are used to select for transmission that part of an arbitrary waveform which lies above or below some particular reference voltage level. A clipping circuit may be used to convert a sinewave into a square wave.

4. A circuit that places either the positive or negative peak of a signal at a desired level is known as a clamping circuit. It simply adds or subtracts a dc component to the input signal and, therefore, it is also called a dc restorer.

The clamping circuit is often referred to as dc reinserter because the dc component introduced may be different from the dc component lost during transmission. The clamping circuit only alters the dc level of the signal but does not affect its shape. A clamper circuit uses sometimes a dc battery in addition to a diode, a capacitor and a resistor.

Clamping circuits may be positive clamping circuits or negative clamping circuits. A positive clamper pushes the signal on the positive side or upward while a negative clamper pushes the signal on the negative side or downward.

The difference between the clippers and clampers is that the clipping circuit removes some portion of an input signal whereas clamping circuit places either the positive or negative peak of the signal at a desired level.

Clamping circuits are often employed in television receivers as dc restorers. Clamping circuits are also used in storage counter analog frequency meter, capacitance meter, divider and staircase waveform generator.

EXERCISES ....

1. What do you mean by a clipper ? Discuss in details diode as a clipper.

2. What is clipping circuit ? Give various configurations of clipping circuits.

3. What is meant by clipping circuit ? Draw and explain different types of diode clipping circuits.

4. Describe with a circuit diagram and waveform function of a transistor clipper.

5. With the help of neat circuit diagram and waveform, explain the working of an emitter-coupled clipper.

6. Explain comparator circuit and its uses.

7. What is meant by comparator ? How do comparator circuits differ from clipping circuits. Give the applications of voltage comparators.

8. What is meant by a clamping circuit ? Give different types of clamping circuits along with their output waveforms and explain operation of any one circuit.

9. What do you mean by (*i*) positive clamping and (*ii*) negative clamping ? List two uses of clamping circuits.

10. What is a comparator circuit ? How does it differ from a clipping circuit ?

SHORT ANSWER TYPE QUESTIONS WITH ANSWERS ....

Q. 1. What is meant by a clipping circuit ?

Ans. A waveshaping circuit which controls the shape of the output waveform by removing or clipping a portion of the input wave is known as the clipping circuit.

Q. 2. What is the difference between a positive clipper and negative clipper ?

Ans. The positive clipper removes the positive half cycles of the input waveform while the negative clipper removes the negative half cycles of the input waveform.

Q. 3. What is the drawback of having a diode as a series element in a clipper ?

Ans. The use of the diode as a series element has a drawback that when the diode is in 'off' condition, there should be no transmission of input signal to the output. But in case of high frequency signals, transmission occurs through diode capacitance which is undesirable.

Q. 4. "Clipping circuit is to suppress amplitude variations in the input signal voltage". Is this statement correct and why ?

Ans. This statement is correct, because sometimes due to noise or other sources input signal may contain variations, which are undesirable and are below certain limits. Such variations can be removed by using clipping circuits.

Q. 5. "Clipping circuit is to remove a certain portion of input signal above or below a certain level". Is this statement correct and why ?

Ans. Yes, the given statement is correct because clipping is a process of removing a portion of the input signal.

Q. 6. Describe the operation of non-saturating transistor clipper.

Ans. The transistor has two pronounced nonlinearities — one occurs when the transistor passes from cutin region to the active region and the second occurs when it passes from active region to the saturation region. If any input signal makes excursion which takes the transistor across the boundary between cutin region and active region, or across the boundary between the active region and saturation region, a portion of the input signal waveform will be clipped off.

Q. 7. What are amplitude selectors ?

Ans. Amplitude selectors are also called clipping circuits. These circuits are used to select for transmission that part of an arbitrary waveform which lies above or below some particular reference voltage level.

Q. 8. What is an emitter coupled clipper ?

Ans. The emitter coupled clipper is a two-level clipper. It is an emitter coupled differential amplifier.

Q. 9. What is meant by a comparator ?

Ans. A circuit used to mark the instant when an arbitrary waveform attains some particular reference level is called a comparator.

Q. 10. How does comparator differ from clipper ?

Ans. The only difference between the comparator and clipper is that, in a clipping circuit part of the signal is required to be reproduced without any distortion whereas in a comparator circuit we have no interest in reproducing any part of the signal waveform.

Q. 11. What is meant by a clamping circuit ?

Ans. A circuit that places either the positive or negative peak of a signal at a desired level is known as a clamping circuit. It simply adds or subtracts a dc component to the input signal and, therefore, it is also called a dc restorer.

Q. 12. What is the difference between a positive clamping circuit and a negative clamping circuit ?

Ans. A positive clamper pushes the signal on the positive side or upward while a negative clamper pushes the signal on the negative side or downward.

Q. 13. A clamper circuit sometimes uses a dc battery in addition to a diode, a capacitor and a resistor, why ?

Ans. A clamper circuit uses sometimes a dc battery in addition to a diode, a capacitor and a resistor to cause an additional shift.

Q. 14. How does a clamper circuit affect the peak-to-peak and rms values of a waveform ?

Ans. No change.

Q. 15. How does a clamper circuit affect the peak and average values of the waveform ?

Ans. The peak and average values of a waveform are changed by a clamper circuit.

Q. 16. What is the difference between clipping and clamping ?

Ans. The clipping circuit removes some portion of an input signal whereas clamping circuit places either the positive or negative peak of the signal at a desired level.

Q. 17. What is the difference between the output from a clipping circuit and a clamping circuit ?

Ans. The output of a clipping circuit appears as if a portion of the input signal were clipped off but a clamper circuit simply clamps (lifts up or down) the input to a different dc level.

Q. 18. List two uses of clamping circuits.

Ans. Clamping circuits are often employed in television receivers as dc restorers. Clamping circuits are also used in storage counter analog frequency meter, capacitance meter, divider and staircase waveform generator.

PROBLEMS

1. Sketch V_{out} for the following network (Fig. 1)

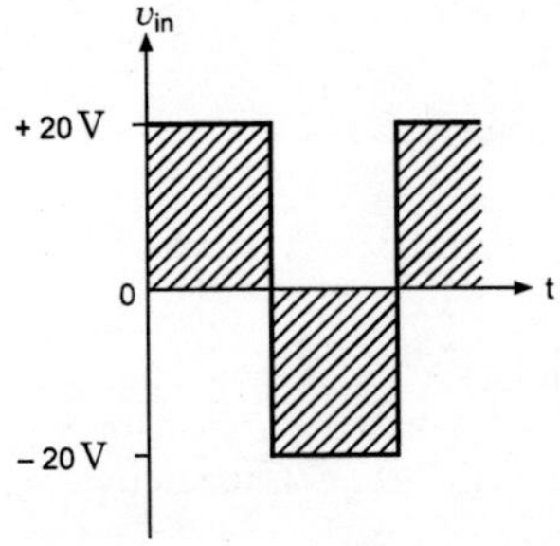

(a)

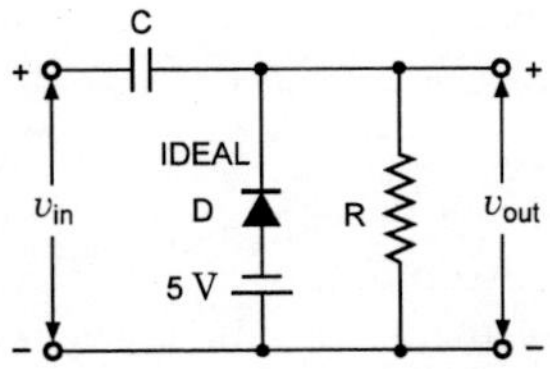

(b)

Fig. 1

2. Determine the peak output voltage for a negative series clipper circuit connected to an input ac sinusoidal signal of peak value 12 V. The barrier potential for silicon diode is 0.7 V. Draw the circuit diagram and output waveform. [**Ans.** 11.3 V]
3. Determine the peak output voltage for a positive shunt clipper circuit connected to an input ac sinusoidal signal of peak value 10 V. The barrier potential for germanium diode is 0.3 V. Series resistor is of 400 Ω and load resistor is of 2 kΩ. [**Ans.** 0.3; –8.33 V]
4. Sketch the output waveform of a clamping circuit shown in Fig. 2.

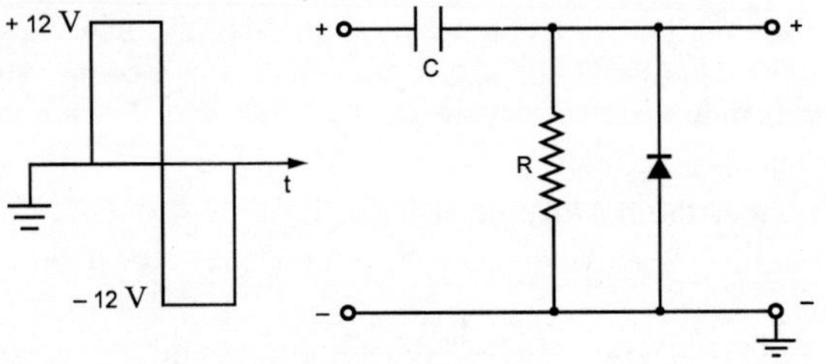

Fig. 2

5. Sketch the output waveform of a clamping circuit shown in Fig. 3.

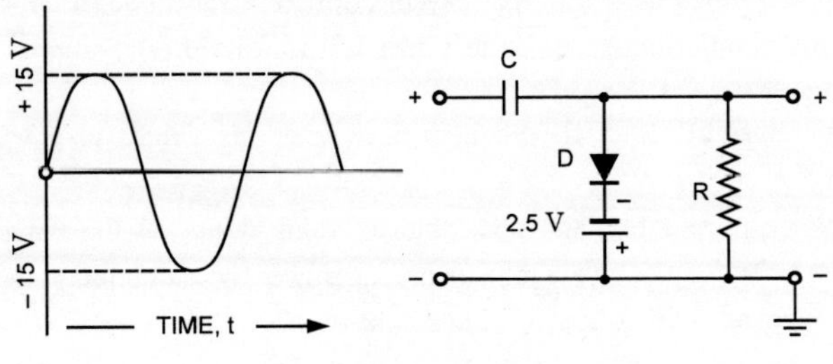

Fig. 3

❑❑❑

CHAPTER 11

Nonlinear Circuits

INSIDE THIS CHAPTER

11.1. INTRODUCTION

As already mentioned in Art. 3.1, op-amp is viewed as a basic building block of linear and nonlinear analog systems. Linear circuits using op-amp have already been discussed in chapter 3. Nonlinear circuits, that include comparators, logarithmic amplifier, antilogarithmic amplifier, logarithmic multiplier, precision rectifiers, clipper and clamper, will be discussed here.

11.2. BASIC OP-AMP COMPARATOR CIRCUIT

Often two voltage signals are to be compared and to be distinguished which is stronger. For such situations, a *comparator* may be the perfect solution. It also forms the basic building block required for *non-sinusoidal waveform generators* or *relaxation oscillators*, so it deserves priority in discussion over relaxation oscillators.

We have studied in Art. 2.11 that when the op-amp is used in open-loop configuration (or without feedback) any input signal (differential or single) which even slightly exceeds zero drives the output into saturation because of very high open-loop voltage gain (nearly infinity) of op-amp. It means that the application of a small differential input signal of appropriate polarity causes the output to switch to its either saturation.

Thus op-amp comparator is a circuit with two inputs and one output. The two inputs can be compared with each other *i.e.,* one of them can be considered a reference voltage, V_{ref}

Figure 11.1 (*a*) shows an op-amp comparator circuit. A fixed reference voltage V_{ref} is applied to the inverting (–) input terminal and sinusoidal signal v_{in} is applied to the noninverting (+) input terminal. When v_{in} exceeds V_{ref} the output voltage goes to positive saturation because the voltage at the (–) input is smaller than at the (+) input. On the other hand, when v_{in} is less than V_{ref} the output voltage goes to negative saturation. Thus output voltage v_{out} changes from one saturation level to another whenever $v_{in} \cong V_{ref}$, as illustrated in Fig. 11.1 (*b*). In short, the comparator is a type of an analog-to-digital converter (ADC). At any given time the output voltage waveform shows whether v_{in} is greater or less than

V_{ref}. The comparator is sometimes referred to as a *voltage level detector* because, for a desired value of V_{ref}, the voltage level of the input voltage v_{in} can be detected.

Diodes D_1 and D_2 are provided in the circuit [Fig. 11.1 (*a*)] to protect the op-amp against damage due to excessive input voltage. Because of these diodes, the differential input voltage v_d is clamped to either + 0.7 V or – 0.7 V, hence the diodes are called *clamp diodes*. There are some op-amps with built-in input protection. Such op-amps need not to be provided with protection diodes. The resistance R_1 in series with v_{in} is used to limit the current through protection diodes D_1 and D_2 while resistance R is connected between the inverting (–) input terminal and V_{ref} to reduce the offset problem.

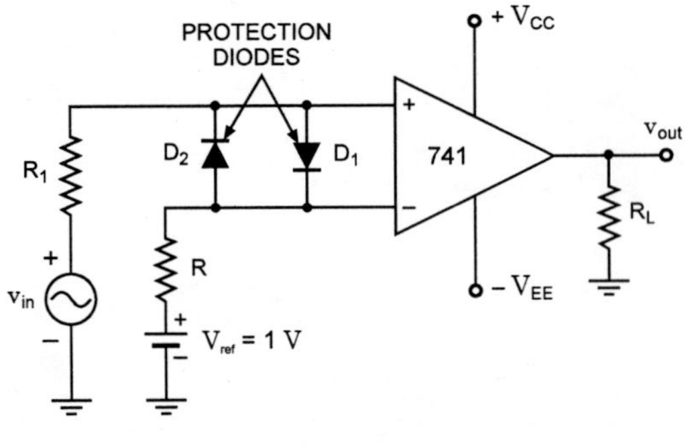

(a) Noninverting Comparator Circuit

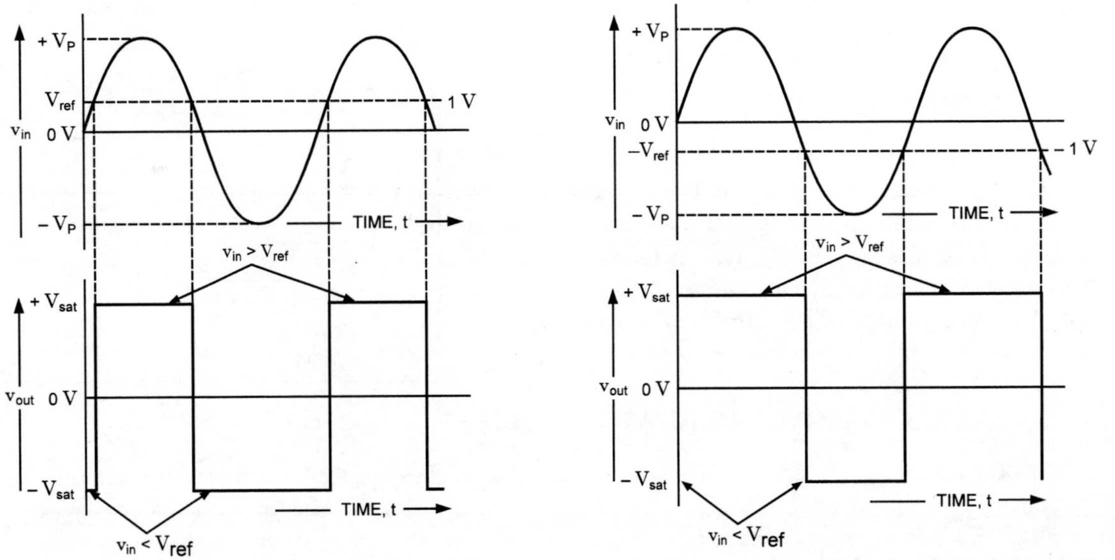

(b) Input and Output Waveforms For Positive V_{ref}

(c) Input and Output Waveforms For Negative V_{ref}

Fig. 11.1

When the reference voltage V_{ref} is negative w.r.t ground, with a sinusoidal signal applied to the noninverting input terminal, the output voltage will be as illustrated in Fig. 11.1 (*c*).

Obviously, the amplitude of v_{in} must be large enough to pass through V_{ref} for switching action to take place.

Since the sinusoidal input signal is applied to the noninverting terminal, this circuit is called the *noninverting op-amp comparator*.

Similarly an *inverting op-amp comparator* can be had by applying the sinusoidal input to the inverting (–) input terminal to the op-amp.

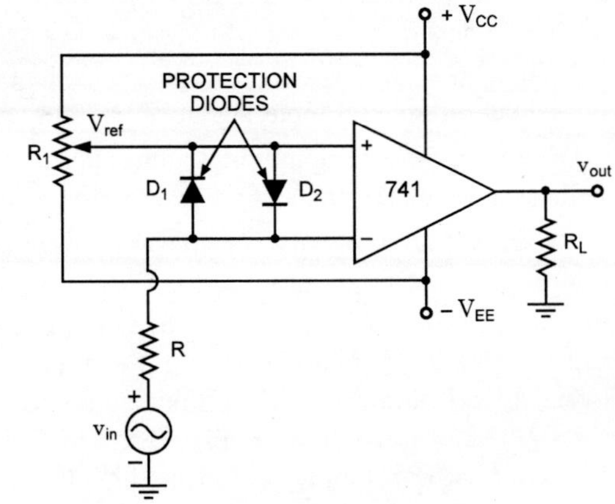

Fig. 11.2 *(a) Inverting Comparator Circuit*

Figure 11.2 (*a*) shows the circuit for an inverting comparator in which the sinusoidal input signal v_{in} is applied to the inverting (–) input terminal while the reference voltage V_{ref} is applied to the noninverting (+) input terminal. In this circuit V_{ref} is obtained by the use of a potentiometer forming a potential divider arrangement with dc supply voltage $+V_{CC}$ and $-V_{EE}$. As the wiper connected to (+) terminal is moved toward $+V_{CC}$, V_{ref} becomes more positive, while if it is moved toward $-V_{EE}$, V_{ref} becomes more negative.

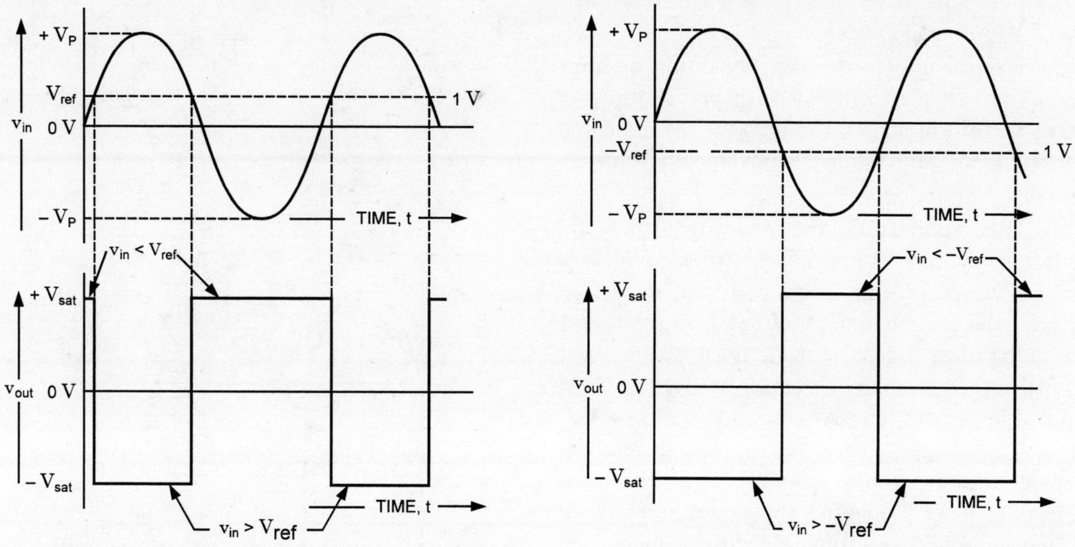

(*b*) *Input and Output Waveforms For Positive* V_{ref} (*c*) *Input and Output Waveforms For Negative* V_{ref}

Fig. 11.2

The input and output waveforms are shown in Figs. 11.2 (*b*) and 11.2 (*c*).

Comparators are used in circuits such as discriminators, voltage-level detectors, oscillators, digital interfacing, Schmitt trigger etc.

Example 11.1. For the comparator shown in Fig. 11.3 plot the transfer curve if the op-amp is an ideal one and $V_{z_1} = V_{z_2} = 9$ V.

Solution : (*a*) $V_{z_1} = V_{z_2} = 9$ V

when v_{in} is positive $v_{out} = V_z + V_\gamma$

when v_{in} is negative $v_{out} = -(V_z + V_\gamma)$

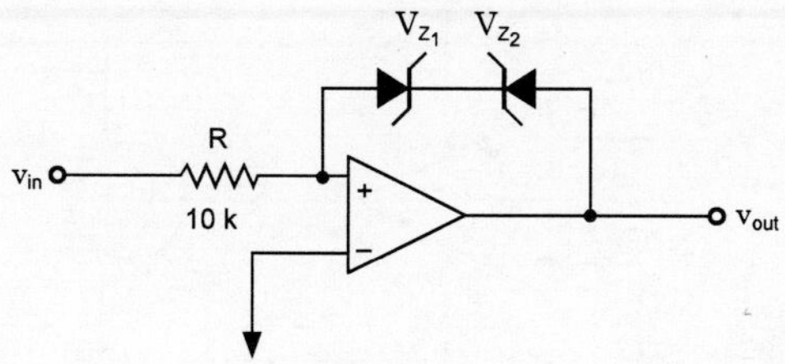

Fig. 11.3

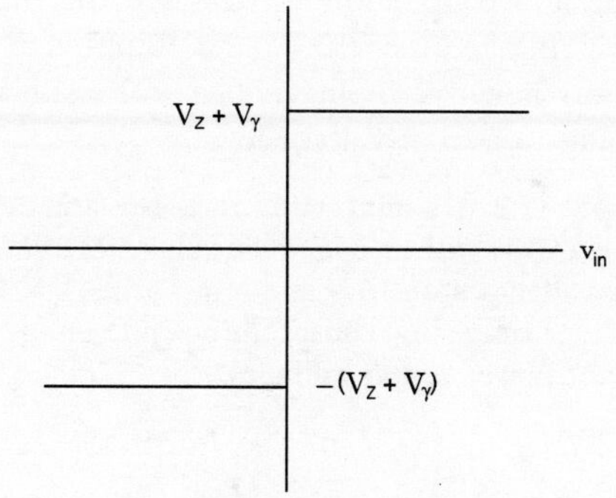

Fig. 11.4

Transfer characteristic is shown in Fig. 11.4.

11.3. ZERO-CROSSING DETECTOR

Zero-crossing detector is an applied form of comparator. Either of the op-amp basic comparator circuits discussed in Art. 11.2 can be employed as the zero-crossing detector provided the reference voltage V_{ref} is made zero.

Zero-crossing detector using inverting op-amp comparator is depicted in Fig. 11.5 (*a*). The output voltage waveform shown in Fig. 11.5 (*b*) indicates when and in what direction an input signal v_{in} crosses zero volt.

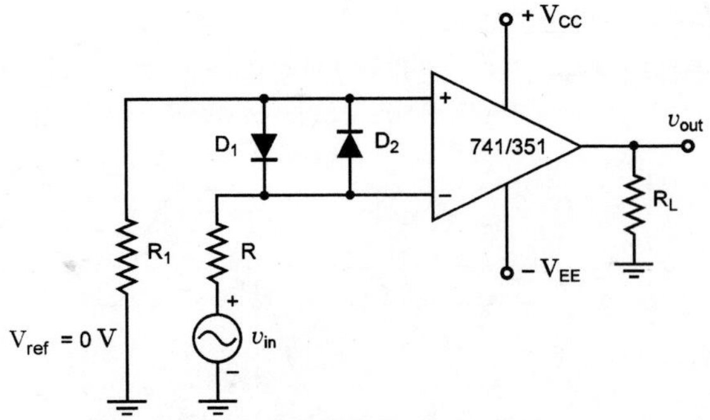

(a) Circuit

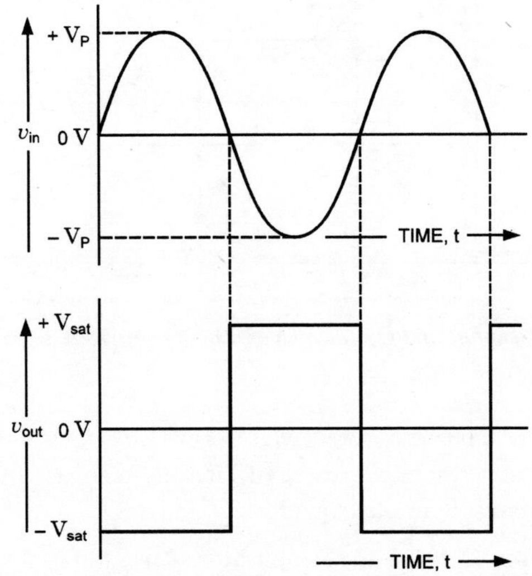

(b) Input and Output Waveforms

Fig. 11.5. ***Zero-Crossing Detector***

In some applications the input signal may be low-frequency one (*i.e.,* input may be a slowly changing waveform). In such a case output voltage v_{out} may not switch quickly from one saturation state to the other. Because of the noise at the input terminals of the op-amp, there may be fluctuation in output voltage between two saturation states ($+ V_{sat}$ and $- V_{sat}$ voltages). Thus zero crossings may be detected for noise voltages as well as input signal v_{in}. Both of these problems can be overcome, if we use *regenerative* or *positive feeding* causing the output voltage v_{out} to change faster and eliminating the false output transitions that may be caused due to noise at the input of the op-amp.

A zero-crossing detector using IC311 (8 pin DIP unit) is illustrated in Fig. 11.6. The positive (above 0 V) going input signal drives the output transistor *on*, the output going *low* (–10 V in this connection). The negative (below 0 V) going input drives the output transistor *off*, the output going *high* (+10 V in this connection).

The output is thus an indication of whether the input is above or below 0 V. When the input is any positive voltage (above 0 V), the output is at *low* level, while any negative input voltage will result in the output going to a *high*-voltage level.

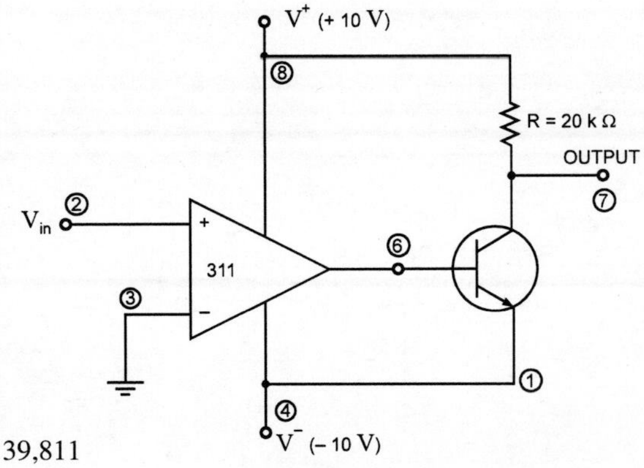

Fig. 11.6

Example 11.2. A noninverting zero-crossing detector uses op-amp μA 741 with an open-loop gain of 92 dB. If the power supply voltages are ±15 V and input offset 0 V, find the input voltage range that causes the output voltage to slew between the power supply voltages.

Solution :

$$\text{Open-loop gain, A} = 92 \text{ dB or Antilog}_{10} \frac{92}{20} = 39{,}811$$

$$\text{Power supply voltages, } V_{CC} = \pm 15 \text{ V}$$

Input offset voltage $= 0$ V

Output voltage, $V_{out} = \pm 15$ V or 30 V

$$\text{Input voltage} = \frac{V_{out}}{A} = \frac{30}{39{,}811} = 7.535 \times 10^{-4}\text{ V or } 0.7535\text{ mV} \quad \textbf{Ans.}$$

11.4. LOGARITHMIC AND ANTILOGARITHMIC AMPLIFIERS

Logarithmic/antilogarithmic amplifiers have several areas of applications. Analog computation may need $\log_e x$, $\log_{10} x$, or sin hx. This can all be performed continuously with analog log amplifiers. It is very convenient to have direct dB displays on digital voltmeters and spectrum analyzers which employ log amplifiers. Signal dynamic range compressors enhance computer signal processing as well as audio recording. Logarithmic amplifiers can be employed to compress a signal's dynamic range.

11.4.1. Logarithmic Amplifiers

The fundamental log op-amp circuit is shown in Fig. 11.7 (*a*) where a feedback resistor is replaced by a diode. Here the output voltage obtained is proportional to logarithm of the input voltage.

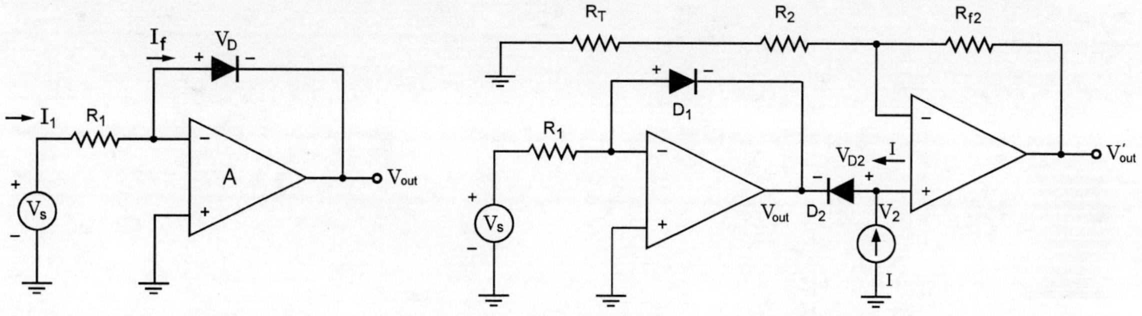

(*a*) (*b*) *Log Op-amp With Temperature Compensation*

Fig. 11.7

The V-I relationship in the feedback path is given by

$$I_f = I_0\, e^{V_D/\eta V_T} - 1 \quad \text{...(11.1)}$$

$$\text{or } I_f \simeq I_0\, e^{V_D/\eta V_T} \quad \text{...(11.2) assuming } \frac{V_D}{\eta V_T} >> 1$$

$$\therefore \quad \frac{I_f}{I_0} = e^{V_D/\eta V_T}$$

$$\text{or } \log\left(\frac{I_f}{I_0}\right) = \frac{V_D}{\eta V_T}$$

$$\text{or } V_D = \eta V_T (\log_e I_f - \log_e I_0) \quad \text{...(11.3)}$$

$$\text{Also } I_f = I_1 = \frac{V_s}{R_1} \quad \text{(due to virtual ground)}$$

$$\text{and } V_{out} = -V_D$$

$$\text{or } V_{out} = -\eta V_T\left(\log_e \frac{V_s}{R_1} - \log_e I_0\right) \quad \text{...(11.4)}$$

The Eq. (11.4) gives us the result that output voltage V_{out} is proportional to logarithm of input voltage V_s. We can also find $\log_{10} x$ by using relation

$$\log_{10} x = 0.4343 \log_e x.$$

The circuit, however, shows that V_{out} is dependent on ηV_T and on saturation current I_0 and both these have temperature dependency.

The temperature effects can be reduced by using circuit shown in Fig. 11.7 (*b*) where diodes D_1 and D_2 are matched, we have used a constant current source I which is independent of temperature, although R_T is a temperature dependent resistor.

Here in Fig. 11.7 (*b*) the input voltage at noninverting terminal of second op-amp is V_2 (say) and diode D_2 voltage say, V_{D_2} so that

$$V_{out} = V_2 - V_{D2} \qquad ...(11.5)$$

$$\text{or } V_2 = V_{out} + V_{D2} \qquad ...(11.6)$$

Now using Eq. (11.3) we can express V_{D2} as

$$V_{D_2} = \eta V_T (\log_e I - \log_e I_0) \qquad ...(11.7)$$

Using Eqs. (11.4), (11.6) and (11.7) we have

$$V_2 = -\eta V_T \left(\log_e \frac{V_s}{R_1} - \log_e I_0\right) + \eta V_T (\log_e I - \log_e I_0) = \eta V_T \left(\log_e I - \log_e \frac{V_s}{R_1}\right)$$

$$= -\eta V_T \log_e \left(\frac{V_s}{I R_1}\right) \qquad ...(11.8)$$

We can express V'_{out} as

$$V'_{out} = \left(1 + \frac{R_{f2}}{R_T + R_2}\right) V_2 \qquad ...(11.9)$$

So that the output voltage

$$V'_{out} = -\left(1 + \frac{R_{f2}}{R_T + R_2}\right) \eta V_T \log_e \left(\frac{V_s}{I R_1}\right) \qquad ...(11.10)$$

Here, the resistance R_T is selected to compensate approximately for the factor ηV_T.

The diode of basic log op-amp circuit of Fig. 11.7 (*a*) can also be replaced by a transistor, as shown in Fig. 11.7 (*c*).

It can be shown that circuit shown in Fig. 11.7 (*c*) gives the same result as with diode

$$I_E = I_0 (e^{V_E/\eta V_T} - 1)$$

As the base of transistor is grounded $I_C \simeq I_E$,

$$I_C = I_0 (e^{V_E/\eta V_T} - 1)$$

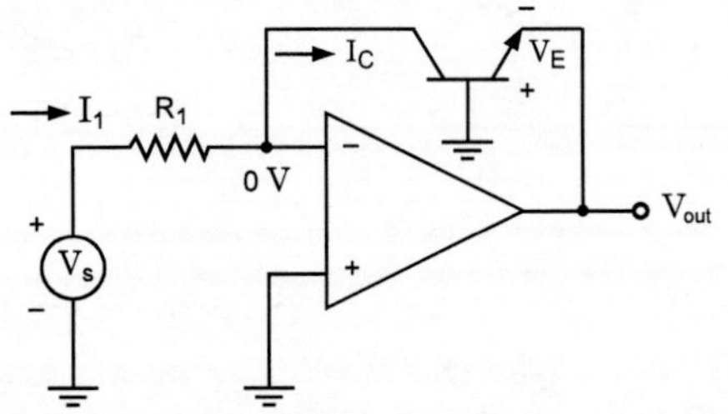

Fig. 11.7 (*c*). *Log Op-amp Using Transistor in Feedback*

Log Amplifier Circuit Using Two Differential Amplifiers

The temperature effects can also be reduced by using difference logarithmic amplifier circuit shown in Fig. 11.8.

The emitter voltage of transistor Q_1 can be had from Eq. (11.4). This is also V_A.

$$V_A = -\eta V_T \log_e \frac{V_s}{I_o R_1} \qquad ...(11.11)$$

The base-emitter voltage for Q_2 is

$$V_{Q_2BE} = \eta V_T \log_e \frac{V_{REF}}{I_o R_1} \quad \text{...(11.12)}$$

The base voltage of Q_2, then is

$$V_{Q_2\text{ base}} = V_{Q_2\text{ emitter}} + V_{Q_2BE}$$

But $V_{Q_2\text{ base}} = V_B$ and $V_{Q_2\text{ emitter}} = V_A$

$$\text{Thus, } V_B = V_A + V_{Q_2BE} \quad \text{...(11.13)}$$

Now from Eqs. (11.11), (11.12) and (11.13) we have

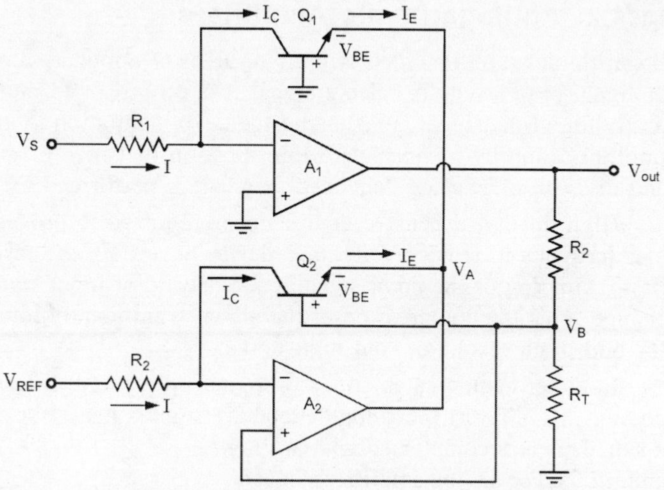

Fig. 11.8. ***Difference Logarithmic Amplifier***

$$V_B = -\eta V_T \log_e \frac{V_s}{I_o R_1} + \eta V_T \log_e \frac{V_{REF}}{I_o R_1}$$

$$= -\eta V_T \left(\log_e \frac{V_s}{I_o R_1} - \log_e \frac{V_{REF}}{I_o R_1} \right)$$

$$= -\eta V_T \log_e \frac{V_s}{V_{REF}} = -\eta V_T \log_e \frac{V_s}{IR_1} \quad \text{...(11.14)}$$

This is the voltage in the centre of the voltage divider

$$V_B = \frac{R_T}{R_2 + R_T} V_{out} \quad \text{...(11.15)}$$

From Eqs. (11.14) and (11.15), we have

$$-\eta V_T \log_e \frac{V_s}{IR_1} = \frac{R_T}{R_2 + R_T} V_{out}$$

$$\text{or} \quad V_{out} = -\frac{R_2 + R_T}{R_T} \eta V_T \log_e \frac{V_s}{IR_1} \quad \text{...(11.16)}$$

The noteworthy point is that the specifications of op-amp need careful consideration. At low levels of input current (and, therefore, input voltage), bias current and drift can cause linearity and log conformance errors. With bias currents as large as 10 percent of the input current, output errors of the order of 2.5 mV may be encountered.

Frequency stability considerations may need external compensation. This is normally had by placing a capacitor across the negative feedback element. This causes lowering in gain with rising frequency. However, the effective resistance in the negative feedback loop is

$$r_e \cong \frac{26 \text{ mV}}{I_C}$$

At high levels of input (collector) current, r_e becomes very small, shorting out the parallel compensation capacitor. The capacitor becomes ineffective. The solution is to include a resistor in series with the emitter, placing a lower limit on the feedback resistance through the transistor.

A typical frequency compensation network is given in Fig. 11.9. Values of R and C are best determined experimentally, as parasitic effects play a vital role.

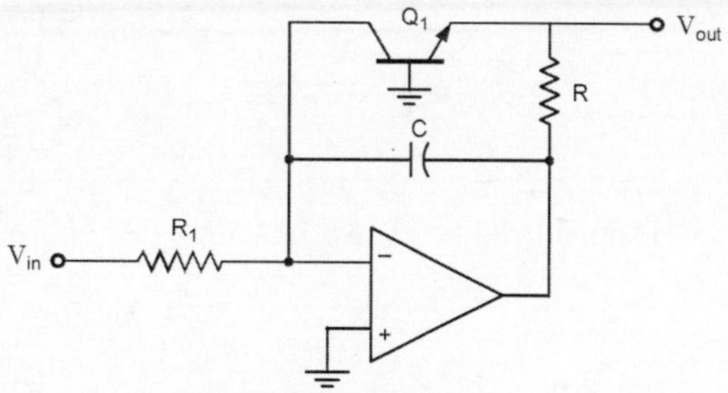

Fig. 11.9. ***Log Amplifier Frequency Compensation***

11.4.2. Antilogarithmic Amplifiers

Logarithmic amplifiers have two major areas of application namely, in complex analog calculations and in signal processing. To raise a signal to a power, or to multiply two signals or to divide two signals, firstly log of the signal is taken and then exponentiation to any power is accomplished by amplification (multiplication by a constant). Multiplication of two log "encoded" signals is performed by an adder, and division of two log "encoded" signals is performed by a subtractor.

High quality, expensive equipment is required to process signals, having range varying from tens of microvolts to tens of volts, in order to have both the resolution (μV) and range (tens of volts). By first taking log of the input signal, each decade of input signal change needs only 1 V of output signal change. This log compression of the signal significantly lowers the demands on the processing circuits for both high resolution and high range.

In either analog calculations or signal processing, the signal after proper conditioning, needs "de-coding". To convert these log "encoded" signals back into real-world terms, antilog is required to be taken. This is accomplished with antilog amplifiers. For the log amplifier a decade change of input signal causes 1 V change on the output. For the antilog amplifier, one volt change of input signals causes a decade's change at the output.

The antilog amplifier gives us the output voltage proportional to antilogarithm of input voltage. The circuit is shown in Fig. 11.10.

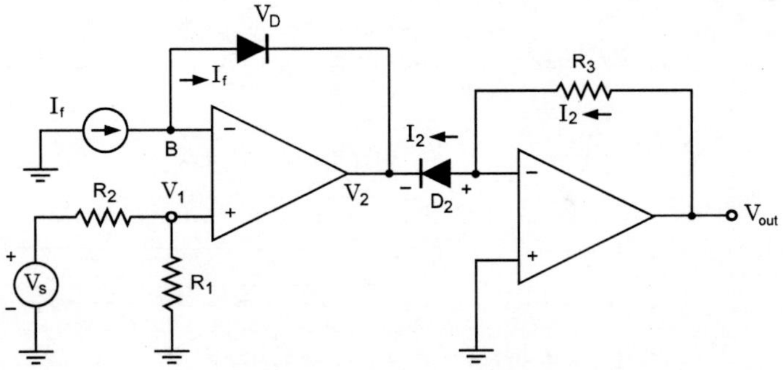

Fig. 11.10. *Antilog Amplifier*

Here we assume that the two op-amps have infinite input resistance and they behave as pure differential amplifier *i.e.,*

$$V_2 = V_1 - V_D \qquad (\because \text{ at node B, voltage drop} = V_1)$$

$$= \frac{R_1}{R_1 + R_2} V_s - \eta V_T (\log_e I_f - \log_e I_0) \qquad ...(11.17) \quad [\text{Using Eq. (11.3)}]$$

$$V_2 = -V_{D2} = -\eta V_T (\log_e I_2 - \log_e I_0) \qquad ...(11.18)$$

$$\text{Also } V_{out} = I_2 R_3 \qquad ...(11.19)$$

From Eqs. (11.17) and (11.18), we have

$$\frac{R_1 V_s}{R_1 + R_2} - \eta V_T \log_e \left(\frac{I_f}{I_0}\right) = -\eta V_T \log_e \left(\frac{I_2}{I_0}\right)$$

$$\text{or } \frac{R_1 V_s}{R_1 + R_2} = -\eta V_T \log_e \left(\frac{I_2}{I_f}\right)$$

$$\text{or } I_2 = I_f \text{ antilog}_e \left[\frac{R_1 V_s}{R_1 + R_2} \left(\frac{-1}{\eta V_T}\right)\right]$$

From Eq. (11.19), we have

$$V_{out} = R_3 I_f \text{ antilog}_e \left[\frac{R_1}{R_1 + R_2} \times \frac{(-V_s)}{\eta V_T}\right] \qquad ...(11.20)$$

$$\text{or } V_{out} = R_3 I_f \text{ antilog}_e \left[\frac{-R_1}{R_1 + R_2} \times \frac{V_s}{\eta V_T}\right] \qquad ...(11.21)$$

Another antilog amplifier circuit is shown in Fig. 11.11. This circuit can be obtained by exchanging input and output connections (more or less) of the difference log amplifier circuit shown in Fig. 11.8. The input to the antilog amplifier is fed to the base of Q_2 through temperature compensating voltage divider network using resistors R_2 and R_T. The output voltage of the antilog amplifier is supplied back to the inverting input terminal of the op-amp A_1 through resistor R_1. Other than this input/output reversal, the only difference between the log and antilog amplifiers is the noninverting input of op-amp A_2. For the log amplifier it is tied to the base of Q_2 while in case of antilog amplifier, noninverting input of op-amp A_2 is tied to ground.

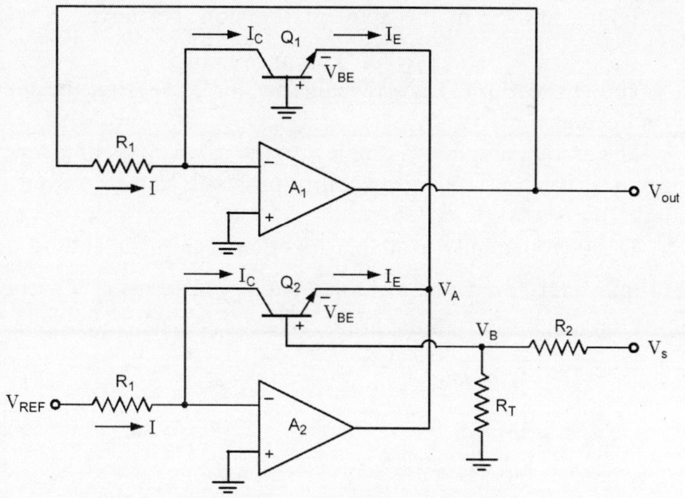

Fig. 11.11. *Antilog Amplifier*

Since the base of Q_1 is tied to ground.

$$V_A = -V_{Q_1BE} = -\eta V_T \log_e \frac{V_{out}}{I_o R_1} \quad ...(11.22)$$

V_B is the base voltage of Q_2, and is the output from the R_2–R_T voltage divider.

$$\text{So } V_B = V_{Q_2\text{ Base}} = \frac{R_T}{R_2 + R_T} V_s \quad ...(11.23)$$

The voltage at the emitter of Q_2 is

$$V_{Q_2E} = V_{Q_2\,base} + V_{Q_2BE} = \frac{R_T}{R_2 + R_T} V_s - \eta V_T \log_e \frac{V_{REF}}{I_o R_1} \quad ...(11.24)$$

But the emitter voltage of Q_2 equals V_A. So by equating Eqs. (11.22) and (11.24) we have

$$-\eta V_T \log_e \frac{V_{out}}{I_o R_1} = \frac{R_T}{R_2 + R_T} V_s - \eta V_T \log_e \frac{V_{REF}}{I_o R_1}$$

$$\text{or } \frac{R_T}{R_2 + R_T} \cdot V_s = -\eta V_T \left(\log_e \frac{V_{out}}{I_o R_1} - \log_e \frac{V_{REF}}{I_o R_1} \right)$$

$$= -\eta V_T \log_e \frac{V_{out}/I_o R_1}{V_{REF}/I_o R_1} = -\eta V_T \log_e \frac{V_{out}}{V_{REF}} \quad ...(11.25)$$

$$\text{or } \frac{-R_T}{R_2 + R_T} \cdot \frac{1}{\eta V_T} V_s = \log_e \frac{V_{out}}{V_{REF}}$$

$$\text{or } \frac{-0.4343}{\eta V_T} \frac{R_T}{R_2 + R_T} V_s = 0.4343 \log_e \frac{V_{out}}{V_{REF}}$$

$$\text{or } \frac{-0.4343}{\eta V_T} \cdot \frac{R_T}{R_2 + R_T} V_s = \log_{10} \frac{V_{out}}{V_{REF}} \qquad \because \log_{10} x = 0.4343 \log_e x$$

$$\text{or } -kV_s = \log_{10} \frac{V_{out}}{V_{REF}}$$

Taking antilog of the above expression, we have

$$V_{out} = V_{REF} 10^{-KV_s} \quad ...(11.26)$$

The above Eq. (11.26) reveals that for increasing the input voltage by unity decreases the output by a decade.

The antilog amplifier is subject to the same types of errors we have seen with the log amplifier. Bias current, drift and frequency stability problems can be solved for the antilog amplifier as in case of log amplifiers.

The commercially available functional chip for antilog amplifier is 755.

Example 11.3. In the Fig. 11.12 find out V_{out} in terms of V_{in} and $V_{ref.}$

Solution : Since in both transistors, collector and base are shorted, they are working as diode.

For diode D_1 $I_{f1} = I_0 \left[e^{v_{f1}/\eta V_T} - 1\right]$

$$I_{f1} \approx I_0 \, e^{V_{f1}/\eta V_T}$$

$$V_{f1} = \eta V_T (\log_e I_f - \log_e I_0)$$

$$V_{out_1} = -V_f$$

$$= -\eta V_T \left(\log_e \frac{V_{in}}{R_1} - \log_e I_0\right)$$

Similarly, $V_{out_2} = -\eta V_T \left(\log_e \frac{V_{ref}}{R_1} - \log_e I_0\right)$

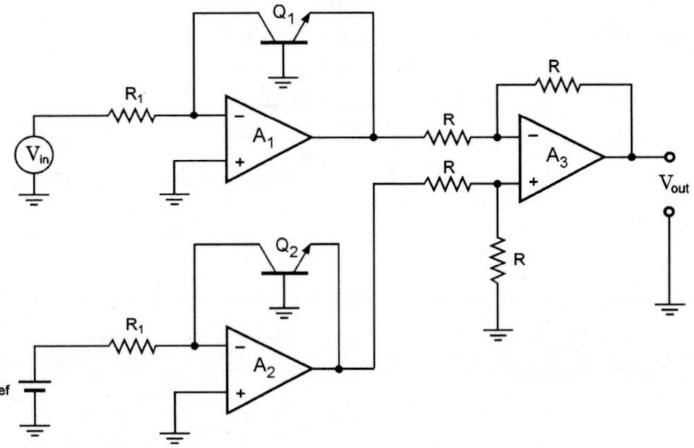

Fig. 11.12

Applying superposition theorem to third op-amp, we have

$$V_{out} = \frac{R}{R}(V_{out_2} - V_{out_1})$$

Substituting the values of V_{out_1} and V_{out_2} and simplifying, we have

$$V_{out} = \eta V_T \log_e \frac{V_{in}}{V_{ref}} \quad \textbf{Ans.}$$

11.5. LOGARITHMIC MULTIPLIER

We can use log amplifiers and antilog amplifiers (described earlier) for multiplication or division of two analog signals V_{in1} and V_{in2}.

The log of two signals are taken and fed to an amplifier and then the antilog of resulting sum is taken to get the multiplication of the two signals, as illustrated in Fig. 11.13.

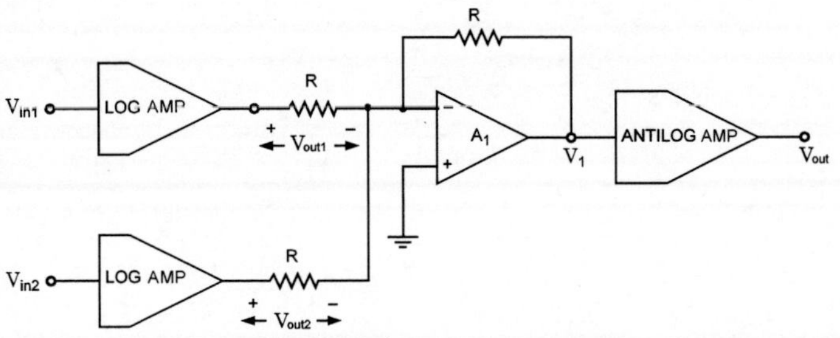

Fig. 11.13. *Logarithmic Multiplier of Two Signals* ($V_{out} = KV_{in1}V_{in2}$)

From Fig. 11.13

$$V_{out1} = K_1 \log_e V_{in1} \quad ...(11.27)$$

$$V_{out2} = K_1 \log_e V_{in2} \quad ...(11.28)$$

where K_1 is some constant

At the output of summing op-amp, we have

$$V_1 = V_{out1} + V_{out2}$$

$$= K_1 \log_e V_{in1} + K_1 \log_e V_{in2}$$

$$= K_1 \log_e V_{in1} V_{in2} \qquad \text{(Since } R_f = R = R_1 = R_2)$$

and hence $V_{out} = K_2 \text{ antilog}_e K_3V_1 = K_2 \text{ antilog}_e [K_3K_1 \log V_{in1}V_{in2}]$

If $K_3K_1 = 1$ (say) $\quad V_{out} = K_2V_{in1} V_{in2}$ ...(11.29)

To use circuit shown in Fig. 11.13 for division of two signals the op-amp A_1 can be used for subtraction of output signals V_{out1} and V_{out2} so that

$$V_1 = K_1 \log_e V_{in1} - K_1 \log_e V_{in2} = K_1 \log_e \frac{V_{in1}}{V_{in2}}$$

$$\text{and } V_{out} = K_2 \text{ antilog}_e K_3V_1 = K_2 \text{ antilog}_e \left(K_3K_1 \log_e \frac{V_{in1}}{V_{in2}} \right)$$

$$\text{if we set } K_1K_3 = 1$$

$$V_{out} = K_2 \frac{V_{in1}}{V_{in2}} \quad \text{...(11.30)}$$

The circuit shown in Fig. 11.13 produces one quadrant multiplication or division as both inputs are required to be positive only.

The device AD 530 manufactured by Analog Devices Inc. is completely a monolithic multiplier/divider with basic accuracy of 1 per cent and bandwidth of 1 MHz. The output is XY/10 as multiplier and as divider $+10 \frac{Z}{X}$. As a divider it works like two quadrant (where X is negative only). Other multiplier IC chips are AD 533 and AD 534.

11.6. PRECISION RECTIFIERS

Precision rectifier is a small signal rectifier which is capable of rectifying signals of very small peaks (of the order of a few millivolts). Since such small ac signals are not able to drive a diode directly due to cutin voltage of the diode, therefore, an op-amp is introduced prior to diode, as illustrated in Fig. 11.14. The circuit is essentially a noninverting voltage follower when the diode is forward biased. When v_{in} is positive, the output of the amplifier, v'_{out} is positive, the diode is forward biased, and a low-resistance path is established between v'_{out} and v^-, as necessary for a voltage follower. The output voltage, v_{out}, then follows the positive variations of $v_{in} = v^+$. Note that even a very small positive value of v_{in} will cause this result because of the large differential voltage gain of the amplfier; *i.e.*, the large voltage gain and the action of feedback cause the usual result that $v^+ = v^-$. Note also that the voltage drop across the diode does not appear in v_{out}.

When the input goes negative, v'_{out} becomes negative, and the diode is reverse biased. This effectively opens the feedback loop, so v_{out} no longer follows v_{in}. The amplifier itself, now operating open-loop, is quickly driven to its maximum negative output, thus holding the diode well into reverse bias. Thus during negative half cycle no output is obtained.

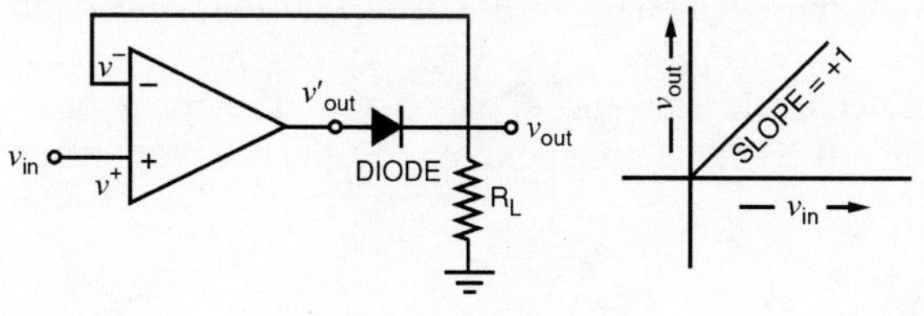

Fig. 11.14

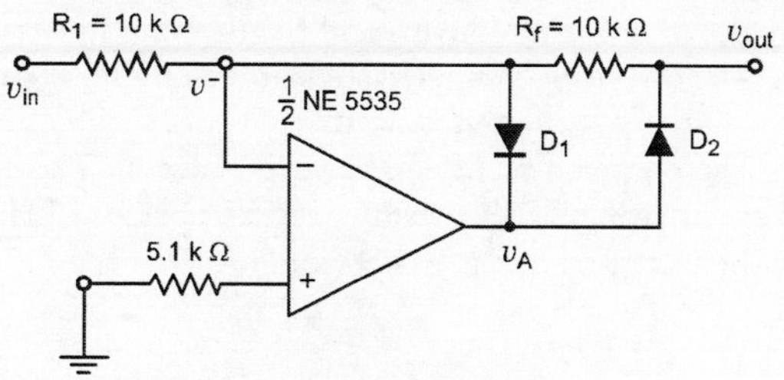

Fig. 11.15. ***Precision Half-Wave Rectifier Using NE 5535 High SR Dual Op-Amp***

Another half-wave rectifier using op-amp is depicted in Fig. 11.15. The op-amp used is high SRNE5535 dual op-amp.

When input voltage v_{in} is positive, output voltage v_A is negative, diode D_1 being forward biased is *on* while diode D_2 being reverse biased is *off*. The diode D_1 shorts the outut v_A to v^-, which is held at virtual ground, so output voltage v_{out} is 0.

When input voltage v_{in} is negative, v_A is positive, diode D_1, being reverse biased, is off (open) and diode D_2, being forward biased, is *on* (closed) and therefore connects the amplifier output v_A through R_f to v^-. Thus, the circuit behaves like an ordinary inverting amplifier with gain $-R_f/R_1$. The output voltage is an amplified and inverted (positive) version of the negative variations in v_{in}. The result is a half-wave rectified output with no diode drops to reduce the signal.

Two half-wave rectifier circuits can be combined, as shown in Fig. 11.16, to provide a precision full-wave rectifier. Note that input connections are reversed for the second op-amp.

Wave shapes of precision half-wave and full-wave rectifiers are shown in Figs. 11.17 (*a*) and 11.17 (*b*) respectively.

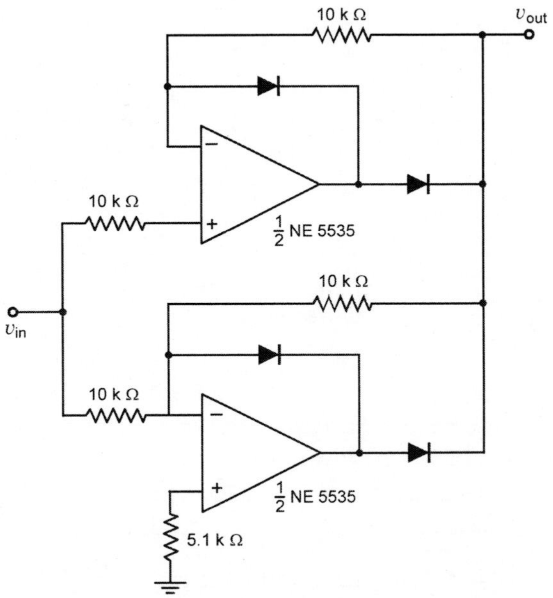

Fig. 11.16. ***Precision Full-Wave Rectifier Using NE 5535 High SR Dual Op-Amps***

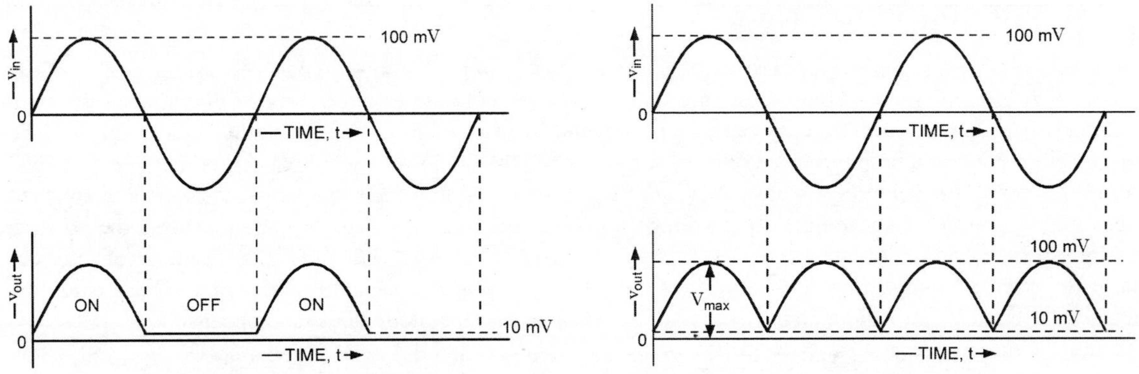

(*a*) *For Half-Wave Rectifier* (*b*) *For Full-Wave Rectifier*

Fig. 11.17. ***Wave Forms of Precision Rectifiers***

The advantages of precision rectifiers are that (*i*) they can rectify signals of very low amplitude (all the way down to the mV level) and (*ii*) they are non-saturating ones.

The inverting characteristics of the output V_{out} can be circumvented by the use of an additional inversion for achieving a positive output.

The precision rectifiers, however, have their speed limited by the slew rate of the op-amp. The operation of the circuit using 741 is usually restricted below 1 kHz. When JFET input op-amps are used, the operating range may be increased to the 10 kHz range.

11.7. CLIPPERS

As already mentioned in Art. 10.2, clippers may be positive clippers or negative clippers. The clipper which removes the positive portion of the input signal, is known as the *positive clipper*. The positive clipper can be formed by using an op-amp and a rectifier diode, as shown in Fig. 11.18. In this circuit the op-amp is basically employed as a voltage follower with the rectifier diode in the feedback path. The clipping level is determined by the reference voltage V_{ref}, which should be less than the input voltage range of the op-amp. Also, since V_{ref} is derived from the positive supply voltage ($+V_{CC}$), the dc supply voltage must be well regulated. The output voltage has portions of the positive half-cycles above V_{ref} clipped off.

The circuit operates as shown in Figs. 11.18 and 11.19.

During positive half cycle of the input signal, the diode D conducts only until $v_{in} = V_{ref}$, because when v_{in} is lower than V_{ref}, the voltage (V_{ref}) at the (–) input terminal is higher than that at the (+) input terminal and consequently the output voltage v'_{out} of the op-amp becomes sufficiently negative to drive diode D into

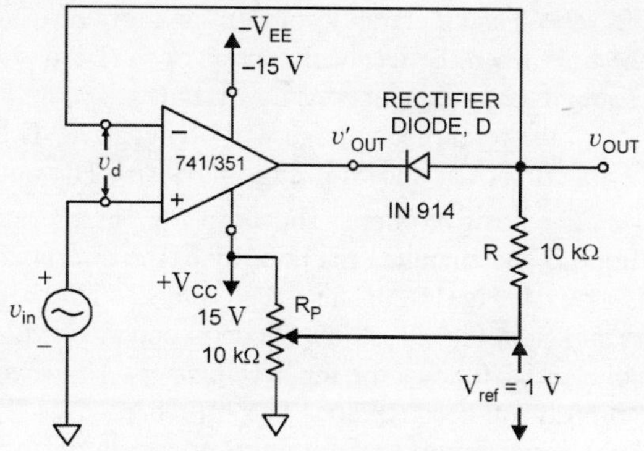

Fig. 11.18. ***Positive Clipping Circuit***

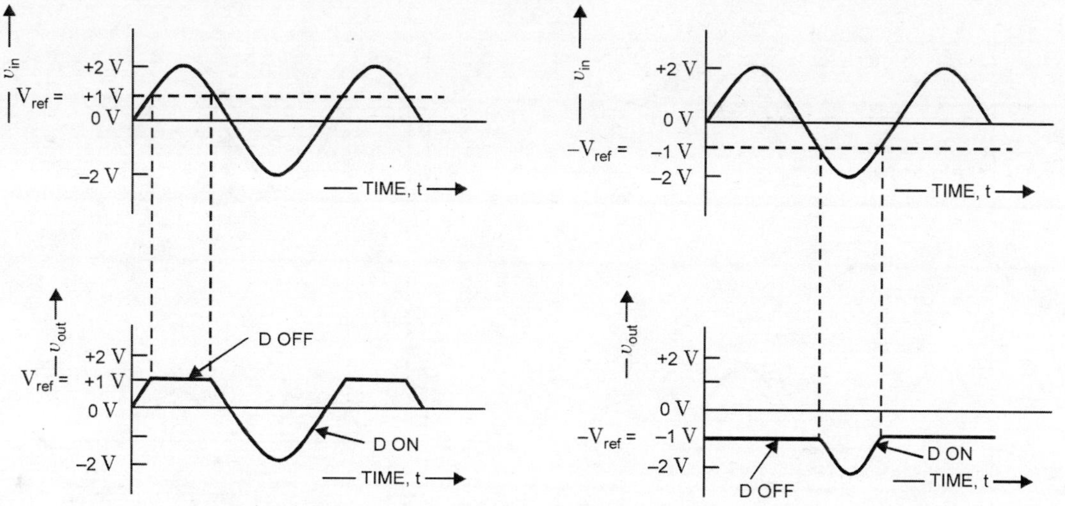

(a) Input and Output Waveforms With $V_{ref} = 1$ V *(b) Input and Output Waveforms With $-V_{ref} = -1$ V*

Fig. 11.19

conduction. When diode D conducts, the feedback loop closes and the op-amp operates as a voltage follower *i.e.*, output voltage v_{out} follows input voltage v_{in} until $v_{in} = V_{ref}$. However, when v_{in} exceeds V_{ref}, slightly, the op-amp output voltage v'_{out} becomes sufficiently positive to drive diode D into cutoff. This makes the feedback loop open, and so the op-amp operates open-loop and output voltage v'_{out} is further driven toward positive saturation ($\cong +V_{CC}$). When diode D is reverse biased the output voltage $v_{out} = V_{ref}$. Thus,

For $v_{in} > V_{ref}$, $v'_{out} \cong +V_{CC}$ and $v_{out} = V_{ref}$, as shown in Fig. 11.19 (*a*).

When input voltage v_{in} becomes less than V_{ref}, the op-amp output voltage v'_{out} becomes sufficiently negative to drive diode D into conduction. This closes the feedback loop and output follows the input. Thus diode D is *on* for $v_{in} < V_{ref}$ and *off* for $v_{in} > V_{ref}$. The output follows the input only when the diode is *on*. The op-amp alternates between the open-loop and closed-loop operations as the diode D is turned *off* and *on*, respectively. That is why, the op-amp used must be

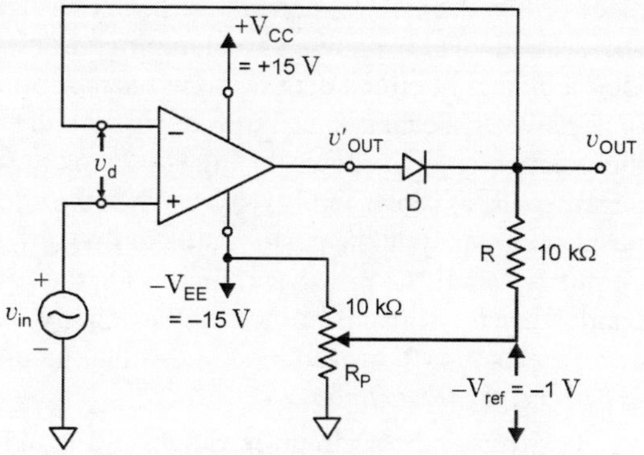

Fig. 11.20. ***Negative Clipping Circuit***

high speed and preferably compensated for unity gain. In addition, the difference input voltage v_d is high during the time when the feedback loop is open (Diode D *off*), so the op-amp with a high difference input voltage is required to avoid input breakdown. In Fig. 11.18, if pot R_P is connected to the negative supply $-V_{EE}$, instead of $+V_{CC}$, the reference voltage will be negative. This will make the entire output waveform above $-V_{ref}$ to be "clipped" off, as shown in Fig. 11.19 (*b*). The output follows the input only when v_{in} is less than $-V_{ref}$.

The positive clipper shown in Fig. 11.18 can be converted into a negative clipper by simply reversing diode D and changing the polarity of the reference voltage V_{ref}. The circuit for a negative clipper so obtained is shown in Fig. 11.20. This circuit "clips" off the negative portions of the input signal below the reference voltage, as shown in Fig. 11.21 (*a*). Diode D conducts when v_{in} exceeds $-V_{ref}$ and, therefore, during this period output voltage v_{out} follows the input voltage v_{in}. However, the negative portion of the output voltage below $-V_{ref}$ is "clipped" off because diode D is off when $v_{in} < -V_{ref}$. In the circuit shown in Fig. 11.20, if $-V_{ref}$ is changed to $+V_{ref}$ by connecting the potentiometer R_P to the $+V_{CC}$, the output voltage below $+V_{ref}$ will be "clipped" off, as shown in Fig. 11.21 (*b*). The diode D must be *on* for $v_{in} > V_{ref}$ and off for $v_{in} < +V_{ref}$.

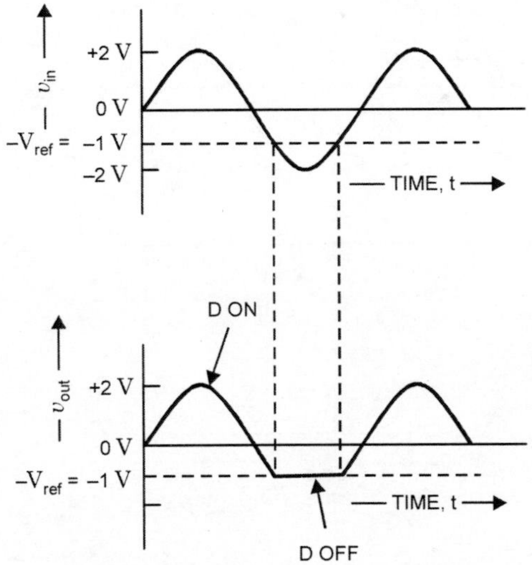

(*a*) *Input and Output Waveform With* $-V_{ref} = -1$ *V*

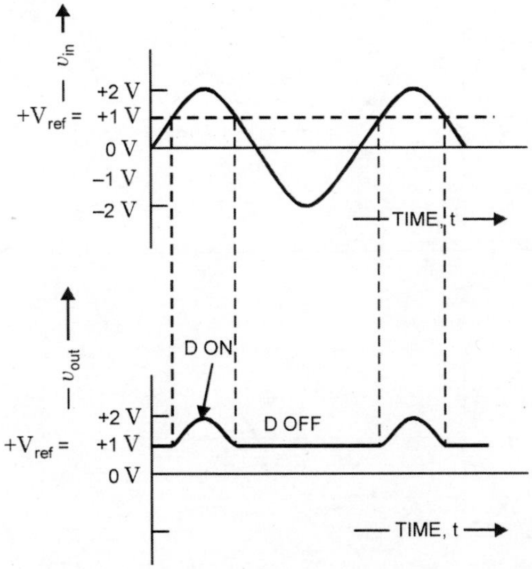

(*b*) *Input and Output Waveforms With* $+V_{ref} = +1$ *V*

Fig. 11.21

11.8. CLAMPERS

As already mentioned in Art. 10.7, a circuit that places either the positive or negative peak of a signal at a desired level is known as a clamping circuit. In fact, a clamping circuit adds dc component (positive or negative) to the input signal so as to push it either on the positive side as shown in Fig. 11.23 or on the negative side as shown in Fig. 11.24. When the circuit pushes the signal on the positive side or upward, the circuit is called the *positive clamper*. On the other hand, when the signal is pushed on the negative side or downward, as shown in Fig. 11.24, the circuit is called the *negative clamper*.

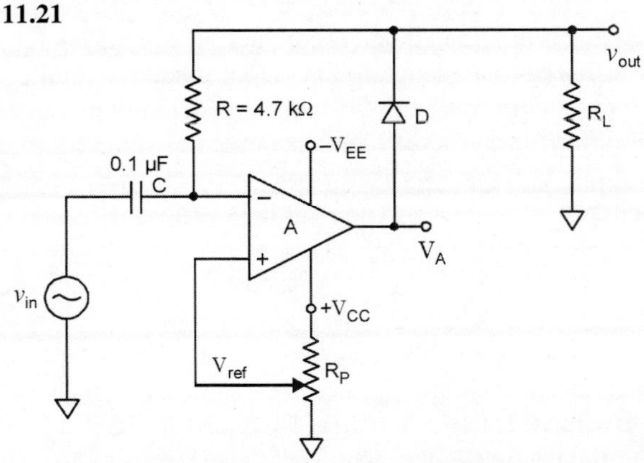

Fig. 11.22. ***Positive Clamper***

Positive clamper circuit is shown in Fig. 11.22. In absence of v_{in}, V_A is positive, diode D is forward biased and conducts. The circuit operates as a voltage follower and its output voltage $v_{out} = V_{ref}$. When v_{in} is applied to the inverting input terminal, the diode conducts in the negative half cycle and the capacitor C gets charged to the peak negative value of v_{in} through the forward biased diode. However, during positive half cycle of v_{in}, the diode D is reverse biased and, therefore, remains *off* and thus capacitor circuit remains open. The capacitor C retains the voltage stored in the negative half cycle V_{max}. Since this V_{max} is in series with the ac input signal, the output voltage will be $V_{max} + v_{in}$. The total output voltage will be equal to $v_{in} + V_{max} + V_{ref}$, as shown in Fig. 11.23.

Positive clamper can be converted into negative clamper by reversing the diode and using a negative reference voltage $-V_{ref}$. The expected waveforms are shown in Fig. 11.24. The resistor R is used for protection of the op-amp against excessive discharge currents from capacitor C especially when the dc supply voltages are switched off.

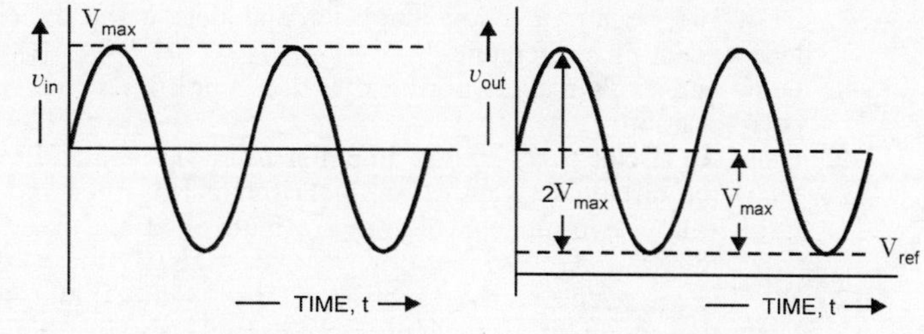

Input Signal *Output Waveform*

Fig. 11.23. ***Input and Output Waveforms With Positive Clamper***

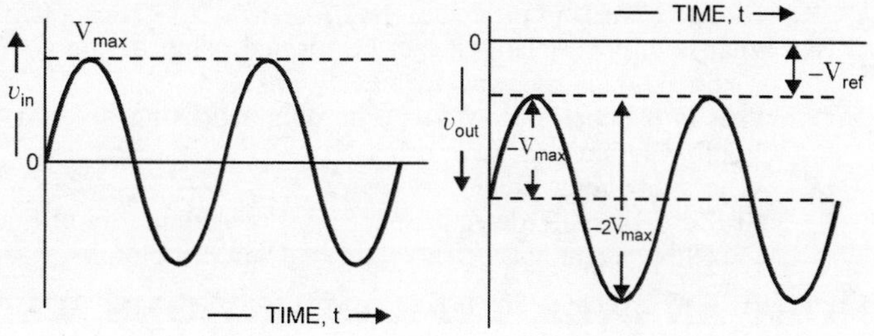

Input Signal *Output Waveform*

Fig. 11.24. ***Input and Output Waveforms With Negative Clamper***

HIGHLIGHTS ....

1. A comparator is essentially an op-amp operated in open-loop configuration. The output signal is either a high or low saturated voltage. It is used in digital interfacing, Schmitt triggers, analog-to-digital converters, oscillators etc.
2. A comparator with a reference voltage of zero volt is called a zero-crossing detector. Diode clampers are often used to protect the comparator against excessively large input voltages.
3. Logarithmic/antilogarithmic amplifiers have several areas of applications. Analog computation may need $\log_e x$, $\log_{10} x$, or sin *hx*. This can all be performed continuously with analog/log amplifiers. It is very convenient to have direct dB displays on digital voltmeters and spectrum analyzers which employ log amplifiers. Signal dynamic range compressors enhance computer signal processing as well as audio recording. Logarithmic amplifiers can be employed to compress a signal's dynamic range.
   The output voltage of a logarithmic amplifier is proportional to the logarithm of the input voltage.
   The antilog amplifier provides the output voltage proportional to the antilogarithm of the input voltage.
4. Precision rectifier is a small signal rectifier which is capable of rectifying signals of very small peaks (of the order of few millivolts).
   Two half-wave rectifier circuits can be combined to provide a precision full-wave rectifier.
5. A clipper is a circuit that removes certain portions of the input waveform. The clipper can be formed by using an op-amp with a rectifier diode.
6. A circuit that places either the positive or negative peak of a signal at the desired level is known as clamping circuit.

EXERCISES ....

1. What is comparator ? What is difference between the inverting op-amp comparator and noninverting op-amp comparator ? Explain the working of an op-amp comparator with circuit diagram and input and output waveforms.
2. Discuss the application of op-amp as comparator circuit.
3. Explain the operation of a zero-crossing-detector.
4. What is zero-crossing comparator ? What is its drawback ? How is this minimised ?
5. How op-amp can be used as logarithmic amplifier ?
6. Draw the circuit of a log amplifier using op-amps and explain its operation. Enlist the possible applications of this circuit.

7. Draw the circuit of a log amplifier and derive the expression for the output voltage. Is the circuit independent of temperature ? **[G.G.S.I.P. Univ Analog Electronics II May 2010]**
8. Draw and explain the working of a log amplifier which provides saturation current and temperature compensation. **[G.G.S.I.P. Univ. Analog Electronics-II May 2011]**
9. Draw the circuit of an antilog amplifier using op-amp and explain its operation.
10. Draw the circuit diagram of antilogarithmic amplifier and explain it.
11. Explain how two analog voltages are multiplied and divided by using log and antilog amplifiers. **[G.G.S.I.P. Univ. Analog Electronics-II May 2011]**
12. Why do we need precision rectifier ? Draw and explain how a precision rectifier works.
13. Draw the circuit of a precision rectifier and explain its working.
14. What do you mean by precision rectifier, explain with circuit. **[G.G.S.I.P. Univ. Analog Electronics-II May 2011]**
15. Make a precision op-amp full-wave rectifier and explain its working. Can this circuit be used as an absolute value circuit ? **[G.G.S.I.P. Univ. Analog Electronics-II May 2011]**
16. What is clipper ? How it can be formed using op-amp ? Discuss negative clipping circuit along with circuit diagram and waveforms.
17. What is clamper ? How it can be obtained using op-amp ? Discuss positive clamper with the help of circuit diagram and waveforms.
18. Write short notes on the following :
    (*i*) Zero-crossing detectors. (*ii*) Logarithmic amplifier. (*iii*) Antilog amplifier.
    (*iv*) Clipper circuits. (*v*) Clamper circuits.

SHORT ANSWER TYPE QUESTIONS WITH ANSWERS ....

Q. 1. What is a comparator and where is it used ?

Ans. A comparator is essentially an op-amp operated in open-loop configuration. The output signal is either a high or low saturated voltage. It is used in digital interfacing, Schmitt triggers, analog-to-digital converters, oscillators etc.

Q. 2. What is zero-crossing detector ?

Ans. A comparator with a reference voltage of zero volt is called a zero-crossing detector. Diode clampers are often used to protect the comparator against excessively large input voltages.

Q. 3. What is logarithmic amplifier ? Name the major applications of log amplifier.

Ans. The circuit, which has output voltage proportional to logarithm of the input voltage, is known as logarithmic amplifier. Log amplifiers have two major areas of applications viz in complex analog calculations and in signal processing.

Q. 4. What is antilog amplifier and for what purpose it is used ?

Ans. The antilog amplifier is circuit that provides the output voltage proportional to antilogarithm of input voltage. It is used to convert log "encoded" signals back into real-world terms.

Q. 5. What is clipper and how can it be formed ?

Ans. A waveshaping circuit that controls the shape of the output waveform by removing or clipping a portion of the input wave is known as the clipping circuit. The clippers may be positive or negative. The positive clipper can be formed by using an op-amp, and a rectifier diode as shown in Fig. 11.18. The positive clipper shown in Fig. 11.18 can be converted into a negative clipper by simply reversing diode D and changing the polarity of reference voltage V_{ref}.

PROBLEMS ....

1. For a basic comparator shown in Fig. 11.1(*a*), v_{in} = 2V_{PP} sine wave at 1 kHz . V_{ref} = 500 mV, R = 100 Ω and supply voltages = ±12 V. Draw output waveform.
2. For an LM 339 comparator given in Fig. 1, R_1 = 10 kΩ and R_2 = 2 kΩ. Determine reference voltage V_{ref}. **[Ans.** 2.5 V**]**

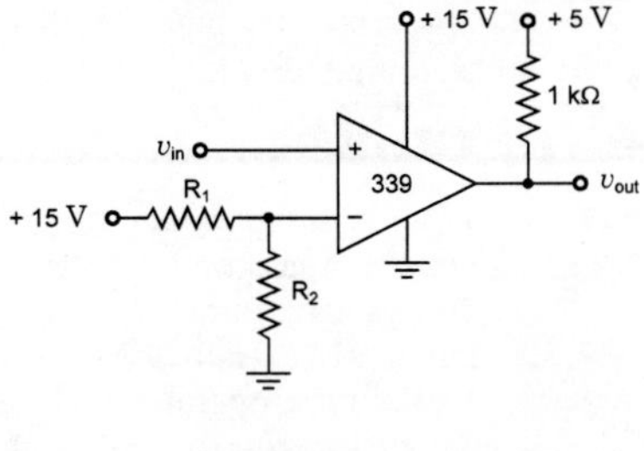

Fig. 1

❑❑❑

Unit-IV

ACTIVE RC FILTERS AND APPLICATIONS OF IC ANALOG MULTIPLIER

CONTENTS OF THIS UNIT

ACTIVE RC FILTERS : Idealistic and Realistic Response of Filters (LP, BP, HP); Butterworth and Chebyshev Approximation Filter Functions; All Pass, Notch Filter; Operational Transconductance Amplifier (OTA)-C Filters.

APPLICATIONS OF IC ANALOG MULTIPLIER : IC Phase-Locked Loops; IC Voltage Regulators; IC Function Generators.

CHAPTER 12

Active Filters

12.1. INTRODUCTION

An electric filter is a network designed to attenuate certain frequencies but pass others without attenuation. A filter circuit, therefore, possesses at least one passband—a band of frequencies in which the output is approximately equal to the input (*i.e.,* attenuation is zero) and an attenuation band in which output is zero (*i.e.,* attenuation is infinite). The frequencies that separate the different pass and attenuation bands are called the *cutoff frequencies*.

Filters may be of any type such as electrical, mechanical, pneumatic, hydraulic, acoustical etc. but the most commonly used filters are of the electrical type.

Electrical filters are used in practically all circuits which require separation of signals according to their frequencies. Applications include (but are certainly not limited to) noise rejection and signal separation in industrial and measurement circuits, feedback of phase and amplitude control in servo loops, smoothing of digitally generated analog signals, audio-signal shaping and sound enhancement, channel separation, and signal enhancement in communication circuits.

Filters are specified, analyzed, and designed somewhat differently than the circuits we have discussed so far. Filter's performance is specified in terms of frequency response, how the gain and phase shift change with frequency. Angular frequency, –3 dB frequency, bands, ripple, rolloff rate, centre frequency, and magnification factor are among the terms unique to filters.

12.2. FREQUENCY RESPONSE

The gain of a filter is a *complex number* that is a function of frequency. It means that at a given frequency, the gain will have a specific magnitude as well as a phase angle. The magnitude and phase shift of a filter varies with the variations in operating frequency. Indeed, this is the purpose of a filter. Performance is often described with a graph of gain (magnitude and phase) versus frequency. The manner in which the gain of the filter responds to different frequencies is called the *frequency response*. A graph of the magnitude of the gain versus frequency is called a *frequency response plot*. Although the magnitude of the gain may be expressed either in decibels (dB) or as a numerical value, the frequency is always plotted on *logarithmic scale*. Remember that to accommodate large frequency ranges the frequency is assigned a logarithmic scale, just as the gain amplitude is assigned a linear scale and is expressed in decibels to accommodate very high gain (of the order of 10^5 or higher).

Logarithmic scale is shown in Fig. 12.1. In Fig. 12.1 (*a*), the distance between 1 Hz and 10 Hz is the same as the distance between 10 and 100 Hz or 100 Hz and 1,000 Hz, or 1,000 Hz and 10,000 Hz or 10,000 Hz and 100,000 Hz. It means, each decade (factor of ten) increase or decrease moves us the same distance along the scale. There are two other noteworthy points. First, the starting point is not at zero. If we move farther left, each increment would lower the scale by 10 (to 0.1, then 0.01, then 0.001, etc.), but we would never reach zero. So log scale is started at the lowest frequency decade of interest, not at zero. Second, the divisions between decades are not uniform. This will be obvious when we look at Fig. 12.1 (*b*), an expansion of a single decade. 5 does not fall halfway between 1 and 10; $3\frac{1}{2}$ does. Be always cautious while interpolating between major scale divisions.

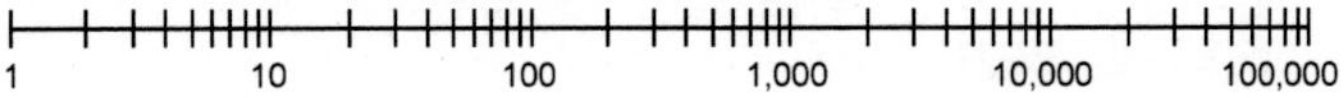

(a) 5 Decade Log Scale

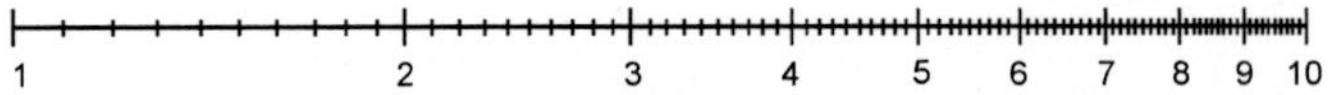

(b) 1 Logarithmic Decade Expanded Logarithmic Scale

Fig. 12.1

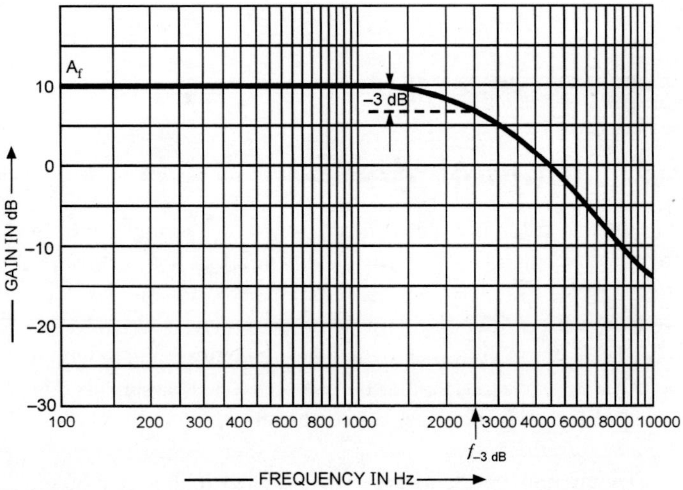

Fig. 12.2. ***Frequency Response Curve For a Typical Low-pass Filter***

Another technique used in the ac analysis of networks is the *Bode plot*, composed of magnitude versus frequency and phase angle versus frequency plots. In magnitude versus frequency plots the magnitude is always taken in decibels. Although both frequency response and Bode plots indicate the effect of variation of frequency on gain. Bode plot is normally used for stability determination and network design.

Figure 12.2 depicts a typical low-pass filter frequency response curve. Gain is taken on vertical axis, which is scaled in decibels. The highest gain is known as the passband gain, A_f. In the figure, it is + 10 dB.

The frequency is taken on X-axis, which is scaled logarithmically. Plotting dB versus log frequency means actual plot of log gain versus log frequency, so any linear relationship will be displayed as a straight line.

A parameter of significance is the –3 dB frequency, $f_{-3\text{ dB}}$. At the –3 dB frequency, the power delivered to the load has been reduced to one-half, so also called the half-power point.

We know that $P = \frac{v^2}{R}$ or $v = \sqrt{PR}$

At the half-power point, *i.e.*, when frequency, $f = f_{-3\text{ dB}}$.

$$v_{f-3\text{ dB}} = \sqrt{\frac{P}{2}R} = \frac{1}{\sqrt{2}}\sqrt{PR} = 0.707\sqrt{PR} = 0.707\, v_{\text{low-frequency}} \qquad ...(12.1)$$

Equation (12.1) reveals that at the –3 dB frequency, the power has been reduced to one-half while the voltage (and the voltage gain) has been reduced to 0.707. In Fig. 12.2 the –3 dB frequency is 2,500 Hz or 2.5 kHz.

The frequency can also be expressed as angular velocity in radians per second. There are 2π radians in a cycle (or circle), so

$$\omega = 2\pi f \text{ radians per second} \qquad ...(12.2)$$

In frequency response, a key parameter is $f_{-3\ dB}$, the frequency at which the power output has been reduced to half and the voltage gain has fallen –3 dB (0.707). But with the drop in gain, the phase relationship between the input voltage and output voltage also shifts. At the critical frequency, f_c, that shift will be a specific multiple of 45° (45°, 90°, 135° etc.). For simpler filters, built with a single RC pair, there will be a phase shift of 45°. For each additional RC pair added to the circuit, the phase shifts another 45° at the critical frequency, f_c. Two R-C pairs cause a phase shift of 90° at the critical frequency while three pairs lead to 135°, etc.

In many filters, the – 3 dB frequency, $f_{-3\ dB}$ and the critical frequency are the same. Most filters built with RC pair(s) without op-amps are that way. However, when an op-amp is used to add gain and positive feedback to a filter these two frequencies often are no longer equal.

Each filter produces its own frequency response. Such responses can be much more easily grouped, identified, analyzed, or circuits designed by normalizing the frequency axis. This is accomplished by dividing all points on the frequency axis by the filter's critical frequency. Thus, the critical frequency of all normalized frequency response plots is

$$f_c / f_c = 1$$

The gain (vertical) axis of a normalized plot is adjusted to set the passband gain, A_f to 0 dB. This may be accomplished by subtracting A_f from each point on the gain axis.

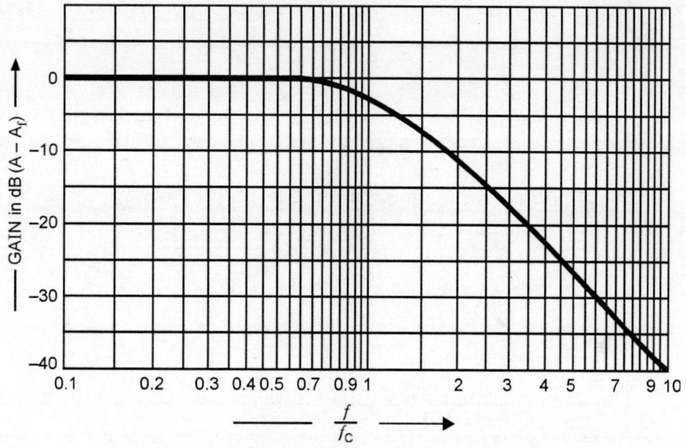

Fig. 12.3. ***Normalized Frequency Response Curve***

Figure 12.3 shows the normalized plot of the plot shown in Fig. 12.2. For denormalization or scaling the frequency axis, all frequency points may be simply multiplied by the desired critical frequency, f_c

$$i.e., \ f_{actual} = f_{normalized} \times f_c \qquad ...(12.3)$$

For denormalization or scaling the gain axis, the desired A_f may be added to all gain points.

Figure 12.4 shows a typical plot of phase shift versus frequency. A complete frequency response plot, shown in Fig. 12.5, is the combination of gain variation and phase shift as a function of frequency. Not only is plotting both graphs on the same axis convenient, but it also facilitates in determining if the filter (or any system) we anticipate building will be stable (*i.e.,* will not oscillate.)

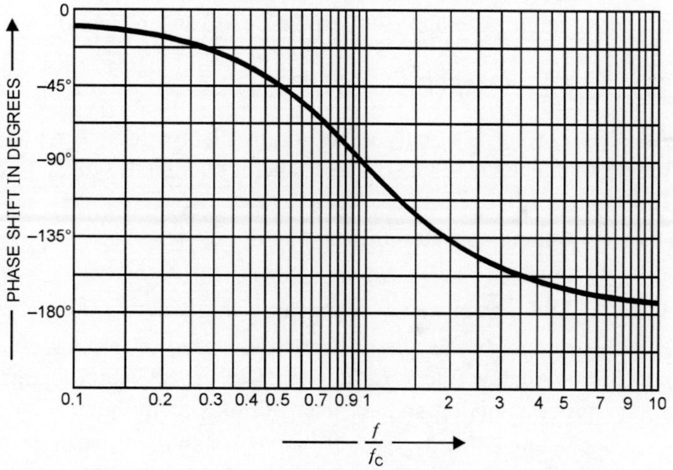

Fig. 12.4. ***Phase Response of a Typical Low-Pass Filter With Normalized Frequency Axis***

Analysis of circuits consisting of several reactive components is most easily handled by using Laplace transforms. It will not be possible to present a rigorous coverage of Laplace transform circuit analysis. However, a very simplified touch is given here, which may prove useful to the readers.

While using the Laplace transform of an equation (or a circuit), differential and integral terms are replaced by s and $\frac{1}{s}$. Then the equation may be manipulated or simplified, using the algebraic rules (rather than those of differential equations).

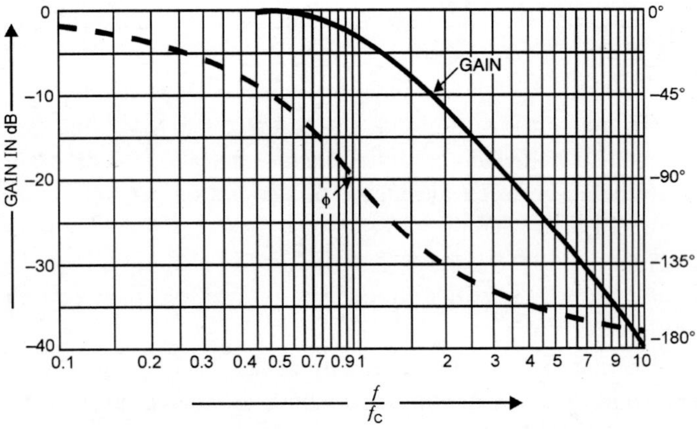

Fig. 12.5. ***Composite Gain and Phase Frequency Response***

As far as active filters are concerned, the primary term is

$$i_C = C\frac{dv_C}{dt}$$

Laplace transform of the above term can be had simply by replacing the derivative with s:

$$I_C = CsV_C$$

The Laplace impedance, then, of a capacitor is

$$Z_C = \frac{V_C}{I_C} = \frac{1}{Cs} \quad ...(12.4)$$

and the Laplace admittance is

$$Y = \frac{I_C}{V_C} = Cs \quad ...(12.5)$$

The term s contains both the amplitude and phase information regarding an equation's (or a circuit's) response.

For obtaining frequency response information from a Laplace equation, substitute $s = j\omega$, where j is the imaginary number and equals $\sqrt{-1}$ and $\omega = 2\pi f$.

12.3. CLASSIFICATION OF FILTERS

An electric filter is usually a *frequency-selective* network that passes a specified band of frequencies and blocks or attenuates signals of frequencies outside this band. Filters may be classified in a number of ways as follows :

1. Analog or digital filters.
2. Passive or active filters.
3. Audio-frequency (AF) or radio-frequency (RF) filters.

Depending on the type of techniques used in the process of analog signals the filters may be analog or digital. Analog filters are designed to process analog signals using analog techniques, while digital filters process analog signals using digital techniques.

Depending on the type of elements used in their construction, filters may be passive or active. A passive filter is built with passive components such as resistors, capacitors and inductors. Active filters, on the other hand, make use of transistors or op-amps (providing voltage amplification, and signal isolation or buffering) in addition to resistors and capacitors. The type of element used dictates the operating frequency range of the filter.

According to the operating frequency range, the filters may be classified as audio-frequency (AF) and radio-frequency (RF) filters.

Filters may also be classified as (*i*) low-pass, (*ii*) high-pass (*iii*) band-pass (*iv*) band-stop and (*v*) all-pass. The filter circuit may be so designed that some frequencies are passed from the input to the output of the filter with very little attenuation while others are greatly attenuated.

Figure 12.6 shows the frequency responses of the five types (mentioned above) of filters. These are ideal responses and cannot be achieved in actual practice.

A filter that provides a constant output from dc up to a cutoff frequency f_c and then passes no signal above that frequency is called an *ideal low-pass filter*. The ideal response of a low-pass filter is illustrated in Fig. 12.6 (*a*). The voltage gain (the ratio of output voltage and input voltage *i.e.,* v_{out}/v_{in}) is constant over a frequency range from zero to cutoff frequency, f_c. The output of any signal having a frequency exceeding f_c will be attenuated *i.e.,* there will be no output voltage for frequencies exceeding cutoff frequency f_c. Hence output will be available faithfully from 0 to f_c with constant gain, and is 0 from f_c onward. The frequencies between 0 and f_c are, therefore, called the *passband frequencies*, while the range of frequencies, those beyond f_c, that are attenuated include the *stopband frequencies*. The bandwidth (BW) is, therefore, f_c.

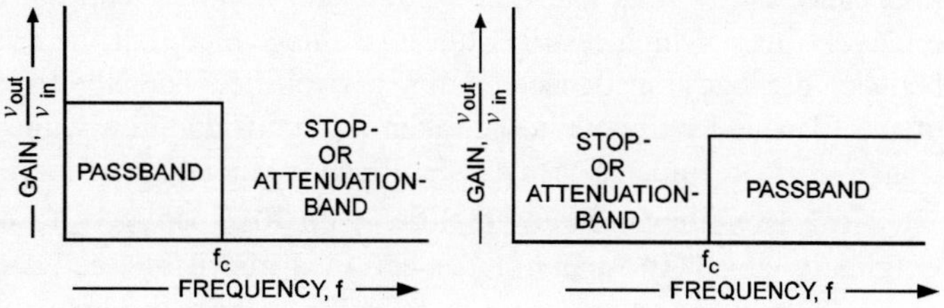

(*a*) *Low-pass Filter* (*b*) *High-pass Filter*

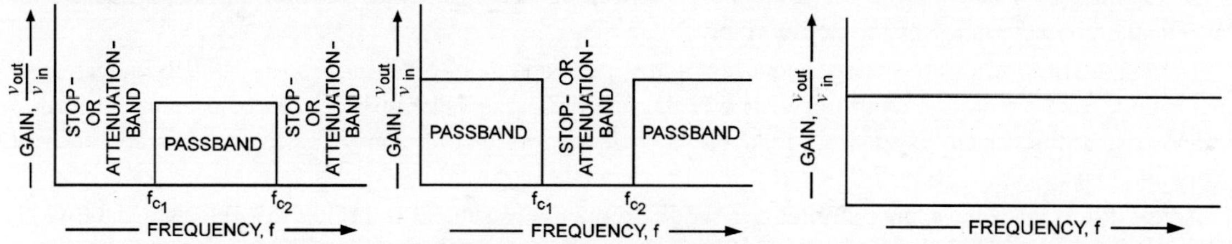

(*c*) *Band-pass Filter* (*d*) *Band-stop (or reject) Filter* (*e*) *All-pass Filter*

Fig. 12.6. ***Ideal Response of Different Types of Filters***

A filter that provides or passes signals above a cutoff frequency f_c is a high-pass filter, as idealized in Fig. 12.6 (*b*). The high-pass filter has a zero gain starting from zero to a frequency f_c, called the cutoff frequency, and above this frequency, the gain is constant, as illustrated in Fig. 12.6 (*b*). Thus signal of any frequency beyond f_c is faithfully reproduced with a constant gain, and frequencies from 0 to f_c will be attenuated.

When the filter circuit passes signals that are above one cutoff frequency and below a second cutoff frequency, it is called a band-pass filter, as idealized in Fig. 12.6 (*c*). Thus a band-pass filter has a pass band between two cutoff frequencies f_{c2} and f_{c1}, where $f_{c2} > f_{c1}$ and two stop-bands : $0 < f < f_{c1}$ and $f > f_{c2}$. The bandwidth of the band-pass filter is, therefore, equal to $f_{c2} - f_{c1}$ where f_{c1} and f_{c2} are lower and higher cutoff frequencies respectively.

The band-stop or band-reject filter performs exactly opposite to the band-pass *i.e.,* it has a bandstop between two cutoff frequencies f_{c2} and f_{c1} and two pass bands : $0 < f < f_{c1}$ and $f > f_{c2}$. The ideal response of a band-stop filter is illustrated in Fig. 12.6 (*d*). This is also called a *band-elimination* or *notch* filter.

The ideal response of an all-pass filter is shown in Fig. 12.6 (*e*). This filter passes all frequencies equally well, *i.e.,* output and input voltages are equal in amplitude for all frequencies. The important feature of this filter is that it provides predictable phase shift for frequencies of different input signals.

The filter discussed above have ideal characteristics and a sharp cutoff but unfortunately, ideal filter response is not practical because linear networks cannot produce the discontinuities. However, it is possible to obtain a practical response that approximates the ideal response by using special design techniques, as well as precision component values and high-speed op-amps.

12.4. ACTIVE VERSUS PASSIVE FILTERS

The simplest approach to build a filter is with passive components (resistors, capacitors, and inductors). In the R-F range it works quite well but with the lower frequencies, inductors create problems. AF inductors are physically larger and heavier, and therefore, expensive. For lower frequencies the inductance is to be increased which needs more turns of wire. It adds to the series resistance which degrades the inductor's performance.

Input and output impedances of passive filters are both a problem, specially below RF. The input impedance is low, that loads the source, and it varies with the frequency. The output impedance is usually relatively high, which restricts the load impedance that the passive filter can drive. There is no isolation between the load impedance and the passive filter. Thus the load will have to be considered as a component of the filter and will have to be taken into consideration while determining filter response or design. Any change in load impedance may significantly alter one or more of the filter response characteristics.

An active filter uses an amplifier with R-C networks to overcome these problems of passive filters. Originally built with vacuum tubes and then transistors, active filters now normally are centred around op-amps. By enclosing a capacitor in a feedback loop, the inductor (with all its low-frequency problems) can be eliminated. By proper configuration input impedance can be increased. The load is driven from the output of the op-amp, giving a very low output impedance. Not only does this improve load drive capability, but the load is now isolated from the frequency determining network. Thus variation in load will have no effect on the characteristics of the active filter.

The amplifier allows us to specify and easily adjust passband gain, passband ripple, cutoff frequency, and initial rolloff. Because of high input impedance of the op-amp, large value resistors can be used and therefore size and cost of the capacitors used are reduced. By selecting a quad op-amp IC, steep rolloffs can be built in very little space and at very little cost.

Active filters also have their limitations. High frequency response is limited by the gain bandwidth (GBW) and slew rate of the op-amps. High frequency op-amps are more expensive, making passive filters a more economical choice for RF applications. Active filter needs a power supply. For op-amps there may be two supplies. Variations in the power supply's, output voltage may affect, to some extent, the signal output from the active filter. In multistage applications, the common power supply provides a bus for high frequency signals. Feedback along the power supply lines may cause oscillations unless decoupling techniques are rigorously applied. Active devices, and therefore active filters, are much more susceptible to RF interference and ionization than are passive R-L-C filters. Practical considerations limit the Q of the band pass and notch filters to less than 50. For circuits requiring very selective (narrow) filtering, a crystal filter, because of its high Q value, will prove to be the best.

Although active filters are most widely employed in the field of communications and signal processing, they are used in one form or another in almost all sophisticated electronic systems. Radio, TV, telephone, radar, space-satellites, and biomedical equipment are but a few systems that make use of active filters.

12.5. CLASSIFICATION OF ACTIVE FILTERS

The most widely used active filters are: (*i*) low-pass (*ii*) high-pass (*iii*) band-pass (*iv*) band-stop or band-reject (also called the band-elimination or notch) and (*v*) all-pass filters. All of these filters use op-amps as the active elements and R-C networks. Although the 741 type op-amp operates satisfactorily in these filter circuits, high-speed op-amps like the LM 318 or ICL 8017 improve the performance of the filter circuits through their increased slew rates and higher unity GBW.

The frequency response characteristics of all of the five types of active filters are depicted in Fig. 12.7. Dashed lines indicate the ideal response while the practical response is depicted by solid curves in the figure.

A low-pass filter has a constant gain from 0 Hz to a high cutoff frequency f_H. Therefore the bandwidth is also f_H. At high cutoff frequency f_H the gain is reduced by 3 dB and for $f > f_H$ it decreases with the increase in input frequency. The frequencies between 0 and f_H are known as *passband* frequencies while the range of frequencies beyond f_H are attenuated and are therefore called the *stopband* frequencies.

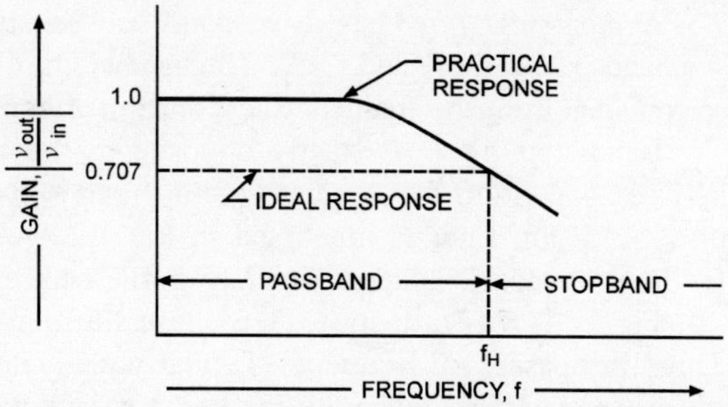

(a) Low-pass Filter

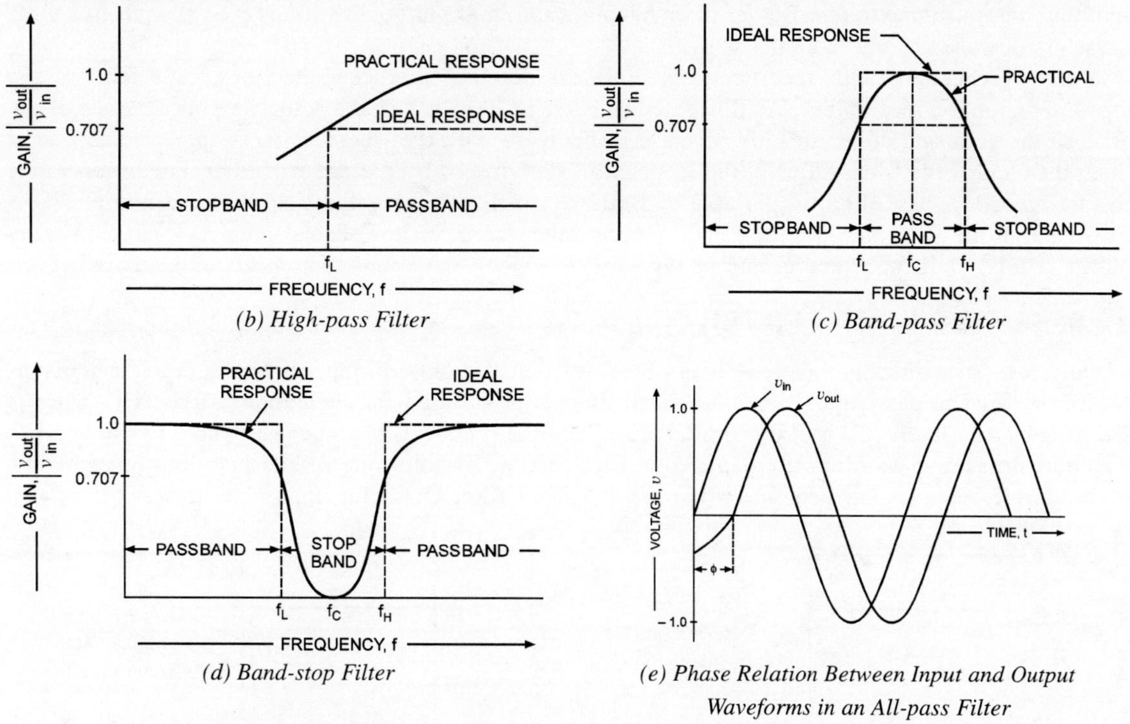

(b) High-pass Filter

(c) Band-pass Filter

(d) Band-stop Filter

(e) Phase Relation Between Input and Output Waveforms in an All-pass Filter

Fig. 12.7. ***Frequency Response of Different Types of Active Filters***

Frequency response of a low-pass filter is illustrated in Fig. 12.7 (*a*). As indicated by the dashed line, an ideal filter has no loss in the passband and infinite loss (or attenuation) in the stopband. But ideal filter response is not practical because linear networks cannot produce the discontinuities. However, it is possible to achieve a practical response that approximates the ideal response by employing special design techniques, precision components and high-speed op-amps such as LM 318 or ICL 8017.

Butterworth, Chebyshev, Bessel and Elliptic filters are some of the most widely employed practical filters for approximating the ideal response. The key characteristic of the *Butterworth filter* is that it has a flat passband as well as a flat stopband. This is the reason that it is sometimes called a flat-flat filter. The *Chebyshev filter* has a ripple passband but a flat stopband. The *Elliptic filter* has a ripple passband as well as a ripple stopband. Generally, the Elliptic filter provides the best stopband response among the three.

A *high-pass filter* with a stopband $0 < f < f_L$ and a passband $f > f_L$ is shown in Fig. 12.7 (*b*). Here f_L is the lower cutoff frequency and f is the operating frequency.

A *band-pass filter* has a passband between two cutoff frequencies f_H and f_L where $f_H > f_L$ and two stopbands at $0 < f < f_L$ and $f > f_H$. The bandwidth of the band-pass filter is, therefore, equal to $f_H - f_L$. All these are obvious from the frequency response of a band-pass filter illustrated in Fig. 12.7 (*c*).

Band-stop filter is exactly opposite to the band-pass filter in performance *i.e.,* it has a bandstop between two cutoff frequencies f_H and f_L and two passbands, $0 < f < f_L$ and $f > f_H$. The frequency response of a band-stop filter is illustrated in Fig. 12.7 (*d*). In Figs. 12.7 (*c*) and 12.7 (*d*), f_C is called the centre frequency, since it is approximately at the centre of the passband or stopband.

Figure 12.7 (*e*) illustrates the phase shift between the input and output voltages of an all-pass filter. This filter passes all frequencies equally well (*i.e.,* the output and input voltages are equal in magnitude for all frequencies) but with the phase shift between the two; phase shift being a function of the input frequency. The highest frequency up to which the magnitudes of the input and output remain equal depends on the unity GBW of the op-amp. At this frequency, however, the phase shift between the input and output is maximum. Bessel filter having a minimal phase shift used, even though its cutoff characteristics are not very sharp.

Before proceeding with specific types of filters, let the filter characteristics, especially in the stopband region, be reexamined. As illustrated in Figs. 12.7 (*a*) to (*d*), the actual response curves of the filters in the stopband either steadily increase or decrease with the increase in frequency. The rate at which the gain of the filter varies in the stopband is determined by the order of filter. For instance, for the first-order low-pass filter, in the case of Butterworth filters, the gain rolls off at the rate of 20 dB per decade in the stopband *i.e.,* for $f > f_H$. On the other hand, for the second-order low-pass filter the roll-off rate is 40 dB per decade, and so on.

12.6. BUTTERWORTH FILTER

In many low-pass filter applications it is necessary that the closed-loop gain is as close to unity as possible within the passband. The Butterworth filter is best suited for such applications. This filter is also called a *maximally flat* or *flat-flat filter*. Ideal and the practical frequency responses for three types of Butterworth low-pass filters are depicted in Fig. 12.8 (*a*) by solid line and dashed lines respectively. As the rolloff becomes steeper, they approach the ideal filter characteristics more closely.

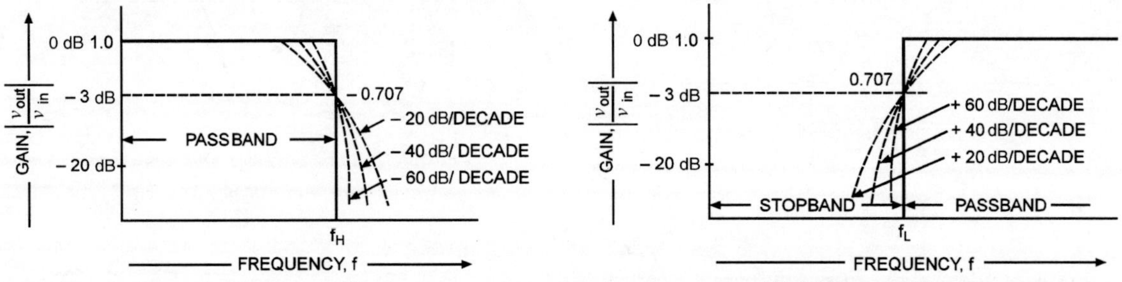

(*a*) *Low-pass Filter* (*b*) *High-pass Filter*

Fig. 12.8. ***Frequency Response For Three Types of Low-pass and High-pass Butterworth Filters***

The frequency responses for three types of high-pass Butterworth filters are shown in Fig. 12.8 (*b*). By contrast, for the first order high-pass filter, the gain increases at the rate of 20 dB per decade in the stopband while increase is 40 dB per decade for the second-order high-pass filter and so on.

Butterworth low-pass and high-pass filters are simple in design, and are the most popular active filters. Here we will discuss only Butterworth filters.

12.7. FIRST-ORDER LOW-PASS BUTTERWORTH FILTER

A first-order low-pass Butterworth filter using an R-C network for filtering is illustrated in Fig. 12.9. The op-amp is used in the noninverting configuration so that it does not load the R-C network. Gain of the filter is determined by resistors R_1 and R_f.

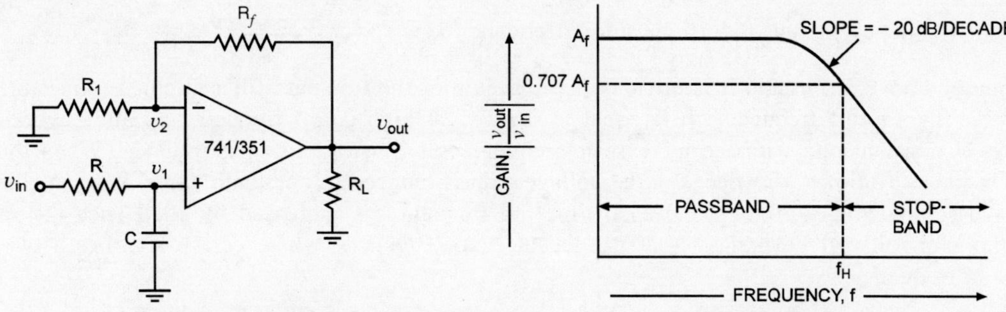

(a) Circuit Diagram *(b) Frequency Response*

Fig. 12.9. ***One-Pole or First Order Low-Pass Butterworth Filter***

Voltage at the noninverting (+) input terminal (voltage across the capacitor C), by voltage divider rule is

$$v_1 = \frac{-jX_C}{R - jX_C}\, v_{in}$$

Substituting $-j\,X_C = \dfrac{1}{j2\pi fC}$ in above equation, we have

$$v_1 = \frac{v_{in}}{j2\pi fCR + 1}$$

and the output voltage, $v_{out} = \left(1 + \dfrac{R_f}{R_1}\right) v_1$

$$\text{So } v_{out} = \left(1 + \frac{R_f}{R_1}\right) \frac{v_{in}}{j2\pi fCR + 1}$$

$$\text{or } \frac{v_{out}}{v_{in}} = \frac{A_f}{1 + j2\pi fCR}$$

where $A_f = \left(1 + \dfrac{R_f}{R_1}\right)$, passband gain of the filter.

Substituting $\dfrac{1}{2\pi RC} = f_H$ (high cutoff frequency of the filter) in the above equation, we have

$$\frac{v_{out}}{v_{in}} = \frac{A_f}{1 + j\dfrac{f}{f_H}} \quad \text{...(12.6)}$$

where $\dfrac{v_{out}}{v_{in}}$ is the gain of the filter as a function of frequency and f is the frequency of input signal in Hz.

The gain magnitude and phase angle can be determined by applying modulus to Eq. (12.6)

$$\left|\frac{v_{out}}{v_{in}}\right| = \frac{A_f}{\sqrt{1 + (f/f_H)^2}} \quad \text{...[12.7 (a)]}$$

$$\text{and phase angle } \phi = -\tan^{-1}\left(\frac{f}{f_H}\right) \quad \text{...[12.7 (b)]}$$

The operation of the low-pass filter can be verified from the gain magnitude Eq. [12.7 (*a*)].

1. For very low frequency *i.e.*, $f << f_H$; $v_{out}/v_{in} \cong A_f$
2. For frequency equal to high cutoff frequency *i.e.*, $f = f_H$;

$$\frac{v_{out}}{v_{in}} = \frac{A_f}{\sqrt{2}} = 0.707\ A_f$$

3. For frequency greater than high cutoff frequency *i.e.,* $f >> f_H$; $\frac{v_{out}}{v_{in}} < A_f$

From the above discussion it is obvious that the gain of the low-pass filter circuit is constant from 0 Hz to the high cutoff frequency. It is equal to $0.707A_f$ at high cutoff frequency f_H and beyond f_H it decreases at constant rate with the increase in frequency, as illustrated in Fig. 12.9 (*b*). When the frequency becomes 10 times (one decade), the voltage gain is reduced to one-tenth value. It means that the gain decreases by 20 dB (20 $\log_{10}$ 10) each time the frequency is increased by 10. Hence the rate at which the gain rolls off beyond f_H is 20 dB/decade or 6 dB/octave, where octave signifies a two fold increase in frequency.

The frequency $f = f_H$ is called the cutoff frequency because the filter gain at this frequency is down by 3 dB (= 20 log 0.707) from 0 Hz. Sometimes it is also called, –3 dB frequency.

The active filter shown in Fig. 12.9 (*a*) passes all frequencies from zero to the cutoff frequency f_H. Such a filter is called a *low-pass filter.* A low-pass filter can be recognized as it has usually one or more bypass circuits. Another way to recognize a low-pass filter is the presence of $\left(1 + \frac{jf}{f_H}\right)$ factor in the denominator of the *transfer function* (the formula for voltage gain).

We often see that word pole is used with active filter. For preliminary discussion, a pole is a bypass circuit that appears in an active filter. For instance, the circuit shown in Fig. 12.9 (*a*) has one pole because it has one bypass circuit. Mathematically, each pole in an active filter produces one *j* factor in the transfer function. Since in Fig. 12.9 (*a*) has only one bypass circuit, Eq. (12.6) has only one *j* factor.

In selection of components for building such a filter there are several small points to be kept in mind. The op-amp, along with R_f and R_1 sets the gain $\left(A_f = 1 + \frac{R_f}{R_1}\right)$. Its gain bandwidth (GBW) and slew rate must allow it to operate well beyond the cutoff frequency ($f_H \times 10$). Otherwise, the filter response may be partially caused by the op-amp's frequency response (which is not entirely predictable). Select C of a value less than or equal to 1 μF. Mylar or tantalum capacitors are recommended for better performance. The filter resistor R strongly affects the input impedance as well as cutoff frequency f_H. So R should be of a value less than or equal to 200 kΩ, unless a precision op-amp is to be used. To minimize the effects of bias currents on the output, R_f and R_1 must be selected of such values so as to give

$$R = R_f \parallel R_1$$

Frequency Scaling. The procedure of converting a given cutoff frequency to a new cutoff frequency is called the *frequency scaling*. Frequency scaling is accomplished as follows.

To obtain a new cutoff frequency, R or C (but not both) is multiplied by the ratio of original cutoff frequency to the new cutoff frequency. In filter design the required values of R and C are usually not standard. Besides, a variable capacitor is not commonly used. Hence the capacitor is taken of a standard value and then the value of resistance R is calculated for a desired cutoff frequency f_H from the relation :

$$R = \frac{1}{2\pi f_H C}$$

This is because for a non-standard value of a resistor, a pot can be used.

Example 12.1. Design a first-order low-pass filter with a high cut-off frequency of 2 kHz and passband gain is 2. [*G.G.S.I.P. Univ. Analog Electronics-II May 2011*]

Solution : Cutoff frequency, f_H = 2 kHz

Let the capacitor C be of 0.01 μF. Then

$$R = \frac{1}{2\pi f_H C} = \frac{1}{2\pi \times 2 \times 10^3 \times 1 \times 10^{-8}} = 7.95 \text{ k}\Omega$$

R may be taken a pot of 10 kΩ

Since the passband gain is 2, so $1+\frac{R_f}{R_1} = 2$ or R_f should be equal to R_1

$$\text{Since } R_f \parallel R_1 = R$$

$$\text{or } \frac{R_f \times R_1}{R_f + R_1} = R$$

$$\text{or } \frac{R_1^2}{2R_1} = R$$

or $R_f = R_1 = 2R = 2 \times 7.95 = 15.9$ kΩ (say, 15 kΩ) **Ans.**

Example 12.2. Design a first order LP Butterworth filter with a cutoff frequency of 1 kHz and a passband gain of 2. Given : C = 0.01 μF, $R_1 = R_f$ = 10 kΩ. Give design steps and draw frequency response.

[G.G.S.I.P. Univ. Analog Electronics II May-June 2009]

Solution :

$$C = 0.01\ \mu F = 1 \times 10^{-8}\ F$$

$$f_H = 1\ kHz$$

$$R = \frac{1}{2\pi f_H C} = \frac{1}{2\pi \times 1 \times 10^3 \times 1 \times 10^{-8}} = 15.9\ k\Omega$$

R may be taken a pot of 18 kΩ

$$R_1 = 10\ k\Omega \text{ Given}$$

$$R_f = 10\ k\Omega \text{ Given}$$

Frequency response is shown in Fig. 12.9(*b*).

12.8. SECOND-ORDER LOW-PASS BUTTERWORTH FILTER

Improved and considerably variable performance is achieved by going to a second-order (rather than first order) active filter.

No doubt a second-order filter can be obtained by coupling two first-order active filters, but this would not be economical because it needs two op-amps. A second-order filter can be obtained simply by introducing an additional R-C network in the first-order filter, as illustrated in Fig. 12.10 (*a*). Second-order filters are important because higher-order filters are designed using them.

The circuit illustrated in Fig. 12.10 (*a*) is also called a two-pole low-pass filter because it has two bypass circuits. The two poles of this circuit modify the cutoff frequency and the response of the circuit. The gain of the second-order filter is set by R_1 and R_f, while the high cutoff frequency f_H is determined by R, R′, C and C′, as given below.

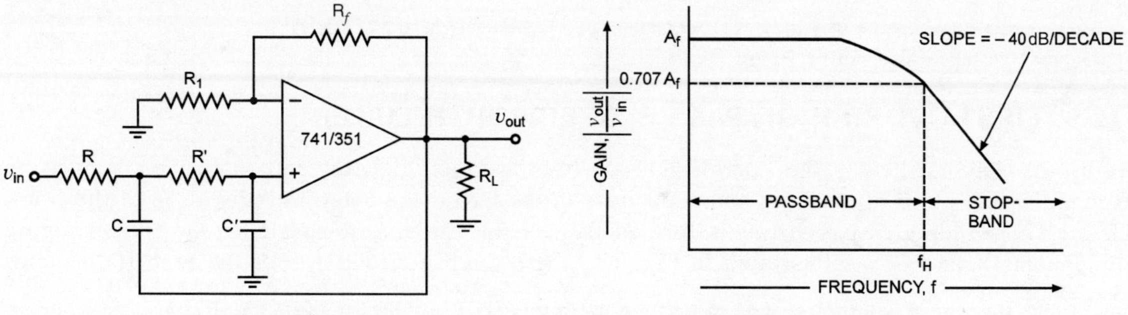

(*a*) *Circuit Diagram* (*b*) *Frequency Response*

Fig. 12.10. ***Two-Pole or Second-Order Low-Pass Active Filter***

High cutoff frequency, $f_H = \frac{1}{2\pi\sqrt{RR'CC'}}$ ...(12.8)

The voltage gain magnitude equation for a second-order low-pass Butterworth filter is given as

$$\left|\frac{v_{out}}{v_{in}}\right| = \frac{A_f}{\sqrt{1+\left(\frac{f}{f_H}\right)^4}} \quad ...(12.9)$$

where $A_f = \left(1+\frac{R_f}{R_1}\right)$, the passband gain of the filter, f is the frequency of the input signal in Hz and f_H is the high cutoff frequency in Hz.

A mathematical analysis reveals that a closed-loop gain A_f of 1.586 is a *critical* value. When the gain is 1.586 flattest possible gain is achieved in the midband. This response is called the *Butterworth* or *maximally* flat response and is one that is most popular.

Since the closed-loop voltage gain A_f must be 1.586 for a Butterworth response,

$$1.586 = 1 + \frac{R_f}{R_1}$$

$$\text{or } R_f = 0.586\, R_1$$

R_1 may be taken of some standard value below 100 kΩ and the value of R_f be worked out.

For simplication of calculations we may take R′ = R and C′ = C. C may be taken of some standard value below 1 μF and then R may be worked out by using equation

$$R = \frac{1}{2\pi f_H C}$$

Example 12.3. Design a second-order low-pass filter at a high cutoff frequency of 2 kHz using op-amp.

Solution : High cutoff frequency, f_H = 2 kHz

Let C′ = C = 0.033 μF

$$\text{Then } R' = R = \frac{1}{2\pi f_H C} = \frac{1}{2\pi \times 2 \times 10^3 \times 0.033 \times 10^{-6}} = 2.4\text{ k}\Omega \text{ (say 2.5 k}\Omega)$$

However, to minimize the effect of offset current (which may be quite important in a low-pass filter), the resistance at the inverting terminal must be equal to the resistance at the inverting terminal.

$$\text{So } 2R = \frac{R_f R_1}{R_f + R_1} = \frac{0.586\, R_1^2}{1.586\, R_1} \qquad \because R_f = 0.586 R_1$$

$$\text{or } 2 \times 2.4\text{ k}\Omega = \frac{0.586\, R_1}{1.586}$$

$$\text{or } R_1 = 13\text{ k}\Omega$$

R_1 may be taken of a value of 15 kΩ and then R_f = 0.586 × 15 = 8.79 kΩ

R_f may be taken as a pot of 10 kΩ **Ans.**

12.9. FIRST-ORDER HIGH-PASS BUTTERWORTH FILTER

High-pass active filter is the *complement* of low-pass active filter and is formed simply by exchanging the place of resistors and capacitors in the frequency determining section of the filter. Thus a first-order high-pass filter is formed from a first-order low-pass filter by interchanging components R and C, as illustrated in Fig. 12.11 (*a*). Such a circuit passes the high frequencies but blocks the low frequencies. The cutoff frequency is still given by $\frac{1}{2\pi RC}$. Below this frequency, called the *low cutoff frequency* f_L, the output voltage decreases at a rate of 20 dB per decade. All frequencies higher than f_L are in the passband with the highest frequency determined by the closed-loop bandwidth of the op-amp.

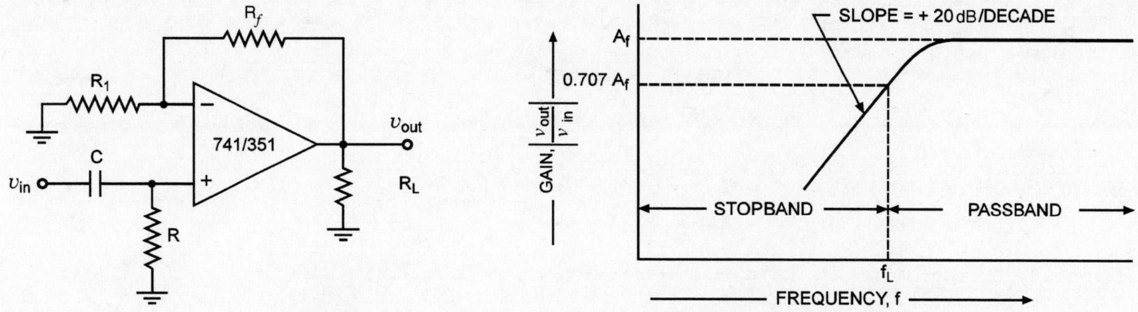

(a) Circuit Diagram *(b) Frequency Response*

Fig. 12.11. ***First-Order High-Pass Butterworth Filter***

The output voltage is given as

$$v_{out} = \left(1 + \frac{R_f}{R_1}\right) \frac{j2\pi f RC}{1 + j2\pi f RC} v_{in}$$

$$\text{or} \quad \frac{v_{out}}{v_{in}} = A_f \left[\frac{jf / f_L}{1 + j(f / f_L)}\right] \qquad \text{...(12.10)}$$

where $A_f = \left(1 + \frac{R_f}{R_1}\right)$, the passband gain of the filter, $f_L = \frac{1}{2\pi RC}$, the low cutoff frequency in Hz, and f is the frequency of the input signal.

The magnitude of the voltage gain is given as

$$\left|\frac{v_{out}}{v_{in}}\right| = \frac{A_f \cdot f / f_L}{\sqrt{1 + (f / f_L)^2}} \qquad \text{...(12.11)}$$

Since high-pass filters are built from low-pass filters simply by exchanging the place of resistors and capacitors in the frequency determining section of the filter, the design and frequency scaling procedures are the same as for the low-pass filters.

Example 12.4. A simple active filter is shown in Fig. 12.12. Assume ideal op-amp. Derive the transfer function $\frac{v_{out}}{v_{in}}$ of the circuit, and state the type of filter (*i.e.,* high-pass, low-pass, band-pass, or band-reject). Determine the required values of R_1, R_2 and C in order for the filter to have a 3-dB frequency of 1 kHz, a high-frequency resistance of 100 kΩ and a high frequency gain magnitude of 10.

Solution : Applying Kirchhoff's current law at inverting node, we have

$$\frac{0 - v_{in}}{R_1 + \frac{1}{sC}} + \frac{0 - v_{out}}{R_2} = 0$$

$$\text{or} \quad \frac{v_{out}}{v_{in}} = \frac{-sCR_2}{1 + sCR_1} = \frac{-j\omega R_2 C}{1 + j\omega R_1 C} \quad \textbf{Ans.}$$

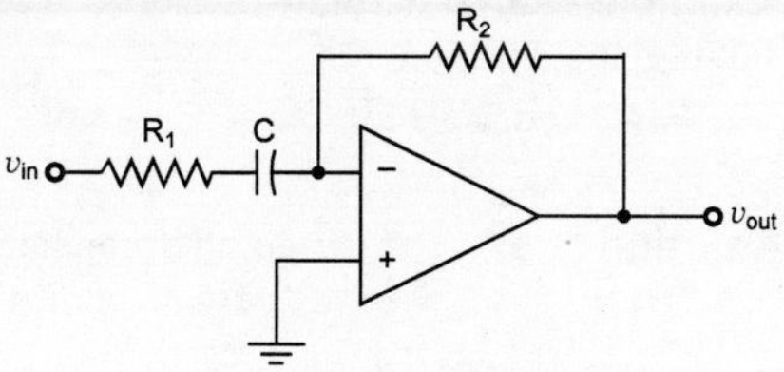

Fig. 12.12

The given filter is high-pass filter **Ans.**

Now $\left|\frac{V_{out}}{V_{in}}\right|$ at a 3-dB frequency of 1 kHz

$$= \frac{1}{\sqrt{2}} = \frac{\omega R_2 C}{\sqrt{1 + \omega^2 R_1^2 C^2}} \qquad \text{...}(i)$$

$$\text{Given } \frac{V_{out}}{V_{in}} = \frac{-R_2}{R_1} = -10$$

$$R_1 = 100 \text{ k}\Omega \quad \textbf{Ans.}$$

$$R_2 = 1{,}000 \text{ k}\Omega = 1 \text{ M}\Omega \quad \textbf{Ans.}$$

From Eq. (*i*)

$$1 + (2\pi \times 1{,}000)^2 \times (100 \times 10^3)^2 C^2 = 2 \times (2\pi \times 1{,}000)^2 \times (10^6)^2 \times C^2$$

$$\text{or } 1 + 4\pi^2 \times 10^{16} C^2 = 8\pi^2 \times 10^{18} C^2$$

$$\text{or } C^2 = \frac{1}{8\pi^2 \times 10^{18} - 4\pi^2 \times 10^{16}} = 1.27288 \times 10^{-20}$$

$$\text{or } C = 1.1282 \times 10^{-10} \text{ F} \quad \textbf{Ans.}$$

12.10. SECOND-ORDER HIGH-PASS BUTTERWORTH FILTER

A second-order high-pass Butterworth filter can also be formed like a first-order high-pass Butterworth filter, from a second-order low-pass Butterworth filter simply by interchanging the frequency-determining resistors and capacitors, as illustrated in Fig. 12.13 (*a*).

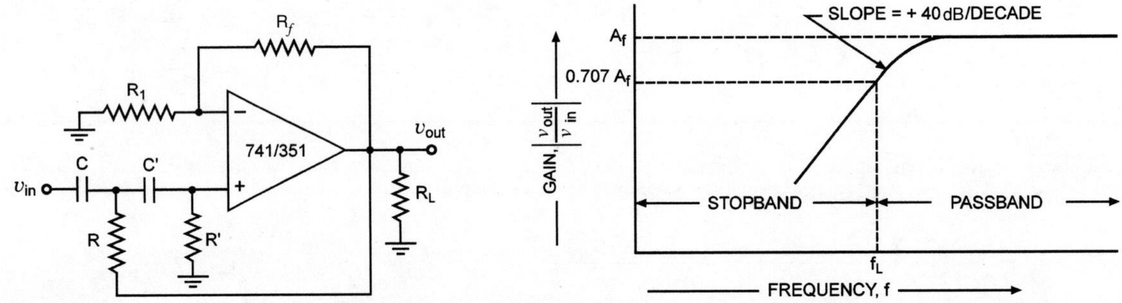

(*a*) *Circuit Diagram* (*b*) *Frequency Response*

Fig. 12.13. ***Second-Order High-Pass Butterworth Filter***

The voltage gain magnitude equation for the second-order high-pass filter is given as

$$\left|\frac{v_{out}}{v_{in}}\right| = \frac{A_f}{\sqrt{1 + \left(\frac{f_L}{f}\right)^4}} \qquad ...(12.12)$$

where $A_f = 1.586$, passband gain for the second-order Butterworth response, f is frequency of the input signal in Hz and f_L is the low cutoff frequency in Hz.

Example 12.5. Determine the cutoff frequency of a second-order high-pass filter [circuit shown in Fig. 12.13 (*a*)] having the components

R = R' = 2.1 kΩ; C = C' = 0.05 μF; R_1 = 20 kΩ and R_f = 60 kΩ.

Solution : Low cutoff frequency, $f_L = \frac{1}{2\pi RC} = \frac{1}{2\pi \times 2.1 \times 10^3 \times 0.05 \times 10^{-6}} = 1.516 \text{ kHz}$ **Ans.**

12.11. HIGHER-ORDER FILTERS

From the discussion made so far on the filters, it may be concluded that in the stopband the gain of the filter changes at the rate of 20 dB/decade for first-order filters and 40 dB/decade for second-order filters. This means that as the order of the filter is increased, the actual stopband response of the filter approaches its ideal stopband characteristics. In general, a third-order filter produces 60 dB/decade, a fourth-order filter produces 80 dB/decade and so on.

Higher-order filters, such as third, fourth, fifth, and so on, are built simply by using the first and second-order filters.

The simplest way to build a third-order low-pass filter is by cascading a first order filter with a second-order. Similarly a fourth-order low-pass filter can be formed by cascading two second-order low-pass filters. Although there is no limit to the order of the filter that can be formed, as the order of the filter increases, so does its size. Also the accuracy declines, in that the difference between the actual stopband response and the theoretical stopband response increases with an increase in the order of the filter.

A third-order low-pass Butterworth filter is illustrated in Fig. 12.14.

The voltage gain of the first section is optional, it can be set, whatever is required. The voltage-gain of the second section, however, affects the flatness of the overall response. If closed-loop gain is kept 1.586, then the overall gain will be down 6 dB (3 dB for each section) at the cutoff frequency. By increasing the voltage gain of the second section slightly, cumulative loss of voltage gain is offset. By using an advanced mathematical derivation, it can be proved that an A_f of 2 is the critical value required for a maximally flat response.

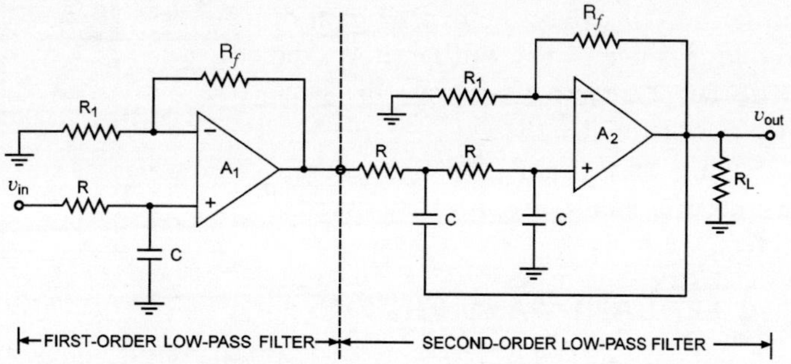

Fig. 12.14. ***Third-Order Low-Pass Butterworth Filter Circuit***

In this case $R_f = R_1$

When $A_f = 2$, the cutoff frequency is given as

$$f_H = \frac{1}{2\pi RC} \qquad ...(12.13)$$

where R and C are the resistance and capacitance of each section. At cutoff frequency, the overall voltage gain is down 3 dB. Above the cutoff frequency, the voltage gain drops at a rate of 60 dB per decade equivalent to 18 dB per octave.

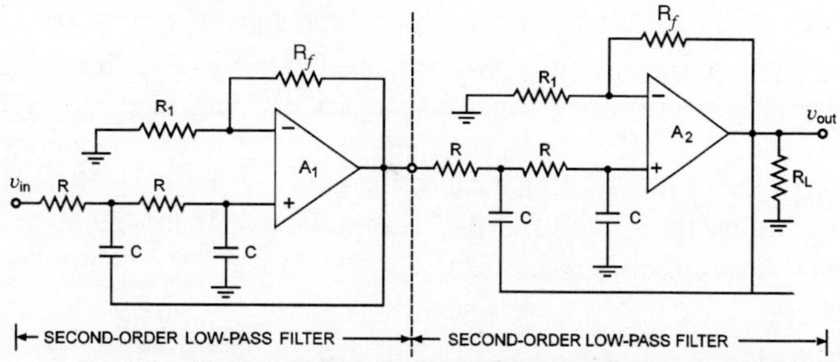

Fig. 12.15. ***Fourth-Order Low-Pass Butterworth Filter Circuit***

A fourth-order low-pass Butterworth filter is illustrated in Fig. 12.15. It is formed by cascading two second-order low-pass filters. If A_f of 1.586 is used for both sections, the voltage gain will be down 6 dB at the cutoff frequency. By using different gain for each section, we can strike a compromise that produces a maximally flat response. An advanced derivation shows that we need to use $A_f = 1.152$ for the first section and $A_f = 2.235$ for the second section.

Also, the overall filter gain is equal to the product of the individual voltage gains of the filter sections. Hence, the overall gain of a fourth-order filter is $1.152 \times 2.235 = 2.575$. In all our Butterworth designs, the cutoff frequency is given as $\frac{1}{2\pi RC}$.

As with the first- and second-order filters, the third- and fourth-order high-pass filters are formed by simply interchanging the positions of the frequency determining resistors and capacitors in the corresponding low-pass filters. The high-order filters can be designed by following the procedures outlined for the first- and second-order filters.

Generally, the minimum-order filter required depends on the application specifications. Although a high-order filter than necessary provides a better stopband response, the high-order filter is more complex, occupies more space and is more expensive.

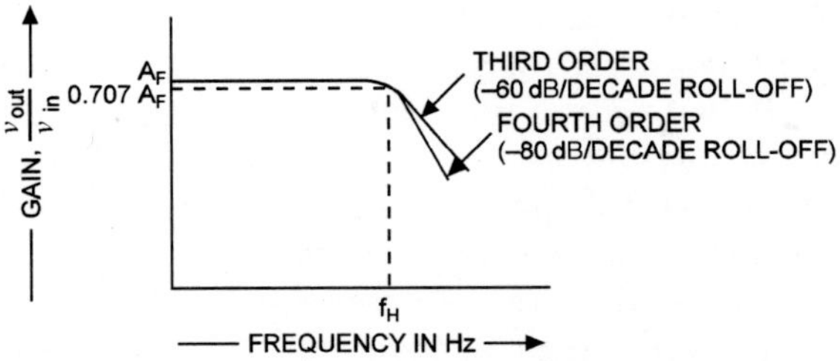

Fig. 12.16. ***Frequency Response of Third-Order and Fourth-Order Low-Pass Butterworth Filters***

It is noteworthy here that in all filters, the same resistance and capacitance values are used in the bypass or R-C networks, a definite convenience in selection of components and ease of construction. This fixes the overall gain of the high-order filters. Furthermore, the 3-dB cutoff frequency is always the same and is equal to $1/2\pi RC$.

Table 12.1 lists the voltage gains we need to build low-pass Butterworth filters.

TABLE 12.1. *Voltage Gains For Butterworth Filters*

Number of Bypass Circuits	*Rolloff in dB/ Decade*	*1st Section 1 or 2 Bypass Circuits*	*Second Section 2 Bypass Circuits*	*Third Section 2 Bypass Section*
1	20	Optional		
2	40	1.586		
3	60	Optional	2	
4	80	1.152	2.235	
5	100	Optional	1.382	2.382
6	120	1.068	1.586	2.482

12.12. BAND-PASS FILTERS

A band-pass filter is a circuit which is designed to pass signals only in a certain band of frequencies while attenuating all signals outside this band. The parameters of importance in a band-pass filter are the high and low cutoff frequencies (f_H and f_L), the bandwidth (BW), the centre frequency f_C, centre-frequency gain, and the selectivity or Q.

There are basically two types of band-pass filters viz. wide band-pass and narrow band-pass filters. Unfortunately, there is no set dividing line between the two. However, a band-pass filter is defined as a wide band-pass if its *figure of merit* or *quality factor* Q is less than 10 while the band-pass filters with Q > 10 are called the narrow band-pass filters. Thus Q is a measure of selectivity, meaning the higher the value of Q the more selective is the filter, or the narrower is the bandwidth (BW). The relationship between Q, 3-dB bandwidth, and the centre frequency f_C is given by an equation

$$Q = \frac{f_C}{BW} = \frac{f_C}{f_H - f_L} \quad ...(12.14)$$

For a wide band-pass filter the centre frequency can be defined as

$$f_C = \sqrt{f_H f_L} \quad ...(12.15)$$

where f_H and f_L are respectively the high and low cutoff frequencies in Hz.

In a narrow band-pass filter, the output voltage peaks at the centre frequency f_C.

12.12.1. Wide Band-Pass Filter

A wide band-pass filter can be formed by simply cascading high-pass and low-pass sections and is generally the choice for simplicity of design and performance though such a circuit can be realized by a number of possible circuits.

To form a ±20 dB/decade band-pass filter, a first-order high-pass and a first-order low-pass sections are cascaded; for a ±40 dB/decade band-pass filter, second-order high-pass filter and a second-order low-pass filter are connected in series, and so on. It means that, the order of the band-pass filter is governed by the order of the high-pass and low-pass filters it consists of.

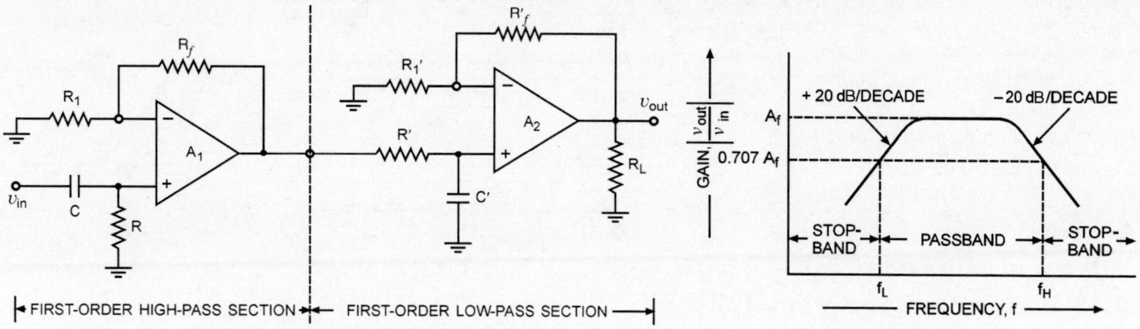

(a) Circuit Diagram *(b) Frequency Response*

Fig. 12.17. ±20 *dB/Decade Wide Band-Pass Filter*

A ± 20 dB/decade wide band-pass filter composed of a first-order high-pass filter and a first-order low-pass filter, is illustrated in Fig. 12.17 (*a*). Its frequency response is illustrated in Fig. 12.17 (*b*).

Example 12.6. Determine the cutoff frequencies and a bandwidth of the band-pass filter circuit shown in Fig. 12.17 (*a*) with R = R′ = 10 kΩ, C = 0.1 μF and C′ = 0.0025 μF.

Solution : Higher cutoff frequency, $f_H = \frac{1}{2\pi R'C'} = \frac{1}{2\pi \times 10 \times 10^3 \times 0.0025 \times 10^{-6}} = 6.366$ kHz **Ans.**

Lower cutoff frequency, $f_L = \frac{1}{2\pi RC} = \frac{1}{2\pi \times 10 \times 10^3 \times 0.1 \times 10^{-6}} = 159.15$ Hz **Ans.**

Bandwidth, BW $= f_H - f_L = (6.366 - 159.15 \times 10^{-3})$ kHz $= 6.2$ kHz **Ans.**

Example 12.7. Design a bandpass filter with f_L = 200 Hz, f_H = 1 kHz and a passband gain = 4.

Solution : *First-Order Low-Pass Filter*

Cutoff frequency, $f_H = 1$ kHz

Let the capacitor C′ be of 0.01 μF then $R' = \frac{1}{2\pi f_H C'} = \frac{1}{2\pi \times 1 \times 10^3 \times 1 \times 10^{-8}} = 15.9\ k\Omega \simeq 20\ k\Omega$

First-Order High-Pass Filter

$f_L = 200$ Hz

Let C = 0.05 μF

then $R = \frac{1}{2\pi f_L C} = \frac{1}{2\pi \times 200 \times 0.05 \times 10^{-6}}$ $\quad \because f_L = \frac{1}{2\pi RC}$

$= 15.9\ k\Omega \simeq 20\ k\Omega$

The bandpass gain is 4. Let the gain of each section (high pass as well as low-pass sections) be set equal to 2. It means, input and feedback resistances must be of same value, say 10 kΩ each.

Thus $R_f' = R_f = R_1' = R_1 = 10\ k\Omega$ [Refer to Fig. 12.17 (*a*)]

$R = R' = 20\ k\Omega$

$C' = 0.01\ \mu F$

$C = 0.05\ \mu F$

Example 12.8. Design a wide band-pass filter having f_l = 500 Hz and f_h = 2 kHz and passband gain of 4. Draw the circuit.

[*G.G.S.I.P. Univ. Analog Electronics-II May 2011*]

Solution : The circuit is shown in Fig. 12.17 (*a*).

The passband gain is 4. Let the gain of each section (high-pass as well as low-pass sections) be set equal to 2. It means that input and feedback resistance must be of same value, say 10 kΩ each.

Thus, $R'_f = R_f = R'_1 = R_1 = 10$ kΩ [Refer to Fig. 12.17 (*a*)]

Let C′ be of 0.01 μF, then
$$R' = \frac{1}{2\pi f_H C'} = \frac{1}{2\pi \times 2 \times 10^3 \times 1 \times 10^{-8}} = 7.95 \text{ k}\Omega$$

R′ may be taken a pot of 10 kΩ

Let C be of 0.02 μF

then
$$R = \frac{1}{2\pi f_L C} = \frac{1}{2\pi \times 500 \times 0.02 \times 10^{-6}} = 15.9 \text{ k}\Omega$$

R may be taken a pot of 20 kΩ. **Ans.**

12.12.2. Narrow Band-Pass Filter

A narrow band-pass filter employing multiple feedback is depicted in Fig. 12.18 (*a*). This filter employs only one op-amp, as shown in the figure. In comparison to all the filters discussed so far, this filter has some unique features that are given below.

1. It has two feedback paths, and this is the reason that it is called a multiple-feedback filter.
2. The op-amp is used in the inverting mode.

The frequency response of a narrow band-pass filter is shown in Fig. 12.18 (*b*).

Generally, the narrow band-pass filter is designed for specific values of centre frequency f_C and Q or f_C and BW. The circuit components are determined from the following relationships. For simplification of design calculations each of C_1 and C_2 may be taken equal to C.

$$R_1 = Q/2\pi f_C C A_f \quad ...(12.16)$$

$$R_2 = Q/2\pi f_C C\,(2Q^2 - A_f) \quad ...(12.17)$$

and
$$R_3 = \frac{Q}{\pi f_C C} \quad ...(12.18)$$

where A_f is the gain at centre frequency and is given as

$$A_f = \frac{R_3}{2R_1} \quad ...(12.19)$$

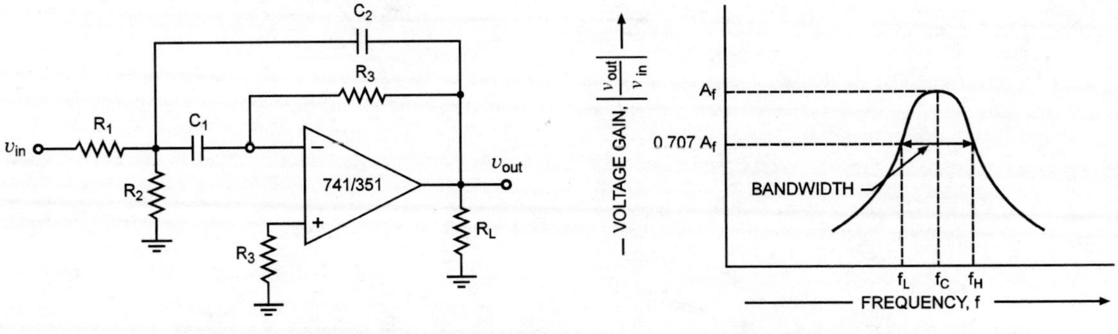

(*a*) *Circuit Diagram* (*b*) *Frequency Response*

Fig. 12.18. ***Multiple Feedback Narrow Band-Pass Filter***

The gain A_f, however, must satisfy the condition $A_f < 2\,Q^2$.

The centre frequency f_C of the multiple feedback filter can be changed to a new frequency f'_C without changing, the gain or bandwidth. This is achieved simply by changing R_2 to R'_2 so that

$$R'_2 = R_2\left(\frac{f_C}{f'_C}\right)^2 \quad ...(12.20)$$

Example 12.9. Design a narrow band-pass filter [circuit shown in Fig. 12.18 (*a*)] with a centre frequency f_C = 1 kHz, Q = 5, A_f = 8.

Change the centre frequency to 2 kHz, keeping A_f and BW constant.

Solution : Let $C_1 = C_2 = 0.01\ \mu F$

From Eqs. (12.16), (12.17) and (12.18), we have

$$R_1 = \frac{Q}{2\pi f_C C A_f} = \frac{5}{2\pi \times 1 \times 10^3 \times 0.01 \times 10^{-6} \times 8} = 9.95\ k\Omega \simeq 10\ k\Omega \quad \textbf{Ans.}$$

$$R_2 = \frac{Q}{2\pi f_C C(2Q^2 - A_f)} = \frac{5}{2\pi \times 1 \times 10^3 \times 0.01 \times 10^{-6}\ (2 \times 5^2 - 8)} = 1.89\ k\Omega \simeq 2\ k\Omega \quad \textbf{Ans.}$$

$$R_3 = \frac{Q}{\pi f_C C} = \frac{5}{\pi \times 1 \times 10^3 \times 0.01 \times 10^{-6}} = 159.2\ k\Omega = 160\ k\Omega \quad \textbf{Ans.}$$

The value of R'_2 required to change the centre frequency from 1 kHz to 2 kHz is

$$R'_2 = R_2\left(\frac{f_C}{f'_C}\right)^2 = 2\ k\Omega \times \left(\frac{1}{2}\right)^2 = 0.5\ k\Omega \text{ or } 500\ \Omega \quad \textbf{Ans.}$$

Example 12.10. Design a band-pass filter so that cut-off frequency f_C = 1 kHz, Q = 20 and A_f = 10. Assume C = 1 μF. [*G.G.S.I.P. Univ. Analog Electronics-II May 2011*]

Solution : Cutoff frequency, f_C = 1 kHz

Gain at centre frequency, A_f = 10

Q = 20

Since Q = 20 (>> 10), the filter is narrow band-pass, which also satisfies the condition $A_f < 2Q^2 < 2 \times 20^2$ *i.e.*, 800

$$C_1 = C_2 = 1\ \mu F = 1 \times 10^{-6}\ F \quad \text{...Given}$$

$$R_1 = \frac{Q}{2\pi f_C C A_f} = \frac{20}{2\pi \times 1 \times 10^3 \times 1 \times 10^{-6} \times 10} = 318.3\ \Omega \text{ (say, 330 } \Omega)$$

$$R_2 = \frac{Q}{2\pi f_C C(2Q^2 - A_f)} = \frac{20}{2\pi \times 1 \times 10^3 \times 1 \times 10^{-6}(2 \times 20^2 - 10)} = 4\ \Omega \quad \textbf{Ans.}$$

$$R_3 = \frac{Q}{\pi f_C C} = \frac{20}{\pi \times 1 \times 10^3 \times 1 \times 10^{-6}} = 6.37\ k\Omega \text{ (say 6.8 k}\Omega) \quad \textbf{Ans.}$$

Example 12.11. (*i*) Design a wide bandpass filter with f_L = 200 Hz, f_H = 1 kHz and a passband gain = 4.

(*ii*) Draw the frequency response of this filter.

(*iii*) Calculate the value of Q for the filter.

Solution : (*i*) Please refer to Solution of Example 12.7.

(*ii*) Since the wide bandpass filter is formed by cascading high-pass and low-pass sections, the magnitude of the voltage gain of the bandpass filter is equal to the product of the magnitudes of voltage gain of the high-pass and low-pass filters.

The magnitudes of voltage gain of high-pass and low-pass filters are given by equations

$$|G_H| = \frac{A_{f_H} \cdot \frac{f}{f_L}}{\sqrt{1 + \left(\frac{f}{f_L}\right)^2}} \quad \text{...(i) Refer to Eq. (12.11)}$$

and

$$|G_L| = \frac{A_{f_H}}{\sqrt{1 + \left(\frac{f}{f_H}\right)^2}} \quad \text{...(ii) Refer to Eq. [12.7(a)]}$$

Thus the voltage gain magnitude of band-pass filter is given by

$$\left|\frac{v_{out}}{v_{in}}\right| = |G_H G_L| = \frac{A_{f_H}^2 \cdot \frac{f}{f_L}}{\sqrt{\left[1+\left(\frac{f}{f_L}\right)^2\right]\left[1+\left(\frac{f}{f_H}\right)^2\right]}} = \frac{A_{f_T}^2 \cdot \frac{f}{f_L}}{\sqrt{\left[1+\left(\frac{f}{f_L}\right)^2\right]\left[1+\left(\frac{f}{f_H}\right)^2\right]}} \quad ...(iii)$$

where A_{fT} = Total passband gain = 4

f_L = Low cutoff frequency in Hz = 200 Hz

f_H = High cutoff frequency in Hz = 1 kHz = 1,000 Hz

f = Frequency of the input signal

Centre frequency, $f_C = \sqrt{f_L f_H} = \sqrt{200 \times 1{,}000} = 447.2$ Hz Refer to Eq. (12.15)

The voltage gain magnitude for the wide bandpass filter with f_L = 200 Hz, f_H = 1,000 Hz and passband gain = 4 is determined for different values of input signal frequency f and is tabulated as shown below :

Input Frequency f in Hz	Magnitude of Voltage Gain $\frac{v_{out}}{v_{in}}$	Voltage Gain Magnitude in dB $= 20 \log_{10} \frac{v_{out}}{v_{in}}$
10	0.19974	–13.99
20	0.39794	–8.004
50	0.96893	–0.274
100	1.77998	5.008
447.2	3.3333	10.456
800	3.0302	9.629
1,000	2.7735	8.861
2,000	1.780	5.008
5,000	0.7838	–2.115
10,000	0.3979	–8.004

The frequency response plot is shown in Fig. 12.19.

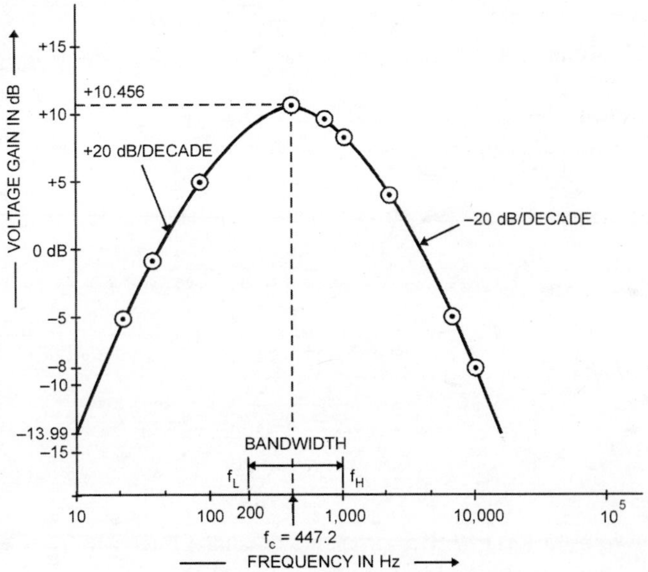

Fig. 12.19. ***Frequency Response***

(*iii*) $Q = \frac{f_C}{f_H - f_L} = \frac{447.2}{1{,}000 - 200} = 0.559$ **Ans.**

Refer to Eq. (12.14)

12.13. BAND-STOP [BAND-REJECT] FILTER

The band-pass filter passes one set of frequencies while rejecting all others. The band-stop filter does just the opposite. It rejects a band of frequencies, while passing all others. This is also called a *band-reject* or *band-elimination filter*. Like band-pass filters, band-stop filters may also be classified as (*i*) wide band and (*ii*) narrow band reject filters.

The narrow band-reject filter is also called a *notch filter*. Because of its higher Q, which exceeds 10, the bandwidth of the narrow band-reject filter is much smaller than that of a wide band-reject filter.

12.13.1. Wide Band-Stop (or Reject) Filter

A wide band-stop filter using a low-pass filter, a high-pass filter and a summing amplifier is shown in Fig. 12.20. For a proper band reject response, the low cutoff frequency f_L of high-pass filter must be larger than the high cutoff frequency f_H of the low-pass filter. In addition, the passband gain of both the high-pass and low-pass sections must be equal.

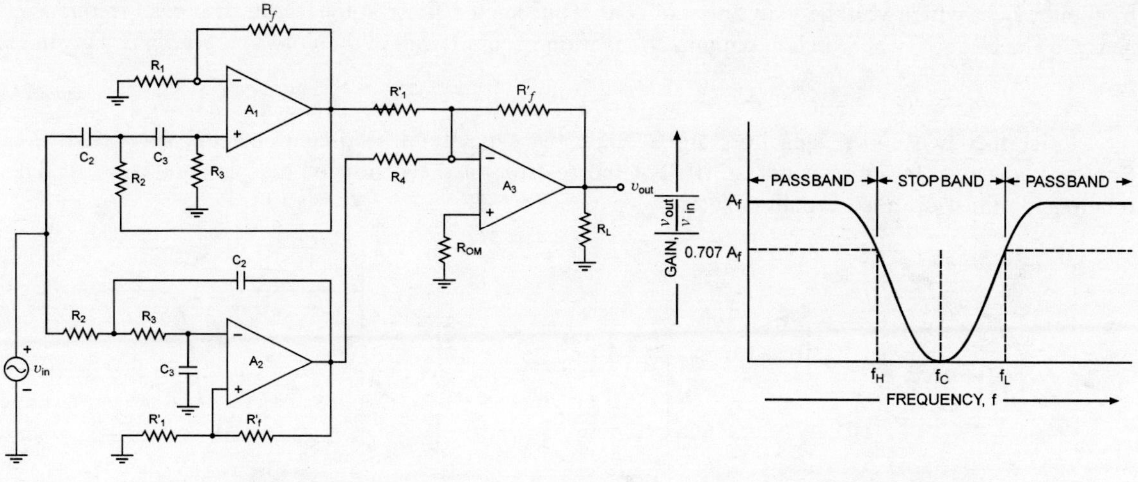

(a) Circuit Diagram *(b) Frequency Response*

Fig. 12.20. ***Wide Band-reject Filter***

Example 12.12. Design a wide band stop filter with f_H = 200 Hz and f_L = 2 kHz.

Solution : Let the capacitance, $C' = C = 0.05\ \mu F$

For low-pass filter, $f_H = 200$ Hz

$$R' = \frac{1}{2\pi f_H C} = \frac{1}{2\pi \times 200 \times 0.05 \times 10^{-6}} = 15.9\ k\Omega \simeq 18\ k\Omega \quad \textbf{Ans.}$$

For high-pass filter, $f_L = 2$ kHz

$$R = \frac{1}{2\pi f_L C} = \frac{1}{2\pi \times 2 \times 1{,}000 \times 0.05 \times 10^{-6}} = 1.59\ k\Omega \simeq 1.8\ k\Omega \quad \textbf{Ans.}$$

12.13.2. Narrow Band-Stop Filter

This is also called a *notch filter*. It is commonly used for attenuation of a single frequency such as 50 Hz power line frequency hum. The most widely used notch filter is the twin-T network illustrated in Fig. 12.21 (*a*). This is a *passive filter* composed of two T-shaped networks. One T-network is made up of two resistors and a capacitor, while the other is made of two capacitors and a resistor.

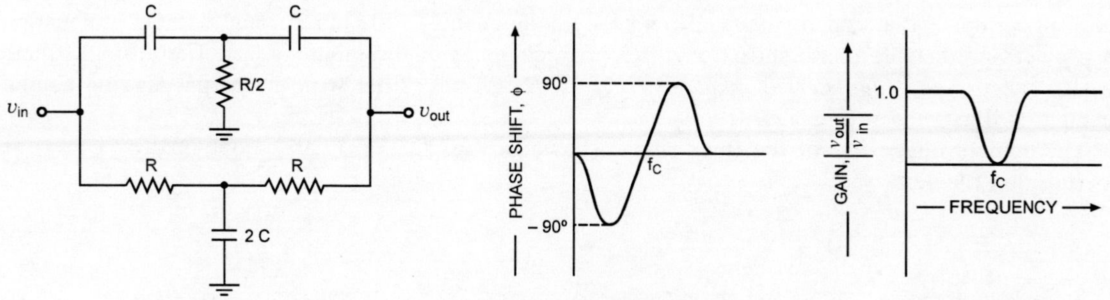

(a) Circuit Diagram *(b) Phase Shift* *(c) Voltage Gain*

Fig. 12.21. ***Twin-T Notch Passive Filter***

A mathematical analysis of this circuit shows that it acts as a lead-lag circuit with a phase angle, shown in Fig. 12.21 (*b*). Again, there is a frequency f_C at which the phase shift is equal to 0°. In Fig. 12.14 (*c*), the voltage gain is equal to 1 at low and high frequencies. In between, there is a frequency f_C at which voltage gain drops to zero. Thus such a filter notches out, or blocks frequencies near f_C. The frequency at which maximum attenuation occurs is called the *notch-out frequency* given by

$$f_N = f_C = \frac{1}{2\pi RC} \quad ...(12.21)$$

Notice that two upper capacitors are C while the capacitor in the centre of the network is 2 C. Similarly, the two lower resistors are R, but the resistor in the centre of the network is 1/2R. This relationship must always be maintained.

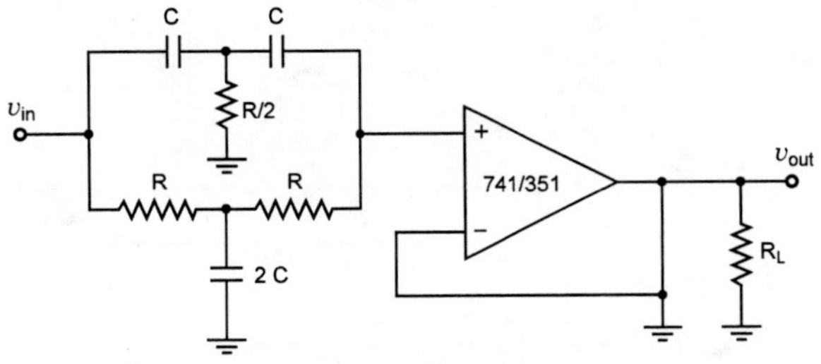

(*a*) *Circuit Diagram*

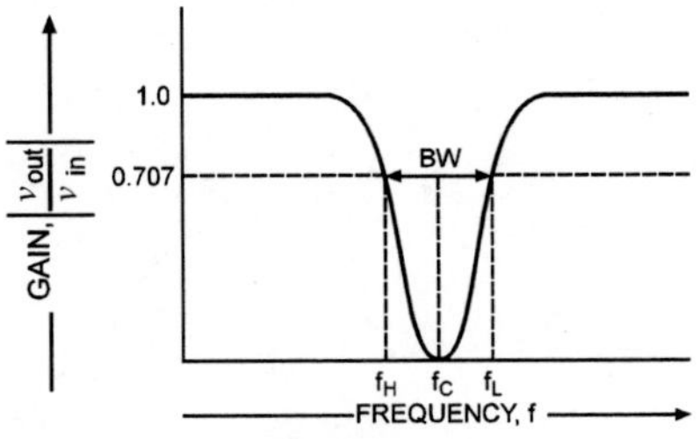

(*b*) *Frequency Response*

Fig. 12.22. ***Twin-T Active Notch Filter***

One drawback of above notch filter (passive twin-T network) is that it has relatively low figure of merit Q. However, Q of the network can be increased significantly if it is used with the voltage follower, as illustrated in Fig. 12.22 (*a*). Here the output of the voltage follower is supplied back to the junction of R/2 and 2 C. The frequency response of the active notch filter is shown in Fig. 12.22 (*b*).

Notch filters are most commonly used in communications and biomedical instruments for eliminating the undesired frequencies.

Example 12.13. Design a 60 Hz active notch filter. [*G.G.S.I.P. Univ. Analog Electronics-II May-June 2009*]

Solution : Let C = 0.06 μF

$$R = \frac{1}{2\pi f_N C} = \frac{1}{2\pi \times 60 \times 0.06 \times 10^{-6}} = 44{,}210\ \Omega \text{ or } 44.21\ k\Omega \cong 47\ k\Omega \quad \textbf{Ans.}$$

[Refer to Eq. (12.21)]

For R/2, two 47 kΩ resistors connected in parallel may be used and for 2C component, two parallel-connected 0.06 μF capacitors may be used.

12.14. ALL-PASS FILTER

An all-pass filter is that which passes all frequency components of the input signal without attenuation but provides predictable phase shifts for different frequencies of the input signals. The all-pass filters are also called *delay equalizers* or *phase correctors*. An all-pass filter with the output lagging behind the input is illustrated in Fig. 12.23.

The output voltage v_{out} of the filter circuit shown in Fig. 12.23 (*a*) can be obtained by using the superposition theorem

$$v_{out} = -v_{in} + \frac{-jX_C}{R - jX_C} v_{in} \times 2 \quad ...(12.22)$$

Substituting $-jX_C = \frac{1}{j2\pi fC}$ in above equation we have

$$v_{out} = v_{in}\left(-1 + \frac{2}{j2\pi fRC + 1}\right)$$

or $$\frac{v_{out}}{v_{in}} = \frac{1 - j2\pi fRC}{1 + j2\pi fRC} \quad ...(12.23)$$

where f is the frequency of the input signal in Hz.

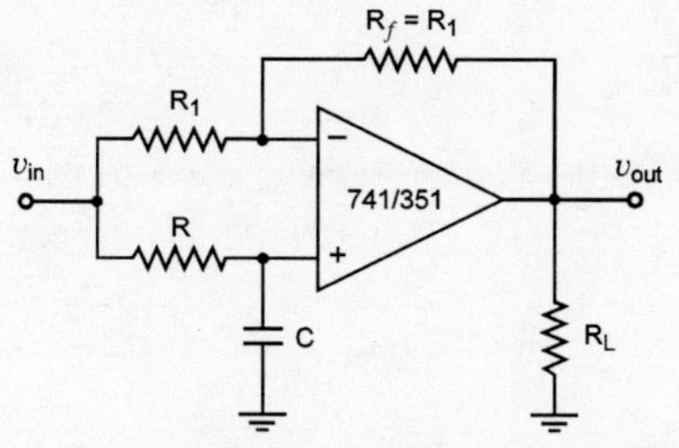

(a) Circuit Diagram

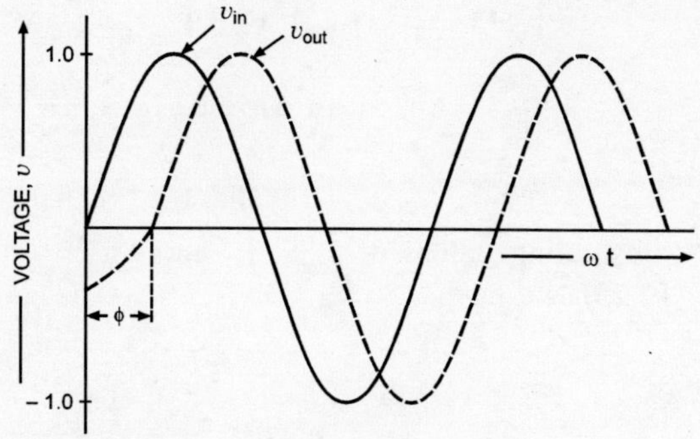

(b) Input and Output Waveforms

Fig. 12.23. ***All Pass Filter***

From Eq. (12.23) it is obvious that the amplitude of $\frac{v_{out}}{v_{in}}$ is unity *i.e.,* $|v_{out}| = |v_{in}|$ throughout the useful frequency range and the phase shift between the input and output voltages is a function of frequency. The phase shift ϕ in degrees is given by the equation

$$\phi = -2 \tan^{-1} \frac{2\pi f RC}{1} \quad ...(12.24)$$ where f is in Hz; R is in ohms and C is in farads.

By interchanging the positions of R and C in the circuit shown in Fig. 12.23 (*a*), the output can be made leading the input.

These filters are most commonly used in communications. For instance, when signals are transmitted over transmission lines (such as telephone wires) from one point to another point, they undergo change in phase. To compensate for such phase changes, all-pass filters are employed.

Example 12.14. For an all-pass filter circuit shown in Fig. 12.23 (*a*), determine the phase shift ϕ if R = 15 kΩ; C = 0.01 μF and frequency of the input signal is 2 kHz.

Solution : Phase shift, $\phi = -2 \tan^{-1} 2\pi f RC = -2 \tan^{-1} (2\pi \times 2 \times 10^3 \times 15 \times 10^3 \times 0.01 \times 10^{-6})$
$= -124°$ *i.e.,* 124° (lagging) **Ans.**

With the advancement in IC technology, a number of manufacturers now offer *universal filters* having simultaneous low-pass, high-pass, and band-pass output responses. Notch and all-pass functions are also available by combining these output responses in the uncommitted op-amp. Because of its versatility, this filter is called the universal filter. It provides the user with easy control of the gain and Q-factor. It is also called a *state-variable filter.*

12.15. OPERATIONAL TRANSCONDUCTANCE AMPLIFIER (OTA)-C FILTERS

The OTAs have been found very useful and attractive for making active filters. It is because of possibility of control of the filter performance by an external voltage V_C or control current I_C. The controlled parameters are the circuit midband gain, or the filter 3-dB frequency. Even the critical frequency of the filter can be changed without any change in passband gain. The type of response from low-pass to all-pass can also be altered by continuous adjustment of the transconductance g_m. A very simple first-order low-pass filter is shown in Fig. 12.24.

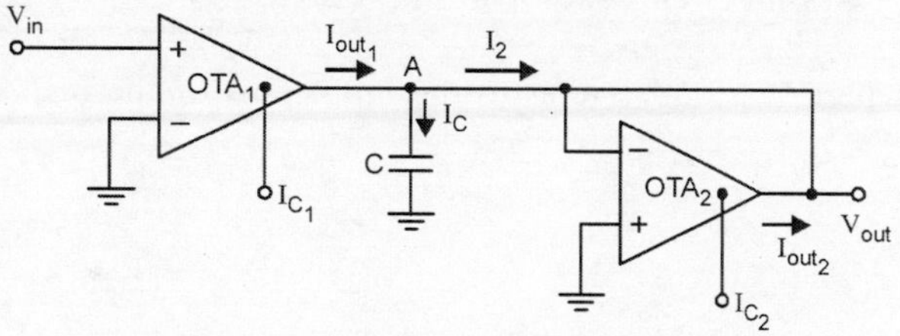

Fig. 12.24. ***First-Order Low-Pass OTA-C Filter***

Applying KCL at node A, we have

$$I_{out_1} = I_C + I_2 = I_C - I_{out_2} \quad ...(12.25)$$

$$\text{Also } I_{out_1} = g_m(V^+ - V^-) = g_m V_{in} \qquad ...(12.26)$$

$$\text{Control current, } I_C = \frac{V_{out}}{X_C} = j\omega C V_{out} \qquad ...(12.27)$$

$$\text{and } I_{out_2} = g_m(V^+ - V^-) = g_m(0 - V_{out}) = -g_m V_{out} \qquad ...(12.28)$$

Substituting values of I_{out_1}, I_C and I_{out_2} from Eqs. (12.26), (12.27) and (12.28) respectively in Eq. (12.25), we have

$$g_m V_{in} = j\omega C V_{out} - (-g_m V_{out}) = V_{out}(j\omega C + g_m)$$

$$\text{or } \frac{V_{out}}{V_{in}} = \frac{g_m}{j\omega C + g_m} \qquad ...(12.29)$$

The 3-dB frequency is given by

$$f_{3\text{-dB}} = \frac{g_m}{2\pi C} \qquad ...(12.30)$$

In the circuit shown in Fig. 12.24, OTA_2 has been configured as a voltage variable resistor which provides the variable cut-off frequency given by Eq. (12.30).

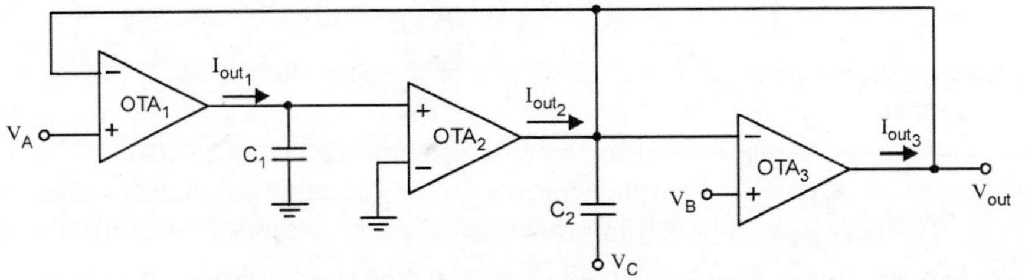

Fig. 12.25. ***A Generalized Second-Order Filter Circuit LPF (V_{out}/V_A); HPF (V_{out}/V_C) and BPF (V_{out}/V_B) Other Terminals Grounded***

A generalized second-order filter circuit with three voltage control terminal is depicted in Fig. 12.25. Low-pass, high-pass, band-pass (or a notch) filter can be realized by using terminals V_{out} and V_A, V_{out} and V_C and V_{out} and V_B respectively with the rest terminals set to ground, as shown in Fig. 12.25. Each of these filters has a centre or critical frequency that can be set by variation of transconductance, g_m of the three OTAs. These filters are also known as adjustable frequency adjustable Q filters because of independently controllable f_c and Q. This independent control makes realization of a constant bandwidth over a wide range of center frequencies possible.

From circuit shown in Fig. 12.25.

$$V_1^- = V_3^- = V_{out} \text{ and } V_1^+ = V_A \qquad ...(12.31)$$

where V_1, V_2, V_3 refer to the input pins of the respective OTAs.

So for OTA_1,
$$I_{out_1} = g_{m_1}(V_1^+ - V_1^-) = g_{m_1}(V_A - V_{out}) \qquad ...(12.32)$$

For OTA_2,
$$V_2^+ = I_{out_1} X_{C_1} = \frac{g_{m_1}(V_A - V_{out})}{sC_1}$$

$$\text{and } V_2^- = 0$$

$$\text{So } I_{out_2} = g_{m_2}(V_2^+ - V_2^-) = \frac{g_{m_1} g_{m_2}(V_A - V_{out})}{sC_1} \qquad ...(12.33)$$

For OTA_3
$$V_3^+ = V_B, \quad V_3^- = V_{out} = I_{out_2} X_{C_2} + V_C$$

$$I_{out_3} = g_{m_3}(V_3^+ - V_3^-) = g_{m_3}(V_B - V_{out}) \qquad ...(12.34)$$

Applying KCL at the output of both OTA_2 and OTA_3, we have

$$I_{out_2} + I_{out_3} = \frac{V_{out} - V_C}{X_{C_2}} = sC_2(V_{out} - V_C)$$

Substituting values of I_{out_2} and I_{out_3} respectively from Eqs. (12.33) and (12.34) in above equation, we have

$$\frac{g_{m_1} g_{m_2}(V_A - V_{out})}{sC_1} + g_{m_3}(V_B - V_{out}) = sC_2(V_{out} - V_C)$$

or $$V_{out}\left[\frac{g_{m_1} g_{m_2} + sC_1 g_{m_3} + s^2 C_1 C_2}{sC_1}\right] = \frac{g_{m_1} g_{m_2} V_A}{sC_1} + g_{m_3} V_B + sC_2 V_C \quad ...(12.35)$$

$$\text{or } V_{out} = \frac{s^2 C_1 C_2 V_C + sC_1 g_{m_3} V_B + g_{m_1} g_{m_2} V_A}{s^2 C_1 C_2 + sC_1 g_{m_3} + g_{m_1} g_{m_2}} \quad ...(12.36)$$

As already mentioned above, a *second-order low-pass* filter can be had by applying input signal to V_A and grounding V_B and V_C. The standard biquadratic form of a low-pass filter can be had by assuming $g_{m_1} = g_{m_2} = g_{m_3}$ and dividing both numerator and denominator by C_1C_2 in Eq. (12.36)

$$\frac{V_{out\,LP}}{V_A} = \frac{g_m^2 / C_1 C_2}{s^2 + \dfrac{s g_m}{C_2} + \dfrac{g_m^2}{C_1 C_2}} \quad ...(12.37)$$

$$= \frac{\omega_h^2}{s^2 + s\dfrac{\omega_h}{Q} + \omega_h^2} \quad ...(12.38)$$

$$\text{where } \omega_h = \frac{g_m}{\sqrt{C_1 C_2}} \text{ and } Q = \sqrt{\frac{C_2}{C_1}} \quad ...(12.39)$$

A *high-pass filter* can be had by applying input signal to V_C and grounding V_A and V_B. The expression for high-pass filter is given as

$$\frac{V_{out\,HP}}{V_C} = \frac{s^2}{s^2 + s\dfrac{g_m}{C_2} + \dfrac{g_m^2}{C_1 C_2}} \quad ...(12.40) \quad \text{if } g_{m_1} = g_{m_2} = g_{m_3} = g_m$$

A *band-pass filter* can be obtained by applying input signal to V_B and grounding V_A and V_C. So

$$\frac{V_{out\,BP}}{V_B} = \frac{s(g_{m_3}/C_2)}{s^2 + s\dfrac{g_{m_3}}{C_2} + \dfrac{g_{m_1} g_{m_2}}{C_1 C_2}} \quad ...(12.41)$$

The above transfer function is of the form

$$\frac{V_{out}}{V_{in}} = \frac{A_1 s}{s^2 + s\left(\dfrac{\omega_0}{Q}\right) + \omega_0^2} \quad ...(12.42)$$

which is second order band-pass filter.

The centre frequency of the filter is given by expression

$$\omega_0 = \sqrt{\frac{g_{m_1} g_{m_2}}{C_1 C_2}} \quad ...(12.43)$$

and 3-dB bandwidth is found directly from the transfer function to be

$$BW = \frac{\omega_0}{Q} = \frac{g_{m3}}{C_2} \text{ radians/second} \qquad ...(12.44)$$

12.16. TYPES OF ACTIVE FILTERS

As already mentioned in Art. 12.5, there are basically four types of active filters viz Butterworth, Chebyshev, Bessel and Elliptic filters.

1. Butterworth Filter. This filter is also called a *maximally flat* or *flat-flat filter*. This class of filters approximates the ideal filter well in the passband. Frequency response curves of different types of filters are shown in Fig. 12.26. The Butterworth filter has an essentially flat amplitude-frequency response up to the cutoff frequency. The sharpness of the cutoff can be seen in the figure. It is to be noted that all the three filters reach a rolloff slope of –40 dB/decade at frequencies much larger than cutoff. Although Butterworth filters achieve the sharpest attenuation, their phase shift as a function of frequency is nonlinear. It has a monotonic drop in gain with frequency in the cutoff region and a maximally flat response below cutoff frequency, as illustrated in Fig. 12.27. The Butterworth filter has characteristics somewhere between those of Chebyshev and Bessel filters. It has a moderate roll-off of the skirt and a slightly nonlinear phase responses.

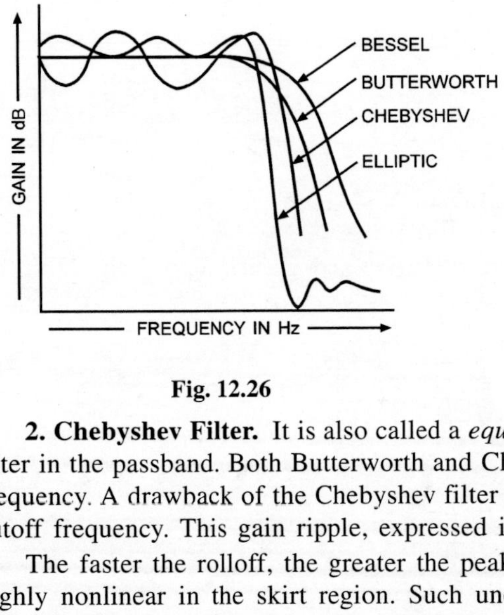

Fig. 12.26

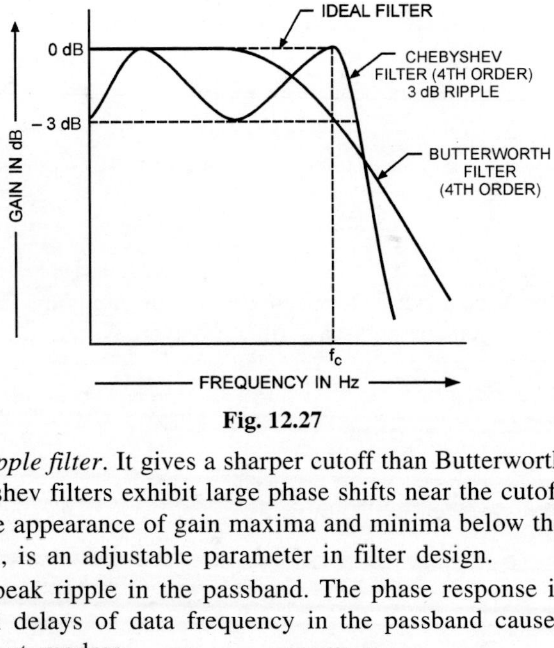

Fig. 12.27

2. Chebyshev Filter. It is also called a *equal ripple filter*. It gives a sharper cutoff than Butterworth filter in the passband. Both Butterworth and Chebyshev filters exhibit large phase shifts near the cutoff frequency. A drawback of the Chebyshev filter is the appearance of gain maxima and minima below the cutoff frequency. This gain ripple, expressed in dB, is an adjustable parameter in filter design.

The faster the rolloff, the greater the peak-to-peak ripple in the passband. The phase response is highly nonlinear in the skirt region. Such unequal delays of data frequency in the passband causes severe pulse distortion and thus increased errors at modern demodulators. This can be overcome somewhat by increasing the BW of the filter so that the phase region is extended.

A Chebyshev filter is used where very sharp rolloff is required. However, this is achieved at the expense of a gain ripple in the lower frequency passband.

3. Bessel Filter. The Bessel filter provides ideal phase characteristics with an approximately linear phase response up to nearly cutoff frequency. Though it has a very linear phase response but a fairly gentle skirt slope, as shown in Fig. 12.28.

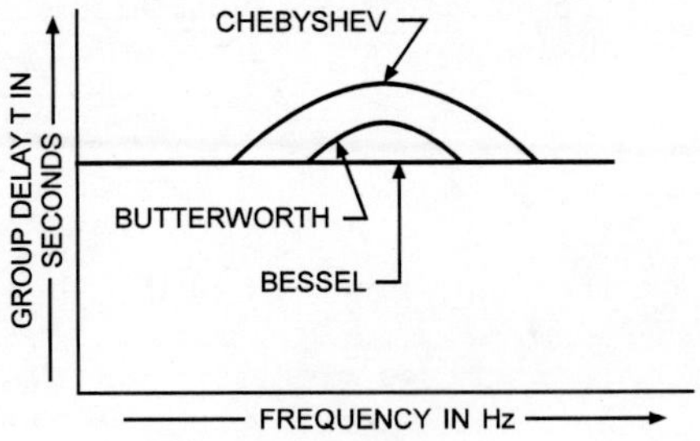

Fig. 12.28

For applications where the phase characteristic is important, the Bessel filter is used. It is a minimal phase shift filter even though its cutoff characteristics are not very sharp. It is well suited for pulse applications.

4. Elliptic Filter. This filter has the sharpest rolloff of all filters in the transition region but has ripples in both the passband and stopband regions, as illustrated in Fig. 12.26. The elliptic filter can be designed to have very high attenuation for certain frequencies in the stopband, which reduces the attenuation for other frequencies in the stopband.

HIGHLIGHTS

1. A filter is often a frequency-selective circuit that passes a band of frequencies and attenuates signals of frequencies outside this band.
2. The gain of a filter is *complex number* that is a function of frequency. Filter performance is often described with a graph of gain (magnitude and phase) versus frequency. The manner in which the gain of the filter responds to different frequencies is called the *frequency response.* A graph of the magnitude of the gain versus frequency is called a *frequency response plot.*
3. Filters are classified in a number of ways : analog or digital, passive or active, audio or radio frequency. Passive filters use inductors. At audio frequencies, inductors are bulky, expensive and have poor electrical characteristics. Active filters use op-amps.
4. The most widely used active filters are: (*i*) low-pass (*ii*) high-pass (*iii*) band-pass (*iv*) band-stop or band-reject (also called the band-elimination or notch) and (*v*) all-pass filters.

   The –3 dB frequency serves as the dividing line between the stopband and the passband. Low-pass filters pass low frequencies and attenuate high frequency signals. High-pass filters do just the opposite. The band-pass filter passes only a certain range of frequencies. All lower and higher frequency signals are attenuated. The notch filter stops the band of frequencies, while passing higher and lower frequencies. Finally, the all-pass filter has input and output amplitudes equal at all frequencies; however, the phase shift between the two is a function of frequency.
5. Passive filters usually have low input impedance and high output impedance, and are easily loaded and frequency shifted by the load. There is poor isolation between load and source. To obtain second-order or higher response, inductors are to be used. At audio frequencies or below these inductors are expensive, bulky and have rather poor electrical characteristics.

   Active filters, especially those built with op-amp, have none of these problems. However, they need power supplies, and may be more expensive than simple passive filters.
6. The order of the filter indicates the rate at which the gain changes while the input frequency is approaching or exceeding the cutoff frequency of the filter.
7. By interchanging the frequency determining components (resistors and capacitors) a low-pass filter can be changed into a high-pass filter or vice versa.
8. Higher-order filters can be built by simply cascading first and second-order filters. For instance, a third-order filter is formed by cascading first and second-order filters; a fourth-order is formed by cascading two second-order filters and so on.
9. OTA-C filters are advantageous in many aspects : possibility of control of (*i*) filter performance parameters, such as midband gain, critical or centre frequency, by an external voltage V_C or control current I_C, (*ii*) type of filter by adjustment of OTA g_m.

EXERCISES

1. What is meant by a filter ? Give the classification of the filters. What is difference between active and passive filters ?
2. What are the advantages of an active filter over a passive filter ?
3. Differentiate between low-pass, high-pass, band-pass and band-reject types of filters, sketching the frequency plots. Give one application for each category.
4. Draw the schematic of an IC low-pass active filter and explain the operation. Also plot the gain-frequency response.

5. Draw the circuit of a second-order low-pass filter and derive an expression for its transfer function for a Butterworth filter. **[G.G.S.I.P. Univ. Analog Electronics-II 2011]**
6. Draw the circuit of a second-order low-pass filter and derive an expression for its transfer function for a maximally flat response in the pass band. **[G.G.S.I.P. Univ. Analog Electronics-II, May 2010]**
7. Write a short note on second order LPF using op-amp. **[G.G.S.I.P. Univ. Analog Electronics December 2009]**
8. Why are active filters preferred over passive filters ? Draw the schematic of a second-order high-pass filter and explain briefly the cutoff characteristics.
9. Draw the schematic of an active second-order high-pass filter and sketch the frequency response.
10. What are the merits of active filters employing op-amps ? Draw the schematic of a band-pass filter and explain the working.
11. Discuss the set-up for realising a band-pass active filter. Sketch its typical frequency response.
12. Draw frequency characteristics of LP, HP and BP filters (first- and second-order). **[G.G.S.I.P. Univ. Analog Electronics II May-June 2009]**
13. Draw the circuit for a wide band-reject active filter.
14. Sketch the response of a notch filter and cite an application for the same.
15. Draw the circuit of a notch filter and derive an expression for its transfer function. **[G.G.S.I.P. Univ. Analog Electronics II May 2010]**
16. Draw the circuit of an all-pass filter and find its transfer function. Give an application of All-pass filter. **[G.G.S.I.P. Univ. Analog Electronics II May 2010]**
17. Explain an all-pass filter. **[G.G.S.I.P. Univ. Analog Electronics-II May 2011]**
18. What is OTA-C filter ? How is it advantageous ? Draw the circuit of a simple first-order low-pass OTA-C filter and derive expressions for transfer function and 3-dB frequency.
19. Draw the circuit of a generalised second-order filter using OTA and derive the transfer functions for second-order (*i*) low-pass and (*ii*) high-pass filters.
20. Write short notes on the following :
    (*i*) Active filters versus passive filters.
    (*ii*) Classification of filters.
    (*iii*) Second order LP Butterworth filter
    (*iv*) High-order filters.
    (*v*) Single op-amp band-pass filter.
    (*vi*) Generalised second-order filter circuit.

SHORT ANSWER TYPE QUESTIONS WITH ANSWERS....................

Q. 1. What is meant by 'frequency response' of a filter ?

Ans. A graph of the magnitude of the gain versus frequency is called a frequency response plot.

Q. 2. The gain of a filter is a complex number. What is meant by it ?

Ans. Because at a given frequency, the gain has a specific magnitude as well as a phase angle, the gain of a filter is said to be a complex number.

Q. 3. What is meant by –3 dB frequency ?

Ans. –3 dB frequency is one at which the power is reduced to one half of the maximum magnitude.

Q. 4. What are the commonly used filters ?

Ans. The most commonly used filters are low-pass, high-pass, band-pass, band-reject and all-pass filters.

Q. 5. Distinguish between active and passive filters ?

Ans. A passive filter is formed with passive components, such as resistors, inductors and capacitors while active filters make use of transistors or op-amps in addition to resistors and capacitors.

Q. 6. Define filter parameters.

Ans. The filter exhibits a wide variety of parameters such as order of filter, cutoff frequencies *i.e.,* upper cutoff frequency and lower cutoff frequency, passband gain, break frequency etc.

Q. 7. Explain why active filters are preferred over passive filters ?

[G.G.S.I.P. Univ. Analog Electronics-II May 2011]

Ans. Active filters are preferred over passive filters because they provide gain and frequency adjustment flexibility. They pose no loading problem because of high input impedance and low output impedance and they are comparatively less expensive.

Q. 8. What is a low-pass filter ?

Ans. A filter that provides a constant output from dc up to a cut-off frequency f_c and then passes no signal above that frequency is called a low-pass filter.

Q. 9. Define first-order low-pass filter ? What does order of a filter signifies ?

Ans. A first order low-pass filter is that in which there is a constant gain from 0 Hz to a high cutoff frequency f_H and beyond f_H the gain decreases at constant rate with the increase in frequency. The rate at which the gain rolls off beyond f_H is 20 dB/decade or 6 dB/octave.

The order of a filter signifies that at what rate stopband response increases or decreases. Higher the order of the filter, better are the attenuation characteristics of the filter. It means that as the order of the filter is increased, the actual stopband response of the filter approaches its ideal stopband characteristics.

Q. 10. Define second order low-pass filter ? What does order of a filter signify.

Ans. A second order low-pass filter is that in which there is a constant gain from 0 Hz to a high cutoff frequency f_H and beyond f_H the gain decreases at constant rate with the increase in frequency. The rate at which the gain rolls off beyond f_H is 40 dB/decade.

The order of a filter signifies that at what rate stop band response increases or decreases. Higher the order of the filter, better are the attenuation characteristics of the filter. It means that as the order of the filter is increased, the actual stopband response of the filter approaches its ideal stopband characteristics.

Q. 11. What is a high-pass filter ?

Ans. A filter that provides or passes signals above a cutoff frequency f_c is called a high-pass filter. It has a zero gain starting from zero to a frequency f_c and above this frequency the gain is constant.

Q. 12. What is the advantage of increasing the order of a filter ?

Ans. With increasing the order of a filter, the actual stopband response of the filter approaches its ideal stopband characteristics.

Q. 13. Why do we use higher order filters ?

Ans. Higher order filters are used to obtain stopband response of the filter approaching ideal stopband characteristics.

Q. 14. What is the wide bandpass filter ?

Ans. A wide bandpass filter has a wide passband between two cutoff frequencies f_H and f_L where $f_H >> f_L$ and two stopbands at $0 < f < f_L$ and $f > f_H$.

Q. 15. What is a notch filter ?

Ans. Notch filter is a narrow band-stop filter. It notches out or blocks frequencies near centre frequency f_c, also called a notch-out frequency.

Q. 16. Fill in the blanks :

(*i*) The manner in which the gain of the filter responds to different frequency is called the ________.

(*ii*) A graph of the magnitude of the gain versus frequency is called a ________.

(*iii*) According to the operating range the filters may be classified as AF and ________ filters.

(*iv*) The rate at which the gain rolls off beyond f_H is ________ in first order low-pass filters.

(*v*) To convert a low-pass filter into a high-pass filter or vice versa simply interchange ________.

Ans. (*i*) Frequency response, (*ii*) Frequency response plot, (*iii*) RF, (*iv*) 20 dB per decade, (*v*) frequency determining components R and C.

PROBLEMS

1. Design a low-pass filter having a cutoff frequency of 1 kHz and has passband gain of 2.
   [**Ans.** C = 0.01 μF, R ≅ 20 kΩ (plot), $R_1 = R_F$ = 10 kΩ]

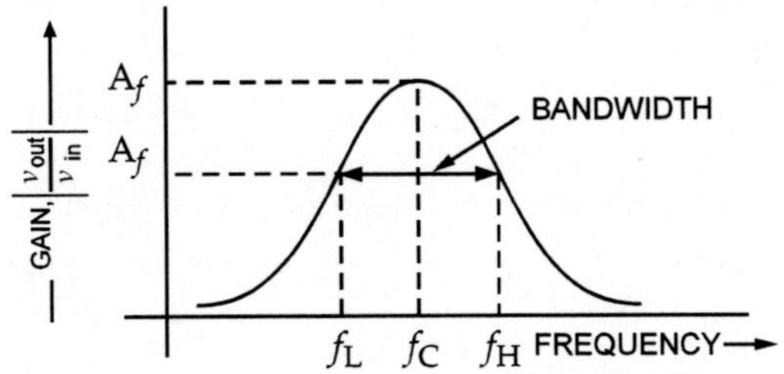

Fig. 1

2. Design a wide band-pass filter with f_H = 1 kHz; f_L = 200 Hz and a passband gain = 4.
   Plot the frequency response also.
   What is value of Q for this filter ? [**Ans.** C = 0.05 μF, R = 15.9 kΩ, f_c = 447.2 Hz, Q = 0.56]
3. Using the op-amp 741 shown below (Fig. 2) design a bandpass filter so that f_c = 1 kHz, Q = 3 and gain A_f = 10. [**Ans.** $C_1 = C_2$ = C = 0.01 μF, R_1 = 4.77 kΩ, R_2 = 5.97 kΩ, R_3 = 95.5 kΩ]

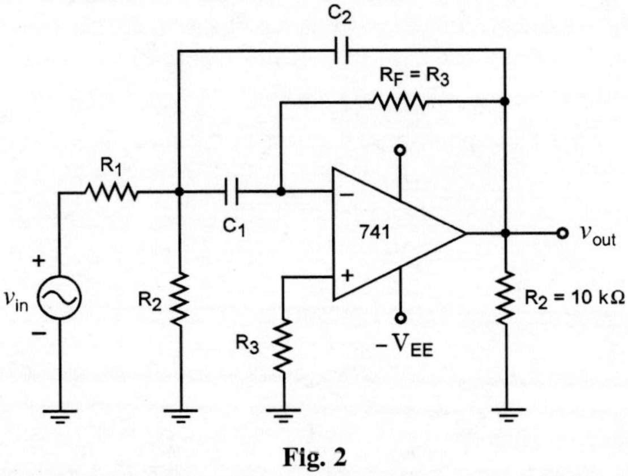

Fig. 2

4. Design a 60 Hz active notch filter. [**Ans.** C = 0.068 μF, R = 39 kΩ]
5. (*a*) Design a LP filter at a cutoff frequency of 1 kHz with passband gain of 2. (*b*) Design a 400 Hz active notch filter.

❑❑❑

CHAPTER 13

Phase-Locked Loops

13.1. INTRODUCTION

INSIDE THIS CHAPTER

Phase-locked loop is a feedback loop consisting of a phase detector, a low-pass filter, amplifier (operational) and a voltage-controlled oscillator (VCO), as illustrated in Fig. 13.1. It plays the same role in the frequency or phase world as the op-amp does in the voltage world. The op-amp has two voltage inputs, noninverting and inverting (normally used for feedback from the output). Similarly, the PLL has two inputs; the PLL's feedback input is normally connected to the circuit's output. Digital frequencies are usually applied. The op-amp changes its output voltage to whatever values it is necessary to drive the difference in voltage between its two inputs to zero. The PLL changes its output phase and frequency to whatever frequency or phase is necessary to make the two input frequencies and phase track. Placing a voltage divider in the feedback loop of an op-amp causes the output voltage to be increased by the amount of the feedback voltage division (amplification). Placing a frequency divider in the feedback of a PLL causes the output frequency to be increased by the amount of the feedback divider. A firm grasp on similarities between the PLL and the op-amp simplifies our analysis and design of circuits containing PLLs.

With the rapid development of IC technology, the phase-locked loop (PLL) has emerged as one of the fundamental building blocks in electronic technology. Common applications of a PLL include (*i*) frequency synthesizers that provide multiples of a reference signal frequency; (*ii*) FM demodulation networks for FM operation with excellent linearity between the input signal frequency and the PLL output voltage; (*iii*) demodulation of the two data transmission or carrier frequencies in digital-data transmission employed in frequency-shift keying (FSK) operation and (*iv*) a wide variety of areas including telemetry receivers and transmitters, tone decoders, AM detectors, tracking filters and motor speed controls.

13.2. OPERATING PRINCIPLE

The block diagram showing operating principle of PLL is given in Fig. 13.1. As illustrated in this figure, the PLL consists of a phase detector, a low-pass filter and a voltage controlled oscillator.

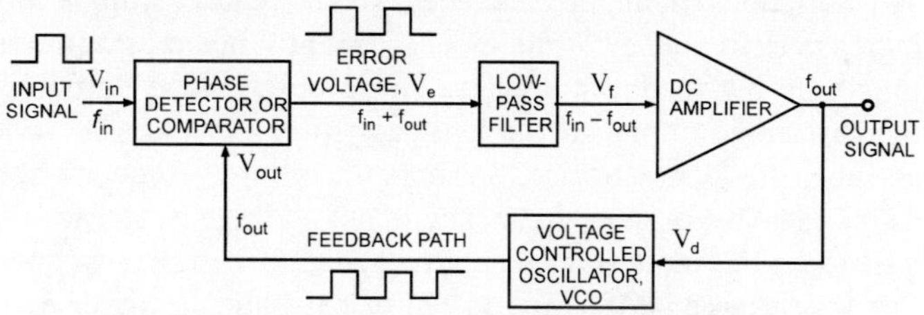

Fig. 13.1. ***Block Diagram of Basic Phase-Locked Loop (PLL)***

A *phase detector* is basically a comparator that compares the input frequency

f_{in} with feedback frequency f_{out}. The phase detector receives two digital signals, one from the input, the other fed back from the output. The output of the phase detector is proportional to the phase difference between f_{in} and f_{out}. The loop is locked when these two signals are of the same frequency and have a fixed phase difference (A locked PLL is analogous to an op-amp not being saturated). The output of a phase detector is a dc voltage and therefore is often referred to as the *error voltage*, V_e. DC output voltage becomes maximum when the phase difference between the two frequencies f_{in} and f_{out} is π radians or 180°. Without input signal, the error voltage V_e is equal to zero and the VCO operates at a set frequency 'f_r' which is also called *free-running frequency* of the VCO. When the input signal frequency is the same as that from the VCO to the phase comparator (PC), the voltage, V_d, taken as output is the value required to hold the VCO in lock with the input signal. If the two input pulses to the PC are of exactly the same frequency and phase, the output of the PC is zero, otherwise there will be an output proportional to their phase difference.

Low-pass filter is used to remove high frequency components and noise from the output of the phase detector. The phase detector is basically a multiplier that produces the sum ($f_{in} + f_{out}$) and difference ($f_{in} - f_{out}$) components of frequency at its output. The low-pass filter removes the high frequency component ($f_{in} + f_{out}$). The difference frequency component of the signal is amplified and then applied as control voltage to voltage-controlled oscillator (VCO). It affects the dynamic characteristics of the PLL including bandwidth, capture and lock ranges and transient response. The low-pass filter accepts the output from the phase detector, removes the high frequency noise and produces a dc level.

Voltage-controlled oscillator generates frequency controlled by input voltage. The dc level output of a low-pass filter is applied as control signal to the voltage-controlled oscillator (VCO). The output frequency of the VCO is directly proportional to the input dc level. The VCO frequency is adjusted till it becomes equal to the frequency of the input signal. During this adjustment, PLL goes through three stages free running, capture and phase lock. Best operation is obtained if the centre frequency of the VCO is set with the dc bias voltage midway in its linear operating range. The amplifier allows this adjustment in dc voltage from that obtained as output of the filter circuit. When the loop is in lock, the two signals to the PC are of the same frequency, although not necessarily in phase. A fixed phase difference between the two signals to the comparator results in a fixed dc voltage to the VCO. Variation in the input signal frequency then causes variation in the dc voltage to the VCO. Within a capture-and-lock frequency range, the dc voltage will drive the VCO frequency to match that of the input.

While the loop is trying to achieve lock, the output of the PC contains frequency components as the sum and difference of the signals compared. A low-pass filter passes only the lower-frequency component of the signal so that the loop can obtain lock between input and VCO signals.

Owing to the limited operating range of the VCO and the feedback connection of the PLL circuit, there are two important frequency bands specified for a PLL. The *capture range* of a PLL is the range of frequencies centred about the VCO free-running frequency f_r, over which the output signal frequency of the VCO can acquire lock with the input signal frequency. Once the PLL has achieved capture, it can maintain lock with the input signal over a somewhat wider frequency range called the *lock range*.

In brief, it can be stated that the VCO gives a digital signal whose frequency and phase are controlled by the dc input from the low-pass filter. Some VCOs also provide a sinusoidal wave.

Variations in input frequency should cause similar variations in the output frequency. A small increase in frequency at the input is initially taken as an increase in the difference in phase between the input and the feedback signals. The average value output of the phase detector increases, causing an increase in dc output of the low-pass filter. An increase in the voltage into the VCO causes its phase, and then frequency to rise. So increase in input frequency causes the output frequency to rise. As long as the increase in input frequency is not so large as to cause the loop to loose lock, the output frequency will track the input. Of course, to generate a higher frequency the VCO must have a larger input dc. For this a larger difference in phase between the input and the feedback signals is required. But the frequencies stand locked, though their phase relationship may change. As a microprocessor can easily change an output digital frequency, this PLL frequency follower allows the control of a sinusoidal frequency by an embedded (microprocessor) controller.

The other major problem with analog signal generators is variation in frequency caused by component value shifts, either because of initial tolerances, or the drift. Such a variation in either the low-pass filter or VCO components may result in increase in output frequency. This is initially taken as increase in phase at the output, and also at the feedback input. But increasing the feedback phase reduces the difference in phase between the input and the feedback. The average output from the phase comparator, therefore, decreases. This reduces the dc output from the low-pass filter, lowering the frequency from the VCO. Since the initial problem was an increase in frequency from VCO, the loop has corrected itself. True, the phase difference at the input of the phase comparator may be a bit changed. But, the output frequency remains locked with the input frequency, even though component values have varied from the nominal.

Now we shall consider the detailed function of each of the blocks.

1. Phase Detector. As mentioned earlier, the phase detector compares the input frequency and the frequency of the VCO and produces a dc voltage that is proportional to the phase difference between the two frequencies. Best performance is often obtained when these signals have a 50% duty cycle.

Depending on the analog or digital phase detector used, the PLL is either called an analog type or digital type respectively. Even though most of the monolithic PLL ICs make use of analog phase detectors, the majority of discrete phase detectors in use are of the digital type chiefly because of its simplicity.

A double-balanced mixer is a classic example of an analog phase detector. On the other hand, examples of digital phase detectors are:

1. Exclusive-OR phase detector.
2. Edge-triggered phase detector.
3. Monolithic phase detector (such as CD 4046A)

The simplest phase detector is an exclusive-OR gate. The output of the exclusive-OR gate is *high* only when f_{in} or f_{out} is high as shown in Truth Table.

TRUTH TABLE FOR EXCLUSIVE-OR GATE

Inputs		*Output*
f_{in}	f_{out}	
LOW	LOW	LOW
LOW	HIGH	HIGH
HIGH	LOW	HIGH
HIGH	HIGH	LOW

Figure 13.2 (*a*) illustrates how phase detection is implemented. In this figure, f_{in} is leading f_{out} by ϕ degrees; *i.e.,* the phase difference between f_{in} and f_{out} is ϕ degrees. The dc output voltage of the exclusive-OR phase detector is a function of phase difference betwen its two inputs. Fig. 13.2 (*b*) shows dc output voltage as a function of the phase difference between f_{in} and f_{out}. This graph indicates that the maximum dc output is obtained when the phase difference is of π radians or 180°. The slope of the curve between 0 and π radians is the conversion gain k_p of the phase detector. For instance, if the exclusive-OR gate uses a supply voltage V_{CC} = 10 V, the conversion gain,

$$k_p = \frac{10\,\text{V}}{\pi} = 3.18 \text{ V per radian}$$

There is a linear relationship between phase angle difference ϕ and the duty cycle of the ouptut waveform

$$\text{Duty cycle, D} = \frac{\phi}{180°}$$

where ϕ = phase difference between the input and the feedback signals in degrees.

The average value of a signal is the area under the curve, divided by the period (*i.e.,* average height = area/length)

$$\text{Average value} = \frac{V_{HIGH} \times t_{on}}{T}$$

But, Duty cycle, $D = \frac{t_{on}}{T}$

Hence, average value = $V_{HIGH} \times D$

or Average value = $V_{HIGH} \times \frac{\phi}{180}$ ...(13.1)

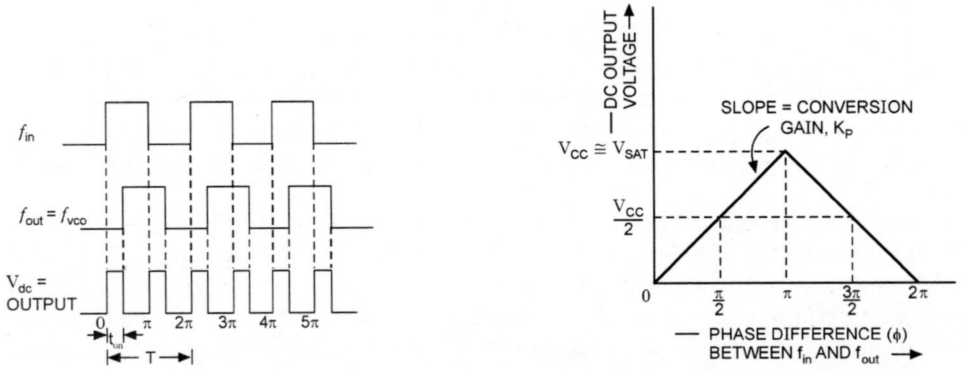

(a) Input and Output Waveforms *(b) Average Output Voltage Versus Phase Difference Between f_{in} and f_{out}*

Fig. 13.2

Of course, any logic IC exclusive-OR gate may be chosen. However, the CD4046A contains all of the active components that are required to form a PLL within its single package, including an exclusive-OR phase detector. The schematic symbol is given in Fig. 13.3. The IC is CMOS, allowing power from 5 V to 15 V. This allows us to match input levels with traditional 5 V logic or to power the IC from the positive side of analog supplies. The V_{DD} pin (pin 16) is for V^+, while V_{SS} (pin 8) is usually grounded. The exclusive-OR phase detector's output is pin 2.

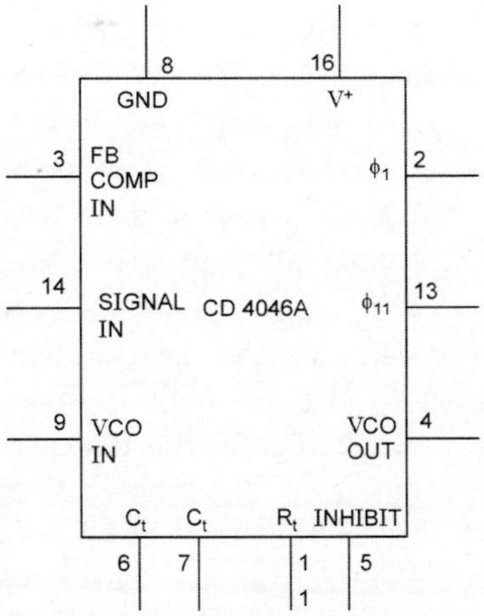

Fig. 13.3. ***Schematic Symbol of CD 4046A PLL IC***

As already mentioned, if there is a significant difference in frequencies of the input signal and the feedback signal, the PLL may not be able to have a lock. An extreme case is illustrated in Fig. 13.4. There are two examples. In Fig. 13.4 (*a*) the input frequency is twice of that being fed back from VCO. As illustrated, $V_{out} = 50\% \ V^+$. When the input frequency becomes three times the feedback signal the result is shown in Fig. 13.4 (*b*). Further increase in input frequency should increase the output from the phase detector to drive the output frequency from VCO even higher. However, the exclusive-OR type phase detector actually outputs only one-third (V^+) when the input frequency is three times too high. Thus increase in input frequency has actually reduced the output of the phase detector. The output frequency from VCO will decrease (not increase). Lock has been lost. It can be reestablished only if the input frequency is close to the output frequency from VCO. Once locked, changes in the input or feedback frequency must be in small steps.

The tendency to loose lock is a major drawback of the exclusive-OR type phase detector. Within the CD4046A is a second phase detector, which shares the input and feedback signals with the ϕ_1 (exclusive-OR) detector but has its own, ϕ_{11}, output (pin 13). It consists of digital logic with flip-flops for memory and a three-state output driver. When the input frequency is more than feedback frequency (*i.e.*, $f_{in} > f_{out}$), the output is at V^+ most of each cycle. During the remainder of each cycle, the output is open. Outputting V^+ allows the filter capacitor to charge, increasing the voltage to the VCO, increasing the output frequency from VCO (increasing the feedback frequency).

When the ϕ_{11} output is open, the filter capacitance holds the charge. In case $f_{in} < f_{out}$, the ϕ_{11} output is at ground most of each cycle, and open when it is not at ground. Ground to filter allows the capacitor to discharge, reducing the voltage to VCO, decreasing the frequency output from VCO, lowering the feedback frequency.

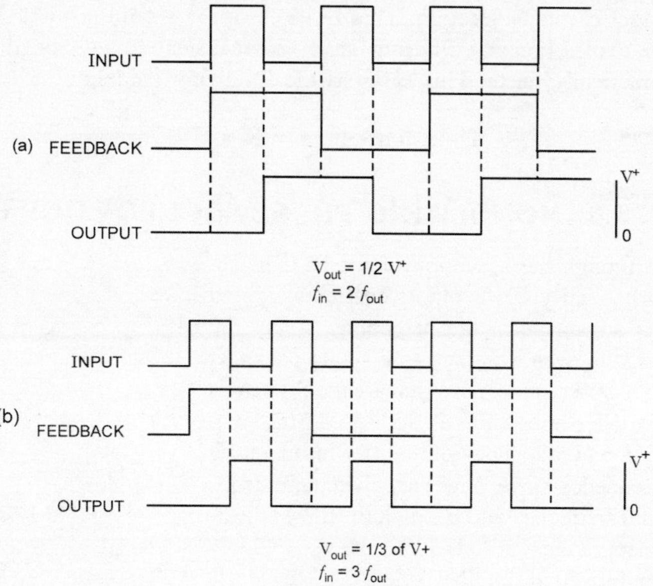

Fig. 13.4

Once the input and feedback frequencies match, the ϕ_{11} output will be high for that part of each cycle that the input signal leads the feedback signal. The remaining of each period, the output is open. Thus the filter capacitor slowly charges a bit more, increasing the phase of the output from VCO. If the phase of feedback signal leads the phase of the input signal, the ϕ_{11} output is ground for that part of cycle when feedback leads input, and open the remaining of the period. This allows the filter capacitor to bleed its voltage down, bringing the VCO's output phase back into the alignment with the input. When both frequency and phase match, the ϕ_{11} output is open.

In brief, it can be stated that, the ϕ_{11} phase detector of the CD4046 A PLL drives the filter capacitor to whatever voltage is necessary to align the feedback frequency and phase with the input frequency and phase.

The second major element in the PLL is the *low-pass filter*. Its function is to extract the average value of the digital output of the signal from the phase detector. It should offer a steady dc level to the input of the VCO which equals the average value of the output of the phase detector.

The requirements of the filter are conflicting. Any variation in the output of the filter produces a change in the frequency of the VCO. So ripple from the filter should be kept the minimum possible. This can be achieved if RC (time constant) is kept larger. Each cycle, the capacitor can only charge or discharge a small amount, giving a low output ripple and a steady VCO output frequency. However, the larger the time constant, the slower the filter is in responding to changes, slowing the overall loops response to a change in input.

The loop filter employed in the PLL may be one of the three types shown in Fig. 13.5. With the passive filters, shown in Fig. 13.5 (*a*) and (*b*), an amplifier is generally used for amplification. On the other hand, the active filter shown in Fig. 13.5 (*c*) includes the amplifier.

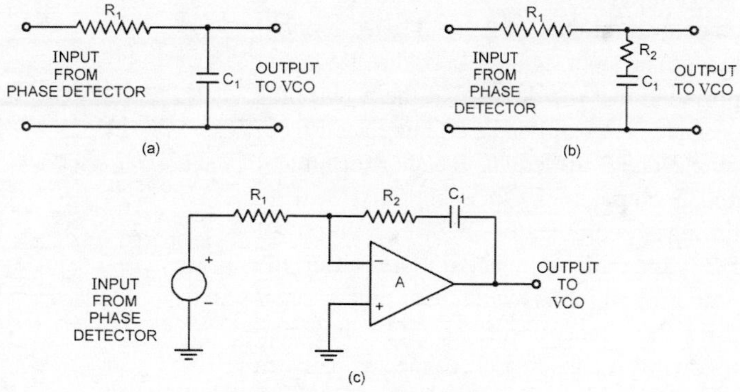

Fig. 13.5. ***Low-pass Filters***

The value of the time constant (R × C) must be selected based on the lowest frequency, the longest period, output from the phase detector. A peak-to-peak ripple out of the filter of 10% of the V_{DC} is certainly the maximum variation of practical consideration. This level of ripple would allow a 10% variation in the frequency, from cycle to cycle. For any filtering at, or more severe than this, the charge and discharge of the capacitor is practically linear. This suggests the following guideline in deciding the time constant of the filter.

$$\text{Time constant, } \tau = R \times C \simeq \frac{50\% \, T_{longest}}{\% \text{ Ripple}} \qquad ...(13.2)$$

13.3. MONOLITHIC PHASE-LOCKED LOOP (PLL) ICs

Although the evolution of the PLL began in the early 1930s but its cost outweighted its advantage in the beginning. Today the PLL is even available as a single package, typical examples of which are the Signetic's SE/NE series such as 560, 561, 562, 564, 565 and 567. They only differ in operating frequency range, power supply requirements, and frequency and bandwidth (BW) adjustment ranges. SE/NE 565 is the most widely employed IC of the series. The device is available as a 14-pin DIP package (Fig. 13.6) and as a 10-pin metal can package. Its important electrical characteristics are given below :

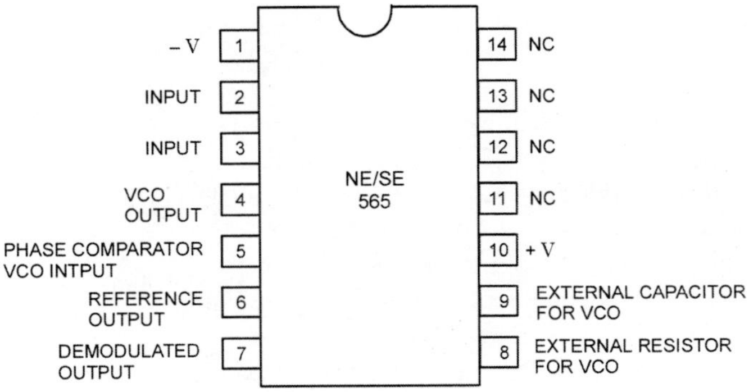

Fig. 13.6. ***14-Pin DIP Package***

1. Operating frequency range : 0.001 Hz to 500 kHz.
2. Operating voltage range : ± 6 to ± 12 V.
3. Input impedance : 10 kΩ typically.
4. Output sink current : 1 mA typically.
5. Output source current : 10 mA typically.
6. Drift in VCO centre frequency with temperature : 300 ppm/ °C typically.
7. Drift in VCO centre frequency with supply voltage : 1.5%/V maximum.
8. Input level required for tracking : 10 mV_{rms} minimum to 3 V peak-to-peak maximum.
9. Bandwidth adjustment range : < ± 1 to > ± 60%.

The block diagram and connection diagram of the SE/NE 565 IC is shown in Fig. 13.7. As shown in the figure, the PLL system consists of a phase detector or comparator (PC), a voltage-controlled oscillator (VCO), an amplifier and R-C combination forming low-pass filter circuit.

The input signals are fed to the phase dctcctor through pins 2 and 3 in differential mode. The input signals can be direct-coupled provided that the dc level at these two pins is made same and dc resistances seen from pins 2 and 3 are equal. By shorting pins 4 and 5 output of VCO is supplied back to the phase comparator (PC). The output of PC is internally connected to amplifier, the output of which is available at pins 6 and 7 through a resistor of 3.6 kΩ connected internally. A capacitor C_2 connected between pins 7 and 10 forms a low-pass filter with 3.6 kΩ resistor. The filter capacitor C_2 should be large enough so as to eliminate the

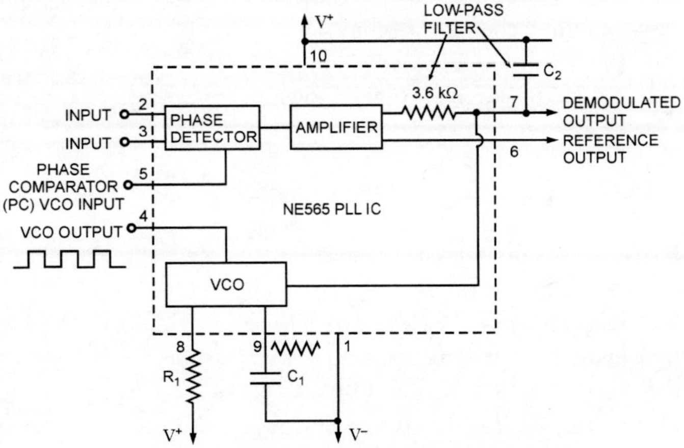

Fig. 13.7. ***NE/SE 565 PLL Block Diagram***

variations in demodulated output and stabilize the VCO frequency. Voltage available at pin 7 is connected internally to VCO as a control signal. At pin 6 a reference voltage nominally equal to voltage at pin 7 is available allowing both the differential stages to be biased. Pins 1 and 10 are supply pins.

The centre frequency of the PLL is determined by the free-running frequency of the VCO which is given as

$$f_{out} \cong \frac{1.2}{4R_1C_1} \text{ Hz} \quad ...(13.3)$$

where R_1 and C_1 are external resistor and capacitor connected to pins 8 and 9 respectively, as illustrated in Fig. 13.7. The free-running frequency f_{out} of the VCO is adjusted, externally with R_1 and C_1, to be at the centre of the input frequency range. Resistor R_1 must have a value between 2 and 20 kΩ. Capacitor C_1 may have any value.

The 565 PLL can lock to and track an input signal typically ± 60% bandwidth with respect to f_{out} as the centre frequency. The lock-range of PLL is given as

$$f_L = \pm \frac{8 f_{out}}{V} \text{ Hz} \quad ...(13.4)$$

where f_{out} is free-running frequency of VCO in Hz and $V = V^+ - V^-$ and capture range is given as

$$f_C = \pm \left[\frac{f_L}{2\pi(3.6)(10^3)C_2}\right]^{1/2} \quad ...(13.5) \text{ where } C_2 \text{ is in farads}$$

The lock range usually increases with an increase in input voltage but falls with an increase in supply voltages.

Example 13.1. The lock range of a certain PLL is specified to be ±15 per cent of the centre frequency. Determine the minimum and maximum frequencies for which the PLL will maintain lock if f_o = 50 kHz.

Solution : Centre frequency, f_o = 50 kHz

Lock range of PLL = ± 15 per cent of f_o

So the minimum frequency for which PLL will maintain lock,

$$f_{min} = f_o - \frac{15}{100} \times f_o = 50 - \frac{15}{100} \times 50 = 42.5 \text{ kHz} \quad \textbf{Ans.}$$

$$f_{max} = f_o + \frac{15}{100} \times f_o = 50 + \frac{15}{100} \times 50 = 57.5 \text{ kHz} \quad \textbf{Ans.}$$

Example 13.2. Determine the free-running frequency, the lock range and capture range for the circuit shown in Fig. 13.7 if V^+ = 12 V; V^- = – 12 V; R_1 = 15 kΩ, C_1 = 0.01 μF and C_2 = 10 μF.

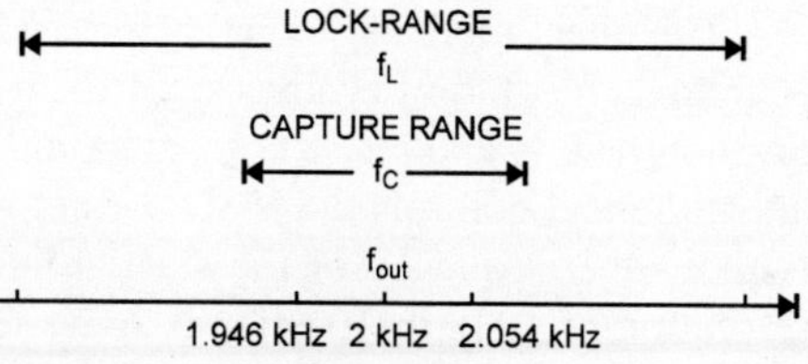

Fig. 13.8

Solution : Free-running frequency,

$$f_{out} = \frac{1.2}{4R_1C_1} = \frac{1.2}{4 \times 15 \times 10^3 \times 0.01 \times 10^{-6}} = 2 \text{ kHz} \quad \textbf{Ans.}$$

$$\text{Lock-range, } f_L = \frac{8 f_{out}}{V} = \frac{8 \times 2 \times 10^3}{+12-(-12)} = \pm\frac{2}{3} \text{ kHz (from 1.33 kHz to 2.67 kHz)} \quad \textbf{Ans.}$$

$$\text{Capture-range, } f_C = \left[\frac{f_L}{2\pi \times 3.6 \times 10^3 \times C_2}\right]^{1/2} = \left[\frac{(2/3) \times 1{,}000}{2\pi \times 3.6 \times 10^3 \times 10 \times 10^{-6}}\right]^{1/2}$$

$$= \pm 54.29 \text{ Hz (from 1.946 kHz to 2.054 kHz)} \quad \textbf{Ans.}$$

The relationship between f_{out}, f_L and f_C is illustrated in Fig. 13.8.

13.4. PLL OPERATION

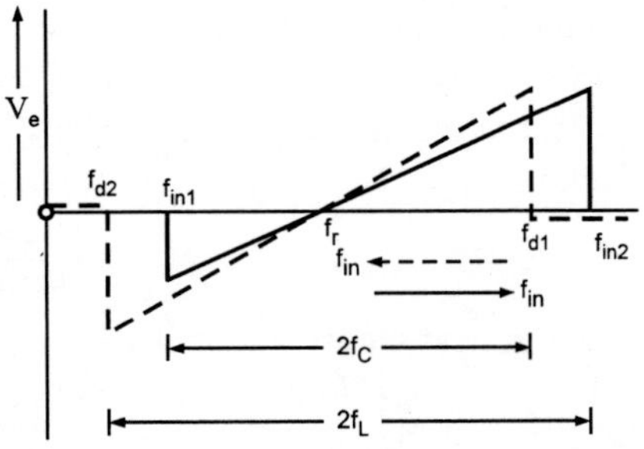

Fig. 13.9

Consider VCO operating without input signal at free-running frequency f_r and input signal of frequency f_{in} increasing from zero is applied to the PC. If the input frequency is less than f_{in1}, then the error voltage V_e is zero as illustrated in Fig. 13.9 and VCO operates at a frequency f_r. When the input signal frequency f_{in} reaches a frequency $f_{in\,1}$ (the lower edge of the capture range), then the output or error voltage V_e jumps from zero to some negative voltage with beat notes of frequency (difference between input signal frequency and actual VCO output signal frequency; $f_{in} - f_{out}$). The error voltage V_e is then filtered, amplified and amplified voltage V_d is applied to the control terminals of the VCO. The instantaneous frequency of VCO decreases because f_{out} falls for negative values of V_d and increases for positive values of V_d. At some instant of time, the decreasing frequency of the VCO equals f_{in1} (lower edge of the capture range), then lock results-in, and the output signal frequency of the VCO may be equal to the input signal frequency (*i.e.*, $f_{out} = f_{in}$). The VCO frequency locks with input signal frequency up to f_{in2} (the upper end of the lock range). If the input signal frequency exceeds f_{in2} then error voltage V_e will fall to zero and the VCO will operate at the free running frequency f_r, as illustrated in Fig. 13.9. If the input signal frequency is now slowly swept back and it attains the value of f_{d1}, then the loop (VCO frequency) locks with the input signal frequency, causing a positive jump of the error voltage V_e. So the VCO output frequency increases from f_r continuously till f_{out} becomes equal to f_{in}. The VCO frequency f_{out} locks with the input signal frequency f_{in} upto f_{d2} (the lower edge of the lock range) as shown in Fig. 13.9 by dotted lines. Now if the frequency of the input signal falls below f_{d2}, then the error voltage V_e will fall to zero and the VCO will operate at the free running frequency.

13.5. PLL APPLICATIONS

13.5.1. Frequency Multiplication or Frequency Synthesis

A *frequency synthesizer* is an electronic device which produces one or more frequencies from a reference source and some control signals. The frequency synthesizer effectively synthesizes either directly or indirectly the desired frequency output from a single frequency obtained from a stable crystal oscillator.

The direct method of synthesis uses harmonic generators and mixers to produce a particular frequency. In indirect method of synthesis, a PLL is normally used, in which the output of VCO is divided by settable division ratio N and phase compared with a standard 10 Hz or 100 Hz signal. The phase comparator output, after filtering, controls the VCO so that when lock has been achieved, the frequency of the VCO is exactly N times the frequency of the standard signal. To enhance the resolution, summing loops are used, and a resolution of 1 Hz in 10 MHz is quite feasible in a synthesizer based on five summing loops.

PLL synthesizer is relatively simple and inexpensive compared to a direct synthesizer. Because of their small size and low power consumption, miniature radio transceivers rely on PLL synthesizers.

The block diagram of a frequency synthesizer is shown in Fig. 13.10 (*a*). In this circuit, a frequency divider is inserted between the output of VCO and the phase comparator. The VCO generates the synthesized frequency according to a dc level at its input. Such an arrangement allows the possibility of slaving a synthesized frequency to a spectrally pure, lower input (or reference) frequency. In the lock condition, the VCO output frequency f_{out} is related to the input frequency f_{in} by $f_{out} = Nf_{in}$ where N is the divider modulus. The desired amount of multiplication can be obtained by selecting a proper divide-by-N network. Fig. 13.10 (*b*) shows this function performed by a 7490 configured as a divide-by-4 circuit. In this figure, transistor Q_1 is used as a driver stage to increase the driving capability of the NE 565.

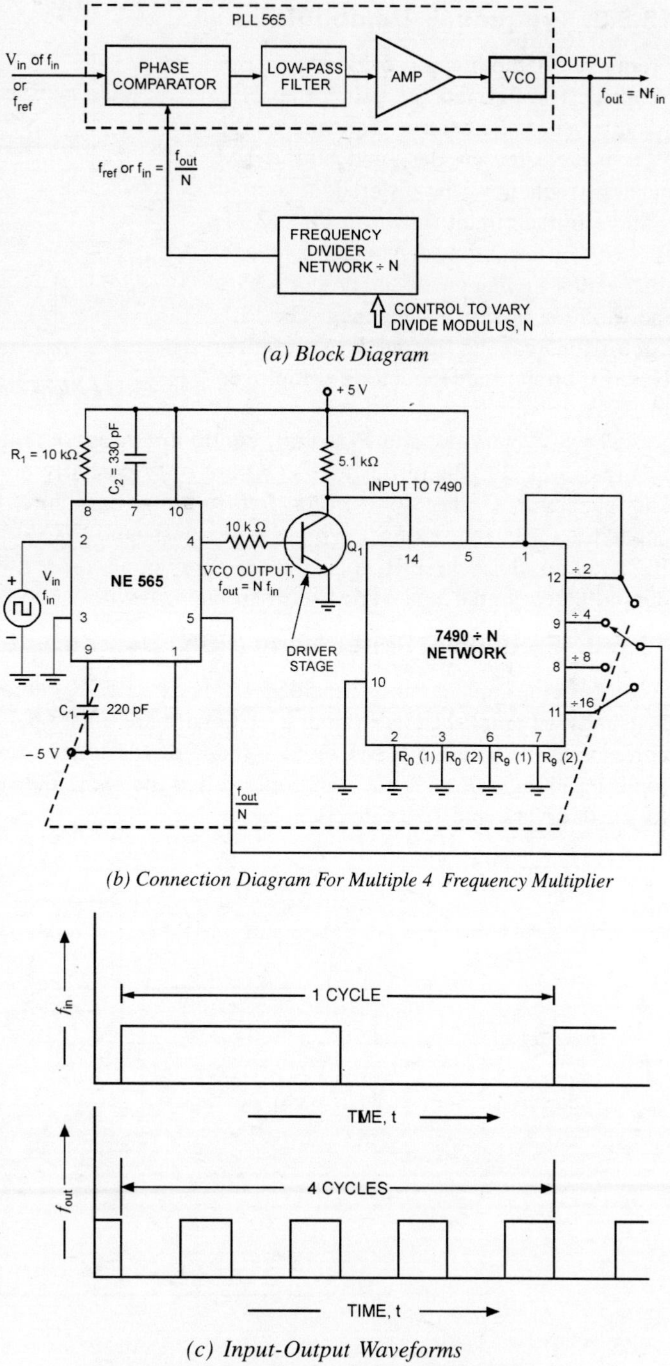

(a) Block Diagram

(b) Connection Diagram For Multiple 4 Frequency Multiplier

(c) Input-Output Waveforms

Fig. 13.10. ***Frequency Multiplier Using The 565***

In this case the input V_{in} at frequency f_{in} is compared with the frequency $\frac{f_{out}}{N}$ at pin 5 of NE565. An output Nf_{in} ($4f_{in}$ in this case) at pin 4 is connected through an inverter circuit to provide an input at the pin 14 of the 7490. The 7490 provides output (one-fourth of its input) at pin 9.

Since the VCO can be adjusted over a limited range from its centre frequency, it may become necessary to change the VCO frequency whenever the divider value is changed.

For verification of the circuit operation, one must determine the input frequency range and then adjust the free-running f_{out} of the VCO by means of R_1 and C_1 so that the output frequency of the 7490 divider is midway within the predetermined input frequency range. The output of VCO should now be equal to 4 f_{in}. In Fig. 13.10 (*b*), the output frequency f_{out} can be adjusted by varying the potentiometer R_1 [see Eq. (13.3)]. In addition, the input waveform can either be sinusoidal or square and may be applied to pin 2 or 3.

Fig. 13.10 (*c*) shows the input-output waveforms for $f_{out} = 4f_{in}$.

The smallest frequency increment generated by the loop is equal to the input or reference frequency (f_{in} or f_{ref}). For instance, if the reference frequency is 1 MHz and N is 10, then the output frequency resolution is 1 MHz and the output frequency is 10 MHz. If the required resolution is, say, 1 Hz, a reference frequency of 1 Hz will be required. For the same frequency output (10 MHz), the value of N should be 10^7.

The switching speed of the PLL is a function of the loop bandwidth and closed-loop frequency response. Since the bandwidth (BW) is always smaller (by about 10 times) than f_{in} or f_{ref}, in order to give a stable correction signal to the VCO, faster switching speeds need higher input frequencies, and therefore, higher loop bandwidths. However, the loop design parameters cannot be chosen to optimize all aspects of the circuit performance. The result is a trade-off between narrow BWs for maximum filtering and wide BWs to provide short rise-times for fast tracking of the input signal. Further, the frequency multiplier loops also multiply noise at the phase comparator, which appears as noise sidebands on the oscillator output.

13.5.2. Frequency Demodulation

Frequency demodulation or detection can be directly obtained by using the PLL circuit. When the centre frequency of the PLL is selected or designed at the FM carrier frequency, the filtered or output voltage in the circuit shown in Fig. 13.11, is obviously the desired demodulated voltage, that varies in magnitude in proportion to the variation of signal frequency. The PLL circuit thus can be operated as complete IF strip, limiter, and detector as employed in FM receivers.

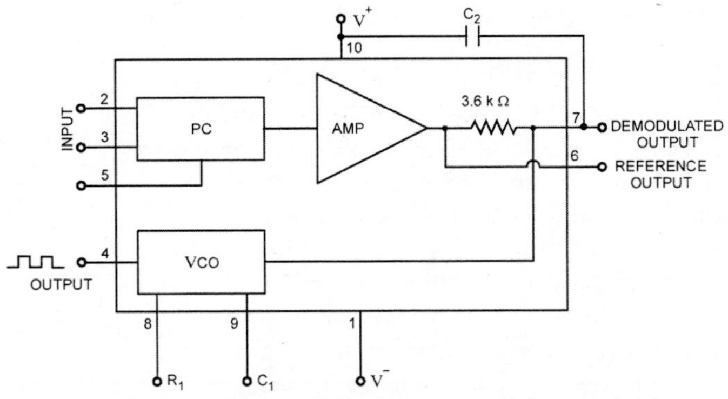

Fig. 13.11. ***Block Diagram of PLL Unit 565 For Use As FM Detector***

The 565, very popular PLL unit, containing a phase comparator PC, amplifier, and VCO, is shown in Fig. 13.11. PC, amplifier and VCO are only partially connected internally. An external resistor R_1 and capacitor C_1 are for setting free-running or centre frequency of VCO and another external capacitor C_2 is for setting low-pass filter passband. VCO output is connected back as the input to the PC so as to close the PLL loop. Power supplies used are two (V^+ and V^-). The output voltage is the demodulated signal that varies with frequency within the operating range specified.

13.5.3. Frequency Shift Keying (FSK) Demodulator

A very useful application of the 565 PLL is as an FSK demodulator. In the 565 PLL the frequency shift is usually accomplished by driving a VCO with the binary data signal so that the two resulting frequencies correspond to the logic 0 and logic 1 states of the binary data signal. The frequencies corresponding to logic 1 and logic 0 states are commonly called the *mark* and *space* frequencies. Several standards are used to set the mark and space frequencies.

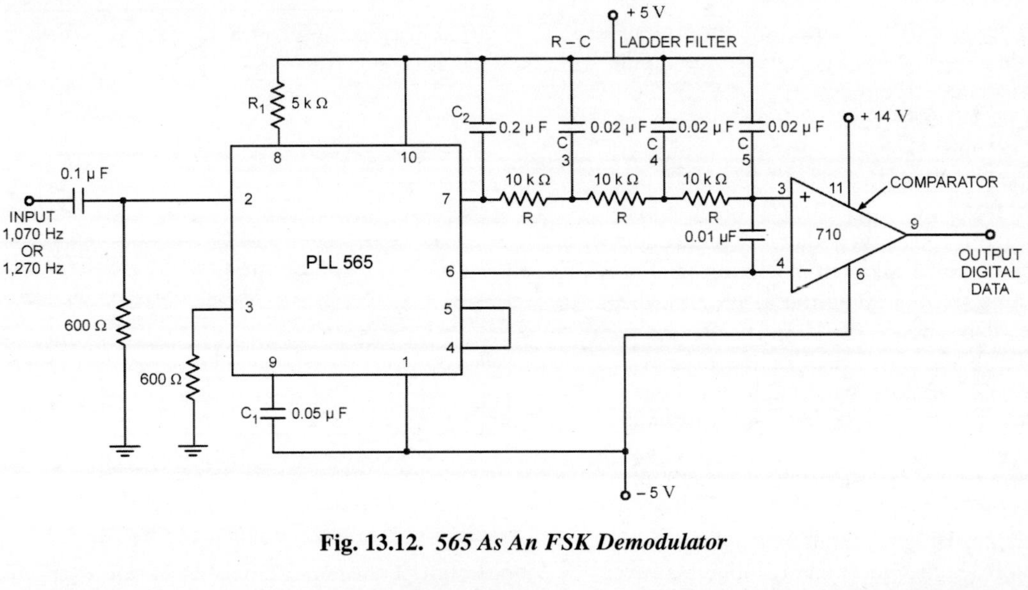

Fig. 13.12. ***565 As An FSK Demodulator***

An FSK signal demodulator can be built as illustrated in Fig. 13.12. The demodulator receives a signal at one of the two distinct carrier frequencies, 1,270 Hz or 1,070 Hz representing the RS-232 C logic levels of mark (– 5 V) or space (+14 V), respectively. Capacitance coupling is used at the input to remove a dc level. As the signal appears at the input of 565 PLL, the PLL locks to the input frequency and tracks it between the two possible frequencies with a corresponding dc shift at the output. Resistor R_1 and capacitor C_1 determine the free-running frequency of the VCO. Capacitor C_2 is a loop filter capacitor that establishes the dynamic characteristics of the demodulator. Capacitor C_2 is chosen smaller than usual one to eliminate overshoot on the output pulse. A three-stage RC ladder filter is employed for removing the sum frequency component from the output. The VCO free-running frequency is adjusted with R_1 so that the dc voltage level at the output (pin 7) is the same as that at pin 6. An input at frequency 1,070 Hz drives the demodulator output voltage to a more positive voltage level, driving the digital output to the high level (space or + 14 V). An input at 1,270 Hz correspondingly drives the 565 dc output less positive with the digital output, which then drops to the low level (mark or –5 V).

HIGHLIGHTS

1. The phase-locked loop (PLL) consists of three elements : a phase detector, an amplifier/low-pass filter and a voltage-controlled oscillator (VCO). The phase detector outputs a signal whose average value is equal to the difference in phase between the two input signals. The low-pass filter accepts the output from the phase detector, removes the high frequency noise and produces a dc level. The output f_{out} of the VCO is fed back to the phase detector, where it is compared with the input frequency f_{in}. When $f_{in} = f_{out}$, the loop is said to be locked. When locked, the PLL will output whatever frequency is necessary to keep the input frequencies the same, and to minimise their difference in phase.
2. The important characteristics of a PLL are : Lock-in range, the capture range and pull-in time. The lock-in range is usually greater than capture range. The capture range depends upon the characteristics of the low-pass filter.
3. The phase detector is basically a multiplier. It may be analog or digital type.
4. The low-pass filter (LPF) may be passive or active. It controls the capture range and lock-in range.
5. Adding a digital divide-by-N counter in the feedback loop means that the output frequency generated by the VCO must be N times above the input frequency. Using the divider allows digital control of the output frequency, with the input signal as the minimum frequency and the resolution.
6. The PLLs are used in applications like FM stereo decoders, motor speed controls, FSK decoders, and FM demodulators.

EXERCISES

1. (*i*) Explain lock-in-range, capture range and pull-in-time of a PLL.
   (*ii*) Draw the block diagram of a PLL and explain its working.
   (*iii*) List the applications of a PLL.
2. Define the terms free-running frequency, capture range and pull in time.
   [G.G.S.I.P. Univ. Analog Electronics-II May 2011]
3. With the help of block diagram, define working of a PLL.
4. Describe the operating principle of PLL. Define capture range and locking range.
5. What is PLL ? Explain its operating principle with the help of block diagram. Describe the functions of each block in detail in block diagram. Discuss at least two applications of PLL.
6. Draw the block schematic of a PLL and explain the function of each block.
   [G.G.S.I.P. Univ. Analog Electronics-II May 2010]
7. Explain operating principle of PL loop in its basic form with block diagram and state function of each block. **[G.G.S.I.P. Univ. Analog Electronics-II May-June 2009]**
8. Draw the block schematic of an IC phase-locked loop and explain its working.
   [G.G.S.I.P. Univ. Analog Electronics-II May 2011]
9. List one application of the PLL and then briefly describe the role of the PLL in that application.
10. Explain how 565 PLL works as (*i*) Frequency multiplier, (*ii*) Frequency shift keying demodulator.

11. Describe PLL with block diagram. Also discuss applications of PLL.
12. Write short note on phase locked loops.
13. Explain how a PLL IC can be employed for FSK demodulation.
14. Describe PLL application as
    (*i*) Frequency translation. (*ii*) FM demodulation.

SHORT ANSWER TYPE QUESTIONS WITH ANSWERS

Q. 1. What is PLL ?

Ans. PLL is basically an electronic circuit that consists of a phase detector, a low-pass filter, a dc amplifier and a voltage controlled oscillator connected, as shown in Fig. 13.1.

Q. 2. What are the basic building blocks of a PLL ?

Ans. The basic building blocks of a PLL are phase detector (or comparator), low-pass filter, dc amplifier and a VCO.

Q. 3. What is the need for a low-pass filter in a PLL ?

Ans. While the loop is trying to achieve lock, the output of the phase detector contains frequency components at the sum and difference of the signals compared. A low-pass filter is provided in a PLL to pass only the low-frequency component of the signal so that the loop can obtain lock between input and VCO signals. Low-pass filter also removes high frequency noise.

Q. 4. What is the role served by the VCO in a PLL chip ?

Ans. The role of the VCO in a PLL is to provide an oscillating output signal (typically of square or triangular waveform), whose frequency can be adjusted over a range controlled by a dc voltage.

Q. 5. In a PLL system define the locked state.

Ans. Before the application of the input, the PLL circuit is in the free-running state. When the input is applied, the VCO output frequency tries to capture the input frequency, so, it is in the capture mode. Once the two frequencies (input signal frequency and VCO output frequency) have matched, the PLL is in the phase-locked state.

Q. 6. Define the lock range of a PLL system.

Ans. The lock range of a PLL system is the frequency range over which it can maintain lock with the input signal frequency after lock has occurred.

Q. 7. Define 'capture range' of a PLL.

Ans. The range of frequencies over which the output signal frequency of the VCO will be able to lock with the input signal frequency is called the capture range. The capture range is always smaller than lock range.

Q. 8. A PLL frequency translator has a centre frequency f_o and the input frequency is f_i. What will be the output frequency ?

Ans. Output frequency, $f_{\text{out}} = f_i \simeq f_o$

Q. 9. What is a low-pass filter ?

Ans. A low-pass filter is a circuit that passes all frequencies from zero to cutoff frequency.

Q. 10. Give the advantages and limitations of PLL synthesizer.

Ans. The PLL synthesizer solves several of the problems associated with purely analog techniques. It is accurate, easily controlled digitally and exhibits good long-term stability.

However, it suffers from limitations of its analog components. The loop must remain locked for reliable performance. Short-term drift is caused by variation in filter and VCO component values. We are to wait for the loop to stabilize when changing frequencies. Frequency is determined by an analog voltage or current. Any noise injected into this point in the circuit will cause jitter.

PROBLEMS

1. The lock range of a certain PLL is specified to be ±12.5% of the centre frequency. Determine the frequency range for which the PLL will maintain lock if f_o = 56 kHz. [**Ans.** 49 kHz to 63 Hz]
2. Determine the free-running frequency f_{out}, the lock range f_L and capture range f_c for the following values : $C_1 = 0.01\ \mu F$, $C_2 = 10\ \mu F$, $R_1 = 12\ k\Omega$, $+V = +10$ V, $-V = -10$ V.
   [**Ans.** f_{out} = 2.5 kHz; f_L = ±1 kHz; f_c = ±66.49 Hz]

❑❑❑

CHAPTER 14

Voltage Regulators

14.1. INTRODUCTION

INSIDE THIS CHAPTER

Almost all electronic devices used in electronic circuits need a dc source of power to operate. The source of dc power is used to establish the dc operating points (Q-points) for the passive and active electronic devices incorporated in the system. The dc power supply is typically connected to each and every stage in an electronic system. It means that the single requirement common to all phases of electronics is the need for a supply of dc power. For portable low-power systems batteries may be used, but their operating period is limited. Thus, for long time operation frequent recharging or replacement of batteries become much costlier and complicated. More frequently, however, electronic equipment is energized by a *power supply*, derived from the standard industrial or domestic ac supply by transformation, rectification, and filtering. The combination of a transformer, a rectifier and a filter constitutes an *ordinary dc power supply*, also called an *unregulated power supply*.

The block diagram of an ordinary power supply is depicted in Fig. 14.1. Usually, a small dc voltage, in the range of 2–24 volts is required for the operation of different electronic circuits, while in India, single-phase ac supply is available at 230 V. So a small step-down transformer is used at the beginning which reduces the voltage level according to the needs. Next block is a rectifier which converts the sinusoidal ac voltage into pulsating dc. In the last there is a filter block which reduces the ripples (ac components) from the rectifier output voltage. The filter is a device which passes dc component to the load and blocks ac components of the rectifier output.

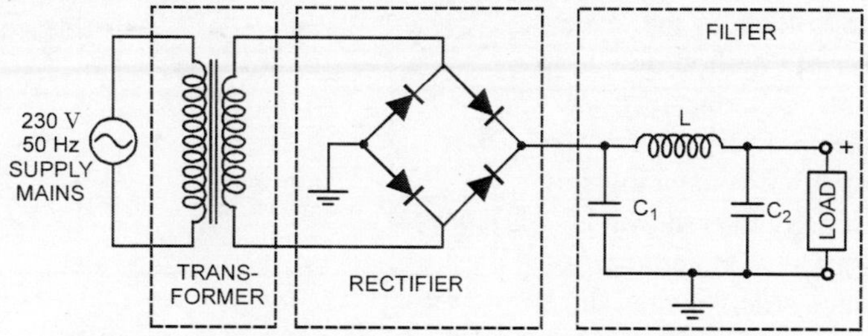

Fig. 14.1. *Block Diagram of an Unregulated Power Supply*

For many applications in electronics, unregulated power supply is not good enough because of the following reasons.

1. *Poor regulation.* The output voltage does not remain constant as the load varies. The internal resistance of an ordinary power supply is relatively large (more than 30 Ω). So output voltage is

significantly affected by the magnitude of current drawn from the supply. The voltage drop in the internal resistance of the supply increases directly with an increase in load current.

2. *Variations in the ac supply mains.* The permissible variation in the ac supply mains voltage as per Indian Electricity Rules is 6% of its rated value. But in India, the variations in ac mains voltage is much more than this (sometimes it may vary from 180 V to 260 V). The dc output voltage being proportional to the input ac voltage, therefore, varies largely.

3. *Variations in temperature.* The dc output voltage varies with temperature, particularly if semi-conductor devices are employed.

These variations in dc output voltage may cause inaccurate or erratic operation or even malfunctioning of many electronic circuits. For instance, in oscillators the frequency will shift, in transmitters output will get distorted, and in amplifiers the operating point will shift causing bias instability.

Some feedback arrangement (acting as a voltage regulator) is employed in conjunction with an unregulated power supply to overcome the above mentioned three shortcomings and also to reduce the ripple voltage. Such a system is called a *regulated power supply.*

In many applications, it is important to protect the power supply output against inadvertant short circuits that might destroy either the circuit under test or operation of the supply itself. Thus, current limiting circuits are often incorporated into the regulator design.

Power supplies are becoming steadily more sophisticated in terms of performance objectives and application strategies. A commercial power supply is typically a complex system that makes use of ICs to reduce ripple, improve regulation, and widen control options. Programmable power supplies are also available to allow remote operation that is useful in many settings.

A *regulated power supply* is an electronic circuit that is designed to provide a constant dc voltage of predetermined value across load terminals irrespective of ac mains fluctuations or load variations.

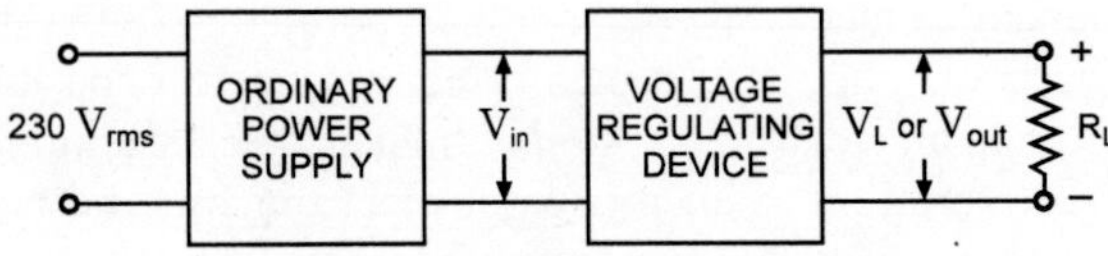

Fig. 14.2. ***Regulated Power Supply***

A regulated power supply essentially consists of an ordinary power supply and a voltage regulating device, as illustrated in Fig. 14.2. The output from an ordinary power supply is fed to the voltage regulating device that provides the final output. The output voltage remains constant irrespective of variations in the ac input voltage or variations in output (or load) current.

Figure 14.3 shows the complete circuit of a regulated power supply with a transistor series regulator as a regulating device. The ac voltage, typically 230 V_{rms} is connected to a transformer which transforms that ac voltage to the level for the desired dc output. A bridge rectifier then provides a fullwave rectified voltage that is initially filtered by a π- (or C-L-C) filter to produce a dc voltage. The resulting dc voltage usually has some ripple or ac voltage variation. A regulating circuit uses this dc input to provide

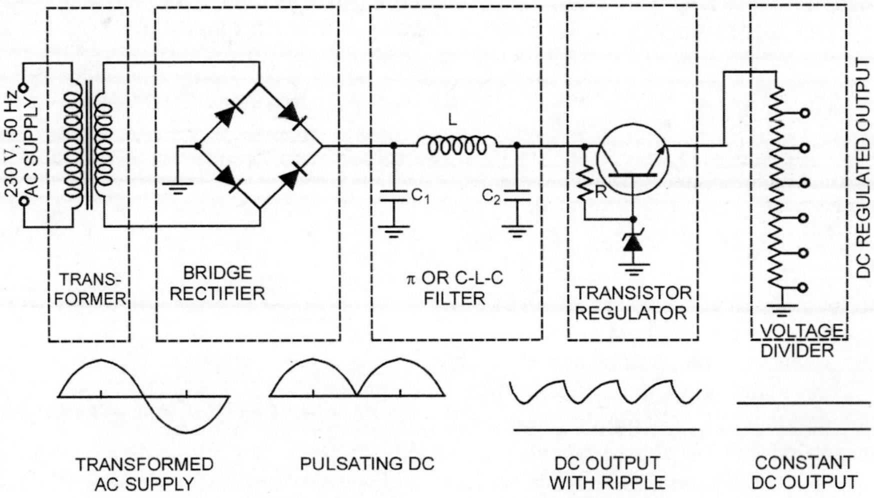

Fig. 14.3. ***Complete Regulated Power Supply Circuit***

a dc voltage that not only has much less ripple voltage but also remains constant even if the input dc voltage varies somewhat or the load connected to the output dc voltage changes. The regulated dc supply is available across a voltage divider.

Often more than one dc voltage is required for the operation of electronic circuits. A single power supply can provide as many voltages as are required by using a voltage (or potential) divider, as illustrated in the figure. As illustrated in Fig. 14.3, a potential divider is a single tapped resistor connected across the output terminals of the supply. The tapped resistor may consist of two or three resistors connected in series across the supply. In fact, bleeder resistor may also be employed as a potential divider.

14.2. POWER SUPPLY CHARACTERISTICS

The quality of power supply depends on different factors such as its load voltage, load current, voltage regulation, source regulation, output impedance, ripple rejection etc. Some of the characteristics of regulated power supplies are discussed below.

1. Load Regulation. The *load regulation*, abbreviated LR (also called the *load effect*), is the change in regulated output voltage when the load current changes from minimum to maximum value *i.e.,*

$$LR = V_{NL} - V_{FL} \qquad ...(14.1)$$

where V_{NL} is load voltage at no load and V_{FL} is the load voltage at full load. In this equation V_{NL} occurs when the load resistance is infinite (*i.e.,* out terminals are open circuited) and V_{FL} occurs when the load resistance is of the minimum value where voltage regulation is lost.

Load regulation is usually expressed in percentage and is given as

$$\%LR = \frac{V_{NL} - V_{FL}}{V_{FL}} \times 100 \qquad ...(14.2)$$

For an ideal power supply output voltage should be independent of dc load current and the percentage load regulation should be equal to zero.

2. Minimum Load Resistance. The load resistance at which a power supply delivers its full-load rated current at rated voltage is referred to as a *minimum load resistance*, $R_{L\,(min)}$.

$$i.e., \quad R_{L\,(min)} = \frac{V_{FL}}{I_{FL}} \qquad ...(14.3)$$

The value of full-load current I_{FL} should never increase beyond that mentioned in the data sheet of the power supply. For instance, if a data sheet specifies that the power supply will provide an output voltage of 24 V at a maximum rated current of 0.2 A, then minimum load resistance that can be connected across the supply is $R_{min} = \frac{24}{0.2} = 120\ \Omega$. If any attempt is made to decrease the value of load resistance R_L below this value, the rated dc output voltage will not be available.

3. Source or Line Regulation. The input line voltage has a nominal value of 230 V but in practice, there are considerable variations in ac supply mains voltage. Since this ac supply mains voltage is the input to the ordinary power supply, the filtered output of the bridge rectifier is almost directly proportional to the ac mains voltage. Filtered output of the bridge rectifier is the input to the voltage regulating device.

Another way of specifying the quality of a regulated power supply is by its *source regulation* (also called the *source effect* or *line regulation*) abbreviated as SR. The source regulation, SR is defined as the change in regulated output voltage for a specified range of line voltage, typically 230 V ± 10 per cent. The defining equation is

$$SR = V_{HL} - V_{LL} \qquad ...(14.4)$$

where V_{HL} is output voltage with high input ac line voltage and V_{LL} is output voltage with low input ac line voltage.

The percentage of source regulation is given as

$$\%SR = \frac{V_{HL} - V_{LL}}{\text{Nominal load voltage}} \times 100 = \frac{SR}{V_{nominal}} \times 100 \quad ...(14.5)$$

The output voltage under specific operating conditions is known as the *nominal load (or output) voltage.*

4. Output Impedance. A regulated power supply is a very stiff dc voltage source. This means that the output resistance is very small (in milliohms). Even though the external load resistance is varied, almost no change is seen in the load voltage. An ideal voltage source has an output impedance of zero. Modern regulated power supplies approach ideal voltage sources.

5. Ripple Rejection. Voltage regulators stabilize the output voltage against variations in input voltage. Ripple is equivalent to a periodic variation in the input voltage. Thus, a voltage regulator attenuates the ripple that comes in with the unregulated input voltage. Since a voltage regulator uses negative feedback, the distortion is reduced by the same factor as the gain. Ripple rejection is a measure of a power supply's ability to reject ripple voltages and is usually expressed in decibels.

14.3. VOLTAGE REGULATORS

The *regulator* may be constructed from a zener diode, and/or discrete transistors, and/or integrated circuits. All voltage regulators must have a stable voltage reference source which is provided by a special type of diode operated in reverse breakdown called a *breakdown diode*, or *zener* diode.

The primary function of a voltage regulator is to maintain a constant dc output voltage. However, it also rejects ac ripple voltage that is not removed by the filter. The regulator may also include protective functions such as short-circuit protection, current limiting, thermal shutdown, or overvoltage protection.

14.3.1. Series Voltage Regulators

The basic connection of a series voltage regulator circuit is shown in the block diagram given in Fig. 14.4.

The series element controls the magnitude of the input voltage that gets to the output. The output voltage is sampled by a circuit that provides a feedback voltage to be compared to a reference voltage.

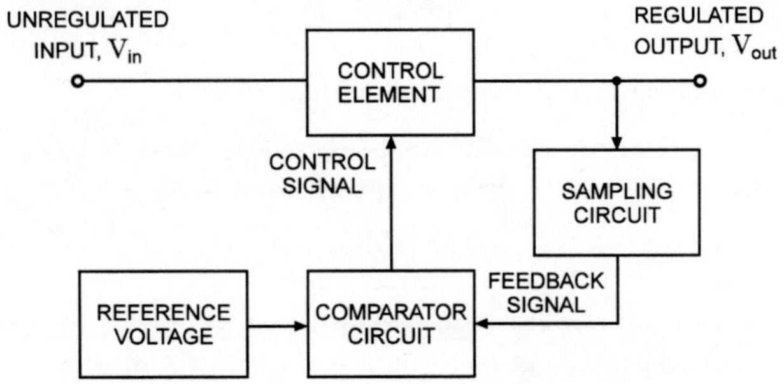

Fig. 14.4. ***Block Diagram of a Series Voltage Regulator***

If the output voltage increases, the comparator circuit provides a control signal to cause the series control element to reduce the magnitude of the output voltage—thereby maintaining the output voltage. On the other hand, if the output voltage falls, the comparator circuit provides a control signal to cause the series control element to increase the magnitude of output voltage. Thus, the output voltage is maintained constant.

14.3.2. Shunt Voltage Regulator

A shunt voltage regulator provides regulation by shunting current away from the load. The block diagram of such a voltage regulator is depicted in Fig. 14.5. The input unregulated voltage provides

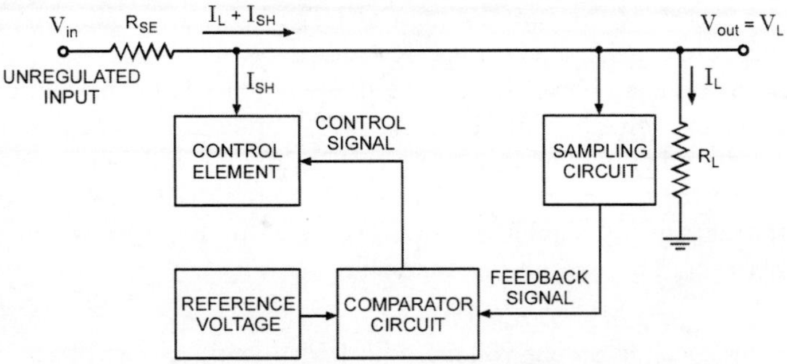

Fig. 14.5. ***Block Diagram of a Shunt Voltage Regulator***

current to the load. Some of this current is shunted away by the control element to maintain the regulated output voltage across the load. If the output voltage tends to change due to change in load, the sampling circuit provides a feedback signal to a comparator circuit which then provides a control signal to vary the magnitude of current shunted away from the load. For example, when the output voltage tends to fall, the sampling circuit provides a feedback signal to the comparator circuit which then provides a control signal to draw lesser shunt current, providing more load current, thereby keeping the regulated voltage constant.

14.4. OP-AMP REGULATORS

14.4.1. Op-Amp Series Voltage Regulator Circuit

Op-amp series voltage regulator circuit is shown in Fig. 14.6. The op-amp compares the zener diode reference voltage with the feedback voltage from sensing resistors R_f and R_1. If the output voltage varies, the conduction of transistor Q_1 is controlled to maintain the output voltage constant. The op-amp places a negligible load on the zener diode, allowing it to operate at a *single fixed* point. This results in a very stable output from the zener diode. Also, the zener diode no longer must absorb large swings in the current. Hence it can be replaced by a low-power, highly precise reference diode.

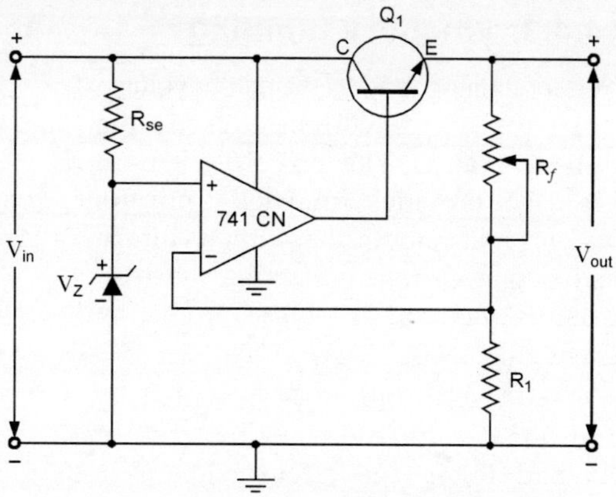

Fig. 14.6. ***Op-Amp Series Voltage Regulator Circuit***

The output voltage will be maintained at a value given by the following equation :

$$V_{out} = V_z\left(1+\frac{R_f}{R_1}\right) \quad ...(14.6)$$

The output voltage can be varied simply by varying R_f in the circuit.

To bias the transistor, the output voltage of the op-amp must exceed the circuit's output voltage V_{out} by 0.7 V.

i.e., $V_{op\text{-}amp} = V_{out} + 0.7\ V$ ...(14.7)

Cautions: 1. The peak value of input voltage (input voltage from the filter) must be kept below the maximum supply voltage for the op-amp (op-amps with rather high supply voltage ratings are available).

2. The voltage from the filter must be at least 2 V above the output voltage V_{out}. This is essential in order to keep the op-amp out of saturation and to keep the zener diode in its breakdown region

i.e., $V_{peak} \leq \pm V_{supply\ rating\ of\ op\text{-}amp}$ ...(14.8)

$V_{min} \geq V_{out} + 2\ V$ ...(14.9)

The noteworthy point is that the minimum output voltage is V_z, when R_f is shorted. Hence it is necessary that the selected zener diode have a voltage smaller than the lowest desired output voltage.

Finally, making of R_f the potentiometer needs care. Making R_f a potentiometer and running the wiper all the way down to ground will remove all negative feedback. The output voltage will be driven to $+V_{sat}$, as the op-amp would be operating as a comparator.

The op-amp only has to provide the base current. This is then multiplied by the transistor β to give load current. The transistor β is determined by equation

$$\beta \geq \frac{I_{output}}{I_{op\text{-}amp\ max}} \quad ...(14.10)$$

Also, the transistor must be able to dissipate the power generated:

$$P_{c\ max} = (V_{dc} - V_{out})I_{output} \quad ...(14.11)$$

where V_{dc} is the dc output voltage from the filter, V_{out} is the output (or load) dc voltage and I_{output} is the output (or load) current.

14.4.2. Current Limiting Circuit

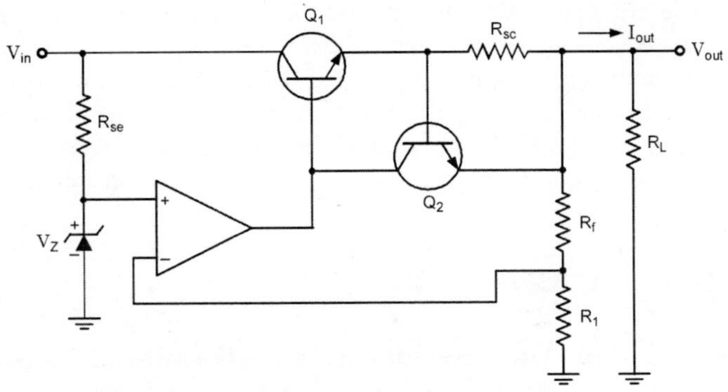

Fig. 14.7. *Current Limiting Voltage Regulator*

One form of short-circuit or overload protection is current limiting, as illustrated in Fig. 14.7. With the increase in output or load current, the voltage drop across the short-circuit sensing resistor R_{sc} increases. When the voltage drop across R_{sc} becomes large enough, it will drive transistor Q_2 *on*, diverting current from the base of transistor Q_1, thereby reducing the load current through transistor Q_1, preventing any additional current to load resistor R_L. The action of components R_{sc} and Q_2 thus provides limiting of the maximum load current.

14.4.3. Foldback Limiting

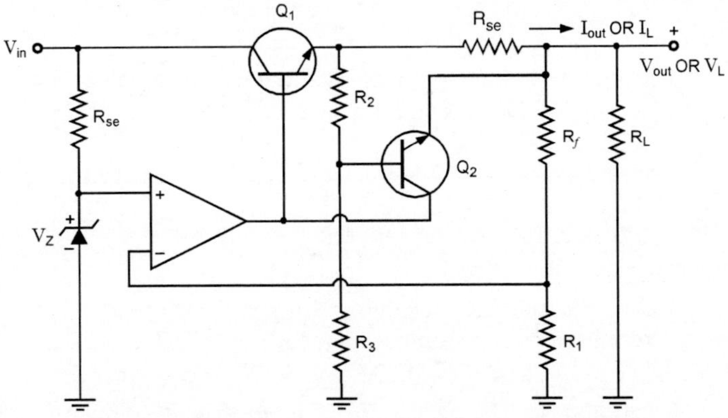

Fig. 14.8. *Foldback-Limiting Series Regulator Circuit*

Current limiting reduces the output voltage when the output current exceeds the limiting value. The circuit given in Fig. 14.8 provides foldback limiting, which reduces both the output voltage and output current protecting the load from overcurrent, as well as protecting the regulator.

Foldback limiting is provided by additional voltage-divider network of R_2 and R_3 in the circuit shown in Fig. 14.8 (over that of Fig. 14.7). The divider circuit senses the voltage at the output (emitter) of Q_1. With the increase in output current to its maximum value, the voltage across R_{se} becomes large enough to drive Q_2 *on*, thereby providing current limiting. If the load resistance is made smaller, the voltage driving Q_2 *on* becomes less, so that output current falls when V_{out} also falls in value—this action being foldback limiting. When the load resistance regains its rated value, the circuit resumes its voltage regulation action.

14.4.4. Op-Amp Shunt Voltage Regulator Circuit

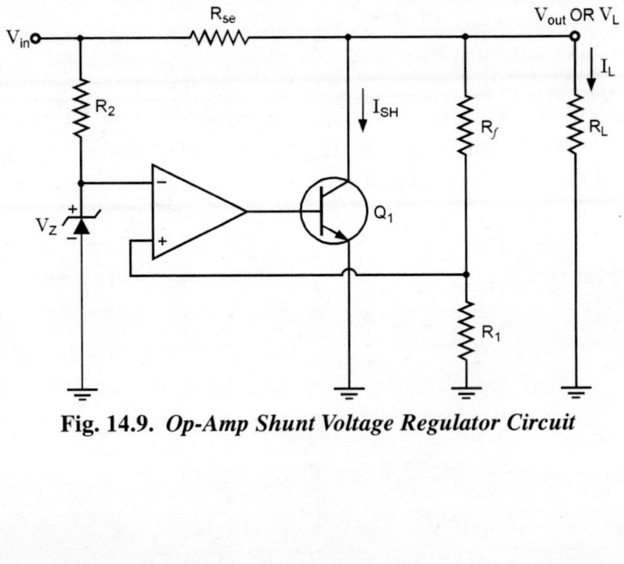

Fig. 14.9. *Op-Amp Shunt Voltage Regulator Circuit*

Op-amp shunt voltage regulator circuit is shown in Fig. 14.9. The zener voltage is compared to the feedback voltage obtained from voltage divider network $R_f - R_1$ to provide the control drive current to shunt element Q_1. The current through series resistor R_{se} is thus controlled to drop a voltage across R_{se} in order to maintain output voltage constant.

TABLE 14.1. *IC Voltage Regulators*

IC Number	V_{out} *in volts*	I_{max} *in amperes*	*LR in mV or %*	*SR in mV or %*	*RR dB*	*Dropout in volts*	*Remarks*
LM309	+ 5	1	15	4	75	2	Fixed positive
LM317	—	1.5	0.1%	0.2%	65	2.5	Adjustable : 1.2 to 32 V
LM320-5	– 5	1.5	50	10	65	2	Fixed negative
LM320-15	– 15	1.5	30	5	80	2	Fixed negative
LM338	—	5	0.1%	0.1%	60	2.7	Adjustable : 1.2 to 32 V
LM340-5	+ 5	1.5	10	3	80	2.3	Fixed positive
LM340-15	+ 15	1.5	12	4	80	2.5	Fixed positive
LM350	—	3	0.1%	0.1%	65	2.5	Adjustable : 1.2 to 33 V
RC4194	—	0.15	0.2%	0.2%	75	3	Dual tracking : 0 to 32 V
RC4195	± 15	0.15	3	2	70	3	Dual tracking

14.5. IC VOLTAGE REGULATORS

IC voltage regulators are versatile and relatively inexpensive and are available with features such as a programmable output, current-voltage boosting, internal short-circuit current limiting, thermal shutdown, and floating operation for high voltage applications.

Voltage regulators comprise a class of widely employed ICs. Regulator IC units contain the circuitry for reference source, comparator amplifier, control device, and overload protection all in a single IC. Although the internal construction of IC is somewhat different from that explained in case of discrete voltage regulator circuits, the external operation is almost the same.

A power supply can be built using a transformer connected to the ac supply line to transform the ac voltage to a desired level, then rectifying the ac voltage, filtering with a capacitor and RC filter, if desired, and finally regulating the dc voltage employing an IC regulator. The regulators can be selected for operation with load currents ranging from hundreds of milliamperes to tens of amperes, corresponding to power ratings from milliwatts to tens of watts.

IC regulators are of the following types :

1. Fixed output voltage regulators : positive and/or negative output voltage.
2. Adjustable output voltage regulators : positive or negative output voltage.
3. Switching regulators.
4. Special regulators.

Except for the switching regulators, all other types of regulators are *linear regulators*. The impedance of the active element of the linear regulator may be continuously varied to provide a desired current to the load. On the other hand, in a *switching regulator* a switch is turned *on* and *off* at a rate such that the regulator provides the desired average current in periodic pulses to the load. The switching regulators are more efficient than the linear regulators. This is because there is negligible power dissipation in switching elements in either the *on* or *off* state. Nevertheless, in switching regulators the power dissipation is substantial during the switching intervals (*on* to *off* or *off* to *on*). Also, most of the loads (devices) cannot accept the average current in periodic pulses. Therefore, most practical voltage regulators are of the linear type.

Voltage regulators, especially the switching type, are employed as control circuits in pulse width modulation, push-pull bridges, and series type switch mode supplies. Almost all power supplies make use of some type of voltage regulator IC because they are simple to use, reliable, cheaper in cost, and, above all, available in a variety of voltage and current ratings.

Table 14.1 lists data for some popular IC regulators. For instance, the LM 309 is a fixed positive regulator with an output of + 5 V, a maximum load current of 1 A, a load regulation of 15 mV, a source regulation of 4 mV, and a ripple rejection of 75 dB. For the adjustable regulators, LR and SR are given in percentage rather than millivolts. The table also includes the dropout voltage, or the minimum permissible difference between the input and output voltages. For example, an LM 309 has a dropout voltage of 2 V. It implies that the input voltage must be at least 2 V greater than output voltage *i.e.,* input voltage must be at least 7 V, because its output voltage is 5 V.

14.6. THREE-TERMINAL IC VOLTAGE REGULATORS

The latest generation of IC voltage regulators has devices with only three pins: one for the unregulated input voltage, one for the regulated output voltage, and one for ground. The new devices can supply load current from 100 mA to more than 5 A. Available in plastic or metal packages, these three-terminal voltage regulators have become extremely popular because they are inexpensive and easy to use. Aside from a couple of bypass capacitors, the new three-terminal IC voltage regulators do not need any external component.

The integrated three-terminal voltage regulators typically incorporate many of the functions discussed so far in a single package, as illustrated in Fig. 14.10.

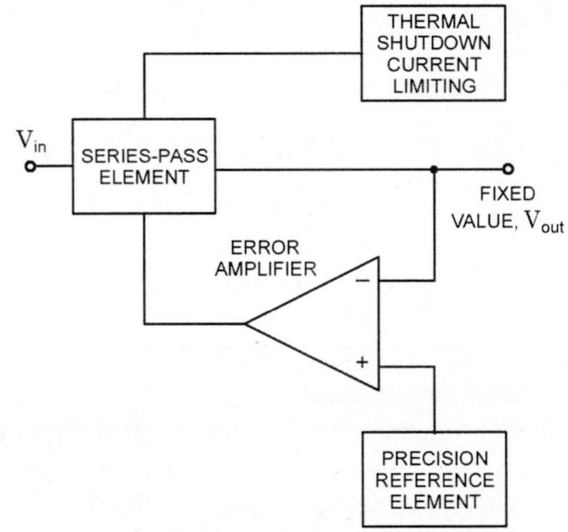

Fig. 14.10. ***Fundamental Block Diagram of a Three-Terminal IC Voltage Regulator***

The error amplifier is used to maintain a constant voltage through a negative feedback. The internal voltage reference is tightly controlled during the fabrication of IC. So, the nominal output voltage of most of the three-terminal voltage regulators has tolerances that range from ±6% to better than ±2%. The series-pass element is driven by the output of the error amplifier. It acts as an automatically controlled variable resistor. Its resistance varies as required for maintaining the output voltage constant. The series-pass element is typically a BJT that is rated to pass the maximum load current.

The basic connection of a three-terminal voltage regulator IC to a load is shown in Fig. 14.11. The fixed voltage regulator has an unregulated dc input voltage V_{in}, applied to one input terminal, a regulated output dc voltage, V_{out} from a second terminal, with the third terminal connected to ground.

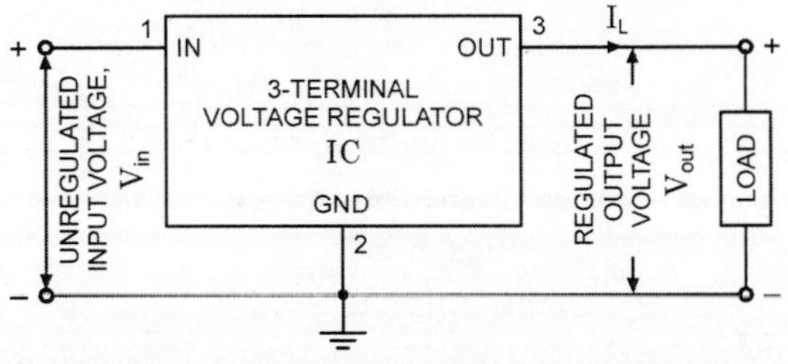

Fig. 14.11. ***Block Representation of a Three-Terminal IC Voltage Regulator***

14.6.1. Fixed Positive Voltage Regulators

The series 7800 regulators provide eight voltage options, ranging from 5 to 24 V, as indicated in Table 14.2. These ICs are designed as fixed voltage regulators and with adequate heat sinking. Although these devices do not require any external component, such components can be employed for providing adjustable voltages and currents. These ICs also have internal thermal overload protection and internal short-circuit current limiting. Fig. 14.12 illustrates how one such IC, a 7815, is connected to provide voltage regulation with output of + 15 V dc from this unit. An unregulated input voltage V_{in} is filtered by capacitor C_1 and connected to the pin 1 (IN terminal) of IC. The pin 3 (OUT terminal) of the IC provides a regulated + 15 V which is filtered by capacitor C_2 (mostly for any high frequency noise). The middle pin (GND terminal) of the IC is connected to ground. While the input voltage may vary over some permissible voltage range, and the output load may vary over some acceptable range, the output voltage remains constant within specified voltage variation limits. These limitations are mentioned in the manufacturer's specification sheet. In addition, the difference between input and output voltages ($V_{in} - V_{out}$), called the *dropout voltage*, must be typically 2.0 V, even during the low point on the input ripple voltage. Furthermore, the capacitor C_1 is required if the regulator is located at an appreciable distance from a power supply filter. Even though C_2 is not required, it may be used to improve the transient response of the regulator. A table of positive voltage regulator ICs is given in Table 14.2.

TABLE 14.2. *Positive Voltage Regulators in 7800 Series*

Device Type	*Output Voltage in Volts*	*Maximum input Voltage in Volts*	*Minimum Input Voltage in Volts*
7805	+ 5		7.3
7806	+ 6		8.3
7808	+ 8		10.5
7810	+ 10	35	12.5
7812	+ 12		14.6
7815	+ 15		17.7
7818	+ 18		21.0
7824	+ 24	40	27.1

The connection of an IC7815 in a complete voltage supply is shown in Fig. 14.13. The ac line voltage is stepped down to 24 V_{rms} across each half of the centre-tapped transformer. A full-wave rectifier and capacitor filter then provides an unregulated dc voltage with ac ripple of a few volts as input to the voltage regulator. The IC7815 then provides an output of + 15 V dc.

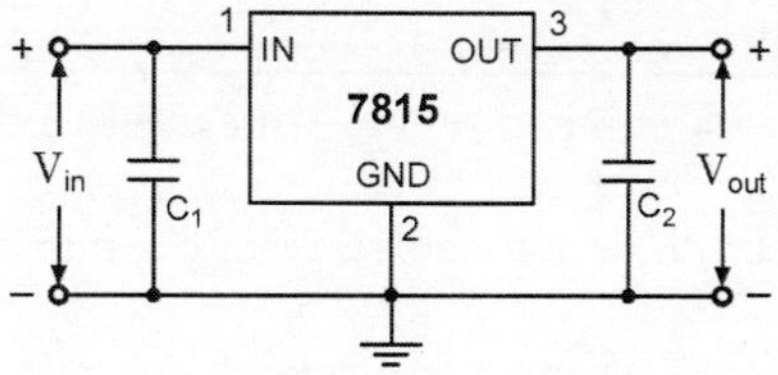

Fig. 14.12. ***Connection of 7815 Voltage Regulator***

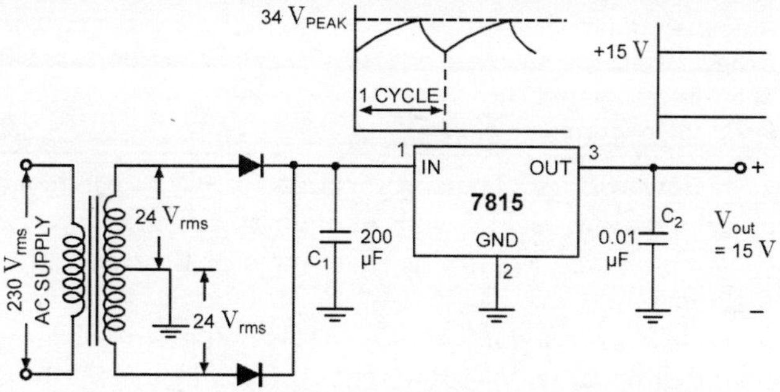

Fig. 14.13. ***+15 V Complete Power Supply***

IC7800 regulators can also be employed as current sources. A typical connection diagram of 7805C as a 0.5 A current source is depicted in Fig. 14.14.

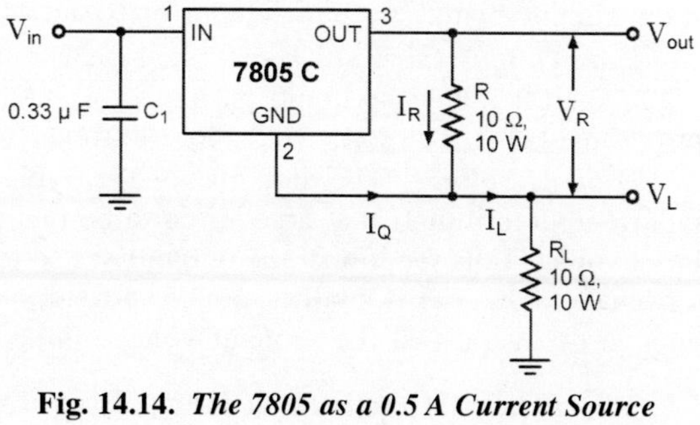

Fig. 14.14. ***The 7805 as a 0.5 A Current Source***

The current supplied to the load is given as

$$I_L = \frac{V_R}{R} + I_Q$$

when I_Q is quiescent current in amperes (4.3 mA typically for IC7805)

In Fig. 14.14, $V_R = V_{32} = 5$ V and $R = 10\ \Omega$

$$\text{So } I_L \cong \frac{5}{10} = 0.5 \text{ A}$$

The output voltage w.r.t. ground is

$$V_{out} = V_R + V_L$$

The load resistance, $R_L = 10\ \Omega$, therefore $V_L = 5$ V

$$\text{Thus } V_{out} = V_R + V_L = 5 + 5 = 10 \text{ V}$$

Minimum input voltage required,

$$V_{in} = V_{out} + \text{dropout voltage} = 10 + 2 = 12 \text{ V}$$

Specifications. The specifications sheet of fixed positive voltage regulators of 7800 series is given below. Some considerations of a few of the more important parameters should be considered.

Device Type	*Nominal Output Voltage*
7805	5 V
7806	6 V
7808	8 V
7810	10 V
7812	12 V
7815	15 V
7818	18 V
7824	24 V

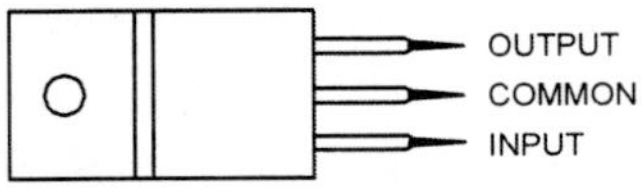

Absolute Maximum Ratings

Input voltage : 40 V

Continuous total power dissipation : 2 W

Operating free-air temperature range : –65 to 150°C

Electrical Characteristics for μA7812C

Parameter	*Minimum*	*Typical*	*Maximum*	*Units*
Output voltage	11.5	12	12.5	V
Input regulation		3	120	mV
Ripple rejection	55	71		dB
Output regulation		4	100	mV
Output resistance		0.018		Ω
Dropout voltage		2.0		V
Short-circuit output current		350		mA
Peak output current		2.2		A

Output Voltage. The specification for IC7812 indicates that its nominal output voltage is 12 V but could be as low as 11.5 V or as high as 12.5 V.

Input or Line Regulation. The input or line regulation is seen to be typically 3 mV, to a maximum of 120 mV.

Output or Load Regulation. The output or load regulation is seen to be typically 4 mV to a maximum of 100 mV (for output currents from 0.25 to 0.75 A). It means that the output voltage can typically vary only 4 mV from the rated 12 V dc.

Short-circuit Output Current. The amount of current is limited to 350 mA if the output were to be short circuited (may be by accident or by another faulty component).

Peak Output Current. The typical peak output current that might be drawn from the supply is 2.2 A against rated maximum current of 1.5 A. It indicates that though the IC is rated as capable of providing 1.5 A, but somewhat more current can be drawn (possibly for a short duration of time).

Dropout Voltage. The dropout voltage, typically 2 V, is the minimum amount of voltage across the input-output terminals that is required to be maintained if the IC is to operate as a regulator. In case the input voltage falls too low or the output rises so that at least 2 V is not maintained across the input-output terminals of IC, the IC will no longer provide voltage regulation. So input voltage is maintained large enough to ensure that the dropout voltage is provided.

14.6.2. Fixed Negative Voltage Regulators

As we have seen many times previously, positive and negative regulated power supply voltages are often required. In response, many manufacturers have developed negative three-terminal regulators. They offer the same types of features and protection as are found in positive voltage regulators.

The 7900 series of fixed output negative voltage regulators are complements to the 7800 series devices. The negative voltage regulators are available in the same eight voltage options as the 7800 devices. In addition, two extra voltage options – 2 V, and –5.2 V are also available in the negative 7900 series, as shown in Table 14.3.

TABLE 14.3. *Negative Voltage Regulators in 7900 Series*

Device Type	*Output Voltage in Volts*	*Maximum Input Voltage in Volts*	*Minimum Input Voltage in Volts*
7902	– 2		– 4.1
7905	– 5		– 7.3
7905.2	– 5.2		– 7.5
7906	– 6		– 8.4
7908	– 8	35	– 10.5
7909	– 9		– 11.5
7912	– 12		– 14.6
7915	– 15		– 17.7
7918	– 18		– 20.8
7924	– 24	– 40	– 27.1

14.6.3. Adjustable Voltage Regulators

Sometimes variable voltage is required and under such a situation adjustable voltage regulator is used. A number of IC regulators, like LM317, LM338, and LM350, are adjustable regulators. They have maximum load currents from 0.10 to 1.5 A.

For example, the LM 317 is a three-terminal **positive** voltage regulator, which can be operated with the output voltage regulated at any setting over the range of 1.2 V to 37 V. Figure 14.15 illustrates how the regulated output voltage of an LM 317 can be set.

Resistors R_1 and R_2 set the output to any desired voltage over the adjustment range (1.2 to 37 V). The output voltage desired can be determined using equation

$$V_{out} = V_{ref}\left(1+\frac{R_2}{R_1}\right) + I_{ADJ}\,R_2 \qquad \text{...(14.12)}$$

where V_{ref} = 1.25 V, the reference voltage between the output and adjustment terminals.

However, the current I_{ADJ} is very small (100 μA) and constant. So the voltage drop across R_2 due to I_{ADJ} is also very small and can be neglected so that Eq. (14.12) can be written as

$$V_{out} = 1.25\left(1+\frac{R_2}{R_1}\right) \qquad \text{...(14.13)}$$

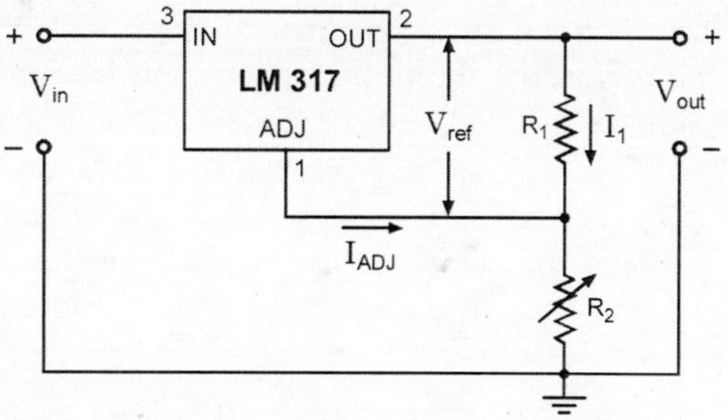

Fig. 14.15. *Connection of LM 317 Adjustable Voltage Regulator*

The current set resistor R_1 is usually 240 Ω, and to achieve good load regulation it should be tied directly to the output of the regulator rather than near the load.

The load regulation is 0.1 per cent while the line regulation is 0.01 per cent per volt. This means that the output voltage varies only 0.01 per cent for each volt of input voltage. The ripple rejection is 80 dB, equivalent to 10,000.

The LM 337 series of adjustable voltage regulators is a complement to the LM317 series devices. The **negative** adjustable voltage regulators are available in the same voltage and current options as the LM317 devices.

Tables 14.4 and 14.5 show the different grades of positive and negative regulators in series 317 and 337 respectively.

14.6.4. LM340 Series

This series is typical of the three-terminal voltage regulators. The block diagram is shown in Fig. 14.16. The built-in reference voltage V_{ref} drives the noninverting input of an op-amp. The operational amplifier consists of several stages of voltage gain. Because of the high gain of the op-amp, the error voltage between the two input terminals of the op-amp approaches zero. So the inverting input voltage is approximately equal to V_{ref}. This means that current flowing through the potential divider is

$$I = \frac{V_{ref}}{R'_2}$$

TABLE 14.4. *Positive Voltage Regulators in LM317 Series*

Device	*Output Voltage in Volts*	*Output Current in Amps*	*Maximum Input Voltage in Volts*	*Ripple Rejection in dB*	*Package*
LM317	1.2 to 37	1.5	40	80	TO-39
LM317H	1.2 to 37	0.5	40	80	TO-39
LM317HV	1.2 to 57	1.5	60	80	TO-3
LM317HVH	1.2 to 37	0.50	40	80	TO-39
LM317L	1.2 to 37	0.10	40	65	TO-92
LM317M	1.2 to 37	0.50	40	80	TO-202

TABLE 14.5. *Negative Voltage Regulators in LM337 Series*

Device	*Output Voltage in Volts*	*Output Current in Amps*	*Maximum Input Voltage in Volts*	*Ripple Rejection in dB*	*Package*
LM337	– 1.2 to – 37	1.5	40	77	TO-39
LM337H	– 1.2 to – 37	0.5	40	77	TO-39
LM337HV	– 1.2 to – 47	1.5	50	77	TO-3
LM337HVH	– 1.2 to – 47	0.5	50	77	TO-39
LM337LZ	– 1.2 to – 37	0.10	40	65	TO-92
LM337M	– 1.2 to – 37	0.50	40	77	TO-202

Incidently, the prime attached to R'_2 indicates that this resistor is inside the IC itself, rather than being an external resistor. Since the same current flows through R'_1, the output voltage is given as

$$V_{out} = \frac{V_{ref}}{R'_2}(R'_1 + R'_2) \qquad \text{...(14.14)}$$

Thus the output voltage of the regulator can be precisely controlled by setting up the desired values of R'_1 and R'_2. The data for LM 340-5 and LM 340-15 are given in Table 14.1.

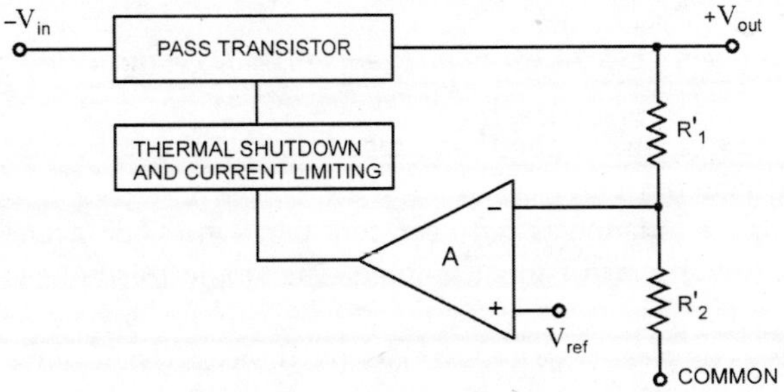

Fig. 14.16. ***Functional Block Diagram of a Typical Three-Terminal IC Voltage Regulator***

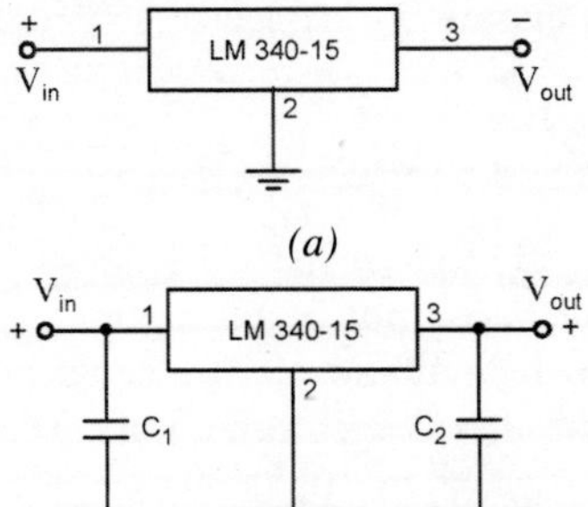

Fig. 14.17

The chip includes a pass transistor that can handle more than 1.5 A of load current provided that adequate heat sinking is used. Also included are *thermal shutdown* and *current limiting*. Thermal shutdown means that the chip will automatically turn itself *off* if the internal temperature becomes excessive, around 175°C. This is a precaution against excessive power dissipation, that depends on the ambient temperature, type of heat sinking, and other variables. Because of thermal shutdown and current limiting, devices in the LM 340 series are almost indestructible.

Figure 14.17 (*a*) shows an LM 340-15 connected as a *fixed voltage regulator*, pin 1 is the input, pin 3 is the output, and pin 2 is ground.

When an IC is more than a few cm from the filter capacitor of the unregulated power supply, the lead inductance may give rise to oscillations within the IC, Capacitor C_1 is required on pin 1 because of only this reason. Capacitor C_2 is sometimes used to improve the transient response of the circuit. Refer to Fig. 14.17 (*b*).

Any device in the LM 340 series needs a minimum input voltage at least 2 to 3 V greater than the regulated output voltage. Otherwise, it will stop regulating. Further more, there is a maximum input voltage because of excessive power dissipation.

Example 14.1. Using the 7812 voltage regulator, design a current source that will deliver a 0.5 A current to a 25 Ω, 10 W load.

Solution : Resistance required R (Fig. 14.14)

$$= \frac{V_R}{I_L} = \frac{12}{0.5} = 24\ \Omega \text{ neglecting quiescent current } I_Q$$

$$\text{Output voltage, } V_{out} = V_R + V_L = V_R + I_L R_L = 12 + 0.5 \times 25 = 24.5 \text{ V}$$

$$\text{Input voltage} = V_{out} + 2 = 24.5 + 2 = 26.5 \text{ V} \textbf{ Ans.}$$

Example 14.2. Determine the regulated output-voltage in the circuit shown in Fig. 14.18 with R_1 = 240 Ω and R_2 = 1.2 kΩ.

Solution : From Eq. (14.13)

$$V_{out} = 1.25\left(1 + \frac{R_2}{R_1}\right) = 1.25\left(1 + \frac{1.2 \times 1{,}000}{240}\right) = 7.5 \text{ V} \textbf{ Ans.}$$

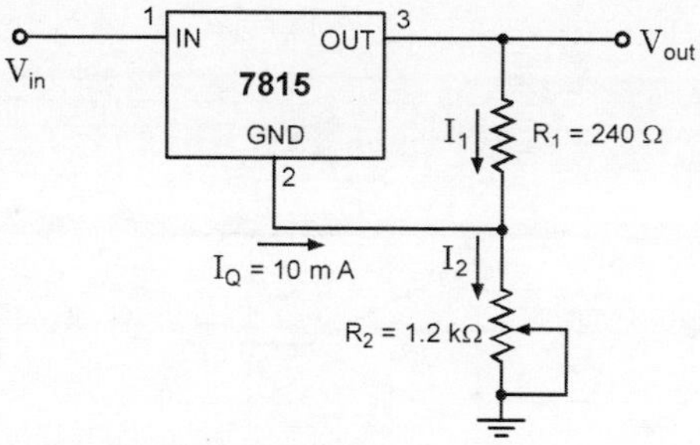

Fig. 14.18

Example 14.3. For the voltage regulator shown below, determine the minimum and maximum output voltages. A constant 1.25 V is maintained by the regulator between the output terminal and the adjustment terminal. The constant current at the adjustment terminal is 50 μA.

Solution : Reference voltage V_{ref} = 1.25 V

Current I_{ADJ} = 50 μA = 50×10^{-6} A

$$\text{From Eq. (14.12) } V_{out} = V_{ref}\left(1 + \frac{R_2}{R_1}\right) + I_{ADJ} R_2$$

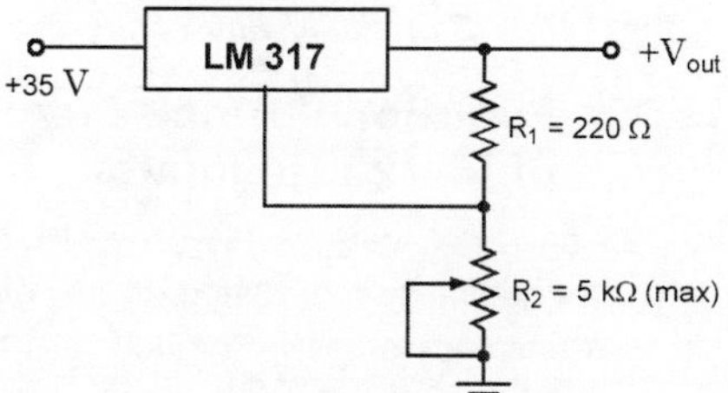

Fig. 14.19

When the potentiometer R_2 is set at minimum *i.e.,* at 0 Ω, output voltage will be minimum and is given by

$$V_{out\,(minimum)} = 1.25\,(1 + 0) + 50 \times 10^{-6} \times 0 = 1.25 \text{ V. } \textbf{Ans.}$$

When potentiometer R_2 is set at maximum *i.e.,* at 5 kΩ, output voltage will be maximum and is given by

$$V_{out\,(maximum)} = 1.25\left(1 + \frac{5 \times 10^3}{220}\right) + 50 \times 10^{-6} \times 5 \times 10^3 = 29.909 \text{ V} \simeq 30 \text{ V } \textbf{Ans.}$$

14.7. IC 723 VOLTAGE REGULATORS

The three terminal voltage regulator ICs have mainly two limitations :

1. No short-circuit protection is there.
2. Output voltage (which may be positive or negative) is fixed.

The above two limitations have been removed in general purpose regulator IC 723 whose output voltage can be adjusted over a wide range of both positive or negative regulated voltage. The IC is inherently a low current device, but can be boosted to provide 5 amps or more current by adding external components.

Yet, the IC also has two limitations :

(*i*) There is no in-built thermal protection.

(*ii*) No limits for short-circuit currents.

The IC is available as 14 pin DIP and 10 pin metal can (Figs. 14.20 and 14.22).

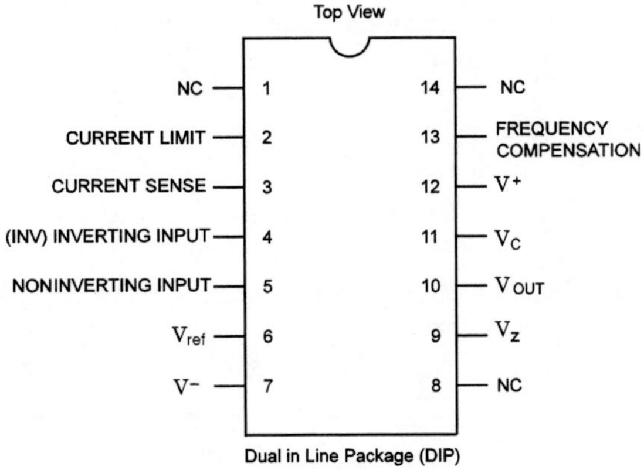

Fig. 14.20. ***Pin Diagram of 14 Pin DIP 723 IC***

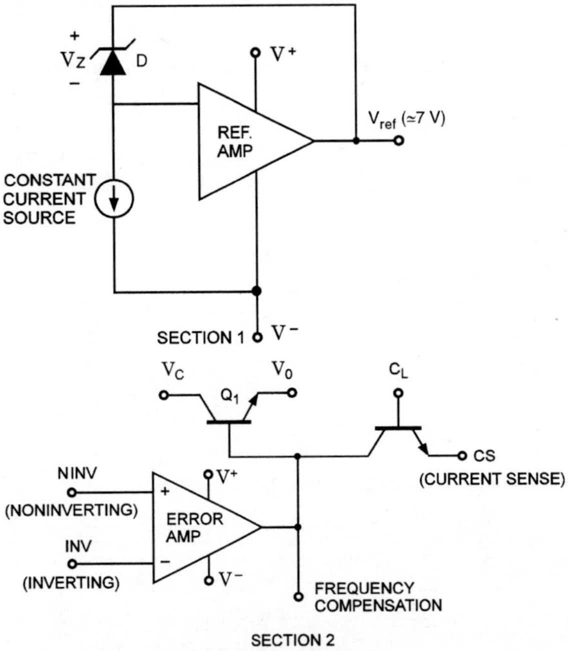

Fig. 14.21. ***Functional Block Diagram of IC 723***

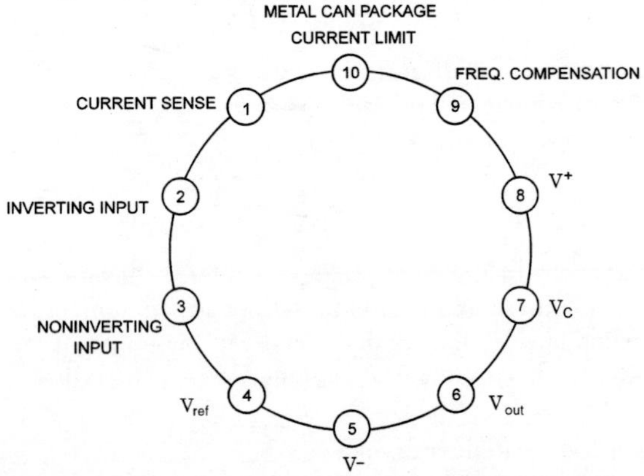

Fig. 14.22. ***Pin Diagram of 10 Pin Metal Can IC 723***

14.7.1. Functional Block Diagram of IC 723 Regulator

IC 723 has two sections (see Fig. 14.23).

(*i*) A constant current source, a zener diode and a reference amplifier producing a constant voltage of about 7 volts at V_{ref} end. The zener diode is being compelled to operate at fixed point (so that zener output is a fixed voltage) by a constant current source.

(*ii*) The second section of IC 723 comprises of an error amplifier, a series pass transistor Q_1 and a current limit transistor Q_2. The error amplifier compares a sample of output voltage applied at INV (Inverting) input terminal to the reference voltage V_{ref} applied at the NINV (Noninverting) input terminal.

The conduction of the transistor Q_1 is controlled by the error signal.

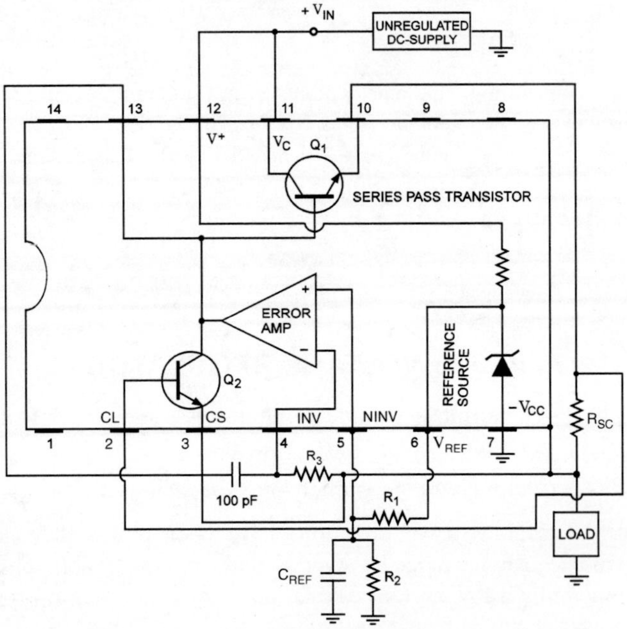

Fig. 14.23. ***Basic Circuit of IC 723 Voltage Regulator For Low Voltage (2 to 7 V)***

The two sections do not have any internal connection and they are brought out on the IC package (Fig. 14.21). IC723 regulator is available in 14 dual-in-line package (DIP) (Fig. 14.20) or 10 pin metal can (Fig. 14.22).

The important features and electrical characteristics are given in Table 14.6.

TABLE 14.6. ***IC723 Regulator Electrical Characteristics***

IC723 Voltage Regulator

Important Features :

- Input voltage 40 V max
- Output voltage adjustable from 2 V to 37 V
- 150 mA output current without external pass transistor
- Output currents in excess of 10 A possible by adding external transistors
- Can be used as either a linear or a switching regulator

PARAMETER	CONDITIONS	LM 723			LM 723C			UNITS
		Min	Typ	Max	Min	Typ	Max	
Line Regulation	V_{IN} = 12 V to V_{IN} = 15 V		0.01	0.1		0.01	0.1	$\%V_{out}$
	$-55°C \le T_A \le +125°C$			0.3				$\%V_{out}$
	$0°C \le T_A \le +70°C$						0.3	$\%V_{out}$
	V_{IN} = 12V to V_{IN} = 40 V		0.02	0.2		0.1	0.5	$\%V_{out}$
Load Regulation	I_L = 1 mA to I_L = 50 mA		0.03	0.15		0.03	0.2	$\%V_{out}$
	$-55°C \le T_A \le +125°C$			0.6				$\%V_{out}$
	$0°C \le T_A \le +70°C$						0.6	$\%V_{out}$
Ripple Rejection	f = 50 Hz to 10 kHz, C_{REF} = 0		74			74		dB
	f = 50 Hz to 10 kHz, C_{REF} = 5 μF		86			86		dB
Average Temperature Coefficient of Output Voltage (Note 8)	$-55°C \le T_A \le +125°C$		0.002	0.015				%/°C
	$0°C \le T_A \le +70°C$					0.003	0.015	%/°C
Short-Circuit Current Limit	R_{SC} = 10 Ω, V_{OUT} = 0		65			65		mA
Reference Voltage		6.95	7.15	7.35	6.80	7.15	7.50	V
Output Noise Voltage	BW = 100 Hz to 10 kHz, C_{REF} = 0		86			86		μV_{rms}
	BW = 100 Hz to 10 kHz, C_{REF} = 5 μF		2.5			2.5		μV_{rms}
Long Term Stability			0.05			0.05		%1000 hrs
Standby Current Drain	I_L = 0, V_{IN} = 30 V		1.7	3.5		1.7	4.0	mA
Input Voltage Range		9.5		40	9.5		40	V
Output Voltage Range		2.0		37	2.0		37	V
Input-Output Voltage Differential		3.0		38	3.0		38	V

The internal diagram of 14 pin DIP is shown in Fig. 14.23 (for low voltage 2 to 7 V).

Circuit Operation. The voltage at noninverting input (NINV) terminal of error amplifier due to R_1-R_2 divider is

$$V_{NINV} = V_{ref} \cdot \frac{R_2}{R_1 + R_2} \qquad \text{...(14.15)}$$

A potentiometer may be included between R_1 and R_2.

The differential voltage between V_{out} and V_{NINV} (which is directly fed back to INV terminal) is amplified by the error amplifier.

The output of error amplifier drives the series pass transistor Q_1 so as to minimize the difference between the INV and NINV inputs of error amplifiers, Q_1 operates as emitter follower.

$$V_{out} = V_{ref}\frac{R_2}{R_1 + R_2} \quad \text{...(14.16)}$$

The voltage at INV terminal also goes down if the output voltage becomes low. This makes the output of error amplifier to become more positive, thereby driving the transistor Q_1 more into conduction. Thus, the V_{Q1} across Q_1 is reduced and it drives more current into the load and thus the voltage across load increases. Thus, initial drop in the load voltage has been compensated.

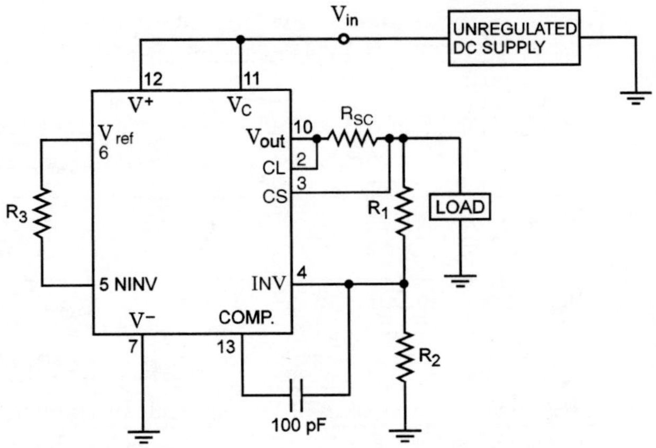

Fig. 14.24. *723 As High Voltage Regulator*

Similarly, any change in input voltage or increase in load voltage is regulated in same manner.

Typically, $V_{ref} = 7.15$ V

$$\therefore \quad V_{out} = 7.15\frac{R_2}{R_1 + R_2} \quad \text{...(14.17)}$$

Thus, the circuit can be used as low voltage (< 7 V) regulator.

If $V_{out} > 7$ volts is desired, circuit of Fig. 14.24 can be used.

Here, the NINV terminal (of error amplifier) is connected directly to V_{ref} via resistance R_3.

Thus, $V_{NINV} = V_{ref}$. The error amplifier produces ouput voltage V_{out} with action of series pass transistor as

$$V_{out} = 7.15\left(1 + \frac{R_1}{R_2}\right) \quad \text{...(14.18)}$$

Since gain of error amplifier [in noninverting mode) is

$$A_V = 1 + \frac{R_1}{R_2} \quad \text{...(14.19)}$$

14.8. CURRENT LIMIT PROTECTION

The circuit arrangement for IC 723 as low voltage or high voltage regulator has no protection. If short circuiting takes place (when load demands more current), regulator IC makes efforts to provide it at a constant output voltage which results in heating that may finally burn the IC.

To prevent the load current from increasing a set value, say I_{LIMIT} is used so that current gets limited. As current increases due to want of load the output voltage falls, and current limit is provided by connecting an internal resistance R_{sc} between pin No. 10 (V_{out}) and pin No. 3 (emitter of Q_2), as shown in Figs. 14.23 and 14.24. The emitter of Q_2 is also connected to the load.

The load current generates a small voltage drop, V_{SENSE} across R_{sc} which is directly applied to base-emitter junction of Q_2. When the transistor becomes ON (when $V_{BE} \cong 0.6$ V), the current from error amplifier reaches collector of Q_2, and I_B of Q_1 decreases, and hence I_E of Q_1 too. Thus, any increase in load current gets cancelled. Similarily, if the load current decreases, V_{BE} of Q_2 drops, the cycle gets repeated such that load current is held constant to produce a voltage drop V_{SENSE} across R_{sc} to turn ON Q_2.

$$\therefore \; I_{Limit} = \frac{V_{SENSE}}{R_2} \simeq \frac{0.6\,V}{R_{SC}} \quad \text{...(14.20)}$$

This method of current limiting is also called as current sensing technique.

14.9. CURRENT FOLDBACK

The current limiting technique maintains a constant load current at some specified value and under overload condition, the V_{out} drops to zero. But, in case of short circuiting under overload conditions the maximum current flows into the regulator.

To protect regulator from overload condition, current foldback method is employed. As shown in the current feedback characteristics [Fig. 14.25 (*a*)], as current requirement increases, the output voltage is held constant till a fixed current level, I_{knee} is reached. If the current requirement reaches above this level, both output voltage and output current decrease. The circuit in Fig. 14.25 (*b*) shows the requirement of current fold back.

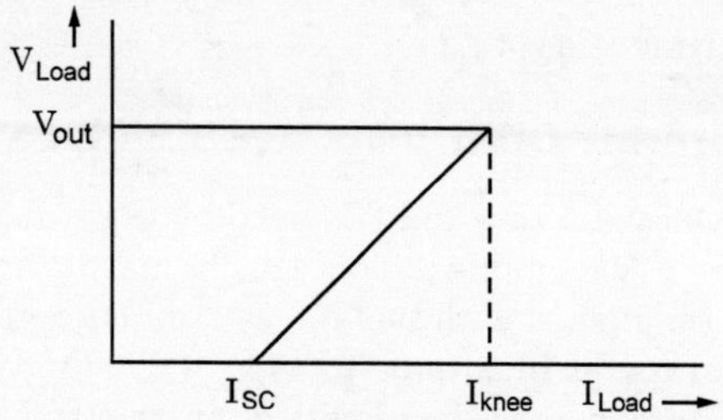

(a) Current Foldback Characteristic Curve

The voltage at pin 2 is divided by $R_3 - R_4$ network. The current limit transistor Q_2 conducts only when the drop across the resistance R_{sc} is large enough to produce a base-emitter voltage of Q_2 to be at least V_{BE}. When Q_2 starts conduction, Q_1 turns *off* and hence I_L decreases. This results in decrease of volt drop at emitter of Q_1, say V_{EQ1} and also the output voltage V_{out}.

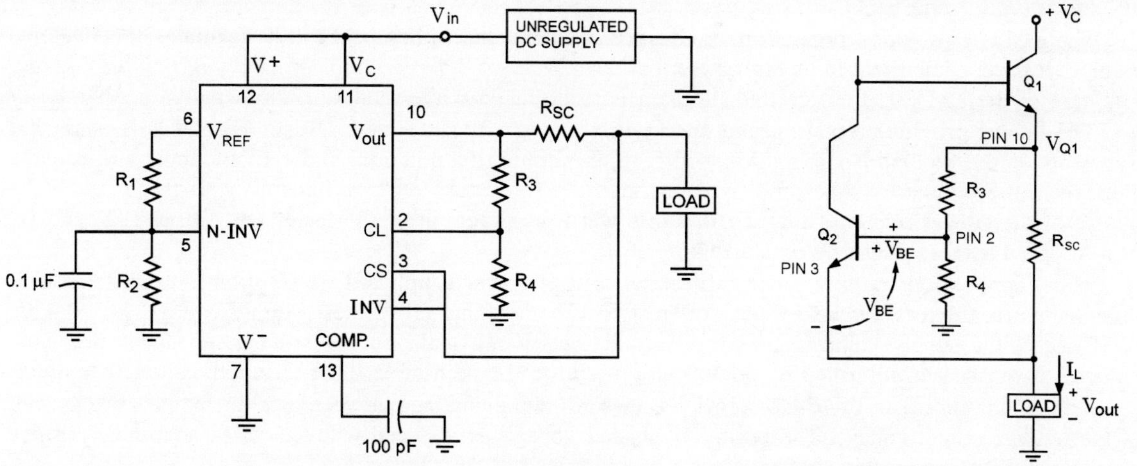

(b) A Low Voltage Regulator Using Current Foldback

(c) Current Foldback

Fig. 14.25

$$\therefore \quad \text{At base of } Q_2 \qquad V_{BQ2} = \frac{V_{EQ1}\, R_4}{R_3 + R_4} \qquad \text{...(14.21)}$$

Thus, the voltage at pin 3 drops by smaller amount as compared to at pin 2. This increase in V_{BE} of Q_2 or increase in conduction of Q_2 decreases conduction of Q_1. Thus, I_L again gets reduced. The process continues till V_{out} = 0 V and V_{EQ1} makes 0.5 V between pin 3 and pin 2. This is I_{sc} and thus has been lowered by both I_L and V_{out}.

14.10. CURRENT BOOSTING

IC723 regulator is limited to provide a maximum current of 140 mA. To increase its current capacity, we add a transistor Q by connecting its collector to pin 11, base to pin 10, and emitter to pin 2, as shown in Fig. 14.26.

Thus, the collector current of transistor Q comes from unregulated dc supply. The pin 10, or say output current from V_{out} is into the base of the transistor, so that

$$I_{load} = \beta_{pass\ transistor} \times I_{out\ (723)} \quad ...(14.22)$$

Thus, the maximum current handling capacity of IC723 gets increased by factor current gain β of the connected transistor [Eq. (14.22)].

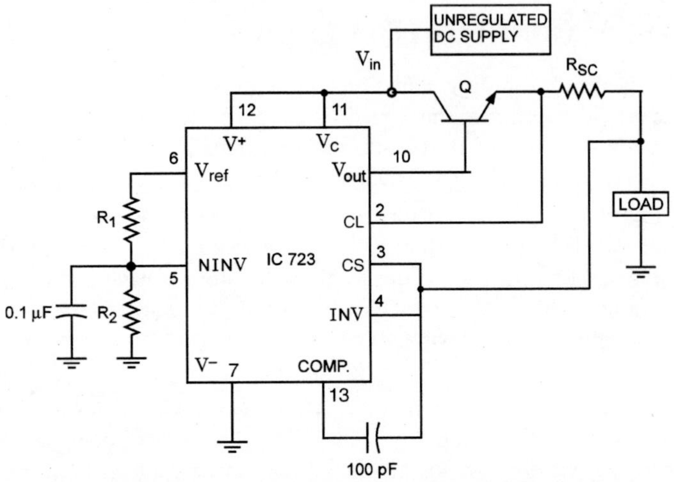

Fig. 14.26. *Current Boosted Low Voltage Regulator*

14.11. DUAL VOLTAGE SUPPLIES

Many discrete and ICs need bipolar (dual or ± V) supplies. This can be easily accomplished with two three-terminal regulators, as illustrated in Fig. 14.27. Opposite-phase ac is provided by the transformer's secondary and a grounded center tap. The single full-wave bridge converts these into +ve and –ve dc voltages (with respect to the grounded center tap). Filtering (with respect to ground) is provided by capacitors C_1 and C_2.

The LM340 provides regulation of the positive voltage, while the LM320 regulates the negative voltage. It is very important to mention here that *LM320 has a different pin configuration than the LM340.* The case of the LM320 is not ground. So care is to be taken while mounting the negative regulator.

The diodes provide protection, but ensure that they are not reversed. Diodes D_1 and D_2 ensure that transients on the regulator outputs do not drive the outputs to a potential above their inputs and cause a damage to the regulators. Also, the two regulators may not turn *on* simultaneously. If this occurs, the output of the slower regulator may be driven toward the potential of the faster one. Diodes D_3 and D_4 prevent these reverse polarities on start up.

Many applications require several different voltage power supplies. One solution is to build several independent regulators. However, very often it is important that all of these supply voltages track. That is, if one of the supply voltages goes up 2%, it is best if all of the supply voltages go up by the same amount. It can be accomplished by adding an op-amp to the adjustable three-terminal voltage regulator.

The circuit shown in Fig. 14.28 provides current limiting and thermal shutdown for the negative as well as the positive output voltage. The positive regulated voltage is produced with a LM317 adjustable positive regulator IC, as it was in Fig. 14.28.

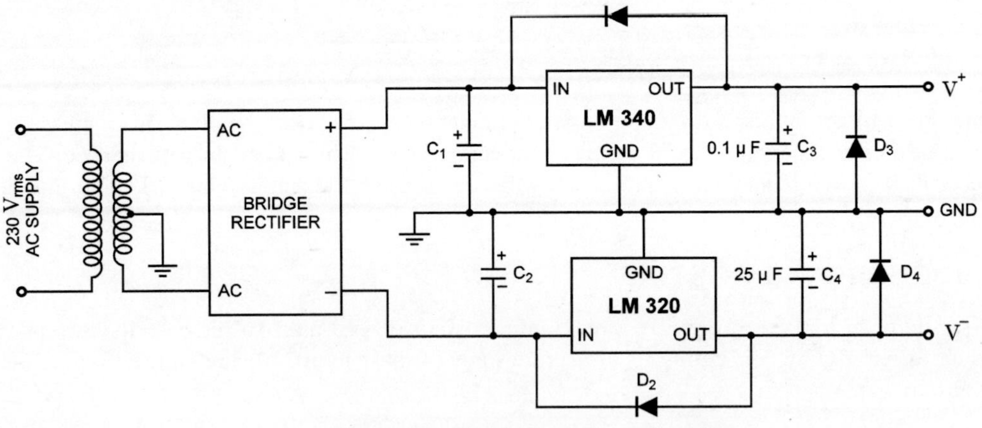

Fig. 14.27. *Simple Dual Power Supply*

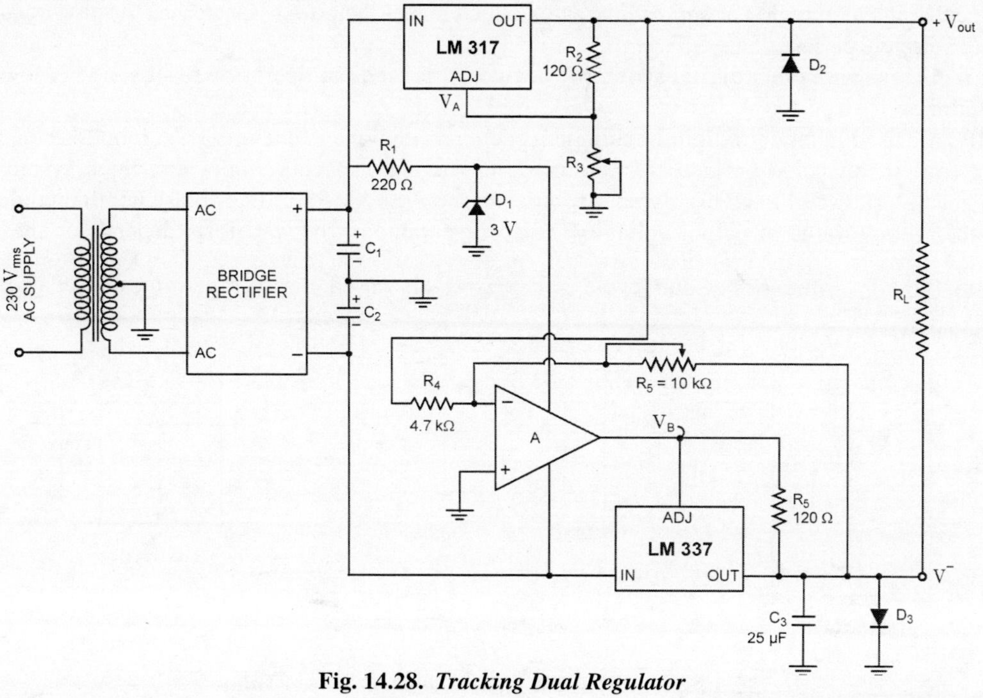

Fig. 14.28. ***Tracking Dual Regulator***

The regulated positive voltage is always 1.2 V more positive than the voltage across R_3 (on the adjust pin). This regulated output voltage is used as the input of an inverting amplifier. The op-amp A is the input stage of this amplifier, and the LM 337 negative voltage regulator is thc power output stage. Consequently, the negative regulated output voltage V^- is an amplified and inverted version of the positive regulated voltage V^+. Current limiting and thermal shutdown are independently provided by the LM 317 (for the +ve output) and the LM 337 (for the –ve output). The op-amp assures that the negative output voltage tracks the positive output by driving the LM 337's adjust pin to +1.2 V above the required negative output. Since the LM 337 is inside the op-amp's negative feedback loop, this +1.2 V offset appears between the op-amp and the regulator, not at the regulator output.

14.12. SWITCH MODE POWER SUPPLIES (SMPS) OR SWITCHING REGULATORS

The regulated power supplies described earlier were basically linear voltage regulators. A linear voltage regulator possesses many limitations:

(*i*) The input transformer is bulky and is a very expensive component in the whole regulated power supply since it is used at low line frequency (50 Hz).

(*ii*) As the ac frequency is 50 Hz, filter capacitor of large values are required to minimize the ripples introduced.

(*iii*) The series regulators exhibit 50% efficiency and hence input voltage should be more than output voltage. If efficiency is less, then that part (difference in input and output voltage) of voltage will be dissipated in the form of power in series pass transistor which is always in active region. A TTL system regulator (V_{out} = 5 V) when operated at 10 V ac input provides 50% efficiency and only 25% for 20 V dc input.

(*iv*) Another limitation is that in a system with one dc supply voltage (such as +5 V for TTL) if there is need ±15 V for op-amp operation, it may not be economically and practically feasible to achieve this.

Switch mode power supply (SMPS) overcomes the limitations of regulated power supplies. Here in SMPS (or sometimes called *switching regulator*), the pass transistor is used as a controlled switch and is operated either in *cutoff* region or *saturation* region which makes the power transfer via the pass transistor in the form of discrete pulses.

As in the case with other digital switching devices, this mode of operation results in very low power consumption.

SMPS are used in modern digital equipment such as telephone exchange, PCs, robotics etc., battery powered equipment, video projectors, measuring instruments, space vehicles etc. Such systems need very compact, light weight and highly energy efficient supplies. SMPS utilizes pulse width modulation to control the average value of output voltage. The average value of the waveform depends on the area of waveform. If the T_{ON} time (or say, duty cycle $\frac{T_{ON}}{T_{ON} + T_{OFF}}$) is varied, the average value of voltage changes proportionately.

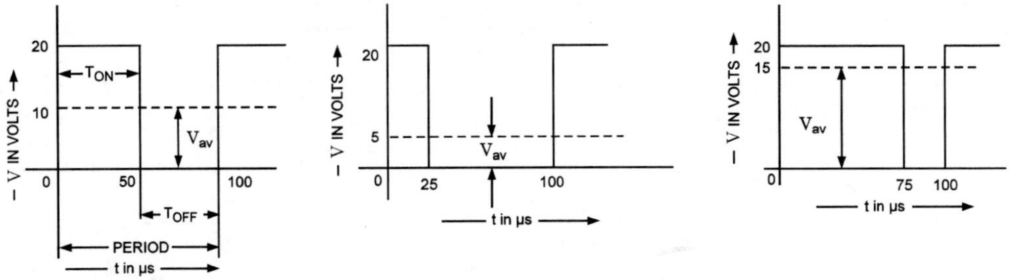

(a) *(b)* *(c)*

Fig. 14.29. ***Pulse Width Modulation and Average Value***

In the PWM switching at constant switching frequency, the switch control signal is generated by comparing $V_{control}$ and repetitive waveform, as shown in Fig. 14.30.

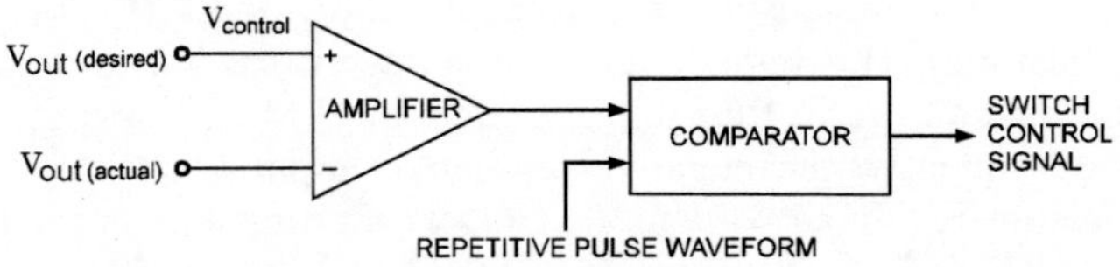

(a) Pulse Width Modulator

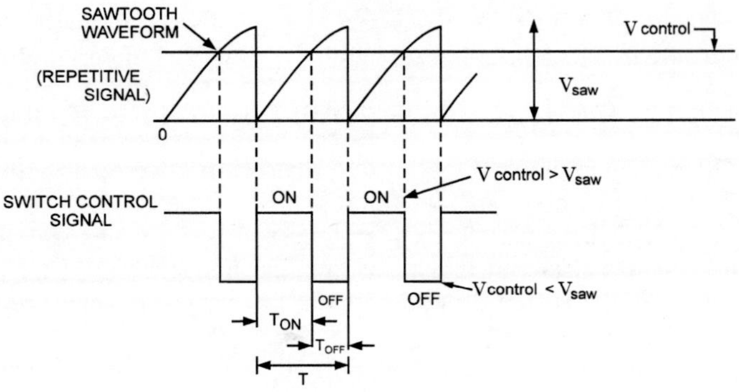

(b) Comparator Signals

Fig. 14.30

SMPS or switching regulators can provide large load currents at low voltage precisely what is required in PCs (personal computers).

Switching regulators are available in three basic configurations viz. step-down, step-up and polarity inverting configurations.

Step-down version is shown in Fig. 14.31(*a*). The rectangular pulses on the base saturate and cut off the pass transistor during each cycle. This generates a rectangular voltage at the input to LC filter. This filter blocks the ac component and allows the dc component to pass to the output. Because of the *on-off* switching, the average value is always less than the input voltage. This is why the circuit is called the step-down version.

Step-up version of the switching regulator is shown in Fig. 14.31(*b*). Again, the transistor is alternately saturated and cut off. When the transistor is saturated, current flows through the inductor. When the transistor suddenly cuts off, the magnetic field around the coil collapses and induces a large voltage across the coil of opposite polarity. This keeps the current flowing in the same direction. Furthermore, the inductive kickback voltage is larger than the input voltage. This is why the circuit is called *step-up configuration*.

Polarity Inverting regulator is shown in Fig. 14.31(*c*). When the transistor is saturated, current flows through the inductor. When the transistor cuts off, the magnetic field collapses, and the inductive kickback keeps current flowing in the same direction. Since the transistor is off, the only path is through the capacitor. If the direction of charging current through the capacitor is checked, output voltage is found to be *negative*.

A *low-power design*, using circuits that we are already familiar with, is illustrated in Fig. 14.32. The relaxation oscillator generates a square wave whose frequency is determined by R_5 and C_3. The square wave is integrated to provide a triangular wave, which is used to drive the noninverting (+) input of a triangular-to-pulse converter. The pulses out of this circuit then drive the pass transistor. The duty cycle of these pulses will determine the output voltage.

The duty cycle D is the ratio of the *on time* T_{ON} to the *time period* T. By controlling the duty cycle of the pulse generator, the duty cycle of the input voltage to the LC filter is controlled. The output of the LC filter is a dc voltage with only a small ripple. This output voltage is directly proportional to the duty cycle and is given as

$$V_{out} = DV_{in} \quad ...(14.23)$$

Since D can vary from 0 to 1, V_{out} can vary from 0 to V_{in}.

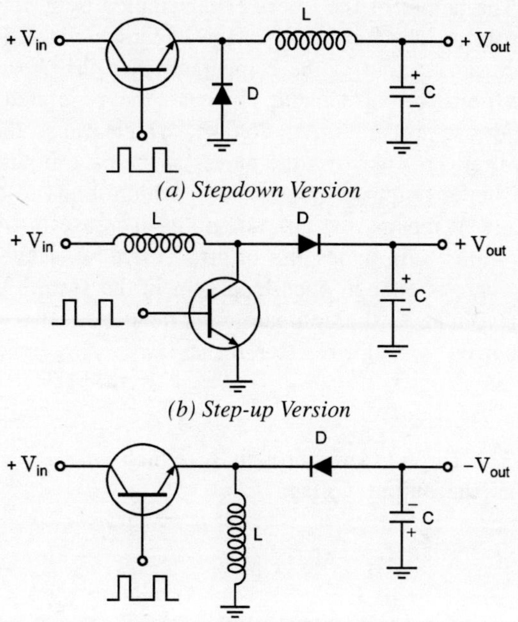

(*c*) *Polarity Inverting Configuration Switching Regulators*

Fig. 14.31

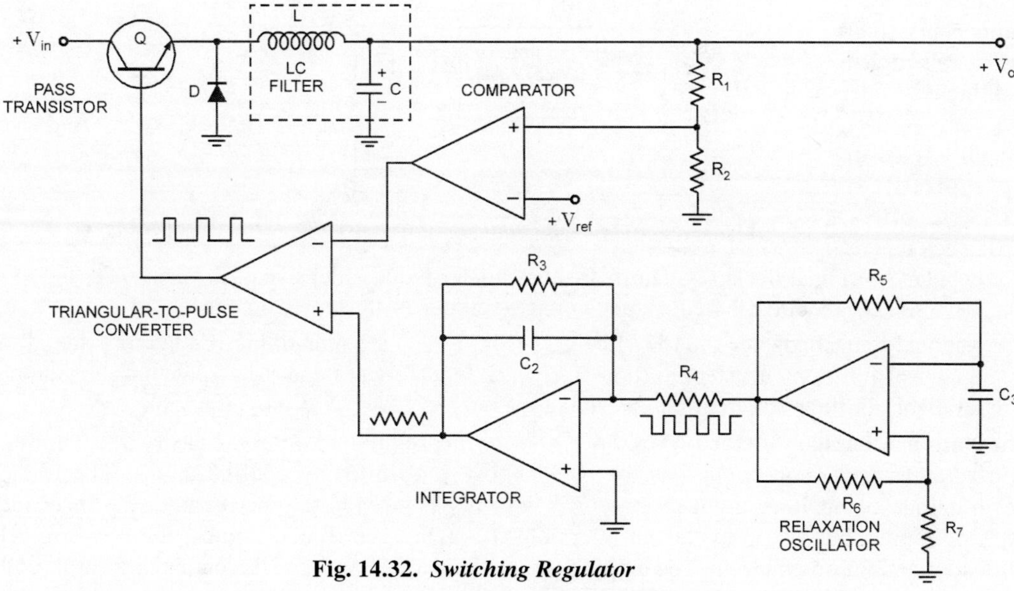

Fig. 14.32. ***Switching Regulator***

The output of the LC filter is sampled by a voltage divider, which returns a feedback voltage to the comparator. The feedback voltage is compared with a reference voltage V_{ref} from a zener diode or other source. The output of the comparator then drives the inverting input of the triangular-to-pulse generator.

Operation of Circuit. In case the regulated output voltage tends to increase, the comparator provides a higher output voltage, which raises the inverting input voltage to the triangular-to-pulse converter. This narrows the pulses at the base input of the pass transistor. Since the duty cycle is lower, the filtered output voltage is less, which tends to cancel almost all the original increase in the output voltage. It means that any attempted increase in output voltage generates a negative feedback voltage that almost eliminates the original increase. Reverse happens should the output voltage fall.

There is enough open-loop gain in the system to ensure a well-regulated output voltage. Since the error voltage to the comparator is near zero, the voltage across R_2 is approximately equal to V_{ref}. So the current through resistor R_2 is

$$I = \frac{V_{ref}}{R_2} \quad \text{...(14.24)}$$

This current flows through R_1, which means the output voltage is

$$V_{out} = \frac{V_{ref}}{R_2}(R_1 + R_2) \quad \text{...(14.25)}$$

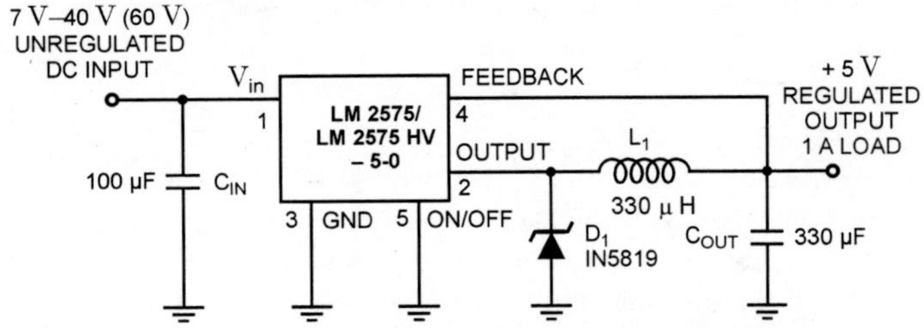

Fig. 14.33. *Note : Pin Numbers are For the TO-220 Package*

Switching regulators are available in different circuit configurations including the flyback, feed-forward, push-pull, and non-isolated single-ended or single-polarity types. Also, the switching regulators can operate in any of three modes step-down, step-up, or polarity inverting.

LM 2575 series of regulators developed by National Semiconductor are monolithic ICs that provide the active functions for *a step-down (buck)* switching regulator, capable of driving a 1 A load with excellent line and load regulation. These devices are available in fixed output voltages of 3.3 V, 5 V, 12 V, 15 V and an adjustable output version.

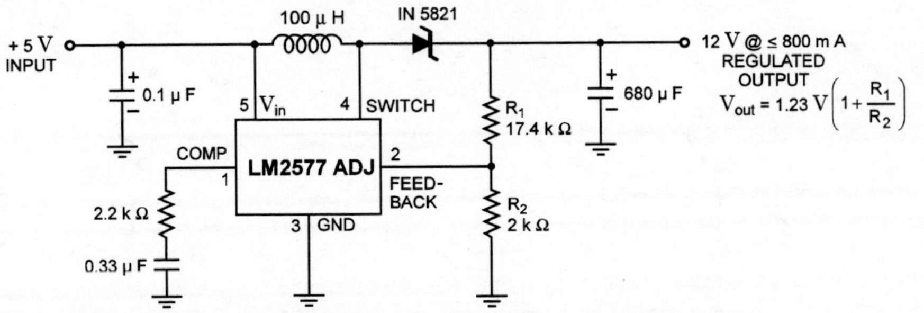

Fig. 14.34. *Note : Pin Numbers Shown are For TO-220(T) Package*

Requiring a minimum number of external components, these regulators are simple to use and include internal frequency compensation and a fixed-frequency oscillator. LM2575 series offers a high-efficiency replacement for popular 3-terminal linear regulators. It substantially reduces the size of the heat sink, and in many cases no heat sink is required. Fixed output voltage version is illustrated in Fig. 14.33.

The National Semiconductor LM1577/LM2577 (Fig. 14.34) are monolithic ICs that provide all of the power and control functions for step-up (boost), flyback, and forward converter switching regulators. The device is available in three different output voltage versions: 12 V, 15 V and adjustable.

The National Semiconductor LM1578A is a switching regulator which can easily be set up for such dc-to-dc voltage conversion circuits as the buck, boost, and inverting configurations. The LM1578A features a unique comparator input stage which not only has separate pins for both the inverting and noninverting inputs, but also provides an internal 1.0 V reference to each input, thereby simplifying circuit design and PC board layout. The output can switch up to 750 mA and has output pins for its collector and emitter to promote design flexibility. An external current limit terminal may be referenced to either the ground or the V_{in} terminal, depending upon the application. In addition, the LM1578A has an on board oscillator, which sets the switching frequency with a single external capacitor from < 1 Hz to 100 kHz (typical). It operates from supply voltages of 2 V to 40 V. It is provided with current limit and thermal shutdown. Duty cycle up to 90 %. Functional diagram is given in Fig. 14.35.

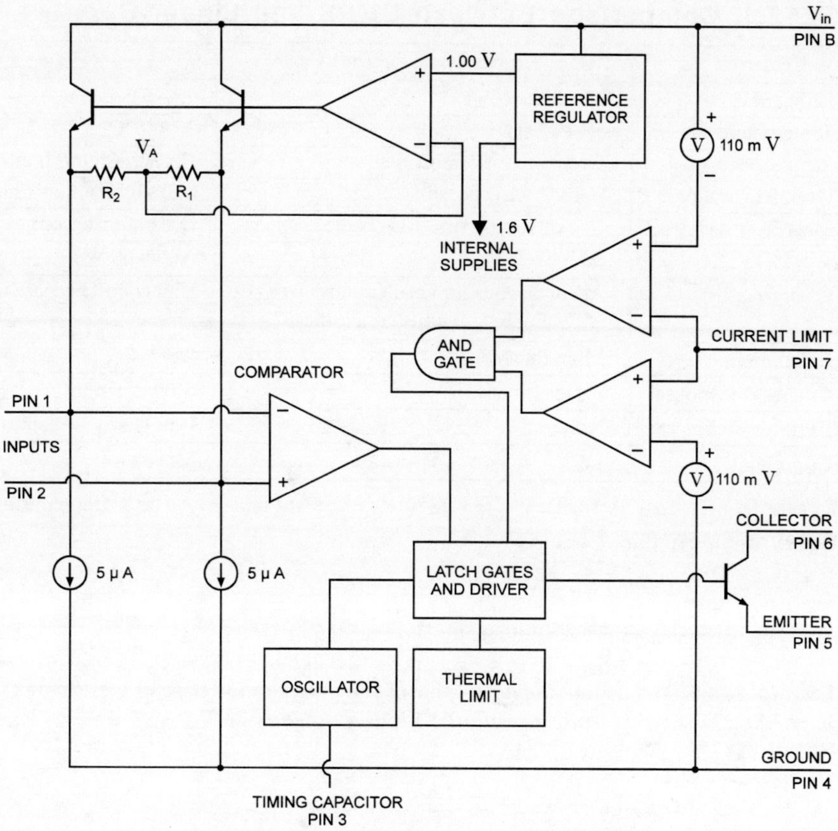

Fig. 14.35. ***Functional Diagram of LM1578A***

The switching regulators have the following advantages over linear regulators:

1. High efficiency because of less heat dissipation.
2. Light in weight and small in size because of reduced size of transformer.
3. Reduced harmonic feedback into the supply mains.
4. Isolation from main supply without the requirement of large mains transformer.
5. Protection against excessive output voltage by quick acting guard circuits.
6. Possibility of conversion of an unregulated input of 24 V into a regulated output of 1,000 V dc.
7. Generation of low and medium voltage supplies are easy.

However, they have the following drawbacks :

1. Because it operates by switching at high frequency, noise may be troublesome on both the output and the ac input lines (due to induction). Switching regulators throw current transients back into the unregulated supply which are somewhat larger than the maximum load current. Thus, switching need heavy filtering.
2. The switching regulator also is limited in its transient response by the switching frequency and is usually considerably slower than a linear regulator (ms versus μs).
3. The switching regulator is more complex than linear type. However, this is often a substitution of electrical complexity for the thermal and mechanical complexity of high power linear regulators.
4. Though RF interference can be a problem in SMPS unless properly shielded. SMPS in TV sets is in synchronisation with the line frequency (15,625 kHz) and thus switching effects are not visible on screen.

For above reasons, the use of switching regulators is usually limited to power levels around 100 W.

14.12.1. Comparison Between SMPS and Linear Supply

Features	Linear Power Supply	SMPS
Efficiency	25-50 per cent	65-75 per cent
Temperature rise	50-100 °C	20-40 °C
Ripple value	Even 5 mV possible	Higher, 25 to 50 mV
Overall regulation	Even 0.1% is possible	0.3% is common, tighter regulation is difficult to have
Magnetic material	Use of stalloy or CRGO core	Use of ferrite core
Weight	20-30 W/kg	About 60 W/kg
Cost	Advantageous for smaller units up to 10-15 W, but bigger units are costlier	Higher cost because of more and special components.
Reliability	More reliable	Depends upon availability of suitable transistors
Transient response	Faster (in μs)	Slower (in ms)
RF interference	None	Can pose problem unless properly shielded
Complexity	Less	More

Example 14.4. In Fig. 14.32, V_{ref} = 1.25 V, R_1 = 2.5 kΩ and R_2 = 1 kΩ. Determine the regulated output voltage.

Solution : According to Eq. (14.24)

$$I = \frac{V_{ref}}{R_2} = \frac{1.25\ \text{V}}{1\ \text{k}\Omega} = 1.25\ \text{mA}$$

This current also flows through R_1. So the output voltage,

$$V_{out} = I(R_1 + R_2) = (1.25\ \text{mA})\ (2.5\ \text{k}\Omega + 1\ \text{k}\Omega) = 4.375\ \text{V} \quad \textbf{Ans.}$$

Example 14.5. Determine the duty cycle of the pulses at the base of transistor Q in Fig. 14.32, if V_{ref} = 1.25 V, R_1 = 3 kΩ, R_2 = 1 kΩ and unregulated input voltage is 20 V.

Solution : According to Eq. (14.25)

$$V_{out} = V_{ref}\frac{R_1 + R_2}{R_2} = 1.25\left(\frac{3+1}{1}\right) = 5\ \text{V}$$

$$\text{Duty cycle, } D = \frac{V_{out}}{V_{in}} = \frac{5}{20} = 0.25 \text{ or } 25\%. \quad \textbf{Ans.}$$

HIGHLIGHTS

1. A regulated power supply is an electronic circuit designed to provide constant dc voltage of predetermined value across load terminals, which is independent of variations in load current, ac mains voltage and temperature.
2. A transformer, a rectifier circuit, a filter circuit, a voltage regulator along with some current limiting arrangement constitute a regulated power supply.
3. Monolithic voltage regulators are available in a variety of different output voltage ratings and are also quicker and easier to use. The 7800 is a fixed positive voltage regulator series with eight voltage options while the 7900 is a fixed negative voltage regulator series with ten voltage options. Adjustable voltage regulators such as the LM317 and LM337 are more popular because of their versality, performance and reliability.
4. Switching regulators are most efficient compared to fixed or adjustable voltage regulators. The efficiency of such regulators may be as high as 95%. In a switching regulator, its efficiency is improved by using the series-pass transistor as a switch rather than as a variable resistor as in the linear mode.

EXERCISES

1. Explain different parameters of power supply.
   Draw the block diagram of regulated power supply and explain the function of each block.
2. Draw shunt voltage regulator and explain its working.
3. Describe the operation of a regulated power supply. Give the standard representation of IC voltage regulator.
4. Draw and explain the circuit diagram of a basic op-amp series voltage regulator.
5. What is the need of voltage regulator in a power supply ? Discuss the applications of voltage regulator IC 7805.
6. Why do we use three-terminal IC regulators ? Explain its circuit and working.
7. How does an IC LM 340 series incorporates thermal shutdown and current limiting ?

8. Discuss how a three-terminal voltage regulator IC be employed to realize an adjustable voltage output.

9. List and explain the characteristics of three-terminal IC regulator. Explain the protection used in 78XX.

10. What type of fixed voltage regulators are available ? Show the standard representation and discuss main characteristics. **[G.G.S.I.P. Univ. Analog Electronics-II May 2010]**

11. Explain adjustable positive voltage regulator. **[G.G.S.I.P. Univ. Analog Electronics-II May-June 2009]**

12. Explain the operation of 723 regulator. **[G.G.S.I.P. Univ. Analog Electronics-II May 2011]**

13. Explain the current foldback characteristics of IC723 voltage regulator.

14. Explain with neat sketch, the current limiting features of IC723 voltage regulator.

15. Describe four different types of voltage regulators and give their applications. **[G.G.S.I.P. Univ. Analog Electronics-II May-June 2009]**

16. What are dual voltage supplies. Give their advantages.

17. What are the advantages of "switching mode" dc voltage regulator over a "linear mode" one. Give the basic block diagrams of the two types and explain their operation. What is the purpose of energy storage elements in "switching mode" one.

18. Draw schematic circuit of a basic switching regulator and explain the function of its for major components. **[G.G.S.I.P. Univ. Analog Electronics-II May-June 2009]**

19. What are switching regulators ? Describe theory of switching regulators.

20. What is the principle of switch mode power supplies ? Discuss its advantages and disadvantages.

21. How SMPS is different from a regulated power supply ? Draw the circuit and explain its working principles.

22. Write short notes in the following:

(*i*) Voltage regulator. (*ii*) Series and shunt voltage regulators.

(*iii*) Three terminal IC regulator. (*iv*) Switching regulators.

SHORT ANSWER TYPE QUESTIONS WITH ANSWERS

Q. 1. What is a regulated power supply ?

Ans. A regulated power supply is an electronic circuit designed to provide constant dc voltage of predetermined value across load terminals, which is independent of variations in load current, ac mains voltage and temperature.

Q. 2. What is practical importance of voltage regulation in power supply ?

Ans. Since the output voltage available from an unregulated power supply varies with the variations in load current, ac mains voltage and temperature and variations in dc output voltage may cause inaccurate or erratic operation or malfunctioning of many electronic circuits. Unregulated power supply is not considered suitable for many of the applications in electronics. That is why voltage regulation is of utmost importance in power supplies.

Q. 3. What is a regulated power supply and what are its parts ?

Ans. A transformer, a rectifier circuit, a filter circuit, a voltage regulator along with some current limiting arrangement constitute a regulated voltage supply.

Q. 4. What do you mean by the parameter 'load regulation' ?

Ans. The load regulation is the change in regulated output voltage when the load current changes from minimum to maximum value *i.e.,*

Load regulation, LR = No-load voltage, V_{NL} – Full-load voltage V_{FL}.

It is usually expressed in percentage and is given as

$$\%LR = \frac{V_{NL} - V_{FL}}{V_{FL}} \times 100$$

Q. 5. Define line and load regulation.

Ans. The source or line regulation is defined as the change in regulated output voltage for a specified range of line voltage, typically 230 V ± 10%.

The defining equation is $SR = V_{HL} - V_{LL}$ ...(*i*)

where V_{HL} is output voltage with high input ac line voltage and V_{LL} is output voltage with low input ac line voltage.

The percentage of source regulation is given as

$$\%SR = \frac{V_{HL} - V_{LL}}{\text{Nominal load voltage}} \times 100 = \frac{SR}{V_{nominal}} \times 100 \quad ...(ii)$$

The output voltage under specific operating conditions is known as the nominal load (or output) voltage.

Q. 6. What is the function of a voltage regulator ?

Ans. The primary function of a voltage regulator is to maintain a dc output voltage constant regardless of variations in load current or ac mains voltage.

Q. 7. What is meant by 'dropout' voltage of an IC voltage regulator ?

Ans. The dropout voltage, typically 2 V, is the minimum amount of voltage across the input-output terminals that is required to be maintained if the IC is to operate as a regulator.

Q. 8. What is the function of error amplifier in IC voltage regulator ?

Ans. An error amplifier is essentially an op-amp with extremely high open-loop voltage gain. Because of high gain of an op-amp, the error voltage between the two input terminals approaches zero.

Q. 9. Why are series regulators called linear voltage regulators ?

Ans. Series regulators are called linear voltage regulators because the pass-transistor used in such regulators operate in the active region (linear portion of its output characteristic).

Q. 10. What are the limitations of transistorized power supplies ?

Ans. The stabilized voltage that can be obtained from transistorized power supplies is limited to about 30 – 40 V only, because maximum safe value of V_{CE} is about 50 V. This puts a limit to the use of transistorized power supplies.

Q. 11. What are the limitations of linear voltage regulators ?

Ans. The main drawback of linear voltage regulator is the power dissipation in the pass-transistor which is operated in its linear mode. Other drawbacks of linear voltage regulators are that regulated power supplies using these regulators require a stepdown transformer (most expensive and bulky component of the power supply) and large sized filter capacitors to reduce the ripple.

Q. 12. What are adjustable voltage regulators ?

Ans. A voltage regulator is a circuit that supplies a constant voltage regardless of variation in load current or ac mains voltage. Sometimes variable voltage is required and under such a situation adjustable voltage regulator is used.

Q. 13. What are advantages of adjustable voltage regulators over fixed voltage regulators ?

Ans. The adjustable voltage regulators offer variety of performance and reliability advantages over fixed voltage regulators *i.e.*, improved system performance by having line and load regulation of a factor of 10 or better. Also it improves the system reliability and performance.

Q. 14. What is switching regulator ? List the four major components of the switching regulator.

Ans. A switching regulator can be thought of as a similar to a linear regulator, but with the series-pass transistor operating as a switch that is either *off* or switched-*on* (in a saturated state). The output voltage from the switch is in a pulse waveform which is smoothed into a dc voltage by the action of an LC filter.

A basic switching regulator consists of four major components—voltage source, transistor (or thyristor) switch, pulse generator and a filter. The block diagram of a basic switching regulator is shown in Fig. 1.

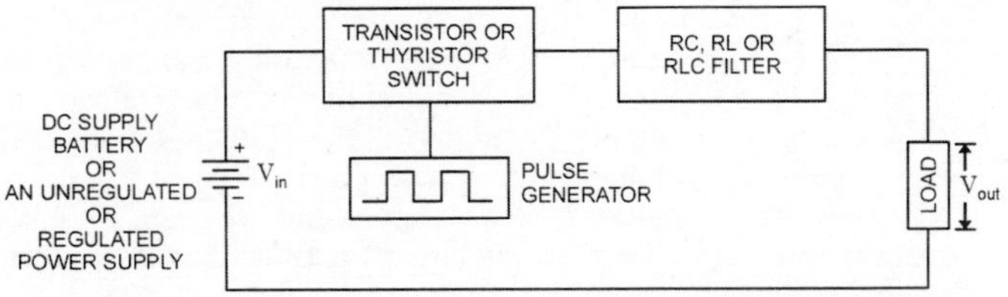

Fig. 1. ***Pulse Generator Produces an Asymmetrical Square***

Pulse generator produces an symmetrical square wave varying in either frequency or pulse width known as frequency modulation (FM) or pulse width modulation (PWM) respectively. The pulse generator output alternately turns the switch *on* and *off.* Filter converts the pulse waveform from the output of the switch into a dc voltage. The basic filters are RC, RL or RLC, the RLC filter is the most commonly used.

Q. 15. What is the working principle of switching regulator ?

Ans. In switching regulators, a series pass transistor is used as a controlled switch and is operated either in cutoff region or saturation region which makes the power transfer via the pass transistor in the form of discrete pulses. SMPS utilizes pulse width modulation to control the average value of output voltage. The output is filtered through a low-pass L-C filter to produce an average dc output voltage. The output of L-C filter is a dc voltage with only a small ripple.

PROBLEMS

1. Using the IC 7805 voltage regulator, design a current source that would deliver a 0.5 A current to the 48 Ω, 10W load. [**Ans.** R = 10 Ω, V_{out} = 29 V, V_{in} = V_0 + dropout voltage = 31 V]
2. Determine the regulated output voltage in the circuit of Fig. 14.15 with R_1 = 540 Ω and R_2 = 2.4 kΩ. [**Ans.** V_{out} = 6.8 volts]
3. Determine the minimum and maximum output voltage of the voltage regulator shown in Fig. 14.18. Assume I_Q of value 20 mA (instead of 10 mA). [**Ans.** 15 V; 94 V]
4. In Fig. 14.32, determine the regulated output voltage if R_1 = R_2 = 2 kΩ and V_{ref} = 2 V. [**Ans.** 4 V]
5. Determine the duty cycle of the pulses at the base of transistor Q in Fig. 14.32 if V_{ref} = 2.5 V, R_1 = 6 kΩ, R_2 = 2 kΩ and regulated input voltage is 20 V. [**Ans.** 50%]

❑❑❑

CHAPTER

15

Function Generators

15.1. INTRODUCTION

INSIDE THIS CHAPTER

A function generator is a signal source that has the capability of producing different types of waveforms as its output signal. The most common output waveforms are sine waves, triangular waves, square waves, and sawtooth waves. The frequencies of such waveforms may be adjusted from a fraction of a hertz to several hundred kHz.

Actually the function generators are very versatile instruments as they are capable of producing a wide variety of waveforms and frequencies. In fact, each of the waveform they generate are particularly suitable for a different group of applications. The uses of sinusoidal outputs and square wave outputs have already been described in the earlier Arts. The triangular wave and sawtooth wave outputs of function generators are commonly used for those applications which need a signal that increases (or reduces) at a specific linear rate. They are also used in driving sweep oscillators in oscilloscopes and the X-axis of X-Y recorders.

Many function generators are also capable of generating two different waveforms simultaneously (from different output terminals, of course). This can be a useful feature when two generated signals are required for particular application. For instance, by providing a square wave for linearity measurements in an audio system, a simultaneous sawtooth output may be used to drive the horizontal deflection amplifier of an oscilloscope, providing a visual display of the measurement result. For another example, a triangular wave and a sine wave of equal frequencies can be produced simultaneously. If the zero crossings of both the waves are made to occur at the same time, a linearly varying waveform is available which can be started at the point of zero phase of a sine wave.

Another important feature of some function generators is their capability of phase locking to an external signal source. One function generator may be used to phase lock a second function generator, and the two output signals can be displaced in phase by an adjustable amount. In addition, one function generator may be phase locked to a harmonic of the sine wave of another function generator. By adjustment of the phase and the amplitude of the harmonics, almost any waveform may be produced by the summation of the fundamental frequency generated by one function generator and the harmonic generated by the other function generator. The function generator can also be phase locked to an accurate frequency standard, and all its output waveforms will have the same frequency, stability, and accuracy as the standard.

The block diagram of a function generator is given in Fig. 15.1. In this instrument the frequency is controlled by varying the magnitude of current that drives the integrator. This instrument provides different types of waveforms (such as sinusoidal, triangular and square waves) as its output signal with a frequency range of 0.01 Hz to 100 kHz.

The frequency controlled voltage regulates two current supply sources. Current supply source 1 supplies constant current to the integrator whose output voltage rises linearly with time according to output signal voltage equation $v_{out} = \frac{-1}{C}\int_0^t i\,dt$. An increase or decrease in the current increases or reduces the slope of the output voltage and thus controls the frequency.

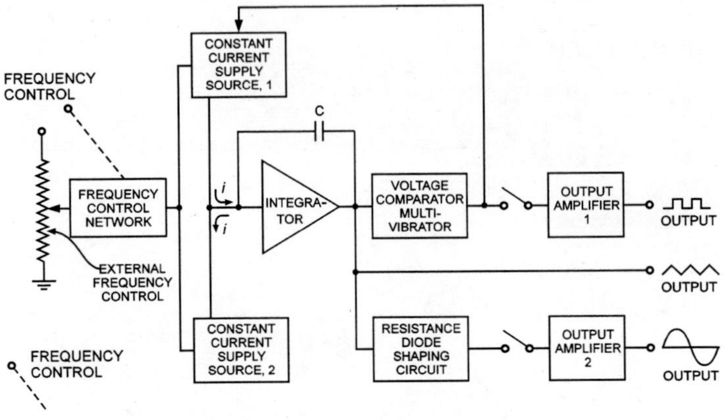

Fig. 15.1. *Block Diagram of Function Generator*

The voltage comparator multivibrator changes state at a predetermined maximum level, of the integrator output voltage. This change cuts off the current supply from supply source 1 and switches to the supply source 2. The current supply source 2 supplies a reverse current to the integrator so that its output drops linearly with time. When the output attains a predetermined level, the voltage comparator again changes state and switches *on* to the current supply source 1.

The output of the integrator is a triangular wave whose frequency depends on the current supplied by the constant current supply sources.

The comparator output provides a square wave of the same frequency as output.

The resistance diode network changes the slope of the triangular wave as its amplitude changes and produces a sinusoidal wave with less than 1% distortion.

15.2. MULTI- OP-AMP FUNCTION GENERATOR

We have seen that the circuit shown in Fig. 15.1 produces both the triangular wave and the square wave (Art. 6.12). This circuit can be converted into a function generator simply by adding a sine-wave shaper. The sine-wave shaper is a circuit that rounds the tips of the triangular wave. Thus, a fairly low-distortion (< 1%) sinusoidal wave can be had from a triangular wave. This may be accomplished with an array of resistors and diodes.

Figure 15.2 shows the schematic of a multi-op-amp function generator. Op-amp A_1 provides

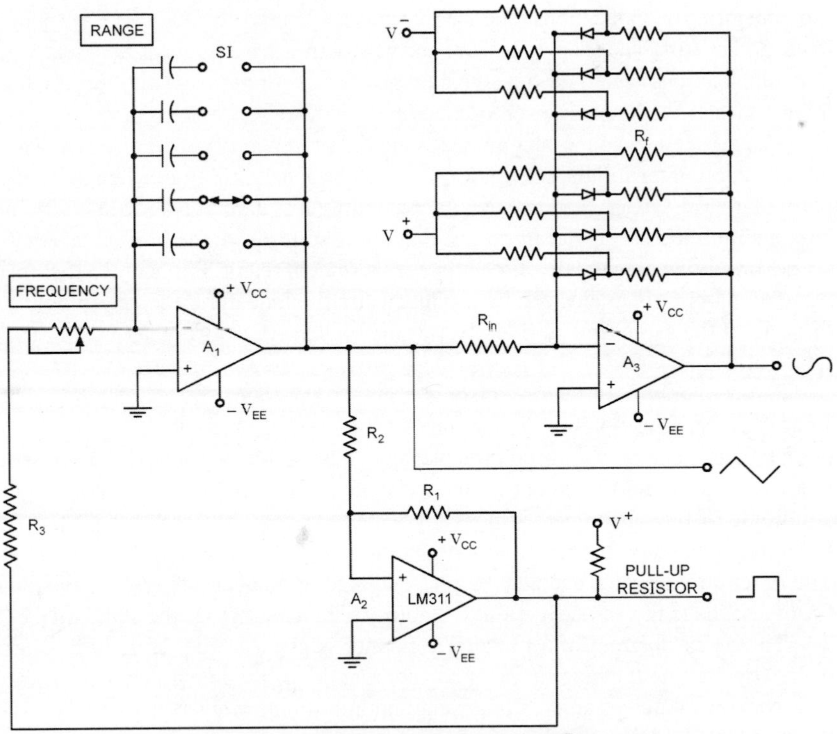

Fig 15.2. *Multi-Op-Amp Function Generator*

the triangular wave while A_2 provides the square wave, as explained in Art. 6.12. The frequency potentiometer is provided to adjust the frequency continuously across a given range. Movable SI selects a frequency range by selecting a capacitor of required capacitance for that particular frequency. The frequency potentiometer should be large (100 kΩ to 1 MΩ), because the capacitors in the negative feedback path cannot be electrolytic.

The two-level comparator A_2 is built with a LM311 high speed comparator. It generates a square wave with much faster rise and fall times than from the less expensive 741C. Resistor R_3 should be of not less than 1 kΩ.

The circuit around A_3 is a nonlinear, sine wave shaper amplifier. It is employed to shape the triangular wave into a sinusoidal wave. When the amplitude of the triangular wave is small, all of the diodes in feedback path of A_3 are off. The gain of the amplifier is $-R_f/R_{in}$. This is illustrated as line segment in Fig. 15.3.

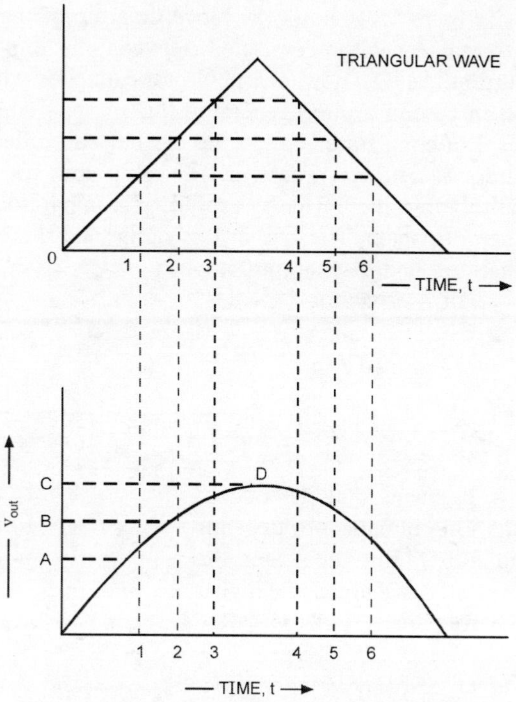

Fig. 15.3. ***Operation of Sinewave Shaper***

When the output attains an amplitude A at time instant 1, a diode in the feedback path gets forward biased and places a resistor in parallel with R_f and thus the feedback resistance is lowered resulting in reduced voltage gain of the amplifier. Consequently, after time instant 1, the output increases at a slower rate (segment B).

When the output attains amplitude B at time instant 2, a second diode in the feedback path is turned on and consequently, the feedback resistance is further reduced, reducing the gain of the amplifier A_3. This reduces the slope of the output wave once more. This process continues for line segments C and D.

The sinusoidal wave shaper converts triangular wave into sinusoidal wave by rounding off triangular wave tips. The resulting sinusoidal wave may have distortion of several per cent, which is much worse than sinusoidal waves generated by R-C oscillators. However, this is the only practical method of generating sinusoidal waves of frequency ranging from very low (0.01 Hz) to very high (exceeding 10 MHz). No significant changes are to be made to the sine wave shaper as the frequency or range controls are adjusted.

For having an instrument quality function generator, function switch, and an output amplifier with amplitude and dc offset controls are to be provided.

15.3. IC FUNCTION GENERATOR

Function generator can be built from a monolithic IC, simply by adding only a few external components. There are several such monolithic ICs available. The XR-2206 is widely available, quite flexible and is easy to use. It can generate triangular wave, square wave, sinusoidal wave, and variable duty cycle pulse. The amplitudes of the output waveforms can be adjusted with external pots (potentiometers), and with an external voltage. This allows true amplitude modulation (AM). The frequency can be adjusted through a range extending from 0.01 Hz to over 1 MHz. The frequency is determined by an external resistor and capacitor, but can be swept over a range of atleast 100 : 1 by an external voltage. This allows frequency modulation (FM). In addition, the output can be stepped between two frequencies with a logic level. Thus, the XR-2206 may be employed for coding a digital pulse train into a series of tones. This is frequency shift keying (FSK).

Figure 15.4 shows the block diagram of the XR-2206. Current controlled oscillator is its heart. An external capacitor connected between the pins 5 and 6, is charged and discharged with a constant current. As we know, charging of a capacitor with a constant current generates a linearly rising voltage. As in case of a ramp generator, this ramp is compared to two levels. When the ramp exceeds the upper level, a comparator within the current controlled oscillator block changes state resulting in reversal of direction of flow of current. The ramp now goes down at the same rate it had gone up. When the ramp falls below the lower threshold of comparator, the current gets reversed again and capacitor starts charging again. The rate of this ramp, and therefore, the frequency of function generator, is set by this charging and discharging.

For a capacitor,

Current,

$$I = \frac{dQ}{dt} = C\frac{dV}{dt} \quad \because Q = CV$$

or $$\frac{dV}{dt} = \frac{I}{C}$$

It means that the rate of charging of capacitor depends upon I/C. But the faster the capacitor is charged, the higher the frequency

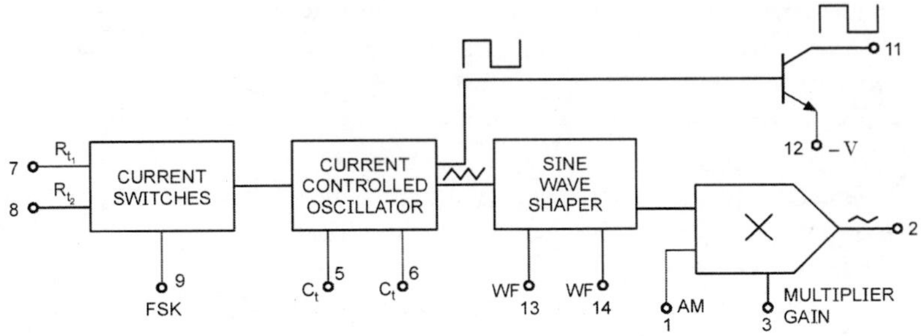

Fig. 15.4. ***Block Diagram of XR-2206***

$$i.e., \quad f \propto \frac{I}{C}$$

The value of C is set by the external capacitor connected between the C_t pins (5 and 6). The current I depends on the current flowing through the external resistors connected to R_{t1} or R_{t2}. Thus the current I is controlled by the value of these external resistors and the voltage level to which they are connected. Which of the resistor is selected depends on the logic level at the FSK pin (pin 9).

In brief it can be said that, the frequency is determined by the rate of charge and discharge of external timing capacitor. This is set by the capacity of the capacitor, which external resistor sets the charging current as selected by the FSK logic level, the value of external resistor, and what voltage is connected to that resistor. Frequency may be controlled by a capacitance, a resistance, a logic level, or a voltage.

The comparator within the current-controlled oscillator generates a 50% duty cycle (square wave). This drives an open-collector buffer. When the output of the comparator is low, the transistor is *off*. A pull-up resistor, $R_{pull\text{-}up}$ is added in order to produce a voltage level at the output. This resistor goes between the square wave output pin and the voltage level desired for the logic high output. The logic low output level is created when the signal from the comparator turns the buffer transistor *on*. At that instant the square wave output pin is shorted to V^-. This may or may not be ground.

Hence a pull-up resistor must be added to the square wave output pin. The 50% duty cycle wave (square wave) goes between the voltage connected to the external pull-up resistor as its high level, and the voltage connected to the V^- of the IC as its low level.

The other signal from the current-controlled oscillator is the triangular wave, which is passed to the sine wave shaper.

The transfer function (input-output relationship) for the sine wave shaper is given in Fig. 15.5. With waveform pins (pins 13 and 14) open, the gain of the sine wave shaper is linear, as illustrated in Fig. 15.5.

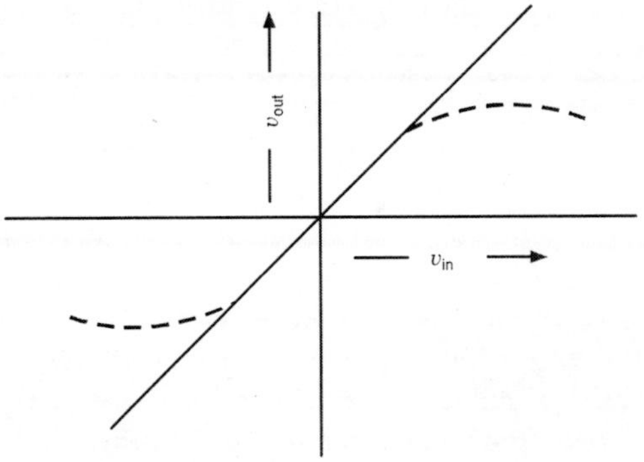

Fig. 15.5. ***Sine Wave Shaper's Transfer Function***

Thus triangular wave from the current-controlled oscillator generates a triangular wave output. However, connecting a 180 Ω resistor between the waveform pins (pins 13 and 14) loads the sine wave shaper, changing its transfer curve as illustrated by the dotted line. Now as the triangular wave from the current controlled oscillator varies the input voltage linearly, the output voltage is curved, having its peaks rounded, giving a sinusoidal wave.

This signal (either a triangular or sinusoidal) is supplied to the multiplier. This forms a true four quadrant multiplier, the output from which is the product of the two inputs, scaled by a multiplication gain factor. Thus varying a dc voltage connected to the AM input (pin 1) varies the output of the multiplier and we have dc voltage control over the output (triangular or sinusoidal) amplitude. Or if the proper low-frequency sinusoidal wave is applied to pin 1, the output's amplitude will vary sinusoidally. Finally, the amplitude of the output can be controlled by changing the multiplier factor. This constant is set by the resistor connected to the multiplier gain pin (pin 3).

Quantitative Details of IC : Power is applied to the inputs (pin 12). The potential difference must be between 10 V and 26 V (*i.e.*, not less than 10 V and not exceeding 26 V).

Power may be single supply, usually with V^- connected to the ground or it may be a split-supply.

$$\pm 5\text{ V} < \pm\text{V} < \pm 13\text{V}$$

Most of the control functions are referenced to negative power pin (pin 12) voltage. Thus the way in which the power is supplied seriously affects how the other signals will be applied for control of performance.

Frequency is controlled by three sets of pins; pins 5 and 6, pin 9 and pins 7 and 8. The timing capacitor, C_t, is connected between pins 5 and 6. The timing capacitor value should be between 1,000 pF and 100 μF.

The voltage at the FSK pin (pin 9) decides which external timing resistor is to be used. When the voltage at pin 9 exceeds 2 V above the negative power pin voltage V^-, then R_{t1} at pin 7 is to be selected. Conversely, if the voltage is within 1 V or less of negative power pin voltage V^-, R_{t2} at pin 8 is to be selected. Leaving the FSK pin open selects R_{t1}. However, leaving a pin open is an invitation to noise-induced problems. If FSK feature is not to be used, pin 9 may be connected to the positive power pin.

For operating XR-2206 from a single supply (such as +12 V or +15 V), the FSK logic levels are TTL compatible. High selects R_{t1} (pin 7) while low selects R_{t2} (pin 8). However, some care is required to be taken in driving the FSK pin (pin 9) if IC is powered from a split-supply (*i.e.*, ± 5 V). The XR-2206 interprets any voltage exceeding 2 V above the voltage at the negative power pin as a logic high. It means that operation with ± 5 V supplies, any voltage greater than – 3 V (such as – 2 V, or even ground) would select R_{t1}. To select R_{t2}, the FSK voltage must be within 1 V of the negative supply (*i.e.*, –4 V to –5 V). This could present a problem. Hence for driving the FSK pin from a digital IC, the function generator must be powered from a single supply.

The final parameter to set the frequency is the current through the selected resistor. This frequency is

$$f = \frac{0.32\text{I}}{\text{C}} \qquad \text{...(15.1)}$$

This current must flow out of the IC and must be kept within the limits

$$1\ \mu\text{A} < \text{I} < 3\text{ mA} \qquad \text{...(15.2)}$$

The XR-2206 regulates the voltage at both R_t pins (pins 7 and 8), keeping 3 V above negative supply voltage V^-. So in the simplest case, a resistor is connected between selected R_t pin and negative power supply pin V^-, as shown in Fig. 15.6 (*a*). Current through this resistor is just

$$\text{I} = \frac{3\text{ V}}{\text{R}_t} \qquad \text{...(15.3)}$$

For Eq. (15.3) to hold true when a negative voltage is tied to the pin 12, R_t must be connected to that same negative supply.

Placement of a voltage on the left end of R_t allows voltage control of the frequency, as shown in Fig. 15.6 (*b*). The current through R_t is now set by

$$\text{I} = \frac{[\text{Negative supply voltage } (\text{V}^-) + 3\text{V}] - \text{V}_{\text{op-amp}}}{\text{R}_t} \qquad \text{...(15.4)}$$

Variation in op-amp output voltage causes variations in the current drawn from the XR-2206, which, in turn varies the frequency.

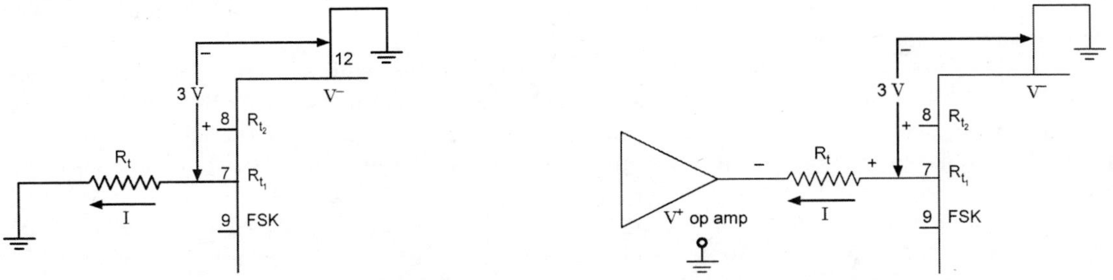

(a) *Simple R_t Control of Frequency* (b) *Voltage Control of Frequency*

Fig. 15.6. ***XR-2206's Timing Current Setting***

An op-amp is employed to set this voltage for two reasons. First, the output impedance of the op-amp is usually much smaller than R_t. Hence only the voltage at its output need be considered. If this voltage is set by a potentiometer (pot) current I will be based on the resistances presented by the pot.

The second reason that an op-amp is used to drive R_t, is its ability to produce an inversion and an offset. To force I (and, therefore, the frequency) to zero, $V_{op\text{-}amp}$ must be set to 3 V. This puts the same voltage on each side of the R_t, leaving no voltage across the resistor. For voltage control of frequency, it is appropriate that 0 V into the circuit should generate 0 Hz. So the op-amp is configured as an inverting adder, with a 3 V offset. Zero volt to that adder produces 3 V to the left end of R_t, which drives the frequency to zero.

This op-amp adder also solves the last problem. For increasing the frequency, I must be increased and for increasing, $V_{op\text{-}amp}$ must be set down, below 3 V. The more negative $V_{op\text{-}amp}$ becomes, the larger the difference in voltage across R_t, the larger I becomes, and therefore, the higher the frequency. This relationship is opposite what is usually needed for voltage control of frequency. Normally driving the voltage down should produce a decrease in frequency. Configuring the op-amp as an inverting adder accomplishes it.

HIGHLIGHTS

1. A function generator is a signal source that has capability of producing different types of waveforms as sine waves, triangular waves and sawtooth waves. The frequency of such waveforms may be adjusted from a fraction of Hz to several kHz. In fact, the function generator is a very versatile instrument as it is capable of producing a wide variety of waveforms and frequencies.

EXERCISES

1. Draw the block diagram of a function generator and explain the method of producing sine waves.
2. What is a function generator? Explain how a multi-op-amp function generator can be had. Explain its operation. Give its few applications.
3. Draw the block diagram of XR-2206. How is it used as a function generator ?
4. Explain the working principle of any one type of function generator with the help of block diagram. **[G.G.S.I.P. Univ. Analog Electronics-II May 2011]**
5. Write technical notes on the following:
   (*i*) Multi-op-amp function generator. (*ii*) IC function generator.

SHORT ANSWER TYPE QUESTIONS WITH ANSWERS

Q. 1. What is function generator ?

Ans A function generator is a signal source that has capability of producing different types of waveforms such as sine waves, triangular waves, square waves, and sawtooth wave. The frequencies of such waveforms may be adjusted from a fraction of Hz to several kHz. In fact, the function generator is a very versatile instrument as it is capable of producing a wide variety of waveforms and frequencies.

Q. 2. What is a sine-wave shaper ?

Ans. The sine-wave shaper is a circuit that rounds the tips of the triangular wave. It makes use of an array of resistors and diodes.

❑❑❑

EXAMINATION PAPERS

B.TECH.

END TERM EXAMINATION
FOURTH SEMESTER MAY-JUNE 2009
ANALOG ELECTRONICS-II

Time : 3 Hours *Total Marks : 75*

Note : *Attempt one question from each unit including Q. No. 1, which is compulsory. Assume data where necessary.*

1. Short answer type questions. (Attempt five questions) **(15)**

(*a*) Explain building blocks of an op-amp and their function.

(*b*) Draw op-amp differentiator and integrator circuits.

(*c*) Define CMRR and slew rate.

(*d*) Explain transformer coupled class B push-pull amplifier.

(*e*) Draw frequency characteristics of LP, HP and BP filters (1st and 2nd order).

(*f*) Explain oscillator principles and frequency stability.

(*g*) Explain quadrature oscillator.

(*h*) Explain adjustable positive voltage regulator.

UNIT-I

2. (*a*) Describe an ideal op-amp and its electrical characteristics. Draw equivalent circuit and ideal voltage transfer curve. **(10)**

(*b*) Explain the working of a practical instrumentation amplifier. **(5)**

3. (*a*) Draw the circuit of an op-amp voltage follower. Explain its working. Why is it called a noninverting buffer ? **(8)**

(*b*) Differentiate between the three closed loop differential amplifier configurations with negative feedback. **(7)**

UNIT-II

4. (*a*) Describe the conditions for sustained oscillations and frequency stability in positive feedback amplifier. Classify the oscillators giving frequency range and the type of waveform generated. **(8)**

(*b*) How is a square-wave signal generated ? Draw circuit diagram and derive equations for the frequency of oscillation. Does it have high frequency limitation ? Discuss. **(7)**

5. (*a*) Explain op-amp Wien bridge oscillator and state conditions for sustained oscillators. **(10)**

(*b*) If resistors and capacitors in the circuit have the values of 8.2 kΩ and 0.05 μF, determine the frequency of oscillation. **(5)**

UNIT-III

6. (*a*) Describe four different types of voltage regulators and give their applications. **(5)**

(*b*) Draw schematic circuit of a basic switching regulator and explain the function of its four major components. **(10)**

7. (*a*) State classification of power amplifiers. Draw circuit diagrams and state their characteristics. **(10)**

(*b*) Design a clipping circuit to clip a sinusoidal wave of 2 V peak voltage at 100 Hz above +1 volt. Draw input and output waveforms. **(5)**

UNIT-IV

8. (*a*) Design a first order L.P. Butterworth filter with a cut-off frequency of 1 kHz and a passband gain of 2 [given C = 0.01 μF, $R_1 = R_f = 10$ kΩ]. Give design steps and draw frequency response. **(8)**

(*b*) Design a 60 Hz active notch filter. State design steps and frequency characteristics (Assume data). **(7)**

9. (*a*) Explain operating principles of P.L. loop in its basic form with block diagrams and state function of each block **(12)**

(*b*) Give application of P.L. loop circuits. **(3)**

B.TECH.

END TERM EXAMINATION
FOURTH SEMESTER MAY 2010
ANALOG ELECTRONICS-II

Time : 3 Hours *Total Marks : 75*

Note : *Attempt one question from each unit including Q. No. 1, which is compulsory. Internal choice is indicated.*

1. (*a*) Explain the function of all the basic building blocks of an op-amp.

(*b*) What is the importance of current mirror ? Draw the circuit of the basic current mirror and explain its working.

(*c*) List the important features of an instrumentation amplifier. Draw the circuit of a practical instrumentation amplifier and find an expression for its output voltage.

(*d*) What types of fixed voltage regulators are available ? Show the standard representation and discuss main characteristics.

(*e*) Draw the circuit of an All-pass filter and find its transfer function. Give an application of all-pass filter.

(5 × 5 = 25)

UNIT-I

2. (*a*) Draw the circuit of a MOSFET source coupled differential amplifier. Derive an expression for its CMRR. **(6.5)**

(*b*) Draw the circuit of a Wilson current source and show that its output current is almost equal to the reference current. **(6)**

OR

3. (*a*) Discuss the external compensating techniques used in op-amps. Compare their performance. **(6.5)**

(*b*) The time constant (RC) of an op-amp integrator circuit is 0.1 ms. Determine the output response for the following input signals : **(6)**

(*i*) 1 V peak sine wave at 5 kHz.

(*ii*) Step voltage $V_1 = 1$ V for $0 \le t \le 0.5$ ms.

(*iii*) 2 V P-P square wave of 5 kHz

Draw the input and output waveforms.

UNIT-II

4. (*a*) Draw the circuit of a multivibrator which can generate a single output pulse of duration T. Explain the working and derive the expression for the pulse width T. **(6.5)**

(*b*) Draw the circuit of an op-amp oscillator to generate radio frequency signals. Determine the expression for the frequency of oscillation and the condition for oscillation. **(6)**

OR

5. (*a*) Explain the working of a square waveform generator and derive an expression for the frequency of oscillation. Suggest a method for obtaining asymmetrical square wave. **(6.5)**

(*b*) For an RC-phase shift oscillator, derive the expression for the frequency of oscillation and condition for sustained oscillation. **(6)**

UNIT-III

6. (*a*) For the circuit shown in Fig. 1, find the output voltage waveform for a sine wave input signal. **(6.5)**

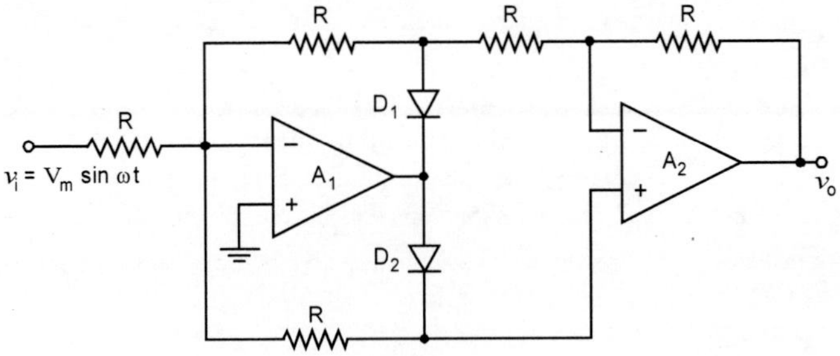

Fig. 1

(*b*) Draw and explain the working of a complementary symmetry push-pull amplifier. Derive an expression for its efficiency. **(6)**

OR

7. (*a*) A comparator circuit using positive feedback is shown in Fig. 2. A sine wave input is applied. Derive the output response. Explain Hysteresis in this circuit. **(6.5)**

(*b*) Draw the circuit of a log amplifier and derive the expression for the output voltage. Is the circuit independent of temperature ? **(6)**

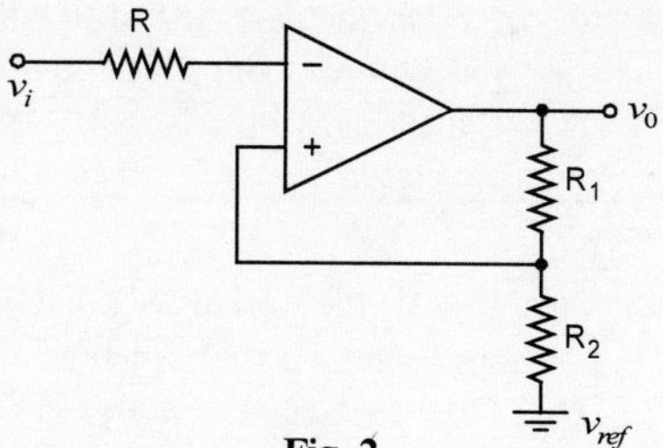

Fig. 2

UNIT-IV

8. (*a*) Draw the circuit of a second-order low-pass filter and derive an expression for its transfer function for a maximally flat response in the pass band. **(6.5)**

(*b*) Draw the block schematic of a PLL and explain the function of each block. **(6)**

9. (*a*) Draw the circuit of a notch filter and derive an expression for its transfer function. **(6)**

(*b*) Draw a block schematic of voltage controlled oscillator (VCO). Explain its working and derive an expression for its frequency. **(6.5)**

B.TECH.

END TERM EXAMINATION
FOURTH SEMESTER MAY 2011

ANALOG ELECTRONICS-II

Time : 3 Hours *Total Marks : 75*

Note : *Attempt five questions including Q. No. 1, which is compulsory. Internal choice is indicated.*

1. (*a*) Define the term "Common mode rejection ratio." Explain how to measure it ?

(*b*) What is an active load ? Where it is used ?

(*c*) Determine the range of variable inductance, if frequency of oscillation is to vary between 950 kHz to 2050 kHz. Assume C_1 = 7500 pF and C_2 = 100 pF.

(*d*) Explain the two conditions for oscillation in feedback amplifier.

(*e*) Explain the difference between a voltage amplifier and a power amplifier by giving an example.

(*f*) Why class B and class C amplifiers have more efficiency than class A amplifier ?

(*g*) What do you mean by precision rectifier, explain with circuit ?

(*h*) Explain why active filter preferred over passive filters ?

(*i*) Explain an all-pass filter.

(*j*) Define the terms free-running frequency, capture range and pull-in time. **(2.5 × 10 = 25)**

2. (*a*) Find the output of a given circuit in Fig. 1. **(6)**

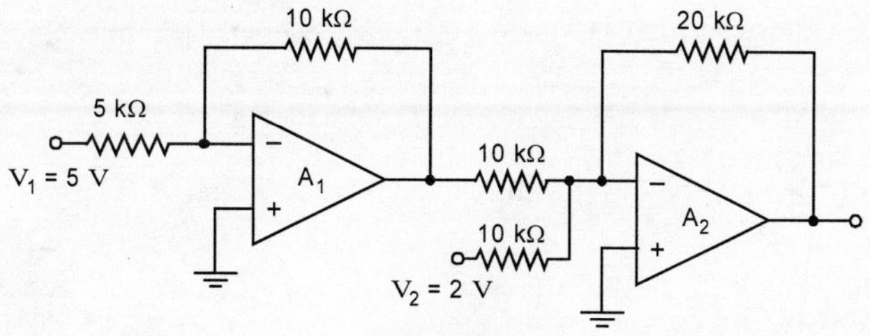

Fig. 1

(*b*) Explain the difference between constant current bias and current mirror. **(6.5)**

OR

3. (*a*) Draw and explain the operationof a voltage to current converter when load is grounded. Is there any limitations on the size of the load when grounded ? **(6.5)**

(*b*) Explain the working of differentiator and it is excited by the input shown in Fig. 2. Find the output voltage if $R_f = 2$ kΩ, $C_1 = 0.1$ μF. **(6)**

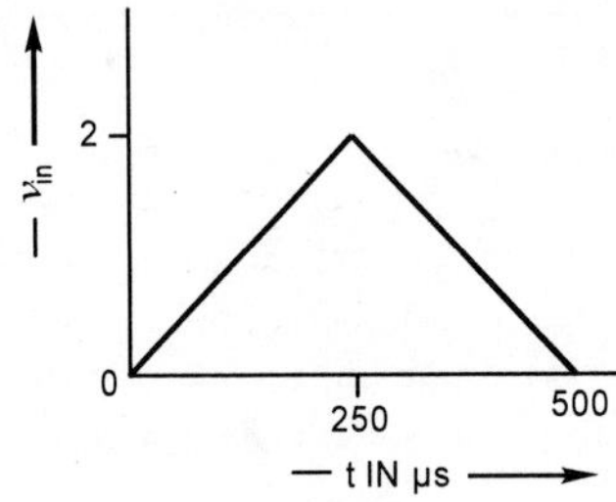

Fig. 2

4. (*a*) Explain the operation of an astable multivibrator and derive the expression for frequency of oscillations. **(6.5)**

(*b*) Explain how the triangular wave generator can be converted into a ramp generator ? **(6)**

OR

5. (*a*) Explain the working of Wein bridge oscillator and derive the expression for the frequency of oscillation. **(6.5)**

(*b*) Discuss the applications of OTA as programmable voltage amplifiers. **(6)**

6. (*a*) Explain how push-pull configuration reduce the harmonic distortion in an amplifier ? **(6)**

(*b*) Explain the advantage of Schmitt Trigger over basic comparator. How it will convert sine wave into square wave ? **(6.5)**

OR

7. (*a*) Define conversion efficiency. Determine the maximum value of conversion efficiency for a series-fed class A power amplifier. **(6)**

(*b*) Explain how two analog voltages are multiplied and divided by using log and antilog amplifier ? **(6.5)**

8. (*a*) Design the band-pass filter so that cutoff frequency $f_c = 1$ kHz, Q = 20 and $A_f = 10$. Assume C = 1 μF. **(6)**

(*b*) Explain the operation of 723 regulator. **(6.5)**

OR

9. (*a*) Design a first-order low-pass filter for a high cut-off frequency of 2 kHz and pass band gain is 2. **(6)**

(*b*) Explain the operation of VCO with the help of block diagram. **(6.5)**

B.TECH.

END TERM EXAMINATION
FOURTH SEMESTER MAY 2011
ANALOG ELECTRONICS-II

Time : 3 Hours *Total Marks : 75*

Note : *Q. No. 1 is compulsory. Attempt One question from each unit.*

1. (*a*) Draw the block schematic of an op-amp and explain the function of each block. **(5 × 5 = 25)**

(*b*) Derive and plot the transfer characteristics of an emitter-coupled differential amplifier and explain the various regions of operation.

(*c*) A class B push-pull amplifier gives cross-over distortion. Explain and suggest a circuit to eliminate it.

(*d*) Explain the working of an ideal op-amp differentiator. What are its disadvantages and how to overcome these ?

(*e*) Draw and explain the internal diagram of an OTA and explain its main features.

UNIT-I

2. (*a*) Derive an expression for the CMRR of an emitter-coupled differential amplifier. Suggest one method to increase CMRR. **(6.5)**

(*b*) Explain the working of a current source suitable for generating low values of constant currents. Design the current source for $I_0 = 10$ μA. Assume $V_{CC} = 10$ V, $V_{BE} = 0.7$ V and β = 100. **(6)**

3. (*a*) Explain the terms :

(*i*) Input bias current.

(*ii*) Input offset current,

(*iii*) Input offset voltage, total output offset voltage

(*iv*) Slew rate.

(*b*) For the circuit shown in Fig. 1, $R_1 = 10$ kΩ, $R_f = 100$ kΩ, $C_f = 10$ nF. Derive and plot the response for the following input signals : **(6)**

(*i*) Sine wave of IV peak amplitude and 5 kHz

(*ii*) Square wave of IV peak amplitude and 5 kHz.

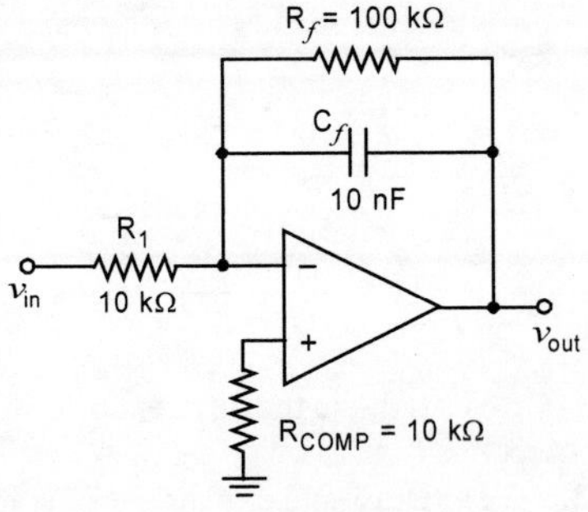

Fig. 1

UNIT-II

4. (*a*) Explain the working of an op-amp square-wave generator. Derive an expression for the frequency of oscillation. What is the maximum amplitude of the square wave and how it can be varied ? **(6.5)**

(*b*) Explain the working of a Wien bridge oscillator. Derive the frequency of oscillation and the condition for oscillation. **(6)**

5. (*a*) Draw and explain the working of a triangular waveform generator using minimum components. Derive an expression for the frequency of oscillation. **(6.5)**

(*b*) Explain the condition for self sustained sinusoidal oscillations. Name an oscillator which gives highly stable fixed frequency oscillations. Explain its working. **(6)**

UNIT-III

6. (*a*) Explain the working of a complementary symmetry push-pull amplifier and derive an expression for its conversion efficiency. **(6.5)**

(*b*) Explain the working of the circuit shown in Fig. 2. Draw the output waveform for a sine wave applied at the input. **(6)**

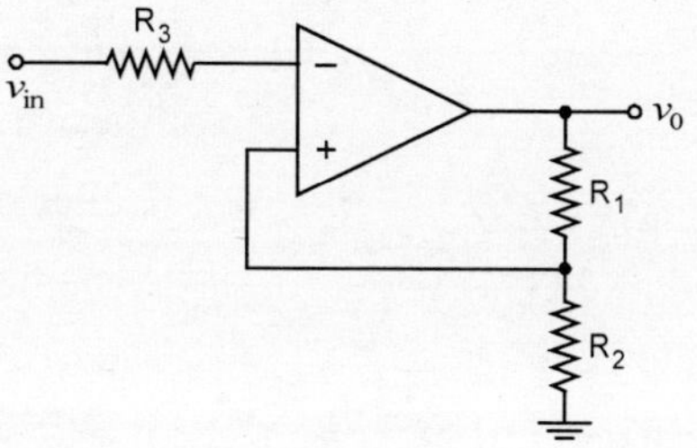

Fig. 2

7. (*a*) Make a precision op-amp full-wave rectifier and explain its working. Can this circuit be used as an absolute value circuit ? **(6.5)**

(*b*) Draw and explain the working of a log-amplifier which provides saturation current and temperature compensation. **(6)**

UNIT-IV

8. (*a*) Draw the circuit of a second order low-pass filter and derive an expression for its transfer function for a Butterworth filter. **(6)**

(*b*) Design a wide-band pass filter having $f_1 = 500$ Hz, $f_h = 2$ kHz and pass band gain of 4. Draw the circuit. **(6)**

9. (*a*) Draw the block schematic of an IC phase-locked loop and explain its working. **(6.5)**

(*b*) Explain the working principle of any one type of function generator with the help of block diagram. **(6)**

❑❑❑

INDEX

A

B

C

H

I

L

M

N

O

T

V

W

Z

❑❑❑

Win Prizes

ATTENTION READERS

Win Prizes

We request for your frank assessment regarding various aspects of the book as given below :

ANALOG ELECTRONICS-II

BY **(GGSIPU)**

J.B. GUPTA, RAJEEV MANGLIK & ROHIT MANGLIK

Please Fill up the Sheet in Neat and Clean Writing

(i) What topics of your syllabus, important for your examination point of view, are not covered in this book ?

....

....

....

....

(ii) What are the chapters wherein the treatment of the subject matter is not systematic or organised or updated ?

....

....

....

....

(iii) Have you come across misprints/mistakes/factual inaccuracies in the book ? Please specify the chapters and the page numbers.

....

....

....

....

(iv) What other books on the same subject have you found/heard better than the present book ? Please specify in terms of price and quality.

....

....

....

....

(v) Further suggestions and comments for the improvement of this book.

....

....

....

....

(PS : If need be please attach a separate sheet to write your views)

OTHER DETAILS

(i) Who recommended you this book ? (Please tick (✓) in the box near the option relevant to you.

☐ Teacher ☐ Friends ☐ Bookseller

....

....

....

....

(ii) Name of the recommending teacher, his designation and address:

....

....

....

....

(iii) Name and address of the bookseller you purchased the book from :

....

....

....

....

(iv) Name and address of your institution (Please mention the University or Board as the case may be)

....

....

....

....

(v) Your name and complete address :

....

....

....

....

(vi) Course you are enrolled in :

....

....

....

....

The best assessment will be awarded each month : The award will be in the form of our publications as decided by the Editorial Board.

Please mail the filled up coupon at the vearliest to :

S.K. KATARIA & SONS®

Publisher of Engineering Books

4885/109, Prakash Mahal, Dr. Subhash Bhargav Lane,

Opposite Delhi Medical Association, Daryaganj, New Delhi–110002 (India)

Phone: +91-11-43551243; *Mobile:* +91-9871775858, 7678425209

e-mail: katariabook@gmail.com, katariabooks@yahoo.com

Website: www.skkatariaandsons.com